T0327614

Engineering Optimization

WILEY-ASME PRESS SERIES LIST

Engineering Optimization

Applications, Methods, and Analysis

R. Russell Rhinehart

This Work is a co-publication between ASME Press and John Wiley & Sons Ltd.

This edition first published 2018

© 2018 R. Russell Rhinehart

This Work is a co-publication between ASME Press and John Wiley & Sons Ltd

Registered Offices

John Wiley & Sons, Inc., 111 River Street, Hoboken, NJ 07030, USA

John Wiley & Sons Ltd, The Atrium, Southern Gate, Chichester, West Sussex, PO19 8SQ, UK

Editorial Office

The Atrium, Southern Gate, Chichester, West Sussex, PO19 8SQ, UK

For details of our global editorial offices, customer services, and more information about Wiley products visit us at www. wiley.com.

Wiley also publishes its books in a variety of electronic formats and by print-on-demand. Some content that appears in standard print versions of this book may not be available in other formats.

Library of Congress Cataloging-in-Publication Data

Names: Rhinehart, R. Russell, 1946– author.

Title: Engineering optimization : applications, methods and analysis / by R. Russell Rhinehart.

Description: First edition. | Hoboken, NJ : John Wiley & Sons, 2018. | Includes index. |

Identifiers: LCCN 2017052555 (print) | LCCN 2017058922 (ebook) | ISBN 9781118936313 (pdf) | ISBN 9781118936320 (epub) | ISBN 9781118936337 (cloth)

Subjects: LCSH: Engineering–Mathematical models. | Mathematical optimization.

Classification: LCC TA342 (ebook) | LCC TA342 .R494 2018 (print) | DDC 620.001/5196–dc23

LC record available at https://lccn.loc.gov/2017052555

Cover design: Wiley

Cover image: © John Wiley & Sons

Set in 10/12pt Warnock by SPi Global, Pondicherry, India

Contents

Preface

Introduction

Optimization means seeking the best outcome or solution. It is an essential component of all human activities. Whether personal or professional, we seek best designs, best choices, best operation, more bang for the buck, and continuous improvement.

Here are some professional examples: Minimize work events that lead to injury while remaining economically competitive. Structure workflow to maximize return on investment. Design an antenna that maximizes signal clarity for a given power. Define a rocket thrust sequence to maximize height. Determine the number of parallel devices to minimize initial cost plus future risk.

Here are some personal examples: Seek the best vacation experience for the lowest cost. Minimize grocery bill, but meet desires for nourishment and joy of eating. Set the family structure for raising children that leads to well-adjusted, happy, productive outcomes, but keep within the limits of personal resources. Create a workout regime that leads to fastest and most attractive muscle development, with no injury, and in balance with other desires in quality of life.

Optimization is not just an intellectual exercise; although often, solving the challenge is as rewarding as completing a Sudoku puzzle. We implement the optimized decision. Accordingly, within any application it is essential to completely and appropriately assess the metrics that quantify "best." If the description of what you want to achieve is not quite right, then the answer will also be wrong, which the implementation will reveal in retrospect. You want to get it right prior to implementation. So, part of this book is about development of the optimization objective.

After the objective is stated, we desire an efficient search logic to find the best solution, with precision and with minimal computational and experimental effort. So, other parts of this book are about the optimizer—the search logic, or algorithm.

Both aspects are essential, and I find that most books on optimization focus on the intellectually stimulating mathematics of the algorithms. So, I offer this book to provide a balance of essential topics to the application to guide user choices in structuring the objective, defining constraints, choosing convergence, choosing initialization, etc. Some will be disappointed that this book is not a compendium of every optimization algorithm conceived by mankind. However, others will value the application perspective.

Also, I find that most people using optimization as a tool did not have a course on it while in school. So, I have written this book in a style that I hope facilitates self-study by those who need to understand optimization applications while keeping it fit for use as a graduate-school course textbook.

Key Points

Here are a few essential aspects of optimization:

Point 1: Although optimization offers the joys of solving an intellectual puzzle, it is not just a stimulating mathematical game. Optimization applications are complicated, and the major challenges are the clear and complete statement of:

1) The objective function (OF—the outcome you wish to minimize or maximize)
2) Constraints (what cannot or should not be violated, or exceeded)
3) The decision variables (DV—what you are free to change to seek a minimum)
4) The model (how DVs relate to OF and constraints)
5) The convergence criterion (the indicator of whether the algorithm has found a close enough proximity to the minimum or maximum and can stop or needs to continue)
6) The DV initialization values
7) The number of starts from randomized locations to be confident that the global optimum has been found
8) The appropriate optimization algorithm (for the function aberrations, for utility, for precision, for efficiency)
9) Computer implementation in code   Oh yes,
10) The mathematics of the optimization algorithm (understanding this is also important)

This book seeks to address all 10 aspects, not just the 10th.

Point 2: Do not study. Learning is most effective if you integrate the techniques into your daily life. You will forget the material that you memorized in order to pass a test. Since this book provides skills that are essential for both personal and career life, I want you to take the techniques with you. I want this book to be useful in your future. Although memorization and high-level mathematical analysis are both elements of the book, understanding the examples and doing of the exercises is more important. To maximize the impact of this material, you need to integrate it into your daily life. You need to practice it.

Oh, I see I omitted a comma in the first sentence of the paragraph above. It should have been "Do, not study." Learn by doing. After you read a section and think you understand it, see if you can implement it. Of course, the comma "error" above was intentional to wake up curiosity about the message.

Point 3: Optimization is universal to all engineering, business, science, computer science, and technology disciplines. Although primarily written for engineering applications, this introductory book is designed to be useful for all those seeking to apply optimization in all fields.

Point 4: The implementation of optimization requires computer programming, which for many is an aggravation. To help the reader, I currently have, and plan to support, a website that offers to any visitor optimization software and examples. Visit www.r3eda.com. The "r3" in the address is my initials, and the appended "eda" means "enabling data analysis." Seeking to maximize ease of use and accessibility, the programs are written as VBA macros for MS Excel. VBA is not the fastest-computing environment, nor does it have the best scientific data processing functions. However, it has been adequately functional for all of my applications, and if you need something better, the code can be translated. This book provides a VBA primer (Appendix F) for those needing the help in accessing and modifying the code. The programs on the r3eda site solve many of the examples in this book.

Book Aspirations

Readers should be pleased with their ability to:

- Understand and use the fundamental mathematical techniques associated with optimization
- Define objective functions, decision variables, models, and constraints for a variety of optimization applications
- Develop, modify, and program simplified versions of the more common optimization algorithms
- Understand and choose appropriate methods for:
  - Constrained optimization
  - Global optimization
  - Convergence criteria
  - Surface aberrations
  - Stochastic applications
- Understand diverse issues related to optimizer desirability
- Explore, contrast, and evaluate the performance of optimization algorithms and user choices of convergence criteria, numerical derivative estimation, threshold, constraint handling method, parameter values, etc. with respect to precision, user convenience, and other measures of optimizer desirability
- Apply optimization algorithms to case studies relevant to the reader's career
- Continue learning optimization methods from texts, reports, Internet postings, and refereed journal articles

Optimization is the name for the procedure for finding the best choices. "Procedure," "best," and "choices" are separate aspects, and the user must understand each to be able to appropriately define the application. And each aspect has a large range of options.

Procedure

This relates to the method used to find the optimum:

- In process or device design, for example, the choices could be the equipment specifications (type, materials, size), and the evaluation of best in the design could be to minimize capital cost with a constraint on reliability. With mixed continuous, discrete, and class variables as the choices, a direct search algorithm might be the best optimizer.
- Alternately, in scheduling a rocket thrust to reach a desired height, the stage choices might be height, best might be evaluated as minimizing either time or fuel use, and the appropriate algorithm might be dynamic programming.
- Another example is characterized as the traveling salesman problem in which the objective is to determine a sequence of locations to visit to minimize travel distance. Here the choice is the sequence, and the best sequence might be impacted by a priority of visits, expenses, wasted time, etc. The procedure might use the random keys method to convert a sorted list of rational numbers into the sequence.
- As a final contrasting example, in model-predictive control, the objective might be to minimize time to move a response to a set point while penalizing excessive manipulated variable moves while avoiding constraints; and the choices might be the future sequence of manipulations. If the penalties are quadratic, the appropriate algorithm might be a gradient-based procedure.

Best

Within optimization terminology, the definition of best for a specific application (and the method for calculating a value to quantify best) is variously termed the cost function or the objective function (OF). It is the function that returns a value representing an assessment of goodness. Best usually means minimize undesirable aspects and/or maximize desirable aspects, and the OF can represent a wide range of metrics related to economics, safety, time, resource conservation, quality, deviation, probability, etc. But best might mean to minimize a worst-case feature (min the max, or min–max), such as finding a path through mountains that minimizes the steepest ascent or finding a process design that minimizes the worst-case outcome (risk).

Defining the appropriate OF is situation specific, and often it is the key challenge in an optimization application. The user needs to clearly understand the complex situation and realize that a first statement of the OF usually embodies a superficial understanding. Subsequent analysis of the results will lead to an evolution of the OF. For example, a challenge might be to choose the best pipe diameter in a process design. A smaller diameter means a less expensive pipe and lower in-pipe inventory cost, but it means a larger pump. An initial OF choice might be to minimize capital. However, reconsideration from a business investment view might reveal that operating costs associated with pumping power and maintenance are also important issues, and perhaps net present value (NPV) is a right way to combine initial capital with future expenses. Then, reconsideration might bring understanding of the sensitivity of the optimum solution to uncertainty in the "givens," which will lead to a refinement of the OF to represent the 95% worst case of the NPV in a Monte Carlo analysis, making it a stochastic function. Risk might then be perceived as an additional issue, and the OF might be split into a multi-objective version (risk and NPV) that provides a non-dominated set of solutions for a user to select a best for the particular situation. Finally, the user might realize that pipe comes in discrete diameter values and that the pipe diameter is not a continuous-valued number. This application might have evolved from an initial simple deterministic (textbook example) case to a complicated application, classified as mixed integer, stochastic, and multi-objective.

This book will address how to develop the OF and will show examples from a wide range of applications.

Choices

The choices a user has (you may call these inputs, decisions, degrees of freedom, or independent variables) to change things toward the best outcome are termed decision variables (DVs).

In regression DVs are the model coefficient values. In product design DVs could be polymer type, blend concentration, operating a process, color, or shape. In process design DVs could be the pipe diameters and pump sizes. In flying aircraft, the DVs would be the stick, throttle, and pedal positions. In control and scheduling, in operating a business, the DVs would be the future plan for both the timing and magnitude of the actions. Alternately, the DVs might be the coefficients in an equation that would define the future schedule for control actions. Again, there are many possibilities for how to choose the DVs; and the user choices impact efficiency of solution, the appropriate optimizer algorithm, and precision of solution. The book will also address such issues.

Organization

As with most books on engineering optimization, this one describes and develops many common algorithms. It starts with simple univariate (line) search approaches and progresses to multivariable and multiplayer approaches. I do not seek to cover every version, or every method. I use archetypical examples of the many approaches, from which readers can grasp the concepts of other methods. Book topics include gradient based, Newton's, and blends such as Levenberg–Marquardt. They include surrogate function methods to characterize the "surface" such as successive quadratic. They include direct searches such as a simple heuristic cyclic, Hooke–Jeeves, and Nelder–Mead. They include multiplayer mimetic approaches of leapfrogging, particle swarm, and genetic algorithms. They include dynamic programming, in which the DVs are the states, and linear programming that takes advantage of certain structures. The book develops the basic techniques and addresses refinements that improve performance, such as quasi-Newton estimates of the Hessian elements, and grid refinement in dynamic programming.

The book provides a guide to match optimization procedures with features of the application such as discontinuities, flat spots, nearly flat spots, constraints, multiple optima, stochastic responses, parameter correlation, etc. Several sections discuss the issues that certain OF features create. Other sections are devoted to the analysis of the optimizers for precision, accuracy, global identity, work to converge, and robustness. Another section reveals sensitivity to user parameters such as contraction and expansion coefficients, thresholds, triggers, etc. A user needs to understand which optimizer is appropriate for which application and how to make the best choice of optimizer parameter values.

The book also addresses choices of convergence criteria that are appropriate for the application and for the optimizer. For example, in choosing thresholds on the DV as the convergence criteria (which is common practice), the user should use propagation of uncertainty to project the DV tolerance on the OF. As a contrasting example, in optimizing results of either experimental outcomes or a Monte Carlo stochastic simulation, the optimizer needs to stop when the noisy response is not making improvement relative to the noise amplitude.

The book is aimed at engineering applications, where optimization is essential for model development, product design, process and device design, dynamic system control, or system operation. However, the applications of optimization extend into all aspects of our lives from purchasing choices to investment choices, to career planning, and to dressing for a desired impact. The reader should be able to extend the guidance of the book to both personal and other professional decisions.

Rationale for the Book

Optimization is ages old. Prior to calculus, optimization was empirical, guided by heuristics and experience. Improvement was by a direct search, one that only uses the OF value and not the derivative information. The mathematics of calculus, however, created a new era, and Simpson (1740) extended Newton's root finding (1685) to the derivative of the function to find the optimum. Cauchy's sequential line search appeared in 1847. Modest technique progress continued through about 1944, at which time the power of the digital computer led to both practical applications and an explosion in the development of diverse techniques. In 1955 Levenberg blended "Newton's" with incremental steepest descent to spawn many approaches to using both the gradient and Hessian to guide sequential improvements in the trial solution. Advances continued to capitalize on computational power. Then the 1960s gave rise to mimetic multiparticle algorithms and multi-objective applications.

In that brief historical overview, gradient-based techniques replaced the precalculus era direct search techniques. Gradient-based techniques remain the mainstay of texts. However, the power of the digital computer is permitting new direct search techniques such as particle swarm, genetic programming, and leapfrogging to outperform gradient-based techniques on nonlinear and stochastic applications with discontinuities—today's relevant problems. One reason for the book is to promote the use of the new direct search techniques.

Most books on engineering optimization focus on the optimization algorithms. However, most users will not write the code; they will buy it. Of more need for a user is instruction on how to create an appropriate OF, how to choose DVs, how to identify and incorporate constraints, how to define convergence, and how to determine the number of independent starts needed to ensure that the global is found. This book seeks to fill in those application essentials.

I developed and used optimization throughout my initial 13-year career in the industry. However, my college preparation for the engineering career did not teach me what I needed to know about how to create and evaluate optimization applications. I recognized that my fellow engineers, regardless of their *alma mater*, were also underprepared. We had to self-learn what was needed. Recognizing the centrality of optimization to engineering analysis, I have continued to explore its application and technique development in my subsequent 30-year academic career.

This book is based on college and professional training courses that I've offered and is a collection of what I consider to be best practices in engineering optimization. It includes the material I wish I had known when starting my engineering career, and I hope the book is useful for the readers.

Target Audience

The examples and discussion presume basic understanding of engineering models, statistics, calculus, and computer programming. This book will have enough details, explicit equation derivations, and examples to be useful either as an introductory course or for self-study.

The book is aimed at a bachelors, or higher, graduate of engineering or a mathematical science (physics, chemistry, statistics, computer science), who has had an undergraduate course in calculus, mathematical models, statistics, and computer programming. However, upper-level undergraduates have been successful in my course. The reader could be either a student or a practicing engineer or scientist.

Presentation Style

In my experience, students cannot grasp the depth of one topic in isolation of the others. Depth in understanding two-dimensional (2-D) OF surface features is required to be able to relate to N-D issues. An initial understanding of the optimization algorithms is required to be able to set up the application OF and DVs. An understanding of the application is required to be able to choose the appropriate convergence criterion and thresholds. Accordingly, I start the book with elementary versions of each of the aspects of optimization in one-dimensional applications, demonstrate the whole of the applications on several case studies to reveal issues, then return to each item in more depth, and demonstrate the improvements of the second-level techniques in 2-D applications, discuss issues, and then extrapolate to N-D implementations.

I offer the reader with software (and access to computer code through my website www.r3eda.com) to execute key operations. Although there are many strong programming environments, the code is written in Excel VBA (Visual Basic for Applications), which is widely accessible. The book includes a listing of the code for the techniques.

A unique feature of the book is the "takeaway" sections associated with the chapters, which summarize the methods of choice using a practical, applications, utility perspective. This is intended as a user's how-to book grounded in fundamentals, not as a math-analysis-of-the-fundamentals book. However, relevant properties of the optimization problems will be mathematically analyzed, the optimization algorithms will be developed from theory, propagation of uncertainty will be related to choices, and the book contains some proofs related to surface analysis and OF transformations.

Acknowledgments

I am especially grateful to Junyi Li, Balashakar (Soundar) Ramchandran, Ming Su, and Scott Essner, who, as graduate students, brought and revealed to me essential optimization concepts. I am also grateful to many other graduate students who worked with me in exploring and developing understanding of optimization. In reverse chronological order, thanks to Anand Govindarajan, Upasana Manimegalai-Sridhar, Thomas Hays, Haoxian Chen, Chetan Chandak, Prithwijit Ghoshal, Judson Wooters, Gaurav Aurora, Samuel O. Owusu, Venkat Padmanabhan, Solomon Gebreyohannes, Preetica Kumar, Naggappan Muthiah, Nitin Sharma, S. Samir Alam, Qing Li, Jing Ou, Siva Natarajan, Sandeep Chandran, Ganesh Narayanswamy, Mahesh S. Iyer, R. Paul Bray, Abrahao Naim Neto, Ganesh Venkataramanan, Songling Cao, Jin Yong Choi, and Vikram Singh.

I consider myself very fortunate to have been granted the health and ability to enjoy, and now to relay, many experiences and a developing understanding related to optimization. I count my industrial application experience to be as valuable as my academic research investigations. Both are essential for the creation of this book.

Other authors have provided books that have been very valuable to my understanding. I recommend these publications: Ravindran, Ragsdell, and Reklaitis, *Engineering Optimization—Methods and Applications*, Wiley, 2006; Beveridge and Schechter, *Optimization: Theory and Practice*, McGraw-Hill, 1970; Edgar, Himmelblau, and Lasdon, *Optimization of Chemical Processes*, McGraw-Hill, 2001; Snyman, *Practical Mathematical Optimization*, Springer, 2005; Hillier and Lieberman, *Introduction to Operations Research*, McGraw-Hill, 2001; Nocedal and Wright, *Numerical Optimization*, Springer-Verlag, 1999; and Rao, *Engineering Optimization: Theory and Practice*, 4th Edition, Wiley, 2009.

As a professor, funding is essential to enable research, investigation, discovery, and the pursuit of creativity. I am grateful to both the Edward E. & Helen Turner Bartlett Foundation and the Amoco Foundation (now BP) for funding endowments for academic chairs. I have been fortunate to be the chair holder for one or the other, which means that I was permitted to use some proceeds from the endowment to attract and support graduate students who could pursue ideas that did not have traditional research support. This book presents many of the techniques explored, developed, or tested by graduate students. Similarly, I am grateful to the National Science Foundation Industry–University Cooperative Research Centers Program and to a number of industrial sponsors of my graduate program who recognized the importance of applied research and its role in workforce development. These include Amoco, Arco Exploration & Production, Aspen Technologies, Cargill, Chevron Phillips, Diamond Shamrock, Dow Chemical, ExxonMobil, Fina, Gensym, Hoechst Celanese, IMC Agrico, Johnson Yokogawa, LAM, Monsanto, Pavilion Tech, Phillips 66, Tennessee Eastman, Texas Instruments, Union Carbide, and Valero.

Career accomplishments of any one person are the result of the many people who nurtured and developed the person. I am of course grateful to my parents, teachers, and friends, but mostly to Donna, who has encouraged and enabled my work initiatives (really just play and hobbies), as well as appropriately guiding my growth.

Nomenclature

Acronyms

ANOFE	average number of function evaluations
ARIMA	autoregressive integrated moving average
ARMA	autoregressive moving average
CDF	cumulative distribution function
CHD	cyclic heuristic direct
CSLS	Cauchy's sequential line search
D	number of decision variables, the optimization dimension
DCFRR	discounted cash flow rate of return
DE	differential evolution
DMC	Dynamic Matrix Control
DV	decision variable(s), or its value(s)
DV^*	optimum value of the decision variable(s)
EC	equal concern factor (the weights relative importance of additive values)
EHS&LP	Environmental, Health, Safety, and Loss Prevention
FL	fuzzy logic
EPA	Environmental Protection Agency
FOPDT	first-order plus deadtime
GA	genetic algorithm
GRG	generalized reduced gradient
HJ	Hooke–Jeeves
ISD	incremental steepest descent
J	objective function
K.I.S.S.	keep it simple and safe
LF	leapfrogging
LHS	left-hand side
LM	Levenberg–Marquardt
LTROA	long-term return on assets
NM	Nelder–Mead simplex
NN	neural network
NOFE	number of function evaluations
NPV	net present value

NR	Newton–Raphson
$O(x)$	on the order of the value of x
ODE	ordinary differential equation
OF	objective function, or its value
OF*	optimal value of the objective function
OSU	Oklahoma State University
PBT	payback time
PDE	partial differential equation
pdf	probability density function
PNOFE	probable number of function evaluations
PS	particle swarm
PSO	particle swarm optimization
RHS	right-hand side
rms	root-mean-square value
SOPDT	second-order plus deadtime
ST	subject to, also S.T.
SQ	successive quadratic
SS	steady state
SSD	sum of squared deviations, alternately just sum of squares
TS	trial solution, a set of DV values
TS	transient state
TSP	traveling salesman problem
w.r.t.	with respect to

Definitions

a posteriori	A Latin term, indicating "after it has been done." A choice made after event outcomes have been observed.
a priori	A Latin term, indicating "before doing it." A choice made before the event, from earlier experience or understanding, not after observation.
Bottom line	The reveal of comprehensive issues.
CDF(OF)*	The cumulative distribution function is a useful tool for visualizing the probability of finding the global optimum and the certainty of its location.
Constraints	These are what should not, or must not, be violated. Some constraints must not be violated because the violation may be catastrophic, like an implosion. Alternately, if other constraints are violated, there might just be a modest penalty. Constraints can be inequality or equality. If equality relations in the DVs, they can be used to reduce the number of DVs if the structure permits solving for one DV given the others. "Hard" constraints are of the "must not" violate category and limit DV TS choices. "Soft" means that the objective function is given a penalty for the

	constraint violation, which softens the base of the cliff with a curve, which converts the surface to one that is analytically tractable.
Convergence and stopping criteria	Something has to indicate that either the optimum is found or the optimizer is hopelessly lost and needs to be stopped. *Convergence* means that the DV* values have been found, that they will be effectively unchanged in sequential trial solutions, and that the optimizer can be stopped, claiming convergence or close enough proximity to the ideal DV*. The tolerance, or precision, or accuracy could be specified on the change in the DVs or the change in the OF value, or on any of many more complex relations such as the maximum impact on the OF value due to range of the DVs, or when the improvement in the OF is inconsequential to the uncertainty on the OF. The optimizer iterations are stopped when convergence is claimed; but if it has been running excessively without convergence, it could be stopped and "no convergence" reported. *Stopping* criteria could be on the number of iterations, run time, or such indicators that more computation will be futile. Both the criteria and the threshold values for both stopping and convergence action are user choices and are critical to the validity of the reported solution.
Convex	The feasible space boundary is always open to, curved to enclose, the feasible DV space. Take any two points in feasible DV space and draw a straight line between them. If every point on the line is within the feasible space, for any pair of feasible points, then the OF is convex. For instance, a rectangle, circle, and ellipse enclose feasible spaces that are convex. However, a kidney bean or a boomerang shape is not. And a circular infeasibility region within a square makes the application non-convex.
Customer	A person, or entity, or group that has a legitimate claim related to desirability or undesirability of your application. Alternately termed a stakeholder.
Derivative evaluation	Gradient-based optimizers require the values of the derivatives and/or second derivatives. If the OF is a relatively simple function, then the derivative formulas can be analytically derived, and values explicitly calculated. But in meaningful applications the OF is usually a procedure, in which case derivatives need to be estimated numerically. Should one use the central difference? Or should one use a forward or backward approximation? The central difference requires an additional OF evaluation but provides a better estimate of the true derivative. But is the work worth the benefit? A forward Δx_i difference might cross over a constraint into an infeasible region. What then? What should be the Δx_i value? Too small and it will cause truncation error. Too large and the numerical procedure will misrepresent the local surface.
DV (decision variables)	These are what you can change to improve the OF value. There may be several or just one. The DVs might be independent or interrelated (reflux must be less than vapor boilup—constrained by each other). They may have rate-of-change constraints. The DVs might be scheduled

with another variable, as in the Goddard problem of choosing rocket thrust to maximize height (thrust could be scheduled with time, height, or remaining fuel). The DVs might be coefficients in an equation that relates the process inputs to the state variable that they are scheduled onto. The DVs might be a continuum variable, an integer or discretized variable, or a class or category. The DV choice is critical. It must match the customers' perception of the flexibility that you have within the application context. Further, your selection of number of stages in scheduling or functional relations in converting DVs to equation coefficients has a substantial impact on both the DV* and OF* values and computational work.

Evaluation of optimizers We want a high probability that a procedure will find the global optima. And we want to find OF* with fewest number of function evaluations, greatest robustness to surface aberrations and least user involvement and dependency on user choices. I'll use number of function evaluations (NOFE) as a measure of work and combine this with probability of finding the global optima as an indicator of success to get the probable NOFE (PNOFE) as a primary evaluator of optimizer performance.

Feasible A DV value that neither violates a hard constraint nor leads to a computer execution error (overflow, divide by zero, log of a negative, subscript out of range, etc.).

Givens These diverse aspects are the basis, assumptions, conditions, procedures, models, etc. in the analysis. Consider these givens in a problem statement, "The glass is half full of water, how long does it take to evaporate?" Is the temperature exactly known and constant over the evaporation period? Is the relative humidity exactly known? How can a glass be exactly half filled? Are there any air currents over the glass? What is the impurity content (salts, dissolved CO_2) of the water? What model is exactly right for the vapor–liquid equilibrium, equation of state, or meniscus effect calculations? There is uncertainty in all of these givens. The givens are not truths. They are just approximations. The uncertainty in the givens has an impact on the application solution, and the consequential uncertainty in the DV* and OF* should be acknowledged.

Greedy algorithm Take the local best action. Only look at the current situation, not the future implications of the action. For example, your car may be in the garage, but you cannot walk directly toward it because of the wall. You need to take a longer constraint free path through the door to the garage, which also includes the stage of opening the door. In the traveling salesman problem, a greedy algorithm is the heuristic of going to the next closest city.

Initialization This refers to both the initial trial solution and the optimizer parameter values. If you start in St. Louis, the downhill path takes you to New Orleans at sea level. But if you start west of Las Vegas, downhill moves you to Death Valley at an elevation below sea level. In a multi-optima surface, the optimizer solution depends on where you start. The exact path an optimizer travels also depends on coefficients in the optimizer that control step size and on other user choices such as whether to use a central difference or backward difference numerical estimate of the derivative. Such coefficients will affect both the number of function evaluations to reach the optimum and the TS value when convergence is claimed.

Number of independent trials

If local optima, constraints, ridges, etc. can trap a trial solution, then run multiple optimizer trials from independent DV initializations. The more independent trials you run, the greater is the probability that you'll find the global.

Maximum or minimum

Since MAX J = OF provides the same DV* as MIN J = –OF, it is irrelevant whether the optimizer is set to minimize (find lower values of the negative OF) or maximize (find higher OF values). Following convention, without loss of generality, this book seeks to minimize.

Model

There must be mechanisms for determining the OF value and constraint conditions from the DV TS values. The models have many choices (ideal gas law or virial equation of state, Hooke's law or a viscoelastic relation). And the models need to include how goodness is evaluated, which means the models reflect human values as well as engineering analysis. The mechanism might include experimental testing, followed by constructing assessment metrics.

Multiple objectives

Many applications seek a balance of opposing ideals (e.g., profit vs. waste, personal image vs. cost, perfection of fit vs. number of coefficients). There are many diverse metrics of application desirability. Most optimizers need a single OF value and seek to optimize it. In such cases the multiple objectives need to be combined into a single function. Usually the separate objective functions are added and weighted to unify dimensions and scale to an appropriate balance. For this approach, I like the equal concern approach for generating weighting factors. But sometimes the diverse OF criteria are multiplied. Alternately, the OF functions are kept separate, and in a Pareto analysis, all the non-dominated TSs are kept as the Pareto optimal set of possibilities for a user to select which is appropriate for a particular situation.

Non-convex

The feasible space boundary is not always open to, curved to enclose, the feasible DV space. Take any two points in feasible DV space and draw a straight line between them. If any point on the line is outside of the feasible space, for any pair of feasible points, then the OF is non-convex. For instance, a rectangle, circle, and ellipse enclose feasible spaces that are convex. However, a kidney bean or a boomerang shape is not. And a circular infeasibility region within a square makes the application non-convex.

OF (objective function)

This is the procedure for determining the measure of goodness, the assessment of desirability. Defining it is a critical step, because the metrics you use to assess value determines the solution you get. You need to be sure that you are defining the right metrics of goodness. For instance, in an economic optimization for a process design, with pipe diameter as the DV, you get different answers if the objective is to minimize capital, expenses, net present value, or probable risk. You also get different answers if the business model is for a 5-year period or for a 20-year period. As another example, in placing surveillance devices, you get different answers if the objective is to maximize observation of the boundary or to maximize the minimum coverage of the internal area. Also, often there are competing issues that need to be balanced with appropriate weighting. In all, it is essential to use metrics, calculation models, and weighting that reflect the customers' values and context for use.

OF transformations For an OF with positive values, the minimum of the log of an OF has the same DV^* values as the minimum of the original OF. DV^* is unchanged with an OF that is transformed with a positive definite function. Also, minimizing the OF has the same DV^* as maximizing the negative of the OF. OF transformations can be conveniences in scaling for display, converting max to min, or simplifying of the concepts. They could convert an OF to a functional form that is more quadratic-like, which may improve speed of convergence for Newton-like algorithms, but they have no impact on speed in direct search algorithms. It does not affect the presence of local optima. Transformations might distort OF-based convergence criteria. As a caution, functional transforming individual elements in the OF or changing the weighting of individual terms will change the DV^* value.

OF value (objective function value) This is the value that your procedure returns.

Optimizer This is the procedure for iteratively improving the TS toward DV^*. If an analytical method is possible, the solution directly calculates DV^*. However, usually, practical applications are analytically intractable, and an iterative approach is needed. There are two general approaches. Gradient-based approaches use the local surface slope to indicate the direction of steepest descent and move TS in that direction. Incremental steepest descent and Cauchy's are of this type. Newton-type approaches use the surface slope and also the second derivatives to adjust the leap-to TS. They are derived by a truncation of a Taylor series model of the surface to quadratic terms. Levenberg–Marquardt combines both incremental steepest descent and Newton's. Successive quadratic generates a local quadratic surrogate model of the surface and then leaps to the DV^* of the surrogate model. Although SQ explicitly uses neither gradient nor Hessian elements, it is equivalent to Newton's in performance. Both are based on a local quadratic model of the surface. Surface aberrations of flat spots or discontinuities are a problem for those approaches that are based on a rational model of a continuous and quadratic surface. Contrasting the gradient-based approaches are the direct search approaches that do not use suppositions of the surface structure. Direct searches only consider the OF value and constraint "PASS" or "FAIL" status and move the TS using heuristic or stochastic rules. Cyclic Heuristic, Hooke–Jeeves, Nelder–Mead simplex, particle swarm, genetic algorithms, differential evolution, and leapfrogging are direct search algorithms. Because these do not leap-to the surrogate model optimum, but approach it gradually, one would think that the gradient-based techniques are faster. However, to numerically evaluate the local surface in a 3-DV problem, SQ and Levenberg–Marquardt each need 10 local OF evaluations; the direct search approaches use many fewer and in many cases converge with fewer NOFE. Further, the direct searches are more robust to surface features. Some older direct search algorithms (HJ, NM) are a single trial solution search (one initial guess moves locally downhill). However, the more recent ones (PSO, GA, DE, LF) use multiple "simultaneous" trial solutions, and the collective evidence of the surface provides a "global" view of where all should go.

Risk   A penalty for a possible undesired event. Risk is the probability of the undesired event times the cost or equivalent impact of the event.

Stakeholder   A person, or entity, or group that has a legitimate claim related to desirability or undesirability of your application. Alternately termed a customer.

Surface aberrations   It would be nice if the optimization application created a smooth TS path to the bottom of the valley, so that gradient-based approaches could guide the DV TS to the optimum. But applications have a variety of surface aberrations that confound a search. It is easiest to visualize surfaces as a 3-D response to 2 DVs and to use familiar terms to describe surface features; but the concepts and issues are scalable to N-D applications. Flat sections in the surface arise because of discretization in DV size or category or OF rank or category. On a flat section the optimizer cannot determine what direction is down. Cliffs and slope discontinuities arise with constraints or with switches in units or modes (turbulent to laminar, three to four parallel units), which may be expressed as IF-THEN conditionals in the model. On either side of a steep valley, the slope on the walls points across the valley, and the optimizer will zigzag across the valley, not move along the ridge. Multiple optima mean that there are multiple local minima; and at any one, all directions seem worse. These often arise in nonlinear models, and you don't want an optimizer to get stuck in a local optimum and report that value without acknowledging the other locations. The derivative is zero at saddle points, maximum, and minimum, and some gradient-based optimizers seek any such point, not just the optima. Stochastic surfaces are the result of experimental or Monte Carlo simulation responses. In a stochastic surface, replicate DV trials do not return the exact same OF value. If a new OF value is better, does that mean that the TS moved in the right direction or simply that the vagaries of noise made it seem better?

Surrogate functions	These are useful for optimizer testing. They are simple to code, quick to compute, and reveal the issues that serious applications would present.
TS (trial solution)	A set of values for the DVs. A "guess" at the DV^* values. Optimizers progressively, iteratively, seek TS values that lead to a better OF.
Uncertainty	There are always "givens" in the application. Givens are the specifications that you are given by the problem statement. Consider this statement: "The weight of the truck is 1.23 tons and the tires are inflated to 55 psig. What is the optimum speed?" Will the tire pressure always be exactly 55 psig? Will other conditions be known with certainty? What about road elevation, barometric pressure, air relative humidity, fuel BTU content, wind, temperature, and so on? The DV^* value may be exactly right for a particular set of givens, but might not be the overall best for the range of possible conditions and influences. One should investigate the impact of uncertainty of the "givens" on the DV^* value.

Symbols

$a, b, c, d, \ldots$	model coefficients, scalar variables
$w, x, y, z, \ldots$	model variables, scalar variables
$\tilde{y}$	modeled response value (as opposed to the measured value)
x	a vector of x-elements
$\mathbf{M}$	a matrix of m elements
$\mathbf{H}$	the Hessian matrix of second-order partial derivatives
$\forall$	for all
Δ	delta, a small increment
∇	gradient, or Jacobean, the $\partial/\partial x$ operator
$:=$	the assignment statement in a computer program, to differentiate from an algebraic $=$
$\lvert x \rvert$	absolute value of a scalar variable
$\lVert x \rVert$	magnitude of a vector $\sqrt{\sum x_i^2}$
$\parallel$	parallel
$\perp$	perpendicular
$!$	factorial
$O(3)$	on the order of three

About the Companion Website

This book is accompanied by a companion website:

www.wiley.com/go/rhinehart/engineeringoptimization

The website includes:

- Techniques
- Excel VBA programs
- Software

Section 1

Introductory Concepts

These three chapters introduce terminology and concepts of optimization and provide an illustration of the diversity of applications and associated issues and a self-validation guide for the person using this book for self-study.

Engineering Optimization: Applications, Methods, and Analysis, First Edition. R. Russell Rhinehart.
© 2018 R. Russell Rhinehart. Published 2018 by John Wiley & Sons Ltd.
Companion website: www.wiley.com/go/rhinehart/engineeringoptimization

1

Optimization

Introduction and Concepts

1.1 Optimization and Terminology

Optimization is a procedure for seeking the best choices, and there are essential elements to the procedure. First, there must be criteria for rating choices. This can be alternately stated as a method for determining the value of a metric of goodness, the assessment value of how a person balances desirables and undesirables. One must have this assessable or quantifiable metric of desirability to be able to choose the best. Also, there must be a relation between choices and the evaluation of the outcome. This relation can be either a model or an experiment. The procedure (equation, algorithm, experimental method) for determining the value of the desirability metric is called the objective function (OF). The choices that you can change to improve the OF value are called the decision variables (DVs).

You might want to minimize (costs, expenses, risk, etc.) or maximize (profit, reliability, probability of success, etc.). Optimization will seek DV values that lead to the optimum, either minimum or maximum.

A trial solution (TS) is a particular DV choice. It might not be the optimum. The optimum is denoted as DV*. For simple, idealized applications, one can determine the exact value of DV*; but, for most applications, this is an ideal concept, like the value of π. You'll get close enough to the true value for a particular need and call the close enough TS the DV*.

There are usually several opposing concepts in the comprehensive statement of an OF. For example, one may wish to minimize cost of food, which leads to the obvious solution—don't consume any. Acknowledging this extreme reveals an opposing concept that needs to be considered as a comprehensive assessment of desirability, which is acquiring adequate nourishment. Further consideration of the situation may also lead one to acknowledge that joy of eating is important and including occasional treats is part of a healthy life. So, the opposing ideals might be to balance cost with nourishment and joy.

Once the equation is obtained, using optimization to determine the optimum is fairly easy. The difficulty in optimization is usually related to obtaining the equation that provides a comprehensive and complete statement of the context.

As another example of opposing ideals, consider how much insulation should be placed in a house. One ideal would be to minimize cost. Calculating the cost of insulation as price per volume times area times thickness results in the model $C = pAt$. If the objective is to determine the thickness that minimizes installed cost, capital, the obvious solution is $t = 0$. Don't install insulation. This extreme, however, might reveal that zero insulation leads to energy losses that are also important. Annual value of energy lost, an expense, might be simply modeled as $E = eA(T_{house} - T_{outside})/(r + kt)$, where k is the specific factor for the insulation, r is the insulation capacity of the walls, and e is the unit cost of energy.

Engineering Optimization: Applications, Methods, and Analysis, First Edition. R. Russell Rhinehart.
© 2018 R. Russell Rhinehart. Published 2018 by John Wiley & Sons Ltd.
Companion website: www.wiley.com/go/rhinehart/engineeringoptimization

Looking at an N-year horizon, the total cost is the initial capital and the N annual expenses. $J = C + NE = pAt + NeA(T_{house} - T_{outside})/(r + kt)$. Now the opposing trends are visible. Increasing t increases the first term but decreases the second.

Optimization is a structured, rational search through the DV "space" from an initial TS toward DV*. Although for simple applications the explicit solution to a simple equation can provide the DV* answer, optimization is usually not a magical single leap to the optimum, but an iterative progression. The locus of trial solutions through DV space is termed the path. Optimization desires to find DV* with minimal computational or experimental burden (minimum number of TS evaluations along the path), minimum *a priori* knowledge or human involvement, and maximum assurance that the global optimum has been found and that DV values are close enough to DV*.

Constraints impact both the DV* value and the path to get there. Some constraints are mathematical (such as divide by zero, or square root of a negative number), some are computational (overflow, or subscript out of range), some are physical (composition cannot be >100% or <0%, or keep temperature above freezing point), some are environmental health safety and loss prevention related (keep composition less than lower explosive limit, or keep pressure above vacuum, or don't let tank overflow), some are legal/political (don't violate contract specifications, or stay beyond the 200-mile offshore limit), etc. Some constraints are "hard," meaning there is no tolerance, and absolutely do not violate the constraint. Some are "soft," wherein small violations are permitted, but with a penalty. Hard constraints complicate the TS search sequence. Soft constraints do not limit the TS values, but the user needs to assess the magnitude of the penalty relative to all OF issues.

The canonical statement of an optimization application is

$$\begin{array}{c} \min \\ \{\mathbf{DV}\} \end{array} J = \text{OF}(\mathbf{DV})$$
$$\text{S.T.:} \quad g(\mathbf{DV}) < 0 \tag{1.1}$$

If the OF is an equation, for example $y = 3 - x + x^2$, and the DV is the variable x, and there are no constraints, the statement becomes

$$\begin{array}{c} \min \\ \{x\} \end{array} J = 3 - x + x^2 \tag{1.2}$$

For this, an analytical solution is simple. Set $dJ/dx|_{x^*} = 0$, which is $0 = 0 - 1 + 2x^*$, and solve. Here, the optimum value of x, $x^* = 1/2$, and with that value of x^*, OF* = 2.75.

For applications, however, the OF is usually not a simple equation, but a complicated procedure (often including loops, look-up tables, and internal root finding), which would incorporate design and economics and other sustainability metrics. Although simple examples like the one aforementioned are instructive (and intellectually stimulating as they permit closed-form mathematical analysis), the practical applications are more complicated. This book is about the practical applications, although often it will use simple relations to facilitate reader understanding.

1.2 Optimization Concepts and Definitions

We usually optimize for the enterprise. This may be the employing organization, a volunteer team, or humankind; and each is an assembly of humans with diverse views on the desirables. Often, the person doing the optimization has a limited view of the context and the perspective that other customers and stakeholders have. Accordingly, clarity of terminology is essential to ensure that the person doing the optimization is making choices and assumptions that are compatible to the enterprise needs. Here is some essential terminology:

Objective: This is the thing or issue that you wish to maximize or minimize—the indicator of goodness or desirability. Usually there are multiple objectives. Perhaps profitability is an objective, but other associated measures of desirability would include resource conservation, safety, sustainability, compliance to contracts, etc. Perhaps health, lifelong joy, bang for the buck, risk, and effort are the objectives. Although those concepts are understood, profitability, health, safety, risk, joy, and effort are vague statements. There are many ways to interpret each of these abstract concepts. The user must provide specifics. Further, it is essential that the objective is comprehensive. If it only represents a portion of the issues that stakeholders will use, the solution will not be quite right.

Objective function (OF, J): This is the procedure to measure desirability (goodness) or undesirability (badness) associated with the thing or issue that you wish to maximize or minimize. If profitability is the objective, then perhaps DCFRR (discounted cash flow rate of return), or PBT (payback time), or NPV (net present value), or LTROA (long-term return on assets) may be the metric used to assess profitability. Each of these balances profit and expenses with investment. Why is there not one measure of profitability? Because, for any one objective there are many possible ways to measure desirability, and depending on the enterprise context, one profitability index may be more appropriate than another. The OF (objective function) is often called a "cost function" representing an understanding that minimizing resource consumption is the only objective, but I prefer to not use that simplistic terminology. Classically, the symbol for the OF is the capital letter *J*. The OF must be concrete, it must provide a value, and it must be quantifiable.

Objective function value (OF, J): This is the value of the objective function, this is what optimization seeks to make as large as possible or as small as possible. OF value examples could include the following: LTROA is 23.7%, or PBT is 1.7 years, or fuel consumption is 42.1 miles/gal.

Decision variables (DV): These are the choices you can make, the values or decisions you can choose to obtain the optimum *J*, or the things that you can change. The DVs are the independent variables that can be changed, and the OF is the response. Some values you cannot change, or so it first seems. However, a second look at the application often will reveal alternate variables that you can change. You need to differentiate between the two and properly identify the degrees of freedom within an application context.

Optimum values (DV*, OF*): These are the DV values that either maximize or minimize the OF. The optimum values are represented as DV* and OF*.

Trial solution (TS): This is a possible value of the DVs. It might not be the optimum value. It is often the next guess as the DV values are being changed to improve the OF.

Model: This is the relation that permits calculating the OF value from a TS of the DV values. The DV may or may not be explicitly shown in the OF. The model is usually not a simple equation. More so it is a procedure. Sometimes experimental results will be used to determine OF values rather than a model.

Constraints: These are limits on the DV values or on other associated variables. For example insulation thickness must be nonnegative, $t \geq 0$. However, constraints can be more complicated, for example, structurally, the ceiling might only be able to hold a certain weight, so $wAt <$ limit. Constraints can be on a rate of change, on a future value, on an executable operation, or on any number of aspects of variables. Consider wanting to minimize the number of trips to carry 5 gal of water across the street. The answer is one trip. But if the bucket has a 3 gal capacity, it imposes a constraint on the amount that can be carried in one trip, and the solution is two trips. Constraints can be as influential on the answer as the objective function.

Method: This is the procedure used to find DV* values. How will you move from an initial TS toward DV*? How will you specify an initial TS? The answers to such questions are part of the method of optimization.

Convergence criterion: Optimization is an iterative sequential procedure to progressively move the TS toward DV*. Probably, it never gets exactly to DV*. When it is close enough, stop the search and claim that DV* has been found. The criterion to determine that the optimizer has found a TS close enough to the optimum to stop the search and call it DV* is termed the convergence criterion.

1.3 Examples

Here are four diverse application examples that reveal the meaning of the several definitions given so far.

Example 1   A one-line function or equation

$$y = 8x - 2x^2 + 4 \tag{1.3}$$

The objective is to find the value of x that maximizes the value of y. This can be graphically illustrated in Figure 1.1.

$y = $ OF (often called dependent response)
$x = $ DV (often called independent input)

Using the analytical procedure of setting the derivative of the function to zero and then solving for the DV* value, we easily obtain

$$x_{\text{optimum}} = 2 = x^* \tag{1.4}$$

$$y(x_{\text{optimum}}) = 12 = y^* \tag{1.5}$$

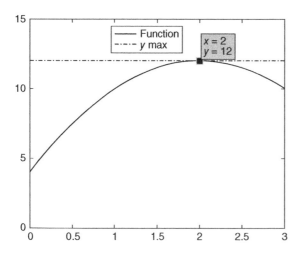

Figure 1.1 Illustrating a function with a maximum.

However, most important applications are not a simple one-line equation that permits an analytical derivative. Although it is an instructive example, the method is limited to ideal applications.

Example 2   Economic Optimization
Find the optimal thickness of insulation in an attic. The initial capital cost is the cost of insulation as price per volume times area times thickness, $C = pAt$, and thickness must be nonnegative, so there is a constraint $t \geq 0$. There may be alternate constraints related to total weight of insulation that the ceiling can support, $\rho At < W$. The annual value of energy lost, an expense, might be simply modeled as $E = eA(T_{house} - T_{outside})/(r + kt)$, where k is the specific factor for the insulation, r is the insulation capacity of the walls, and e is the unit cost of energy. Ignoring the time value of money and looking at an N-year horizon, the total cost is the initial capital and the N annual expenses. $J = C + NE = pAt + NeA(T_{house} - T_{outside})/(r + kt)$. The optimization statement then becomes

$$
\begin{aligned}
&\min_{\{t\}} J = pAt + \frac{NeA(T_{house} - T_{outside})}{r + kt} \\
&\text{S.T.:} \quad \begin{aligned} &t \geq 0 \\ &\rho At < W \end{aligned}
\end{aligned}
\tag{1.6}
$$

Analytically, this admits a solution. Set $dJ/dt = 0$, and solve for insulation thickness.

$$
t^* = \frac{1}{k}\left[\sqrt{\frac{p}{kNe(T_{house} - T_{outside})}} - r\right]
\tag{1.7}
$$

One needs to check that the value of t^* is within constraints.

The role of the coefficients in the equation could be questioned. Are they "givens" or are they DVs? The parameters associated with the insulation are p and k, but an alternate insulation material choice could permit alternate values. Further, the value for r is related to other construction choices such as ceiling material or roofing material. Is the designer free to change these or not? Finally, the equations reveal that the value of energy lost is related to the house temperature setting. Is an option for minimizing costs to change the house temperature setting? If yes, then perhaps there needs to be a comfort penalty for deviations from the nominally ideal 72°F (22°C). Characteristically, the penalty scales with the square of the deviation from desired, but there needs to be a weighting factor that makes the temperature deviation equivalent to the cost. With such, the optimization statement has evolved to

$$
\begin{aligned}
&\min_{\{t, T_{house}\}} J = pAt + \frac{NeA(T_{house} - T_{outside})}{r + kt} + \lambda(T_{house} - 72)^2 \\
&\text{S.T.:} \quad \begin{aligned} &t \geq 0 \\ &\rho At < W \end{aligned}
\end{aligned}
\tag{1.8}
$$

Example 3   Least Squares Regression
Find best line that goes through data. In Figure 1.2 the data are represented as circles, and a not-good model is indicated by the dashed line. The best model is represented as the solid line. How is the best model chosen? In least squares regression, it is the line that minimizes the sum of squared deviations between data and model-dependent variable (vertical deviations as represented by the vertical lines).

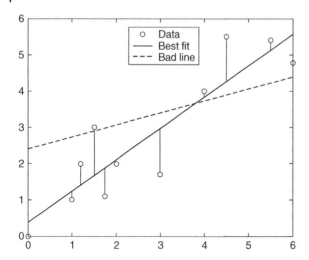

Figure 1.2 Regression illustration.

In classic regression, we evaluate best as minimizing sum of squared deviation:

$$\text{OF} = \text{SSD} = \sum_{i=1}^{N}(y_i - \tilde{y}(x_i))^2 \tag{1.9}$$

where y_i and x_i are data pairs, N is the number of data points, and $\tilde{y}$ is a model value.

The equation of a line is $\tilde{y} = a + bx$, which is linear in both of the coefficients. The objective is to find the values for a and b that minimize the OF. Here, the decision variables are the model coefficients $\{a, b\}$, and the optimization application can be stated as

$$\min_{\{a,b\}} J = \sum_{i=1}^{N}(y_i - \tilde{y}(x_i))^2 \tag{1.10}$$
$$\text{S.T.:} \quad \tilde{y}_i = \tilde{y}(x_i) = a + bx_i$$

The statement previously is not the answer; it is a canonical statement of the problem. Any set of model coefficients $\{a, b\}$ are feasible. The set that results in the not best dashed line in Figure 1.2 is a trial solution, a possible DV choice. The optimum DV set that results in the best line in the figure is indicated as DV*, $\{a^*, b^*\}$. The classical analytical method to solve a least squares regression application with a model that is linear in coefficients is to take the derivative of the OF w.r.t. each model coefficient, set each derivative to zero (the normal equations), and solve the simultaneous linear equations.

Substituting the model into the OF, $J = \sum_{i=1}^{N}(y_i - (a + bx_i))^2$, taking the derivatives, and setting them to zero,

$$\frac{\partial J}{\partial a} = 0 = -2\sum_{i=1}^{N}(y_i - a - bx_i) \tag{1.11a}$$

$$\frac{\partial J}{\partial b} = 0 = -2\sum_{i=1}^{N}x_i(y_i - a - bx_i) \tag{1.11b}$$

Rearranging, the normal equations are

$$(N)a + \left(\sum x_i\right)b = \left(\sum y_i\right) \tag{1.12a}$$

$$\left(\sum x_i\right)a + \left(\sum x_i^2\right)b = \left(\sum x_i y_i\right) \tag{1.12b}$$

Since each term in parentheses has known values from the data, the normal equations result in two linear equations in two unknowns, variables a and b.

This was a classic least squares approach with a model that is linear in coefficients. A two-coefficient model results in two linear equations. This procedure is scalable. An M coefficient model that is linear in the coefficients results in M normal equations, linear in the M coefficient unknowns, requiring linear algebra techniques to obtain the model coefficient values.

However, nonlinear models are not always amenable to this analytical approach. The derivative might not have a tractable analytical expression, and if it does, the solution to the resulting equations will likely require iterative nonlinear root-finding techniques.

Example 4 Best Path or Sequence
Find the best path to visit major baseball parks in the Northeastern United States. Should this be least miles, least time, lowest cost, or most scenic? Start at home, and return to home. One possible sequence is to go to Chicago (C), then Boston (Bo), then New York (N), and so on.

Give a number to each city in sequence. Assign home as points 1 and 8, start and finish. For example, here is one plan, one sequence, one path:

H	C	Bo	N	Pit	Phi	Bal	H
1	2	3	4	5	6	7	8

But this sequence might have fewer total miles:

H	C	Bo	N	Pit	Phi	Bal	H
1	7	6	5	2	3	4	8

If minimizing distance for the total trip is the objective, using l_{ij} = distance between city i and j, the optimization application can be stated as

$$\min_{\text{sequence}\{2,3,\ldots,7\}} J = \sum_{i=1}^{N} l_{ij} \tag{1.13}$$

This is a classic example of the traveling salesman problem (TSP), which is a key element in scheduling planning, sequencing, and logistics. The example reveals several issues. One is that the decision variables are nominal (or category variables), indicated here as integers for the sequence. Accordingly, there is no concept for a derivative, revealing that many applications are not amenable to the classic analytical approach to determining DV*.

The other issue is an additional concept, the constraint. Here the sequence is constrained so that (i) each city number is used, (ii) each is used only once in the sequence, and (iii) the initial and final cities are constrained to H.

1.4 Terminology Continued

Returning to concepts and nomenclature, constraints are a key issue in optimization.

1.4.1 Constraint

These are values or issues that cannot be violated. These may include physical "laws," man-made laws or procedures, infeasible math operations, procedural steps, or many other features. Constraints can be on the DV, OF, or auxiliary (other related, but secondary) variables. Constraints can be on values, rates of change, or transitions. They can be on current action or on future implications of today's action. Constraints often are the most important influence on determining the DV* values.

Here is an introduction to the diversity of constraints:

- Constraint on transition: When visiting major league ballparks, don't go to NYC immediately after Boston or vice versa because of fan loyalty. When operating a mixer, you must fill and blend material in the mixer before dumping.
- Constraint on rate of change: Don't immediately "floor" the automobile accelerator pedal position, because it startles passengers. Gradually move it from one position to another, no faster than 5° of arc per second.
- Constraint on DV: A composition must be between 0 and 100%. A flow rate must be nonnegative.
- Constraint on secondary variables (auxiliary variables): Don't let a downstream tank overflow or run dry. The sum of all compositions must be equal to unity. Keep the fire temperature below the melting point of the nozzle material. An automobile engine temperature cannot exceed 180°F.
- Constraints may relate to natural laws: Human bodies cannot float in the Earth's atmosphere because of the density difference between mostly water and air. Heat does not naturally flow from cold to hot, because entropy must increase. As close as cellulose is to sugar, and even though grazing animals can digest cellulose, humans cannot live on cellulose.
- Constraints may relate to human laws or procedures: You cannot build your house on your neighbor's property. You cannot freely travel from any one country to any other. Wear your uniform and follow this procedure to clock your work hours.
- Constraints may occur in the future: Today you can make the purchases on your credit card; but the constraint happens in the future when all of your income is paying accumulated interest. You can drive the race car now and pass a pit stop; but the worn tires will fail in half a lap.
- Constraints may be soft or hard: It is OK to violate soft constraints. The speed limit is 60 mph, but enforcement permits you to drive 4 mph higher. Other soft constraint examples, in which mild violation is permissible might include, "Don't yell at your children." Wear white hats from Memorial Day to Labor Day. By contrast, hard constraints may not be violated. "Don't let the mixture composition get into the explosive limits." "Pay your taxes." Don't try to take the logarithm of a negative number.

1.4.2 Feasible Solutions

Some ideas, such as perpetual motion, are infeasible. They violate hard constraints, such as the first or second laws of conservation. Feasible solutions are the DV values that do not violate any of the constraints such as those listed previously.

1.4.3 Minimize or Maximize

Minimizing seeks the lowest point, or smallest value. Maximizing seeks the highest or largest. But these are mathematically equivalent when the OF is multiplied by negative unity. If you wish to maximize, you could minimize the negative of the OF and find the same DV* value. Turn the graph of Figure 1.3a upside down and compare it to the Figure 1.3b to intuitively confirm this. Chapter 16 will provide mathematical analysis of this and other related useful OF transformations.

This book will focus on minimization. With only one optimization objective, the concepts are less confusing, and there is no loss of generality.

1.4.4 Canonical Form of the Optimization Statement

There is a conventional style of presenting the optimization application and associated elements. Equation (1.14) is an example. To begin the statement, in the upper left, state whether the objective is to minimize (min) or maximize (max) the OF. Below the min or max statement explicitly, list the decision variables within braces. Define the objective function, J. It might be a simple formula, or something that represents a procedure such as SSD, DCFRR, or the calculation of risk. The objective function might even be its own optimization, for instance, to identify the worst or maximum possible event, and the optimization might be to identify DV values that minimize a worst-case outcome. In designing a product, one might wish to assess the worst possible failure situation and design to minimize the maximum risk. List the constraints in lines that follow the S.T.: (subject to) label. Illustrated here are three types of constraints—one on the decision variables, one on a function of the DVs, and one on a rate of change:

$$\begin{aligned}
&\min_{\{DV\}} J = \text{you define this relation}\\
&\text{S.T.:} \quad P_1 = f(DV) \leq P_{10}\\
&\qquad\quad a \leq DV_1 \leq b\\
&\qquad\quad \text{rate} = \frac{dP_2}{dt} \leq \text{rate}_0\\
&\qquad\qquad \vdots
\end{aligned} \tag{1.14}$$

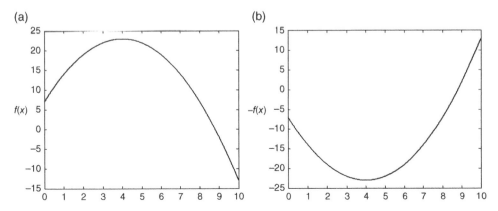

Figure 1.3 Equivalency of maximizing and minimizing the negative: (a) seeking the maximum and (b) seeking the minimum of the negative of the function.

This canonical form is just a communication tool. It is neither the solution nor the method for finding the solution. It may also be incomplete; and if it is, include the necessary supporting information. For instance, is the OF being evaluated experimentally or by a simulator? In either case the reader may wish to know essential details about the experimental design or the idealizations and methodology of the simulator model.

The optimization math is relatively easy, standard, and widely available. The application problems are typically related to:

- Characterizing the valid and appropriate and comprehensive measure of goodness, how to decide best, the OF
- Identifying the DVs (are they values, timing between values, sequence of operations, which variables?)
- Identifying the constraints ("Oh, I didn't know shortest path required me to stay on the roads!")
- Generating the models so the OF value or constraint conditions can be calculated given the DVs
- Choosing the optimization algorithm that is right for the application
- Choosing thresholds and parameters related to the algorithm initialization, operation, and convergence

Although this book describes many fundamental optimization algorithms, it is dedicated to a focus on the application issues listed previously.

1.5 Optimization Procedure

There are eight basic stages in an optimization procedure:

1) *State the OF, or objective function.*
2) *State whether you wish to minimize or maximize.*
3) *State the DVs, or decision variables, and what you are free to change to effect the OF value.*
4) *Use model relationships to relate the DV to all parts of the OF.*
5) *Define constraints.*
6) *State the method used for solving the optimization application. This is the optimization approach, algorithm, procedure, or logic.*
7) *Execute the procedure.*
8) *Reflect on it all.*

A few examples will reveal the stages in developing and solving an optimization application and reveal the often overlooked importance of the eighth.

Example 5 Minimize the perimeter of a rectangle that provides a desired area by choosing the height and width. Figure 1.4 represents a mental construct of a rectangle. We need the mental construct to derive models that represent the interaction of DVs and OF. The model of the mental construct is not the reality, but it is a mathematical approximation of reality.

1) State the OF, or objective function: J = perimeter = $2h + 2w$.
2) State whether you wish to minimize or maximize: Minimize $J = 2h + 2w$.
3) State the DVs, or decision variables, and what you are free to change to effect the OF value. In this case it is just one item, either the height or the width. Given one value, the other is

Figure 1.4 Analysis of a rectangle.

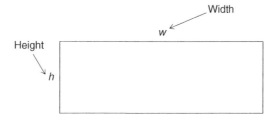

calculated from the desired area. $A = hw$. I'll choose height as the DV. The basic optimization statement is

$$\min_{\{h\}} J = 2h + 2w \tag{1.15}$$

4) Use model relationships to relate the DV to all parts of the OF. In general we need a model of the device or process to relate state variables to influences. In this case h is explicit in the OF. However, h also affects w, which is not explicitly revealed in the OF. We need the complete relation. These models are often termed constitutive relations. From the definition of area in a rectangle,

$$w = \frac{A}{h} \tag{1.16}$$

Inserting it into the OF completely indicates how the choice of h-value affects the OF value:

$$\min_{\{h\}} J = 2h + \frac{2A}{h} \tag{1.17}$$

5) Define constraints

$$h > 0 \tag{1.18}$$

6) State the method used for solving the optimization application and for adjusting h to find the value h^* that minimizes the OF. This is the optimization approach, algorithm, method, or logic. The analytical method is one algorithm. Since the OF permits its use, I'll use it. Assume J is a function with one minimum, as illustrated in Figure 1.5.

Figure 1.5 A function with one minimum.

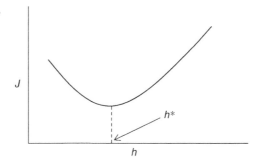

At the minimum, $dJ/dh|_{h^*} = 0$. Elsewhere, the slope is not zero. So, get the analytical derivative of J w.r.t. h, equate it to zero, and solve for the value of h^*.

7) Execute the procedure:

$$\frac{dJ}{dh}\bigg|_{h^*} = 0 = 2 - \frac{2A}{(h^*)^2} \tag{1.19}$$

The solution is

$$h^* = \sqrt{A} \tag{1.20}$$

which also means that $w^* = h^*$ and the rectangle is a square.

8) Reflect on it all (the model, the assumptions, the answer, the method, the acceptance of your solution by others in the enterprise). Evaluate your procedure and results. The first application and solution provides progressive insight. What would you do differently in OF, DV, constraints, model, and method? What idealizations limit validity of result? When others look at your analysis, what might they find incomplete or overlooked? Redo the work with this added depth of insight. This was an ideal initial analysis.

For instance: Is the item being modeled an ideal rectangle of lines of zero width? Other than mental constructs and intellectual exercises, where do you find such things? An exercise like this might be the idealization of an application to choose a window pane shape to minimize the consumption of frame material. But the frame has thickness, and it must both overlap the window a bit and extend beyond. So, perhaps the volume of frame material, not the perimeter, is the real OF. Also, is it the window pane area or the open area inside the frame that is important? The answer to the analysis above was a square, but how does this fit with your personal experience? Are window panes square? What considerations seem to override the analysis to give an alternate (not $h = w$) solution?

Example 6 Maximize power delivered to a resistor from a battery with voltage V. Figure 1.6 represents a mental construct of a battery within the dashed lines as an ideal battery followed by an internal

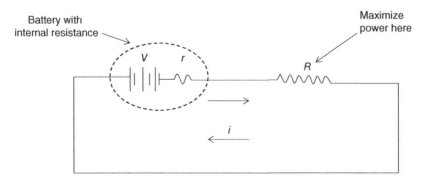

Figure 1.6 Mental construct of a battery in a resistance circuit.

resistor. We need the mental construct to derive a model that represents the interaction of DVs and OF. The model of the mental construct is not the reality, but it is a mathematical approximation of reality.

1) State the OF, or objective function: $J = \text{Power} = i^2 R$.
2) State whether you wish to minimize or maximize: Maximize $J = i^2 R$.
3) State the DVs, or decision variables, and what you are free to change to effect the OF value. In this case it is just one item, the external resistor resistance, $DV = R$. The basic optimization statement is

$$\underset{\{R\}}{\max} J = i^2 R \tag{1.21}$$

4) Use model relationships to relate the DV to all parts of the OF. In general we need a model of the device or process to relate state variables to influences. In this case R is explicit in the OF. However, R also affects i, and the model is not explicitly revealed in the OF. We need the complete relation. These models are often termed constitutive relations. From elementary circuit analysis there is a relation between current and resistance:

$$V = i(R + r) \rightarrow i = \frac{V}{R + r} \tag{1.22}$$

Inserting it into the OF completely indicates how the choice of R value affects the OF value:

$$\underset{\{R\}}{\max} J = \frac{V^2 R}{(R + r)^2} \tag{1.23}$$

5) Define constraints

$$R > 0 \tag{1.24}$$

6) State the method used for solving the optimization application and for adjusting R to find the value R^* that maximizes OF. This is the optimization approach, algorithm, method, or logic. The analytical method is one algorithm and applicable to this example. Assume J is a function with one maximum as illustrated in Figure 1.7.

At the maximum, $dJ/dR|_{R^*} = 0$. Elsewhere, the slope is not zero. So, get the analytical derivative of J w.r.t. R, equate it to zero, and solve for the value of R.

Figure 1.7 Illustration of a function with a single maximum.

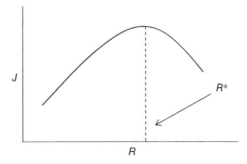

7) Execute the procedure:

$$\frac{dJ}{dR}\bigg|_{R^*} = 0 = \frac{V^2}{(R+r)^2} - 2\frac{V^2 R}{(R+r)^3} \tag{1.25}$$

The solution is

$$R^* = r \tag{1.26}$$

8) Reflect on it all (the model, the assumptions, the answer, the method, the acceptance of your solution by others in the enterprise). Evaluate your procedure and results. The first application and solution provides progressive insight. What would you do differently in OF, DV, constraints, model, and method? What idealizations limit validity of result? When others look at your analysis, what might they find incomplete or overlooked? Redo the work with this added depth of insight. This was an ideal initial analysis.

For instance, is the model of the battery correct (is the internal resistance in series with the ideal battery or is it more complicated)? Is the circuit model correct (zero resistance in the wire)? Is there an impedance effect on start-up? Does the battery voltage remain at the initial "V" while it is discharging? If you buy a resistor of $R^* = r$ what value will R actually be? Are the "given" V and r values the absolute truth for all circuits to be made? This OF looked at maximizing delivered power, but are there other measures of performance or cost that have been overlooked?

Returning to the importance of stage 8, the simplicity of examples and solutions, which is typical of the classroom teaching style, may provide a pedagogical method that facilitates initial understanding; but it falls short of the comprehensive issues that affect reality and so misguides a student. Optimization is not about the idealized, intellectual mathematical analysis. It is done to improve human choices in design, operations, and procedures, and the application needs to align with the comprehensive reality, not an isolated idealization. Often the step-back reflection on the outcomes that reshapes the statement objective and assumptions is the most important aspect to optimization.

1.6 Issues That Shape Optimization Procedures

As mentioned in the prior examples, the classical analytical optimization concept (set the derivative to zero and solve for the DV*) is often not feasible. As a result, a diversity of optimization procedures have been developed to surmount the issues. There are many reasons that make the ideal approach inapplicable. Here is a preview listing of difficulties that real applications present:

- Nonlinearities—The objective function may not have an analytically tractable derivative. If it does, the derivative may require iterative nonlinear root-finding procedures to solve an implicit nonlinear relation.
- Discontinuities—The OF or its derivatives may have discontinuities.
- Discretization—The decision variable might represent integer values (number of queuing lines, number of parallel devices, number of samplings for a delay) or discretized sizes (pipes, resistors,

and shirts come in discrete sizes). Further, the numerical time or space discretization in the model used for the OF will generate steps or ridges (striations) on what might be considered as a smooth OF. Convergence criterion on root-finding techniques used within the model can also cause striations. For these, analytical optimization techniques, which were developed for continuum-valued OFs and DVs, will be confounded.

- Multiple optima—Many objective functions have local minima that would trap the optimization procedure within a local, not the global optimum.
- Flat spots—Some OFs have zero, or effectively zero, response to the DV in regions of saturation or inconsequence, and if the derivative is zero, there is no guidance as to how to improve the DV solution.
- Stochastic response—When OF data is being generated experimentally, there is noise (uncertainty, experimental error) on the OF value. A replicate experiment (an attempt to implement the same DV values) will not produce exactly the same OF value. Here, because of experimental vagaries, moving the TS toward the true DV* value might return a worse, not better OF value, which would indicate that the optimum is in the false direction. When Monte Carlo simulations are used as a surrogate for experiments, the impact is the same.
- Uncertainty—There is uncertainty in the givens. Air pressure, temperature, humidity, and wind velocity continually change. If you design an airplane wing for one set of conditions, what is the outcome at other realizations of the givens? Should the design be to minimize the worst possible outcome over all possible realizations?
- Constraints—Types of constraints were described previously. When constraints are hard (cannot be violated), the optimization procedure needs to be modified to choose an alternate path. If you encounter a high wall blocking your downhill walk, but want to get to the other side, perhaps walk along the wall, not directly downhill.

The useful applications of optimization encounter those sorts of issues. Accordingly, the useful methods for optimization need to work in spite of such issues.

1.7   Opposing Trends

If the OF has a monotonic response to the DV, then the optimum is at an extreme. Consider this example:

$$\min_{\{x\}} J = 7e^{-x/10} \tag{1.27}$$

The function decreases monotonically with the x-value, as shown in Figure 1.8, and the minimum value for J occurs at $x^* = +\infty$, at an extreme value.

Since a value of infinity is not practical, the solution is to use the largest x-value as possible. Some optimization applications legitimately send the DV to a constraint. For example, "What speed should you travel to get there in minimum time?" The answer is drive at the maximum speed possible. However, in my experience, usually such an outcome indicates that the user has not properly included all relevant features in the OF. In a car speed example, these additional considerations may include fuel consumption, wear on the vehicle, and the risks associated with an accident, being arrested, or setting an example for others.

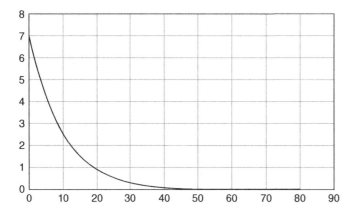

Figure 1.8 Illustrating a monotonic OF response to a DV.

Usually there are at least two opposing responses affecting the OF, and optimization seeks to find the best balance of the opposing ideals. One response would get better with increasing the DV value, and the other would get worse. This can be illustrated as a product of functions; a simple example building on the OF above is

$$\begin{matrix} \min \\ \{x\} \end{matrix} J = 7xe^{-x/10} \tag{1.28}$$

Here the first function, $7x$, rises with the value of x, and the second function, $e^{-x/10}$, diminishes with the value of x. The solution is an intermediate value $x^* = 10$, as illustrated in Figure 1.9.

Alternately the opposing functionalities may be additive, as with this example and Figure 1.10 illustration:

$$\begin{matrix} \min \\ \{x\} \end{matrix} J = 7e^{-x/10} - 5e^{-x/5} \tag{1.29}$$

Here the $-5e^{-x/5}$ functionality rises with x, and the $7e^{-x/10}$ falls with x-value. And $x^* = 10^*\ln(10/7) = 3.566749...$

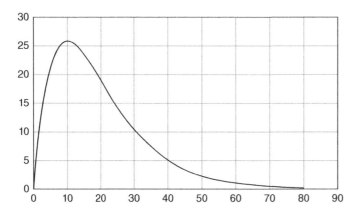

Figure 1.9 Illustrating a dual OF response to a DV—multiplicative functionality.

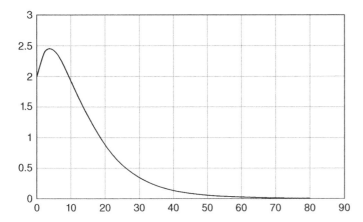

Figure 1.10 Illustrating a dual OF response to a DV—additive functionality.

In a real application, the opposing influences may be much more complicated functions, such as iterative computational procedures; but for an optimum to exist at a non-extreme x-value, at an intermediate x-value, the OF must represent a balance of opposing ideals.

Further, one ideal must make larger OF changes than the other when x changes in the $x < x^*$ range, and vice versa in the $x > x^*$. To illustrate this consider the slight perturbation to the application represented by Equation (1.29) in which the x-value in the argument of the second term is scaled by 8 instead of 5:

$$\begin{array}{l} \underset{\{x\}}{\min} J = 7e^{-x/10} - 5e^{-x/8} \\ \text{S.T.:} \quad x > 0 \end{array} \tag{1.30}$$

The opposing idealities of the functionalities are the same (one increasing with x, the other decreasing), but the relative magnitude of the two influences does not change, and the OF is monotonic, as illustrated in Figure 1.11.

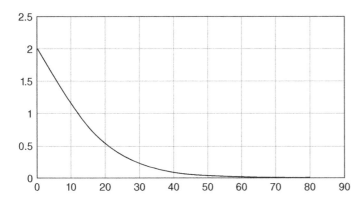

Figure 1.11 Illustrating a monotonic OF response to a DV—additive functionality.

1.8 Uncertainty

A typical optimization application is confounded with multiple aspects of uncertainty. Consider this simple application: "What insulation thickness on a pipe is optimal?" The OF will be related to annualized cost associated with energy losses and a 5-year life of the insulation. Model the cost of insulation as proportional to the volume of material coating a pipe of length L and radius of r_p. The insulation outer thickness is r. A classic model for the rate of heat transfer per surface area from an insulated pipe (with zero resistance from the metal pipe or internal liquid to pipe convection) is

$$\dot{q}'' = \frac{T_{surrounding} - T_{internal}}{\left[\ln(r/r_p)/2\pi k\right] + (1/2\pi rh)} \tag{1.31}$$

where k represents the thermal conductivity of the insulation and h is the heat transfer coefficient between the insulation surface and the surrounding air.

The objective is to minimize annualized costs over a 5-year period:

$$\min_{\{r\}} J = c_{insulation} \pi \left(r^2 - r_p^2\right) L + 5 c_{energy} Y 2\pi r L \frac{T_{surrounding} - T_{internal}}{\left[\ln(r/r_p)/2\pi k\right] + (1/2\pi rh)} \tag{1.32}$$

Here c-values represent insulation and energy costs per unit, Y is the number of time units in a year, and L is the piping length.

While the formula might look complicated, and the pipe insulation application might not be familiar to many, it is still a relatively simple application. Input the values for the "givens" ($c_{insulation}$, r_p, c_{energy}, $T_{surrounding}$, $T_{internal}$, k, h, etc.), and it is relatively easy to let a computer determine the optimal value, r^*.

If the "givens" remain unchanged, then any computer that solves Equation (1.32) will get the same value for r^* regardless of the day, or who is the data-entering user. However, that answer might not be right. Consider two aspects. One is uncertainty in the givens, and the other is associated with the model.

Givens: Will the surrounding temperature remain at the value of $T_{surrounding}$ for a 5-year period? Will the energy cost remain unchanged from the value of c_{energy}? Might not age and environmental effects change the insulation k-value from the installed value? What will the programmer choose to use for the value of π? Since none of the givens have values that are known with certainty, changing those values to other reasonable values will change the calculated t^* (insulation thickness) $(t = r - r_p)$ value. For instance, if the optimal value at nominal conditions is $t^* = 2.54$ cm, it might range between 2 and 3 cm for combinations of alternate reasonable values of the givens.

In any case, the same optimization procedure would evaluate the same method of OF calculation. The optimization procedure is not dependent on the values of the givens, but the resulting OF* or DV* outcomes are.

Model: Is there no internal resistance to heat transmission (fluid to pipe)? Is the outside heat transfer coefficient, h, uniform along the length of the pipe and at the bottom, top, and sides? What is the right model to use to estimate the h-value? Alternate models for the process may lead to t^* values that range between 1.7 and 3.5 cm, and compounded with uncertainty in the givens, the conclusion might not be the ideal $t^* = 2.54$, but a more realistic $1.3 < t^* < 3.8$ cm.

The business decision is not based on $t^* = 2.54$ cm. For instance, if the capital supply is ample, then the decision might be to use the 3.8 cm thickness. But if capital is scarce, use the 1.7 thickness;

and if it is found to be inadequate, then add more insulation later. Recognizing uncertainty in the DV^* value is as important as the DV^* value.

1.9 Over- and Under-specification in Linear Equations

Over-specified means that there are more independent equations than independent variables, more constraints than variables—in this case the degrees of freedom <0. A straight line can go through any two points, but if you add a third point that is not on the line, the single line cannot simultaneously go through all three points.

Under-specified means that there are fewer equations (constraints) than variables, degrees of freedom >0. For example, if there is only one point, then you could choose the slope of the line and still find a line to go through the point. You are free to choose a variable value. You have a degree of freedom.

Balanced means that there are the same number of equations as variables: DoF = 0.

From a linear equation understanding, consider three equations with three unknowns $\{x, y, z\}$:

$$a_1 x + b_1 y + c_1 z = q_1$$
$$a_2 x + b_2 y + c_2 z = q_2 \tag{1.33}$$
$$a_3 x + b_3 y + c_3 z = q_3$$

If the equations are linearly independent, there is a unique solution for the $\{x, y, z\}$ values. The DoF = 0. The three might be considered equations, or specifications, or active constraints, and the $\{x, y, z\}$ set might be considered to be the DVs. The equations can be written in vector-matrix form:

$$\begin{bmatrix} a_1 & b_1 & c_1 \\ a_2 & b_2 & c_2 \\ a_3 & b_3 & c_3 \end{bmatrix} \begin{bmatrix} x \\ y \\ z \end{bmatrix} = \begin{bmatrix} q_1 \\ q_2 \\ q_3 \end{bmatrix} \tag{1.34}$$

In this familiar representation, the number of equations (which appears to be just one equation when counting the number of equal signs) may not necessarily match the number of unknown variables. Here degrees of freedom = number of DVs − number of specifications (active constraints).

Under-specified: Consider 2 equations and 5 unknowns. Here DOF = 5 − 2 = 3. This means that you can choose any value for three (any three) unknowns and then solve for the others. The solution is not unique. The problem is under-specified. If you can choose which variable and its value, then which is best? This is an optimization. Would your choices be based on the simplest remaining solution, or a conventional value, or something that pleases your boss?

Balanced: Consider 3 equations and 3 unknowns. Here DOF = 0. This means you can find an exact and unique solution.

Over-specified: Consider 4 equations and 3 unknowns. Here DOF = −1. This means that you cannot fit all four conditions simultaneously. The problem is constrained. In this case you could (i) relax (eliminate) one condition and solve for others or (ii) find a best compromise solution with a weighted objective function that includes a penalty for each equation deviation from zero as an OF term. For instance, square each deviation, sum the weighted squares, and seek a DV set of values that minimizes the objective function. (The choice you make is an optimization. Which might be best for an application? How would you assess best?)

1.10   Over- and Under-specification in Optimization

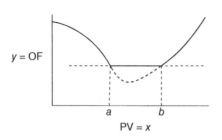

y = OF

a b

PV = x

Figure 1.12 Illustrating an under-specified application.

There are parallels in optimization. Figure 1.12 illustrates an under-specified application, for example:

$$\min_{\{x\}} J = y \begin{cases} = c & \text{if } a \leq x \leq b \\ = 1 - e^{-(x-d)^2} & \text{otherwise} \end{cases} \tag{1.35}$$

Any x in the $[a, b]$ range is equivalent to any other x-value. There is flexibility in the solution. There are extra choices you can make. This may also arise if there are distinct but identical solutions. For instance,

$$\min_{\{x\}} J = (x^2 - 9)^2 \tag{1.36}$$

Here there are two identical solutions: $x = (3, -3)$. If you can choose one, which is preferred? In this case there are probably reasons other than that expressed in the OF as to why one x-value is better than another. Consider resource conservation, future flexibility, reliability, political capital, etc., and include this additional concept into the OF.

Alternately, under-specified may be the result of redundant parameters. Here is an example of a hyper-elastic model of stress versus strain:

$$\sigma = A\left(e^{B\epsilon} - 1\right) \tag{1.37}$$

It appears to have independent model coefficients A and B. However, if either B or ϵ is very small, then $e^{B\epsilon} \cong 1 + B\epsilon$ and the model effectively is equivalent to

$$\sigma \cong AB\epsilon = k\epsilon \tag{1.38}$$

In this view, stress is proportional to a constant, k, times strain. The model is dependent on the value of k, and it does not matter what the individual values for A and B are, as long as their product is equal to k. A contour plot of the least squares solution of

$$\min_{\{A,B\}} J = \sum (\sigma - \tilde{\sigma})^2 \tag{1.39}$$

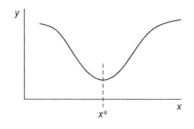

y

x^* x

Figure 1.13 Illustrating a balanced application.

would show a nearly common (nearly constant valued) minimum (valley) along the curve $B = k/A$, or DV_1 proportional to the reciprocal of DV_2.

If you notice an under-specified application, multiple DV^* values giving essentially identical OF^* values, then use this finding to trigger reassessment of the models and description.

Figure 1.13 reveals a balanced application with a unique minimum solution.

Figure 1.14 reveals an over-specified or constrained application, in which the minimum of $y(x)$ occurs in the constrained x-region:

$$\begin{aligned} \min_{\{x\}} J &= y(x) \\ \text{S.T.}: \quad x &< a \end{aligned} \tag{1.40}$$

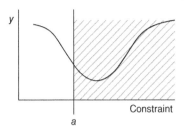

Figure 1.14 Illustrating an over-specified application.

If over-specified or constrained, you could either (i) use the hard constraint to limit optimizer trial solution values or (ii) add a penalty (soft constraint) for the constraint violation to the OF.

If the solution is on a constraint, and if the constraint defines the DV* value, then the constraint is the critical aspect. Then move your focus from solving the original optimization statement and return focus to the application and do these: (i) Be sure that the constraint is properly specified. (ii) Seek alternate designs (of the process, device, procedure) that would remove the constraint. The real solution may be outside of the mathematical optimization exercise that would seek how to include the constraint in the problem.

1.11 Test Functions

Any practical application will not have a simple one-line equation as the OF. However, for instruction and for convenient testing, many people have created simple one-line equations as easy-to-implement, rapidly computed surrogates to reveal fundamental issues. Many of these test functions have become the standard benchmark problems that people use to develop and compare optimizers. The problem is that when these relations constitute the majority of the examples for a novice, it misrepresents the true complexity of optimization.

Appendix E presents a set of applications that are grounded in physical reality illustrating diverse application issues but that are still relatively simple for exploring optimization techniques.

1.12 Significant Dates in Optimization

I find that a review of significant dates related to optimization provides revealing issues about why we have certain procedures that seem to dominate our legacy and instructional material. Here are selected dates for publications of algorithms:

- 1669: Newton—Root finding (this is not optimization, but it is a common element in many algorithms) (not published until 1711).
- 1690: Raphson simplified Newton's method for root finding.
- 1740: Simpson extended Newton's method to root finding on the derivative (analytical optimization). Perhaps the optimization application should be named Simpson's method.
- 1807: Legendre presented the least squares method (there were parallel developments by Gauss).
- 1847: Cauchy—Sequential line search on a steepest descent line.
- 1939: Kantorovich—Linear programming (LP) model and an algorithm for solving it.
- 1944: von Neumann and O. Morgenstern solved sequential decision problems by using the idea of dynamic programming (DP). A. Wald (1947) did related research. Another early application of DP is presented by P. Massé (1944) for reservoir management.

- 1948: Dantzig—Linear programming and the simplex tableau, now a common algorithm.
- 1952: Bellman's key publication on the theory of dynamic programming.
- 1952: Hestenes and Stiefel—Conjugate gradient approach.
- 1955: Levenberg—Blending of incremental steepest descent and Newton's methods (seemingly lost in the literature).
- 1956: Frank and Wolfe—Reduced gradient algorithm for constrained optimization.
- 1961: Hooke and Jeeves—Pattern search.
- 1962: Spendley, Hext, and Himsworth—Simplex search.
- 1963—Marquardt's rediscovery and popularizing of Levenberg's approach (now known as Levenberg–Marquardt).
- 1963: Wilson—Sequential quadratic programming.
- 1960s: Genetic algorithms and in general the start of mimetic approaches (algorithms that attempt to mimic nature's way of finding the optimum).
- 1965: Nelder and Mead's heuristic rules for expansion and contraction of SH&H's simplex.
- 1969: Broyden—Quasi-Newton procedures.
- 1974: Lasdon, Fox, and Ratner—Generalized reduced gradient method.
- 1979: Khachaturyan, Semenovskaya, and Vainshtein—Simulated annealing.
- 1980: Beginning of randomized search methods.
- 1989: Glover—Tabu search.
- 1990: Beginning of diverse mimetic algorithms—ant colony and bee colony.
- 1995: Kennedy and Eberhart—Particle swarm.
- 1997: Storn and Price—Differential evolution.
- 2003: Raphael and Smith—Probabilistic Global Search Lausanne.
- 2012: Leapfrogging.

Notable about this timeline is that prior to 1950, prior to widespread computer access, optimization was based on manual calculations. This generally limited applications to linear, quadratic, unconstrained, and well-behaved (mathematically tractable, low dimension) functionalities. And appropriate for these were second-order techniques such as Newton–Raphson.

In the 1950–1960 period, mainframe computers permitted inclusion of algorithms to handle surface aberrations. During this period, direct search (function evaluation only) approaches began to invade what was formerly exclusively optimization based on gradient-based (sensitivity of the OF to the DVs) search algorithms. This period began the rise of the more complicated second-order methods (LM, GRG), direct search (HJ, SHH-NM), and numerical approaches. But, there was limited access to mainframes for day-to-day applications with low economic or nonmilitary impact.

During 1960–1980 mainframes had widening access and began to gain acceptance in process design and optimization, and diverse and innovative ways to use this tool for formerly difficult optimization applications blossomed.

In the 1980s personal desktop computers and widespread convenient access to powerful machines shifted the algorithm focus from the number of function evaluations and single trial solution optimizers to multiplayer methods and their likelihood of finding the global minimum. This also shifted the algorithm development focus from deterministic logical/mathematical algorithms to mimetic stochastic algorithms that mimic nature (genetic algorithms, evolutionary algorithms, particle swarm, ant farm, simulated annealing, etc.).

Today, I sense that methods to solve stochastic applications are rising in importance, even at the day-to-day business level.

Although gradient-based and Newton-type approaches are still important, they were a product of the pre-computer era, and my bias is that direct search approaches are more compatible with computer techniques and solving difficult applications. This book does address archetypical gradient-based approaches but provides greater coverage to direct search approaches. Further, although this book will address classical formulation of optimization applications, it will also rise above the deterministic and quadratic limitations that characterize last-century methods to reveal today's solutions. Finally, as much fun as the mathematical analysis of optimization can be, the application issues have much greater practical importance. So, although this book will reveal some of the underlying analysis, I will limit it to the essential for understanding and emphasize the practical aspects of applications.

The pace of progress has been accelerating. Will it continue, or are we about to the limit of novel optimization ideas that take advantage of the computer? I have to think it will continue.

Nearly all of the techniques aforementioned were developed for deterministic applications. If you repeat the DV values, you get exactly the same OF value as a response. But life is not like that. It is stochastic. Here are three examples: First, if you are seeking to use experimental data as the OF, then experimental variability makes replicate runs not have the exact same response. Second, simulations of what life might impose are also stochastic. For instance, when deciding how much to invest now, for retirement 40 years in the future, one has to forecast inflation rates, interest rates, personal health, career outcomes, etc. There is great uncertainty in these "givens." Any "given" could have this value, or that value, and the value will change from year to year in unpredictable patterns. With the same DV (investment portion of salary) different values for the "givens," different possible situations, different realizations of what is possible, lead to different OF values. Finally, as the third example, I'll mention bootstrapping, a technique for random re-sampling of data to estimate the uncertainty in models as a result of variability in data. I believe that optimization of stochastic functions (Monte Carlo simulations) is now widely possible, and it is gaining acceptance and appreciation within the business–technical community. It appears that we have only seen the tip of the iceberg of applications and techniques for optimization of stochastic functions.

Also, nearly all of the techniques aforementioned were developed for continuum-valued functions. By contrast, when the model contains either variables that are discretized or conditionals, the OF response to the DV will have ridge/valley/cliff discontinuities. Often a magnified view of the surface is required to visualize these. First, consider discretization striations with a familiar example: a cloth fabric (for a shirt, or curtain, or socks) seems to be a continuum material from afar, but a microscopic view reveals the ridges in the weave or knit. The local striations on a surface will lead many optimizers downhill along the striation, not downhill globally. Alternately, to visualize a sharp valley, consider draping a piece of paper that contains a crease. Many optimizers will tend to jump back and forth across the crease, not follow the slope discontinuity downhill. Again, I believe that there is much opportunity for development of both applications and solutions for non-continuum objective functions.

What tools will be next in the progression of tools for optimization? How might they reshape algorithms? It is a wonder to me how easily water finds the downhill path in a steep valley. It remains difficult for all optimizers, but seems easy for water. Maybe the magic that water uses is that it is composed of individual molecules, and when billions of them are in the downhill side of the valley, the probability of one of them finding the downhill path is high.

All in all, it appears that the near-term future promises much in continued development and analysis of optimization.

1.13 Iterative Procedures

The several examples previously have simple equations that permit taking the analytical derivative and then using it to explicitly solve for DV*. However, applications that I've encountered are analytically intractable. The equation that relates OF value to TS value may be too complicated to be confident that the derivative is correctly formulated. Or the OF value might be computed from a black-box simulator that does not reveal internal equations. Or the OF value might be the result of physical experimentation. In these cases, we need another procedure to determine DV*.

There are many that will be revealed in subsequent chapters, but here, I'll introduce a heuristic direct search. The term direct search contrasts procedures that use gradient (derivative) information. A direct search only uses the function evaluation. The term heuristic means that it uses intuitive or tried-and-true human experience, as opposed to a mathematically formulated rule. To understand the heuristic direct search, imagine how a blindfolded tightrope walker would find the bottom of the rope span if he were placed randomly along the rope. The rule is take a step in one direction. If you go downhill, the step was the right direction; so, from this best-so-far spot, make the next step in the same direction. If you go uphill, the step is probably the wrong direction; so, return to the best-so-far spot, and make the next step in the other direction. If the reverse step had the same size, then you return to a prior spot, so when reversing direction, also cut the step size in half. By contrast, if going in the right direction, make the next step a bit larger. Near the minimum the alternating steps are progressively cut in half. Stop, and claim success when the step distance is small enough.

Here is an algorithm outline for the heuristic direct search

1) Initialize
   Choose the initial feasible TS, the base case DV
   Evaluate the OF-value, OF_{base} at the DV_{base}
   Choose a step size, DV_{delta}
   Choose a convergence threshold for DV_{delta}
2) Test a new TS
         Set the new TS=$DV_{base}+DV_{delta}$
         Evaluate the OF-value, OF_{TS} at the TS
         IF OF_{TS} is better than OF_{base} THEN
             Set $OF_{base} = OF_{TS}$
             Set $DV_{base} = TS$
             Set $DV_{delta} = 1.2*DV_{delta}$
         ELSE (meaning that the OF is worse or a constraint is hit)
             Set $DV_{delta} = -0.5*DV_{delta}$
         ENDIF
         IF $|DV_{delta}|<$ convergence threshold EXIT
      Return to a new TS (Go to Point 2)

A range of DV_{delta} attenuation factors seem to work. The 1.2 expansion factor could be from 1.05 to 1.25. If larger than 1.25, it seems to excessively enlarge the step size when going in the right direction, overstepping the minimum, and requiring additional back steps to temper the excess. Alternately, an expansion factor that is closer to unity does not accelerate the progress after a contraction. A value of 1.1–1.2 seems to work well. Similarly, there is a range of good values for the contraction factor. If the new TS is bad, then the contraction should be less than 1/expansion so that the step does not go on the other side of the prior best. However, if the contraction factor is too small, for instance, 0.1, then after a reversal, subsequent steps in the right direction are very small. Contraction values in the 0.5–0.8/expansion range seem good.

In addition to not requiring analytical derivatives, this simple procedure is easy to understand and implement (either in code or intuitively), is robust to hard constraints (or execution errors) and surface discontinuities (such as those related to numerical approximation of DV discretization), and executes rapidly.

Notice that contrasting the analytical method of the prior examples, this procedure does not intelligently jump to the zero derivative spot, where DV^* is calculated from the analytical derivative. This iterative procedure incrementally moves the TS toward the vicinity of DV^* and stops, and claims convergence, when the incremental changes are less than a user-defined convergence value. Like incrementally calculating the next number in the irrational value of π, this procedure gets progressively closer and stops when the user has defined that the TS is close enough—when $|DV_{delta}| <$ convergence threshold.

The user's choice of the convergence threshold defines the precision of the solution. Somehow the user must relate the impact of the threshold to the undesirables. If too large a threshold, the solution is not very precise. If too small, the procedure requires excessive iterations.

1.14 Takeaway

Optimization is essential in all aspects of your personal and professional life.

Understand the terminology in this chapter. The concepts are essential to doing.

Do, not study: Apply the terminology and canonical form of the statement to everything you can. Practice seeing your reality from the structure of this optimization perspective.

The first solution is probably inadequate. Sketch the procedure and solution, evaluate its elements and outcome with respect to your developing understanding and other's input, erase your solution, re-sketch, evaluate, and erase, sketch, eval., erase, s, e, e, s, e, e, … See. See? Eventually you'll have a comprehensive solution. The solution is not the intellectual exercise that produces a value. It is the right solution as defined by all constituents.

1.15 Exercises

1 Qualitatively discuss issues that would relate to these optimization applications:
   A How wide should a hat brim be?
   B What is the best temperature for a freezer?
   C What is the optimum length for a vacuum cleaner cord?
   D Where should you cross the street?
   E Should you password protect your cell phone?
   F How thick should be a pane of glass for a window?
   G What are the criteria to open your umbrella?

2 State OF, DV, and constraints for the following purposely vague situation statements:
   A Increasing the coating thickness on pills inhibits O_2 diffusion in, which increases shelf-life.
   B The larger the pipe diameter, the smaller the ΔP required to achieve target flow rate—hence the smaller the pump capital and energy cost and expense.

C A set of y versus x data, in general, shows curvature, but it could be that separate linear fits in the large x and small x regions will give as good a fit as a comprehensive cubic model.

D If a cab breaks, the driver can't work. If the company has extra cabs, then when one or several are in the shop the drivers can still generate revenue.

E There is a temperature control knob in the refrigerator.

F How much to eat at each meal, what time to have meals, and how many meals per day?

G Boiling campsite water to purify it.

H The number of tasks to accept in a job (or courses in a semester).

I Layout of a resume.

J Length of an electric cord for a home-use vacuum cleaner.

K Dimension of the packaging of a multi-pack of soft drink cans (12, 15, 18, 24, 36, ... cans in a pack?) (side by side, layered, hexagonally or rectangularly packed?)

L Thickness of newspaper.

M Mathematical model of a power series type.

N Game strategy.

O Path to get from house to class.

P Method to raise children.

Q Orientation of a house.

R Type of computer.

S Menu for a week of home-prepared meals.

T Green-on time of a traffic light at an intersection.

U Speed limit on a road.

V Number of elevators in a building.

W Gasoline contains ethanol.

X Passenger jet flying speed.

Y Designing a bicycle.

Z Driving a bicycle.

3 Use the analytical technique, $dJ/dDV|_{DV^*} = 0$, to determine the value of x that maximizes $y = 8x - 2x^2 + 4$.

4 Use the analytical technique, $dJ/dDV|_{DV^*} = 0$, to determine the value of x that maximizes $y = \cos(2\pi x/10)$ over the range $0 \le x \le 10$.

5 Use the analytical technique, $dJ/dDV|_{DV^*} = 0$, to determine the value of x that maximizes $y = xe^{-x}$.

6 Minimize this function of D. You choose numerical values for coefficients ($a > 0$ and $b > 0$), $J = aD^{-4} + bD^{+2}$. Use the analytical technique, $dJ/dDV|_{DV^*} = 0$, to solve for D^*. Since you can specify coefficient values, there are an infinite number of right answers. How do you know that your D^* is the right value?

7 Choose three (x, y) data pairs, and generate a linear model that best fits the data in a least squares sense. For each of the $i = 1,2,3$ pairs, the equation $y_i = a + bx_i$ is true. Use the analytical technique $dJ/dDV|_{DV^*} = 0$ to solve for the $\{a, b\}$ coefficient values. Here, the OF is the sum of squared deviations between model and data. Because there are two decision variables, there are two $dJ/dDV|_{DV^*} = 0$ equations, termed the normal equations. They will be linear in coefficients

{a, b}. Solve them simultaneously. Show each stage of your process. To verify that you have gotten the right answer, compare your answer to any automated method of curve fitting (such as the trend line function in Excel charts).

8   A person goes to bed at midnight and has to be on the job at $8:00$ am the next day. Before going to bed she makes a decision about the time, T (hour), for the alarm clock to ring. It takes 30 min to travel to work. The longer she sleeps, the more sleep benefit she gets. Benefit $= 1 - e^{-T/3}$. However, the longer she sleeps the greater is the anxiety about, and potential cost associated with, being late to work. Badness $= 1/(7.5 - T)$. Graph these to show a qualitative relation of benefit versus hours of sleep and the cost of wake-up time. Additively combining these multiple and competing objectives as a single OF, $J = \left(1 - e^{-T/3}\right) - 1/(8 - T)$. Use the analytical technique, $dJ/dDV|_{DV^*} = 0$, to solve for the best wake-up time.

9   As a batch process ages over several days, it makes a higher yield of desired product, but there is a diminishing returns approach of product made to an asymptotic limit. $M(t)$ is the mass of material produced after a batch time t, in days, and a simple diminishing returns equation is $M(t) = k\left(1 - e^{-(t/\tau)}\right)$. When the batch process is terminated, the vessel is emptied to recover product, filled with fresh feed, and a new batch is restarted. The turnaround time for emptying and recharging is θ. The process owner wants to maximize annual production of the desired product. Letting the batch run longer increases the yield per batch, but it means fewer batches per year. How long should each batch run to maximize annual productivity of the vessel? Set up the optimization statement. Clearly indicate OF, DV, and the equations (models) that relate DV to other elements. Use the analytical technique, $dJ/dDV|_{DV^*} = 0$, to solve for the best batch duration.

10 A right circular open tank of height, h, and radius, r, has volume $\pi r^2 h$ and surface area $\pi r^2 + 2\pi rh$. The cost of a tank is related to the amount of material in the surface of thickness, t, density, ρ, and cost per unit mass, c. Cost $= ct\rho(\pi r^2 + 2\pi rh)$. What dimensions, r and h, minimize cost while meeting a specified volume, V? State the optimization in standard form. Explain simplifications that you are using. Use the analytical technique, $dJ/dDV|_{DV^*} = 0$, to solve the simplified application for {r^*, h^*}.

11 Use the analytical technique, $dJ/dDV|_{DV^*} = 0$, to show that if x^* is the optimum DV value for $f(x)$, it is also the optimum value for $a + f(x)$.

12 Implement the heuristic direct search method on any of Exercises 3–11.

13 Provide an example of each for the following terms from your personal life in the past day, or so, sufficiently and explicitly, but briefly explain so that a reader can clearly understand. Reveal that you understand the terms. But don't reveal anything improper! (i) State an OF and one or more DVs. (ii) State your algorithm for determining the Optimum. (iii) State your stopping criterion (or criteria).

14 State whether each of these is, or is not, a DV. In case it could be either, clearly explain your reason. Use normal situations: (i) outside temperature, (ii) oven temperature, (iii) fever temperature, (iv) car temperature, and (v) house temperature.

15 For each model, state whether it is linear or nonlinear in coefficients (a, b, c, etc. that need to be valued in regression modeling) *and* linear or nonlinear in $y(x)$ functionality (for finding x^* to optimize y). $y = a*\sin(x)$, $y = a*\sin(x + b)$, $y = a + bx_1 + cx_1x_2$, $y = a + bx_1 + cx_2^d$, $y = a + b/x$.

16 If you were to optimize a written sentence, what would you choose as the DVs, OF, and Constraints? Just list the terms or concepts, not equations.

17 The volume of a rectangular block of length (l), width (w), and height (h) is $V = lwh$. Its surface area is $S = 2(lw + lh + hw)$. Set up the optimization statement to determine the block dimensions (l, w, h) that meet a particular volume requirement (V_o) yet minimize surface area.

18 Consider optimization of the number of holes in a shoe for the shoelaces and state an OF, the DV, and a constraint.

19 Consider optimization of a parasol (an umbrella to keep sunlight off a person). Is the handle color an OF, a DV, or a constraint?

20 Why use optimization algorithms to find the minimum when you could plot the function and see directly where it is?

21 Here is an equation relating an independent variable, x, to the dependent variable, y. $y = a + bx + cx^2$. In an optimization application, when would: "b" be a DV? "x" be a DV?

22 Reveal that you understand optimization concepts by describing an OF, an associated DV, and a constraint for each of the following. Provide brief and elementary answers.
   A Consider that you are *designing* a bicycle. State an OF, an associated DV, and a constraint.
   B Consider that you are *driving* a bicycle. State an OF, an associated DV, and a constraint.
   C Consider that you are *designing* a better leaf for a tree. State an OF, an associated DV, and a constraint.
   D Consider that you are *driving* a car. State an OF, an associated DV, and a constraint.

23 Something decided the thickness of newspaper. It is thicker than tissue paper, but thinner than writing paper. List several of the issues that would guide selecting an optimal thickness value.

24 Show that in minimizing the sum of squared deviations of data from a common value, $\min\limits_{\{a\}} J = \sum_{i=1}^{N}(x_i - a)^2$, the common value becomes the average, $a = (1/N)\sum x_i$. Use the analytical method to solve the optimization application for the DV value.

25 True or false? Explain.
   A ___ The optimizer calculates the OF value.
   B ___ The convergence threshold affects proximity to the true DV*.
   C ___ If three optimizer runs from randomized starts find the same DV*, that is the global.
   D ___ Education should have students work in the knowledge, but not the evaluation level.

E ___ The development of the OF models for an application requires engineering skill.

F ___ The convergence type must be related to the DV to find DV*.

26 For two optimization examples, one from your personal life and another from work/research, describe with precision the OF, DV(s), and constraint(s). In this exercise you will describe two optimization applications representing diverse aspects of your experience. For each, state a plausible choice for the OF, DV(s), and constraint(s). Explain why the ones you chose were "right" for your situation. "Describe with precision" means you should provide equations that would quantify how the OF value is calculated, unambiguous statements of quantifying the DV(s) and constraint(s), and units on the OF and DV. For example, "amount of food" as a DV is not as precise as "calorie intake per day." Also, for example, "quietness" of the automobile ride is not a quantifiable OF, but "microphone dB could be." And, for example, "don't place too much in the tank" is not as precise a constraint statement as "the tank contents cannot exceed 55 gal." In a text statement of your objective, you may use value-laden words like good, best, works, smooth, large, fits, pleases, etc. But your OF, DV, and constraint descriptions should not have any subjective, intuitive, or ambiguous meaning.

27 Return benefit from effort has a diminishing returns response. There is also a threshold effort before there is any return. Consider building a piece of furniture. Effort may start with buying the wood, stain and varnish, and sharpening and aligning the cutting tools. This effort is necessary, but so far there is nothing to sell; there is no benefit to the carpenter. After assembly, light sanding and one coat of finish may make a salable item; but more sanding and another coat of finish makes it an excellent item, increases perceived value to the buyer, and willingness to pay a higher price. Additional carpenter effort beyond excellence moves the item toward perfection, but doesn't increase value over something already excellent. Often an S-shaped relation is used to relate value to effort. The logistic relation is one, $V = a/[1 + e^{-b(E-c)}]$, where "E" represents the time or effort invested and "V" the value. You probably have 20 things that you are working on and only 100% effort to give. If you work totally, you don't eat or sleep, which are necessary functions to be able to invest energy in any project. As you personally seek to optimize life outcome, what are DVs, OF, and constraints? For simplicity and clarity, only consider three projects.

2

Optimization Application Diversity and Complexity

2.1 Optimization

Optimization is a procedure that seeks the best and is relevant to a wide range of applications, but uses a common statement:

$$\begin{matrix} \min \\ \{DV\} \end{matrix} J = OF \tag{2.1}$$

However, there is a wide diversity of applications presenting a variety of characteristics that should be considered when choosing an optimizer.

Many algorithms have a common approach to determining the DV^* values: Set the derivative of the OF w.r.t. each DV equal to zero, and solve the set of equations for the DV^* value. However, characteristics of the OF and DV often create difficulties (multiple solutions, nonlinear nature, no analytic methods to get derivatives, discrete values) that make this approach impossible to implement.

This chapter provides an overview of the diversity of applications and reveals the unique issues and difficulties associated with them.

2.2 Nonlinearity

Consider the simple example of determining the dimensions of rectangular box to meet a desired volume while minimizing surface area. If the sides have dimensions of height, width, and length, but no thickness, and if the sides can be constructed from ideal planes, then the volume is $V = hwl$ and the surface area is $S = 2hw + 2lw + 2hl$. Since choosing two dimensions constrains the third to meet the volume constraint, the optimization statement is

$$\begin{matrix} \min \\ \{h,w\} \end{matrix} J = S = 2hw + \frac{2V}{h} + \frac{2V}{w} \tag{2.2}$$

Solving analytically, by setting the derivatives of the OF w.r.t. each DV to zero, results in the following set of equations:

$$\begin{aligned} 0 &= 2w - \frac{2V}{h^2} \\ 0 &= 2h - \frac{2V}{w^2} \end{aligned} \tag{2.3}$$

Engineering Optimization: Applications, Methods, and Analysis, First Edition. R. Russell Rinehart.
© 2018 R. Russell Rinehart. Published 2018 by John Wiley & Sons Ltd.
Companion website: www.wiley.com/go/rhinehart/engineeringoptimization

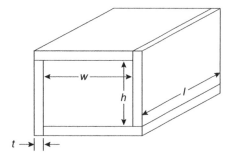

Figure 2.1 A box.

These equations are nonlinear but can be easily rearranged and solved analytically to give $h = w = l = \sqrt[3]{V}$.

However, boxes need to be constructed of material of finite thickness, t. Perhaps the box is assembled in the manner shown in Figure 2.1 in which the edges of the front-facing and rear-facing ends are surrounded by the four other sides, and each of the four sides overlaps one edge of the other. The front and rear ends have cut dimensions of h and w. But the height and width dimensions of the other sides are $h + t$ and $w + t$. So that the internal volume is as desired, the length of those other sides must each be $l = (V/hw) + 2t$. The surface area must now include the faces of the six sides as well as the eight exposed edges on the front and back ends and the four exposed edges on the sides.

The optimization statement is

$$\min_{\{h,w\}} J = S = 2hw + 2(h + w + 4t)\left(\frac{V}{hw} + 2t\right) + 4t(h + w + 2t) \tag{2.4}$$

Now, taking the calculus derivatives without error will be a challenge for many, and if done, then attempting to solve the system of nonlinear relations is another challenge.

Whether mathematical complexity is the result of nonlinear relations or not, for realistic versions of even simple applications, the mathematics often become intractable—or effectively so for an engineer who is not also a hobbyist mathematician.

2.3   Min, Max, Min–Max, Max–Min, …

Most often optimization seeks to find the DV value that minimizes or maximizes an OF value. But it is not unusual to seek the DV value to minimize a maximum value. Consider the location of a hiking path through the mountains. If the objective is to get from point A to point B in the shortest distance, that particular path might travel up (or down) steep slopes, making it difficult for the hiker. In such a case, the best path might be the one that minimizes the steepest slope in addition to the overall path length. The procedure would be to look at all of the slopes along the path, find the maximum, and seek a path that minimizes the maximum.

In devising a business model, the nominal objective is to maximize some measure of profitability. But the "givens" related to forecasts of sales, tax rates, energy costs, raw material availability, tariffs, rainfall, maintenance, etc., are all uncertain. Rather than choose DV values to maximize the profitability index at the nominal situation forecast, a better objective would be to search over the range of the external influence values, then to see what the worst possible (the minimum possible) outcome is, and then to maximize the minimum over all situations.

A similar max–min example from my personal life relates to the allocation of investments in my retirement account. When I was working and investing in a financial plan, I chose to place nearly all of monthly contributions in high growth potential investments. The logic is that over a 40-year investment program, the risk of a few years' loss of investment value to me is inconsequential, knowing that, in the long term, the investment will be maximized. However, now that I am withdrawing from the plan, if the value drops and I continue to withdraw retirement income, the value loss is compounded. Forecasting diverse but realistic possible trends in the investment in a high growth portfolio, if the market has a significant downturn, there is actually a chance that I could consume all of the investment prior to my life expectancy. Running out of income is not my end-game philosophy. By contrast, if I now invest in a balance of steady and growth investments, then, although the end-game inheritance I leave for my children will likely be lower, the chance of it being zero or less is minimized. I choose a balance of investments to minimize the probability of the worst-case situation given the vagaries of the market. I choose to minimize the minimum.

As a last example, in standard least squares regression, the objective is to determine the model coefficient values to minimize the sum of squared deviations between model and data. This is a case limiting the broader objective, which is to determine the model coefficient values that maximize the likelihood that the model could have generated the data. Consider (y, x) data pairs and a $y = f(x)$ model. If the uncertainty in the x- and y-values is similar, then maximizing likelihood that the data could have been generated by the model is the same as finding the model coefficients that make the model come closest to the data points. Choose model coefficient values, and then search along the model line to find the point on the model that has the minimum distance to each experimental data point. Sum the squares of the minimum distances, and seek model coefficient values that minimize the squares of the minimum distances—minimize the minimum.

2.4   Integers and Other Discretization

Often the decision variable can only have integer values. Here are a few examples: Your business can hire 1, 2, 3, or 4 people, or it can buy 1, 2, 3, or more delivery trucks. You can install 1, 2, 3, or more light bulbs in a room. True, one could consider illumination wattage as a continuum-valued variable to replace the number of lights or a shared truck or part-time employees as providing any fractional portion of a unit capacity. But one still has to install an integer number of light fixtures, and although a business can use one truck 37% of the time, it cannot send out 37% of a truck or provide parking for 37% of a truck.

When you purchase material, it comes in units of cans, barrels, sacks, tank cars, rolls, etc. Packages contain an integer number of items. True, at the engineering/business scale, weight is a continuum-valued variable, and you could be supplied with sacks of any weight, for instance, 26.83 lbs. But standard supplies come in sacks of 30 or 40 lbs. Of course, it is almost true that quantity is continuum valued. When one gets to the molecular limit, half of a molecule is a different material. And many optimization applications within particle physics and chemistry need to consider the single molecule and quantized energy aspects.

Those were integer examples. Additionally, variables that could be considered to be continuum valued are often discretized. As an example, on the human macroscale, time is a continuum. However, in a dynamic simulation, time is discretized into Δt increments for numerical integration. If you are seeking the optimal time for the simulated process, the DV values can only have integer in multiples of Δt. If you are seeking to optimize some other variable, then the time discretization will create ridge

Figure 2.2 Illustrating striations.

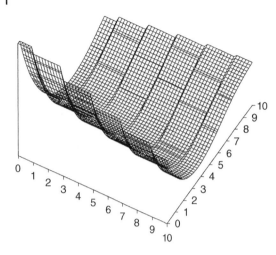

discontinuities (striations) on the response surface. As another example, length is a continuum variable, but when you seek to order a pipe, the diameter only comes in standard sizes, perhaps discretized to an integer number of millimeters. Or is it discretized in 10ths of a centimeter or 1000ths of a meter?

Whether the discretization is on integer numbers of whole items or on incremental portions of some unit of mass, length, or time, the impact is equivalent. First, if the DV is limited to discretized values, then there is a constraint on optimizer choices for the trial solution. The constraint may be that values are limited to an integer number of increments, $DV = N\Delta$. Or it could be that the DV is limited to a value from a set, $DV = \{DV_1, DV_2, DV_3, ..., DV_N\}$. Second, if another variable in the model is discretized, then a magnified view of the OF response will reveal irrational ridges or flat spots that will confound an optimizer search.

Figure 2.2 is a close-up of a 2-DV surface with striations due to time discretization, Function 15 (Retirement). It is nominally a continuum application, and over 70 years, the surface appears smooth, but when zoomed into a local section, the discontinuities are visible.

2.5 Conditionals and Discontinuities: Cliffs Ridges/Valleys

Models often use IF–THEN rules, which are conditional statements in programming. For instance, if the Reynolds number is greater than 2100, use a turbulent flow model; otherwise use a laminar flow model. As another instance, if a delivery load exceeds 1 ton, use the next larger truck. We use conditionals in the choice of thermodynamic models, tax structures, equipment choices, safety or compliance choices, and many other aspects of models that we are using in an optimization.

Conditionals lead to either cliff or sharp valley/ridge discontinuities in the OF response to the DV. These invalidate local derivative evaluation and invalidate the basis of gradient-based, Newton-type, and surrogate-model optimizers that presume continuum-valued surfaces.

Whether gradient-based or direct search algorithms, sharp valleys tend to make the trial solution jump back and forth from side to side across the valley, rather than to follow it downhill.

Figure 2.3 Illustration of sharp valleys.

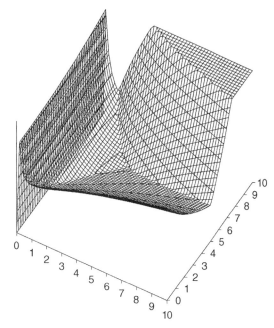

Figure 2.3 reveals an application with a sharp valley, Function 27 (Design of a Liquid/Vapor separator). The discontinuities are due to a conditional that takes the larger of several methods to calculate the tank diameter. Near to the valley, but not at the optimum, local gradient searches will tend to direct the next trial solution across the valley, not down the middle. Further, direct search methods will only find a better trial solution if chance places the next TS in the "V" of the local contour.

2.6 Procedures, Not Equations

Contrasting most textbook examples, applications of utility are usually not convenient, one-line, mathematical statements of either the OF or the constraints. Although simple examples are important to understanding, in my experience what we seek to optimize for either career or personal life does not lead to simple one-line equations. What I encounter are procedures—computer simulations that are solved by numerical techniques.

One issue is constraints. Procedures often have a myriad of internal variables, which can lead to execution errors such as divide by zero, log of a negative number, or subscript out of range. There may be path integrals and unknowable constraints until the procedure finds it on the path. It may not be possible to express such constraints as a simple function of the DVs.

Another issue is obtaining derivatives and second derivatives. Without a closed-form, analytically tractable relation between OF and DV, optimizers need to estimate derivatives numerically. Here, choices of increment size and whether forward, backward, or central difference can affect the solution.

Finally, often, nonlinear functions are locally linearized to obtain rational direction for the optimizer next TS value. Then the functions are re-linearized at the next TS. The sequential re-linearization is a procedure.

2.7   Static and Dynamic Models

Dynamic models are time-dependent and reveal how a variable will evolve over time. They are often called transient models. By contrast, static models, often called steady-state models, reveal what the end value will be once all transients settle. In batch processes, not continually influenced externally, the steady state is the thermodynamic equilibrium. However, if the batch is continually influenced, the steady-state result might be the result of the influence rate being balanced by a relaxation rate, such as in a continuously fed flow-through batch reactor, often termed a continuous stirred tank reactor (CSTR).

Static models are most often used in process design and process analysis. But dynamic models are essential to process control and scheduling, in which understanding and shaping the path from one state to another is the objective. Since a dynamic model will settle to the steady-state value, the dynamic model must contain the functionality of the static model. Additionally, dynamic models include the time dependence. The issues related to how something evolves over time adds several aspects to optimization.

First, one may wish that the transient path between the initial and final states minimizes time, energy, or waste, or maximizes impact. This reveals the need to accumulate the OF value over a path, which is termed a path (line) integral and discussed in Section 2.8.

Second, one may wish that the transient path between the initial and final states minimizes proximity to constraints or minimizes the worst potential case. This would be a min–max application, as discussed in Section 2.3.

Third, since most dynamic models are computed with numerical techniques, as opposed to analytical techniques, the incremental time discretization of the numerical technique creates time jumps, a discontinuous path. The discontinuity may not be visible to the eye when graphed, as a good numerical method will provide, but at the level of the optimizer convergence criterion, the striation discontinuities may have an impact. Issues related to discretization discontinuity are discussed in Section 2.4.

2.8   Path Integrals

When the objective is to maximize or minimize the accumulation or consumption along a path, one has an integral along a curve. Perhaps it is to choose a car accelerator position schedule that minimizes the consumption of fuel from point A to point B along a road. Perhaps it is to choose a time of day for travel that minimizes the number of bug splats accumulated on a windshield along a particular road. The path does not need to be a road or a physical path through Cartesian space. The objective might be to determine control action for a process to transition from one product to another with minimum total waste. Here the path is how the process state (temperature, composition, etc.) changes in time. Chapter 5 shows how to analyze along a path.

2.9   Economic Optimization and Other Nonadditive Cost Functions

In optimization there must be opposing functionalities, some aspects must get better, while the other get worse as the DV value changes. In economic optimization this typically seeks to balance benefit with initial capital cost and future expenses. Typically, one wishes to minimize capital investment, c, and maximize annual cash flow, F. For instance, purchasing an inexpensive car today may permit a schoolboy to get paid to deliver pizza today, but it may become a maintenance or operating expense burden in the future. Instead, purchasing a more expensive car today may have higher resale value, and lower operating expenses, but it may take longer to pay off and realize the delivery income. There are many methods for combining fixed initial capital and the annual cash flow (income less expenses). A simple profitability index is payback time (PBT), where the initial capital investment is divided by the expected annual income less expenses, $\text{PBT} = C/(I-E)$, to determine the number of years that earnings will pay for the investment. This does not account for either taxes or depreciation (an annual accumulation of \$ to regenerate the capital at its end of life). Long-term return on assets (LTROAs) includes working capital and the average value of the fixed capital, $\text{LTROA} = (1-T)(I-E-D)/(0.5C+W)$. But neither LTROA or PBT includes scrap value at the end-of-useful-life or the time-value of money. Net present value (NPV) discounts the cash flow in each future year to the value you would have to have in the current year, which would be compounded to equal the future cash flow. NPV is the sum of all present and future discounted cash flows. $\text{NPV} = -c_0 + \sum \left(F_i/(1+r)^i \right) + c_N$, where r is the inflation rate, F is the cash flow in the ith year, and c represents the capital or scrap value after N years.

In all, the concept is to balance capital and cash flow, but none of these are simple additive relations of the two opposing objective functions. One could compose the OF as $J = F - c$. But since F and c have different units, there needs to be a weighting factor to permit combination, as $J = F - \lambda c$. And here, the value of λ needs to reflect the appropriate balance of initial capital and annual cash flow. In economics the opposing OFs are not added, but combined in a more complicated manner that reflects the economic reality.

There are other nonadditive ways to combine opposing objective functions. For example, in choosing the number of terms in a regression relation, the user seeks to minimize complexity (have fewer number of terms and coefficients) while improving the goodness of fit between model and data. Of several measures with similar bottom-line impact, the final prediction error (originally proposed by Akaike and subsequently promoted by Ljung) is defined as $\text{FPE}_m = ((N+m)/(N-m))\text{SSD}_m$, where N is the number of data, m is the number of coefficients in the model, and SSD is the sum of squared deviations between model and data. The complexity factor, $(N+m)/(N-m)$, a measure of undesirability, multiplies the SSD, another measure of undesirability.

2.10 Reliability

A light bulb might fail. If there is only one bulb in the room, when the bulb fails, the room goes dark. So, place two bulbs in the light fixture, and when one bulb fails, one bulb is left to provide light to keep the room functional and to provide light so that the bad bulb can be changed. But the second bulb might fail before the first can be fixed. There are different questions that can be asked. A simple one is "How many bulbs are needed to have the room functional 99.999% of the instances of use?" The situation may be that the chance of failure of each bulb is independent and that when one of more bulbs is operating, the room is functional. Then the probability of at least one out of N bulbs being operable is the same as the probability of either 1 or 2 or 3 or 4 or … N out of N being operational, which is the same as the complement the probability that none is operable. Using $p_{\text{individual}}$ as the probability that a bulb works on demand,

$$P_{\text{desired system reliability}} = P(n > 1 | N) = 1 - \left(1 - p_{\text{individual}}\right)^N \tag{2.5}$$

Solving for the number of bulbs,

$$N = \frac{\ln\left(1 - P_{\text{desired for system}}\right)}{\ln\left(1 - p_{\text{individual}}\right)} \tag{2.6}$$

This is a simple introductory probability example, which does not need optimization to solve for the DV value, the number of bulbs.

Optimization might aim to balance system reliability with cost. If c_1 is the price of one bulb, then $c_1 N$ is the cost of N bulbs. If c_2 is the opportunity cost of a nonfunctional room, then $c_2 \left(1 - P_{\text{system functionality}}\right)$ is the risk. The optimization might be to determine N to minimize the initial and opportunity cost:

$$\min_{\{N\}} J = c_1 N + c_2 \left(1 - P_{\text{system functionality}}\right) = c_1 N + c_2 \left(1 - p_{\text{individual}}\right)^N \tag{2.7}$$

Although this is a simple application, it reveals that the decision variable is an integer, not a continuum-valued number, which is a difficulty for many optimization algorithms.

Another more complicated question might recognize that the room is less and less functional with each bulb failure, and the question might be "How many bulbs are required to ensure that the room is at least 50% functional, 99.99% of the time?" The next level of complication might recognize that the failure probability for a bulb is not a constant, but that it is a "bathtub" curve, high initially because there might be defects, dropping too low if there were no defects, and then rising after many uses due to age. Other complications might include common cause failure modes—if one bulb fails, then it might have been due to an external/environmental feature that will increase the chance that another bulb fails. These conditions do not lead to simple probability equations that might admit an analytical solution. The probability can still be calculated, but from numerical, Monte Carlo techniques, that add stochastic uncertainty to the calculated OF value.

2.11 Regression

The objective is to determine model coefficient values that make a model to best fit the data. Here the DVs are the model coefficients. By contrast, in optimization for design or control of processes, devices, or products, we have the model, we have the mathematical description of how the thing behaves, and optimization aims the design choices (such as number of elements, size, insulation type, valve characteristic) or influences (such as flow rate, operating temperature, and pressure) to make the process or device behave as desired. The influences are the design choices, and the consequences are the measures of goodness related to a process or product (cost, reliability, function, etc.). In process or product optimization, the model coefficients are "given," and the inputs (influences) and outputs (consequences, outcomes, results) are the variables that an optimizer adjusts and observes.

However, in regression the model coefficients are the unknowns, and the influences and results are known. In regression, an experiment has revealed how the process or product responds to the influences, and these inputs and outputs are the givens. These sets of inputs and outputs are called data. The model coefficients are adjusted to make the modeled response to the input data values the best match to the experimental output data values.

A steady-state model can be expressed as $\tilde{y} = f(c,x)$, in which the response is represented by the variable y and the vectors c and x represent model coefficients and the influences. The squiggle over the variable y explicitly acknowledges that it is the modeled response, not the experimental data response. For a particular set of data, the ith of N total number of data sets, the model is $\tilde{y}_i = f(c,x_i)$. Note that the subscript on the model response $\tilde{y}_i$ and on the experimental influences, x_i, indicates a particular data set. Note that there is no subscript on the model coefficient vector, c, indicating that the c-values are universal across all of the data range.

A dynamic model acknowledges how some variables change in time. It is a time-dependent or transient model, which often can be represented as $d\tilde{y}/dt = g(\tilde{y},c,x)$. Here, the rate of change of a response may be dependent on the response. For example, if a reactor temperature is high, then the rate of reaction and the rate of change of temperature will be high. Also, if the level in a draining tank is high, then the rate of change of level will be high. So, the response variable is explicitly indicated in the RHS function. Likely, the response and influences are not recorded continuously in time, but are digitally recorded at discrete time samplings. The model is valid at any particular time sampling, j, indicated by $(d\tilde{y}/dt)_j = g(\tilde{y},c,x)_j$. Likely, the dynamic response cannot be solved by analytical methods, but can be approximated with a finite difference method, $d\tilde{y}/dt = \Delta\tilde{y}/\Delta t = (\tilde{y}_{j+1} - \tilde{y}_j)/\Delta t$, in which the subscript j represents a time-counting index. Then the model can be used to solve for the new y-value given the influences and y-value from the past time sampling. Using Euler's explicit method (forward finite difference),

$$\tilde{y}_{j+1} = \tilde{y}_j + \Delta t\, g\left(\tilde{y}_j, c, x_j\right) \tag{2.8}$$

With a steady-state model and data, the order of listing the experimental data is irrelevant. However, for a time-dependent model, any one value depends on the prior value, so any listing of data and model must preserve the chronological order.

Note: At steady state the rate of change of the response variable is zero, $(d\tilde{y}/dt)_{SS} = 0 = g(\tilde{y},c,x)_{SS}$. Adding $\tilde{y}_{SS}$ to both sides produces $\tilde{y}_{SS} = \tilde{y}_{SS} + g(\tilde{y},c,x)_{SS}$. Then, explicitly acknowledging that the static model is valid at steady state, $\tilde{y}_{SS} = f(c,x_{SS})$, reveals that the dynamic model, $g(\tilde{y},c,x)$, must contain the same functionality as the steady-state model, $f(c,x)$, $f(c,x_{SS}) = \tilde{y}_{SS} + g(\tilde{y},c,x)_{SS}$.

In regression the objective is to make the model match the data. Desirably, $\tilde{y}_i = y_i$ for all i (or use subscript j if a dynamic model). However, two aspects make this impossible. First is experimental uncertainty (error) in both the y and x data. Second is process–model mismatch (the model will not exactly capture all nuances of the process). Accordingly, a realistic desire is to minimize the deviation between modeled and data responses, $d_i = y_i - \tilde{y}_i$. However, if the modeled values were absurdly large (bad values), making them larger (worse) would make the d-values largely negative, minimizing them. So, the concept is not to make them as negative as possible, but to make their magnitudes as small as possible. Some interpret this as to minimize the absolute values of all of the deviations. However, the absolute value has discontinuous derivatives at the $d_i = 0$ value, which leads to nonanalytic responses of the OF. Further, human experience imposes a penalty for a deviation from that desired, which roughly with the square of the deviation, not proportional to the magnitude. Lastly, theoretical analysis of normal statistical variation, likelihood analysis, reveals that the idealized penalty should be proportional to the square of the deviation. So in regression, the commonly accepted objective is to minimize the sum of squared deviations (SSD). For a static model,

$$\min_{\{c\}} J = \text{SSD} = \sum_{i=1}^{N}(d_i)^2 = \sum_{i=1}^{N}(y_i - \tilde{y}_i)^2 = \sum_{i=1}^{N}(y_i - f(c,x_i))^2 \tag{2.9}$$

For a dynamic model,

$$\min_{\{c\}} J = \text{SSD} = \sum_{j=1}^{N}(d_j)^2 = \sum_{j=1}^{N}\left(y_j - \tilde{y}_j\right)^2 = \sum_{j=1}^{N}\left(y_j - \left(\tilde{y}_{j-1} + \Delta t g\left(\tilde{y}_{j-1}, c, x_{j-1}\right)\right)\right)^2 \tag{2.10}$$

2.12 Deterministic and Stochastic

A deterministic model returns one value for a set of givens (coefficients and influences). For instance, when you use the ideal gas law to calculate the pressure of a gas, $P = nRT/V$. Given a quantity, temperature, and volume, the calculated value of P is exactly the same regardless of who does it, what calculator we use, or when we decide to do the calculation.

By contrast, stochastic models do not return the exact same value each time they are asked "What is the answer?" Consider the rolling standard (cubical) of game playing dice. The outcome value on any one roll is either a 1, 2, ..., or 6. The average value might be 3.5, but the average of 10 rolls might be 3.3 or 3.7 or 3.4. Each possible outcome value is a realization, a possible response. This variation is generated in a model in a technique referred to as Monte Carlo, in honor of its application origins in the gaming casino. There are many ways to have a computer return a stochastic value, most often based on a random number generator that provides a uniformly distributed value, r, in the 0–1 interval $0 \leq r < 1$. There is a one-sixth chance that the value of r is in the $0 \leq r < 1/6$ interval, or in the $1/6 \leq r < 2/6$ interval, so, the number that appears on a die can be simulated as

$$n = 1 + \text{INT}(6r) \tag{2.11}$$

Such variation of the result is a problem for an optimizer. If, for instance, it is moving the DV in the right direction, but a statistical perturbation provides the appearance of a worse value, the optimizer may think best is in the other (wrong) direction. In a stochastic process, optimizers will tend to seek a phantom minimum, one that appears to be best simply because of the vagaries of the statistical perturbations.

Sometimes, in very idealized situations we can derive the average or statistical expectation. However, in complicated applications (such as in a game or business projection in which the outcome is subject to many rules, influences, and conditional inputs) an analytical average is not possible to derive. Another approach to identify the expectation is to increase the number of realizations and average the results. But, although variation reduces, it is not eliminated. The central limit theorem indicates that the range on the variable drops with the square root of the number of realizations.

In structuring a business, for instance, we might forecast possible outcomes considering the vagaries of possible future tax changes, raw material and energy costs, customer demand, etc., to see the result of our business choices. We might have a thousand possible realizations and look at either the average or 95% possible worst outcome. In any case, the value for 1000 simulations is not the certain truth, it has statistical vagaries. In designing a reservoir, for instance, we might consider the vagaries of rainfall and water use over a 100-year simulation to see the impact of choices on location and dam height. In designing an airplane, for instance, one might model the response to 1000 possible arrangements of

Figure 2.4 Illustration of a stochastic function.

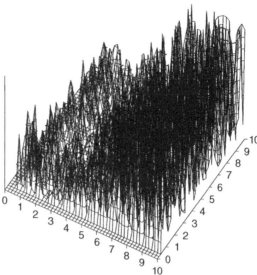

baggage distribution and weight, passenger seating, fuel weight, air properties (humidity, temperature, density), wind, flight path changes, and pilot response.

If considering minimizing the 95% worst case (not the worst possible, but of all possible outcomes the one for which a worse outcome only happens 5% of the instances) and optimizing a design to max-imize its value, the optimization statement might appear as

$$\begin{matrix} \max \\ \{\mathbf{DV}\} \end{matrix} J = OF_{0.05} \tag{2.12}$$

Figure 2.4 shows a stochastic surface, from Case Study 9, an *in-vitro* fertilization (IVF) study. The ver-tical axis represents several undesirability aspects, and the lower left axis is the number of eggs to be fertilized in the first procedure, and if it is not a success, the lower right axis represents the number to be fertilized in the follow-on procedure. The stochastic nature is the result of the unknown probability of procedure success.

2.13 Experimental w.r.t. Modeled OF

Models have various degrees of fidelity to nature. Some match (predict, forecast) very well, but in my experience, none is perfect. Improved sensors, better experimental techniques, and more data always seem to come along and reveal some inadequacy in the model. If there is strong belief that the model is an adequate representation of nature, then models are used as a surrogate for experimentation to guide optimization.

However, often the models are not available, or when a model exists, it does not match the data with desired fidelity to make results from its use credible. In such cases, we can use experimental results to provide the OF response to DVs.

However, experimentation is time-consuming and expensive and generates risk. So, minimizing the number of trial solutions to find a reasonable approximation to DV^* is even more important in experimental work than when there is a model that can be used in computer simulation.

Additionally, experimental data includes uncertainty (experimental error) on both the input and response values. The statistical vagaries of experimentation return a stochastic OF response to the TS. And, like Monte Carlo simulators, the stochastic nature of the OF is important to choosing an optimization algorithm.

A visualization of the result is similar to the stochastic response in Figure 2.4.

2.14 Single and Multiple Optima

Figure 2.5a and b reveals objective functions of one variable with several optima.

Each figure illustrates three local minima. The left-hand figure might illustrate delivery costs for a shipping company as cost changes with business size. With small size, buy a small truck. The left portion of Figure 2.5a illustrates this. The more you ship, the less is the truck cost allocated per delivery. However, when shipping volume exceeds the small truck capacity, you need to either trade it for a larger truck or buy a second small truck. This extra capital increases cost per delivery but permits you toward larger volumes.

These are local optima with discontinuities. In contrast Figure 2.5b shows an OF as a continuous function with multiple optima. In each case, one optimum is clearly the global best. This is called the global optima. The others are local optima. We want our optimizer to find the global overall best.

The issue this presents is that depending on where one starts, on the initial trial solution, the optimizer search may lead to a local optimum, not the global best. For certain classes of objective functions, a particular optimization technique will find the global best, but, in general, there is no guarantee that an optimizer will find it. However, there are things a user can do to improve the probability that the global best will be found. Some techniques include multiplayer algorithms and multiple optimization trials from randomized initialization.

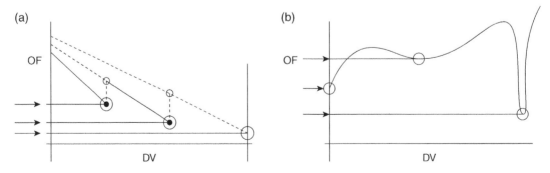

Figure 2.5 Illustrations of multiple optima: (a) discontinuous, (b) continuum. *Source:* Rhinehart 2016. Reproduced with permission of Wiley.

2.15   Saddle Points

Figure 2.6 illustrates a saddle point in a 2-D application, named after the shape of a saddle for a rider on a horse. Values of the two DV values are labeled on the floor of the illustration, and the OF is the vertical dimension. Contour lines (of constant OF value) are shown on the surface net (grid lines). The lowest point is illustrated as the circular contour at the right, near the DV values of (9, 8). This would be the optimum.

The saddle point is near the center of the illustration. Moving from the left to the right, from the DV pair (0, 0) toward location (10, 10), the OF value rises and then falls. However, moving the DV pair from the front (10, 0) to the back (0, 10), the OF falls and then rises. The saddle point is at DV pair (5, 5), where the horse rider would sit. Going across the saddle (left to right in the illustration), the OF rises and then falls; however going front to back, the OF falls and then rises.

At the saddle point, the derivatives of the OF w.r.t. each DV is zero. This creates a problem for optimizers that seek to find the point at which the derivatives are zero. Whether it is a one-dimensional search with a single DV, a 2-D search, or one with many more DVs, optimizers that use a "find the point where the derivatives are all zero" logic will mistakenly indicate that a saddle point is the solution. This logic is mathematically represented by the elements of gradient vector, all having a zero value:

$$\nabla f = \mathbf{0} \tag{2.13}$$

Figure 2.6 Illustration of a saddle point.

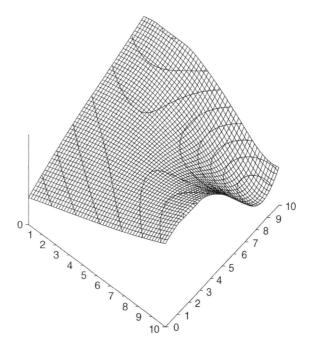

2.16 Inflections

An inflection is the point at which a second derivative changes sign, when the curve changes from convex to concave. As with saddle points, this relates to any dimension, but a single variable visualization is a convenient relation.

Figure 2.7a and b reveals a Lennard–Jones 6–12 energy potential well as a function of atomic separation. The minimum is at a distance value of about 0.75. It would seem that any simple optimizer can find this point. But some optimizers use derivative information and can get misdirected by it. The derivative of the function is illustrated in Figure 2.7b. Note that the derivative is zero at a DV value of 0.75 where the OF is minimum. Note also that the derivative has a maximum at a DV value of about 1 and that this corresponds to the inflection point on the OF curve. In the vicinity of DV*, the OF function is convex (upward curved). In this region the second derivative has a positive value (the slope of the derivative curve and the derivative of the derivative are positive). However, to the right of the inflection point, the OF is concave (downward curve). In this region the second derivative has a negative value (the slope of the derivative curve and the derivative of the derivative are negative).

Any of the several Newton's method optimizers seek to root-find on the derivative. Start at a DV value of 1.5, and root finding on the derivative will send a search toward higher DV values, not back to the optimum. In general for Newton-type algorithms, if the trial solution is on the other side of an inflection point from the optimum, then the search will move further away from the optimum, not

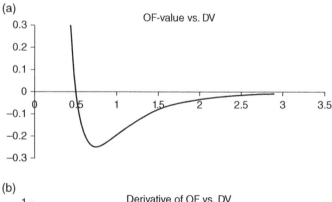

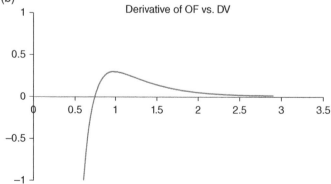

Figure 2.7 (a) A function and (b) its derivative.

toward it. Convergence will not necessarily mean that the optimum has been found if an optimizer wandered off in the wrong direction.

2.17 Continuum and Discontinuous DVs

Conceptually, decision variable values can be continuum valued. For instance, you could choose the dimensions of a cube to be 1.2345678910111213… meters. Or an optimization such as what box dimensions minimize surface area while permitting a volume of $5\,\mathrm{ft}^3$,

$$\min_{\{h,w\}} J = S = 2hw + \frac{10}{h} + \frac{10}{w} \tag{2.14}$$

indicates $h = w = l = \sqrt[3]{5} = 1\,\mathrm{ft}$ and 8.5197113…in. Although the answer is continuum valued, the reality is that dimensions will be discretized to some reasonable manageable increment, such as 1/32nd inch, 1 ft, 8 and 17/32 in.

Or perhaps an optimization indicates that the optimum pipe diameter is 3.718293… in., but when you go to buy one, you see that the sizes are discretized and that exact value is not a choice.

In those examples, realizable DV values are discretized, even though the concept is continuum valued. However, in many applications the DV concept is an integer number. For instance, how many screws should be used to fasten a picket fence? The more used, the better it will resist boards being blown off by high wind or knocked off when bumped into by playing children. But an excessive number of fasteners increases material cost and assembly effort, and too many holes can actually weaken the wood. Another example could be reliability, in which increasing the number of spares improves the likelihood that the device will work. In these examples the DV value can only be an integer, a form of discretization.

The optimizer cannot choose any possible value, but only those that are valid for the discretization.

In those examples the DV values preserve a ratio property. Six meters is twice as long as 3 m; 10 fasteners are twice as many as 5. Those were ordinal numbers. More broadly, however, the DV may be a class or category variable. It might be a choice between a centrifugal or turbine pump, between a propeller or a water-jet propulsion for a boat, or between visiting city A before or after city B. In these cases the DV has a nominal value; it is a string variable (or text). And a derivative concept is not possible.

2.18 Continuum and Discontinuous Models

Models useful for optimization applications are usually not simple one-line equations. While such may be instructive as learning tools, applications are usually from complicated models, in which there are several aspects that create discontinuities.

OFs that require solution to differential equations or other forms of function integration are likely to be solved with a numerical technique. Whether it be a rectangle rule, trapezoid rule, Simpson's rule, Richardson extrapolation, Euler's explicit, Runge–Kutta, or orthogonal collocation, numerical discretization can lead to striation-type discontinuities that affect the fine scale of a DV that is related to the integration time or the spatial dimension of the model.

Many OFs require some sort of root-finding method to solve for internal variables such as those related to thermodynamics. Whether it be a Newton's method or interval halving, the numerical method must have a convergence criterion to determine when the trial solution for the root is adequately close to the true value. The method always stops prior to finding the ultimate, true value, and the numerical error can cause fine-scale disruptions to the OF response to DV choices.

Some models switch from one rule to another. Perhaps it is gain scheduling controllers or fuzzy logic type rules (heuristic rules or a Takagi–Sugeno–Kang rule) or switches in a thermodynamic model as states make one more appropriate than another. Again, rule transitions can lead to local discontinuities to the OF value.

In each of those situations, the conceptual expectation might be that the OF is a continuum and smooth response to the DV, but the actuality is that the surface may have discontinuities, cliffs, or ridges from the numerical method. And these will misdirect many optimizers that might follow a local ridge, the artifact of a numerical method, to a false minimum.

2.19   Constraints and Penalty Functions

Hard constraints are values or events that cannot be accepted. They may be on a DV value, which for physical implementation must be a positive, non-complex value. Or the constraint might be on a secondary consequence of a DV choice that happens nearly immediately (it violates an explosive limit) or in the future (such as the tank cannot overflow or the cash flow must remain positive). These are illustrated as the subject to (ST) portion of this generic optimization statement:

$$\min_{\{x\}} J = f(x)$$
$$\text{S.T.:} \quad a < x_3 < b$$
$$g(x) > c$$

(2.15)

Alternately, the constraint might be on a calculation within the model that might lead to a divide by zero, log of a negative number, subscript out of range, etc. Feasible DV choices cannot lead to computer execution errors. Usually, there is no simple relation from the diverse constraint situations that would reveal a simple range of feasibility on the DVs. So, what happens if the optimizer does its calculations to determine the "TS" values and then discovers that the solution violates constraints? Any retry of the calculations would result in the same constraint-violating "TS."

There are many solutions. If the constraints can be mathematically expressed as an equality relation between DVs, then each equality constraint can be used to reduce the number of DVs by one. The optimizer chooses trial solutions for the $N-1$ DVs, and the constraint is used to calculate the Nth. Optimizers with such logic are called reduced gradient types. The implementation logic is not simple and gets confounded if other constraints are simultaneously encountered. And there might not be a feasible way to get the Nth DV value from constraints that do not permit analytic inversion.

There are other possible solutions to handle hard constraints, but a common approach is to convert them into a penalty function and add the penalty for a constraint violation to the OF. First, consider the STs as desirables, but not absolute requirements. Examples of rules that are desirable are "Wash hands prior to eating." "Obey the speed limit." "Floss daily." "Keep the tank level between 25 and 85%." "No more than two fertilizer bags on any single pallet will leak."

A soft constraint acknowledges that there is a penalty for the rule violation, but that violating the rule is not a catastrophic disaster. Normally, the penalty value is proportional to the square of the constraint violation and is scaled by a coefficient to make it a reasonable balance with the other elements in the objective function.

In the example earlier, a constraint is $g(x) > c$. If the constraint is violated then the amount of violation is $\varepsilon = c - g(x)$; otherwise the magnitude of the violation is zero:

$$\begin{aligned} &\text{if } g(x) \leq c \quad \text{then } \varepsilon = c - g(x) \\ &\text{if } g(x) > c \quad \text{then } \varepsilon = 0 \end{aligned} \tag{2.16}$$

The penalty is added to the OF, where λ is the factor for equilibrating the importance of the constraint violation to the original OF:

$$\begin{aligned} &\underset{\{x\}}{\min} J = f(x) + \lambda \varepsilon^2 \\ &\text{S.T.}: \quad a < x_3 < b \end{aligned} \tag{2.17}$$

Note that the $g(x) > c$ constraint is no longer indicated in the S.T. list. It is now part of the OF calculation.

Soft penalties often permit a bit of constraint violation and are termed "soft constraints" because the OF may include a little constraint violation at the minimum. Figure 2.8 illustrates this. There is a constraint on the DV that should keep it out of the hatched region, but this is where the OF is a minimum, as the dashed line indicates. If there were no constraint, DV* would be the x-value at the minimum of the dashed curve. Increasing λ makes the penalty larger, and the three solid curves reveal three λ choices. If λ is too small, the penalty addition permits a substantial violation of the constraint. Larger λ values will move the optimum back toward the feasible region but still permit the violation, on a lesser amount.

If there are multiple constraints, each can be converted to additional penalty functions.

Choosing the right λ-values that provide an appropriate balance of penalty for violation w.r.t. desirability of the OF can be a challenge to a user, especially when there are several soft constraints included. Often users seek to balance the $-value of the violation with the $-value of the OF, but as often, interpreting the $-value is difficult. My recommended approach is to consider a level of concern that is associated with the OF and penalty and scale both by equal concern factors (see Chapter 21).

The advantages of this approach are as follows: (i) The OF w.r.t. DV surface remains analytic, continuous, and differentiable. It converts the discontinuity of the hard constraint into something that can be handled by gradient-based optimizers. (ii) Hard constraints block optimizers, and this permission

Figure 2.8 Illustration of a penalty function, a soft constraint.

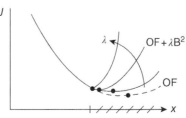

of a bit of violation accelerates the search for DV*. (iii) When there are multiple objectives, they can be combined in this manner.

The disadvantages of this approach are as follows: (i) A bit of violation is permitted. (ii) It is dependent on the user's choice of scaling factors. (iii) It cannot be performed unless there is a measure of the magnitude of the constraint violation.

2.20 Ranks and Categorization: Discontinuous OFs

If runners come in 1st, 2nd, and 3rd in a race, they have ranks of 1, 2, and 3. Rank 1 is better than rank 2. But rank values (or category values or such) do not indicate a sense of how superior one was from another. The 1st could have been faster than the 2nd by a fraction of a second, and both could be a minute faster than the distant 3rd. In this case, 1st and 2nd are effectively identical, but well differentiated from 3rd. Or it could be 1st was much faster than either 2nd or 3rd who may have been essentially identical. The race time is a continuum variable, but the rank is a discrete category, a classification, whether expressed as an integer number (1, 2, or 3) or as a nominal classification (first, second, or third). If the time for runner A was 3 min and 24 s (represented as $3'24''$) and C was $4'17''$, and if the time for B is any number between $3'24''$ and $4'17''$, then the rank for B is unchanged. If the finish time for C was $3'24.1''$, C is 2nd. But if the C-time drops a fraction of a second to $3'23.9''$, then the C-rank abruptly changes to 1st. Rank is a discontinuous response to performance.

Ranking is a significant category of optimization applications. For example, an application could be to develop a model to predict how well a tennis player might perform in competition (or any individual-event athlete such as golfer, sprinter, diver, skater, gymnast, race horse). Why do this? Perhaps you need a method to invite people to be junior members of your school team and need to know who has the best potential to develop into a winner at the varsity level. Perhaps you want to know whether your 6-year-old child has talent and whether to invest time and money in developing him/her for a possible place on the Olympics team. Perhaps you want to bet on tennis matches. Perhaps you want to know if you should join the tennis ladder at work.

The hope is that there are fundamental attributes of the individuals that can indicate their competitive fitness. For instance, reflex rate could be easily measured in a lab test of the individual's delay in hand response to visual signals, and such reflex rate seems to be a relevant metric to imply fast responses on the court, which seems important to winning. Smaller would be better. Another measureable attribute might be how far a person can reach when standing. Larger would be better. There are several such attributes, and the hope is that a model of overall fitness for the game can be developed from measureable attributes.

Perhaps overall fitness can be predicted with a classic power law function $F = f_1^a f_2^b f_3^c...$, where F is the overall player fitness for the game, f_i are the lab-measured attributes, and coefficients a, b, c, etc., are adjustable model parameters (these are the DVs in the optimization). This model indicates that if one factor is very low (perhaps $f_2 = 0.02$) and two are very high (perhaps $f_1 = 0.93$ and $f_3 = 0.89$), the player is not as fit for the game as the one with modest values for all attributes (perhaps $f_1 = f_2 = f_3 = 0.7$). As a note there are many types of models used in ranking, which include fuzzy logic, neural networks, and others. In any case, however, the models must process the attribute values to arrive at an overall fitness and have coefficients that define the processing.

The model coefficients could be determined by taking N number of high-level players, testing them for their attribute values, and having them play in tournaments under various conditions (indoor, outdoor, clay court, morning, evening, etc.) to determine their ranks. The model will be adjusted so that the rank predicted by the overall fitness best matches the average rank found from the tournaments. Perhaps the objective is to minimize the squared deviation between modeled rank and player rank, over all players, and all tournaments (there are many other measures of assessing goodness of ranking):

$$\underset{\{a,b,c,d,\dots\}}{\min} J = \sum_{\text{all tournaments}} \left[\sum_{\text{all players}} (r_{i,j} - \tilde{r}_i)^2 \right] \tag{2.18}$$

Regardless of the details of a model or assessing the goodness of ranking, rank is a discontinuous response to the continuum-valued model coefficient values. So, the OF will have flat spots, DV regions where the OF value will not change as the optimizer changes the DV values. On a flat spot, gradient information is meaningless.

Rank is a discontinuous response, and it is one of a general class of categorization that leads to flat spots on the OF response. In many applications we are seeking to develop a categorization or selection algorithm that uses basic attributes of an image, or situation, or pattern and seeks to return the category. Optical character recognition looks at features of symbols to determine which letter or number it represents. Speech recognition aims to translate human vocal patterns to text. Defense algorithms are tasked to identify the airplane from its shape projection. In all of these applications, the OF makes discontinuous jumps to flat spots as the DV values change.

2.21 Underspecified OFs

If the optimization application is underspecified, there will be multiple DV values with the same OF. This means that you have a choice in the DV value. Should you use the large one or the small one? If you can make the choice, then something makes one choice better than the other. Identify that rationale and include the new relation in the OF so that the optimizer finds it.

A common instance in regression modeling is to structure a model with redundant coefficients. For instance, it might seem that these three $\{a, b, c\}$ coefficients are independent in a model of the form $y = f(x) = (a + bx)/c$. However, dividing each term by coefficient c reveals that there are only two coefficients: $y = f(x) = (a/c) + (b/c)x = \alpha + \beta x$. If there is a unique $\{\alpha, \beta\}$ set, then there are an infinite number of $\{a, b, c\}$ sets that are equivalent, and values for these will be correlated. In the event of parameter correlation, remove a parameter and reformulate the model.

Some applications are effectively underspecified. Instead of the minimum being identical for a range of DV values, the OF appears as a gently sloped valley between steep walls.

2.22 Takeaway

Understanding the reality of the application is critical to the selection of the optimization procedure.

2.23 Exercises

1 Describe your own example to represent these diverse applications:
   A Nonlinear
   B Min–max (or so)
   C Integer or other discretization

- **D** Stochastic
- **E** Underspecified
- **F** Rank or Categorization of OF
- **G** Constraints
- **H** Discontinuous Models
- **I** Saddle Points
- **J** Multiple Optima
- **K** Economic
- **L** Regression
- **M** Reliability
- **N** Path Analysis
- **O** Dynamic Model
- **P** Conditionals

3

Validation

Knowing That the Answer Is Right

3.1 Introduction

This book is focused on people who will be self-learning. I think that optimization is an essential application skill for all who are in business and engineering, but is rarely offered as a course in degree programs. So, I think that there is a need to help those in the practice to acquire the techniques.

Self-learning is different from the school days of teacher-directed learning where the teacher chooses the topics, the students read and attempt to understand, and then the teacher creates tests and grades the student's knowledge or skill. The teacher decides whether the student has adequately acquired the desired skill. In the survival-of-the-fittest environment that is school, graduates have been prepared for passive learning that self-stops when the student thinks they've memorized enough to pass a classroom test. By contrast, in self-learning, the student must create the tests and exercises that affirm that the material is properly understood and that skill is adequate.

If someone says, "Trust me. I validated it. Use these calculations to build the bridge." Would you?

"Yes, I trust you." The boss will say and politely add, "So, I know that you have validated it in a way that the senior engineers will agree is a correct and complete analysis. Show me the evidence."

How does one validate the claim, "My program gives the right results" or "I have adequate skill"? How does one create and provide the evidence? Unfortunately, we usually inadequately develop this ability or perspective in the school experience of students.

3.2 Validation

I don't have a theory on how to self-validate understanding and skill. But here are things that I do:

1) Have expectations about an equation, rule, theory, or procedure. If it is right, what do you expect to see, and not to see? Know how to test, assess, and evaluate the knowledge.
2) Don't just accept the procedure, recipe, formula, or rule. Understand the basis, assumptions, and context. Know the why about it. Be able to explicitly and quantitatively express the cause-and-effect mechanism.
3) Don't just accept computer output. Test it over a range of inputs and givens, and be confident that the output is consistent with your expectations.
4) Test your understanding by creating your own exercises. Explore alternate examples, values, assumptions, and the inverse relation. Be sure that trends are as expected.

Engineering Optimization: Applications, Methods, and Analysis, First Edition. R. Russell Rhinehart.
© 2018 R. Russell Rhinehart. Published 2018 by John Wiley & Sons Ltd.
Companion website: www.wiley.com/go/rhinehart/engineeringoptimization

5) What else does it apply to? What if you extrapolate it? Does the application make sense?

6) Compare a next-step better model to ideal calculations. Does it approach the ideal in the limit? Does it fit expected trends and homologous trends? When you adjust a parameter value, does the result behave as expected?

7) Compare to alternate methods such as an old-style handbook graphical method, software product A, software product B, or prior work.

8) Seek knowledge from product bulletins, handbooks, trade magazines, vendor's white papers, and the Internet. Textbooks are often an acceptable source of fundamentals and procedures. However, in spite of its importance to the academic community, avoid the scientific journal literature when you are seeking practical knowledge.

9) Think of analysis, synthesis, and evaluation in terms of Bloom's taxonomy of cognitive skill (see Section 3.3). Critically question the basis and assumptions in your knowledge. See how to apply it and how to integrate it in context. Consider how stakeholders will see the outcomes. Reveal how you know that it is correct by providing assessments of multiple, comprehensive, and competing criteria.

10) Be your own devil's advocate. Take the perspective of those who could claim to have an alternate opinion about the thing (maintenance, purchasing, labor, community, politicians, scientists, operator, opposition, etc.), and consider what aspects they could find and claim are undesired.

11) Learn by doing, not by studying. Don't just read or follow. Extrapolate on your own. Prepare by doing, not by intellectualizing. I tell my students the secret to success is this: "Do not study." To be sure they get the message, I write it on the board. Then, I pretend to be surprised at their not understanding, look at the statement, feign puzzlement, and agree that it doesn't make sense. Then pretend to have overlooked the comma when writing the secret to learning and add it, converting the sentence to say, "Do, not study."

12) Test and evaluate your own learning. Make your own quiz problems. If you can't, you don't understand it yet. Solve them. Implement your procedures in a spread sheet or structured code, and explore the validity of the outcomes.

13) Guide learning by what is needed to be able to do some task, not by what is interesting, or by the basic body of knowledge or by what everyone else knows.

14) Test it on simple ideal cases. Show that it gives the right answer. Then test it on more complicated cases, and show that it gives the answers that several experts agree on. Be sure to challenge it. Don't choose one or two cases that are simple to implement because that can provide a false affirmation. For example, my favorite pretend claim is that addition is the same as multiplication. I use $2 \times 2 = 2 + 2$ and $0 + 0 = 0 \times 0$ as examples to defend that claim. There are an infinite number of supporting examples. Further, the claim has a sophisticated name "Theory of Positional Invariance," which states that regardless of the observer's viewpoint the object retains its properties. There are many examples: Whether observed from the north or south poles, the moon has the same mass, although the moon appears upside down to one observer. Whether you look at a person from top or back or front, it is still that person with the same personality and color of eyes. Applying the principle, observe that except for their $45°$ rotation, the $\times$ and the $+$ symbols are the same, so the theory claims that $2 + 2 = 2 \times 2$. There you are! The theory is intuitively logical, has a sophisticated name, and is confirmed by data, which has an infinite number of cases. (Given any x-value, the constraint $x + y = xy$ can be solved for $y = x/(x-1)$. This even works for complex numbers.) So, the claim must be true. Challenge your knowledge and understanding with situations that might reveal the error.

15) Test it on real data, not just by calculations. It should provide a good enough match to the real data.

16) Accept your new knowledge on a tentative basis until you come to know better. You will have a tendency to want to accept your self-learned knowledge. You created it. It is your progeny. It may be difficult for you to see its inadequacy. Realize that even mankind's best theories have been proven false. We once thought that the magical substance called the ether transmitted electromagnetic waves. That led to Maxwell's equations, which were affirmed by data of that era. The caloric theory of heat led to the diffusion equations, again, affirmed by data. At one time, data and logic seemed to support the flat Earth concept as the center of the universe. Perfection in knowledge is elusive.

17) If it is a numerical procedure, see if smaller step sizes or convergence criteria change the answer. If right, calculated values should not change.

18) Extrapolate the application to large chronological time or large sizes and to initial values or very small dimensions. Test parameter and coefficient value extremes of 0, 1, or ∞. Take variable values to extreme conditions (dilute or concentrated, hot or cold, high or low flow rate, short or long tube, early or long time) and look at asymptotic limits of the model terms. Do they reduce to ideal conditions? Does it still make sense?

19) When the data is functionally transformed, is the trend as expected? For instance, when a power law model is log transformed, the data trend should be linear.

20) Consider and report the uncertainty in any value. There is uncertainty in the givens, in the coefficients, and in the models. How do these sources affect the output? How does uncertainty on the application impact a decision?

21) Seek challenges to your skill or knowledge that are sufficient, meaning that all relevant cases considered, testing is complete w.r.t. your context.

22) Also, seek challenges that are credible, meaning that it is tested with meaningful, known cases and returns "right" answers.

3.3 Advice on Becoming Proficient

The question is "How does one acquire skill?" Partly, the answer is to understand cognitive levels. Table 3.1 presents Bloom's taxonomy of cognitive skill. There are six levels. The lower level, number 1, is termed "Knowledge" and means memorization. The upper level, number 6, is Evaluation.

Education nearly exclusively has students work at the lower three levels of Knowledge, Understanding, and Application. By contrast, professional partners in human enterprises are expected to add value to employer and to human welfare and, therefore, must work in the upper three levels of Analysis, Synthesis, and Evaluation. Those are the "doing" levels. In our personal lives the upper levels are essential to success. Although the lower levels are important and need to constitute the introduction to optimization applications, I am attempting to explicitly tie this book experience to all six levels.

Note the parallels of Bloom's upper levels and optimization. Analysis is required to develop the models and determine the metrics for desirability. Synthesis is required to implement a solution. And Evaluation is required to determine whether the solution is right or whether some portion of the procedure needs to be fine-tuned.

Table 3.1 Bloom's taxonomy of cognitive skill.

Level	Label	Function—a person does	Examples
6	Evaluation (E)	Judge goodness, sufficiency, and completeness of something; choose the best among options; know when to stop improving. Must consider all aspects	Decide that a design, report, research project, or event planning is finished when considering all issues (technical completeness, needs of all stakeholders, ethical standards, safety, economics, impact, etc.)
5	Synthesis (S)	Create something new: purposefully integrate parts or concepts to design something new that meets a function	Design a device to meet all stakeholders' approvals within constraints. Create a new homework problem integrating all relevant technology, design a procedure to meet multiple objectives, create a model, create a written report, and design experiments to generate useful data
4	Analysis (An)	Two aspects related to context:	
		One: Separate into parts or stages, and define and classify the mechanistic relationships of something within the whole	*One*: Describe and model the sequence of cause-and-effect mechanisms; tray-to-tray model that relates vapor boil-up to distillate purity, impact of transformer start-up on the entire grid, impact of an infection on the entire body and a person's health
		Two: Critique, assess goodness, determine functionality of something within the whole	*Two*: Define and compute metrics that quantify measures of utility, desirability, or goodness
3	Application (Ap)	Independently apply skills to fulfill a purpose within a structured set of "givens"	Properly follow procedures to calculate bubble point, size equipment, use the Excel features to properly present data, and solve classroom "word problems"
2	Understanding/ comprehension (U/C)	Understand the relation of facts and connection of abstract to concrete	Find the diameter of a $1''$ dia. pipe, convert units, qualitatively describe staged equilibrium separation phenomena, explain the equations that describe an RC circuit, and understand what Excel cell equations do
1	Knowledge (K)	Memorize facts and categorization	Spell words, recite equations, name parts of a valve, read resistance from color code, and recite the six Bloom's levels

3.4 Takeaway

Do, not study.

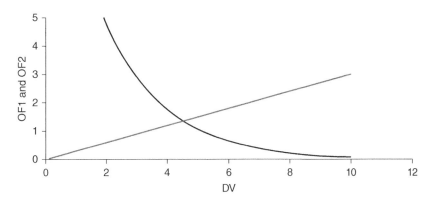

Figure 3.1 Illustration of two functions to be added.

3.5 Exercises

1 As you are doing something (such as cooking, reading, conversing, playing, or driving), reflect on the cognitive levels in which you are working. Explain your classification.

2 When you are learning something, test yourself using several of the 22 items listed in Section 3.2. Report the results.

3 Figure 3.1 is a generic illustration of an optimization situation. As the DV increases, one consequence gets better, while the other gets worse. The OF is the sum of the two features (or some other composite such as the product). Consider the sum in this exercise. Often, people illustrate the optimum of the OF as occurring at the crossover (about DV = 4.5 in this illustration). However, this is wrong. For a generic case of OF = f(DV) + g(DV), use mathematical analysis to show that the optimum happens when the slope of f is the negative of the slope of g. What conditions would make DV* equal to the DV at the point of the crossover of the two curves?

Section 2

Univariate Search Techniques

Univariate means one variable. In this section, there is only one DV, and all optimizers seek to find the minimum. The single DV applications make it easy to understand optimizer concepts, procedures, convergence criteria, path integrals, and the issues that confound them.

Engineering Optimization: Applications, Methods, and Analysis, First Edition. R. Russell Rhinehart.
© 2018 R. Russell Rhinehart. Published 2018 by John Wiley & Sons Ltd.
Companion website: www.wiley.com/go/rhinehart/engineeringoptimization

4

Univariate (Single DV) Search Techniques

4.1 Univariate (Single DV)

The term "line search" is often used to mean that the optimization application has a single decision variable. I will use the term univariate, one variable, to indicate a single DV:

$$\min_{\{x\}} J = f(x) \tag{4.1}$$

The variable x could be a primary, fundamental variable, such as what might be graphed in an $f(x)$ w.r.t. x presentation.

However, the decision variable might be distance along a line that goes through space, and the objective might be to find the point along the line to maximize proximity (minimize distance) to a planet. Using a parametric notation for a line to calculate the (x, y, z) position as a function of parameter "s" from the origin:

$$x = x_0 + as$$
$$y = y_0 + bs \tag{4.2}$$
$$z = z_0 + cs$$

Note: At some position (x, y, z) distance along the line from the origin (x_0, y_0, z_0) is S, not s:

$$S = \sqrt{(x - x_0)^2 + (y - y_0)^2 + (z - z_0)^2} = \sqrt{(as)^2 + (bs)^2 + (cs)^2} = s\sqrt{a^2 + b^2 + c^2} \tag{4.3}$$

However, here the objective is not to minimize distance along the line, but the distance from the line to the point. Then, with the OF to minimize distance to a planet, a point object, at (x_p, y_p, z_p), the optimization statement becomes

$$\min_{\{s\}} J = \sqrt{(x(s) - x_p)^2 + (y(s) - y_p)^2 + (z(s) - z_p)^2} \tag{4.4}$$

Here the DV is not distance along an axis in Cartesian space, or even distance along the line, but a parameter representing distance along a line.

Although this is a 3-D application, there is a single DV, which is related to distance along a line, supporting the "line search" terminology.

Engineering Optimization: Applications, Methods, and Analysis, First Edition. R. Russell Rinehart.
© 2018 R. Russell Rinehart. Published 2018 by John Wiley & Sons Ltd.
Companion website: www.wiley.com/go/rhinehart/engineeringoptimization

Alternately, the path might not be a straight line; it might be a curve in space. Consider a path through mountains on a Cartesian map in (x, y) space, where elevation is a function of the x–y position, $z(x, y)$. The decision variable might be the distance along a path, and the objective might be to find it from the point of origin to the point where the maximum slope occurs:

$$\min_{\{S\}} J = \frac{dz(x,y)}{dS} \tag{4.5}$$

Here the rate of change of the z-elevation w.r.t. path distance, S, is the objective, and the elevation depends on the (x, y) position, which depends on distance along the path. Although this path through (x, y, z) space might be curved (not a straight line), and although the path may change in three dimensions, it is still a single DV search.

The term line search means a single DV search, whether the search path is along an axis, along a line, along a curved path in physical space, along a straight or curved path in N-dimensional DV space, or described by a related parameter. However, the single path must have unique model influence values for the single DV value.

4.2 Analytical Method of Optimization

Consider a continuously valued function of one variable, which also has continuous derivatives. At the minimum point, any DV deviation from DV* in either direction increases the OF. To one side of DV*, the slope is positive, and to the other side, the slope is negative. At DV*, the slope is zero. The classic optimization method seeks the DV* value that makes the derivative of the OF to be zero. This is termed a stationary point. To solve the optimization statement,

$$\min_{\{x\}} J = f(x) \tag{4.6}$$

Determine the x^* that makes $df/dx|_{x^*} = 0$.

Example 1   Quadratic Function

What is the value of x^*, if $y = a + bx + cx^2$? $(c = 5)$
Set the derivative to zero, $0 = b + 2cx^*$, and solve for $x^* = -b/2c$.

However, this technique might determine a maximum, not a minimum. Consider the shape of the function if $c = -5$. So, either before or after using the analytical method, check to see that the function is at a minimum. This could be by a graph of the function, or it could be by evaluating the second derivative at the DV* value. If the function has continuous second derivative values, then it is a minimum if $d^2f/dx^2|_{x^*} > 0$. Alternately, if $d^2f/dx^2|_{x^*} < 0$ the point is at a maximum, and if $d^2f/dx^2|_{x^*} = 0$, the point is either a saddle point or a flat spot in the function.

Also, consider this radial basis function example.

Example 2   Radial Basis Function

What is the value of x^*, if $y = e^{-((x-c)/s)^2}$?
Set the derivative to zero, $0 = -2((x^*-c)/s)e^{-((x^*-c)/s)^2}$, and solve for $x^* = c$.

The reader should look at the second derivative (or graph the function) to see that $x^* = c$ defines a maximum, not a minimum.

Further, this analytical technique might find a saddle point, a local point at which the derivative is zero in an otherwise monotonically increasing function.

Example 3 Saddle Point Function

What is the value of x^*, if $y = a + bx + cx^2 + dx^3$? $(b = 12, c = -6, d = 1)$
Set the derivative to zero, $0 = b + 2cx^* + 3dx^{*2}$, and solve for $x^* = 2$.

The reader should verify: Perhaps graph the function and also use the second derivative to see that the answer is not the minimum, but a saddle point.

Accordingly, there needs to be a method to determine whether the stationary point, the solution to $df/dx|_{x^*} = 0$, is a minimum, maximum, or saddle point. The analytical method is to use the value of the second derivative. If the function has a minimum, then starting at a DV value less than DV* has a greater OF value than OF*, and increasing the DV value toward DV* decreases the OF to OF*. Further increasing the DV value above the DV* value increases the OF value. So, below DV* the derivative of the OF w.r.t. DV is negative, and above DV* the derivative is positive. The change of the derivative is the second derivative:

$$\frac{\text{slope}(DV^+ + \Delta) - \text{slope}(DV^+ - \Delta)}{2\Delta} \cong \frac{d}{dDV}\frac{dOF}{dDV} = \frac{d^2OF}{dDV^2} \tag{4.7}$$

If $d^2f/dx^2|_{x^*} > 0$ then a minimum has been found. Using similar logic, if $d^2f/dx^2|_{x^*} < 0$, then a maximum has been found. Similarly, if $d^2f/dx^2|_{x^*} = 0$, then a saddle point has been found.

The reader can easily test this with the prior examples.

4.2.1 Issues with the Analytical Approach

Unfortunately, the analytical method may lead to an equation for the stationary point that is nonlinear and does not permit an analytical solution.

Example 4 Nonlinear Function

What is the value of x^* if $y = a + b\sqrt{x} + c^x$?
Set the derivative to zero, and after rearrangement to isolate DV*, $\sqrt{x^*}c^{x^*} = -b/(2\ln(c))$. Solve for x^*?

A method to solve this nonlinear relation is to use some form of root finding such as Newton's method. Define

$$g(x^*) = \sqrt{x^*}c^{x^*} + \frac{b}{2\ln(c)} = 0 \tag{4.8}$$

Then iteratively estimate the value of x^* from

$$x_{k+1}^* = x_k^* - \frac{g(x_k^*)}{dg/dx|_{x_k^*}} \tag{4.9}$$

where k represents the iteration number and x_{k+1}^* indicates the sequential estimate of x^*.

Although possibly feasible, if this analytical procedure to determine the stationary point requires an imbedded numerical iterative solution (which must have an initial value for the $k = 0$ trial and both a convergence criterion and corresponding threshold value selected to stop the iterations), then perhaps it is best to just start with a numerical procedure to determine the optimum.

Further, since Newton's method root-finding techniques can send the solution in the wrong direction if the initial guess is beyond an inflection point, or it can require excessive iterations for s-shaped functions, a more primitive method, such as interval halving, may be more robust, reliable, and have a guaranteed number of iterations to converge.

Finally, there might not be a unique optimum. A fourth-order polynomial could have two minima (and a maximum). If you find the value of DV*, it might not be the globally best value.

Those issues are each of concern if you can determine an analytical expression for the derivative of the OF w.r.t. the DV. Where the OF is a simple one-line equation, such is often possible, but in my experience in real applications, the function is a procedure with no possibility of calculating an analytical derivative.

Accordingly, my preference is to use numerical optimization procedures except, of course, for the rare-to-encounter or pedagogically contrived simple relations that permit the analytical approach.

4.3 Numerical Iterative Procedures

Most practical optimization techniques are iterative numerical procedures, as opposed to an analytical solution. There are a wide variety of such iterative numerical techniques.

4.3.1 Newton's Methods

There are several ways to develop a Newton's iterative method for optimization with a single DV. The methods provide a common functionality but reveal the possibility for flexible (human choice in) coefficient values. As a first derivation, choose a point x_0, hopefully near x^*, and use a Taylor series quadratic expansion for the f value at x^*. Eliminate high-order terms as having negligible impact when x-values are near x^*. (This step is grounded in the value of x_0 being sufficiently near x^*, so that the quadratic approximation is locally valid.)

$$f(x^*) \cong f(x_0) + \frac{df}{dx}\bigg|_{x_0} (x^* - x_0) + \frac{1}{2}\frac{d^2f}{dx^2}\bigg|_{x_0} (x^* - x_0)^2 \qquad (4.10)$$

At x^*, at the minimum, the slope $df/dx|_{x^*}$ is zero. (Note: This is an analytical concept and assumes that the function has a continuous value and first and second derivatives in the region of x^* and x_0.) Rearrange the truncated Taylor series model of the function to approximate the derivative at x_0 with a backward finite difference. (Subtract $f(x_0)$ from both sides of the equation, and then divide by $(x^* - x_0)$.)

$$0 = \frac{df}{dx}\bigg|_{x^*} \cong \frac{f(x^*) - f(x_0)}{x^* - x_0} \cong \frac{df}{dx}\bigg|_{x_0} + \frac{1}{2}\frac{d^2f}{dx^2}\bigg|_{x_0} (x^* - x_0) \qquad (4.11)$$

Note the coefficient on the second derivative is ½, which generates 2 when Equation (4.11) is rear-ranged to solve for x^*:

$$x^* \cong x_0 - 2\frac{df/dx|_{x_0}}{d^2f/dx^2|_{x_0}} \tag{4.12}$$

Recognize that x^* is just the next approximation. So, use the formula recursively:

$$x_{k+1} = x_k - 2\frac{df/dx|_{x_k}}{d^2f/dx^2|_{x_k}} \tag{4.13}$$

This is one version of an analytical Newton's method. An alternate legitimate relation can be generated from a similar analysis. Repeat the Taylor series quadratic expansion about x_0, but have it predict $f(x)$, not $f(x^*)$:

$$f(x) \cong f(x_0) + \left.\frac{df}{dx}\right|_{x_0}(x-x_0) + \left.\frac{1}{2}\frac{d^2f}{dx^2}\right|_{x_0}(x-x_0)^2 \tag{4.14}$$

Take the derivative of $f(x)$

$$\frac{df}{dx} \cong \frac{d}{dx}\left(f(x_0) + \left.\frac{df}{dx}\right|_{x_0}(x-x_0) + \left.\frac{1}{2}\frac{d^2f}{dx^2}\right|_{x_0}(x-x_0)^2\right) = 0 + \left.\frac{df}{dx}\right|_{x_0} + \frac{1}{2}\cdot 2\left.\frac{d^2f}{dx^2}\right|_{x_0}(x-x_0) \tag{4.15}$$

Note that the coefficient of ½ is normalized by the coefficient of 2.
At the minimum, at $x = x^*$, the derivative is zero:

$$\left.\frac{df}{dx}\right|_{x^*} \cong 0 = \left.\frac{df}{dx}\right|_{x_0} + \left.\frac{d^2f}{dx^2}\right|_{x_0}(x^* - x_0) \tag{4.16}$$

which can be rearranged to solve for x^*

$$x^* \cong x_0 - \frac{df/dx|_{x_0}}{d^2f/dx^2|_{x_0}} \tag{4.17}$$

The recursion formula is

$$x_{k+1} = x_k - \frac{df/dx|_{x_k}}{d^2f/dx^2|_{x_k}} \tag{4.18}$$

This is the standard Newton's method for optimization. Again, it is a recursive, successive approximation method.

Note: In this second derivation, the coefficient value is 1, not 2, the consequence of the choice of base point. Both Equations (4.13) and (4.18) are valid choices. Both are mathematically perfect in their derivation. But both are inexact techniques because they truncate the Taylor series expansion to the quadratic terms and model the function as a quadratic. This can be termed a surrogate model, a reduced-order local approximation. So, the coefficient value of 2 is as right as the value of 1. Further, both techniques are just approximations. They do not jump to the true answer, but iteratively approach it. To plant the seed of user flexibility in choosing the coefficient value, maybe a 1.5 is a valid coefficient value, or maybe it should be a value of 0.732648.... In Chapter 9, I will give implementer's permission to appropriate the mathematics, to keep the essence of Newton's method, and to keep the functional relation, but to change the coefficient values to enhance algorithm robustness.

Note some aspects related to initialization, computer work, and convergence:

1) The user must initiate the first x-value trial solution, the value of x_0.
2) Then, at each iteration the computer must evaluate two functions, the first and second analytical derivatives. However, I also want to observe the function value as it evolves with iteration, so as a preference, at each iteration the computer must perform three function evaluations.
3) Some criteria must be used to decide when the iterative procedure should be stopped. Nominally, one might claim convergence and stop the iterative procedure when trial-to-trial estimates of x^* are close together. The user must also define the threshold. However, if the procedure is diverging or converging on a maximum or saddle point, then the procedure should be stopped. Further, if the procedure is approaching an infeasible value for x^*, perhaps one that leads to an execution error, then the procedure should be stopped. In all, there needs to be supervision that determines whether the procedure should stop or continue.

The Newton's method for optimization can be considered root finding on the derivative, which adds insight about why it might diverge to a wrong estimate for x^*. Figure 4.1a and b indicates a function $f(x)$ with a minimum and the derivative of the function, $g(x)$. The derivative has a value of zero at the optimum. So, searching for the root value of the derivative determines the same value as searching for the x^* to optimize $f(x)$.

Note that the function and derivative sketches in Figure 4.1 are internally consistent. They are not just abstract squiggles. The derivative has a zero value at the same point as the optimum of the function. Where the slope of the function is negative, the derivative value is negative, and vice versa. Where the slope of the function is high, the derivative function is far from zero. I think that preservation of such relations is important to grasping concepts.

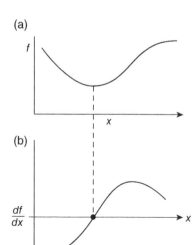

(a)

f

x

(b)

$\frac{df}{dx}$

x

Figure 4.1 (a) A continuous function of a single variable and (b) its derivative.

The analytical version of Newton's root-finding technique on the function $g(x)$ is

$$x^* \cong x_0 - \left.\frac{g}{dg/dx}\right|_{x_0} \tag{4.19}$$

With df/dx substituted for $g(x)$, it is identical to Equation (4.17):

$$x^* \cong x_0 - \left.\frac{g}{dg/dx}\right|_{x_0} = x_0 - \left.\frac{df/dx}{d^2f/dx^2}\right|_{x_0} \tag{4.20}$$

Note: In the illustration in Figure 4.1, there is an inflection point on the function, which corresponds to the maximum value of the derivative. At an inflection point the second derivative changes sign, and the function transitions from convex to concave. This is the point of maximum rate of change and slope of the function. With an initial trial solution, within the x^* proximity of the inflection point, Newton's method of root finding on the derivative, the desire to get the derivative closer to zero, will move toward x^*. However, on the other side of the inflection point, the downward trend of the derivative indicates that moving further away leads to g-values that might approach zero.

Each step farther to the right indicates that the solution is farther still. Newton's method for optimization is equivalent to Newton's method of root finding on the function derivative, and this discussion reveals that if the initial trial solution is beyond the inflection point from x^*, Newton's method iterations will diverge away from the optimum.

Alternately a geometric, or secant, version of root finding can be applied to the derivative, using Figure 4.2 as a guide to nomenclature.

Any two points on the function can be used to define the line through them, and from the equation of the line, any g-value along the line:

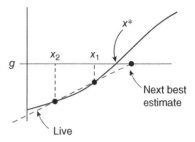

Figure 4.2 A secant method for root finding on a derivative.

$$g(x) = g_1 + \frac{g_2 - g_1}{x_2 - x_1}(x - x_1) \tag{4.21}$$

The objective is to determine the x-value that makes $g = 0$. Rearranging Equation (4.21)

$$x^* \cong x_1 - \frac{g_1}{(g_2 - g_1)/(x_2 - x_1)} \tag{4.22}$$

This produces a finite difference representation. In a recursion formula

$$x_{k+1} = x_k - \frac{g_k}{(g_k - g_{k-1})/(x_k - x_{k-1})} \tag{4.23}$$

I like this. It does not require an analytical second derivative, and it is relatively robust. Here, the user must specify the first two x-values, but the computer only performs two function evaluations at each iteration (the function and the analytical derivative).

Better yet, g_k can be estimated numerically, using either a forward difference approximation:

$$g_k \cong \frac{f(x_k + \delta) - f_k}{\delta} \tag{4.24}$$

Or a backward difference approximation:

$$g_k \cong \frac{f_k - f(x_k - \delta)}{\delta} \tag{4.25}$$

where δ represents a small incremental change in the x-value. The advantage of this approach is that an analytical expression for the derivative is not needed. The disadvantage of this method is that at each x_k value, both $f_k = f(x_k)$ and $f(x_k - \delta)$ need to be calculated—two function evaluations at each iteration.

Alternately, if the sequential changes in x_k are small enough for the derivative to be estimated by the backward difference:

$$g_k \cong \frac{f_k - f_{k-1}}{x_k - x_{k-1}} \tag{4.26}$$

Then the recursion relation for x_k becomes

$$x_{k+1} = x_k - \frac{f_k - f_{k-1}}{[(f_k - f_{k-1})/(x_k - x_{k-1})] - [(f_{k-1} - f_{k-2})/(x_{k-1} - x_{k-2})]} \tag{4.27}$$

Initiating this requires three initial x-values, but once initiated, this only requires one function evaluation per iteration.

These several variations on Newton's method provide sequential estimates. Hopefully, each new estimate is better than the prior estimate and provides a better base point for the next estimate. They are successive, or iterative, or recursive. A recursion formula gives the next value from past values, and while the values may change, the formula is invariant.

Each of these Newton-type techniques is predicated on the trial solution being sufficiently close to the x^*-value so that the second-order (quadratic) Taylor series truncation is valid; alternately, the linear extrapolation of the derivative successively leads to better x_k-values. It presumes that the function and its derivative are continuous in the region of x^*. If these conditions are not valid, then the method might send the sequence of trial solutions toward absurd values. Just because a method is grounded in selected mathematical principles does not mean it will work. Nature is not required to comply with human contrivances.

How many iterations to apply? One method to claim convergence is to decide on a Δx value (ε, a precision) for which any x-value is close enough to the true (but unknowable) value of x^*. If the iterative procedure is alternating back and forth over the true, but unknowable, x^*, then, when iteration-to-iteration differences in x-values are less than ε, the most recent x_k is within ε of x^*. However, the procedure may be incrementally creeping up on the x^* value from one side. If this is the situation, then even if $|\Delta x| < \varepsilon$, the most recent x-value might not be in the desired proximity of x^*. Accordingly, when sequential estimates are closer than a tenth of that desired precision, claim that the last calculated value is close enough to x^*. If $|x_k - x_{k-1}| < 0.1\varepsilon$, then $x_k \cong x^*$. This is a convergence criterion. Note that it is just one of many criteria to be discussed.

Is there a bound on the number of iterations to converge? Usually there are not too many, if the method is converging on x^*. But some s-shaped functions make it exceedingly difficult for a Newton's method to converge. And many functions have inflection features that make Newton's methods diverge.

If it converges, did it find a minimum, maximum, or saddle point? To be sure, investigate the second derivative of the function w.r.t. DV at x^*.

Newton-type methods are termed second-order methods, a view that arises in several embodiments of the model basis for the method: The Taylor series is truncated to quadratic terms. Equation (4.27) needs three initial guesses, which could be used to define a quadratic surrogate model of the function.

4.3.2 Successive Quadratic (A Surrogate Model or Approximating Model Method)

This technique seeks to approximate the local surface region with a quadratic model, a surrogate model, which approximates the true $f(x)$ relation. This is also a second-order model-based approach.

The concept is illustrated in Figure 4.3, in which x_1, x_2, and x_3 are trial solutions, initial user-defined values, hopefully near the x^* value.

The method starts with three $(x, f(x))$ pairs and fits them with a quadratic estimate of the function. The model, illustrated as a dashed line, should perfectly fit the three data points, but the model is just an approximation to the function, $f(x) \cong \widetilde{f}(x) = a + bx + cx^2$.

How does one obtain the model a, b, c values? The surrogate model is valid for each of the three data pairs, which yields three equations linear in the three model coefficients of unknown values (a, b, c):

$$f_1 = a + bx_1 + cx_1^2$$
$$f_2 = a + bx_2 + cx_2^2 \qquad (4.28)$$
$$f_3 = a + bx_3 + cx_3^2$$

Figure 4.3 Illustration of the quadratic surrogate model.

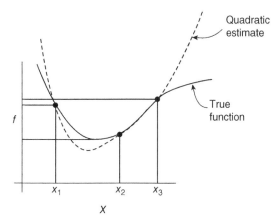

This linear algebra problem of determining coefficient values can be placed in matrix–vector form:

$$\mathbf{Mc} = \begin{bmatrix} 1 & x_1 & x_1^2 \\ 1 & x_2 & x_2^2 \\ 1 & x_3 & x_3^2 \end{bmatrix} \begin{bmatrix} a \\ b \\ c \end{bmatrix} = \begin{bmatrix} f_1 \\ f_2 \\ f_3 \end{bmatrix} = f \tag{4.29}$$

After solving for the values of coefficients (a, b, c) (Gaussian elimination is one method), use the analytical approach to get x^* for the approximating function (the surrogate model):

$$\left.\frac{df}{dx}\right|_{x^*_{\text{function}}} \cong \left.\frac{d\widetilde{f}}{dx}\right|_{x^*_{\text{estimate}}} = 0 = \left.\frac{d}{dx}\left(a + bx + cx^2\right)\right|_{x^*} = b + 2cx^* \tag{4.30}$$

which defines

$$x^* \cong -\frac{b}{2c} \tag{4.31}$$

Since the value of coefficient a is not needed in estimating x^*, the final step in the backward solution stage of the Gaussian elimination algorithm is not needed.

Since this gives the value of x^* for the surrogate model, which is not necessarily x^* for the function, the procedure needs to be repeated. The recursion relation is

$$x_{k+1} = -\frac{b_k}{2c_k} \tag{4.32}$$

One initializes the procedure with three trial solutions. The successive quadratic (SQ) procedure leads to a 4th x-value, then a 5th, and so on. Which three of the past $N (x, f(x))$ pairs should be used in defining the next surrogate model? One good rule is to use the new point and most recent past two points. It is claimed to be a good rule because it works and is simple to implement. Alternately, rules that are a bit more burdensome, but which may converge faster, are to use the new point and (i) the past two points with closest x-values to new or (ii) the past two points with lowest OF values.

Regardless of the selection rule, each iteration requires (i) a new function evaluation and (ii) solution of the linear equations.

The method is predicated on the local shape of the function being adequately approximated by a quadratic model. There are many cases in which this is not valid. Such as a discontinuous function, which might arise with discretized DVs, constraints, conditionals, etc.

The successive quadratic method provides sequential estimates. Hopefully, each new estimate is better than the prior estimate, and provides a better base point for the next estimate. They are successive, or iterative, or recursive. A recursion formula gives the next value from past values, and while the values may change, the formula is invariant.

How many iterations to apply? One method to claim convergence is to decide on a Δx value (ε, a precision) for which any x-value is close enough to the true (but unknowable) value of x^*. Then, when sequential estimates are closer than a 10th of that precision, claim that the last calculated value is close enough to x^*. If $|x_k - x_{k-1}| < 0.1\varepsilon$, then $x_k \cong x^*$.

Is there a bound on the number of iterations to converge? Usually there are not too many, but like Newton's methods some functions make it exceedingly difficult for SQ to converge, and it might diverge if initial trial solutions are beyond the inflection from x^*.

Did it find a minimum, maximum, or saddle point? Also, like Newton's methods, SQ seeks the point of zero slope in the derivative, which may be a minimum, maximum, or saddle point. To be sure, investigate the second derivative of the function w.r.t. DV at x^*.

4.4 Direct Search Approaches

The analytical method and the iterative Newton's and SQ methods all rely on the function being continuous and either analytically differentiable or analytically quadratic-ish in the vicinity of the optimum. Some functions are, but some functions with cliffs, flat spots, or ridges are not. Also, the optimization procedure outcome is dependent on the initial trial solution(s) being placed within the inflection region of the DV*, and since they may seek min, max, or saddle points, the solution type must be verified after convergence to the proximity to x^*. Contrasting the methods that rely on derivatives or surrogate models, direct search methods only use the function value.

4.4.1 Bisection Method

The bisection method does not rely on models or presumed surface behavior. As a result it is more generally applicable than are the analytical or second-order methods, but it is not necessarily as fast or efficient in jumping to the optimum.

The bisection method is a two-stage procedure. First, bound the location of the optimum by a marching method, and then hone in on the optimum with interval halving. The marching method stage is illustrated in Figure 4.4.

First, start at one trial solution value that might be an extreme, perhaps the lower extreme, with a value of $x = x_1$. Evaluate $df/dx|_{x_1}$. If $df/dx|_{x_1}$ is negative, then increasing the value of x is a move toward the minimum. Increment to x_2, where

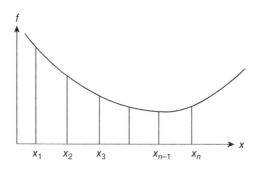

Figure 4.4 Bisection search stage 1—marching.

$x_2 = x_1 + \Delta x$. Evaluate $df/dx|_{x_2}$. Continue until df/dx changes sign; if $df/dx|_{x_{n-1}} df/dx|_{x_n} < 0$, then the values of x_n and x_{n-1} bound the minimum.

Note: If an analytical equation for the derivative is not available, the derivative can be estimated numerically, $df/dx|_{x_k} = g_k \cong [f(x_k + \delta) - f(x_k)]/\delta$. However, precise derivative values are unnecessary. Just the sign is needed to direct the search. So, a coarse estimate of the derivative is all that is justified. Using $g_k \cong [f(x_k) - f(x_{k-1})]/\Delta x$ is fully adequate and avoids an additional function evaluation.

Note: Choose Δx, the marching increment, to have the opposite sign of the slope, or else this algorithm will find the maximum.

Note: If the value for Δx is too small, it may take an excessive number of marching iterations (and function evaluations hence computational time) to bound the optimum. But, if too large a value for Δx, the marching method might skip over an important feature. The right value for Δx typically has a large range but does require user understanding of the application.

Note: The marching method could be stopped by observing the function evaluations, not the derivative. This cuts the number of function evaluations in half. Observe $f(x_j)$. As long as $f(x_j) < f(x_{j-1})$, the search is approaching the minimum. But in this case, when $f(x_n) > f(x_{n-1})$, the minimum is not necessarily between x_n and x_{n-1}. It might be between x_{n-1} and x_{n-2}. So start the next stage with bounds of x_n and x_{n-2}.

The second stage is to use an interval halving approach in the region between x_n and x_{n-1} to find x^*. The values of x_n and x_{n-1} define the initial left and right (low and high) values as the range that includes the x^* value. Figure 4.5 illustrates the procedure.

Select a midpoint value in the range. $x_M = (x_L + x_R)/2$. If the slope at x_M has the same sign as slope at x_L, then reject the x_L to x_M region and set x_M as x_L. Else, do the complementary. With new x-values of the right-most and left-most bounds, recalculate the new midpoint $x_M = (x_L + x_R)/2$. Repeat until the x-interval is small enough to claim that the midpoint is close enough to the optimum, and claim that x^* is near the midpoint of the interval. If $|x_L - x_R| < \varepsilon$, then $x^* \cong (x_L + x_R)/2$.

Interval halving discards 50% of the DV range at each stage in the iterations. If there is an analytical equation for the derivative, then each iteration only requires one function evaluation, but I like to also observe the function value as well as the derivative, so I like two function evaluations per iteration. But, if there is not an analytical derivative, it takes two function evaluations at each x_M to determine the slope. Since two function calculations reduce the x-range by 50%, this implies a 25% range reduction per function evaluation.

If the marching method Δx defines the initial range for interval halving, and each iteration cuts the range in half, then after N interval halving iterations $|x_L - x_R| = \Delta x (1/2)^N$. If convergence is based on $|x_L - x_R| < \varepsilon$, then $\Delta x (1/2)^N < \varepsilon$, and solving for N, $N \cong \ln(\varepsilon/\Delta x)/\ln(1/2)$. The approximately equal sign indicates that N must be an integer, the rounded-up value of $\ln(\varepsilon/\Delta x)/\ln(1/2)$. This indicates that the bisection stage of the optimization is guaranteed to find the optimum in at least N iterations, but the number of trials in the marching stage cannot be predicted.

For the interval halving method, the user must specify (i) the initial x-value, (ii) the Δx value for the marching method (the sign on Δx can be determined by the computer by the derivative evaluation at the initial x-value), and (iii) the convergence ε (or the equivalent N for the interval halving stage).

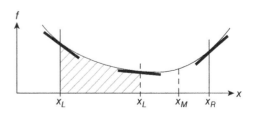

Figure 4.5 Illustration of the bisection method.

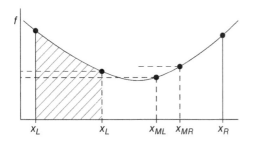

Figure 4.6 Golden Section method.

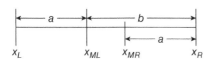

Figure 4.7 Golden
Section apportionment.

This method works if there is a single minimum. If the global minimum is elsewhere, the local search might not find it. This method also is predicted on the derivative evaluation providing valid information, precluding function features such as a discontinuity.

4.4.2 Golden Section Method

There are strong similarities of the Golden Section (GS) method to the bisection method, but GS is more efficient when considering (i) the portion of the x-range discarded per function evaluation and the robustness to discontinuities. GS is also simpler to program because it is not a two-stage procedure.

If the user knows the x-range that bounds the optimum, start with those values. Otherwise use a marching method to find the upper and lower x-values that bound the optimum. Figure 4.6 illustrates a function with a minimum and the upper and lower x-values that bracket the minimum.

The Golden Section method starts with four x-values, a left extreme, a right extreme, and two interior locations mid-left and mid-right. (The interior points do not equally divide the DV range; they are not at one-third and two-thirds values. The rule for interior point placement will be revealed in a bit.) Evaluate f at each point. Keep the exterior region with the lowest of all four f-values. In this case x_{MR} has the lowest OF value, so the right-hand extreme is kept. Discard the other exterior region, the one that is shaded. In this case point x_L is discarded, and three points x_{ML}, x_{MR}, and x_R are retained. Place the new fourth point, x_{new}, to preserve the original spacing pattern (the trick is to be revealed), relabel the two points that changed relative positions, and label the new point appropriately. In this case x_{ML} becomes x_L, x_{MR} becomes x_{ML}, and x_{New} becomes x_{MR}. Repeat, until the range remaining satisfies convergence. If $|x_L - x_R| < \varepsilon$, then stop and report $x^* = (x_L - x_R)/2$.

Note: The rule is to keep the extreme region that contains the lowest of the four OF values, not to discard the region that contains the worst OF value.

The trick with GS is to make each new point preserve the pattern within the spacing of the four points. Figure 4.7 illustrates the points and uses $a + b$ to indicate the total range and the relative positions of the points.

The desire is that when the left portion of the DV range is excluded and a new point is placed in the remaining DV range that the new a–b proportionality is preserved:

$$\frac{x_R - x_{ML}}{x_R - x_L} = \frac{x_R - x_{MR}}{x_R - x_{ML}} \tag{4.33}$$

which means $b/range = b/(a + b) = a/b$, which can be rearranged to the quadratic form, $a^2 + ab - b^2 = 0$. Then, applying the quadratic formula, the ratio of b to *range* or the ratio of a to b is

$$\frac{a}{b} = \frac{b}{a + b} = \frac{-1 \pm \sqrt{5}}{2} \tag{4.34}$$

Since only the "+" root is meaningful

$$\frac{a}{b} = 0.6180339888... = \gamma \tag{4.35}$$

Gamma, γ, is called the golden ratio.

Once x_R and x_L values are chosen, determine initial x_{MR} and x_{MR} values from these equivalent relations:

$$x_{MR} = x_L + \gamma(x_R - x_L) = x_L(1 - \gamma) + \gamma x_R \tag{4.36a}$$

$$x_{ML} = x_R - \gamma(x_R - x_L) = x_R(1 - \gamma) + \gamma x_L \tag{4.36b}$$

Then, each time a region is discarded and new x_R and x_L values are updated, determine the new point:

$$x_{MR} = x_L + \gamma(x_R - x_L) \quad \text{if left is discarded, or} \tag{4.37a}$$

$$x_{ML} = x_R - \gamma(x_R - x_L) \quad \text{if right is discarded} \tag{4.37b}$$

This procedure preserves the pattern in the points with only one point added at each iteration.

Each new point requires one new function evaluation (the function f, not its derivative df/dx) and discards $(1 - \gamma) = 0.381966...$ fraction of area. This is a 38.2% range reduction per function evaluation. GS (at 38.2% DV range reduction per function evaluation) is more efficient in discarding space than interval halving (at 25% per function evaluation).

Further, there is only one iterative procedure in the GS approach, as contrasting the two (marching to bound the root, then interval halving) in the interval halving approach. GS is simpler to program.

Additionally, GS does not depend on derivative information, making it more robust to surface aberrations.

After N iterations GS reduces the DV region by a factor of γ^N. If the convergence threshold, ε, is based on the remaining DV range, then the relation between initial range, R, N, and convergence threshold ε is

$$N = 1 + \text{RoundUp}\left[\frac{\ln(\varepsilon/R)}{\ln(\gamma)}\right] \tag{4.38}$$

The 1 represents the first iteration to initialize the four DVs. The RoundUp function takes a non-integer value to the next higher integer. If such a function is not available, add 1 and take the INT value. Since there are four function evaluations on the first initialization, and one for each remaining iteration, the number of function evaluations for GS is

$$\text{NOFE} = 4 + \text{Int}\left[1 + \frac{\ln(\varepsilon/R)}{\ln(\gamma)}\right] \tag{4.39}$$

Golden Section is efficient, and the number of iterations or function evaluations to converge in this one-rule one-stage procedure is guaranteed.

But, GS is not perfect. If the initial x_R to x_L range does not include the optimum, then it will converge on either the initial x_R or the initial x_L value. Further, if the range includes a local optimum, not the global, it will find the local. Also, if the optimum is a pinhole in the extreme high function side, then it could be in the discarded region. Finally, GS does not scale to higher dimensions; it is for single DV searches only.

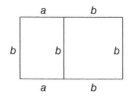

Figure 4.8 Squaring a rectangle.

4.4.2.1 Side Note on γ

The golden ratio is a value that frequently arises in geometry, math, art, and folklore. One classic origin is the question, "What is the aspect ratio of a rectangle such that when the large side is squared, the new rectangle preserves the original aspect ratio?" This is illustrated in Figure 4.8. The original rectangle has short and long side dimensions of a and b. When a square is placed on the long side b, the new rectangle dimensions are long side $(a + b)$ and short side b.

Specify that $a/b = b/(a + b)$ leads to $a/b = 0.6180339888\ldots$

Continuing the side note, and mystique, about the golden ratio, consider a Fibonacci series: 0, 1, 1, 2, 3, 5, 8, 13, 21, ... in which the recursion formula is

$$\text{Term}_N = \text{Term}_{N-1} + \text{Term}_{N-2} \tag{4.40}$$

In the limit of a large N $\text{Term}_{N-1}/\text{Term}_N = \gamma$.

I've often been entertained by much more of the mystique related to how the golden ratio appears throughout nature, math, art, folklore, and architecture.

4.4.3 Perspective at This Point

There are several issues with the prior techniques: The analytical technique is predicated on the derivative of the function being analytically tractable. Often it is not, and the procedure may identify min, max, or a saddle point.

On the other side of the inflection, SQ and Newton-type methods send the trial solution in the wrong direction. Often the analytical derivative is not possible, requiring some form of numerical estimate. They move to a min, max, or saddle, so the solution must be tested. If the trial solution violates a constraint, then what? A recalculation gives the same bad answer. The methods presume that there are no discontinuity or flat spots.

GS is a great technique, but is not scalable to higher DV dimensions. Interval halving is less efficient.

We'd like a direct search technique (function value only, not derivatives) that is efficient (finds the optimum reasonably rapidly), robust (to discontinuities, flat spots, surprise constraints, or stochastic features in the function or its derivatives), and scalable (the same algorithm is applicable to higher DV dimensions).

The two algorithms that follow heuristic direct search and leapfrogging meet those desired attributes.

4.4.4 Heuristic Direct Search

A direct search only uses function values. It does not use derivatives. "FAIL" can be a function value. This aspect makes handling hard constraints easy.

The heuristic search only goes downhill (or uphill if seeking to maximize). It is not seeking a zero derivative, so it is not confounded by an inflection and is not drawn to a saddle or maximum.

Figure 4.9 shows a univariate search (sometimes termed a line search) example. The plot is J (OF or y) w.r.t. DV (or x). The initial trial solution is x_{base}, and the shaded area represents a constrained or forbidden range of x-values.

For the heuristic direct search, initialize by choosing x_{initial} (the starting value) and Δx (the initial search direction and size). Set $x_{\text{base}} = x_{\text{initial}}$, and evaluate $J(x_{\text{base}})$ and assign that value to J_{base}. Then

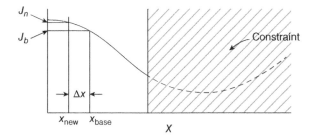

Figure 4.9 Illustration of the beginning of a heuristic direct search.

define a new trial solution as $x_{new} = x_{base} + \Delta x$. Evaluate $J(x_{new})$ and assign the value to J_{new}. If J_{new} or x_{new} or some other aspect of the calculation violates a constraint, set the value of J_{new} as "Fail."

Now there are two possible outcomes: either J_{new} indicates that x_{new} is a better trial solution, or it is not better. Not better could mean that the OF is either worse or equivalent, or the TS is constraint violating. If the trial solution is not better, it could mean that the search was in the wrong direction (uphill) and that it led to a constraint violation, or it was too large a step in the right direction (which crossed over the minimum and started up the other side) or that it is on a flat spot and made no OF improvement. If so, contract and reverse the search direction. IF J_{new} = "Fail" or $J_{new} \geq J_{base}$, THEN $\Delta x = -$contract$\cdot\Delta x$. Otherwise, it was not worse; x_{new} was a step in the right direction and found a better spot. If so, assign the new conditions to the base case, $J_{base} = J_{new}$, $x_{base} = x_{new}$, and with confidence that you are moving in the right direction expand the search step size $\Delta x = \Delta x\cdot$expand.

Repeat until convergence, perhaps based on the step size factor IF $|\Delta x| < \varepsilon$ THEN stop and claim that $x^* = x_{base}$.

Figure 4.10a and b indicates the sequence for the first few iterations of unconstrained and constrained searches.

The expansion factor makes Δx larger as long as J keeps improving. Perhaps make "expand" = 1.2, 20% larger. (1.05 works and creates a more cautious expansion, 1.25 works and makes a more aggressive search, ...) Within this reasonable range, the value of the expansion factor is not critical to the search. Too small a value, and the search does not accelerate when going in the right direction. With too large an expansion value, Δx can become large, and once in the vicinity of the optimum, it takes

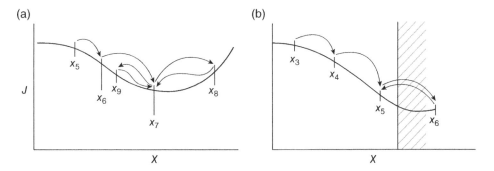

Figure 4.10 Illustration of a heuristic direct search: (a) unconstrained and (b) constrained.

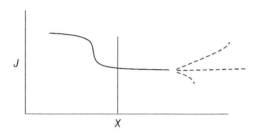

J

X

Figure 4.11 Representation of the quandary over flat spots in the OF.

many iterations to reduce it to a small value for the fine-tuning local search. A value of 1.1 seems to work well for me, balancing the issues.

The contraction factor makes Δx smaller. Perhaps Δx should be 20% smaller if a trial solution was not better. But, if the prior step was a success, and this step was not, that means that the Δx for this unsuccessful step had been expanded. So, if you want a 20% reduction in search step size, use contract = 0.8 ÷ expand. If the value of the contraction factor is too small, a few not better OF values will significantly reduce the search step size, taking longer to move to the optimum. Alternately, a contraction factor near to unity takes many iterations to reduce Δx for the local end-game search. Again, the value is not critical, and values within the reasonable range $0.5 \leq$ contract = 0.9 ÷ expand work. A value of 0.5 for the contraction factor seems to work well for me.

This is a simple concept, and it is simple to implement either in code or as a qualitatively directed (human directed) search. The search algorithm is heuristic, meaning it follows human intuitive rules rather than mathematical derivations—it is understandable and comfortable. It is robust to constraints and to surface aberrations (discontinuities). It is scalable to higher dimensions. Although the heuristic direct search is not as efficient as gradient-based approaches (on well-behaved functions), for simplicity, effectiveness, and robustness, it is one of my favorites as a general optimizer.

Leapfrogging (described in the next section) is my favorite direct search method. It is a bit more complicated, but it performs better overall.

Should the criteria for contracting the search be based on whether J_{new} is worse or not better? If the criteria is worse, the logic is IF J_{new} = "Fail" OR $J_{new} > J_{base}$, THEN contract. Alternately, if the criteria is not better, the logic is IF J_{new} = "Fail" OR $J_{new} \geq J_{base}$, THEN contract. Consider what happens on a flat "floor," illustrated in Figure 4.11, and your current trial solution is as shown and the search came from the left. The OF values to the right are yet unknown, as indicated by dashed lines. If the OF rises or remains flat, then the local point is at an equivalent minimum to any other place and the optimizer should stop. If the criterion for contraction is "worse" (IF J_{new} = "Fail" OR $J_{new} > J_{base}$, THEN contract), then the search accelerates along the flat spot never finding a better spot, and when it encounters an eventual rise in the OF, the search reverses and races back to the point shown, where it will encounter a worse OF, reverse and race forward again. But, if the function value drops, this logic will move along the flat spot to the drop and seek the optimum off to the right.

However, if the criterion for contraction is "not better" (IF J_{new} = "Fail" OR $J_{new} \geq J_{base}$, THEN contract), then along the flat spot the search alternates direction, progressively contracting and rapidly converging on the flat spot. But, if far to the right, the function improves, and this logic will converge on the flat spot and not find a better spot.

Do you want to march across forever flat, to find if it goes up, or to find a new slope down? It is your choice. Which is best, $J_{new} > J_{base}$ or $J_{new} \geq J_{base}$?

4.4.5 Leapfrogging

Most optimizers start with an initial trial solution, or perhaps a range, within which points are placed to estimate a surrogate model, derivatives, or region to eliminate. By contrast, many of the more recently developed optimization algorithms are multi-particle or multiplayer and only use function values (not derivatives). Multiplayer optimizers have a higher probability of finding the global and of not getting stuck in a local optimum.

Leapfrogging is a simple and effective multiplayer direct search. Start by choosing several (about 10) independent trial solutions, randomly scattered within the DV feasible range. Consider them as players on a surface. There is the worst player (the highest OF value if seeking a minimum) and the best (the lowest OF value). If there is a tie for the best (perhaps there are flat spots on the surface), just take the first in the player index list to represent best. In Leapfrogging, the worst leaps over the best (like the children's game of leapfrog), and the leaping player lands in a random spot on the other side of the best. For simplicity, the leap-into window is equal to the leap-from distance between worst and best. The placement for the leaping player, the new DV value, is calculated by

$$x_{w,\text{new}} = x_b + r(x_b - x_{w,\text{old}}) \tag{4.41}$$

where r is a random number (uniformly and independently distributed on the interval or 0–1, UID [0,1]) and subscripts b and w represent best and worst player x-values.

If the new player position is better than the prior best, IF $J(x_{w,\text{new}}) < J(x_b)$, and no constraints were encountered with the $x_{w,\text{new}}$-value, that player becomes the best, and the remaining worst of the other players leaps over it. If the leap-to position is not the best, the worst is found among the remaining team of players, and it leaps over the best. If the leap-to position violates a constraint, then it is the worst, and it leaps from that infeasible spot back over the best.

Figure 4.12 illustrates several situations. For simplicity only five players are shown. The initialization is indicated in Figure 4.12a, where the player DV values are indicated on the x-axis as $x_1, x_2, \ldots, x_5$, and

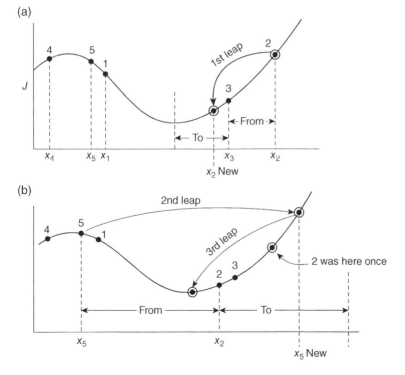

Figure 4.12 Illustration of leapfrogging: (a) first leap and (b) second and third leaps.

the player location on the surface is indicated by the player index 1,2, ..., 5. Note that the index is not the order in either DV or OF value, it is the chronological order of initial randomized player placement in the game. Also note that Player 4 is on a portion of the function that slopes in the direction away from the global optimum toward a local optimum at the left-side of the feasible range.

After initialization, Player 3 is the best, because it has the lowest OF value, while Player 2 is the worst. So, 2 leaps over 3. Figure 4.12a also indicates the leap-from distance and the leap-into window. Player 2 moves to a random location in the leap-to DV location and has an OF value indicated by the open circle in the figure. This location will make Player 2 the best.

Figure 4.12b illustrates the second and third leaps. For the second leap, Player 2 is the best, while 5 is the worst, so 5 leaps over 2, and 5 happens to land on a DV value that has a worse yet OF value. So, Player 5 remains the worst, and on Leap 3 it again leaps over Player 2. As illustrated, 5 becomes the new best.

Note: Not every leap goes to a better spot, but in moving toward the local best, there is an overall tendency to find the optimum.

Note: Eventually Player 4 will be identified as the worst, and it will leap out of the vicinity of the local optimum into the region of the global optimum.

Note: On average each leap cuts the distance between the worst and best in half. But the illustration reveals that some leaps are into the far side of the leap-to window and some to the near side. The stochastic nature of the UID r overrides a deterministic path.

Note: Randomization of the initial player placement may seem less efficient than uniform placement to characterize the surface. In random placement some areas will be relatively unexplored and others more densely characterized. Further, simple placement on a grid would be computationally more efficient than calling the random number generator function. However, nature often devises patterns that would surprise a human planner or human convenience and evade discovery. I like using the random initialization.

Note: For similar reasons, I find that leaping into a randomized location is better than a computationally simpler deterministically leaping a set portion of the distance, such as replacing Equation (4.41) with $x_{w,\text{new}} = x_b + 0.7(x_b - x_{w,\text{old}})$ or so.

Each leap-over is a complete cycle of the procedure. There is only one new function evaluation at each leap-over.

However, I'll define one iteration as N number of leap-overs, where N is the DV dimension. This makes one LF iteration have similar surface exploration to other techniques.

If initially all players were placed on the right-hand portion of the DV space, on the side of a hill, the DV range of the team cluster will expand as the trailing player leaps a long distance over the lead. On the side of a hill, the team cluster expands accelerating the path to the bottom. By contrast, at an optimum, the best gathers all players about it, cutting the distance in half (on average) at each leap, leading to team player convergence in the vicinity of the optimum.

Convergence could be based on the DV range of the cluster. Alternately, convergence could be based on the OF range between best and worst. At convergence the DV of the best player should be reported as the DV*.

Surface aberrations (cliffs, slope discontinuities, flat spots, and infeasible areas) are inconsequential. The feasible area might be separated by an infeasible space or otherwise non-convex.

The procedure is scalable to N-D situations, and as Section 4.4.6 and Chapter 24 reveal, LF can handle stochastic functions.

If initially all players are placed in the local vicinity of the local optimum (in the far left region of Figure 4.12a where Player 4 is), then it is likely that the local best will draw all players toward the local minimum at the far left. So, players should be initialized throughout the entire feasible DV space.

An outline (pseudo code) of the procedure is as follows:

1) Initialize 10 players in feasible locations—randomly assign TS values from within the range of possible DV values, and calculate the OF for each.
2) Find the best and worst player from the 10.
3) Leap the worst over the best into a random location using $x_{w,\text{new}} = x_b + r(x_b - x_{w,\text{old}})$.
4) If the DV position is constrained, then repeat Step 3, leaping from the new constrained DV value back over the best player.
5) If the new position is not constrained, evaluate the OF at the new DV location.
6) If the calculation of the new OF reveals a constraint, then repeat Step 3, leaping from the new position back over the best player.
7) If the new OF is better than the former best, then the new player becomes the best. Otherwise the best remains unchanged. (In either case there is no need to search through all of the players for the best.)
8) Else, if the new OF is equal or worse than the old worse, then it remains the worse. (There is no need to search through all of the players for the worst.)
9) However, if the new OF is in between the former best and worst, then search through the players to identify the new worst.
10) Test for convergence. If not converged, return to Step 3.

4.4.6 LF for Stochastic Functions

A deterministic function returns the same OF value when given the same DV value. Regardless of when, where, how, who, or what adds 7 + 3, the answer is 10 (in base 10 numbers). However, some functions are stochastic, and do not return exactly the same value. Here is a visual example: Consider a pier over a lake. There are waves on the water surface, and the ripples from waves that are reflected by the shore and the pier. If you were measuring the height of the water at a pier, you would have a continually fluctuating value, even if you never change the measurement location, and even if you take the measurements over a small time interval, when nominal level has not changed. This is a stochastic response. There is a nominal average OF value, but the value of any one measurement is perturbed by many random fluctuations.

Standard deviation could be used to estimate the variability of the OF value. One could seek to determine the mean water level by averaging a sequence of measurements, but averaging does not eliminate variability, it just reduces it. The central limit theorem indicates that the uncertainty in the average scales with the square root of the number of samples:

$$\sigma_{\text{average of } N} = \frac{\sigma_{\text{individual measurements}}}{\sqrt{N}} \tag{4.42}$$

Stochastic functions do not return a consistent identical value as a response to any particular DV. Instead, they return an OF value that is perturbed by random fluctuations. The uncertainty of the true (nominal, mean, background) OF value can be reduced by averaging, but variability cannot be eliminated.

Although nearly 100% of undergraduate training is related to deterministic functions, important optimization applications regularly deal with stochastic phenomena. Gaming and economic forecasting are two commonly understood applications. In gaming, someone set the rules of the game such that the house wins on average, but the player wins enough to be encouraged to continue playing. Such games depend on the random outcome of a roll of dice, spin of a wheel, or shuffle of cards. The outcome of any one play is stochastic; it depends on the numbers that chance reveals on that one play.

Because of the origins of stochastic functions in the simulation and optimization of games, these are often termed Monte Carlo methods. But the application of stochastic processes is much broader than game analysis.

In business forecasting, we attempt to evaluate profitability of a decision over the next N years and base the future income and expenses on forecast tax rates, raw material and labor costs, product pricing and demand, etc. Although we might know those values today, we cannot be certain of any of those values over the N-year future. So, one technique in economic forecasting in business is to use a Monte Carlo technique. Randomly sample a possible value from each uncertain variable, and use that as the "given" in the OF to return a response to a particular DV. Alternately, let each value take a random walk path through its N-year future. This is termed one realization and results in one possible OF outcome. It is not the average, or, mean, or median. Because fluctuations may tend to make high values farther from the average than are low values, the average OF value is not the value at the average value of the givens. Further, the objective might not be to maximize some average profitability metric, because this may make the venture vulnerable to catastrophic failure under a particular confluence of future "givens," under a range of realizations. So, the objective might be to maximize the 95% worst possible outcome. In any case, the OF value is stochastic.

When experimental trials are used to determine the OF, the result is stochastic. Each trial outcome is subject to random uncontrolled variation.

The situation can be modeled in any number of ways. A simple way to convert an ideal deterministic OF to a stochastic one, which might represent experimental results, is to add Gaussian distributed noise. If using the Box–Muller technique,

$$\text{OF}_{\text{stochastic}} = \text{OF}_{\text{deterministic}} + \sigma\sqrt{-2\ln(r_1)}\sin(2\pi r_2) \tag{4.43}$$

Here the variables r_1 and r_2 represent uniformly distributed random numbers on the range of zero to unity, $0 < r \le 1$, the complement to the standard pseudo-random number. Their values would be determined anew each time the OF function is evaluated. And σ represents the standard deviation on the OF noise.

If you wish to make coefficients in a model, represent a uniform uncertainty and then perturb the coefficient values within some expected range. If model coefficient a has a nominal value of a_0, and an expected range of $\pm\varepsilon_a$, then at each function call, the coefficient value can be a realization, uniformly distributed, sampled within the range:

$$a = a_0 + 2\varepsilon_a(r-0.5) \tag{4.44}$$

If, for example, the deterministic OF is a simple quadratic, $\text{OF}_{\text{deterministic}} = a + bx + cx^2$, then the stochastic model, due to uncertainty on coefficients could be calculated as

$$\min_{\{x\}} J = \text{OF}_{\text{stochastic}} = a + bx + cx^2 \tag{4.45}$$

$$a = a_0 + 2\varepsilon_a(r-0.5)$$
$$b = b_0 + 2\varepsilon_b(r-0.5)$$
$$c = c_0 + 2\varepsilon_c(r-0.5)$$

This would make the realizations of the coefficient values uniformly distributed. Alternately, your application might indicate that normally distributed, Poisson distributed, log-normally distributed, or other model of variation is more appropriate. Further, your application might indicate that cross correlation between coefficient values, or autocorrelation between successive values of one coefficient are appropriate.

The stochastic nature of an OF, the continual fluctuation, the misrepresentation of the OF, is a problem for optimizers. An optimizer might get stuck in an off-optimum DV value, for which chance returned a one-time, improbable, extreme, phantom good-appearing OF value. To visualize this reconsider measuring the height of the pier over the lake surface, with the objective to find the location on the pier that is farthest from the water. The pier is relatively horizontal with perhaps a few inches difference in elevation. The water has level fluctuations that are on the order of 2 ft. Nominally, the pier is 5 ft above the water. If at one point on the pier, a confluence of random waves and reflections returns an elevation of 8 ft, a 99.9% extreme value and then an optimizer might claim that is definitively the best spot.

Alternately, chance fluctuations can totally misdirect the search, making it appear that the DV is moving in the wrong direction. Consider that the optimizer is at a point on the pier that is sloping up and leading to a more desirable spot, but a wave makes the elevation measurement smaller. Then the optimizer might reverse the search direction and look in the wrong direction.

One approach to finding the optimum in stochastic functions is to use a multiplayer algorithm, such as leapfrogging. One player claims it is the best. Replicate the best player: Resample the OF of the best player, and use the worst of several replicate values. If the best player remains the best, leap the worst over it. The more times the best remains the best, the greater is the chance of its OF value representing a worst possible extreme. Alternately, if the replicate value reveals that best does not remain the best, choose the second best as the lead and leap over it. This changes the location of the leap-over base. After the team has converged on the proximity of the optimum, the stochastic lead continually changes, which will make the cluster randomly move to and fro over that optimum region. The cluster will never converge to a single point.

One method of testing for convergence of stochastic process is to identify when things settle to a noisy steady state. Observe the OF value of the worst player. In the beginning of the iterations, as the cluster is randomly scattered throughout the DV range, the OF of the worst player will be high. As the team converges in the vicinity of the optimum, the OF of the worst player will drop. The player identity with the worst OF value will keep changing, as it leaps to a better spot, making another player the worst. At the end, when the cluster is randomly moving within the region of the optimum, the OF of the worst player will be a fluctuating value with a consistent average. Observing the OF of the worst player with iteration will reveal a noisy signal that relaxes to a noisy steady state. At steady state, there is no improvement of the OF w.r.t. iteration, so claim convergence. There are many methods to identify steady state in a noisy signal.

I like the method of Cao and Rhinehart (Appendix D) because it is computationally simple and works appropriately. The method looks at a ratio of two variances. The numerator variance is that conventionally calculated by deviation from the average, and the denominator variance is estimated from differences in sequential data. At SS, when the average is unchanging, the expectation of the ratio of the variances is unity. However, if not at SS, some data deviate substantially from the average, the numerator variance is large, and the ratio will be greater than unity.

For computational efficiency, the method uses an exponentially weighted moving variance and an exponentially weighted moving average. These are alternately termed first-order filters.

Since the ratio statistic is a stochastic variable, at SS it does not hold a deterministic value of unity, but has a range from about 0.75 to 1.5. Also, when a process is close to, but not quite at SS, the nominal ratio value might be 1.4, but have a value within the range of 0.9–2.3. Accordingly, claim SS when the ratio has a value that would not come from a near SS situation, such as 0.9, but has a value that would confidently represent one that could only be encountered if at SS, perhaps 0.85.

The equations are developed in Appendix D, and the resulting VBA code for the procedure is

```
nu2f = l2 * (OF_worst - xf ) ^ 2 + cl2 * nu2f        'estimate of numerator
variance
xf = l1 * OF_worst + cl1 * xf                        'estimate of average value
delta2f = l3 * (OF_worst - OF_worst_old) ^ 2 + cl3 * delta2f
'denominator
 OF_worst_old = OF_worst                             'reset just past value
 R = (2 - l1) * nu2f / delta2f            'calculate the ratio of variances
 IF R<0.85 THEN SS = 'YES'
```

The coefficients l1, l2, and l3 represent the three filter lambda values. Recommended values for each is 0.1, essentially indicating that the most recent 10 samples are included in the analysis. The coefficients cl1, cl2, and cl3 represent the complementary values, $(1 - \lambda) = 0.9$.

A ratio of variances the statistic is scaled by the inherent noise level in the data. It is also independent of the dimensions chosen for the OF.

Note: At convergence, one player in the team will have the best OF value. but this is not the unique solution, it is the fortuitous best in the cluster on that iteration. The range of lead players in the vicinity of the optimum indicates the general vicinity and its range of uncertainty. Reporting the single best player value will misrepresent the application. Accordingly, report the range of DV* values.

4.5 Perspectives on Univariate Search Methods

The analytical method is simple conceptually, but it requires an analytically tractable expression—a formula that permits obtaining an analytical derivative. Subsequently, hopefully, the derivative equation permits explicit solution for DV*. An advantage of such is that the equation for DV* can be used to reveal the function dependency of the DV* due to other variables. However, depending on complications, the solution for x^* may need a numerical root-finding algorithm. Further, the procedure may find a maximum or a saddle point, since $df/dx = 0$.

Use the second derivative to determine whether the zero-slope point is a maximum, minimum, or saddle point. If a minimum at x^*, then $d^2f/dx^2|_{x^*} > 0$, which is termed "positive definite." If a maximum at x^*, then $d^2f/dx^2|_{x^*} < 0$, which is termed "negative definite." If $d^2f/dx^2|_{x^*} = 0$, it is a "saddle point" or locally "flat."

All other methods are iterative.

Region elimination methods (interval halving and Golden Search) are robust to many function shapes and give guaranteed DV range after N iterations:

$$\text{DV Final Range} = \text{DV Initial Range} \begin{Bmatrix} 0.5^N \\ 0.618...^N \end{Bmatrix} \tag{4.46}$$

According to the rules, they find the either the minimum or the maximum as desired by the user, not a point that might be either minimum, maximum, or saddle. But they each have a somewhat complicated logic.

Newton-type methods and successive quadratic (second-order, iterative approaches) may send the next guess to extremes when slope of the first derivative (the value of the second derivative) is close to zero as illustrated in Figure 4.13.

You can identify this situation by observing the Δx values. They will alternate sign, sending the trial solution too far to the right and then too far to the left. You can temper this potential problem by limiting Δx to a max size or adding a damping coefficient, α, to the iterative rule. Perhaps $\alpha = 0.1$. If a Newton-type algorithm, the damping coefficient can multiply the Δx calculation:

$$x_N = x_{N-1} - \alpha \frac{dy/dx|_{x_{N-1}}}{d^2f/dx^2|_{x_{N-1}}} \tag{4.47}$$

Alternately, Newton-type iterative approaches may very slowly creep up to the optimum where the slope of a function is as illustrated in Figure 4.14.

You can identify this situation by observing the Δx values. They will keep the same sign. You can accelerate the search for the optimum by making the coefficient; α has a larger value, perhaps $\alpha = 2$:

$$x_N = x_{N-1} - \alpha \frac{dy/dx|_{x_{N-1}}}{d^2f/dx^2|_{x_{N-1}}} \tag{4.48}$$

With analytical expressions for the first and second derivatives, Newton-type algorithms are easy to program. With numerical estimates of the first and second derivatives, requiring user choices for the method (forward, backward, central) and for the Δx increment used to estimate the derivatives, they are a bit more complicated than Golden Section.

For well-behaved functions (continuous-valued with continuous derivatives), especially ones with a quadratic-like shape, Newton-type methods will rapidly jump to the vicinity of the optimum and rapidly hone in on the DV* value.

Newton-type algorithms will move to a maximum, minimum, or saddle point, requiring a check of the second derivative at a solution.

They seek a local optimum, the attractor at the initial TS, not a global optimum from a range. Congratulations to you, if you know, from either prior experience or the functionality of the equations, that there is only one optimum for the application. If you do not know such, then graph the function to reveal its shape, or initialize the optimizer from diverse points to see if all initializations lead to the same DV* value.

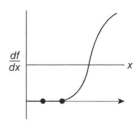

Figure 4.13 Illustration of an overstep when the second derivative has a low value.

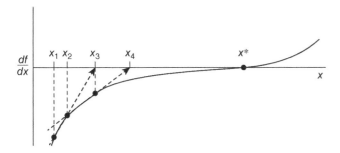

Figure 4.14 Illustration of an understep when the derivative asymptotically approaches zero.

Successive quadratic is somewhat more complicated to program, but like Newton-type algorithms, it is fast and relatively robust in the vicinity of x^* for well-behaved functions. The convergence criterion needs to stop it before the most recent three DV points are very close to x^*, because essentially identical successive values of x^* and f lead to an ill-conditioned matrix (numerical instability). Like Newton's method, successive quadratic will seek a maximum or saddle point as well as a minimum. Outside of the inflection points from a minimum, they seek the maximum. They are only useful near (within the concave vicinity of) the minimum. Again use the second derivative of the function to determine whether the stationary point represents a maximum, minimum, or saddle.

All methods presume well-behaved functions not situations illustrated in Figure 4.15 of a surface with multiple optima (alternately one that is stochastic) or one with likely to be missed pinhole optima.

Although analytical methods (analytical, surrogate model, Newton's types) cannot cope with discontinuities such as illustrated in Figure 4.16, direct search algorithms can.

The direct search methods seek a minimum. They do not get drawn to maxima or saddle points. However they can be programmed to find a maximum, for instance, if Δx has same sign of slope in a marching stage of interval halving, if save the extreme region with the "lowest" is replaced with save "highest" in GS, or if the other direct searches define better if $f_{\text{new}} > f_{\text{base}}$ as opposed to using $f_{\text{new}} < f_{\text{base}}$.

None of the methods are universally perfect. In Figure 4.17a, for example, if a function has local optima, the Golden Section may discard the region with the global minimum. In Figure 4.17b, for example, if $y = 10 + [2/(x-5)]^2$ Newton's, Newton-secant, and successive quadratic all diverge away from the minimum. Further, even the assumption that the derivative of y w.r.t. x is zero at the minimum is incorrect.

(a) (b)

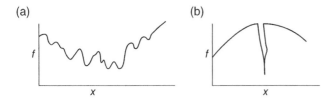

Figure 4.15 Illustration of ill-behaved OF responses: (a) stochastic and (b) pinhole.

(a) (b)

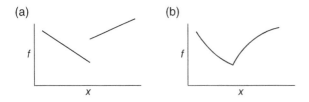

Figure 4.16 Illustration of (a) function discontinuity and (b) slope discontinuity.

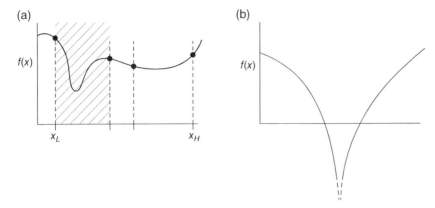

Figure 4.17 Two difficult optimization cases: (a) minimum overstepped and (b) causes divergence in second-order methods.

4.6 Evaluating Optimizers

Optimization seeks the best. Can it be applied to determine which optimization algorithm is best? Here is the statement:

$$\max_{\{AM,\,N,\,SQ,\,HD,\,LF\}} J = \text{Desirable Attributes} \tag{4.49}$$

To do this, one must decide what to include as the desirable and undesirable attributes of an optimizer. Here is a list of attributes that has guided my choices in deciding what to present in this chapter:

- NOFE—number of function evaluations to identify DV*. This would include the functions needed to calculate the derivatives, whether analytical or numerical.
- p-Global—the probability of finding the global optimum in an application with multiple optima.
- Does not get redirected to max or saddle, if seeking min.
- Simplicity of algorithm—for understanding and trouble shooting.
- Simplicity of use—such as initialization of TS and choices for algorithm coefficients.
- Robustness of algorithm—to execution errors and to misdirected excursions.
- Robustness of the algorithm to application aberrations—discontinuities in OF value or derivatives, flat spots, discretization effects, inflections, multiple optima, constrained regions, etc.
- Computational speed—this includes all algorithm-specific aspects, but it would exclude user choices for data output, convergence testing, and procedures to determine function values.
- Broad acceptance—the optimization algorithm is popular, familiar, used, etc.

4.7 Summary of Techniques

I have selected these six optimizer algorithms as the most important to present in the univariate search chapter for the following reasons.

4.7.1 Analytical Method

- Provides exact solution for DV*
- Provides an equation that reveals how DV* is functionally related to auxiliary variables or application coefficients
- Scalable up to N DV dimensions
- Seeks min, max, or saddle point
- May be analytically intractable, or often one step too difficult for a normal STEM person applying optimization

4.7.2 Newton's (and Variants Like Secant)

- A fundamental technique within many N-Dimension algorithms.
- It gets a local second-order model of the OF vs. DV then jumps to the model minimum.
- Scalable up to N DV dimensions.
- It presumes a convex function in which $d^2OF/dDV^2 > 0$ everywhere (when seeking a minimum).
- It presumes a continuous OF with continuous first derivatives everywhere.
- It can jump to ridiculous places if the conditions are not valid, like inflections.
- It needs to have values for the first and second (mixed) derivatives, which may need to be done numerically, requiring additional function evaluations for each jump to iteration.
- Will move to min, max, or saddle point.
- Finds local minimum, not global.
- Cannot handle "FAIL" as a hard constraint.
- Fast and efficient when the surface is quadratic-like.

4.7.3 Successive Quadratic

- Gets a local quadratic model (surrogate model) of the OF w.r.t. DV surface then jumps to the model minimum.
- Scalable up to N DV dimensions.
- It presumes a convex function in which $d^2OF/dDV^2 > 0$ everywhere.
- It presumes a continuous OF with continuous first derivatives everywhere.
- It can jump to ridiculous places if the conditions are not valid.
- Does not need to evaluate derivatives or second derivatives, but needs to solve the linear $N \times N$ equation set for the model coefficients. (Gaussian elimination or such.)
- Requires several function evaluations at each iteration to get the simplest type of model and more for a model with cross product terms.
- Will move to min, max, or saddle point.
- Finds local minimum, not global.
- Cannot handle "FAIL" as a hard constraint.
- Fast and efficient when the surface is quadratic-like.
- Reveals the application of surrogate models; there are many alternate models that could be used.

4.7.4 Golden Section Method

- Limited to a univariate search. Cannot extend to N-D.
- The "line" search may be along a curve in N-D space.

- Direct (as opposed to gradient based)—no models, no derivatives.
- Presumes one minimum within the range.
- Allows discontinuity of OF or its derivative.
- Most efficient of the region elimination algorithms (range reduction per function evaluation).
- Finds local minimum, not global, within the initialization range.
- Will not move to max or saddle point, only to min.
- Cannot handle "FAIL" as a hard constraint.
- Relatively robust to surface discontinuities.

4.7.5 Heuristic Direct

- Uses heuristic rules. You can change them to suit your logic.
- Scalable to *N*-D.
- Direct (as opposed to gradient based)—no models, no derivatives.
- Presumes no jump over of the best (narrow valley) on a general downhill path.
- Allows functions and slope discontinuity.
- Can handle "FAIL" as a hard constraint of a function or response to error trapping in code execution.
- Simple to understand and implement.
- Finds local minimum, downhill from the initialization, not global.
- Will not move to max or saddle point, only to min.
- Intuitive.
- Relatively robust to surface aberrations.

4.7.6 Leapfrogging

- Scalable to *N*-D
- Direct (as opposed to gradient based)—no models, no derivatives
- Allows functions and slope discontinuity
- Allows flat spots
- Can handle "FAIL" as a hard constraint of function response
- Simple to understand and implement
- Tendency to find global minimum, but not guaranteed
- Will not move to max or saddle point, only to min
- One OF evaluation per iteration (in 1-D)
- Can be applied to stochastic applications
- Multiplayer

4.8 Takeaway

If the application is analytically tractable, then use the analytical technique. If not, but the function is unconstrained and quadratic-like, then use Newton's secant (my preference over SQ). If the function has constraints or is discontinuous, then use the heuristic direct method. If the function has either multiple optima or a stochastic response, then use leapfrogging.

4.9 Exercises

1 Explain why successive quadratic could send the search in the wrong direction.

2 Explain why Newton's method could send the search in the wrong direction.

3 If the univariate search "line" is actually a curve, explain how you could use the Golden Section to find the optimum.

4 List one advantage and one disadvantage for each of the following optimization algorithms. Be brief as well as specific and understandable. Don't just copy keywords from the chapter. Reveal that you understand the advantages and disadvantages. (i) Successive quadratic. (ii) Heuristic direct search. (iii) Golden Section. (iv) Newton's (secant). (v) Leapfrogging.

5 Analytically based searches (those using equations to characterize the surface and determine trial solutions) seek a point of zero slope, but this may be either a minimum, maximum, or saddle point. What analytical method (equation based, not testing adjacent points) will indicate whether the optimum is a minimum, maximum, or saddle point?

6 In a univariate search, how many function evaluations are required for (1) initialization and (2) for each successive each stage (iteration) for the following optimization algorithms. Explain your two numbers for each algorithm. Be brief as well as specific and understandable. (i) Heuristic direct search, (ii) Golden Section, (iii) Newton's (secant), and (iv) leapfrogging.

7 Figure 4.18 shows a circle on an (x, y) plane centered at the dot and contour lines for an objective function, $f(x, y)$. In polar coordinates, (r, θ), $x = x_{center} + r \cos(\theta)$, $y = y_{center} + r \sin(\theta)$. Show and explain the first three iterations of the heuristic direct search to perform a univariate search for the minimum on the circle perimeter (keep r fixed, use θ, the angle from the horizontal, as the single DV, use an initial $\Delta\theta$ of +30° of arc).

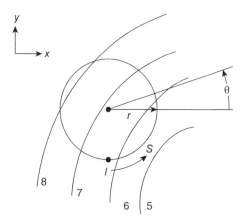

Figure 4.18 Illustration for Exercise 7.

8 Figure 4.19 is a sketch of a function with low and high DV expectations marked. Sketch and explain the first two iterations of the Golden Section method to find the minimum.

Figure 4.19 Illustration for Exercises 8 and 9.

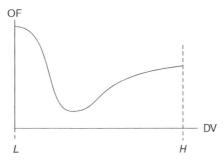

9 Use Figure 4.19 and create your own sketch of the derivative of the OF over the entire DV range, and starting near the high DV value, sketch and explain the first move of a Newton-type search for the optimum.

10 Write the computer code to perform univariate searches to find the minimum of $J = y = 4 + 8x - 2x^2$. Use x as the DV. Alternately, use $J = a + bx + cx^2$, and you pick values for the coefficients a, b, c. Use each of these methods: successive quadratic, Golden Section, secant (or Newton's), and heuristic direct.

  a) Graph OF versus DV.
  b) For each algorithm, reveal the path to the optimum, so that you can "see" the algorithm logic and know that your code is working correctly. You choose the starting point.
  c) Show your computer code for each search method.
  d) Choose the method you liked best. Explain and quantify your evaluation criteria.

11 Repeat Exercise 10 with $J = y = (x_1 - 8)^2 + (x_2 - 6)^2 + 15\,\mathrm{Abs}((x_1 - 2)(x_2 - 4)) - 300\,\mathrm{Exp}(-((x_1 - 9)^2 + (x_2 - 9)^2))$ search along the line defined by $x_2 = a + bx_1$, and use distance along the line as the DV. You pick values for "a" and "b" and for the line origin.

12 Repeat 10 with $J = \cos(2\pi x/10)$.

13 Repeat 10 with $J = [5 + (x - 5)^2][1 - \exp(-10(x - 9)^2)]$.

14 Repeat 10 with $J = 5 - x + 0.45x^2 - 0.08x^3 + 0.005x^4$.

15 Repeat 10 with $J = (1/x) - 5e^{-x}$

16 Repeat 10 with $J = \left[1 + 0.2*(x - 0.5)^2\right] * \left[1 - e^{-200(x-4)^2}\right]$.

17 Repeat 10 with $J = \left(2 - 2x^2 e^{-(x/2)}\right)^2$.

18 Repeat 10 with $J = (50 + \ln(x) - 5\sqrt{x} - x)^2$.

19 Minimize this function of D. You choose numerical values for coefficients ($a > 0$ and $b > 0$). $J = aD^{-5} + bD^{1.2}$.

20 Here are three functions. The DV is x, the OF is y.

a) $y = 5 - x + 0.45x^2 - 0.08x^3 + 0.005x^4$

b) $y = e^{-\frac{1}{2}\left(\frac{x-5}{2}\right)^2}$

c) $y = -2\left(\frac{1}{x-5}\right)^2$

For each, the DV range is 0–6. Function a) has a minimum and a saddle point. Function b) has two minima. The minimum for Function c) has a y-value of $-\infty$. Minimize each using four techniques: analytical (set the derivative equal to zero, and then root-find for the value of x^*), either Newton's using numerical first and second derivatives or secant, successive quadratic, and golden section. This is a 3 * 4 = 12-part exercise. However, depending on the starting location, the initial trial solution, you will get different results for the first two functions. So, for each of the 12 combinations, start at places which reveal the different results. Explain why the optimizer outcomes are what they are. A printout of the sequential trial solutions is not an explanation. Describe why each optimizer method created that DV sequence. I think that you can best explain this by hand illustrations on a printout of the graph of y w.r.t. x for each function.

21 Which is the better optimizer for an unconstrained univariate search—Newton's secant, successive quadratic, Golden Section, or heuristic direct? An evaluation of "better" would include simplicity of understanding, computational burden, probability of finding the minimum from random starts, and what else? How would you measure these aspects of better? Choose your functions and assessment methods, run trials to generate data, interpret the results, and state your opinion as to which search method is better.

22 State this version of an undergraduate physics projectile problem in the standard form of an optimization problem. "A canon is at the same elevation as the distant target. What angle permits the lowest muzzle velocity for the canon ball?" First, ignore air resistance and any other confounding effects (surface curvature, Earth spin). Choose reasonable values for the coefficients of the situation, and solve for angle theta using any of the univariate search techniques.

23 Add air resistance to the model in Exercise 22, with a simple model that the drag force is opposite to the direction of motion and proportional to the velocity squared. Then resolve. For this case, you might not be able to create a one-line function. The projectile with air resistance is a bit challenging— numerical solution to nonlinear ODEs, embedded with root finding, embedded with optimization.

24 Choose three (x, y) data pairs, and generate a quadratic model that fits the data. For each of the $i = 1, 2, 3$ pairs, the equation $y_i = a + bx_i + cx_i^2$ is true. Solve for the coefficient values by the method of Gaussian elimination. Show each stage of your process. Compare your answer to any automated method of curve fitting (such as the trend line function in Excel charts).

25 One wishes to determine the w-value (root) of the implicit relation $w = 50 + \ln(w) - 5\sqrt{w}$. Convert this to an optimization application, to minimize the square of the deviation, and implement the solution. Show the stages on a graph. Explain your justification for stopping the iterations. Compare the result to that from a conventional root-finding procedure.

26 Lectures are effective in the beginning when students are paying attention. But as time drags on, student attention wanders, and fewer and fewer students are sufficiently intently focused to make the continued lecture be of any use. A model that represents the data is $f = 1/\left(1 + e^{(t-15)}\right)$, where f is the fraction of class attention and t is the lecture time in minutes. Graph this, and you'll see that after 15 min $f = 0.5$, half the class is not attentive. After about 20 min, effectively no one is attentive to the lecturer. The average f over a 55-min period is about 0.26. Students only get 26% of the material. That is not very efficient. A teacher wants to maximize learning, as measured by maximizing the average f over a 55-min lecture. She plans on doing this by taking 3-min breaks in which the students stand up, stretch, and slide over to the adjacent seat, and then resume the lecture with renewed student attention ($t = 0$ again). If a teacher lectured 2 min and took a 3-min break, on a 5-min cycle, then the value of f would be nearly 1.00 for the 2 min, but there would only be $2 * (55/5) = 22$ min of lecture per session. But 22 min at an f of 0.9999 over the 55 min is an average f of 0.40. This is an improvement in student focus on the lecture material. Is there a better lecture-stretch cycle period? Express the optimization statement—clearly state the OF and DV and equation(s) relating them. Use a trial-and-error (human intuitive) approach to find an optimum. Use any method you wish to compute how the OF value changes with DV.

27 Choose a function of a single DV, initializations, convergence criterion, and threshold. Run (i) your choice of one of the Newton's optimization approaches, (ii) successive quadratic, and (iii) heuristic direct search to find the minimum. For each of the three methods, explain what the algorithm is doing. Perhaps display and annotate the iteration-to-iteration data. (iv) Find a function that sends Newton's or SQ (choose one) to the wrong place (maximum or saddle, or to extreme values if the initial TS is on the wrong side of the inflection point), and compare how the heuristic direct does not get misdirected. (v) Find a function that is quadratic-ish and show that Newton's (or SQ) jumps right to the answer, while the heuristic direct takes many function evaluations to creep up to it.

28 Choose the best diameter for a sample line to convey a liquid from the main process line to the analyzer. A smaller diameter line means less cost of the line, less cost of the material in the line, and less waste to flush old material through to get the new material to the analyzer. However, flow rate through the line is its response to the diameter and to system properties such as fluid viscosity and line pressure. A small diameter means low flow rate and a longer wait to purge the line of old material to get the new to the analyzer. If the flow is in the laminar regime, then about two flush-throughs of the volume are needed. But, if turbulent, then one flush-through is adequate. The laminar-turbulent transition creates a discontinuity in the OF w.r.t. DV plot. The weighting of the several desirables and undesirables and process attributes (viscosity, pressure, distance of the sample line) all affect the optimum. For a simple, but adequately complex, exercise: Use the Hagen–Poiseuille relation for laminar flow, and calculate average velocity as $u = D^2 \Delta P/(32\mu L)$. If the Reynolds number is less than 2300, then accept that the flow rate is in the laminar regime, $Re = Du\rho/\mu$. If $Re > 2300$, then the flow is turbulent, and use the Darcy law with the Swami–Jain friction factor relation to determine average velocity. $u = [1.14 - 2\log 10(\varepsilon/D + 21.25\mathrm{Re}^{-0.9})]$ $\sqrt{2\Delta PD/(L\rho)}$. This is iterative, the velocity is needed to determine the Re value, but the Re value is needed to calculate velocity. The mass of material in the sample line is $M = \rho L\pi D^2/4$. The cost of the line could be scaled to capacity with the 6/10ths power law, and a simple embodiment could be $C = 5L(D/0.01)^{1.2}$. In SI units for a nominal system, $L = 5\,\mathrm{m}$, $\rho = 1000\,\mathrm{kg/m^2}$, $\mu = 0.002\,\mathrm{Pa\,s}$, $\Delta P = 5\,\mathrm{kPa}$, $\varepsilon = 0.0001\,\mathrm{m}$. The units on u are m/s and on C is \$. Determine the optimum diameter to balance concerns related to transport time, material in the line, flush-through waste, and sample line cost.

29 Numbers S and L are the smaller or larger of two numbers. What number, ε, when added to S and subtracted from L minimizes the product?

5

Path Analysis

5.1 Introduction

We often have to evaluate events of a situation along a path and use either a summation of events or a rate of change as the objective function. This chapter reveals some examples and the mathematical methods for assessing the OF.

Many people use the term line integral, but to help eliminate confusion over the term line search and to admit that the path for accumulation does not need to be a straight line, I'll use path integral.

5.2 Path Examples

Following are five examples to reveal the concepts of path analysis:

Example 1 What path should you walk along to get from point A to point B? Figure 5.1 shows two paths: one walking directly and one weaving among the trees. If it has just started lightly raining and getting wet is undesired, then the longer path under the protection of the trees might be best. How does one calculate the amount of rain encountered on each path to determine which is better? If a path was described by a function $y = f(x)$ such as $y = a + bx + cx^2$, where x and y represent the ground map coordinates, and the walking speed was $v = dS/dt$ where S is the distance along the path, then path analysis provides the time in the path section ΔS at any particular spot $[x(t), y(t)]$. If the rain rate is known, $r = g(x,y)$, then a path integral technique will provide the answer.

In this example, the points A and B constrain the path equation. When $x = x_A$, then the relation $y = a + bx_A + cx_A^2$ must return the value of y_A. Similarly, $y_B = a + bx_B + cx_B^2$. These constraints mean that of the three coefficients (a, b, c), only one can be chosen by the optimizer to determine the best path. Once one is chosen by the optimizer, the other two are determined from the two equality relations. As a result, this is a single DV optimization.

$$\min_{\{c\}} J = \int_A^B r(x,y)\, dS \tag{5.1}$$

In optimization, this is often termed a line search, because it has a single DV. However, the integral is also over a line (a path). The common terminology of "line" has two very distinct meanings. In one it is

Engineering Optimization: Applications, Methods, and Analysis, First Edition. R. Russell Rhinehart.
© 2018 R. Russell Rhinehart. Published 2018 by John Wiley & Sons Ltd.
Companion website: www.wiley.com/go/rhinehart/engineeringoptimization

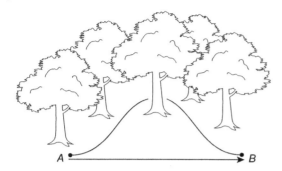

Figure 5.1 Getting from *A* to *B*.

the variable *c*, a coefficient in the path relation. In the other it is the distance from point *A* to point *B*. As mentioned earlier, to avoid the confusion, I will use the terminology path integral or path function to relate to the process along a path (or line). I prefer the use of the term univariate (a single DV) search to describe the DV dimension, instead of the term line search.

Example 2 Here is another example of a path integral: Where do you aim, and what initial velocity of a projectile will maximize accumulated points in a coast through space game (see Figure 5.2)? The best combination will lead to the most accumulated points along the path and at a target on the planet. The equations of motion of a solid particle can be used to describe the path (x, y) position in time. But how is the accumulation of points determined? Again in an integral, an accumulation of points along a path technique is the answer. This is an initial value problem. Once angle and velocity are decided, the dynamics of the path (position and speed, time in a vicinity) are determined by the free-fall equations of motion. The optimizer has two DVs, but the objective function value is still determined by a path integral.

Example 3 In this third path integral example, a driver is seeking to drive the car from point *A* to point *B* within a time constraint, to minimize fuel consumption. Illustrated in Figure 5.3, there are flat and uphill portions of the road and a rough section. The logic might be to floor the accelerator until a certain time and then coast to the finish line:

$$p = 100\%, \quad \text{if } t < \tau$$
$$p = 0\%, \quad \text{if } t \geq \tau$$

$$(5.2)$$

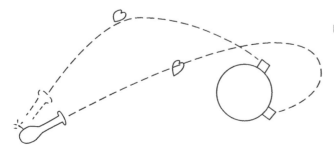

Figure 5.2 Coast into space game.

Figure 5.3 Best speed profile from *A* to *B*.

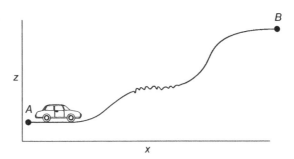

Here the optimizer would search for the value of τ to minimize fuel consumption with the constraint of getting to point *B* within the time limit. Alternately, the pedal position could be scheduled to change from 100 to 0% with elevation or distance along the road, instead of time. Then, given a particular τ trial solution, the engine thrust model (as a function of speed and pedal position) and the equations of motion for the car can give the road position, velocity, and fuel consumption at any point in time. Integrating fuel consumption along the 100% on-time will give the fuel consumption. Here the single decision variable is the time to start coasting, τ, and the path integral might be over time, from zero to τ:

$$\min_{\{\tau\}} J = \int_{t=0}^{t=\tau} F(t,v,z)dt \tag{5.3}$$

Here the single DV is the same as the integration limit, adding confusion when distinguishing the line search DV from the line integral argument. So, don't use the terminology line search; use a univariate (singe DV) search. And don't use the terminology line integral; use path integral.

Example 4   In the maximum likelihood method of fitting a curve to data, the objective is to determine model coefficient values that make the model closest to the data points. Least squares regression, minimizing the sum of vertical squared deviations, is an idealized variant. Consider a simple single-input–single-output situation, $\tilde{y} = f(x)$. The classic sum of squared deviations, SSD, objective function is $J - \sum(y_i - \tilde{y}_i)^2$. In maximum likelihood, however, the "distance" between model and data is related to both *y*- and *x*-deviation, and each is scaled by its uncertainty. The likelihood is related to the probability that the model could have generated the data. Assuming a normal (Gaussian) probability distribution, for the *i*th data point, $L_i = (1/2\pi\sqrt{\sigma_x\sigma_y})e^{-(1/2)\left\{[(x_i-\tilde{x})/\sigma_x]^2 + [(y_i-f(\tilde{x}))/\sigma_y]^2\right\}}$. One stage in the procedure is to find the point on the curve that is closest to the data point, where close is related to the maximum of the joint probability that the curve could have generated the data. Given the trial solution of coefficient values in a modeled curve, for each data point, search along the model to find the *x*-value that maximizes L_i:

$$\max_{\{\tilde{x}\}} J = L_i = \frac{1}{2\pi\sqrt{\sigma_x\sigma_y}}e^{-\frac{1}{2}\left[\left(\frac{x_i-\tilde{x}}{\sigma_x}\right)^2 + \left(\frac{y_i-f(\tilde{x})}{\sigma_y}\right)^2\right]} \tag{5.4}$$

This is a single variable search. Here, the DV was represented by the x-variable, $\tilde{x}$, but it could as well have been the response variable, $\tilde{y}$, or distance along the curve, or some parametric relation such as how position might change in time. Although this is a single DV optimization, it is one stage of optimization imbedded in a multivariable model coefficient optimization. The objective is to search along the model line, a path defined by $\tilde{y} = f(\tilde{x})$, to find each point $(\tilde{x}_i, \tilde{y}_i)$ that maximizes L_i.

Example 5 Rather than integrating along a path determined by a single variable (time or position), here is an example that uses a derivative of the path function. The objective is to determine a walking trail path from point A to point B in a mountain range. Perhaps the x–y path is described by a model from A to C to B with quadratic segments. From A to C the path is $y = f_1(x)$, such as $y = a_1 + b_1 x + c_1 x^2$; and from C to B, $y = f_2(x)$, such as $y = a_2 + b_2 x + c_2 x^2$. The objective is not only to determine the trail with minimum distance (which would be a path integral) but also to keep the trail slope at any point less than some maximum value specified for safety and comfort of the walkers and rain erosion of the trail stones. Here, the slope is the rate of change of elevation w.r.t. distance along the trail. If the x-position is chosen, then the path model gives the y-position. If elevation were modeled as a function, $z = h(x,y)$, then the slope of the path is dz/dS where S is distance along the path (in all three spatial directions).

The optimization statement could be to determine values for the location of point C and the path equations from A to C and then C to B that minimize the steepest slope anywhere along the path:

$$\min_{\{x_C, y_C, c_1, c_2\}} J = \max \left[\frac{dz}{dS} \right], \forall S \tag{5.5}$$

5.3 Perspective About Variables

In many of those examples, the path position was spatial, with (x, y, z) coordinates. The position could be the consequence of one of any number of independent variables such as the x-position, time, S-distance along the path, or accelerator pedal position that is scheduled with distance or time. Both the position response and the local accumulation of desirable aspects (to be maximized) or undesirable aspects (to be minimized) are the dependent variables, and these are in response to the chosen independent variable for the path calculation:

Note: Neither the dependent or independent variable in the path calculation are the optimizer decision variables. The DVs would be the coefficient values (a, b, c, τ, etc.) in the scheduling relations, the initial values (angle and initial projectile velocity) in a dynamic response system, or model coefficient values (as in maximum likelihood curve fitting).

Note: The path does not have to be a spatial path. It could be, for instance, how temperature changes in time or how financial decisions and outcomes change in time.

Note: The independent variable that defines the path position does not have to be either time or a spatial position. It could be an alternate parametric relation.

Commonly in a model, there is a dependent variable that is a response of an independent variable. For example, $v_{\text{dependent}} = f(v_{\text{independent}})$. And the accumulation is the integral of some characteristic, c, over the dependent variable. $J = \int_A^B c(v_d)dv_d$, where A and B are the dependent variable limits. This of course can be transformed to be an integral w.r.t. the independent variable. $J = \int_C^D c(f(v_i))(df/dv_i)dv_i$, where C and D are the associated limits for the independent variable. Further, although the optimization concept is represented by an integral, probably there is not an analytic solution, and the OF value will be calculated by a numerical procedure (such as the rectangle rule of integration, Gaussian quadrature, Simpson's rule, etc.)

Similarly, the rate of change of some function, h, can be expressed as the derivative. If h is a function of the independent variable, then the rate of change w.r.t. the independent variable is $r = dh/dv_i$. However this can be mathematically transformed in a variety of ways. For instance, if h is a function of v_d and the rate of change is w.r.t. a variable $S(v_i)$, then $r = (dh/dv_d)(dv_d/dv_i)/(dS/dv_i)$. Again, it is unlikely that the functions are convenient for determining analytical derivatives, and numerical procedures would be used to estimate the derivative values (e.g., forward, backward, or central finite difference).

5.4 Path Distance Integral

An elementary question to ask is "What is the distance along a curve, $y = f(x)$, from x_1 to x_2?" Figure 5.4a illustrates a curve and indicates that S is the distance along the path from the initial point at x_0. Often this is termed a line integral, although the curve is not a straight line.

This analysis presumes that x and y have the same units. If they do not, then you could scale y and x by their respective ranges, to represent the concept in dimensionless, scaled variable notation.

Using $S =$ distance, then $S = \sum \Delta s$ elements along curve. For small increments, Δs, the incremental distance along curve can be approximated using the Pythagorean relation: $\Delta s = \sqrt{\Delta x^2 + \Delta y^2}$. This is illustrated in Figure 5.4b.

However, as illustrated in Figure 5.4c, for small Δx, the Δs element can also be approximated as distance along the tangent to curve at x_i; so, the slope of the Δs element is approximately $dy/dx|_{x_i}$. Then, $\Delta y \cong dy/dx|_{x_i}\Delta x$. Using the Pythagorean relation, $\Delta s = \sqrt{\Delta x^2 + \left(dy/dx|_{x_i}\right)^2 \Delta x^2}$, and factoring out the common Δx, $\Delta s = \Delta x\sqrt{1 + \left(dy/dx|_{x_i}\right)^2}$.

Then, summing the elements,

$$S = \sum \Delta s = \sum \left(1 + \left(\frac{dy}{dx}\Big|_{x_i}\right)^2\right)^{1/2} \Delta x \qquad (5.6)$$

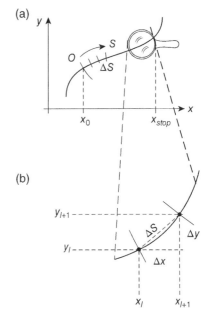

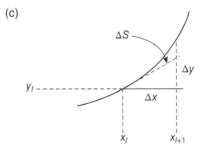

Figure 5.4 Path distance: (a) overall, (b) Pythagorean, and (c) tangent.

This is the rectangle rule of integration. In the limit as $\Delta x \rightarrow 0$, it is represented as an integral:

$$S = \int_{x_0}^{x} \left[1 + \left(\frac{dy}{dx} \right)^2 \right]^{1/2} dx \tag{5.7}$$

Given x_0 and x, and a relation for the function $y = f(x)$, explicitly calculate S from the integral. If lucky, you will be able to use a simple analytical expression. If not, use a numerical method.

The approach can be extended to a multivariate situation, if each variable can be related to one. For instance, if the path goes through the mountains, then the z-elevation depends on the x-latitude and y-longitude positions of the path on a map. If $y = f(x)$ and $z = h(x,y)$, and y and z have unique values for all x-values and the derivatives exist, then

$$S = \int_{x_0}^{x} \left[1 + \left(\frac{dy}{dx} \right)^2 + \left(\frac{\partial h}{\partial x} + \frac{\partial h}{\partial y}\frac{dy}{dx} \right)^2 \right]^{1/2} dx \tag{5.8}$$

Example 6   What (x,y) location is 37.2 units along the path, described by the relation $y = 2x^3$ and starting from $x = 3$ units?

This example uses the inverse of Equation (5.7). Rather than determining the value of S given the value for x, the procedure needs to determine x from S. The derivative of the path relation is $\dot{y} = dy/dx = 6x^2$, and with the specific conditions, the path integral relation becomes $37.2 = \int_{3}^{x} \sqrt{(1 + 36x^4)}dx$. The unknown is the upper limit of integration x. Perhaps an analytical procedure can be used in this example; but in general, the relations might not be so simple. Accordingly, I recommend numerical approaches. Here, it would be a root-finding procedure: Find the value of x to make the value of $g(x) = 37.2 - \int_{3}^{x} \sqrt{(1 + 36x^4)}dx = 0$. Using the rectangle rule of integration, the procedure is to find the value of x to make the value of $g(x) = 37.2 - \Delta x \sum_{i-1}^{N} \sqrt{(1 + 36x_i^4)} = 0$. Given a value for x, and a choice of N increments, $\Delta x = (x - x_0)/N$, and $x_i = x_0 + i\Delta x$, where $x_0 = 3$.

Numerically, in VBA code the integral is as follows:

```
X0=3
Xi=X0
dX=(Xf-X0)/N
S=0
For i = 1 to N
   Xi = Xi + dX
   S = S + dX * ( 1 + 36 * X^4 )^0.5
Next i
```

The root-finding procedure will guess at an x-value and then numerically calculate the resulting S-value and re-guess x-values until the consequential S-value is close enough to the target S. The user needs to decide a root-finding procedure—I think interval halving or Newton's secant version are good. The user also needs to specify a convergence criterion to determine when S is close enough to the target. Since the target is 37.2, implying that there are three significant digits desired, a conventional convergence will be one to two orders of magnitude smaller, perhaps claim convergence when $\text{Abs}(S - 37.2) < .001$. As with any numerical integration, the accuracy improves with smaller Δ values as the finite element estimations in the increment better approximate the true infinitesimal increments. (This of course is true until the values approach the numerical limit of the computer variable storage.) How does one choose N? A rule of thumb is to initially choose 30 for each inflection section on the curve. For this function, for $x > 3$ with $y = 2x^3$, there is only a convex portion, so start with $N = 30$ and get the x-value. Then double the value of N and recalculate the x-value. If the result is effectively the same, then the N value is sufficient. If the $2N$ value gives a noticeably different x-value, then progressively double the N value until there is no change in the resulting x-value.

5.5 Accumulation along a Path

If points are accumulated along a path (or raindrops, or joy, or fuel consumed) and the rate of accumulation (or consumption) depends on the path location and perhaps duration at any location along the path, how is the net accumulation calculated? The rate of accumulation may be dependent not only on the position but also be on the time at a particular location and the velocity, rate $= r(x, y, z, v, t)$. The rate could be expressed as accumulation per time or as accumulation per path distance. If there is a path relation $y = f(x)$ and a z-position dependence on x- and y-positions (such as elevation depends on longitude and latitude) $z = g(x, y)$ and the rate of accumulation is based on distance along the path, then the amount accumulated in a Δs element is $\Delta Q = r\Delta s$. And the total amount accumulated over the entire path is

$$Q = \sum \Delta Q = \sum_{S=0}^{S} r\Delta s = \sum_{i=1}^{N} r_i \Delta s_i \tag{5.9}$$

In the limit of infinitesimal Δs values, the rectangle rule of integration can be expressed as an integral:

$$Q = \int_0^S r(x, y, z, v, t) ds \tag{5.10}$$

There are dozens of possible transformations of this relation, which can indicate how the accumulation depends on distance along the path. However, for instance, if incremental distance is calculated by velocity and incremental time, then the equivalent relation indicates how accumulation is time dependent:

$$Q = \int_0^t r(x, y, z, v, t) v(x, y, z, t) dt \tag{5.11}$$

One can either use mathematical variable transformations or return to the initial incremental accumulation sum to generate the desired representation. For example, accumulation along a path would depend on the density, or probability, or rate of consumption, and the duration in that Δs interval. If it is a simple sweeping (vacuuming) model in which all of the accumulated item is accumulated while moving along the path and the density of item per linear distance is $\rho(x, y)$, then $\Delta\rho = \rho\Delta s$. Then the total accumulated, m is

$$m = \int_{x_0}^{x} \rho(x, f(x)) \left[1 + \left(\frac{dy}{dx} \right)^2 \right]^{1/2} dx \qquad (5.12)$$

Example 7 Path Accumulation

This path accumulation example is a simple Goddard problem. A single rocket is being launched from sea level, and thrust starts at 100%, and then thrust is scheduled linearly with elevation. When all the fuel is consumed, the rocket coasts, until gravity and air resistance slows it to a stop, and it then free-falls back toward the Earth (hopefully an empty ocean or with a parachute to soft land in a vacated area). A set of simple models are used for concept illustration.

Newton's law for a solid body: $\sum F = ma/g_c = (m/g_c)(dv/dt)$. Quadratic wind drag $F_D = A c_D \rho (1/2) v^2$ with air density exponentially dropping with elevation $\rho(h) = \rho_0 e^{-ah}$. Gravity force $F_G = GMm/r^2$, where $r = r_0 + h$, and the rocket mass is that of the rocket and fuel, $m = m_R + m_F$. The fuel is consumed at a rate that is proportional to the thrust, which is proportional to elevation $dm_F/dt = \beta T = \beta(1 - \gamma h)$. And finally, elevation depends on velocity $dh/dt = v$. The maximum thrust is equivalent to 2g's under the initial conditions $T_{max} = 2(GMm_0/r_0^2)$.

The optimization objective is to determine the value of coefficient γ, which determines how thrust is scheduled with elevation, and to maximize height given an initial fuel quantity, making this is a single DV (univariate) optimization problem:

$$\max_{\{\gamma\}} J = h(t|v = 0, m_F = 0) \qquad (5.13)$$

And the model requires a path integral to account for the consumption of fuel.

I have described the variables w.r.t. time. Numerically, differential equations for velocity, position, and fuel quantity can be solved with a simple Euler's finite difference explicit approach, and then thrust, gravity pull, and air drag will be calculated for each Δt increment. After appropriate variable initializations, a VBA routine for the height achieved might be as follows:

```
Do Until v<=0 and mF=0
  T=Tmax*(1-gamma*h)
  mF=mF-dt*beta*T   'fuel mass at the end of the time step
  if mF<0 then
        mF=0
          T=0
  End If
  mR=m0+mF+dt*beta*T/2   'average mass over time step
  dens=rho0*exp(-alpha*h)
  Force=T-G*M*mR/(r0+h)^2 - alpha*A*dens*v^2
  v=v+gc*Force*dt/mR
  h=h+v*dt
  simtime=simtime + dt
Loop
```

Note 1: The code shown earlier is the model that returns the OF value. The optimizer code, which might be Golden Section, Newton's, leapfrogging, etc., is not shown. The optimizer chooses a value of γ and calls the aforementioned model to determine the elevation, the OF value. Which optimizer would you choose?

Note 2: This is a very simplified set of models representing gravity, thrust/fuel, air drag, etc., which is not meant to be a definitive Goddard problem. It is meant to use a not-too-complex concept to reveal the procedure. Seeing the assembly of pieces, a person can make each appropriately complex.

Note 3: Scheduling thrust linearly with elevation is not a best strategy. A commonly accepted, idealized solution to the problem of scheduling thrust to minimize fuel given a desired height (or to maximize height given fuel) has the rocket at full thrust until it reaches terminal velocity and then to schedule the thrust to maintain terminal velocity.

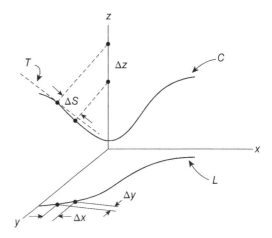

Figure 5.5 Illustrating slope along a path.

5.6 Slope along a Path

Here the question is "What is the surface slope in the direction of a path?" Consider a walking trail along a mountain path and a desire to keep the slope of the path within a limit. On a map, the path will have x and y coordinates, representing latitude and longitude coordinates on the Earth surface. However, the path also has elevation z; so, the incremental distance along the path, ΔS, depends on the incremental Δx, Δy, and Δz elements. The slope of the path will be the change in elevation w.r.t. path distance, $\Delta z/\Delta S$.

Figure 5.5 reveals these concepts. The vertical axis, z, represents elevation. The path is the curve labeled C, and the projection of the path on the x–y surface (the view from outer space of the curve on the flat Earth surface) is line L. The path starts in the upper left and goes down in elevation as it moves off to the back right of the view. The path is the intersection of the $z(x, y)$ surface above x–y curve on a map. The slope of the path at an intermediate point is indicated as the tangent line, T, to the path. The slope is $\Delta z/\Delta S$.

Note: In finite increments the distance along the x–y line L (the projection in the x–y plane) is related to the changes in the two dimensions. $\Delta s = \sqrt{\Delta x^2 + \Delta y^2}$. However, considering elevation, the on-path increment is

$$\Delta s = \sqrt{\Delta x^2 + \Delta y^2 + \Delta z^2} \qquad (5.14)$$

Defining the x–y projection relation as $y = f(x)$ and the elevation response as $z = g(x, y)$ permits Δy to be estimated from Δx, $\Delta y = (dy/dx)\Delta x = (df/dx)\Delta x$. Further, Δz can also be estimated from Δx: $\Delta z = (\partial z/\partial x)\Delta x + (\partial z/\partial y)\Delta y = ((\partial g/\partial x) + (\partial g/\partial y)(df/dx))\Delta x$. Then the path increment can be related to the x increment. In the limit of small Δx,

$$\frac{dS}{dx} = \sqrt{1 + \left(\frac{df}{dx}\right)^2 + \left(\frac{\partial g}{\partial x} + \frac{\partial g}{\partial y}\frac{df}{dx}\right)^2} \tag{5.15}$$

The path rate of change in elevation is the slope along the 3-D path length:

$$\text{slope} = \frac{dz}{dS} = \frac{\partial z}{\partial x}\frac{dx}{dS} + \frac{\partial z}{\partial y}\frac{dy}{dS} = \left(\frac{\partial g}{\partial x} + \frac{\partial g}{\partial y}\frac{dy}{dx}\right)\frac{dx}{dS} = \frac{(\partial g/\partial x) + (\partial g/\partial y)(df/dx)}{dS/dx} \tag{5.16}$$

$$\text{slope} = \frac{(\partial g/\partial x) + (\partial g/\partial y)(df/dx)}{\sqrt{1 + (df/dx)^2 + [(\partial g/\partial x) + (\partial g/\partial y)(df/dx)]^2}} \tag{5.17}$$

If it is not possible to obtain analytical derivatives, this can be evaluated numerically.

Example 8   Path Gradient and Integral

For a road, find the (x,y) path (the east–west and north–south paths on a 2-D map) through the mountains that minimizes distance between points A and B and also avoids constraints on steepness of the road (dz/ds). A quadratic path, $y = a + bx + cx^2$, is a convenience for this illustration, not a necessity. The application statement is

$$\min_{\{a,b,c\}} J = \int_A^B ds$$
$$\text{S.T.:} \quad \left|\frac{dz}{ds}\right| \leq \varepsilon \tag{5.18}$$

Since the path must go through points A and B, there are additional constraints on the $\{a, b, c\}$ values:

$$y_A = a + bx_A + cx_A^2 \tag{5.19a}$$

$$y_B = a + bx_B + cx_B^2 \tag{5.19b}$$

So, with the value of coefficient a chosen by the optimizer, the values for b and c are fixed by the path constraints of its origin and terminal locations.

The optimizer is to find the value of coefficient a that minimizes distance, subject to the constraints that the path goes through points A and B, and is not too steep:

$$\min_{\{a\}} J = S_A^B = \sum_{i=0}^{i=N} \Delta s_i \tag{5.20}$$

$$\text{S.T.:} \quad y_A = a + bx_A + cx_A^2$$
$$y_B = a + bx_B + cx_B^2$$
$$\frac{\partial z}{\partial x_i} < \varepsilon \ \forall i$$

Note 1: This is a single DV application not only with a path integral but also with a gradient analysis along the path.

Note 2: You could use either a, b, or c as the DV. After choosing a value for any one of the coefficients, the equality constraints fix the others. You might consider which choice makes the solution for the others simplest, or most precise. Which choice minimizes sensitivity of the TS to the OF? This is an optimization of the choices within the mathematical optimization application.

Note 3: If the path model was higher order, for example, $y = a + bx + cx^2 + dx^3$, then there would be two DVs. Although it would not be a 1-DV search (a line search), it would still be a path integral (line integral) application.

Note 4: More choices: Which univariate search algorithm is best? What should you use as the convergence criterion and threshold? If one choice is better than another, then the optimization procedure can be optimized. Later chapters reveal how to optimize the procedure and choices.

5.7   Parametric Path Notation

A projectile moves through 3-D space in time, and the x, y, and z positions are often stated as functions of time. In the classic Physics 1 model of a cannon ball or shot-put projectile, on a flat Earth with no air drag, the horizontal and vertical positions are mathematically described as

$$x(t) = x_0 + v_{x,0}t \tag{5.21a}$$

$$y(t) = y_0 + v_{y,0}t - \frac{1}{2}gt^2 \tag{5.21b}$$

In general if three dimensions are considered and the geometry included curved planet and air drag that was dependent on both projectile velocity and air density, the relations would be a bit more complicated but still could be represented as functions of time:

$$x(t)$$

$$y(t)$$

$$z(t)$$

Alternately, a polar coordinate system could be used:

$$r(t)$$

$$\theta(t)$$

$$\phi(t)$$

This is termed a parametric model, where each response variable value depends on the same parameter. In this case the parameter is time. However, one could change the perspective and relate time to x-position, $t = f^{-1}(x)$, and then represent the model as

$$x$$

$$y(x)$$

$$z(x)$$

It is still a parametric model.

The parameter might be distance along the curve, air pressure, temperature, quantity of material produced, etc. Further, the response variables do not need to represent spatial positions. The response variables could be temperature and the composition of several variables in a reactor, responding to the intensity of catalyzing light on the reactants:

$$T(t,I)$$
$$c_1(t,I)$$
$$c_2(t,I)$$
$$c_3(t,I)$$
$$c_4(t,I)$$
$$\vdots$$

It is still a parametric model. If the light intensity represents the single DV and the objective is to maximize the concentration of the third compound after 20 min, max $c_3(t = 20,I)$ while minimizing the cost of supplied energy min $I^* 20$, then the statement is

$$\min_{\{I\}} J = \lambda 20I - c_3(t = 20,I) \tag{5.22}$$

In Equation (5.22), λ balances the relative importance of the two terms.

5.8 Takeaway

Accumulation, proximity, response rate, etc., can be functions of a path. The path can be described in many ways and does not have to be a path through spatial dimensions. Except for trivial example exercises, numerical procedures are usually required.

5.9 Exercises

1 Write the computer code to perform a univariate search to find the minimum of this function. $y = (x_1 - 8)^2 + (x_2 - 6)^2 + 15\text{Abs}[(x_1 - 2)(x_2 - 4)] - 300\text{Exp}\{-[(x_1 - 9)^2 + (x_2 - 9)^2]\}$. Search along the curve defined by $x_2 = 0.1x_1^2$, and use the constraint $S < 4$. Use x_1 as the DV, and [0, 10] as the search range, but report the optimum as "S," distance along the curve from the origin. Use mathematical analysis to show that $S = \int_0^x \sqrt{1 + 0.04x_1^2}\, dx_1$, from which you can either integrate numerically to get a value for "S" or use a table of integrals to find an explicit equation for the value of "S." Use a univariate algorithm from each category: second-order (successive quadratic, or secant/Newton's) and direct (Golden Section, heuristic, LF). For each optimization algorithm, implement the constraint as "soft" (as a penalty added to the OF) and as "hard." This is a total of four investigations. Show the effect of too large and too small values of the multiplier in the constraint penalty. Could you implement a hard constraint on successive quadratic, secant/Newton's, or Golden Section?

2   Set up all of the equations to define an optimization application to determine the shortest path along a surface from point A to point B. You can imagine this as building a road along the surface between two locations in a hilly region. Some function $z = f(x,y)$ defines land height w.r.t. the two distance variables—x (north–south) and y (east–west) distances. The straight-line path between points A and B on the x–y plane (the projection of the Earth surface) might lead up a high mountain and then down. The shortest road might be around the mountain. For simplicity, define the path as a polynomial $y = a + bx + cx^2 + dx^3$, and then the algorithm would have to find the values of coefficients a, b, c, and d that minimize the distance along the path (not the distance in the x–y plane, but the distance along the x–y–z path). Note: Points A and B must be on the path, which provides two equality constraints, which reduces the number of DVs from 4 to 2. You do not need to solve this (but it might be fun).

3   Derive a procedure to perform a univariate search along distance S of a curved line in x–y–z space. The DV is distance S. The y and z positions of the line can be described as functions of the x-position, $y = f(x)$, and $z = g(x)$. The value of the OF along the line is a function of position OF $= v(x,y,z)$. The optimizer is not the issue. Show how can you find the OF value, from the trial solution, S, value. Choose nonlinear functions for f, g, and v, and implement your search for the optimum.

4   The objective is to find the path from point A to point B that minimizes undesirables. You won't find the numerical values of the path coefficients in this exercise; you'll just set up the optimization statement. The cubic relation for the y–x path projected on the flat surface of the Earth is of this relation: $y = y_A + a(x-x_A) + b(x-x_A)(x-x_B) + c(x-x_A)(x-x_B)(x-x_C)$ in which (x_A, y_A) are the longitude and latitude coordinates of point A in Cartesian (flat, planar) space and points B and C have similar notation. This function is termed a Newton interpolating polynomial, here truncated to the cubic functionality. Point C is an intermediate point on the path, but its coordinates are yet to be determined. It's a clever equation: at $x = x_A$, all the x-terms are zero, leaving the intercept as y_A. At x_B, y must be y_B, because the path must end at point B. This requires the value of coefficient a to be $a = (y_B - y_A)/(x_B - x_A)$. The path goes through mountains where elevation, z, depends on (x,y) location: $z = f(x,y)$. So the path distance, S, is the distance in three dimensions. Along the path in (x,y,z) space, there are bothersome flying bugs. The bug density (number per m^3 of air, essentially the number encountered in each meter of distance) is $\rho = g(x,y,z)$. We wish to have our optimizer find the path coefficient values to minimize path distance and to minimize the number of bugs encountered along the path. "I'd walk an extra 50 m just to avoid a bug," one friend says. The answers to this exercise will be equations in terms of the coordinates of points A and B, path coefficients, penalty weighting values, and functions f and g. (i) Show how to derive the aforementioned relation for the value of coefficient a. (ii) Derive a similar relation for the b coefficient value in terms of the coordinates for points A, B, and C. (iii) Clearly indicate the DVs for the optimization. (iv) Use calculus to represent how to obtain the OF value. This will use the path integral techniques. (v) State the optimization application in standard form. (vi) Weight the two OF terms to match the friend's statement about bugs. (Feel free to implement your solution.)

5   Set up the optimization statement that would show that the shortest path between points "A" and "B" on a plane is a straight line. Use a Newton's interpolating polynomial to define the curve between A and B: $(y-y_A) = a(x-x_A) + b(x-x_A)(x-x_B)$. First, show that the point (x_A, y_A) is satisfied by the equation and then define coefficient "a" by the (x_B, y_B) pair. Then, set up the

optimization statement clearly indicating OF, DV(s), constraint(s) (if applicable), and the equations (models) that relate DV(s) to other elements. Solve it for the value of coefficient b.

6   What distance, s, along this quadratic path (x is the E–W direction and y is the N–S direction, and $s = 0$ at the x,y pair 0,5) $y = 5 + 2x - 3x^2$ leads to the highest elevation, h, over this land feature $h = 1 + 2x^2 y - 3y^2$? You can use meters for x, y, s, and h. You will need to (i) define the optimization statement; (ii) define models relating DV to OF; (iii) code the function; (iv) decide on the optimizer, its initialization, and convergence criterion; and (v) comment on the result.

6

Stopping and Convergence Criteria: 1-D Applications

6.1 Stopping versus Convergence Criteria

The optimization searches are iterative. Hopefully, they progressively move the trial solution to the vicinity of the optimum, and when close enough to x^*, they stop and report their x^* value.

However, it may happen that the algorithm diverges from the optimum, oscillates about it, creeps upon it very slowly, and after excessive iterations still has not come close enough. Or it may be that the convergence criteria have been set too "tight" by a user expecting too much precision, and it may take excessive iterations to come adequately close enough to x^*. Or it may be that each trial solution is infeasible and repeated tries to find a better solution just continue to cross over into an infeasible region. Or it could be that the algorithm has encountered an execution error and has not been given an alternate logic to remedy it. In such cases the iterative procedure should stop and report, "Possible failure to find a solution. Giving up on the search." This would acknowledge that continuing the search for an optimum seems futile, with those particular choices for optimizer and its initialization, coefficients, thresholds, etc.

By contrast, if the search was successful and the proximity to x^* is adequately close, then stop and claim, "A solution has been found. Hooray! It has converged."

The optimizer needs both convergence criteria to indicate the trial solution is in desirable proximity to an optimum and stopping criteria to prevent excessive iterations when the search seems futile. The criteria for stopping and convergence are distinct, so the metrics for each are different.

6.2 Determining Convergence

Optimization is often an iterative procedure that uses an incremental approach in adjusting trial solutions toward finding a minimum OF value. In general the approach can be stated as

$$x_k = x_{k-1} + \Delta x_{k-1} \tag{6.1}$$

where x_k represents the kth trial and k is the iteration number (alternately termed stage number or epoch number).

The question is "When to claim convergence and stop iterations?" Here are some standard, conventional, popular criteria for univariate (single trial solution) procedures.

Engineering Optimization: Applications, Methods, and Analysis, First Edition. R. Russell Rhinehart.
© 2018 R. Russell Rhinehart. Published 2018 by John Wiley & Sons Ltd.
Companion website: www.wiley.com/go/rhinehart/engineeringoptimization

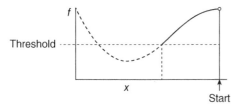

Figure 6.1 Illustration of convergence criterion on OF threshold.

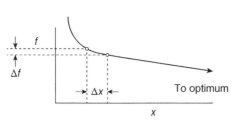

Figure 6.2 Illustration of issues with convergence on Δf.

6.2.1 Threshold on the OF

Stop when the objective function value is low enough.

$$f_k < \text{threshold} \qquad (6.2)$$

Figure 6.1 illustrates this. The problems are as follows: If the threshold is too high, you'll stop not near optimum. If threshold is too low, you'll never stop.

However, in some applications one knows that the minimum OF value is zero and can define a threshold that represents when the OF value is adequately close to zero.

In any case, the user needs to know what can be reasonably expected from the OF* value to be able to define an appropriate threshold value on the OF.

6.2.2 Threshold on the Change in the OF

This criterion claims convergence and stops execution when the incremental change in OF is small.

$$|f_k - f_{k-1}| < \text{threshold} \qquad (6.3)$$

Ideally, the absolute value sign is not needed, because each OF value should be smaller than the proceeding value, but if maximizing, the signs may be reversed.

This metric is also imperfect: (i) A small iteration value of the trial solution Δx relative to Δf could stop far from optimum (as Figure 6.2 illustrates). (ii) If OF asymptotically approaches the minimum as x goes to infinity, the DV will approach a non-implementable extreme value. (iii) It is scale dependent. For the same ΔTemperature, $1°C \cong 2°F$. If you have ΔT set for 0.002, it may need to change to 0.001 if the user changes units.

A common solution is to set the threshold to be two orders of magnitude smaller than what might be perceived to be a permissible deviation from optimum.

6.2.3 Threshold on the Change in the DV

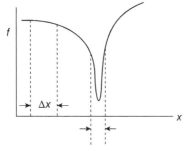

Figure 6.3 Illustration of issues with convergence on Δx.

This criterion claims convergence and stops execution when Δx is small. Note that the absolute value is required because the DV could be either incrementally increasing or decreasing.

$$|\Delta x| < \text{threshold} \qquad (6.4)$$

This has also an issue: A steep OF response (one that is very sensitive to the value of the DV near the optimum) could mean that the solution is still far from optimum (in an OF view) when the Δx criterion is satisfied, as illustrated in Figure 6.3.

Again, a common solution is to set the threshold to be two orders of magnitude smaller than what might be perceived to be a deviation from optimum.

6.2.4 Threshold on the Relative Change in the DV

This criterion claims convergence based on the relative change in the DV:

$$\frac{|\Delta x|}{x} < \text{threshold} \tag{6.5}$$

   This determines all DV thresholds, but in scaled terms. Still, this has the problem of the choice of threshold. Too large, and it misses optimum. Too small, and it causes excessive iterations. In the chance that x could have a value near zero, reformulate the criterion to avoid a possible divide by zero and use IF $|\Delta x| < x^*$threshold THEN STOP.

6.2.5 Threshold on the Relative Change in the OF

This criterion claims convergence based on the relative change in the OF:

$$\frac{|\Delta f|}{f} < \text{threshold} \tag{6.6}$$

   This format is often an aid to a user who may not know what OF value to expect but would know that changes of 1/1000th of the value is inconsequential. In the chance that f could have a value near zero, reformulate the criterion to avoid a possible divide by zero and use IF $|\Delta f| < f^*$threshold THEN STOP.

6.2.6 Threshold on the Impact of the DV on the OF

This criterion claims convergence and stops execution when the impact that the Δx has on the OF value is small:

$$\left|\frac{df}{dx}\Delta x\right| < \text{threshold} \tag{6.7}$$

   This multiplies the local sensitivity of the OF to the DV by the iteration change in the DV to estimate the impact that Δx has on the OF. Again, specify a threshold that is two orders of magnitude smaller than what might be minimally undesirable (or user detectable) as a deviation from OF*.

6.2.7 Convergence Based on Uncertainty Caused by the Givens

This criterion acknowledges the uncertainty of the "givens" (the model coefficients and other influences) in an application and claims convergence when the change in the OF value is small relative to the root-mean-square propagation of uncertainty that the "givens" have on the OF.

$$|\Delta f| < 0.01\sqrt{\frac{1}{N}\sum_c \left(\frac{\partial f}{\partial c}\varepsilon_c\right)^2} \tag{6.8}$$

   "Givens" are the basis for the optimization and include influences such as tax rate, humidity, operating hours per year, cost of electricity, cleanliness factor, emissivity, density, and model coefficients. These are represented by the symbol c. This method multiplies the sensitivity of the OF to the given by the uncertainty in c to estimate a probable uncertainty these have on the OF. It stops the optimization iterations when the change in the OF is two orders of magnitude smaller than the expected certainty of the application.

6.2.8 Multiplayer Range

The previous convergence criteria were suitable for optimizers with a single TS. However, multiplayer algorithms have many "simultaneous" trial solutions (DV sets for each player) and the corresponding OF values for each. The convergence test could be related to the variation in DV values or that of the OF values. The variation could be simply the ranges

$$R_{DV} = DV_{highest} - DV_{lowest} \tag{6.9}$$

$$R_{OF} = OF_{highest} - OF_{lowest} \tag{6.10}$$

where the subscripts *highest* and *lowest* represent the extreme values on the DV or the OF axis. I could have used the terminology maximum and minimum values, but this might be confused with optimization.

In leapfrogging there is a best player and a worst player, as indicated by the OF values. Here the OF range is the same as $R_{OF} = OF_{worst\ player} - OF_{best\ player}$. However, the worst player may be near (in DV space) to the best player, so the DV range is not the DV difference between the worst and best players $R_{DV} \neq DV_{worst\ player} - DV_{best\ player}$.

Although relatively simple, when the range between all players is small, the range between players in the vicinity of the optimum is even smaller. Accordingly, an alternate indication of the range between players might be better for a user to specify. The root-mean-square (rms) deviation from the best player is such a metric that uses all players, not just the two best as an overall indication.

$$rms_{DV^*} = \sqrt{\frac{1}{M-1}\sum_{i=1}^{M}(DV^* - DV_i)^2} \tag{6.11}$$

$$rms_{OF^*} = \sqrt{\frac{1}{M-1}\sum_{i=1}^{M}(OF^* - OF_i)^2} \tag{6.12}$$

Here M is the number of players. Since one player is the best, there are only $M - 1$ terms in the sum. Accordingly, the rms value is scaled by $M - 1$ not M. However, it is inconsequential whether the rms is scaled by M or $M - 1$. With $M = 10$, the difference is 5%, which is probably much smaller than the uncertainty associated with the human choice in specifying a convergence threshold.

6.2.9 Steady-State Convergence

There are two separate categories of optimization that lead to randomly perturbed OF values. In one, the search algorithm is stochastic, which continually perturbs trial solutions. One example is particle swarm optimization. Even if the OF response is deterministic, and even in the vicinity of the optimum, the position of the particles (individuals, players) are randomly perturbed at each iteration. This means that the resulting OF of each player is continually varying in a random nature.

In the contrasting category, the OF is stochastic, and even with a deterministic search algorithm, replicate trial solutions (at the same DV value) will return randomly perturbed OF values about a nominal value. I recommend multiplayer algorithms when optimizing stochastic functions because single TS algorithms can become wholly confounded.

In either category, or in the combined case of stochastic search with a stochastic function, the OF value of any particular player will fluctuate even if the TS values are unchanged. Consider a player that

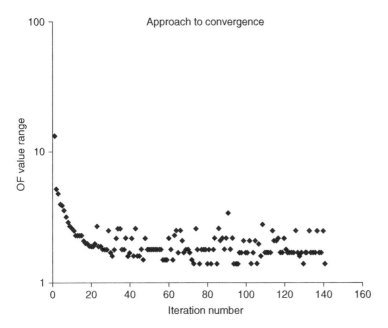

Figure 6.4 Illustration of steady-state convergence criterion.

does not start at the optimum and graph its OF value w.r.t. iteration. The OF value will take a noisy path from the initial high, off-optimum value to a noisy steady state at the optimum value.

The initial worst player may happen to converge early, leaving other players "in the hills" and late to join the team at the optimum. Accordingly, one should not watch one particular player. Instead, observe how the worst OF of all players relaxes to steady state w.r.t. iteration. The graph will be similar to Figure 6.4, which displays how the OF range (worst to best) relaxes w.r.t. iteration. The figure reveals the OF on a log scale for convenience when there are very large changes in value. Steady state was confidently accepted at iteration number 143, at which time convergence was claimed.

Many techniques can be used to indicate when a signal is at a noisy steady state. A conceptually simple one is, at each iteration, to look at data in the most recent past N data, perhaps $N = 10$. This view is termed a moving window of length N. If the slope of the worst OF (or the team OF range) w.r.t. iteration in the window is statistically not distinguishable from zero, then accept that the trend is at SS. My objection to this, and such conventional approaches, is the computational work required at each iteration.

I prefer one of the incremental, iterative approaches. A simple one is described in Section 4.4.6 and Appendix D. Being a ratio of variances, the statistics is scaled by the inherent noise level in the data. It is also independent of the dimensions chosen for the OF.

6.3 Combinations of Convergence Criteria

No single metric is perfect, comprehensive, or universal. Accordingly, you may choose to use a simple combination of both. For instance, stop iterations and claim convergence when $|\Delta f|/f < \varepsilon_f$ and $|\Delta x|_i/x_i < \varepsilon_{DV}$ for each DV.

6.4   Choosing Convergence Threshold Values

An optimizer does not stop exactly at x^*, but instead stops in the proximity of it and claims that the most recent best trial solution is the value of x^*. If the thresholds in the previous criteria were chosen to be zero, the optimizer would stop and claim convergence when the incremental changes in the DV were less than the numerical discretization element of the computer storage capacity for the variable. But that is still not the ideal value. And it would take excessive iterations. And probably it would have much greater precision than justified by the application. So the user needs to balance perfection of the solution (a measure of desirability) by the work required to get that solution (a measure of undesirability). Conceptually, this is stated as

$$\min_{\{\text{threshold}\}} J = \text{undesirability} - \text{desirability} \tag{6.13}$$

The recommended way to judge whether the stopping criteria are too lenient, or too severe, is to observe the results. For instance, plot J w.r.t. k (the iteration counter) and x_i w.r.t. k to observe progress or the process and to see if both converge with k and choose a threshold value for convergence with these objectives in mind:

1) It makes OF and DV both appear to converge.
2) Both x_i and J effectively stay unchanged for enough iterations to establish credibility that it has converged.
3) The value of k does not indicate excessive iterations after confidence in convergence has been established.

Selection of the right values for any of the threshold or limits depends on scale (2°F is about the same interval as 1°C), features of the function topography, and the utility or user need of the result (propagation of uncertainty defines tolerance of J and x intervals). Selection of right values requires user *a priori* knowledge.

Note: Thresholds on ΔOF and on ΔDV are most common. Unfortunately, terms like precision and tolerance do not reveal which is on which, and incremental change is not necessarily related to precision. Be certain you understand how your software defines terms.

6.5   Precision

When convergence is claimed, the optimizer is close enough to the true DV^* value to satisfy the user's convergence criterion. On replicate trials, on repeated optimizer trials with randomized TS or player initializations, and perhaps even on randomized optimizer coefficients, the optimizer should converge in the vicinity of the same optimum, but it will not end at exactly the same spot. Trial-to-trial deviations could be observed on either the DV^* or OF^* values. Both will vary from trial to trial.

If the function is deterministic, then the best OF^* (and associated DV^*) of the replicate trials will represent the best optimized value in that set of N trials. Although it will be close to the true optimum, it will likely not be the absolute true best. However, for a stochastic function, the best of N trials will just be the fortuitous best of N stochastic trials and not represent the true overall or long-term optimum. In any case we would like to quantify and report the value and associated uncertainty of the solution.

Commonly, we use average and 95% confidence interval to relay a value and associated uncertainty. One could run N trials, then calculate the average and the standard deviation for the N DV* values and the N OF* values, and then report the average value and the 95% confidence interval based on $t_{0.05, N-1} \sigma / \sqrt{N}$.

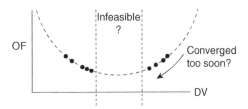

Figure 6.5 Illustration of convergence precision.

If the optimizer sends the trial solution on a path that jumps to and fro over the optimum or if it is a multi-player algorithm in which end-game players surround the optimum, then reporting the average and confidence interval of the solution is appropriate.

However, as a caution, such techniques presume that the data are normally distributed and that the average is feasible. If, for instance, the optimizer approaches the optimum from either side but convergence is claimed prior to getting to the ultimate best, the location of the N solutions might appear as in Figure 6.5. This situation might also arise if the in-between region is constrained and infeasible. As illustrated the data is bimodal; there are two clusters of DV* solutions.

In one case, the gap of data (where the average might be) could reflect the true optimum, and the situation could be that convergence was claimed too soon. In this case the standard deviation of the range will not reflect uncertainty in the average. In this case the user should consider tightening the convergence criteria to converge closer to the optimum. Alternately, if the bimodal clusters are the result of an infeasible interior region, then the average DV value from the solutions is infeasible.

Know the situation. Observe the distribution of solutions. Test that the average DV* is a feasible solution.

Note: The convergence threshold on the DV or on the OF, referred to in this book as ε, is not the 95% confidence limit. The range of the DV data in Figure 6.5 is much greater than the convergence criterion threshold ε. In unconstrained applications, it seems reasonable that smaller values for convergence thresholds will result in smaller variation in the DV* values (in better precision). In constrained applications, there may not be any relation. Unfortunately, in some software the terms precision and tolerance are often used to mean the value of the convergence threshold ε. The user needs to be aware that entering a value in a field labeled "precision" does not necessarily mean statistical uncertainty.

6.6 Other Convergence Criteria

After this book progresses into multi-DV applications, Chapter 17 revisits convergence criteria that I think are better, but they are a bit more complicated. Here "better" is related to robustness, interaction, relation to the problem, surface features, or multiplayer algorithm.

6.7 Stopping Criteria to End a Futile Search

Something must stop the optimizer if the progress seems futile.

6.7.1 *N* Iteration Threshold

Stop after *N* iterations. Report "no convergence after *N* iterations." Problems are as follows: The search may need more iterations to find optimum; it may be on the right path, but too small a user-specified *N* prevents finding x^*. By contrast, too large an *N* uses excessive time. You need to choose the value of *N*.

6.7.2 Execution Error

Stop upon experiencing an execution error. The solution progress may iteratively move the DV into a place that will give a divide by zero, square root of a negative number, log of a nonpositive number, subscript out of range, etc. Include error trapping logic in the algorithm and report "Search led DV to a point where an execution error would occur."

6.7.3 Constraint Violation

Stop upon experiencing persistent constraint violation. The optimizer may be on the feasible boundary of a constraint, and each calculation indicates that the new trial solution should be on the other side of the constraint. Report "Search led DV to a point where a constraint violation occurred."

6.8 Choices!

Any optimizer can use any of the convergence criteria. If there are 7 optimization algorithms to choose from and 10 convergence criteria, then there are 70 choices. But you must also choose a threshold value for the convergence criterion. If you consider three values, large, medium, and small thresholds (for a coarse, intermediate, or precise search result), then the 70 becomes 210 possibilities. It is not uncommon to explore an application with a large threshold, for a coarse initial understanding, with fewer function evaluations, which means lesser computational or experimental time and cost. Then once assured that the right OF has been selected for the application situation and the right procedures selected for the application optimization, use smaller threshold values to get a DV* with appropriate precision.

Further, consider that the choice of DVs and the OF are also user choices. In an economic optimization, for example, (see Chapter 20) the OF could be capital expenses, some combination such as payback time or net present worth, or long-term return on investment.

6.9 Takeaway

Convergence and stopping are distinctly separate concepts, and both must be specified. These concepts and their role within optimization are also distinctly separate from the optimizer, the model, the OF value, the constraints, or the application initialization. Keep the concepts separate. Don't code one concept in the subroutine for another.

The user specifies convergence criterion and threshold values to balance the desires for solution precision (close proximity to the true optimum) with minimal computer work (number of function evaluations to converge). This is a human judgment.

The resulting precision of solution is different from the convergence threshold.

6.10 Exercises

1 Compare the effect of stopping criteria choices. Write the computer code to perform a univariate search to find the optimum for $y = 100 - A * \text{EXP}(-((x - 10)/s)^{\wedge}2)$. The minimum is at $x = 10$. Use $5 < x < 15$ for the DV search range. Use any one (if you want you can use more than one) of the univariate algorithms. You choose positive values for "A" and "s" in the function. Large "A" values make the minimum distinct. In the limit of $A = 0$, there is no minimum. Large "s" values make the minimum very broad, and small "s" values make it very localized. Include graphs of y w.r.t. x for the four cases: large and small values each for "A" and "s" to reveal the four characterizations of the minimum: broad or narrow, for each deep or shallow. For each of the four cases, experiment with each of these two stopping criteria: (i) threshold on the change in DV and (ii) threshold on the change in OF. This is a total of eight investigations. Use one stopping criterion at a time so that you can see results. You should find that smaller threshold values on either stopping criteria stop the optimization closer to the true optimum. You should also see that small "s" values require very small threshold on the DV to get the best OF and that for large "s" values you don't need to stop near the true DV* value to get essentially the same best OF value. There will be similar findings for the range of "A." Report and interpret your findings.

2 On a test case of your choice, calculate precision of the optimizer (use data from many independent initializations) and see how precision trends with convergence threshold value.

3 Compare the value of the convergence threshold on DV to the precision of the OF. Note that the units will be different. Show how some measure of the variability of one (precision, range, etc.) is related to the other by some measure of the sensitivity of the OF to the DV.

4 Compare the DV and OF precision with the same threshold value for two separate convergence criteria on the OF. Use $|f_k - f_{k-1}| < \text{threshold}$ and $|(df/dx)\Delta x| < \text{threshold}$.

5 State a situation when Δx might be preferred over $\Delta x/x$ as the convergence criterion and vice versa.

Section 3

Multivariate Search Techniques

This section contains 12 chapters. After an introduction to multivariable concepts and terminology, eight chapters present and analyze fundamental optimization techniques. Then additional chapters provide proofs and utility related to functional transformations of the objective function, more on convergence criteria, and common enhancements to optimizers.

Engineering Optimization: Applications, Methods, and Analysis, First Edition. R. Russell Rhinehart.
© 2018 R. Russell Rhinehart. Published 2018 by John Wiley & Sons Ltd.
Companion website: www.wiley.com/go/rhinehart/engineeringoptimization

7

Multidimension Application Introduction and the Gradient

7.1 Introduction

Many optimization applications have two or more decision variables. The previous chapters focused on univariate, one-DV, one-dimensional (1-D) applications. Although N-DV applications are more frequently encountered, univariate searches are important, and they reveal key issues in relatively uncomplicated situations. There are new issues and techniques associated with 2-D and higher-dimension applications, and this chapter begins a section related to 2-D, two-decision variable, and N-D applications. Each section of this book on optimization algorithms introduces techniques with 2-D examples and then extends the optimization techniques to N-D applications. In 2-D, we still have the generic optimization concept for maximization:

$$\max_{\{x_1, x_2\}} J = \sum \text{desirables} - \sum \text{undesirables} \tag{7.1}$$

in which the two DVs, x_1, x_2, are explicitly revealed. Alternately, it could be stated as a minimization of the negative of the OF and/or by using the vector symbol for the DV set:

$$\min_{\{x\}} J = \sum \text{undesirables} - \sum \text{desirables} \tag{7.2}$$

In a two-DV situation, the OF is a response of two variables, $J = f(x_1, x_2)$, which means that the OF value can be plotted as the third dimensional response to the two DV values.

Figure 7.1 illustrates Function 11 in the 2-D Optimization Examples file, an exploration of the perpendicular (or normal, or total) least squares method of best fitting a straight line to data. The two DVs are the slope and intercept of the line. The floor represents the two DVs, and the height of the function, the vertical axis; the third dimension represents the response. This reveals surface topography, an aspect of the mathematical analysis that has strong connectivity to the human experience in 3-D space, of land or of other surfaces. The familiarity and comfort of this 3-D view facilitates the mathematical analysis needed for optimization and understanding in the N-D applications.

Note two aspects of this 3-D illustration. The "net lines" run parallel to each DV axis. They make it seem like a woven net, a fish net, or some sort of coarse weave netting was draped over items. The net lines would represent the rollercoaster ride if one traveled the surface in one-DV direction, keeping the other DV constant. If you are relating this to hills and valleys on land, the net lines would be longitude and latitude on a local (Cartesian) section; however, in this rotated view they appear on the diagonal,

Engineering Optimization: Applications, Methods, and Analysis, First Edition. R. Russell Rhinehart.
© 2018 R. Russell Rhinehart. Published 2018 by John Wiley & Sons Ltd.
Companion website: www.wiley.com/go/rhinehart/engineeringoptimization

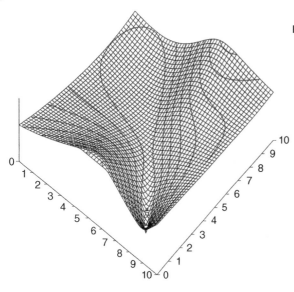

Figure 7.1 3-D surface illustration.

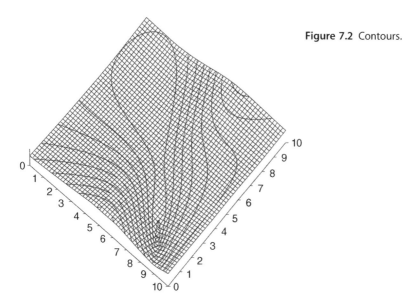

Figure 7.2 Contours.

not vertical or horizontal. In this panoramic view from an airplane, the net lines are curved. However, looking straight down on the surface from far above on the OF axis, like from a satellite, the net lines are straight, like quadrille-lined paper. In the almost directly above view in Figure 7.2, the net lines are nearly straight.

The other aspect to note is the contour lines. These are the curved lines that indicate constant surface height, or OF value. These could be imagined as the cuts of a plane parallel to the x–y plane that intersect the vertical OF axis at some value. Looking straight down on the surface from far above on the OF axis, one sees the contour lines curve around features. This is equivalent to a contour map of a local section of the Earth. Figure 7.3 is a view of my locale.

Notice that W 44th Avenue runs directly east–west, a straight line, with no curves on the map. However, if you drove along W 44th Ave., you would not be at sea level, but you would be averaging about 950 ft above sea level and would be going up- and downhill from 1000 to 900 ft and back.

In optimization we need to understand the change in the OF, represented here as the surface level, as the DV location changes. This chapter begins the requisite supporting mathematical analysis.

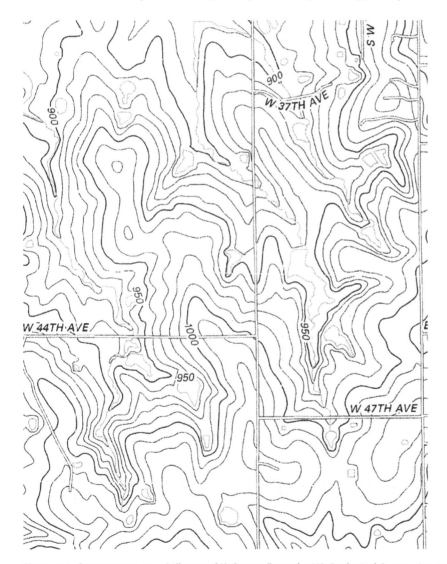

Figure 7.3 Contour map near Stillwater, OK. *Source:* From the US Geological Survey site, www.usgs.gov.

Further, consider that you stop your car on the road near the "A" in the "W 44th Ave" label at the 950 ft elevation and then place the vehicle in neutral and let it drift down the road. It will move east on the road, downhill to the 920 ft valley, then coast up a bit, stop, coast back, and eventually come to rest at the bottom of the dip in the road. By contrast, if you released a large heavy ball near the same initial location, it would not stay on the road. Gravity would pull it down the hillside moving it in the northeast direction to the bottom of the valley, and then it would continue to roll down the bottom of the valley, northwest toward the 900 ft contour, and then on to lower elevations, perhaps like how rainwater might make a temporary stream in the valley. The reader may want to sketch their own view of how a ball would travel and self-validate understanding.

There are two views here. One is the projection of the position on the 2-D Cartesian map, and the other is the trace of the position in the 3-D space. For understanding optimization, mathematical analysis of both viewpoints is important.

7.2 Illustration of Surface and Terms

Figure 7.4 illustrates a surface in x, y, z space. The left portion is the net view of the surface, and the right portion is the contour line view. The variable $z = f(x, y)$ is the response, the OF, and the variables x and y are the independent variables, the DVs. The net lines are those of a constant DV value when the other DV value changes. They reveal how vertical planes cut the surface. The contour lines are of constant OF value. They reveal how horizontal planes cut the surface.

Although it is convenience to think of x, y, z as the familiar special distances, they could represent wholly disparate concepts. For example, in economic optimization of a batch reaction, DV-x could be temperature with units of centigrade, DV-y could be cook time with units of minutes, and OF-z could be a measure of profitability such as $/yr.

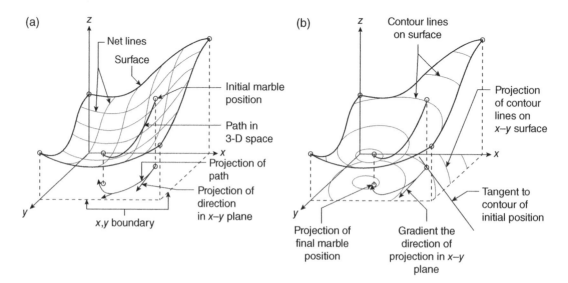

Figure 7.4 (a) Net Lines and (b) contour lines.

The projection of the net and contour lines on the x–y plane are also illustrated. The solid dot on the surface illustrates the initial location of a ball or marble placed on the surface, and the line that spirals around to the bottom of the surface, the minimum elevation, represents the path on which it might roll. The marble initially travels in the local steepest descent direction of its starting point; then as it travels, the surface slope changes, which changes the steepest descent direction. Consequently, the marble curves toward the minimum as it travels. The projection of the path on the 2-D DV plane is also revealed.

As will be revealed the steepest descent direction in the 2-D projection is determined by the negative gradient of the surface:

$$-\nabla z, \quad \text{where } \nabla = \begin{bmatrix} \dfrac{\partial}{\partial x} \\ \dfrac{\partial}{\partial y} \end{bmatrix}$$

Also, it will be revealed that the direction of steepest descent, $-\nabla z$, is $\perp$ (perpendicular) to the contour line projection in the DV plane.

7.3 Some Surface Analysis

Consider $z = f(x, y)$, for example, elevation $= f(NS$ and EW distance); and imagine a contour map projection of the surface. Figure 7.5 illustrates both. The contour map shows lines of constant z-value corresponding to the intersection of the surface and planes parallel to the x–y plane.

At any point (x_0, y_0) there are a corresponding z-value and a contour line at that z-value. The tangent to the contour can be determined in at least two ways.

Method 1: Start with a Taylor series approximation of the $z(x, y)$ function:

$$z(x,y) = z(x_0, y_0) + \left.\frac{\partial z}{\partial x}\right|_0 (x - x_0) + \left.\frac{\partial z}{\partial y}\right|_0 (y - y_0) + 0(2) \tag{7.3}$$

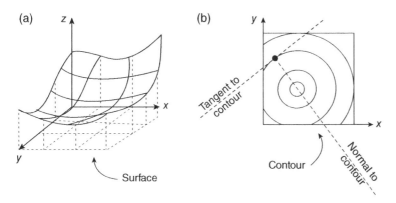

Figure 7.5 (a) 3-D view and (b) contour map.

Discard terms of order 2 and higher to create a linear approximation, which is only a valid representation of the function near the (x_0, y_0) point. On the contour $z(x,y) = z_0 = z(x_0, y_0)$; so, on the contour near to the point (x_0, y_0),

$$0 = \left.\frac{\partial z}{\partial x}\right|_0 (x - x_0) + \left.\frac{\partial z}{\partial y}\right|_0 (y - y_0) \tag{7.4}$$

The tangent line to the contour is a locally valid linear relation between x and y, which is revealed by solving for y as a function of x:

$$y = y_0 - \frac{\partial z/\partial x|_0}{\partial z/\partial y|_0}(x - x_0) = a + bx \tag{7.5}$$

Method 2: Start with a total derivative definition from calculus:

$$dz = \left.\frac{\partial z}{\partial x}\right|_0 dx + \left.\frac{\partial z}{\partial y}\right|_0 dy \tag{7.6}$$

(which is just a Taylor series first-order approximation).

On the contour $dz = 0$. Approximating the differentials with finite differences $dx = \Delta x = (x - x_0)$, $dy = \Delta y = (y - y_0)$, then the total derivative becomes

$$y = y_0 - \frac{\partial z/\partial x|_0}{\partial z/\partial y|_0}(x - x_0) \tag{7.7}$$

which is the same relation for the tangent to the contour.

The tangent line L_{T_0} through (x_0, y_0) can be described as a vector of (x, y) deviations from the base point (x_0, y_0):

$$L_{T_0} = \begin{bmatrix} x' \\ y' \end{bmatrix} = \begin{bmatrix} x - x_0 \\ y - y_0 \end{bmatrix} = \begin{bmatrix} x - x_0 \\ y_0 - \frac{\partial z/\partial x|_0}{\partial z/\partial y|_0}(x - x_0) - y_0 \end{bmatrix} = \begin{bmatrix} 1 \\ -\frac{\partial z/\partial x|_0}{\partial z/\partial y|_0} \end{bmatrix}(x - x_0) \tag{7.8}$$

The subscript T represents tangent. The prime superscripts indicate deviation variables.

In 2-D space in which both dimensions have the same units, a perpendicular to a line has the negative reciprocal slope. A line perpendicular to the tangent at (x_0, y_0) has the negative reciprocal slope:

$$(y_\perp - y_0) = + \frac{\partial z/\partial y|_0}{\partial z/\partial x|_0}(x_\perp - x_0) \tag{7.9}$$

The symbol $\perp$ represents the perpendicular, in this case perpendicular to, or normal to, or at a right angle from the contour.

Note: For pure mathematics with dimensionless variables, this is OK, but for engineers the DV units will often not be identical. Then, this relation for the line perpendicular to the contour will not make dimensional sense. However, engineers and scientists often overlook dimensional inconsistency in many applications such as $F = ma$, $.Q = C_v\sqrt{\Delta P/G}$, or ln $[x]$. There are two solutions to this problem: (i) include the unit-valued dimensional unifier (such as g_c), or (ii) use scaled (dimensionless) variables for x and y.

The line perpendicular to the contour at the base point can also be stated in vector notation, the normal vector:

$$L_{\perp_0} = \begin{bmatrix} x - x_0 \\ y_0 + \dfrac{\partial z/\partial y|_0}{\partial z/\partial x|_0}(x - x_0) - y_0 \end{bmatrix} = \begin{bmatrix} 1 \\ \dfrac{\partial z/\partial y|_0}{\partial z/\partial x|_0} \end{bmatrix}(x - x_0) \tag{7.10}$$

The gradient of z is the vector of partial derivatives of z w.r.t. each DV. It is the list of local sensitivities:

$$\nabla z = \begin{bmatrix} \dfrac{\partial z}{\partial x} \\ \dfrac{\partial z}{\partial y} \end{bmatrix} \tag{7.11}$$

If the gradient at a point $(x_0\, y_0)$, ∇z_0, is normal to the tangent line L_{T_0}, then the dot product of the two vectors is zero:

$$\nabla z|_0 \cdot \Delta L_{T_0} = 0 \tag{7.12}$$

Substituting terms

$$\begin{bmatrix} \dfrac{\partial z}{\partial x}\Big|_0 \\ \dfrac{\partial z}{\partial y}\Big|_0 \end{bmatrix} \cdot \begin{bmatrix} 1 \\ -\dfrac{\partial z/\partial x|_0}{\partial z/\partial y|_0} \end{bmatrix} \cdot (x - x_0) = \left(\dfrac{\partial z}{\partial x}\Big|_0 \quad -\dfrac{\partial z}{\partial x}\Big|_0 \right) \cdot (x - x_0) = 0 \tag{7.13}$$

we find that it is zero. Good!

Also, if the gradient at the base point is parallel to $L_{\perp_0}$, then their elements are the same, or they are linearly dependent (can be scaled by the same quantity to be the same). Here the gradient is multiplied by a scalar:

$$\left\{ \nabla z|_0 = \begin{bmatrix} \dfrac{\partial z}{\partial x}\Big|_0 \\ \dfrac{\partial z}{\partial y}\Big|_0 \end{bmatrix} \right\} \cdot \left(\dfrac{x - x_0}{\partial z/\partial x|_0} \right) \tag{7.14}$$

And when expanded it is revealed as the perpendicular line:

$$\nabla z|_0 \left(\dfrac{x - x_0}{\partial z/\partial x|_0} \right) = \begin{bmatrix} 1 \\ \dfrac{\partial z/\partial y|_0}{\partial z/\partial x|_0} \end{bmatrix} \cdot (x - x_0) = L_{\perp_0} \tag{7.15}$$

Good!

So $-\nabla z$ is a direction of descent and perpendicular to the tangent to the contour, but not yet shown as the direction of steepest descent. What is direction of steepest descent?

Again, consider a function of two variables. The equation $z = f(x, y)$ is the surface response to independent variables (x, y). As Figure 7.6 illustrates, S is a line on the x–y plane, passing through the base point (x_0, y_0). It has the relation

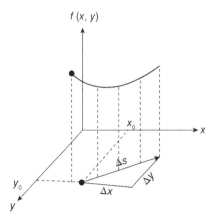

Figure 7.6 Illustration of Equation (7.16).

$$(y - y_0) = a(x - x_0) \tag{7.16}$$

The question is, "What value of a (what line slope, what line direction in the x–y plane) maximizes df/dS the slope (rate of descent or ascent) along the 2-D line in the (x, y) plane (not necessarily along either individual (x, y) direction)?" This optimization statement is

$$\begin{array}{c} \max \\ \{a\} \end{array} J = \text{slope along } s = \frac{df}{dS} \tag{7.17}$$

Apply the calculus chain rule to the slope of the function along S in three dimensions to relate it to the DVs:

$$\frac{df}{dS} = \frac{\partial f}{\partial x}\frac{dx}{dS} + \frac{\partial f}{\partial y}\frac{dy}{dS} = f_x\frac{dx}{dS} + f_y\frac{dy}{dS} \tag{7.18}$$

where f_x is shorthand notation for $\partial f/\partial x$.

The projection of the 3-D descent line in the x–y plane in two dimensions is

$$y = y_0 + a(x - x_0) \tag{7.19}$$

from which the slope is $dy/dx = a$.

Combining several prior relations results in

$$\frac{df}{dS} = f_x\frac{dx}{dS} + f_y\frac{dy}{dx}\frac{dx}{dS} = (f_x + af_y)\frac{dx}{dS} \tag{7.20}$$

Applying the Pythagorean theorem to relate distance along the line to the incremental x and y changes and relating the change in y to the change in x,

$$\Delta S^2 = \Delta x^2 + \Delta y^2 = \Delta x^2 + a^2\Delta x^2 = (1 + a^2)\Delta x^2 \tag{7.21}$$

Taking the square root to convert to distances,

$$\Delta S = (1 + a^2)^{1/2}\Delta x \tag{7.22}$$

In the limit of very small changes, the differential representation is

$$dS = (1 + a^2)^{1/2}dx \tag{7.23}$$

or

$$\frac{dS}{dx} = (1 + a^2)^{1/2} \tag{7.24}$$

Combining Equations (7.20) and (7.24) provides a relation for the rate of descent (change in f w.r.t. distance along the search direction) as a function of the direction coefficient, a:

$$\frac{df}{dS} = (f_x + af_y)(1 + a^2)^{-1/2} \tag{7.25}$$

That relation is the model in the optimization application and the objective is to find the value of a to maximize $J = df/dS$. The optimization algorithm will be the analytical method: set $dJ/da = 0$, and solve for the value of a:

$$\frac{d}{da}\left(\frac{df}{dS}\right) = 0 = \left[f_y - (af_x + a^2f_y)(1 + a^2)^{-1}\right]\left[(1 + a^2)^{-1/2}\right] \tag{7.26}$$

This is a product of two terms. If either is 0, we have a solution. Can the second term have a value of zero? Yes. But only if $a = \infty$, which is not realizable. Can the first term have a value of zero? If so,

$$0 = \left[f_y - \left(a f_x + a^2 f_y \right) \left(1 + a^2 \right)^{-1} \right] \tag{7.27}$$

Since the term $(1 + a^2)^{-1}$ cannot be zero, multiply by it and rearrange

$$f_y \left(1 + a^2 \right) = a f_x + a^2 f_y \tag{7.28}$$

which reveals that the value of the slope is

$$a = \frac{f_y}{f_x} \tag{7.29}$$

Yea! $a = f_y/f_x$ maximizes df/ds. This is the same direction defined by $+\nabla f$ and by the line $\perp$ to tangent to contour.

Therefore, $-\nabla f$ defines the projection in the 2-DV plane of the local line of steepest descent, which is also the line $\perp$ to the local projection of the contour.

In summary, to find the equation of a line of steepest descent given $z = f(x, y)$, a line is defined by a linear relation between x and y:

$$L = \begin{vmatrix} \Delta x \\ \Delta y \end{vmatrix} = \begin{vmatrix} x - x_0 \\ y - y_0 \end{vmatrix} = \begin{vmatrix} x - x_0 \\ a(x - x_0) \end{vmatrix} \tag{7.30}$$

The negative gradient is

$$-\nabla f = - \begin{vmatrix} \dfrac{\partial z}{\partial x} \\ \dfrac{\partial z}{\partial y} \end{vmatrix} \tag{7.31}$$

For the line to be colinear with the negative gradient,

$$L = \begin{vmatrix} \Delta x \\ \Delta y \end{vmatrix} = -\alpha \begin{vmatrix} \dfrac{\partial z}{\partial x} \\ \dfrac{\partial z}{\partial y} \end{vmatrix} \tag{7.32}$$

This represents two equations, one for x and one for y. The x equation can be used to define alpha from delta x, and then that used for alpha in the delta y equation to define the x–y line:

$$\Delta y = \frac{\dfrac{\partial z}{\partial x}}{\dfrac{\partial z}{\partial y}} \Delta x \tag{7.33}$$

or

$$y = y_0 + \frac{\dfrac{\partial z}{\partial x_0}}{\dfrac{\partial z}{\partial y_0}} (x - x_0) \tag{7.34}$$

Example 1 For the simple function $J = 1 + 2x_1 + 3x_2 + 4x_1^2 + 5x_2^2 + 6x_1x_2$. What is the tangent to the contour and the line of steepest descent at the point $x_1 = 2$, $x_2 - 2$?

$$\frac{\partial J}{\partial x_1} = J_1 = 2 + 8x_1 + 6x_2 \qquad (7.35a)$$

$$\frac{\partial J}{\partial x_2} = J_2 = 3 + 10x_1 + 6x_1 \qquad (7.35b)$$

At the base point $(2, -2)$

$$J_1 = 6 \qquad (7.36a)$$

$$J_2 = 35 \qquad (7.36b)$$

The projection of the tangent to the contour, in the x–y plane, at the base point is

$$x_2 = (-2) - \frac{6}{35}(x_1 - 2) \cong -0.171x_1 - 1.657 \qquad (7.37)$$

The projection of the line of steepest descent, in the x–y plane, at the base point is

$$x_2 = (-2) + \frac{35}{6}(x_1 - 2) \cong 5.83x_1 - 13.67 \qquad (7.38)$$

7.4 Parametric Notation

Consider the function represented as $y = f(x_1, x_2)$, as illustrated in Figure 7.7.

Consider a plane $\perp$ to the (x_1, x_2) plane with a direction of steepest descent of the function at the point (x_{10}, x_{20}). The curve on the $\perp$ plane shows how the f changes with that particular (x_1, x_2) relation. It is not a constant f contour.

In parametric notation, the location along the line, S, in the (x_1, x_2) plane can be expressed in terms of a common "distance" parameter, α:

$$\left.\begin{cases} x_1(\alpha) = x_{1_0} - \left.\dfrac{\partial f}{\partial x_1}\right|_0 \cdot \alpha \\[2mm] x_2(\alpha) = x_{2_0} - \left.\dfrac{\partial f}{\partial x_2}\right|_0 \cdot \alpha \end{cases}\right\} = x(\alpha) = x_0 - \alpha\nabla f|_0$$

$$(7.39)$$

In deviation variables

$$x'(\alpha) = -\alpha\nabla f|_0 \qquad (7.40)$$

This is termed a parametric model. The values of the decision variables (x_1, x_2) depend on the value of the parameter α. The value of α is not the incremental distance along the line, but is related to it:

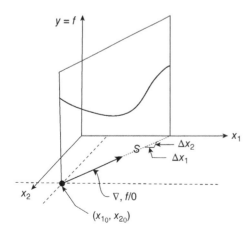

Figure 7.7 Illustration of a function trend w.r.t. a DV line of steepest descent.

$$\Delta S = \sqrt{\sum \Delta x_i^2} = \sqrt{\sum x_i'^2} = \alpha \sqrt{\sum \left.\frac{\partial f}{\partial x_i}\right|_0^2} = \alpha \sqrt{\nabla f|_0 \cdot \nabla f|_0} \tag{7.41}$$

Note that this distance calculation presumes that the units on each x_i are additive.

Using time as a parametric coefficient is a familiar concept. So, here is a bit of a diversion to the classic projectile example of a cannon ball trajectory. We model the height and distance of the projectile as $x(t) = x_0 + v_{x,0}t$ and $y(t) = y_0 + v_{y,0}t - (1/2)gt^2$. Here t, time, is the parameter, and both variables x and y are functions of time. However, they can be related to each other by solving one equation for time and inserting into the other:

$$t = \frac{x - x_0}{v_{x,0}} \tag{7.42}$$

$$y(x) = y_0 + \frac{v_{y,0}(x - x_0)}{v_{x,0}} - \frac{1}{2}g\left(\frac{x - x_0}{v_{x,0}}\right)^2 \tag{7.43}$$

Returning to the parametric representation in 2D, in the aforementioned, what are the units on α? Note the slope of the line in the $(x_1\, x_2)$ plane is

$$\frac{\Delta x_2}{\Delta x_1} = \frac{\partial f/\partial x_2|_0}{\partial f/\partial x_1|_0} \tag{7.44}$$

By substituting for Δx and rearranging,

$$x_2 = x_{2_0} + \frac{\partial f/\partial x_2|_0}{\partial f/\partial x_1|_0}(x_1 - x_{1_0}) \tag{7.45}$$

So, along direction of descent from $(x_{1_0}\, x_{2_0})$,

$$y = f(x_1, x_2) = f\left(x_1, x_{2_0} + \frac{\partial f/\partial x_2|_0}{\partial f/\partial x_1|_0}(x_1 - x_{1_0})\right) = f(x_1) \tag{7.46}$$

The parametric model results in the same relation as the vector analysis.

Alternately, x_1 and x_2 can be considered as functions of distance S along the line of descent:

$$S = \sqrt{\Delta x_1^2 + \Delta x_2^2} = \sqrt{\left[\alpha\frac{\partial f}{\partial x_1}\right]^2 + \left[\alpha\frac{\partial f}{\partial x_2}\right]^2} = \alpha\sqrt{\left[\frac{\partial f}{\partial x_1}\right]^2 + \left[\frac{\partial f}{\partial x_2}\right]^2} \tag{7.47}$$

Then

$$y = f(x_1, x_2) = f\left(x_{1_0} + \alpha\left.\frac{\partial f}{\partial x_1}\right|_0, x_{2_0} + \alpha\left.\frac{\partial f}{\partial x_2}\right|_0\right) = f(S) \tag{7.48}$$

where

$$\alpha = \frac{S}{\sqrt{(\partial f/\partial x_1)^2 + (\partial f/\partial x_2)^2}} \tag{7.49}$$

Whether the application is 2-D, 3-D, ..., N-D or whether specifying distance from the initial point as S or the parameter α or whether specifying one DV and calculating the other(s) from that value, the steepest descent is the DV direction indicated by the negative gradient, $-\nabla f$.

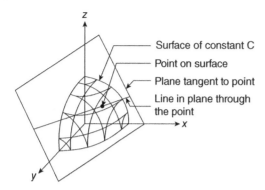

Surface of constant C
Point on surface
Plane tangent to point
Line in plane through the point

Figure 7.8 Illustrating a line in the tangent plane through the point of contact.

7.5 Extension to Higher Dimension

That was for a 2-D DV problem. A 3-D function of spatial variables could represent any number of familiar situations:

- Fog density in x–y–z space
- Droplet size within a rain cloud
- Temperature distribution with a room
- Odor distribution around a dead skunk
- Charge in an electron cloud
- Gravitational pull in space

Further, the decision variables do not have to represent spatial dimension. If one is seeking to manage career and investments to maximize personal joy in a lifetime, one might use these three DVs: portion of salary to invest in a retirement fund, the portion of investment allocated to growth or to secure investments, and retirement age as the third. Here $J = f\left(p_{salary}, p_{portfolio}, age_{retirement}\right)$, revealing both nonspatial DVs and DVs with disparate dimensions.

Now, the function $C = f(x, y, z)$ has constant-value contour surfaces in 3-DV space, not lines in the 2-DV plane. An onion has a skin layer of constant color. Inside are layers of varying moisture. Here too, $-\nabla z$ is descent (toward lower T, %RH, odor, dust, etc. values), and it is perpendicular to the tangent of the contour.

Following the 2-D analysis procedure above, but in three dimensions, with $C = f(x, y, z)$, the total derivative is

$$dC = \frac{\partial f}{\partial x}\bigg|_0 dx + \frac{\partial f}{\partial y}\bigg|_0 dy + \frac{\partial f}{\partial z}\bigg|_0 dz \tag{7.50}$$

The subscript 0 means at the point (x_0, y_0, z_0).

On a surface of constant C value, $dC = 0$. With this, and using finite differences for the calculus differentials (which is locally valid for small enough deviations),

$$0 = \frac{\partial f}{\partial x}\bigg|_0 (x - x_0) + \frac{\partial f}{\partial y}\bigg|_0 (y - y_0) + \frac{\partial f}{\partial z}\bigg|_0 (z - z_0) \tag{7.51}$$

Rearranging to solve for the z deviation

$$(z - z_0) = -\frac{\partial f / \partial x|_0}{\partial f / \partial z|_0}(x - x_0) - \frac{\partial f / \partial y|_0}{\partial f / \partial z|_0}(y - y_0) \tag{7.52}$$

And combining constant terms results in

$$z = a_0 + b_0 x + c_0 y \tag{7.53}$$

here the subscripts on the coefficients indicate that they are evaluated at the point (x_0, y_0, z_0).

This is a linear relation of z to x and y. It is the equation of a plane tangent to the surface of constant C at the point (x_0, y_0, z_0).

The three dimensions can be illustrated as in Figure 7.8 in which the spherical constant C surface is represented by the net illustration. Note that the net lines are now in all three dimensions. The (x_0, y_0, z_0) point is indicated, as is the plane tangent to the surface at the point.

Any line can be drawn in the tangent plane though the base point, and that line has a projection in the x–y plane. $(y - y_0) = a(x - x_0)$ defines any line through the (x_0, y_0) pair in the x–y plane:

$$L_{\text{Tangent}} = \begin{bmatrix} x - x_0 \\ y - y_0 \\ z - z_0 \end{bmatrix} = \begin{bmatrix} 1 \\ a \\ -\dfrac{\partial f / \partial x|_0}{\partial f / \partial z|_0} - a \dfrac{\partial f / \partial y|_0}{\partial f / \partial z|_0} \end{bmatrix} (x - x_0) \tag{7.54}$$

Since $\nabla C \cdot L_T = 0$, again $-\nabla C$ is in the direction of $\perp$ to contour, and a descent toward lower C values.

$$\nabla C = \begin{bmatrix} \dfrac{\partial C}{\partial x} \\[2ex] \dfrac{\partial C}{\partial y} \\[2ex] \dfrac{\partial C}{\partial z} \end{bmatrix} \tag{7.55}$$

7.6 Takeaway

As I was teaching the optimization course, one professor told me, "The most important concept students need to get is that of the gradient."

The negative gradient is the direction of steepest descent. It is a line whether the optimization is 2-D, 3-D, or N-D. The line of steepest descent can be stated as a vector, or as an algebraic relation in which one DV is a function of another, or in parametric notation. In the parametric description, the parameter might represent distance along the line in the projection on the N-DV space, time of flight, a scalar that multiplies each gradient element, or any number of alternatives.

When the OF response is included in 2-DV applications, it produces a 3-D representation, which can be visualized because of our familiarity with 3-D space. The impact of surface features such as cliffs, ridges, boundaries, and flat spots in our 3-D life can be understood, and they translate directly to 2-DV optimization. Even though we may not be able to visualize such surface features in N-D applications, the issues and solutions can be understood in 2-DV applications.

Although you may mostly encounter N-DV applications, developing understanding on 2-DV examples is important.

7.7 Exercises

1 Draw vectors at the points (4, 6) and (8, 3) on the contour map of a function in Figure 7.9, $f(x, y)$ to illustrate $-\nabla f$. Be sure the directions and relative lengths of your vectors are consistent with the x and y elements. The minima are located at about (6, 4) and (8.5, 8.5).

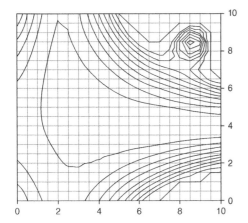

Figure 7.9 Illustration for Exercise 1.

2 If you want a step size ΔS in the steepest descent direction, what must be the value for α in the equation $\Delta x = -\alpha \nabla f$? Derive the formula to determine the value for alpha when the x-vector and the gradient have two elements.

3 In a 2-DV optimization application, the negative gradient defines a line in the DV plane that indicates direction of steepest descent. (i) What geometric shape does it define in a 1-DV application, and (ii) what does it indicate? Use the definition of the gradient to support your answer.

4 A surface is modeled as $z = 3 + 4(x - y)^2 + x^3 + y$. At the x–y pair $(x_0, y_0) = (2, 1)$, what is the equation of the line of steepest descent?

5 If the function is $f(x,y) = 1 + 2x + 3xy + 4y^2$, derive the x–y relation that represents the line of steepest descent from the (x, y) pair $(2, 2)$.

6 Circle T for true, F for false, of both if it could be either. Explain your choice.

T	F	The gradient at a point is parallel to the contour at that point
T	F	In 3-DV dimensions, the gradient is a plane
T	F	The gradient is a direction and only applicable to DVs with length dimension (e.g., not temperature, or color, or power units)
T	F	"Steepest descent" means the direct path, shortest line to the function min (or max)
T	F	The gradient might point to a saddle point instead of a min or max

7 Determine the element values of the gradient for OF $= 3 + (x - 2y)(z - 1)$ at the point $(x, y, z) = (4, 2, 1)$.

8 Figure 7.10 shows a contour of a 2-DV application, and the x-marks indicate the local (upper left) and global (lower right) optima. At the $DV_1 = 4$ and $DV_2 = 5$ point, sketch the tangent to the contour and the negative gradient. Repeat for $DV_1 = 1$ and $DV_2 = 4$ point, and sketch the tangent to the contour and the negative gradient.

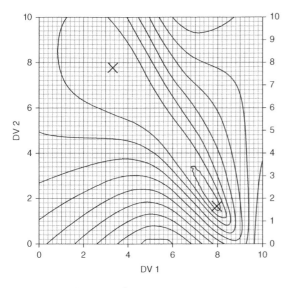

Figure 7.10 Illustration for Exercise 9.

9 Determine and illustrate the line of steepest descent at point (x_0, y_0) on the following parabolic surface (which has elliptic contours): $z = (x-a)^2 + b(y-c)^2$. You choose values for a, b, c, x_0, and y_0. Use a contour plot to illustrate your version of the function. (You could add your function to the Optimization Examples VBA file.) Show the derivation of the line of steepest descent. Overlay the line on the contour map. Describe features that make you know you have the right answer.

10 Determine the analytical expression for the gradient of this function: $J = 3x_1 + bx_1x_2 - x_2^{-1}$

11 Create or find a 2-DV or higher-dimension function, and demonstrate that the negative gradient is the direction of steepest descent. Instead of a mathematical proof such as performed in the chapter, provide an experimental verification. If you choose a function that is symmetric about the minimum (contours are circles, spheres, or hyperspheres), then the steepest descent line will point directly to the minimum. You could show that. Alternately, you could choose any 2-DV function and graph the local contours about a random DV set and show that negative gradient is also the most direct line to the contour with the next lower value.

12 A surface is modeled as $z = 0.1[(x-y)^2 + xy]$. Show the contour over the 0–10 DV range. Choose your own x–y pair $(x_0, y_0) = (2 < x < 8, 2 < y < 8)$, and mathematically derive the vector

equation for the line of steepest descent in terms of a "distance" parameter, s. This asks for a parametric relation in which $x = f(s)$ and $y = g(s)$. Hand draw your x and y line w.r.t. distance parameter s on the contour. (If your line is not perpendicular to the contour, you made a mistake.)

13 What is the best direction from a point to a plane, if the point is not on the plane? Consider Cartesian space, rectangular x–y–z coordinates. Use "best" as minimizing the distance from the point (x_0, y_0, z_0) to the plane $z = a + bx + cy$. Analytically find either the closest point on the plane (x_p, y_p, z_p) or the equation of the line from (x_0, y_0, z_0) to (x_p, y_p, z_p).

14 Show that $-\nabla c$ is the direction of steepest descent for a generic three-variable application, $c = f(x,y,z)$. On a line through DV space the y- and z-positions can be linearly related to the x-position. For example, $\Delta y = a\Delta x$, $\Delta z = b\Delta x$. The objective is to reveal that a and b values that minimize (maximize) the negative of dc/dS match the $-\nabla c$ elements.

15 A surface is modeled as $z = (x-y)^2 + xy$. At the x–y pair $(x_0, y_0) = (5, 1)$, provide the vector equation for the line of steepest descent in terms of a "distance" parameter, s. This asks for a parametric relation in which $x = f(s)$ and $y = g(s)$.

8

Elementary Gradient-Based Optimizers

CSLS and ISD

8.1 Introduction

Regardless of the DV dimension, the negative gradient indicates the direction of steepest decent, which is a logical direction to begin the search. This chapter presents two gradient-based searches: Cauchy's sequential line search (CSLS) and incremental steepest descent (ISD). Although the exercises will primarily be in 2-D applications, these are applicable to N-D situations. The optimization statement is

$$\begin{array}{c}\min\\ \{x_1 x_2, x_3,...\}\end{array} J = f(x_1, x_2, x_3,...) \tag{8.1}$$

8.2 Cauchy's Sequential Line Search

Although the objective is to find the true minimum in a multi-DV application, the Cauchy sequential line search (CSLS) will be applied to univariate single DV searches along a straight line of steepest descent through the multidimensional DV space. At the minimum along one line, the gradient is reevaluated, and a new direction of steepest descent is defined. This sequential univariate search along a line is repeated until convergence.

This is a nested optimization. At each stage, search along the line to find the minimum along the line. That is an optimization. But those local line searches continue until the new search directions do not improve the OF.

First, start at an initial trial solution, and use the negative gradient of the function to define the line of steepest descent. Use any of the several 1-D algorithms to search for the optimum along the line. This requires a convergence criterion for a 1-DV search along the line. At the minimum, a new base point, a pivot point, reevaluate the gradient to define a new line of steepest descent and apply the univariate optimization until it converges. Ultimately, in the vicinity of the optimum, sequential 1-D searches will make very small changes in the sequential pivot points and in this case claim convergence.

Figure 8.1 illustrates a 2-DV application. Start at point A. Search along steepest descent line to the minimum along the line, point B. Restart at B and move to the next minimum, point C.

Engineering Optimization: Applications, Methods, and Analysis, First Edition. R. Russell Rhinehart.
© 2018 R. Russell Rhinehart. Published 2018 by John Wiley & Sons Ltd.
Companion website: www.wiley.com/go/rhinehart/engineeringoptimization

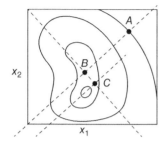

Figure 8.1 Illustrating Cauchy's sequential line search.

Note: All search lines are mutually perpendicular when DVs have the same graph scale (physical distance on the graph per DV unit). The minimum along a line must stop at the tangent point to the minimum contour. If, for instance, a point on the line, which is hypothesized to be a minimum, is not tangent to a contour of that value, then that contour crosses the search line. Then moving a bit along the line crosses the contour to a lower OF value. So, the local point was not the minimum. Accordingly, the current search line of steepest descent is the tangent to the minimum contour that it touches. Then, because the new line of steepest descent is perpendicular to the tangent of the contour, it is perpendicular to the prior steepest descent line. This expectation can be used as a check that the technique is properly implemented.

Note: If a numerical algorithm is used for the line search (GS, NR, HD, SQ, etc.), the search will not stop exactly at the minimum, but in its proximity. So, the sequential lines will not be exactly perpendicular. If noticeably off of perpendicular, consider tightening the convergence criterion threshold for the univariate search. But balance work with perfection. The tighter the convergence criterion, the more function evaluations are required to fine-tune the optimum. But the ideal true path from a random initial trial solution to the optimum is not what is needed, and coarse paths can lead to the optimum with lesser computational work.

Each univariate search stage (the search along the sequential lines of steepest descent) can be defined by a parametric relation:

$$\underset{\{\alpha\}}{\min}\, J = f(\alpha, \text{line defined by gradient}) \tag{8.2}$$

where in 2-D the steepest descent line is defined as

$$\begin{bmatrix} x_{1,\,\text{new}} \\ x_{2,\,\text{new}} \end{bmatrix} = x = x_0 - \alpha \nabla f|_0 = x_0 - \alpha \begin{bmatrix} \dfrac{\partial f}{\partial x_1} \\[2mm] \dfrac{\partial f}{\partial x_2} \end{bmatrix}_0 \tag{8.3}$$

This scales to N-D. The x-vector is the list of DV elements. Alpha is the factor that defines distance along the line, but α is not the distance. Using ΔS as the distance in the DV projection,

$$\Delta S = \sqrt{\Delta x_1^2 + \Delta x_2^2 + \Delta x_3^2 + \cdots + \Delta x_N^2} \tag{8.4}$$

$$\Delta S = \sqrt{\left(\alpha \frac{\partial f}{\partial x_1}\Big|_0\right)^2 + \left(\alpha \frac{\partial f}{\partial x_2}\Big|_0\right)^2 + \cdots + \left(\alpha \frac{\partial f}{\partial x_N}\Big|_0\right)^2} \tag{8.5}$$

$$\Delta S = \alpha \sqrt{\sum \left(\frac{\partial f}{\partial x_i}\Big|_0\right)^2} = \alpha \sqrt{\sum (g_i|_0)^2} = \alpha \sqrt{\sum (f_i'|_0)^2} = \alpha \sqrt{\nabla f|_0 \cdot \nabla f|_0} \tag{8.6}$$

So, using gradient elements at the base point, ΔS can define α, or vice versa.

Note: In this concept both the parametric relation and ΔS are dimensionally consistent only if the units on each DV are the same. Mathematically, there is no issue, and whether the DV units are the same or not, the method works. However, if the DVs do not have identical units, to satisfy engineering sensibility, a solution would be to convert the DVs to scaled variables and then to dimensionless variables.

If the DVs were scaled by a range for each, for example, $x_i' = x_i/R_i$, then the partial derivatives in the gradient would become $(1/R_i)(\partial f/\partial x_i')$ and the incremental change in the DV would become $\Delta x_i' = \Delta x_i/R_i$. Then the parametric search would determine scaled DV values from

$$\begin{bmatrix} \Delta x_1' \\ \Delta x_2' \end{bmatrix} = \begin{bmatrix} R_1^2 & 0 \\ 0 & R_2^2 \end{bmatrix} \alpha \begin{bmatrix} \dfrac{\partial f}{\partial x_1'} \\ \dfrac{\partial f}{\partial x_2'} \end{bmatrix}_0 \tag{8.7}$$

And then the actual DV values for the OF calculation would be the unscaled values:

$$x_i = x_{i0} + R_i \Delta x_i' \tag{8.8}$$

In this univariate search along the steepest descent line through N-dimensional DV space, the single DV is ΔS (or equivalently the scale factor α). The univariate search can choose the position along the line (the value of α, or ΔS, or any one DV) in any of many ways.

Likely, the gradient element values will be determined numerically, with a finite difference approximation. Here, a gradient element is represented as a forward finite difference:

$$\frac{\partial f}{\partial x} \cong \frac{f(x + \Delta x) - f(x)}{\Delta x} \tag{8.9}$$

But what direction is forward? If the search is moving toward larger x values, then use a positive Δx value to explore the surface in the forward direction. However, if the search is moving toward smaller x values, then use a negative Δx value to explore the surface in the forward direction.

In either case, one has to choose the value of Δx. It needs to be large enough to avoid truncation error in the gradient element calculation. However, at the end game, when very near the optimum a one-sided evaluation of the slope may misrepresent the local surface. Accordingly, the value of Δx for estimating the derivative should be smaller than the value of Δx used for convergence. Also, perhaps a central difference will be more appropriate than a one-sided estimate as a measure of the local slope when in the vicinity of the optimum, but it doubles the number of function evaluations to get the gradient.

Any number of univariate searches could be used within the CSLS method.

8.2.1 CSLS with Successive Quadratic

For each sequential line search, choose three alpha values and evaluate the function at each:

$$\begin{aligned} \alpha_0 &= 0 \rightarrow x = x_0 \rightarrow f_0 \\ \alpha_1 &= \text{small} \rightarrow x_1 \rightarrow f_1 \\ \alpha_2 &= \text{another} \rightarrow x_2 \rightarrow f_2 \end{aligned} \tag{8.10}$$

Then fit the f-values to a quadratic model in α:

$$f = a + b\alpha + c\alpha^2 \tag{8.11}$$

Then find $\alpha^* = -b/2c$.

Claim convergence when sequential α^* values meet the threshold criteria. The initial values of α that determine search distance along the line α_1 and α_2 should be large enough initially to explore some distance along the line. But, then at the end game, these may represent large steps away from the optimum, large steps up the hill, which would not represent the local surface near the optimum. So, there should be some rule that adjusts the α_1 and α_2 values as the search progresses. Perhaps, make the sequential values a fraction of the distance between the past two pivot points. Refer to Figure 8.1. In the search from pivot point B to find C, a reasonable distance for the interval in the SQ points is ½ of the distance from points A to B.

8.2.2 CSLS with Newton/Secant

This will use the rate of change of the function w.r.t. the search DV, but along the line of steepest descent. If the search DV is the distance parameter α, choose two values for α. From these calculate the independent variable values, x:

$$\alpha_1 \to x_1 \to g_1 = \text{slope} = \left.\frac{df}{d\alpha}\right|_1 \cong \frac{f(\alpha_1 + \Delta\alpha) - f(\alpha_1)}{\Delta\alpha}$$

$$\alpha_2 \to x_2 \to g_2 \qquad (8.12)$$

$$\alpha_3 = \alpha_2 - \frac{g_2}{(g_2 - g_1)/(\alpha_2 - \alpha_1)}$$

Then apply this recursively:

$$\alpha_n \to x_n \to g_n$$

$$\alpha_{n+1} = \alpha_{n-1} - \frac{g_n}{(g_n - g_{n-1})/(\alpha_n - \alpha_{n-1})} \qquad (8.13)$$

Once convergence is decided on one line, reevaluate the gradient and repeat.

There are, of course, many variations on this Newton-type technique. Some were revealed in Chapter 4.

Again, as the optimum is approached, there will be small differences between sequential pivot points, and the values for α that are used to initialize the search should be scaled as a fraction of recent past distance between pivot points.

8.2.3 CSLS with Golden Section

Consider a search along α. What to use as limits? If the line indicates the direction of descent, then the base point, $\alpha = 0$, will be high and could be used as one end of the GS search. But how far down the line should the other end be placed? Since GS is predicated on bounding the optimum, and that information is probably not available, perhaps it is not the most appropriate search algorithm for CSLS. The situation is the same whether the search parameter is α, ΔS, or a DV.

8.2.4 CSLS with Leapfrogging

Initialize players within some reasonable interval along the new line. For the first univariate search along a line, this might be half of the DV range. Subsequently, it might be equivalent to the distance between most recent pivot points. The advantage of LF over the SQ and Newton-type searches is that it will not be misdirected by inflection points and has a higher probability of finding the global in a multi-optimal search. The advantage of LF over GS is that the LF exploration is not bounded by the initialization range.

8.2.5 CSLS with Heuristic Direct Search

This is my choice in the CSLS search and is what I coded in the 2-D Optimization Examples VBA programs. It does not become misdirected by inflection points along the line, it can handle constraints, it does not require a range initialization, and it is very simple to understand. Although HD will move downhill to the first minimum, and perhaps miss the global in a multi-optimum search, I prefer it over LF within CLSL due to simplicity of implementation.

Once the minimum is found along the line, set that as new trial solution (the base point), then reevaluate the gradient at the new base point, and redo the univariate search on the new line to seek the new minimum. Repeat until stopping criteria is met.

If the univariate DV is α, then you need to specify an initial $\Delta\alpha$ value. As the search along a line converges, that $\Delta\alpha$ increment will progressively contract. Starting the next search with the converged value will cause it to make vary small search steps. So, reinitialize the $\Delta\alpha$ value. There is no sense in using an excessively large value. So, I initialize it using Equation (8.6) as a portion of the ΔS represented by the most recent two base points.

8.2.6 CSLS Commentary

The steepest descent line is perpendicular to the tangent of the contour. From a trial solution, the search line moves perpendicular to the contour in the direction of steepest descent. At the minimum along the line, the line is tangent to the new contour. If not so, the search line would cross the lowest contour, meaning that there is a lower OF value in front or behind. So, since the new steepest descent direction is perpendicular to the new contour, and the new contour is tangent to the old line, each sequential search line is perpendicular to the previous line. To observe this on a display, the physical scale of the plot axes (distance per DV unit) needs to be the same for all DVs.

Regardless of the DV dimension of the OF, CSLS is a search along a line. There is only one DV for the univariate search, which is related to distance along the line.

Cauchy's sequential line search approach identifies the direction of steepest descent. But this is not necessarily aimed at the minimum. And it does not indicate how far to travel along the direction.

It can zigzag from wall to wall when in a long narrow valley. This makes progress down the valley slow, taking excessive function evaluations. Conjugate gradient (Chapter 10) is a fix for such zigzag motion.

The value of α in the formulation is not the step size. Looking at each element in the vector notation,

$$\Delta x_i = \alpha \frac{\partial f}{\partial x_i} = \alpha g_i \tag{8.14}$$

Applying the Pythagorean theorem (presuming scaled variables),

$$\text{step size} = \alpha \sqrt{\sum g_i^2} \tag{8.15}$$

Cauchy's sequential line search permits the optimum to be on a boundary. This is good.

Any univariate search procedure can be used within the CLSL method. If the analytical approach is feasible, then probably it should be used on the original application instead of CLSL. SQ and Newton-type searches are efficient for well-behaved functions; but they can be substantially misdirected if there are inflections, confounded if constraints, and will seek the local optimum. GS does not have those issues, but will only search within its initialization range. LF works well, but I like HD for its overall performance—simplicity, effectiveness, speed, robustness, etc.

What should be used as the ΔS value beginning each line search? The first line searches travel a substantial distance. The last line searches hone in about the minimum and only travel small lengths. If you use a big ΔS, the movement along the initial lines will be fast, but the initial step will be way too large for the end-game lines. Alternately, if you use small ΔS values for the end-game lines, they won't waste time starting along the line, but then it will take many function evaluations to make progress along the initial lines. Here is a description of what I included in the VBA code. Make the first line search ΔS be a relatively large value, $\Delta S = 1$ (which is 1/10th of the DV range). Then on each successive line, make the ΔS be related to the last distance traveled along a line. I chose the initial ΔS to be one-third of the last line length. I found that this enhancement nearly halves the NOFE on a variety of functions. I also found that the portion is relatively unimportant instead of one-third and numbers of 0.2–0.5 were equivalent.

8.2.7 CSLS Pseudocode

Pseudocode, an outline of the CSLS with HD search, is:

1) Initialize DV with a feasible trial solution, evaluate OF, and set DV as the first pivot point
2) Choose a convergence criterion and corresponding threshold for both the univariate search and for the sequential line searches. These could be the same.
3) Initialize ΔS to be used for the first line search. Perhaps make it 1/10th of the DV.
4) Determine the local gradient. This can be either analytical or numerical. If numerical, the ΔDV needed to be initialized, and should be small relative to ΔS.
5) Begin a univariate search along the line defined by the negative gradient at the pivot point
   5.1) Use ΔS and the gradient elements to define α
   5.2) Use α to define the next trial solution along the line
   5.3) Evaluate the OF value at the TS.
       5.3.1) If the OF value is worse or constrained, contract ΔS
       5.3.2) If the OF value is better, expand ΔS and set the trial solution as the base point
   5.4) Test to end the univariate search. If excessive iterations, exit and report
   5.5) Test for convergence of the univariate search.
       5.5.1) If convergence is not met, return to Step 5.1
       5.5.2) If convergence is met, continue to Step 6
6) Calculate ΔS as a fraction of the past pivot to pivot points. Perhaps, $\Delta S = \sqrt{\sum \left(x_{i,j} - x_{i,j-1}\right)^2}$, where the subscript i represents the DV dimension and j represents the past pivot value.
7) Test for stopping criteria if there have been excessive number of sequential lines without convergence, Exit with report.
8) Test for convergence of sequential lines.
   8.1) If converged, exit with message.
   8.2) Otherwise continue, and return to Step 4.

8.2.8 VBA Code for a 2-DV Application

```
Sub Cauchy_Heuristic()
'  Cauchy Sequential Line Searches search along line of steepest descent
'  Here, each line search in steepest direction uses R3 Direct search rules
'  At each iteration, counter i, determine gradient, then search along
'    the same line for minimum.  kk is the line search counter.
'  I enhanced the efficiency of the search by having the initial delta-S
'    based on the past iteration to iteration distance.  In the beginning,
'    the lines take the solution a long way.  At the end game, the lines
'    only make small changes to the DVs.  The Q is what should be the delta-S
'    to start each line search.  I found that .2 times the past iteration-to-
iteration
'    DV change is better than using a fixed initial value.

If i = 0 Then      'initialization
  constraint = "Not Assessed"
  Do Until constraint = "OK"     'Find a feasible initial trial solution for
single player mathods
    x1o = 10* Rnd()          'Randomize initial DV1 value
    x2o = 10* Rnd()          'Randomize initial DV2 value
    OF_Value_Old = f_of_x(Fchoice, x1o, x2o)  'Determine the inital
function value
  Loop
  x1prior = x1o
  x2prior = x2o
  Worksheets("Path_Data").Cells(3, 1) = i
  Worksheets("Path_Data").Cells(3, 2) = x1o
  Worksheets("Path_Data").Cells(3, 3) = x2o
  OF_Value_Old = f_of_x(Fchoice, x1o, x2o)
  Cells(33, 12) = OF_Value_Old
  OF_Value = OF_Value_Old
  expansion - Cells(6, 12)
  contraction = Cells(7, 12)               ' Start line search iteration
  ds = 0.01
Else            'beginning iterations i=1,2,3,4, etc.
  Cells(25, 12) = i
' Define steepest descent direction
  g1v = g1(Fchoice, x1o, x2o)                  'central difference
  g2v = g2(Fchoice, x1o, x2o)
' Start line search iteration
  so = 0        'distance from base
  If i = 1 Then    'initialize the initial step size based on prior iteration
changes
    ds = 1       'initial distance increment, large, but not excessive
  Else
```

```
      ds = 0.3 * Sqr((x1o - x1prior) ^ 2 + (x2o - x2prior) ^ 2)  'any number between
.2 and .5 seems equivalent
   End If
   x1prior = x1o
   x2prior = x2o
   For KK = 1 To 100       'limit of 100 points along the line
' set trial solution
      s = so + ds
      If (g1v ^ 2 + g2v ^ 2) > 0 Then
         x1 = x1o - g1v * s / Sqr(g1v ^ 2 + g2v ^ 2)
         x2 = x2o - g2v * s / Sqr(g1v ^ 2 + g2v ^ 2)
      Else
         x1 = x1o
         x2 = x2o
      End If
      dx1 = x1 - x1o
      dx2 = x2 - x2o
      OF_Value = f_of_x(Fchoice, x1, x2)
      '  search logic for line search - heuristic direct
      If OF_Value > OF_Value_Old Or constraint = "FAIL" Then
         ds = -contraction * ds
      Else
         ds = expansion * ds
         OF_Value_Old = OF_Value
         so = s
      End If
      ' Convergence in line search portion, needs to be smaller than line-to-
line test
      If Abs(ds) < Cells(14, 12) / 10 Then Exit For
   Next KK
   ' update x1 and x2 with best values
   If (g1v ^ 2 + g2v ^ 2) > 0 Then
      x1 = x1o - g1v * so / Sqr(g1v ^ 2 + g2v ^ 2)
      x2 = x2o - g2v * so / Sqr(g1v ^ 2 + g2v ^ 2)
   Else
      x1 = x1o
      x2 = x2o
   End If
   x1o = x1
   x2o = x2
End If

End Sub
```

Example 1   CSLS

Here is a simple 2-DV application:

$$\underset{\{x_1,x_2\}}{\min} J = 15 - 2x_1 - 2x_2 + 0.1{x_1}^2 + 0.1{x_2}^2 + x_1 x_2 \qquad (8.16)$$

The contour and CSLS path are implemented in the Excel VBA program "2-D Optimization Examples" on the www.r3eda.com site. Figure 8.2 reveals both. The contour is a mild ellipse. The initial trial solution is about the point (5.2, 0.2). The perpendicular to the local contour directed the first search along a line to the upper right. The univariate search along that line found the minimum at about the point (8.5, 5). Note that it is tangent to a contour that you might imagine to be between the contours illustrated. Sequential searches along local lines led to the optimum. Pivot points are illustrated as the dots at the line discontinuities. Note that sequential lines are mutually perpendicular.

Here convergence of the univariate search along each line is based on $\Delta S < 0.001$, and the convergence of the overall method is based on the distance between successive pivot points also as $\Delta S < 0.001$. The central finite difference is used to obtain numerical estimates of the derivatives.

Note that only about four lines are visible. In actuality there are seven lines from the initial point to the converged X. Each line represents an iteration. A primitive analysis might think that each dot on the graph represents one function evaluation and using the central difference to estimate derivatives would require two for each DV for each line. So, if each dot represents 5 function evaluations, and there are 7 lines, one might conclude that the method converged with 35 function evaluations. However, between the dots, each step in the univariate line search requires another function evaluation. In all, the search illustrated required 145 function evaluations, roughly 16 for each univariate search.

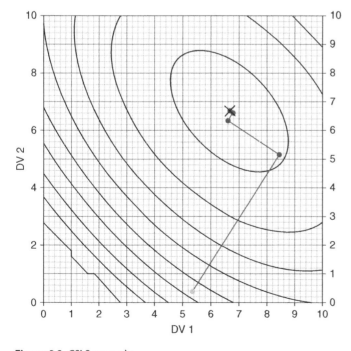

Figure 8.2 CSLS example.

Figure 8.3 Water following an incremental steepest descent path. *Source:* Rhinehart (2016b). Reproduced with permission of Wiley.

8.3 Incremental Steepest Descent

Cauchy's sequential line search initializes the line of steepest descent and then searches along that line. But even if the surface slope changes along the line, CSLS continues along the original line regardless. Perhaps a better method would be to make an incremental step along the line, then reevaluate the local gradient, and make the next step in the local steepest descent direction.

This logic for continually moving downhill is similar to water finding a trickle path downhill, as illustrated in Figure 8.3. I had watered seedlings in pots on my driveway. The excess water first flowed toward the left front of the photo, and then as it encountered a new slope to the driveway, the water eventually changed direction to move toward the right front. As such, incremental steepest descent (ISD) could be labeled a mimetic algorithm, one that employs a "logic" that humans might attribute to some natural behavior, a logic that mimics an "intelligence" within nature. Water seems smart enough to find its way downhill.

8.3.1 Pseudocode for the ISD Method

1) Initialize
    1.1) Choose the first trial solution and set as the base point
    1.2) Choose convergence criterion
    1.3) Choose the initial incremental step size ΔS
2) Evaluate the OF at the initial TS
3) Evaluate the gradient, ∇f. I chose to use numerical methods to accommodate intractable functions. As long as the ΔS step is large relative to the DV range I use a forward finite difference; but when the ΔS is relatively small, suggesting that ISD is in the vicinity of a optimum, I use a central difference.
4) Set α to a value using the gradient elements and ΔS, $\alpha = \Delta S / \sqrt{\sum g_i^2}$
5) Move an incremental ΔS along the line of steepest descent, $x = x_0 - \alpha \nabla f|_0$
6) Evaluate the OF
    6.1) If the new OF value is better than the prior, you are moving in the right direction, expand ΔS (or α) by 10% (or so). $\Delta S_{new} = 1.1 \Delta S_{prior}$. And set the current trial solution (DV and OF) to the base values.
    6.2) Alternately, if the new OF value is worse, or if a constraint is violated, then contract ΔS (or α) by 50% or so. $\Delta S_{new} = 0.5 \Delta S_{prior}$. Note that the contraction does not change the sign. One always wants to search in the steepest descent direction.
7) Test for stopping criteria, such as excessive iterations, and exit with report of a failed search

8) Test for convergence
    8.1) If converged, exit with report
    8.2) Else return to Step 3

Note that I use heuristic rules for adjusting ΔS (or α), which are similar to those I've found effective in the heuristic direct search.

This algorithm defines a path that is similar to the water trickle flowing downhill, or to a marble rolling down the hill, or to an ice cube sliding downhill (but without momentum or viscous damping). Although it has been enjoyable for me to add momentum and friction-like energy dissipation to the path such a particle might take and fun to watch it speed up when going downhill then coast and slow down when moving up the other side of the hill, such mechanisms are less efficient than the simple incremental downhill steps.

The ISD technique must evaluate the gradient as well as the function value at each step. In a 2-DV application this means three function evaluations at each step. By contrast, CSLS only uses one function evaluation at each trial solution once the line has been set, which would seem better, but CSLS follows the original line in spite of the redirection information in the local slope. CSLS can move in the wrong direction.

I like ISD. In my explorations it generally had fewer function evaluations than CSLS.

Example 2   ISD
Here is the same simple 2-DV application illustrated for CSLS:

$$\min_{\{x_1,x_2\}} J = 15 - 2x_1 - 2x_2 + 0.1x_1^{\,2} + 0.1x_2^{\,2} + x_1x_2 \tag{8.17}$$

The contour and IDS path are implemented in the Excel VBA program "2-D Optimization Examples" on the www.r3eda.com site. Figure 8.4 reveals both. Again, the contour is a mild ellipse, and the initial trial solution is about the point (5.2, 0.2). The first step is a length of about 2-DV units and moves perpendicular to the local contour toward the upper right. The step was a success so the next step size is a bit larger and moves in the local steepest descent direction. It was a success. The fourth trial solution, inside the lowest contour illustrated, indicates that the new direction is toward the left and the past success would indicate a larger step size. However, a large step size to the left moved the trial solution up the other side of the hill, so the step size was halved, leading to the fifth successful trial solution. Eventually, the sequence finds the minimum.

Here convergence is based on $\Delta S < 0.001$.

Note that only about eight trial solutions are visible. In actuality there are 28 successful trial solutions and a good number of unsuccessful ones. Each trial solution requires a new function evaluation, and each successful one requires the gradient evaluation (two additional function evaluations if forward difference, four if central). In all, the search illustrated required 101 function evaluations, roughly two-thirds of that required in CSLS.

8.3.2 Enhanced ISD

But nothing is perfect, as it seems. In a valley with a gently sloping floor and steep walls, ISD zigzags across the floor, from wall to wall, as it seeks a path downhill. The issue is that unless it is at the very

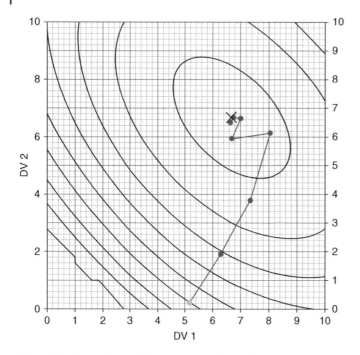

Figure 8.4 Illustrating an ISD approach to the optimum.

local minimum of the valley (if it is near the valley floor but on a wall), the direction of steepest descent is across the valley, not down along the bottom. Figure 8.5 illustrates this on Function 11 in the 2-D Optimization Examples program. The initial trial solution is at the point (4, 3) and moves downhill toward the local optimum at about (3.5, 8). But this optimum is in a long valley with steep walls relative to the slope of the floor. So, the ISD search oscillates to and fro, and the path zigzags across the valley.

In an enhancement (grounded in the concept that led to conjugate gradient techniques), I decided to average recent gradient elements. If one gradient points the search toward east by southeast, and upon going there the gradient points west by southwest, the average gradient points south. Instead of a classic average, for convenience I decided to use an exponentially weighted moving average, alternately known as a first-order filter:

$$g_{f,i} = \lambda g_{f,i-1} + (1-\lambda)g_i \tag{8.18}$$

Here g represents the gradient element, f means filtered value, i is the iteration counter, and λ is a filter factor. Without extensively exploring what might be a right value, $\lambda = 0.5$ gives good results. With $\lambda = 0.5$, old values of the gradient do not have a long-persisting influence, the directional reversals are greatly reduced, yet the filtered gradient rapidly adjusts to new topology. Figure 8.6 illustrates the ISD search with filtered gradient values on the same function and from the same starting point as illustrated in Figure 8.5. Now, the path does not zigzag across the valley, and the optimum is reached with fewer function evaluations.

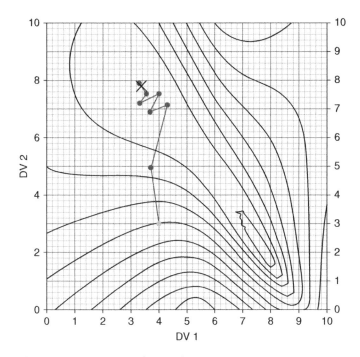

Figure 8.5 Zigzag approach to optimum.

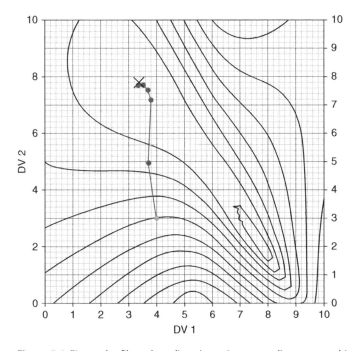

Figure 8.6 First-order filtered gradient (a conjugate gradient approach).

8.3.3 ISD Code

```
Sub ISD()
'  Incremental steepest descent with heuristic step-size rule
'  And filtered gradient values to temper zig-zagging paths

   If i = 0 Then  'initialization prior to iterations
      constraint = "Not Assessed"
      Do Until constraint = "OK"    'Find a feasible initial trial solution for
single player mathods
         x1o = 10 * Rnd()          'Randomize initial DV1 value
         x2o = 10 * Rnd()          'Randomize initial DV2 value
         OF_Value_Old = f_of_x(Fchoice, x1o, x2o)  'Determine the inital
function value
      Loop
      Worksheets("Path_Data").Cells(3, 1) = i
      Worksheets("Path_Data").Cells(3, 2) = x1o
      Worksheets("Path_Data").Cells(3, 3) = x2o
      Cells(33, 12) = OF_Value_Old
      OF_Value = OF_Value_Old
      alpha = Cells(6, 12)
      beta = Cells(7, 12)
      expansion = alpha
      contraction = beta
      ds = 2
      eISD = -g1forward(Fchoice, x1o, x2o)    'negative Gradient element,
      first partial of OF_Value w.r.t. x1
      fISD = -g2forward(Fchoice, x1o, x2o)
   Else   'iterations i=1,2,3,4, ... etc.
      'filtering the gradient elements greatly reduces the zig-zag when in
steep valleys
      'too much filtering makes it rely on way past derivative values.
      'the 0.5 is a good value.  0.75 and 0.25 seem to hold on to past values
too long.
      '0.25 and 0.75 seem to not temper the fluctuations
      If ds < 10 * Cells(14, 12) Then 'near converging, use central finite
difference
         eISD = 0.5 * eISD + 0.5 * (-g1(Fchoice, x1o, x2o)) 'negative Gradient
element, first partial of OF_Value w.r.t. x1
         fISD = 0.5 * fISD + 0.5 * (-g2(Fchoice, x1o, x2o))
      Else
         eISD = 0.5 * eISD + 0.5 * (-g1forward(Fchoice, x1o, x2o)) 'negative
Gradient element, first partial of OF_Value w.r.t. x1
         fISD = 0.5 * fISD + 0.5 * (-g2forward(Fchoice, x1o, x2o))
      End If
      Gradient = Sqr(eISD ^ 2 + fISD ^ 2)
      If Gradient = 0 Then
```

```
      dx1 = 0
      dx2 = 0
      Exit Sub
    End If
    coeff = ds / Sqr(eISD ^ 2 + fISD ^ 2)
    x1 = x1o + coeff * eISD
    x2 = x2o + coeff * fISD
    OF_Value = f_of_x(Fchoice, x1, x2)
    dx1 = x1 - x1o
    dx2 = x2 - x2o
    If OF_Value > OF_Value_Old Or constraint = "FAIL" Then     'Cauchy step
too big?
        ds = contraction * ds          'Make ds smaller
    Else
        ds = expansion * ds            'make step larger
        OF_Value_Old = OF_Value
        x1o = x1
        x2o = x2
    End If
  End If
End Sub
```

8.4 Takeaway

Many options and choices are associated with any optimizer. It is much more than just convergence criterion, convergence threshold, and initial trial solution. There are choices for step-size initialization and reinitialization that can improve performance as the search progresses from initial to end game. There are rules for overriding the base calculation of a gradient or switching to central difference or changing the Δx values in numerical estimates.

The heuristic search techniques ISD and CLSL (with heuristic direct) are more robust to surface aberrations (inflections, saddle, discontinuities, constraints, etc.) than the other CLSL options (Newton's, successive quadratic) and do not presume knowledge of the range of the search DV (as does CSLS with a GS search).

Spilled water is smart enough to find its way downhill.

8.5 Exercises

1 In Figure 8.3, why does the water not flow from the plants on the bricks directly to the lowest point in the driveway, in the lower right? Sketch the elevation contour lines on the driveway.

2 In ISD, what would you use for the logic to set the magnitude of the ΔS jump distance? Would you change the jump distance as the search progresses? If the incremental move goes to a higher OF value, should you search in the other direction from the base point, or a smaller distance in the same direction from the base point?

3 Use the 2-D Optimization Examples file to compare CSLS and ISD looking at ANOFE.

Explore the impact of:
- Convergence criterion threshold value
- Expansion and contraction factors
- Choice of numerical method—forward or central finite difference

Run 100 trials and investigate these outcomes. (Recognizing data variability in a sampling of the population, compare using either a t-test or F-test as appropriate.)

- NOFE
- Precision of OF* (sigma after many runs)
- Precision of DV* (sigma after many runs)

4 What are the implications for the choice of the expansion factor 1.05–1.25 or so, and the contraction factor or 0.5–0.25 or so?

$$\min_{\{\text{contract value}\}} J = \text{number of function evaluations} + \text{robustness} + \cdots$$

5 What if the units on each DV are not the same? What are the implications for the units on ΔS, or variables in the recursive equation $x_n = x_0 - \alpha \nabla f|_0$?

6 Start at a point near the middle left side on the contour in Figure 8.7, and show three iterations of Cauchy's sequential line search toward the optimum. Repeat from a point in the middle right side.

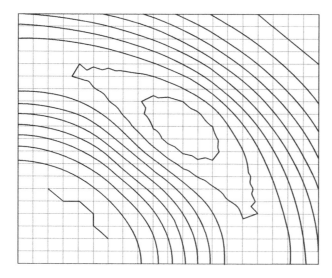

Figure 8.7 For Exercises 6 and 7.

7 Start at a point near the middle left side on the contour in Figure 8.7, and show three iterations of ISD toward the optimum. Repeat from a point in the middle right side.

8 For a two-dimensional (2-D) situation, for example, $z = f(x, y)$, CSLS is a series of one-dimensional (1-D, univariate) line searches. In a 3-D situation, for example, $c = f(x, y, z)$, is it a sequence of 2-D searches? Show the analysis that defends your answer.

9 Figure 8.8 shows contours of a 2-D function. The minimum is near the (x, y) point (5.5, 5). Start at the (x, y) position (9, 7.5), and illustrate the first two iterations of a Cauchy's sequential line search.

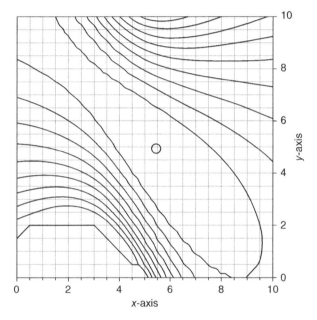

Figure 8.8 For Exercises 9 and 10.

10 From position (1.5, 3) in Figure 8.8, illustrate the first two iterations of an incremental steepest descent search with an initial step size of 2.

11 Explain why Cauchy's sequential line search method is a three-level nested optimization (optimization within an optimization within an optimization).

12 The "line" of a line search, a single DV optimization application can be a curve. Provide an example.

13 The following equation defines how the OF value is dependent on two DVs. $f = 4 + 3x + 3x^2 + 4y + 5y^2 + 6xy$. State the negative gradient, $-\nabla f$, in terms of x and y. Determine the (x, y) value that a line search will move to using $\alpha = 1/5$, the relation $\Delta x = -\alpha \nabla f$, and starting at the location $(x_0, y_0) = (1, 1)$.

14 The following equation defines how the OF value is dependent on two DVs. $f = 1 + 4x + 2x^2 + 3y + 5xy$. The DVs are already scaled. Determine the numerical values of Δx that

ISD would specify, if the line search starts at the location $(x_0, y_0) = (0,0)$ and has an initial step size $S = 0.5$. Use $\alpha = S/\sqrt{\sum(\partial f/\partial x_i)^2}$ in the ISD relation $\Delta x = -\alpha \nabla f$.

15 Provide a flow chart and supporting text explanation for CSLS.

16 Provide a flow chart and supporting text explanation for ISD.

17 Which is best for either CSLS or ISD optimization (you choose which), the forward finite difference or the central finite difference as a numerical approximation to the derivative? The OF is to be a combination of NOFE and some other aspect. You define what the other OF issue is and how to measure it. Choose the functions you think will provide a balanced and comprehensive, but not redundant testing ground. Create experimental results to gather sufficient data. Describe the results and defend your choice of best. This is sketch, consider, erase, sketch, consider erase, …, ink. Don't fix your first impression of your action as the "way." As you follow some path, as you see developing results, you will be smarter about what you should have done. Change your plan accordingly.

18 In my version of CSLS, I use the heuristic direct search for each sequential line search. Any line search algorithm could be used. Change the code to use SQ, or GS, or a Newton's. Reveal the results. Was it better? How do you define "better?"

19 In Cauchy's sequential line search, there is an optimization for each sequential line; and each optimization stops with a convergence criterion. In my program, each line search uses the same convergence criterion. But the initial line searches do not take a path directly to the optimum; so, whether initial searches converge at the exact perfect, or at a "nearly there" spot, is inconsequential. On early lines, continuing the search until exactly perfect requires extra function evaluations that seem unwarranted. Devise a new rule that uses "loose" convergence when the search is not near the optimum, but switches to "tight" convergence at the end game where precision is warranted.

20 Describe the procedure to perform a univariate search along distance S of a path that has a slope discontinuity, as in Figure 8.9. Instead of one straight line, there is a kink in the line. $y = a + bx$ in one segment, and $y = c + dx$ in the other segment. The optimizer is not the issue. Show how can you find the OF value, from the trial solution, S, value.

21 Derive a procedure to perform a univariate search along distance S of a curved path in x–y–z space. The DV is distance S. The y and z positions of the line can be described as functions of the x position, $y = f(x)$, and $z = g(x)$. The value of the OF along the path is a function of position $OF = v(x, y, z)$. The optimizer is not the issue. Show how can you find the OF value, from the trial solution, S, value.

22 A right circular, open tank of height, h, and radius, r, has volume $\pi r^2 h$ and surface area $\pi r^2 + 2\pi rh$. The cost of a tank is related to the amount of material in the surface of thickness, t, density, ρ, and cost per unit mass, c. Cost $= ct\rho(\pi r^2 + 2\pi rh)$. What dimensions, r and h, minimize cost while meeting a specified volume, V? (i) State the optimization in standard form. (ii) Explain simplifications that can be used. (iii) Solve the simplified application.

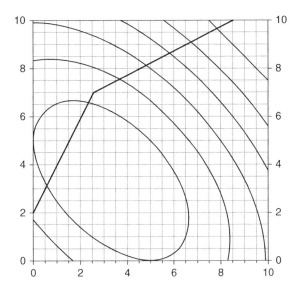

Figure 8.9 For Exercise 20.

23 For a two-dimensional situation, show that the equation to determine the change in DV, Δx, for an incremental steepest descent method is not dimensionally correct if the units on the two DVs, x and y, are not the same.

9

Second-Order Model-Based Optimizers

SQ and NR

9.1 Introduction

This chapter presents two search algorithms that use second-order models of the surface: successive quadratic (SQ) and Newton–Raphson (NR). They presume more about the surface than the gradient-based optimizers. Consequently these are faster when the surface is compatible with the algorithm concepts, which include continuum deterministic surfaces, no flat spots, and an initial trial solution in the vicinity of the optimum. These optimization approaches often are accepted as the premier optimization methods, and they are components in next-level gradient-based optimizers. So, they need to be presented, even though I find that they are wholly inappropriate for many applications, which have features that are inconsistent with the concepts on which these algorithms are predicated.

Although the examples and analysis will primarily be in 2-D applications, SQ and NR are applicable to N-D situations. The objective is

$$\min_{\{x_1 x_2, x_3, \dots\}} J = f(x_1, x_2, x_3, \dots) \tag{9.1}$$

9.2 Successive Quadratic

This is a direct extension of the SQ presentation in Chapter 4 and is often termed a surrogate model method. In this case the surrogate model of how the function depends on the DV values is a quadratic relation. The model is used to approximate the function, and the next estimate of DV* is based on the model, not the function. It is very easy to obtain DV* values from a quadratic model; and if the model is a good fit to the function, the modeled DV* values will be a close approximation to the true DV* values. At each estimate of the DV* values, the function is used to generate data to obtain a new surrogate model, which, hopefully, is progressively more representative of the local surface at the optimum. When successive DV* values meet the convergence criteria, convergence is claimed.

There are a variety of model types that people use for surrogate models. Some are radial basis functions that are progressively adjusted as each new trial solution is added. In many embodiments, the old

Engineering Optimization: Applications, Methods, and Analysis, First Edition. R. Russell Rhinehart.
© 2018 R. Russell Rhinehart. Published 2018 by John Wiley & Sons Ltd.
Companion website: www.wiley.com/go/rhinehart/engineeringoptimization

points are not completely discarded, but continue to shape the global model. Here the concept will be presented with quadratic models.

Regardless of either the surrogate model type or the algorithm to determine the optimum of the surrogate model, this is an iterative procedure with two stages in each iteration. In the first stage, generate function data in a region of interest and generate a model to match the data. In the second stage, determine the DV* values of the surrogate model.

9.2.1 Multivariable SQ

Model the OF surface with a quadratic model, a surrogate model. Here is one example:

$$\text{Form 1}\begin{cases} Y = a + b_1X_1 + b_2X_2 + b_3X_3 + \cdots \\ \quad + c_{11}X_1^2 + c_{22}X_2^2 + c_{33}X_3^2 + \cdots \\ \quad + c_{12}X_1X_2 + c_{13}X_1X_3 + \cdots + c_{23}X_2X_3 + \cdots \end{cases} \tag{9.2}$$

The first line in Form 1 has linear terms only, and the second line has quadratic terms, but no interaction. The third line adds the interacting quadratic terms. The model coefficients are the $\{a, b, c\}$ values and the application decision variables are the $\{x_i\}$ terms.

The same relation can be mathematically stated in any of a variety of ways. Form 2 presents the quadratic model in vector–matrix notation:

$$\text{Form 2}\begin{cases} Y = a + \boldsymbol{b}^T X + X^T(\boldsymbol{C}X) \\ X = \begin{bmatrix} X_1 \\ X_2 \\ X_3 \\ \vdots \end{bmatrix} \quad \boldsymbol{b} = \begin{bmatrix} b_1 \\ b_2 \\ b_3 \\ \vdots \end{bmatrix} \quad \boldsymbol{C} = \begin{bmatrix} c11 & c12 & c13 & \dots \\ 0 & c22 & c23 & \dots \\ 0 & 0 & c33 & \dots \\ \vdots & \vdots & \vdots & \ddots \end{bmatrix} \end{cases} \tag{9.3}$$

And Form 3 presents the quadratic model in a series notation:

$$\text{Form 3}\left\{ Y = a + \sum_{i=1}^{N} b_i X_i + \sum_{i=1}^{N}\sum_{j=1}^{N} c_{ij}X_iX_j \right. \tag{9.4}$$

First, presume the surrogate model has been identified—presume that the coefficient values have been determined. Stage 2 in SQ is to determine the optimum DV* values of the surrogate model. If model coefficient values are known, then the analytical method provides N linear equations to solve for the N x_i^* element values.

At the optimum the derivative of the function w.r.t. each DV is zero. Applying this to determine the optimum of the surrogate model, for each DV,

$$\left.\frac{\partial Y}{\partial X_i}\right|_{X^*} = 0 \tag{9.5}$$

Using Form 3 and setting the derivative w.r.t. each x_i to zero,

$$0 = b_1 + 2c_{11}X_1^* + c_{12}X_2^* + c_{13}X_3^* + c_{14}X_4^* + \cdots$$
$$0 = b_2 + c_{12}X_1^* + 2c_{22}X_2^* + c_{23}X_3^* + c_{24}X_4^* + \cdots$$
$$0 = b_3 + c_{13}X_1^* + c_{23}X_2^* + 2c_{33}X_3^* + c_{34}X_4^* + \cdots \tag{9.6}$$
$$0 = \cdots$$

Note that this results in N equations, each of which is linear in the N unknown x_i^* values. The coefficient values are known from the initial fitting of the surrogate model to the function. Note that $c_{ij} = c_{ji}$.

This system of linear equations can be stated in classic vector–matrix form:

$$\mathbf{D}X^* = -\boldsymbol{b} \tag{9.7}$$

where

$$\mathbf{D} = \begin{bmatrix} 2c_{1,1} & c_{1,2} & c_{1,3} & c_{1,4} & \cdots \\ c_{2,1} & 2c_{2,2} & c_{2,3} & c_{2,4} & \cdots \\ c_{3,1} & c_{3,2} & 2c_{3,3} & c_{3,4} & \cdots \\ c_{4,1} & c_{4,2} & c_{4,3} & 2c_{4,4} & \cdots \\ \vdots & \vdots & \vdots & \vdots & \ddots \end{bmatrix}$$

There are many linear algebra algorithms that can be used to solve for the unknown x_i^* values. Gauss elimination is one.

This development uses the analytical method to find the optimum of a quadratic surrogate model. There are other methods. One could use ISD or CSLS, for example, or any of the algorithms that will be developed in subsequent chapters. But the quadratic model is amenable to the analytical method, and the linear algebra solution to the set of derivative equations is relatively computationally fast and is a deterministic explicit procedure with a guaranteed solution within a deterministic number of computer operations.

But first, the model must be identified, which is stage 1. There are many choices, including model structure; and once it is selected, there are additional user choices related to how to obtain the coefficient values.

You could use a structure that does not model the interaction terms. This simpler model is easier to identify and use, but it is not as true to the surface as a model with interaction terms. However, except in some ideal case, no surrogate model is exactly true to the entire surface:

$$Y = a + b_1X_1 + b_2X_2 + b_3X_3 + \cdots$$
$$+ c_1X_1^2 + c_2X_2^2 + c_3X_3^2 + \cdots \tag{9.8}$$

Alternately

$$Y = a + \boldsymbol{b}^T X + X^T(\mathbf{C}X) \tag{9.9}$$

where

$$
C = \begin{bmatrix}
c_1 & 0 & 0 & \cdots \\
0 & c_2 & 0 & \cdots \\
0 & 0 & c_3 & \cdots \\
\vdots & \vdots & \vdots & \ddots
\end{bmatrix}
$$

There are $1 + N + (N + 1)N/2 = (N + 1)(N + 2)/2$ number of coefficients in the first interactive model of Equation (9.2). There are $1 + N + N = 2N + 1$ coefficients in the simpler model of Equation (9.8). To determine the $(N + 1)(N + 2)/2$ (or $2N + 1$) coefficient values, you have to evaluate that many function values and solve a linear algebra set of equations that large:

$$
\begin{bmatrix}
1 & {}_1X_1 & {}_1X_2 & \cdots & {}_1X_1^{\,2} & {}_1X_1X_2 & {}_1X_1X_3 & \cdots \\
1 & {}_2X_1 & {}_2X_2 & \cdots & {}_2X_1^{\,2} & {}_2X_1X_2 & {}_2X_1X_3 & \cdots \\
1 & {}_3X_1 & {}_3X_2 & \cdots & {}_3X_1^{\,2} & {}_3X_1X_2 & {}_3X_1X_3 & \cdots \\
\vdots & & & & & & & \\
1 & {}_{i-1}X_1 & {}_{i-1}X_2 & \cdots & {}_{i-1}X_1^{\,2} & {}_{i-1}X_1X_2 & {}_{i-1}X_1X_3 & \cdots \\
1 & {}_iX_1 & {}_1X_2 & \cdots & {}_iX_1^{\,2} & {}_iX_1X_2 & {}_iX_1X_3 & \cdots \\
1 & & & & & &
\end{bmatrix}
\begin{bmatrix}
a \\ b_1 \\ b_2 \\ \\ c_{11} \\ c_{12} \\ c_{13} \\ \vdots
\end{bmatrix}
=
\begin{bmatrix}
{}_1Y \\ {}_2Y \\ {}_3Y \\ \\ {}_{i-1}Y \\ {}_iY \\ {}_{i+1}Y \\ \vdots
\end{bmatrix}
\tag{9.10}
$$

The placement of the $(X_1 X_2 X_3 \ldots)$ sets needs to ensure that the matrix is invertible and that all rows are linearly independent. To do so a "cube" or "star" pattern, such as those in classic design of experiments, is often used. Figures 9.1 and 9.2 reveal possible patterns in a 2-DV and 3-DV application. The solid dots represent essential points for the model with either the interacting or non-interacting terms. The open circles are additional points if the interacting terms are used. Other patterns can be used, such as rotated cubes, circular, star, or even random patterns. For developing computer code, the simpler aligned-with-axis patterns are most common. In any case, the user must choose the range for each DV or some metric related to the exploration range.

Not all corner points are needed for the cross-product model. Any number of alternate patterns is used. However, the pattern must permit the linear algebra evaluation of the surrogate model coefficient values.

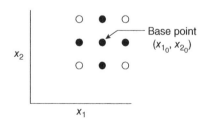

Figure 9.1 Illustrating a 2-D placement of points to evaluate the surrogate quadratic surface.

You need to determine the range of each $X_1 X_2 X_3 \ldots$ cube dimensions. Too small a range and the model can't understand the entire surface. Since it projects the local trend to the entire surface, this would be a concern in the initial iterative stages. However, too large a range of sampling and the model may not see important local features. This is an end-game concern.

But unless the surface is ideally quadratic and is identical to the model, the model will be wrong, and X^* calculated will be wrong; but, hopefully, X^* will be close to the optimum of the function.

If the surface is convex, successive quadratic will send X^* toward a maximum. If the surface model is developed beyond the inflection point from a minimum, SQ will move in the wrong direction, as illustrated in Figure 9.3.

Successive quadratic seeks the point at which the derivative of the surrogate model is zero, which could be a minimum, maximum, or saddle point. Accordingly, at the optimum check second derivatives to see what it has found. If all are second derivatives are greater than zero, then it has found a minimum; if all are less than zero, it found a maximum; if mixed, it found a saddle point. The N number of second derivatives of the non-interactive surrogate quadratic model is

$$\frac{\partial^2 y}{\partial x_i{}^2} = 2c_{ii} \qquad (9.11)$$

The additional $N(N-1)/2$ number of second derivative terms in the interacting surrogate model is

$$\frac{\partial^2 y}{\partial x_i \partial x_j} = c_{ij} \qquad (9.12)$$

With the analytical method of solution, the matrix must be invertible. If the surface has flat sections, then coefficients with zero values may prevent linear algebra from determining DV^* solutions.

The method presumes that the surface is reasonably approximated by the quadratic model. If the surface has discontinuities, flat spots, or stochastic behavior, then the SQ approach may not work.

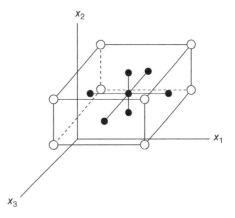

Figure 9.2 Illustrating a 3-D placement of points to evaluate the surrogate quadratic surface.

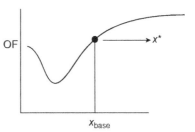

Figure 9.3 SQ diversion when outside of the inflection point.

9.2.2 SQ Pseudocode

Here is pseudocode for successive quadratic:

1) Select an initial trial solution, and evaluate OF.
2) Evaluate OF at sufficient surrounding points to determine model parameters. $\mathbf{X}p = \mathbf{y}$. Initially use the entire DV range for sampling. As the X^* values converge reduce the sampling range to the local region.
3) If $\mathbf{D}$ is positive definite continue. Else restart at Step 1.
4) Use $\mathbf{D}x^* = -\boldsymbol{b}$ to calculate the next trial solution, the new base point.
5) Check for convergence/stopping criteria. Stop or repeat from Step 2.

9.3 Newton–Raphson

The Newton–Raphson method of optimization is Newton's method of root finding applied to the gradient of the function. It is often termed Newton's method. It is a second-order procedure that presumes that the function is effectively modeled as a quadratic response to the DVs or equivalently that the gradient elements are essentially linearly dependent on the DV values. The objective is

$$\min_{\{x\}} J = f(x) \tag{9.13}$$

At the minimum, at x^*, the derivative of the function w.r.t. each DV is zero: $\partial f / \partial x_1 = 0$, $\partial f / \partial x_2 = 0$, … *and* $\partial f / \partial x_n = 0$.

For convenience define a g-vector to represent the gradient elements:

$$g = \nabla f = \begin{bmatrix} g_1 \\ g_2 \\ g_3 \\ \vdots \end{bmatrix} = \begin{bmatrix} \dfrac{\partial f(x_1, x_2, x_3)}{\partial x_1} \\[2ex] \dfrac{\partial f}{\partial x_2} \\[2ex] \vdots \end{bmatrix} \tag{9.14}$$

Recognize that g_i is a function of the local x_1, x_2, x_3 values.

The method is to determine the x-values that make all g-elements have a zero value. Following Newton's method of root finding, use a calculus total derivative or Taylor Series linear approximation for g:

$$g = \begin{bmatrix} g_{1_0} + \dfrac{\partial g_1}{\partial x_1}\bigg|_0 dx_1 + \dfrac{\partial g_1}{\partial x_2}\bigg|_0 dx_2 + \dfrac{\partial g_1}{\partial x_3}\bigg|_0 dx_3 + \cdots \\[3ex] g_{2_0} + \dfrac{\partial g_2}{\partial x_1}\bigg|_0 dx_1 + \dfrac{\partial g_2}{\partial x_2}\bigg|_0 dx_2 + \dfrac{\partial g_2}{\partial x_3}\bigg|_0 dx_3 + \cdots \\[3ex] \vdots \end{bmatrix} \tag{9.15}$$

Rearranging terms and representing the differentials with a finite difference,

$$g = \begin{bmatrix} g_{1_0} \\ g_{2_0} \\ g_{3_0} \\ \vdots \end{bmatrix} + \begin{bmatrix} \dfrac{\partial g_1}{\partial x_1}\bigg|_0 & \dfrac{\partial g_1}{\partial x_2}\bigg|_0 & \dfrac{\partial g_1}{\partial x_3}\bigg|_0 & \cdots \\[3ex] \dfrac{\partial g_2}{\partial x_1}\bigg|_0 & \dfrac{\partial g_2}{\partial x_2}\bigg|_0 & \dfrac{\partial g_2}{\partial x_3}\bigg|_0 & \cdots \\[3ex] \dfrac{\partial g_3}{\partial x_1}\bigg|_0 & \vdots & \vdots & \ddots \\[3ex] \vdots & & & \end{bmatrix} \begin{bmatrix} \Delta x_1 \\ \Delta x_2 \\ \Delta x_3 \\ \vdots \end{bmatrix} \tag{9.16}$$

At the minimum $g = 0$:

$$0 = g_0 + \begin{bmatrix} \dfrac{\partial g_1}{\partial x_1}\bigg|_0 & \dfrac{\partial g_1}{\partial x_2}\bigg|_0 & \dfrac{\partial g_1}{\partial x_3}\bigg|_0 & \cdots \\[3ex] \dfrac{\partial g_2}{\partial x_1}\bigg|_0 & \dfrac{\partial g_2}{\partial x_2}\bigg|_0 & \dfrac{\partial g_2}{\partial x_3}\bigg|_0 & \cdots \\[3ex] \dfrac{\partial g_3}{\partial x_1}\bigg|_0 & \vdots & \vdots & \ddots \\[3ex] \vdots & & & \end{bmatrix} \Delta x \tag{9.17}$$

Solve for Δx, the local estimate of where to move to reach the optimum. I'll first replace the g symbol with its derivative of the function:

$$0 = \nabla f|_0 + \begin{bmatrix} \dfrac{\partial}{\partial x_1}\dfrac{\partial f}{\partial x_1}\Big|_0 & \dfrac{\partial}{\partial x_2}\dfrac{\partial f}{\partial x_1}\Big|_0 & \dfrac{\partial}{\partial x_3}\dfrac{\partial f}{\partial x_1}\Big|_0 & \cdots \\[2ex] \dfrac{\partial}{\partial x_1}\dfrac{\partial f}{\partial x_2}\Big|_0 & \dfrac{\partial}{\partial x_2}\dfrac{\partial f}{\partial x_2}\Big|_0 & \dfrac{\partial}{\partial x_3}\dfrac{\partial f}{\partial x_2}\Big|_0 & \cdots \\[2ex] \dfrac{\partial}{\partial x_1}\dfrac{\partial g_3}{\partial x_3}\Big|_0 & \vdots & \vdots & \\[2ex] & & & \\ & \vdots & & \end{bmatrix} \Delta x \tag{9.18}$$

The g-vector is the familiar gradient, and regardless of the number of DVs, the matrix in Equation (9.18) becomes a 2-D array of second partial derivatives, termed the Hessian matrix. Along the main diagonal the elements are the classic second derivative w.r.t. one variable; however off-diagonal elements are the mixed second derivatives. The matrix is square, and its dimension, the number of rows and columns, is that of the number of DVs. Regardless of the number of DVs, the Hessian matrix is always two-dimensional.

As with the gradient operator of first partial derivatives, which needs to have the function it operates on revealed, the Hessian is an operator of second partial derivatives and also needs to have the operand revealed. Notationally, the statement is presented as

$$0 = \nabla f|_0 + Hf|_0 \Delta x \tag{9.19}$$

This should be recognized as a standard linear algebra application. If the values of $\partial f/\partial x_i|_0$ and $(\partial/\partial x_i)(\partial f/\partial x_j)\big|_0$ have been determined, then these are simply coefficient values in the equation, which can be solved for Δx, and the next trial solution will be

$$x_{k+1} = x_k + \Delta x_k \tag{9.20}$$

Note that if the function is continuously differentiable, then $(\partial/\partial x_2)(\partial f/\partial x_1)|_0 = (\partial/\partial x_1)(\partial f/\partial x_2)|_0$ and the matrix is symmetric about the main diagonal. Then instead of needing to solve for N^2 Hessian elements, only $N + N(N-1)/2 = (N^2 + N)/2$ elements are needed. If analytical expressions are available, then each Hessian evaluation requires that many functions. However, likely one will use numerical procedures to handle the generally intractable derivatives. Central finite difference approaches for the homogeneous second derivatives require two additional function evaluations, each. And the mixed second-order derivatives require four each. There are points in common, so perhaps each Hessian evaluation requires a minimum of $(N^2 + N)$ additional function evaluations.

The user (or the algorithm) needs to define the range of deviations used to determine the derivative values. If too large a deviation, the gradients and Hessian elements could misrepresent the local surface near the optimum. Too small a deviation could lead to truncation error. Perhaps the deviations for evaluating derivatives could be an order of magnitude smaller than either DV convergence criteria, or the last DV changes between iterations.

Equation (9.19) can be rearranged to a standard linear algebra structure to represent the procedure. To solve for the new trial solution, you would use a linear algebra procedure such as Gaussian elimination:

$$\mathbf{H}f|_0 \cdot \Delta x = -\nabla f|_0 \tag{9.21}$$

Note that the RHS is the negative gradient, the direction of steepest descent.

We can represent the method by explicitly solving for Δx:

$$\Delta x = -\mathbf{H}f|_0^{-1} \cdot \nabla f|_0 \tag{9.22}$$

Alternately

$$x = x_0 - \mathbf{H}f|_0^{-1} \cdot \nabla f|_0 \tag{9.23}$$

However, although this seems like an explicit calculation, inversion of the Hessian and the dot product with the gradient are linear algebra procedures.

As a comparison to NR, ISD solves for $\Delta x = -\alpha \nabla f|_0$. Contrasting this to Equation (9.22) reveals that the inverse of the Hessian plays a role in tempering the negative gradient. The NR $\mathbf{H}$ has a role similar to α in ISD. However, $\mathbf{H}$ is not a single scalar, but a matrix of elements that use second-order information to individually adjust each Δx element.

Newton-type methods can proceed slowly or can oscillate about the DV* (in this case the root of the gradient, the DV* value). So, it is permissible to temper or accelerate the move magnitude with a factor, α (not the same as the univariate search alpha):

$$x = x_0 - \alpha \mathbf{H}f|_0^{-1} \cdot \nabla f|_0 \tag{9.24}$$

Conventionally, $\alpha = 1$. However, if the surface slope is not linear, then sequential trial solutions could sequentially slowly approach the optimum, or they could overshoot and oscillate about the minimum. The value of α could be adjusted upon recognition of the pattern in the locus of past trial solutions. If oscillating, reduce α to 0.5. If moving in the generally same direction, increase α to about 2.

This is a recursion relation. From the initial trial solution, estimate where the optimum is. Move there, and redo:

$$x_{k+1} = x_k - \alpha \mathbf{H}f|_k^{-1} \cdot \nabla f|_k \tag{9.25}$$

9.3.1 NR Pseudocode

1) Choose an attenuation factor, alpha.
2) Initialize with a trial solution, an initial value for each DV.
3) Evaluate the gradient elements at the trial solution.
4) Evaluate the Hessian elements at the trial solution.
5) Calculate the next trial solution by $x_{k+1} = x_k - \alpha \mathbf{H}f|_k^{-1} \cdot \nabla f|_k$.
6) If x_k values appear to be oscillating, reduce the value of α. Alternately, if x_k values appear to be continuing in the same direction, increase the value of α.
7) Determine the OF value at x_{k+1}.
8) Test for convergence.
9) If not converged return to Step 3, else exit.

Example 1 Makes NR Oscillate

This is a contrivance, a function that represents an extreme condition that makes NR oscillate. At the same time it is entertaining as well as instructive about an issue with NR. It is Function 32 in the 2-D Optimization Examples file, a sum of two identical functions. As a univariate search, it also makes NR oscillate:

$$f(x) = b + \frac{a}{2}x^2 + (ac + 1)x + \frac{2}{s}\ln\left[1 + e^{s(x-c)}\right] \tag{9.26}$$

The VBA code is

```
ca32 = 1
cb33 = 4
cc32 = 5
cs32 = 10
fofx321 = cb32 + (ca32 / 2) * x1 ^ 2 - (ca32 * cc32 + 1) * x1 + (2 / cs32) * Log(1 +
Exp(cs32 * (x1 - cc32)))
fofx322 = cb32 + (ca32 / 2) * x2 ^ 2 - (ca32 * cc32 + 1) * x2 + (2 / cs32) * Log(1 +
Exp(cs32 * (x2 - cc32)))
f_of_x = fofx321 + fofx322
f_of_x = 10 * (f_of_x + 37) / 35
```

In a 3-D view of Figure 9.4, the function seems relatively well behaved.

But regardless of starting position, NR will oscillate between solutions. In Figure 9.5 the initial trial solution was near the top center, and NR directed the next trial solution to about (4.5, 4), which sent it to about (6, 6), which sent it to about (4, 4) and to then (6, 6), and the oscillations continued.

9.3.2 Attenuate NR

The attenuation factor, α, in Equation (9.25) can be adjusted to either accelerate convergence or temper the oscillations that NR can acquire.

Figure 9.6a reveals a univariate function that seems well behaved, with an easy-to-find minimum, and Figure 9.6b shows its derivative. Recall that NR optimization is equivalent to Newton's method of root finding on the derivative. From the leftmost x-values, the function value drops, and its slope is negative, but the slope progressively diminishes to zero. The left side of Figure 9.6b reveals this. The right side of the function progressively rises, and as the derivative indicates, the slope is positive, but approaching a constant value. Note also that the derivative graph is not linear, but is curved, and recall that Newton's method extrapolates the local slope from a trial solution to project the next trial solution on the zero of the y-axis. Trial solutions far to the left make incremental changes and creep toward the root; this is labeled as Shape A on the figure. By contrast, a TS on the right, in Shape B, jumps far to the left.

Figure 9.7 reveals four possible combinations of Shapes A and B. If B–B combination such as characteristic of pH optimization applications, the solution oscillates. Here, an attenuation factor of 0.5 might be desired to temper the moves. By contrast, if A–A combination, regardless of the TS, the solution creeps up on the solution. Here, an attenuation factor of 2 might be desired to accelerate the moves.

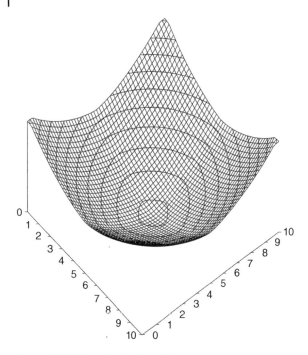

Figure 9.4 Makes Newton's oscillate.

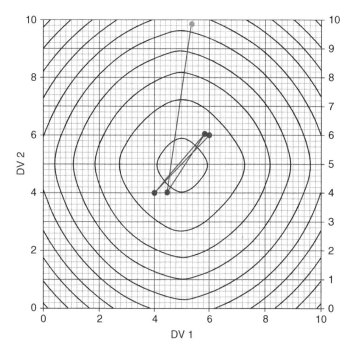

Figure 9.5 An NR path on Function 32.

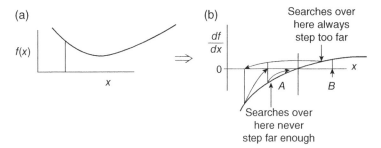

Figure 9.6 (a) A function and (b) its derivative.

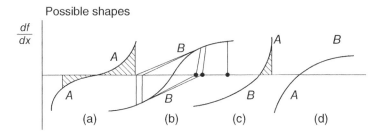

Figure 9.7 Combinations of derivative shapes near the root: (a) A–A, (b) B–B, (c) B–A, and (d) A–B.

A rule that I have found useful in tempering NR is to look at the incremental changes in the DV to indicate what the shape is:

If $\Delta X \cdot \Delta X_{\text{past}} > 0$, you are likely in an "A" shape. Make alpha larger.
If $\Delta X \cdot \Delta X_{\text{past}} < 0$, you are likely in a "B" shape. Make alpha smaller.

For a multivariable application, this would need to be done for each DV:

$$\mathbf{H}\Delta X = -(\mathbf{I}A)\nabla f \tag{9.27}$$

If a 3-DV application

$$A = \begin{bmatrix} a_1 \\ a_2 \\ a_3 \end{bmatrix} \quad \mathbf{I} = \begin{bmatrix} 1 & 0 & 0 \\ 0 & 1 & 0 \\ 0 & 0 & 1 \end{bmatrix} \tag{9.28}$$

Perhaps 0.25 is as small a partial step as is necessary, but perhaps not. Figure 9.8 illustrates a somewhat extreme A–A shape.

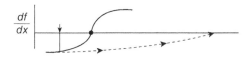

Figure 9.8 An extreme A shape.

Perhaps 4 is as big as an acceleration factor needs to be, but perhaps not. Figure 9.9 illustrates a somewhat extreme B shape.

Rather than moving to the extreme alpha value based on one iteration, I prefer to asymptotically approach the limits. Here is a simple rule to move alpha toward 0.25 or 4.

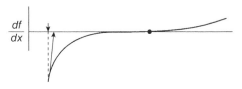

Figure 9.9 An extreme B shape.

```
If ΔX• ΔX_old > 0 Then    'Shape A, approaching slowly
      If α < 1 Then   'formerly thought to be in Shape B
            α = 1  'return to the nominal α -value
      Else            'still in Shape A
            α = 0.2 * 4 + 0.8 * α   'incrementally move toward α =4
      End if
Else                      'Shape B, oscillating
      If α > 1 Then   'formerly thought to be in Shape A
            α = 1  'return to the nominal α -value
      Else            'still in Shape B
            α = 0.2 * 0.25 + 0.8 * α 'incrementally move toward α =0.25
      End if
End if
```

Use this rule for each DV. (What other rules make sense?)

9.3.3 Quasi-Newton

An issue with the Hessian is the calculation of the second derivatives. To do it with analytical expressions, you, the user, must take the mixed partial derivatives of the functions, which leads to a high chance of error. Then the computer must calculate the value for each Hessian element, which may be as much computational burden as calculating the OF. And, to do it analytically, there must be an analytical expression for the OF, which would not be possible if determining this OF value requires a procedure for root finding, Monte Carlo simulation, or experimental data.

To determine Hessian element values numerically requires additional OF evaluations. For an N-dimensional application, the Hessian is an $N \times N$ matrix. There are N homogeneous second derivatives along the main diagonal and N^2-N off-diagonal mixed second derivatives. But if the OF is continuous, then the order of the derivative is inconsequential, the symmetric elements are identical, and there are only half as many $N(N-1)/2$ to evaluate.

If central difference for a mixed derivative, the numerical estimate is

$$\frac{\partial^2 f}{\partial x_i \partial x_j} \cong \frac{f_{+,+} - f_{+,-} - f_{-,+} + f_{-,-}}{4\Delta x_i \Delta x_j} \tag{9.29}$$

Here $f_{+,+}$ indicates that the OF value is evaluated at the base case perturbed by the $+\Delta x_i$ and $+\Delta x_j$ values.

The base case, f_0, will have been calculated as the trial solution. If central differences are used for the Jacobean (gradient) elements, then the derivative is estimated by

$$\frac{\partial f}{\partial x_i} \cong \frac{f_+ - f_-}{2\Delta x_i} \tag{9.30}$$

And then the homogeneous second derivatives can be estimated by

$$\frac{\partial^2 f}{\partial x_i^2} \cong \frac{f_+ - 2f_0 + f_-}{\Delta x_i^2} \tag{9.31}$$

So the base case requires 1 OF evaluation, and the gradient elements require $2N$ more. These OF values can provide the main diagonal Hessian elements, but the off-diagonal elements require 4 OF evaluations for each of the $N(N-1)/2$ elements.

In total, an N-dimensional Newton-type iteration requires $1 + 2N + 4N(N-1)/2 = 1 + 2N^2$ OF evaluations.

It would be nice to reduce the computational burden.

The secant approach to Newton's method of root finding reduces computational burden by using the difference between past and present iterations (possibly not very small incremental perturbations) to estimate the slope of the derivative:

$$g = \frac{df}{dx} \cong \frac{f_k - f_{k-1}}{x_k - x_{k-1}} \tag{9.32}$$

where k is the iteration counter.

The advantage is that the f-values are all known from past searches. It does not need new function values to estimate the derivative. The disadvantage is the coarse estimation of the derivative. But how important is precision or fidelity, when the basis for the method is the supposition that the surface is a simple quadratic?

Applying this sequential secant approach to the Hessian element changes w.r.t. the ith DV, first with a backward difference estimate based on iteration-to-iteration changes:

$$\frac{\partial^2 f}{\partial x_i \partial x_j} \cong \frac{\partial f / \partial x_j|_k - \partial f / \partial x_j|_{k-1}}{x_i|_k - x_i|_{k-1}} \tag{9.33}$$

Recognize that $\partial f / \partial x_j|_k$ is $g_j|_k$, the gradient element for j already calculated for the kth iteration. So don't recalculate the f derivative values. Use the known gradient elements:

$$\frac{\partial^2 f}{\partial x_i \partial x_j} \cong \frac{\partial f / \partial x_j|_k - \partial f / \partial x_j|_{k-1}}{x_i|_k - x_i|_{k-1}} \cong \frac{g_j|_k - g_j|_{k-1}}{x_i|_k - x_i|_{k-1}} \tag{9.34}$$

Or it could be based on the jth DV, also as a backward difference and the already calculated gradient elements for x_i:

$$\frac{\partial^2 f}{\partial x_i \partial x_j} \cong \frac{\partial f / \partial x_i|_k - \partial f / \partial x_i|_{k-1}}{x_j|_k - x_j|_{k-1}} = \frac{g_i|_k - g_i|_{k-1}}{x_j|_k - x_j|_{k-1}} \tag{9.35}$$

Since the expectation is that the **H** elements are symmetric, use an average of both estimates:

$$\frac{\partial^2 f}{\partial x_i \partial x_j} \cong \frac{1}{2} \left[\frac{g_j|_k - g_j|_{k-1}}{x_i|_k - x_i|_{k-1}} + \frac{g_i|_k - g_i|_{k-1}}{x_j|_k - x_j|_{k-1}} \right] \tag{9.36}$$

In my 2D Optimization Examples program, the Jacobean elements are "a" and "b." The past changes in $x1$ and $x2$ are "$dx1$" and "$dx2$." So the one quasi-Newton Hessian element is

$$\frac{\partial^2 f}{\partial x_1 \partial x_2} \cong \frac{1}{2} \left[\frac{b_{\text{new}} - b_{\text{old}}}{dx1} + \frac{a_{\text{new}} - a_{\text{old}}}{dx2} \right] \tag{9.37}$$

For the Hessian element values, this uses only the numerical elements from the current Jacobean (gradient) and those from the prior trial solution evaluation. This eliminates the need for $4N(N-1)/2$

function evaluations. Only $1 + 2N$ function evaluations are required at each iteration, reducing the iteration-to-iteration NOFE by a ratio of $(1 + 2N)/(1 + 2N^2)$, which approaches a reduction in work by a factor of $1/N$ for large dimension. This can be significant when the OF evaluation is complicated.

Some commercial packages only update the Hessian elements every second or third iteration. This saves additional computational work. If the trial solution is near the optimum, then the Hessian elements should not be changing and would not need frequent updates. If the trial solution is far from the optimum, the weighting should be on the incremental steepest descent portion of the algorithm, and the **H** values are not that important.

Are these good practices? A person doing optimization would consider the desirables and undesirables associated with the method and choose the best method.

It makes the Hessian elements one step more approximate, but Newton's approach is a second-order approximation that might not be wholly valid. It could make the new trial solution estimates better or worse!

It makes the algorithm one step more obscure.

This quasi-Newton approach to the Hessian approximation leads to less precise estimates of the second derivative and actually adds iterations. However, the reduction in NOFE per iteration more than makes up for the precision, and the actual NOFE to the optimum is reduced.

So, it is less precise and one step more obscure, but lower computational burden.

I think the benefit in computational burden and speed is worth the additional approximation and human confusion.

9.4 Perspective on CSLS, ISD, SQ, and NR

Cauchy's sequential line search method is a line search in the direction of steepest local descent. Neither where to go when you get way out there on the line nor how far to go is defined in the CSLS method.

Incremental steepest descent reevaluates the surface slope after each incremental downhill step. It does not stick to the original direction.

Although both CSLS and ISD have the conflict of units, when the DVs have disparate units, NR and SQ are dimensionally consistent.

Both successive quadratic (SQ) and Newton–Raphson (NR) methods jump to the hoped-for minimum. They do not incrementally search along a path. Further, the jump-to spot might not be on the local steepest descent line. Both NR and SQ use local information (near the trial solution) to characterize the surface and then use that characterization to determine the optimum.

CSLS "fumbles around" with sequential line searches when near an optimum (when the surface is not a simple quadratic). But CSLS and ISD point in the right direction when far away. They always go downhill. By contrast, SQ and NR will be attracted to a maximum or saddle and diverge from the minimum when initialized beyond an inflection point.

The NR method can make excessive and poor Δx jumps when $\det(\mathbf{H})\tilde{\,}0$ when $dg/dx\tilde{\,}0$, which may frequently happen far from an optimum. But it has very rapid convergence near to an optimum.

NR method jumps out of bounds when outside of the inflection region about the minimum. Further, depending on the surface features, NR may oscillate across the optimum or approach it very slowly (as Newton's method does in root finding).

Like successive quadratic, the Hessian matrix must be positive definite for Newton's to provide a rational new trial solution.

NR method requires first and second derivatives and inverting the **H** matrix for calculating Δx.

In NR, the required number of function evaluations that are needed to calculate all Jacobean (gradient, first derivative) and Hessian (second derivative) matrix elements for an N-dimensional search may be excessive. And NR presumes that second derivatives exist.

In incremental steepest descent, $\Delta x = -\alpha \nabla f_0$, the right scalar "$\alpha$" value is not known, but $-\nabla f$ points to steepest descent. In Newton–Raphson, $\Delta x = -\mathbf{H}_{f_0}^{-1} \cdot \nabla f_0$, the $\mathbf{H}^{-1}$ does two things: it adjusts steepest descent direction to aim at minimum (assuming a quadratic order function), and it determines the right step size, Δs, to jump to min.

Either SQ or NR will want to jump to a new position; if it is in an infeasible region, as illustrated in

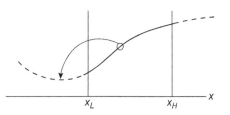

Figure 9.10 NR or SQ encountering a constraint.

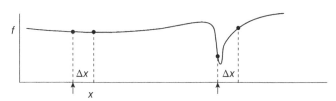

Figure 9.11 Illustrating too large and too small increments for numerical derivative evaluation.

Figure 9.10, what then? A redo will calculate the same spot. One could include heuristic rule overrides to the calculated value when such is encountered.

9.5   Choosing Step Size for Numerical Estimate of Derivatives

How to determine the value of ΔX for calculating numerical derivatives? Figure 9.11 illustrates two situations—one where the function is relatively flat in the vicinity of the trial solution and the other where the trial solution is in the vicinity of a steep optimum.

The value of ΔX is the same in both cases in the illustration. In the left-hand example, Δx is too small because Δf is so small; truncation error may be significant. For example, if $f_1 = 5.00000001$ and $f_2 = 5.00000002$ (differing in the ninth digit), but the values are truncated to five digits, then $f_1 = 5.0000$ and $f_2 = 5.0000$, making the difference appear to be zero $\Delta f = 0$.

By contrast, in the right-hand example, Δx is too large, because the numerical approximation to slope has the wrong sign.

Although these were illustrated with a forward finite difference, central or backward differences can give the wrong approximations also. However, the central finite difference is the best near an optimum, as illustrated in Figure 9.12, although it doubles the number of function evaluations in estimating the derivative.

Knowledge of the function response to each X_i is required for a choice of ΔX_i that is appropriate for evaluating ∇f and $\mathbf{H}f$ elements.

However, if you knew what the "surface" looked like, you would know where the optimum was and would not need an optimizer.

I would offer this as a solution to determine numerical derivatives.

Start with $\Delta X_i' = O(0.01)$ (on the order of 0.01), which means that the unscaled $\Delta X_i = 0.01$ range on the ith DV. Also use a forward difference approximation. At the "end game," close to the optimum,

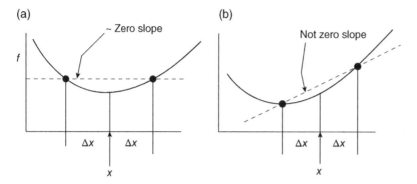

Figure 9.12 Illustrating the central difference numerical approximation to the local slope: (a) near the optimum and (b) away from the optimum.

select ΔX for finite difference $\approx O(0.1)$ (on the order of a tenth) of the ΔX trial solution, and use a central difference approximation. This keeps estimate of ∇ and $\mathbf{H}$ on scale with jumps toward optimum. The end game (near the optimum) might be recognized when the ΔDV values are small (nearing convergence).

I think there should also be an override in case the optimizer jump is very large or very small:

```
dX_i = (0.1) ΔX_i
If dX_i < ε then dX_i = ε
If dX_i > 0.01   Range_i then dX_i = 0.01 R_i
```

But the choices for ε and 0.01 would require some understanding of the problem and the effect of numerical truncation error on estimates for ∇ and $\mathbf{H}$ elements.

Probably you have no control over what choices the software programmer decided to use in the algorithm embodiment to calculate derivative and second derivative values.

9.6 Takeaway

SQ and NR are powerful jump-near-to-the-minimum algorithms when the surface is quadratic-like and continuous and the trial solution is in the vicinity of the minimum. Choices within the algorithm such as where to place points for surface modeling and how to solve the linear algebra procedures are well developed. However, they can provide absurd jump-to locations (entertainingly absurd if the application is for study and has no career or personal relevance) and often need some tempering overrides. Further, they have computational burden in numerical estimates of the surrogate model or the gradient and Hessian elements. Quasi-Newton seeks alternate estimations of the Hessian element values to alleviate the burden.

9.7 Exercises

1   What metrics would you use to compare optimizers? (Bloom's Level 4.) What experiments/investigations would you use to generate the metrics to have a comprehensive and unbiased set of data? (Bloom's Level 5.) How would you prioritize the diverse metrics to make an evaluation? (Bloom's Level 6.)

2 Show that NR moves to a new point, but not necessarily in the direction of local steepest descent.

3 Show a standard numerical formula to provide a value for the mixed second derivative $\partial^2 f / \partial x_1 \partial x_2$.

4 The DVs are x_1 and x_2, and the OF is $J = 5x_1 + 4x_2 - 3x_1 x_2 + 2x_2^2$. What is the Hessian?

5 Write your own code to implement a 2-D successive quadratic optimization in the Excel VBA program optimization applications. Use this function (Function #9, or some other).

```
z   = 0.02 * (((x - 8) ^ 2 + (y - 6) ^ 2) +
      + 1.5 * ((x - 2 - 0.001 * y ^ 3) * (y - 4 + 0.001 * x ^ 3)) ^ 2 +
      - 1500 * Exp(-0.5 * ((x - 9) ^ 2 + (y - 9) ^ 2))) +
      - 0.25 * Exp(-2 * ((x - 3) ^ 2 + (y - 6) ^ 2))
```

For a 2-D quadratic surface, the local quadratic model will be z model $= a + b * x + c * x \char`\^2 + d * y + e * y \char`\^2 + f * x * y$. Since there are six coefficients in this local model of the surface, you will need six data points to determine the coefficient values. Use your trial solution as the central data point and five surrounding data points, perhaps placed in a pentagon shape. Use linear algebra (perhaps Gaussian elimination) to determine the model coefficient values. To determine the next trial solution, take the derivative of the local surface model w.r.t. x and y, set the derivatives to zero, and solve two linear equations for the next trial solutions for x and y. Repeat this iterative procedure until convergence. Note: Depending on the surface shape in the vicinity of your trial solution, successive quadratic can send the next trial solution outside of the limits on the DVs. So, if it does, restart at a new randomly chosen initial trial solution. Note: As the algorithm approaches the optimum value, the size of the exploratory pattern of data points around the trial solution used to determine the local surface model should shrink. Perhaps make the characteristic length of the pentagon proportional to the last change in trial solution location.

A Explain: the 2-D SQ logic, your convergence criterion (or criteria), your override approach if DV is outside of the range, and your choice of pattern shrinking.
B Show your derivation of the equations used in determining local model coefficients.
C Show your derivation of the equations used in determining the next trial solution.
D Show your code that implements this SQ algorithm.
E Provide results that reveal your code works as desired.

6 Provide a flowchart and supporting text explanation for SQ.

7 Provide a flowchart and supporting text explanation for NR.

8 Use the 2-D Optimization Examples software to compare NR, CSLS, and ISD on a variety of functions. Clearly state what criteria you will use to judge optimizer performance, and explain how your choice of functions and associated algorithm initialization leads to a comprehensive unbiased evaluation.

10

Gradient-Based Optimizer Solutions

LM, RLM, CG, BFGS, RG, and GRG

10.1 Introduction

Features of a particular application (constraints, inflections, steep valleys) provide difficulty for SQ and NR, which are second-order model-based optimizers. This chapter covers several solutions that typify a myriad of similar adaptations of the basic incremental steepest descent and Newton-type approaches. The topics of this chapter are Levenberg–Marquardt (LM), conjugate gradient (CG), Broyden–Fletcher–Goldfarb–Shanno (BFGS), and generalized reduced gradient (GRG). LM blends ISD and NR to ensure it moves downhill, not to a saddle, maximum, or jump to absurd locations. CG and BFGS are for unconstrained applications with CG being a variant of Cauchy's sequential line search method and BFGS a quasi-Newton approach. GRG is for handling constraints.

Although these are established and intellectually clever solutions, they are limited to certain application classes and add algorithm complexity. Although they are often considered best-in-class solutions, I favor the direct search approaches of Chapter 11 for simplicity, generality, and effectiveness.

10.2 Levenberg–Marquardt (LM)

The concept is that for the early iterations, use incremental steepest descent. This ensures that the TS moves in a downhill sequence. As you approach the optimum and are more likely to be in a locally quadratic convex region of the function, switch to Newton–Raphson. The switch will be gradual, not abrupt. As a note there are many versions of the incremental gradient search and Newton's (for instance, Gauss–Newton). So, there are many versions of the LM concept. Here is one.

To develop the LM method, structure the ISD relation similarly to NR. In NR

$$[\mathbf{H}_k(f)]\cdot\Delta\mathbf{x}_k = -\nabla f_k \tag{10.1}$$

Divide both sides of the ISD relation by the step-size parameter, and show ISD as

$$\left[\frac{1}{\alpha}\right]\cdot\Delta\mathbf{x}_k = [\lambda_k\mathbf{I}]\cdot\Delta\mathbf{x}_k = -\nabla f_k \tag{10.2}$$

where $\lambda = 1/\alpha$. For both representations, the subscript k indicates iteration. Adding the subscript k to λ_k reveals that the value of the ISD step-size controlling parameter will change with iteration. The scalar λ times the identity matrix does not alter its function, but does present the $\Delta\mathbf{x}_k$ coefficient

Engineering Optimization: Applications, Methods, and Analysis, First Edition. R. Russell Rhinehart.
© 2018 R. Russell Rhinehart. Published 2018 by John Wiley & Sons Ltd.
Companion website: www.wiley.com/go/rhinehart/engineeringoptimization

in a matrix notation that parallels NR. Except for the term in brackets and except for the coefficient values in $\mathbf{H}_k(f)$ and $\lambda_k\mathbf{I}$, the NR and ISD equations are identical.

Add the bracketed coefficients and the method can be stated either as

$$[\mathbf{H}_k(f) + \lambda_k\mathbf{I}]\cdot\Delta\mathbf{x}_k = -\nabla f_k \tag{10.3}$$

or as

$$\Delta\mathbf{x}_k = -[\mathbf{H}_k(f) + \lambda_k\mathbf{I}]^{-1}\cdot\nabla f_k \tag{10.4}$$

The aforementioned are just linear algebra representations of a combination of ISD and Newton's. But the $\Delta\mathbf{x}_k$ solution needs numerical procedures to evaluate the Hessian elements, and then linear algebra procedures to solve for the delta-x element values.

The value of parameter λ defines a smooth, gradual switch between ISD and NR. If the value of λ is much larger than those of the H elements, then the rule is essentially ISD with the step-size parameter $\alpha = 1/\lambda$. Alternately, if λ has a small value relative to the H elements, then the rule is essentially NR. The method starts with a large value of λ and incrementally reduces it at each iteration, shifting the search from ISD initially to NR at the end game.

Wikipedia reports this as Levenberg's original 1955 idea, which seemed to have gotten overlooked and brought back to visibility by Marquardt in 1963. I find it remarkable that it took humans so long to recognize this blend of approaches. But, in retrospect, I imagine it was the power and availability of digital computers that made this invention even feasible.

In a 2-DV application, representing the Hessian elements as coefficients a, b, c, and d and the Jacobean elements as e and f, the method is

$$\begin{vmatrix} a+\lambda & c \\ b & d+\lambda \end{vmatrix}\begin{vmatrix} \Delta x_1 \\ \Delta x_2 \end{vmatrix} = -\begin{vmatrix} e \\ f \end{vmatrix} \tag{10.5}$$

which is easily solved analytically in 2-D as

$$\Delta x_1 = \frac{(d+\lambda)e - cf}{(a+\lambda)(d+\lambda) - bc} \tag{10.6}$$

and

$$\Delta x_2 = \frac{(a+\lambda)f - be}{(a+\lambda)(d+\lambda) - bc} \tag{10.7}$$

In this method, start with a very large value of λ. But what is large? A conventional value is $\lambda \sim 10^4$. Then if the Hessian element values are small by comparison, $\mathbf{H} + 10^4\mathbf{I} \sim 10^4\mathbf{I}$ and

$$\Delta\mathbf{x} \sim -10^{-4}\mathbf{I}^{-1}\nabla f = -10^{-4}\nabla f = -\frac{1}{\lambda}\nabla f \tag{10.8}$$

This is incremental steepest descent, but with a very small step size. Maybe $\alpha = 0.0001$ is small. However, what is classified as small depends on the units used in OF and DV values and the steepness of the surface.

Here $\alpha = 1/\lambda$; so effectively, the initial ISD steps size is $\Delta s = (1/\lambda)\sqrt{\sum(df/dx_i)^2}$.

As information develops, increase or decrease λ as appropriate. If $f_{\text{new}} < f_{\text{old}}$, you moved in the right direction, with confidence shift toward NR, and set $\lambda_{k+1} = (1/2)\lambda_k$, and then continue. Otherwise step size was too big, or the Newton influence was too much. Then set $\lambda_{k+1} = 2\lambda_k$, and repeat from the previous best $x = x_{\text{old}}$. There are many variations on the heuristic rules to adjust λ. Many work with equivalent functionality.

When λ is small, the other extreme after a sequence of successful trial solutions is

$$\mathbf{H} + \lambda \mathbf{I} \sim \mathbf{H} \tag{10.9}$$

and

$$\Delta x \cong -\mathbf{H}_f^{-1} \nabla f \tag{10.10}$$

which is the Newton–Raphson technique.

Levenberg–Marquardt is one of the most accepted, widely used, and balanced (efficient and robust) optimizers.

10.2.1 LM VBA Code for a 2-DV Case

```
Sub LM()
' Levenberg-Marquardt blend of Newton-Raphson and Incremental Gradient
' Originally coded by Mazdak Shokrian, Fall 2011, revised by RRR since
        If i = 0 Then
            constraint = "Not Assessed"
            Do Until constraint = "OK"    'Find a feasible initial trial
solution for single player mathods
                x1o = 10 * Rnd()            'Randomize initial DV1 value
                x2o = 10 * Rnd()            'Randomize initial DV2 value
                OF_Value_Old = f_of_x(Fchoice, x1o, x2o) 'Determine the inital
function value
            Loop
            Worksheets("Path_Data").Cells(3, 1) = i
            Worksheets("Path_Data").Cells(3, 2) = x1o
            Worksheets("Path_Data").Cells(3, 3) = x2o
            OF_Value_Old = f_of_x(Fchoice, x1o, x2o)
            Cells(33, 12) = OF_Value_Old
            LMlambda = 100                  '10,000 was original recommendation,
but this makes initial step size so small that convergence is supposed.
        Else
            a = LMlambda + h11(Fchoice, x1o, x2o)    'Tempered Hessian element,
second partial of OF_Value w.r.t. x1 after x1
            b = h12(Fchoice, x1o, x2o)
            c = h21(Fchoice, x1o, x2o)      'Hessian element, second partial of
OF_Value w.r.t. x1 after x2
            d = LMlambda + h22(Fchoice, x1o, x2o)
            e = -g1(Fchoice, x1o, x2o)      'negative Gradient element, first
partial of OF_Value w.r.t. x1
            f = -g2(Fchoice, x1o, x2o)
            x1 = x1o + (d * e - f * b) / (a * d - b * c)  'Newton's jump to place
            x2 = x2o + (a * f - c * e) / (a * d - b * c)
            dx1 = x1 - x1o
```

```
          dx2 = x2 - x2o
          OF_Value = f_of_x(Fchoice, x1, x2)
          If constraint = "OK" And OF_Value < OF_Value_Old Then
              x1o = x1
              x2o = x2
              OF_Value_Old = OF_Value
              LMlambda = LMlambda / 2
          Else
              LMlambda = 2 * LMlambda
          End If
      End If
End Sub
```

10.2.2 Modified LM (RLM)

The Levenberg–Marquardt method is not perfect. The choice of $\lambda_{k=1} = 10^4$ is scale dependent. It may be too small if **H** elements are $O(10^6)$ such as 4.2×10^8 ft/h^2 being used for 32.2 ft/s^2. Or it may be orders of magnitude too large, if, for instance, choice of units produce an identical value of 6.1×10^{-3} miles/s^2.

If λ is excessively large, initial steps in the incremental steepest descent search will be excessively small, and downhill progress to the minimum will be slow. The search might even meet the convergence criteria on the first TS and stop. By contrast, if λ is not large enough, then Newton's approach will have too strong an initial influence, which could lead the search toward maxima, saddles, or way-off places.

The choice of a right initial value for λ in LM depends on magnitude of the second derivatives, which are not easily seen.

Here is a possible solution to those issues, a modification to the Levenberg–Marquardt approach. Again, recognize the mathematical parallels in the incremental steepest descent and Newton–Raphson structures:

$$\left\{ \begin{array}{l} \Delta x = -\alpha \nabla f \\ \Delta x = -\mathbf{H}^{-1} \cdot \nabla f \end{array} \right\} \tag{10.11}$$

As before, formulate both in a parallel manner:

$$\left\{ \begin{array}{l} \dfrac{1}{\alpha} \mathbf{I} \Delta x = -\nabla f \\ \mathbf{H} \Delta x = -\nabla f \end{array} \right\} \tag{10.12}$$

Here $\alpha = \Delta s / \sqrt{\sum (\partial f / \partial x_i)^2}$ where Δs is step distance along line of steepest descent.

Weight the incremental gradient (ig) by λ (this is not the LM lambda) and Newton–Raphson (nr) by the complementary value $(1 - \lambda)$ where $0 \leq \lambda \leq 1$. Multiply both sides of the incremental gradient equation by λ and the Newton–Raphson equation by $(1 - \lambda)$:

$$\lambda \frac{1}{\alpha} \mathbf{I} \Delta x = -\lambda \nabla f \tag{10.13a}$$

$$(1-\lambda)\mathbf{H}\,\Delta\boldsymbol{x} = -(1-\lambda)\,\nabla f \tag{10.13b}$$

Sum the two equations, and the RHS becomes the negative gradient

$$\left((1-\lambda_k)\mathbf{H}_k + \lambda_k \frac{1}{\alpha_k}\mathbf{I}\right)\Delta\boldsymbol{x}_{k+1} = -\nabla f_k \tag{10.14}$$

Although this was presented with the ISD coefficient α_k, I like using ΔS_k because I can relate to the composite DV step size and can select an initial value for ΔS_0 that is compatible with the convergence threshold. Determine the value for α_k by $\alpha_k = \Delta S_k / \sqrt{\sum (\partial f / \partial x_i)^2}$:

$$\left((1-\lambda_k)\mathbf{H}_k + \lambda_k \frac{\sqrt{\sum (\partial f / \partial x_i)^2}}{\Delta S_k}\mathbf{I}\right)\Delta\boldsymbol{x}_{k+1} = -\nabla f_k \tag{10.15}$$

Now there are two algorithm coefficients, (α, λ) or equivalently $(\Delta S, \lambda)$. The iteration subscript acknowledges that their values will change with iteration, and we need a logic, rule, to guide the evolution of values. I explored various rule versions and like the one that follows as best.

Use Equation (10.15), ignr, to determine the TS and determine the resulting OF value. If better or equal to the base case (better or equal OF value and no constraint), then accept the ignr result as the new base, and make lambda smaller to shift toward nr. Note that this is an LM logic. However, if ignr is worse, then (i) increase lambda to shift the algorithm more toward ig. Also, (ii) test the ig trial solution. If the ig TS is better than the base, expand ΔS; if worse or constraint violation, contract ΔS. Note that this is the ISD logic. Also note that ig is only tested if ignr was not good. The expansion and contraction values for ΔS are those used in ISD. The shift in λ is a heuristic version of an asymptotic approach to either extreme value of $\lambda = 0$ or $\lambda = 1$.

10.2.3 RLM Pseudocode

1) Choose a trial solution that is feasible. Choose Δs, an initial ig stepsize, as appropriate for the DV range. Set $\lambda = 1$ to start with incremental gradient. Evaluate the OF at this initial trial solution.
2) Evaluate gradient and Hessian elements.
3) Use the ignr relation to determine the next trial solution and evaluate OF.
4) If the result is better (better OF, and no constraints) then decrease λ to move toward nr. $\lambda_{k+1} = 0.85\lambda_k$ Accept the new trial solution and OF as the base case. Go to Step 6.
5) However, if the result is worse (worse OF value or a constraint violation) then do two things:
   a) Increase λ to move toward ig. $\lambda_{k+1} = 0.5 + 0.5\lambda_k$
   b) Explore the ISD solution. Determine a trial solution from ISD, and evaluate the OF
      i) If better then accept the new trial solution as the base and enlarge the ISD step size $\Delta S_{k+1} = 1.1\Delta S_k$.
      ii) If worse then reduce the ISD step size $\Delta S_{k+1} = 0.5\Delta S_k$
6) Test for stopping criteria (excessive iterations or convergence). Stop or continue.
7) Go to Step 2.

The heuristic rules for adjusting $\Delta S, \lambda$ values seem to be robust to a variety of rule types and coefficient values.

This method does a bit more work per iteration than the traditional Levenberg–Marquardt because it calculates α from ΔS and the gradient values and may possibly evaluate the ISD solution independently. Since the NR method needs the gradient and Hessian elements, there is no extra work in getting the gradient information for the α calculation. The occasional ISD trial in step 5b will add one new function evaluation. However, compared with the linear algebra solution of the LM trial solution and the multiple OF evaluations needed to determine the gradient and Hessian elements in traditional LM, this extra work is small.

In my explorations the speed of finding the optimum (computational time, number of stages in progressing to the optimum, or number of function evaluations) is faster than traditional LM, and therefore the extra computational work per iteration is well compensated by fewer iterations.

Certainly quasi-Newton techniques could also be used to ease the computational burden of calculating the Hessian element values.

10.2.4   RLM VBA Code for a 2-DV Case

```
Sub RLM_1()
' Rhinehart twist on Levenberg-Marquardt lambda blend of Incremental
Gradient and Newton-Raphson (ignr)
' First test ignr, if better than base case decrease lambda to shift more
toward nr from ig,
' and take new ignr solution as new base. No test for ig.
' If however, ignr was not better than base, then shift lambda toward 1 to move
toward ig, and
' then test the ig solution. If ig is better than base then increase step size
and take new ig
' solution as base, otherwise decrease stepsize.
    If i = 0 Then
      constraint = "Not Assessed"
      Do Until constraint = "OK"     'Find a feasible initial trial solution
for single player mathods
        x1o = 10 * Rnd()            'Randomize initial DV1 value
        x2o = 10 * Rnd()            'Randomize initial DV2 value
        OF_Value = f_of_x(Fchoice, x1o, x2o) 'Determine the inital
function value
        OF_Value_Old = OF_Value
      Loop
      Worksheets("Path_Data").Cells(3, 1) = i
      Worksheets("Path_Data").Cells(3, 2) = x1o
      Worksheets("Path_Data").Cells(3, 3) = x2o
      OF_Value_Old = f_of_x(Fchoice, x1o, x2o)
      Cells(33, 12) = OF_Value_Old
      alpha = Cells(6, 12)
      beta = Cells(7, 12)
      expansion = alpha
      contraction = beta
      Lambda = 1    'initially ISD only, no NR action, a conservative choice
      ds = 0.5
      dx1 = 10    'to prevent false convergence on the first trial if both fail
      dx2 = 10
```

```
      ElSE
         e = -g1(Fchoice, x1o, x2o)        'negative Gradient element, first
partial of OF_Value w.r.t. x1
         f = -g2(Fchoice, x1o, x2o)
         If ds > 10 ^ -12 Then rcoeff = Sqr(e ^ 2 + f ^ 2) / ds
         a = Lambda * rcoeff + (1 - Lambda) * h11(Fchoice, x1o, x2o) 'Tempered
Hessian element, second partial of OF_Value w.r.t. x1 after x1
         b = (1 - Lambda) * h21(Fchoice, x1o, x2o)
         c = (1 - Lambda) * h12(Fchoice, x1o, x2o)     'Hessian element, second
partial of OF_Value w.r.t. x1 after x2
         d = Lambda * rcoeff + (1 - Lambda) * h22(Fchoice, x1o, x2o)
         ' Calculate new trial solution
         x1 = x1o + (d * e - f * b) / (a * d - b * c) 'one-step Incremental Gradient
lambda balanced with Newtons jump to place
         x2 = x2o + (a * f - c * e) / (a * d - b * c)
         OF_Value = f_of_x(Fchoice, x1, x2)
         If OF_Value > OF_Value_Old Or constraint = "FAIL" Then 'Newton's sent
in wrong direction, or did not add value
            Lambda = 0.5 * (1) + 0.5 * Lambda     'make Lambda larger to shift to
Incremental Gradient
            Worksheets("Path_Data").Cells(3 + i, 6) = "ignr-Fail"
            Worksheets("Path_Data").Cells(3 + i, 7) = Lambda
            x1 = x1o + e / rcoeff 'See where incremental gradient wants to go
            x2 = x2o + f / rcoeff
            OF_Value = f_of_x(Fchoice, x1, x2)
            If OF_Value > OF_Value_Old Or constraint = "FAIL" Then 'Gradient
step too big
               ds = ds * contraction        'Make ds smaller
               Worksheets("Path_Data").Cells(3 + i, 4) = "ig-Fail"
               Worksheets("Path_Data").Cells(3 + i, 5) = ds
            Else
               ds = ds * expansion          'Make ds larger
               dx1 = x1 - x1o
               dx2 = x2 - x2o
               x1o = x1 'Update Base Point with ig trial solution
               x2o = x2
               OF_Value_Old = OF_Value
               Worksheets("Path_Data").Cells(3 + i, 4) = "ig-Success"
               Worksheets("Path_Data").Cells(3 + i, 5) = ds
            End If
         Else                    'Lambda-blended ignr was good
            Lambda = 0.8 * Lambda     'make lambda smaller to shift to Newton's
            dx1 = x1 - x1o
            dx2 = x2 - x2o
            OF_Value_Old = OF_Value
            x1o = x1
            x2o = x2
```

```
        Worksheets("Path_Data").Cells(3 + i, 6) = "ignr-Success"
        Worksheets("Path_Data").Cells(3 + i, 7) = Lambda
    End If
    ' If both are fails then this returns with the same OF and xo-values
    ' which will make it appear to have met convergence.
    End IF
    Cells(33, 12) = OF_Value_Old
End Sub
```

10.3   Scaled Variables

In an optimizer with the logic $\Delta x = -\alpha \nabla f$, an issue is that the units may not match when the x-elements have different units. Consider creating a food recipe with these nominal DV values: 1 TSP salt, 2 oz of vanilla, 3 eggs, 1 cup milk, etc. If you seek to optimize them, then the DVs do not have consistent units.

The units can be made dimensionless by scaling the DVs by a nominal, expected range. Here is one of many structures for scaled variables:

$$x' = \frac{(x - x_{min})}{(x_{max} - x_{min})} = \frac{(x - x_{min})}{R} \tag{10.16}$$

Here the variable x is scaled by its range $R = x_{max} - x_{min}$. The notation x_{min} does not mean DV^* or the x-value for which the objective is the minimum. The subscripts refer to the feasible, reasonable, or investigated range for the DV. The scaled variable x' is dimensionless. The scaling relation defines a relation between differentials

$$\frac{dx'}{dx} = \frac{1}{R} \tag{10.17}$$

or alternately

$$\frac{dx}{dx'} = R \tag{10.18}$$

The ISD algorithm in scaled variables is written as

$$\Delta X' = -\alpha' \nabla' f \tag{10.19}$$

Note that the objective function, f, is not scaled; however, the DVs are, which means that the gradient operator is, and then necessarily so is alpha. Revealing detail of the elements in Equation (10.19),

$$\begin{bmatrix} \Delta X'_1 \\ \Delta X'_2 \\ \cdot \\ \cdot \end{bmatrix} = -\alpha' \begin{bmatrix} \partial f/\partial X'_1 \\ \partial f/\partial X'_2 \\ \cdot \\ \cdot \end{bmatrix} = -\alpha' \begin{bmatrix} (\partial f/\partial X_1)(dX_1/dX'_1) \\ \cdot \\ \cdot \\ \cdot \end{bmatrix} \tag{10.20}$$

Converting to unscaled variables,

$$
\begin{bmatrix} \Delta X_1/R_1 \\ \Delta X_2/R_2 \\ . \\ . \\ . \end{bmatrix} = -\alpha' \begin{bmatrix} (\partial f/\partial X_1)*R_1 \\ (\partial f/\partial X_2)*R_2 \\ . \\ . \\ . \end{bmatrix}
\tag{10.21}
$$

and multiplying each individual equation by R_i,

$$
\begin{bmatrix} \Delta X_1 \\ \Delta X_2 \\ . \\ . \\ . \end{bmatrix} = -\alpha' \begin{bmatrix} (\partial f/\partial X_1)*R_1^2 \\ (\partial f/\partial X_2)*R_2^2 \\ . \\ . \\ . \end{bmatrix}
\tag{10.22}
$$

or

$$
\Delta x = -\alpha'(\mathbf{I} \cdot \mathbf{R2}) \cdot \nabla f
\tag{10.23}
$$

where $\mathbf{I}$ is identity matrix and $\mathbf{R2}$ is the vector of squared scaling ranges.

$$
\mathbf{R2} = \begin{bmatrix} R_1^2 \\ R_2^2 \\ . \\ . \\ . \end{bmatrix}, \quad \mathbf{I} = \begin{bmatrix} 1 & 0 & 0 & . \\ 0 & 1 & 0 & . \\ 0 & 0 & 1 & . \\ . & . & . & . \end{bmatrix}
$$

or, in the form for blending incremental gradient with Newton–Raphson,

$$
\left(\frac{1}{\alpha'}\right)(\mathbf{I} \cdot \mathbf{RR2}) \cdot \Delta X = -\nabla f
\tag{10.24}
$$

where

$$
\mathbf{RR2} = \begin{bmatrix} 1/R_1^2 \\ 1/R_2^2 \\ . \end{bmatrix}
$$

λ-Blending the scaled ig + nr

$$
\left(\frac{\lambda}{\alpha'}\right)(\mathbf{I} \cdot \mathbf{RR2}) \cdot \Delta X = -\lambda \nabla f
\tag{10.25a}
$$

$$
(1-\lambda)\mathbf{H} \cdot \Delta X = -(1-\lambda)\nabla f
\tag{10.25b}
$$

which when added become

$$\left[\left(\frac{\lambda}{\alpha'} \right)(\mathbf{I} \cdot \mathbf{RR2}) + (1 - \lambda)\mathbf{H} \right] \cdot \mathbf{\Delta X} = -\nabla f \tag{10.26}$$

The scale factors permit compatible units in step size and similar gradient sensitivities in ig direction and keep the relation for scaled variables dimensionally correct.

Step-size definition:

$$\Delta S' = \sqrt{\left(\Delta X_1'^2 + \Delta X_2'^2 + \cdots\right)} = \alpha' \sqrt{\left(\sum \left(\frac{\partial f}{\partial Xi} Ri\right)\right)^2} \tag{10.27}$$

Rearranging

$$\frac{1}{\alpha'} = \frac{\sqrt{\left(\sum \left((\partial f / \partial Xi) Ri\right)\right)^2}}{\Delta S'} \tag{10.28}$$

10.4 Conjugate Gradient (CG)

Conjugate gradient (CG) is a modification of Cauchy's sequential line search algorithm. Like CSLS, at each iteration, CG searches along a line to find the minimum along that search direction. At the minimum along the line, a new search direction is determined. Although a search along the line of steepest descent is often a very good choice, with some surfaces, steepest descent direction leads to zigzagging across a steep valley or ridge. Figure 10.1 shows a 2-DV example. The initial trial solution is in the

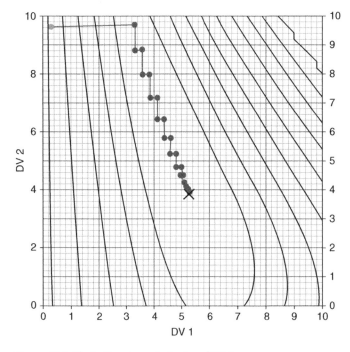

Figure 10.1 Illustrating zigzag behavior of CSLS.

upper left, and it takes a zigzagging path toward the optimum. At each iteration the local steepest descent direction switches from nearly vertical to nearly horizontal.

However, an intelligent observer might see that the alternating directions are not the best and recognize that there is an overall movement in the south-by-southeast direction. An intelligent algorithm might move the trial solution in that general direction, as opposed to zigzagging from valley wall to opposite wall.

The conjugate direction tempers the local steepest descent direction with past directions.

Letting u represent the search direction, in steepest descent $u = -\nabla f$, and the new trial solution is determined by the scalar α, $x = x_0 - \alpha \nabla f$, which is equivalent to $x_{i,k} = x_{i-1}^* + \alpha_k u_i$. The subscript i represents the line-to-line iteration, and x_{i-1}^* is the optimum on the prior line search, the pivot point. The subscript k represents the within-line iteration on the scalar α. For each line there are iterative trial solutions for α to find the minimum along the line.

In a conjugate gradient search, the search direction, $u = -\nabla f$, is tempered by past search directions:

$$u_i = -\nabla f\left(x_{i-1}^*\right) + \beta_i u_{i-1} \tag{10.29}$$

Then

$$x_{i,k} = x_{i-1}^* + \alpha_k u_i \tag{10.30}$$

If β is very small, then the direction for the line search is essentially the conventional steepest descent. If, however, β makes the magnitude of the past direction, u_{i-1}, equivalent to the local steepest descent, then the search direction is an average of the two. Where the function would lead to a zigzagging search across a valley, blending the past and new direction will lead to a search along the average direction, hopefully along the valley bottom. One formulation for β is to equate it to the magnitude of the current gradient scaled by the magnitude of the prior gradient:

$$\beta_i = \frac{\left\|\nabla f\left(x_{i-1}^*\right)\right\|}{\left\|\nabla f\left(x_{i-2}^*\right)\right\|} \tag{10.31}$$

Note the following:

- Like the CSLS or ISD approaches, the algorithm is dimensionally inconsistent if the DV values have disparate units.
- In CSLS, the sequential search directions are mutually perpendicular. Here, however, the directions are the vector addition of local and past directions and are not necessarily perpendicular.
- This method presumes a continuously differentiable function. Its effectiveness degrades with a function that has either slope or level discontinuities.
- This method is not developed to handle constraints.
- The direction u_i is not a unit vector. But it could be normalized to be a unit vector, $u_i' = u_i/\|u_i\|$, which may be a convenience in initializing α on each line iteration.

10.5 Broyden–Fletcher–Goldfarb–Shanno (BFGS)

BFGS is a particularly well-accepted quasi-Newton approach for updating the Hessian elements. There are many alternate approaches, however.

Here, B_k is the approximation to the Hessian matrix at the kth iteration. In BFGS

$$B_{k+1} = B_k - \frac{B_k s_k\, s_k^T B_k}{s_k^T B_k s_k} + \frac{y_k\, y_k^T}{y_k^T s_k} \qquad (10.32)$$

where s_k is the iteration-to-iteration change in the trial solution and y_k is the change in the gradient:

$$s_k = x_{k+1} - x_k \qquad (10.33)$$
$$y_k = \nabla f_{k+1} - \nabla f_k \qquad (10.34)$$

This is reported to be stable and efficient with a guaranteed positive definite character of the Hessian estimate. It can replace the Hessian matrix evaluation in any of the Newton-type algorithms. And satisfyingly, it is dimensionally consistent.

Note, however, the following:

- The method does presume a continuous function with continuous first derivatives.
- It was not devised to handle constraints, but to substantially diminish the work in evaluating Hessian elements.
- For applications that are not quadratic-ish, the Newton logic may misdirect the search, so this should be used in conjunction with a steepest descent search (like LM).

10.6   Generalized Reduced Gradient (GRG)

GRG was devised to handle equality and inequality constraints. The objective is

$$\begin{aligned} \min_{\{x\}}\ & J = f(x) \\ \text{S.T.}:\ & h(x) = 0 \\ & g(x) \le 0 \end{aligned} \qquad (10.35)$$

Of course the $\le$ is equivalent to $\ge$ with a sign change of the inequality g-function. Consider that there are l (the letter ell) number of equality constraints, m number of inequality constraints, and n number of DVs.

I have seen two basic approaches, one reduced gradient and another generalized reduced gradient (GRG).

First consider the RG concept. Start with an unconstrained search from a feasible TS. When a new TS violates an inequality constraint, then convert that inequality to an equality constraint, and use the equality to reduce the DV dimension by one, from n to $n-1$. Continue the $n-1$ DV search until the search direction would either (i) send the TS in a direction that is free of the constraint or (ii) add another constraint. If the constraint is no longer active, then revert to the n DV dimension. If a new constraint becomes active encountered, then convert that inequality to an equality constraint and reduce the DV dimension again.

This approach requires some fairly extensive management logic. One question is, "Which DVs remain in the optimizer set of $n-1$ variables, and which DV has its value calculated by the others?" If a variable is linear in the constraint, then convenience and avoidance of execution errors and the possible instability of solving a nonlinear relation would choose it. But what if none appear linearly?

What nonlinear root-finding method and convergence criterion value are appropriate? Would two active constraints lead to an alternate choice of which are the best variables to use as the dependent ones? Further, if the equality constraints are also nonlinear, how are all nonlinear relations evaluated?

The nonlinearity problem can be alleviated by locally linearizing the constraint. For the *i*th equality constraint,

$$\nabla h_i(\boldsymbol{x}_k) \cdot \Delta \boldsymbol{x}_k = 0 \tag{10.36}$$

When an inequality constraint is active, set it to its constrained value; then the active *i*th inequality constraint being linearized becomes

$$\nabla g_i(\boldsymbol{x}_k) \cdot \Delta \boldsymbol{x}_k = 0 \tag{10.37}$$

Then any one Δx value can be calculated from the others. Perhaps choose the *j*th Δx element with the largest gradient value to minimize the impact of numerical truncation error:

$$\Delta x_j = \frac{\sum_{l=1,\,\text{not}\,j}^{n} (\partial g_i / \partial x_l) \Delta x_l}{\partial g_i / \partial x_j} \tag{10.38}$$

This would be valid if the incremental changes are small enough to make local linearization valid.

The alternate approach, more generically applicable, is to linearize all of the constraints, locally, about the trial solution:

$$\begin{aligned} &\min_{\{\boldsymbol{x}\}} J = f(\boldsymbol{x}) \\ &\text{S.T.:} \quad \nabla h_1(\boldsymbol{x}) \cdot \Delta \boldsymbol{x} = 0 \\ &\qquad\quad \nabla h_2(\boldsymbol{x}) \cdot \Delta \boldsymbol{x} = 0 \\ &\qquad\quad \ldots \\ &\qquad\quad \nabla g_1(\boldsymbol{x}) \cdot \Delta \boldsymbol{x} \le 0 \\ &\qquad\quad \nabla g_2(\boldsymbol{x}) \cdot \Delta \boldsymbol{x} \le 0 \end{aligned} \tag{10.39}$$

Then add a slack variable to each inequality constraint, making the inequality an equality. Here the subscript *i* indicates the constraint index:

$$\begin{aligned} &\min_{\{\boldsymbol{x}\}} J = f(\boldsymbol{x}) \\ &\text{S.T.:} \quad \nabla h_i(\boldsymbol{x}) \cdot \Delta \boldsymbol{x} = 0 \\ &\qquad\quad \nabla g_i(\boldsymbol{x}_k) \cdot \Delta \boldsymbol{x}_k + s_i^2 = 0 \end{aligned} \tag{10.40}$$

This would add *m* number of slack variables, one for each inequality constraint, but the *m* number of equality constraints can be used to make *m* number of the *x* and *s* elements dependent on the original *n*-number of DVs. The constraints are locally linear in *x* values. By making the slack variable squared in the constraint, there are no limits on the s_i value. So, it is convenient to use the *m* number of *s* values and *n-m-l* number of *x* values as the optimizer DVs and to calculate the remaining *n-l* number of *x*-values from the two sets of linear equality constraint equations. Here, the subscript *k* represents iteration, and $\boldsymbol{x}'$ is the set of *n-m-l* DVs:

$$\begin{array}{c} \min\limits_{\{x',s\}} J = f(x) \\[6pt] \text{S.T.:} \quad \nabla h_i(x) \cdot \Delta x = 0 \\[6pt] \nabla g_i(x_k) \cdot \Delta x_k + s_i^2 = 0 \end{array} \qquad (10.41)$$

Although any sort of optimizer could be used, Lasden's GRG technique employs a steepest descent sequential line search. I understand that this is the algorithm that the Excel Solver Add-in uses

The constraints are locally linear, meaning that the Jacobean elements ∇h_i are updated at each iteration. Although this keeps them locally valid, (i) it does add computational burden, (ii) it is predicated on the constraint functions being continuous with continuous derivatives, and (iii) the constraints can be explicitly stated as single functions of the DV (not implicit functions, not equivalent discontinuities resulting from conditionals in the OF calculation, not on non-executable surprises within the OF). Since a linear algebra technique is required to solve the linearized equality constraints for the m number of dependent DVs, this adds computational work. Finally, linearizing is an approximation; accordingly, the converged solution will slightly violate nonlinear constraints.

Choosing initial DV values and the right subset selection to ensure that the matrix inversion is executable needs a bit of management. Small incremental changes are required to ensure that the linearized relations are reasonably valid. Choices can evolve as the solution places DVs on constrained values.

10.7 Takeaway

These algorithms are improvements over the basic ISD or NR. The improvements are related to minimizing function evaluations to convergence and robustness. However, they are designed for continuum-valued surfaces with continuum-valued derivatives. The procedures are also getting complicated. Although many applications fall in the category that make LM, CG, or GRG appropriate choices, many other application types will confound them.

10.8 Exercises

1   Use text to describe the Levenberg–Marquardt method for optimization. Show your understanding of the concept not the equations.

2   Provide a flowchart and supporting text explanation for LM.

3   Use the 2-D Optimization Examples file to compare LM and RLM with CLSL and IDS on diverse applications. Clearly indicate the rationale for your choice of criteria to use as a comparison and your choice of experiments to generate data.

4   Choose a relatively simple nonlinear optimization example with one relatively simple nonlinear inequality constraint and show
    A   How GRG would replace a DV with an unconstrained slack variable
    B   The locally linearized statement

11

Direct Search Techniques

11.1 Introduction

Direct searches do not use gradient or second-derivative information. They do not use models of the surface. Direct searches only use function evaluation, and the trial solution sequence is directed either by human logical or stochastic rules. Typically they creep up to optima, as opposed to understanding the surface and jumping to, or near to, the perfect answer. One might think, then, that direct searches are inferior. Well, they are inferior to second-order methods but only for the limited class of applications that meet the ideal conditions of deterministic functions with continuum variables and derivatives, no constraints, no flat spots, quadratic-ish surfaces, etc. However, most of the applications that I've had to consider are not from that ideal, trivial category. For example, time and space discretization in the models that are used to calculate the OF will generate surface discontinuities, which misdirect second-order optimizers. In general, I find that direct search algorithms beat the best of gradient and second-order optimizers when considering application versatility, speed to find the answer, simplicity of code, robustness, etc.

This section provides a description of several direct search algorithms that I think are best in class and that also reveal diverse search logic categories. The optimizers can be divided into single trial solution algorithms or multiplayer algorithms. Multiplayer algorithms initially scatter trial solutions throughout the DV space and look at the OF value of each. They then move the worst player, or they create new players into a DV region that human logic might want to explore. Single trial solution algorithms start at one DV spot and seek to move downhill from that point. Single trial solution algorithms can get stuck in local optima. Because multiplayer algorithms have an initial sampling of the entire surface, they have a higher probability of finding the global optimum. Although often termed global optimizers, no optimizer is guaranteed to find the global on any one run.

The single TS algorithms in this chapter are cyclic heuristic direct (CHD), Hooke–Jeeves (HJ), and Nelder–Mead (NM). The multiplayer algorithms are the complex method (CM), leapfrogging (LF), and particle swarm optimization (PSO).

There are many other interesting direct search algorithms, and it appears that the bulk of research and development in this era is presently about such algorithms.

Engineering Optimization: Applications, Methods, and Analysis, First Edition. R. Russell Rhinehart.
© 2018 R. Russell Rhinehart. Published 2018 by John Wiley & Sons Ltd.
Companion website: www.wiley.com/go/rhinehart/engineeringoptimization

11.2   Cyclic Heuristic Direct (CHD) Search

The term "direct search" means that it uses information from function evaluations and constraint violations only. There is no derivative or second derivative (no gradient, no Hessian) in the algorithm. No mathematical model of the surface is formulated. The term "cyclic" means that it cycles through the list of DVs, making an incremental change in each, one at a time. Then it repeats a cycle. Each complete cycle of exploring a change in each DV ends a stage, or defines an iteration. The term "heuristic" means that human logical rules guide the search increment expansion or its reversal and contraction.

This is my second favorite algorithm for moving DV values toward the minimum (or maximum). It is high on my list of favorites because of its simplicity, robustness, broad applicability, and effectiveness (in scoring performance points over other optimizers). It can be implemented in a human mentally directed trial-and-error search and does not need a computer to implement it. It is my second favorite algorithm because it is a single trial solution approach that can get stuck in a local optimum.

11.2.1   CHD Pseudocode

The procedure pseudocode is

1) Initialize
    i) DV base (the initial trial solution in a feasible region) and evaluate the OF
    ii) ΔDV_i to be used for exploratory changes
    iii) Convergence method and threshold
2) Start the cycle for one iteration (alternately this may be called an epoch, stage,…).
3) Taking each DV, one at a time, set the new trial DV value,
$$DV_{i,\text{trial}} := DV_{i,\text{base}} + \Delta DV_i$$
    Note DV_{trial} and DV_{base} are vectors. Only add ΔDV_i to the ith element.
    All other DV_{trial} elements remain at the base value.
    i) Evaluate the function at the trial solution value.
        1) If worse or "FAIL" keep DV base and set ΔDV_i to a smaller change in the opposite direction: $\Delta DV_i := -\text{contract}^* \Delta DV_i$. This means that each dimension has its own ΔDV_i value.
        2) Otherwise, keep the better solution (make it the base point) and accelerate future moves in what seems to be the right search direction
$$OF_{\text{base}} := OF_{\text{trial}}$$
$$DV_{i,\text{base}} := DV_{i,\text{trial}}$$
$$\Delta DV_i := \text{expand}^* \Delta DV_i$$
4) At the end of the cycle, check for stopping criteria (excessive iterations, or other indication of futility).
    i) If stopping, exit with a "Did not Converge" message.
    ii) Otherwise go to Step 5
5) Check for convergence criteria
    i) If converged, exit with appropriate message.
    ii) Otherwise return to Step 3

This algorithm is simple to understand and simple to implement. Direct searches can accommodate hard constraints. Direct searches do not have ∇ or ∇^2 operators that (i) have difficulty coping with nonanalytic surfaces (ridges, discontinuities in function or slope, noise or variability on an OF value) and (ii) add computational burden and programming complexity. Although other search algorithms may be more efficient (use fewer iterations or fewer function evaluations), to my taste, the simplicity, robustness, and constraint accommodation of this algorithm are substantial advantages.

This approach started with an idea from one of my graduate students, Junyi Li. I continued the investigation with another graduate student, Sandeep Chandran, and evolved the search to use multiple starts and a "coarse" search to find the most promising DV location to return to for a fine search. Subsequently, I decided that the random perturbations on each ΔDV added complexity with little compensating benefit and dropped that concept. More recently, I added a feature that is suggested by a Hooke–Jeeves pattern search. If the cyclic search directions are all good (constraint free and lead to better OF values) then interpret this as moving down a long slope. Then, rather than searching each DV independently, make them all move together. If searching independently, each iteration takes N (=# DVs) OF evaluations. If moving all together, it only takes one OF evaluation and ends at the same trial solution if the direction is correct. Another option I've explored is to alternate backward and forward sweeps through the cycle, an idea from numerical method techniques for boundary value problems. There is not only one way. Many options can lead to improvements. If you are exploring improvements, clearly state how you will assess the impact.

11.2.2 CHD VBA Code

This CHD version is the basic algorithm.

```
Sub RRR_Cyclic()
' RRR cyclic direct search
' Cyclic meaning that trial investigations for exploring new x1 and x2
  alternate
'    one at a time.
' Direct search meaning only objective function value is considered,
  no derivatives.
' RRR likes the rule, it is simple and generally very good.
'    If the tial DV is better than the base DV, make it the base and
'    increase the delta-x value for the next turn. Alternately, if the trial DV
'    leads to a worse position, then return to the base DV and change direction
'    for the search and use a smaller delta-x.
        If i = 0 Then
          constraint = "Not Assessed"
          Do Until constraint = "OK"   'Find a feasible initial trial solution
for single player methods
            x1o = 10 * Rnd()        'Randomize initial DV1 value
            x2o = 10 * Rnd()        'Randomize initial DV2 value
            OF_Value_Old = f_of_x(Fchoice, x1o, x2o) 'Determine the initial
function value
          Loop
        x1 = x1o
        x2 = x2o
        dx1 = 2 * (Rnd() - 0.5)
        If Abs(dx1) < 0.05 Then dx1 = 0.05
        dx2 = 2 * (Rnd() - 0.5)
```

```
            If Abs(dx2) < 0.05 Then dx2 = 0.05
            Worksheets("Path_Data").Cells(3, 1) = i
            Worksheets("Path_Data").Cells(3, 2) = x1o
            Worksheets("Path_Data").Cells(3, 3) = x2o
            OF_Value_Old = f_of_x(Fchoice, x1o, x2o)
            If Nfevals > 1 Then
               foworst = OF_Value_Old
            For kfeval = 1 To Nfevals - 1
               OF_Value_Old = f_of_x(Fchoice, x1o, x2o)
               If OF_Value_Old > foworst Then foworst = OF_Value_Old
            Next kfeval
            OF_Value_Old = OF_Value_Oldworst
         End If
         Cells(33, 12) = OF_Value_Old
         alpha = Cells(6, 12)
         beta = Cells(7, 12)
         expansion = alpha
         contraction = -beta / expansion
         NGoodTrials = 0
       Else
         Cells(25, 12) = i
         x1 = x1o + dx1
         OF_Value = f_of_x(Fchoice, x1, x2o)
         If OF_Value >= OF_Value_Old Or constraint = "FAIL" Then 'if better
    or equal to past f
          ' If OF_Value > OF_Value_Old Or constraint = "FAIL" Then 'if better
    thatn past f
               dx1 = contraction * dx1
               x1 = x1o
            Else
               dx1 = expansion * dx1
               OF_Value_Old = OF_Value
                x1o = x1
            End If
            x2 = x2o + dx2
            OF_Value = f_of_x(Fchoice, x1o, x2)
            If OF_Value >= OF_Value_Old Or constraint = "FAIL" Then
           ' If OF_Value > OF_Value_Old Or constraint = "FAIL" Then
               dx2 = contraction * dx2
               x2 = x2o
            Else
               dx2 = expansion * dx2
               OF_Value_Old = OF_Value
               x2o = x2
            End If
          End If
      End Sub
```

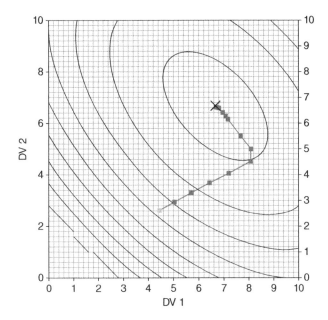

Figure 11.1 Illustration of CHD in 2-D.

Example 1 Function 13 in the 2-D Optimization Examples code is a nonlinear ellipse. Figure 11.1 shows the contours and the path that CHD took when randomly initialized at a point near (4, 3) and with initial step sizes of +1 DV unit. The second dot on the path is the result after exploring changes in both DVs. Both changes were successful, led to better OF values and no constraints, and were accepted. Accordingly, the distance between the initial and the second dot is $1.41... = \sqrt{1^2 + 1^2}$. Since both DV explorations were successful, the exploration step is expanded by 10%. Accordingly, on the successful downhill path between the initial and fifth dots, each distance between dots is progressively larger. At dot #5, the DV1 exploration to the right led to an uphill position; accordingly the search kept the fifth DV1 value and reversed and contracted the DV1 increment it will use in the next cycle. Meanwhile, the continued exploration of DV2 led to a success, so the new DV2 value was kept. Dot #6 is the result. Dot #7 is the result of continued success in the upward direction of DV2, and success in the new leftward direction in DV1. After iteration #8, large steps either to the left or upward led to not as good an OF value. Then contraction and reversal of the search to the right and downward also did not find a better location, so with subsequent reversal and contraction, the ΔDV is back toward the left and upward, but now ¼ the size. These explorations led to successes. The convergence criterion is on each ΔDV < 0.001. The search converged at the optimum with 60 function evaluations.

Note that the initial downhill search is not exactly aligned with the direction of steepest descent.

Example 2 Figure 11.2 shows a CHD search path on a function with a broad valley with a single minimum. Again note that the initial path is not exactly in the steepest descent direction, but it is the direction that the initial ΔDV successes indicated leads to a better spot. The search continues in that direction, expanding the steps by 10% until an exploration is not a success.

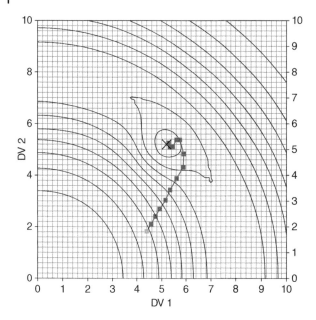

Figure 11.2 CHD on a 2-D Function.

11.3 Hooke–Jeeves (HJ)

Hooke–Jeeves is termed a pattern search. There is a local exploratory pattern used to explore the local surface trends, and the jump to a new trial solution is a "pattern" move. It moves the center of the pattern to the new trial solution. This is a two-step procedure. Jump to a new location, and then explore the local features to determine where the next jump-to TS is to be placed. Like CHD, in the local search, HJ uses one variable at a time during local exploration.

This illustration will be for a 2-D application, but HJ scales to N dimension.

First, jump to a new trial solution. Move the pattern center to a point that is the same distance and direction between prior base points. Base points with superscripts (k) and $(k-1)$ and subscript b are the prior two iteration best locations, and it makes intuitive sense to continue moving in that direction. Here, the expansion factor is explicitly indicated as unity, and the jump-to position has the subscript p, indicating it is the center of the pattern search, a tentative location, not a new base point.

$$x_p^{(k+1)} = x_b^{(k)} + 1\left(x_b^{(k)} - x_b^{(k-1)}\right) \tag{11.1}$$

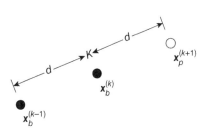

Figure 11.3 Pattern move in HJ.

This is illustrated in Figure 11.3. The prior two base points are indicated by solid dots, and the jump-to trial solution is indicated by an open circle.

Presently, assume that the OF value at the jump-to point $x_p^{(k+1)}$ is better than the OF value at the prior base DV point $x_b^{(k)}$. To be accepted as the new pattern point, $f\left(x_p^{(k+1)}\right) < f\left(x_b^{(k)}\right)$.

Step 2 is to perform a pattern search about the pattern center. Figure 11.4 illustrates this for a 2-D application. The pattern will explore some of the points in a rectangle around the pattern center. Possible exploration points are indicated by open circles, but not all will be explored.

Possible points to explore are labeled with the subscript "*e*," $x_e^{(k+1)}$, and indicated as open circles on Figure 11.4. One of these will become the new base point.

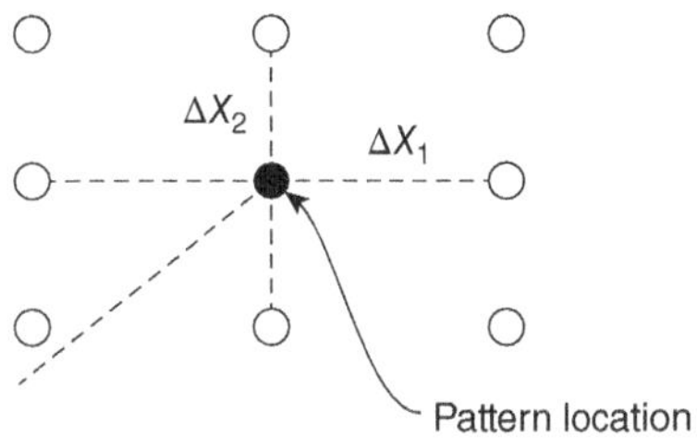

The Δx_i values are initially user chosen and do not have to be identical for each DV. They can have negative values.

Figure 11.4 Illustration of HJ possible exploration points in 2-D.

The local exploration is not exhaustive. Like CHD, keep all of the other DV values in the pattern search the same and just explore the ΔDV of one. Do this in a cyclic manner.

$$x_e^{(k+1)} = \left[x_{1_p} + \Delta x_1, x_{2_p} \right]^T \tag{11.2}$$

If $f(x_e) < f(x_p)$, then the pattern exploration was good. Accept that DV value in the pattern point, and move on to explore the impact of changing another DV. However, and only if, an exploration is not good (worse of or constraint), reverse direction.

$$x_e^{(k+1)} = \left[x_{1_p} - \Delta x_1, x_{2_p} \right]^T \tag{11.3}$$

If $f(x_e) < f(x_p)$, then the pattern exploration was good. Accept that DV value in the pattern point, and move on to explore the impact of changing another DV. However, and only if, an exploration is not good (worse of or constraint), return to the center pattern value for that DV. Then search with another DV.

Note: If the first exploration was a success, the second is not performed. This is not an exhaustive local search.

Then, explore changes in the value of x_2 in a similar logic.

$$x_e^{(k+1)} = \left[x_{1_p} + \Delta x_1, x_{2_p} + \Delta x_2 \right]^T \tag{11.4}$$

However, if an exploration is not good reverse direction

$$x_e^{(k+1)} = \left[x_{1_p} + \Delta x_1, x_{2_p} - \Delta x_2 \right]^T \tag{11.5}$$

Note: If the first exploration for each DV results in OF improvement, then there are only three OF evaluations, one at the pattern point and one each at the ΔDV1 and ΔDV2. Alternately, the highest number of function evaluations would arise if both of the first ΔDV explorations were not better than the pattern point. In this case there would be a maximum of five function evaluations, the pattern point and the $+\Delta$DV and $-\Delta$DV for each DV. Although nine possible exploration locations are illustrated for the 2-DV case on Figure 11.4, only three to five, perhaps averaging four, are evaluated on each iteration.

A decision tree representing the exploration stage would get complicated, attempting to represent the three possible outcomes for explorations in each dimension. However, the logic for each is identical, so as nested conditionals within loop through each DV, the exploration is relatively simple to program.

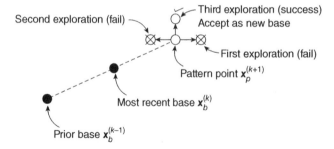

Second exploration (fail)

Third exploration (success)
Accept as new base

First exploration (fail)

Pattern point $\mathbf{x}_p^{(k+1)}$

Most recent base $\mathbf{x}_b^{(k)}$

Prior base $\mathbf{x}_b^{(k-1)}$

Figure 11.5 Illustration of a HJ search in 2-D.

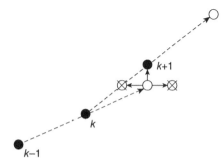

$k+1$

k

$k-1$

Figure 11.6 Illustration of a HJ pattern move in 2-D.

Figure 11.5 illustrates a possible outcome of the exploration. The jump to the new pattern point was successful. Then each of the pattern explorations for the first DV failed; these are indicated by the X^{ed}-out points. Then, the first exploration of the second DV was a success, so the checked point is accepted as the new base point.

Once the local explorations are complete, the two steps in HJ iteration are complete. The procedure returns to the Step 1 movement of the pattern search as illustrated in Figure 11.6.

Note two features: The pattern search direction shifted, and the distance of the pattern shift has expanded. Both are the logical results of local information at the last pattern exploration.

Suppose, however, that the new pattern point is a "fail," meaning either the search encountered a constraint violation or a worse OF value. Then contract and reverse the pattern search and repeat. This is illustrated in Figure 11.7.

The pattern shift equation is

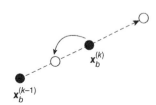

$\mathbf{x}_b^{(k)}$

$\mathbf{x}_b^{(k-1)}$

Figure 11.7 Illustration of a pattern move change if the pattern point is not better.

$$x_p^{(k+1)} = x_b^{(k)} + \lambda\left(x_b^{(k)} - x_b^{(k-1)}\right) \tag{11.6}$$

Nominally, in Equation (11.6) $\lambda = 1$. However, if the new pattern point is not better, contract and reverse the search, making $\lambda = -(1/2)\lambda$. More generally with a contraction factor of α, $\lambda = -\alpha\lambda$

The user chooses initial trial solution and ΔDV values. The ΔDV values do not have to be equal. The user also chooses α the pattern contraction factor to be used if new pattern point is a "fail."

The pattern progressively moves downhill, roughly in an ISD approach. If moving in the right direction, each pattern jump is larger than the prior.

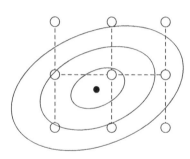

Figure 11.8 Illustration of exploration fails in HJ.

Near a minimum ΔDV values need to be contracted, otherwise all exploration will be in uphill directions, as Figure 11.8 illustrates such in a 2-D situation.

If a pattern center is a success but all exploratory points fail, contract all ΔDV values, and repeat the local exploration. The user needs to choose the contraction factor.

Convergence could be based on any number of criteria, the pattern-to-pattern distance, the ΔDV values, or the associated OF values.

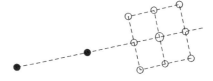

Figure 11.9 Illustration of an alternate pattern exploration layout in HJ.

In all, the basic HJ method has a pattern shift followed by a local search that is used to accelerate or contract and shift the search direction. There are two override conditions for the end game. If the jump-to pattern position center is not good, contract and reverse the pattern. If all local explorations are worse than the pattern center, then contract the local search.

As with any algorithm, one can devise many possible enhancements to improve efficiency. For example, the local exploration does not have to be aligned with DV axes. It could be aligned with the most recent pattern move, as shown in Figure 11.9.

As another, if the pattern is moving in a negative DV direction, then instead of making the first exploration in the positive direction, make the first in the negative ΔDV direction to increase the probability that the first explorations will be successes, minimizing the NOFE per iteration.

Further, there are many possible heuristics that one could use to modify the trigger to contract the pattern jump or to contract the exploration range or how much for either is dependent on the end-game situation.

11.3.1 HJ Code in VBA

```
Sub Hooke_Jeeves()
' With RRR features of
' 1 - expanding or contracting the search area depending on whether success
  of not
' 2 - changing the sign of the first search direction based on last search
'     direction to accelerate progress when the right direction was not the
original.
        If i = 0 Then
          constraint = "Not Assessed"
          Do Until constraint = "OK"   'Find a feasible initial trial
          solution for single player mathods
            x1o = 10 * Rnd()      'Randomize initial DV1 value
            x2o = 10 * Rnd()      'Randomize initial DV2 value
            OF_Value_Old = f_of_x(Fchoice, x1o, x2o) 'Determine the inital
function value
          Loop
          dx1 = 0.1
          dx2 = 0.1
          expansion = Cells(6, 12)
          contraction = Cells(7, 12)
          x1base = x1o
          x2base = x2o
          x1pattern = x1o
          x2pattern = x2o
```

```
          OF_Value_Old = f_of_x(Fchoice, x1base, x2base)
          Worksheets("Path_Data").Cells(3, 1) = i
          Worksheets("Path_Data").Cells(3, 2) = x1o
          Worksheets("Path_Data").Cells(3, 3) = x2o
          Worksheets("Path_Data").Cells(3, 8) = OF_Value_Old
     Else
          x1ExploreSuccess = "YES"    'explore in x1 direction
          x1explore = x1pattern + dx1
          OF_Value = f_of_x(Fchoice, x1explore, x2pattern)
       If OF_Value >= OF_Value_Old Or constraint = "FAIL" Then      'if bad or
equivalent reverse direction
              dx1 = -dx1
              x1explore = x1pattern + dx1          'and explore in other direction
              OF_Value = f_of_x(Fchoice, x1explore, x2pattern)
              If OF_Value >= OF_Value_Old Or constraint = "FAIL" Then 'if also
bad then record it
                  x1ExploreSuccess = "NO"
                  x1explore = x1pattern
              Else
                  OF_Value_Old = OF_Value          'if good reestablish pattern and
OV value
                  x1pattern = x1explore
              End If
          Else
              OF_Value_Old = OF_Value              'if good reestablish pattern and
OV value
              x1pattern = x1explore
          End IF
          x2ExploreSuccess = "YES"
          x2explore = x2pattern + dx2
          OF_Value = f_of_x(Fchoice, x1pattern, x2explore)
          If OF_Value >= OF_Value_Old Or constraint = "FAIL" Then
              dx2 = -dx2
              x2explore = x2pattern + dx2
              OF_Value = f_of_x(Fchoice, x1explore, x2explore)
              If OF_Value >= OF_Value_Old Or constraint = "FAIL" Then
                  x2ExploreSuccess = "NO"
                  x2explore = x2pattern
              Else
                  OF_Value_Old = OF_Value
                  x2pattern = x2explore
              End If
          Else
              OF_Value_Old = OF_Value
              x2pattern = x2explore
          End If
          movex1 = x1explore - x1base      'decide new pattern move
          movex2 = x2explore - x2base
```

```
        x1base = x1explore          'set base at best of explored points
        x2base = x2explore
        x1pattern = x1base + movex1     'move pattern point
        x2pattern = x2base + movex2
        OF_Value = f_of_x(Fchoice, x1pattern, x2pattern)     'test
pattern point
        If OF_Value > OF_Value_Old Or constraint = "FAIL" Then     'return to
new base
           x1pattern = x1base
           x2pattern = x2base
           dx1 = contraction * dx1     'contract exploration
           dx2 = contraction * dx2
        End If
        If x1ExploreSuccess = "NO" Then
           dx1 = contraction * dx1
        Else
           dx1 = expansion * dx1
        End If
        If x2ExploreSuccess = "NO" Then
           dx2 = contraction * dx2
        Else
           dx2 = expansion * dx2
        End If
        x1 = x1pattern
        x2 = x2pattern
        x1o = x1pattern
        x2o = x2pattern
        Worksheets("Path_Data").Cells(3 + i, 1) = i
        Worksheets("Path_Data").Cells(3 + i, 2) = x1pattern
        Worksheets("Path_Data").Cells(3 + i, 3) = x2pattern
        Worksheets("Path_Data").Cells(3 + i, 8) = OF_Value_Old
     End If
End Sub
```

Example 3   Figure 11.10 shows a Hooke-Jeeves path on the nonlinear ellipse of Function 13.
  The initial search is generally downhill, but not exactly along the steepest descent. The acceleration of pattern point moves, the expansion, is obvious along that initial path; the expansion magnitude is the result of the ΔDV choices. The path redirection and jump-to distance reduction are obvious in the top center of the figure. And then, once it finds a new long downhill path, it begins to expand the jump-to pattern moves. Eventually, surrounding the optimum, contraction of the exploration ΔDVs leads to convergence.

11.4   Compare and Contrast CHD and HJ Features: A Summary

Jud Wooters, ChE MS 2008, compared Hooke–Jeeves with CHD. Hooke–Jeeves was generally the winner in the key performance metric of the number of function evaluations. Key to its success was better ability to move around a constraint. However, at what programming complexity? And with no constraint, HJ was only slightly better than CHD.

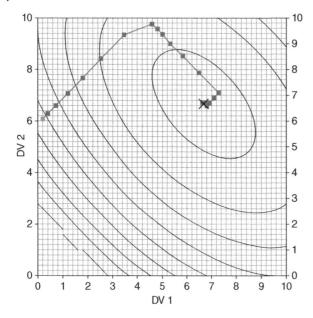

Figure 11.10 HJ path on Function 13.

CHD

1) One, simple logic for both search and end game.
2) Can be intuitively, manually applied.
3) Does not follow steepest descent, but does go downhill.
4) Will not get distracted by saddles or maxima.
5) Can handle surface discontinuities.
6) Basically N (number of DVs) OF evaluations per iteration, but can become 1 OF evaluation with the switch to a line search if all OF individual trial solutions are successful. Each OF evaluation is used to make progress.
7) User needs to choose expansion and contraction factors.

Hooke–Jeeves

1) More complicated logic.
2) Turns toward steepest descent direction.
3) Will not get distracted by saddles or maxima.
4) Better than CHD for circumventing some constraints.
5) 1 pattern move OF evaluation then, for a 2-DV application, 2–4 local search OF evaluations per iteration. Averaging $1 + 3 = 4$ NOFE per iteration.
6) Has slope evaluation features providing some similarity to the ISD approach, some functional benefit, but without the burden of derivative evaluations.
7) Can handle surface discontinuities.

8) Automatic expansion and contraction during search.
9) User needs to choose pattern contraction for end-game convergence.

11.5   Nelder–Mead (NM) Simplex: Spendley, Hext, and Himsworth

Spendley, Hext, and Himsworth originally developed and published the search logic using a simplex, the simplest $N+1$ point object in the N-DV space. In 2-D space, a plane, the three-point simplex object is an equilateral triangle. In 3-D space the four-point simplex structure is a tetrahedron.

The initial feasible trial solution is chosen, and then the simplex is "built" on that spot. The user needs to choose the TS and the DV distance that the simplex takes. The initial TS and the additional N simplex points provide local surface exploration.

Instead of a classic heuristics of either progressively moving the best trial solution to a better spot or moving toward the best spot, this clever approach moves the worst of several trial solutions away from the worst region.

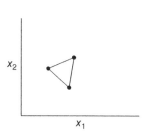

Figure 11.11 illustrates a simplex, an equilateral triangle, in a 2-DV application. Note that the construct is traditionally not aligned with either axis. Evaluate three OF values, one at each simplex vertex, and identify the worst one.

Figure 11.11 A simplex in a 2-DV case.

The worst vertex will be moved by projecting the worst location through the opposite side to invert the triangle. Often this is termed "flip the triangle," as illustrated in Figure 11.12.

Mathematically, one could determine the center of gravity of the two stationary points of the triangle, the center of the side opposite the point to be relocated, and relocate the worst point to an equal position opposite the center of gravity of the other points. Measure the OF. This is one iteration. In higher dimensions, the concept of projecting the one point through the center of gravity of all the other points is simpler than the concept of flipping the structure.

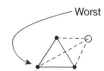

Figure 11.12 An NM or SHH simplex move.

Notably, rather than moving toward the best, this search moves the trial solution away from the worst position. There are classes of optimization called "Tabu" (taboo) that moves away from the worst.

Although there are three points in the 2-DV application, each move only relocates one point and requires one new OF evaluation.

After several iterations (identify worst point on the simplex, flip the triangle, determine the new OF value, repeat), the pattern might appear as shown in Figure 11.13. As the triangle flips along the OF surface seeking the minimum, it curves toward the downhill direction.

Ideally, each "jump-to" point is an improvement over the "from" point and an improvement over the other stationary points on the simplex. But, upon bounding a minimum, the triangle will flip back and forth, or cycle about, repeating the same point. This oscillation may also happen in long valleys prior to

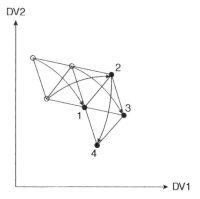

Figure 11.13 The NM-SHH simplex moving toward the optimum.

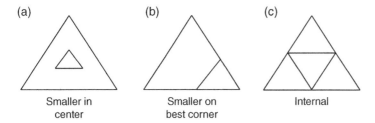

Figure 11.14 Options for placement of the smaller simplex: (a) on center, (b) on the best corner, and (c) on COG of the other points.

going all the way down the bottom. There are several actions, depending on the situation: (i) Keep the "to" point if it was better than the "from" point. Or (ii) if the "to" point was worse than the "from" point, return to the prior configuration, and then flip on the second-worst point. (iii) If there is no better flip or if one vertex remains un-moved in $N + 1$ iterations, then shrink the size of the simplex.

Continue until size of simplex is small enough to meet convergence criteria. This could be based on the DV range that it covers or on the resulting OF values.

You can scale each axis on a 0–1 basis so that simplex dimensions have similar lengths. You should do this; otherwise 12 inch may be flipped to 12 ft or 12°C.

Of course there are many options as Figure 11.14 reveals. You can trigger size reduction of the simplex on many criteria, and you can make size reduction in any of many manners.

Which reduction is easier to calculate? Which minimizes the number of function evaluations?

Suppose two points on the simplex tie for the worst OF value? What logic would you use to choose which to flip?

Nelder and Mead modified this simplex method to allow for expansion and contraction during the reflection or inverting or flipping stage.

The Nelder–Mead version of the simplex with the expansion and contraction of the simplex means that the object does not keep the equal-sided geometry. It is better for speed of finding the optimum and has become the default approach and household name. Regretfully, in my opinion, the Spendley, Hext, and Himsworth names are hardly used, and it has generically become known as the Nelder–Mead simplex search.

Note: The expansion and contraction rules of the move, the choice to initialize the simplex so that no edge is aligned with an axis, and the end-game relocation of the simplex are all choices, heuristics.

Note: This search does not move the best to a better spot. It moves the worst in the general direction that others indicate is better—an approach that is kin to tabu, swarm-like, and evolutionary-like searches.

Note: Simplex is also the name for a linear programming (LP) approach, the simple table or "Simplex Tableau," a very different technique that we'll consider in Chapter 12. Be sure to clarify simplex as either LP simplex or NM simplex.

In my opinion this NM performs equivalently to Hooke–Jeeves.

11.6 Multiplayer Direct Search Algorithms

The direct search techniques presented up to this point in the chapter move from a single trial solution locale. The HJ pattern and the SSH-NM simplex would be appropriately considered as having more than one trial solution, but all of the points are local (in DV space) to the nominal trial solution. Similarly, although incremental steepest descent, successive quadratic, and Newton-type methods are each initialized at one trial solution, the associated local surface exploration to obtain the surrogate model or the gradient and Hessian require multiple local surface evaluations. However, these multiple surface evaluations are local to the single trial solution, and the techniques are considered single trial solution techniques.

By contrast, the techniques in subsequent sections of this chapter are multiplayer direct search algorithms. Rather than starting from one single trial solution in DV space, they start with a randomly placed team of players (swarm of particles, many trial solutions) in feasible spots throughout the entire DV ranges thought to be of interest.

Single TS algorithms only "see" the local portion of the OF response and move in the best downhill direction from the single point. If they move to a local optimum, there is no knowledge that a better optimum might have been reached if it was initialized at a different point.

The benefit of a multiplayer algorithm is that it has a high probability of finding the global optimum. This is related to the broad surface exploration, initially, and subsequent exploration as the players attempt to gather at a best spot.

11.7 Leapfrogging

Initially place many players (trial solutions) on the surface throughout feasible DV space. Determine the players with best and worst OF values. Then move the worst player to the other side of the best player in DV space. The worst player "jumps" over the best player, like the children's game of leapfrog, and lands in a random spot in the projected DV "area" on the other side. Hence the term leapfrogging.

Figure 11.15 an illustration on a 2-D contour, with a minimum at about $x = 7.5$ and $y = 2$. Five trial solutions, five players, are illustrated with markers and are randomly placed over feasible DV space. Note: For low dimension, 10 players per dimension are recommended. Here, there should be 20, but only 5 are shown for simplicity of the presentation.

Calculate the OF for each player, and identify the worst and best. These are labeled W and B in Figure 11.16, which also shows the leap-from DV area (the window with the solid border) and the leap-into area (dashed border). The leap-to area is illustrated with the same size in DV space but is reflected on the other side of B from W. Both window borders are aligned with the DV axes. W leaps to a random spot inside of the reflected DV space window as the arrow indicates.

The calculation of the new DV position is very simple. If there are two DVs, x and y, the leap-to location is calculated as

$$\left.\begin{array}{l} x_{w,\,new} = x_b - r_x\left(x_w - x_b\right) \\ y_{w,new} = y_b - r_y\left(y_w - y_b\right) \end{array}\right\} \tag{11.7}$$

Here, r_x and r_y are independent random numbers, with a uniform distribution and range of $0 < r \le 1$. These random perturbations are independent for x and y. Note: The procedure scales to higher order or to a univariate search.

There are many variations on the concept. A user certainly could choose window amplification or attenuation, or alternate distributions to randomize the leap-to location. Equation set (11.7) can be

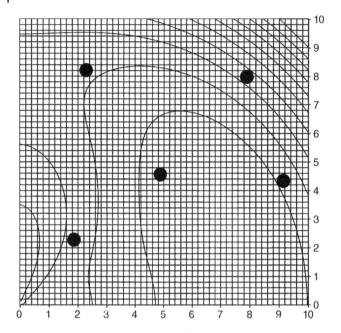

Figure 11.15 LF explanation: (1) Initialization.

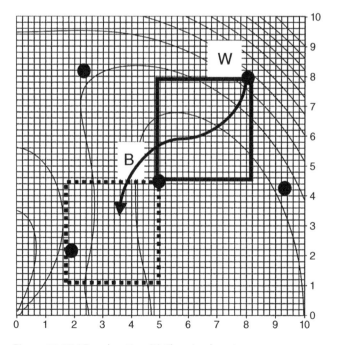

Figure 11.16 LF explanation: (2) Choosing leap-to area.

modified to move the player to the other side of the centroid of all particles, to blend moving to the other side of the centroid and the best, to move randomly either between or beyond the best or the centroid, or to jump into a window that is either larger or smaller than the reflected window. If the leap-to window is smaller than the leap-from window, the team of players converges faster, but does not explore as much. Alternately, if the leap-to window is larger than the leap-from window, it explores the entire space more and has a higher probability of finding the global, but converges slower. The team will converge if the leap-to window size amplification is less than $e = 2.718...$ (see Chapter 30). The leap-to window size can be adjusted as belief increases that the proximity of the global has been found. The leap-into window can be shifted toward W so that the DV region on either side of B is explored. The leap-to window can be circular, or its shape aligned with the path between W and B. Although there are some beneficial features to such enhancements, I find the leap-to rule above to be best—simple and most or equivalently effective on a variety of test applications. The K.I.S.S. principle applies.

As Figure 11.17 reveals, W moved to new position, and its prior location is vacated. The algorithm now must determine the OF value at the new player position. Note several features: The new position is not in the center of the leap-into window. There are still five players. There is one function evaluation per leap-over.

Test for convergence. As Figure 11.18 indicates the players are still far apart in DV space, as indicated by the size of the ellipse that contains them. They are not converged, yet. Of course, many metrics can be used to determine when the players are all close to the optimum, but for presentation simplicity, I often use either average or rms deviation of all players from the best as represented in

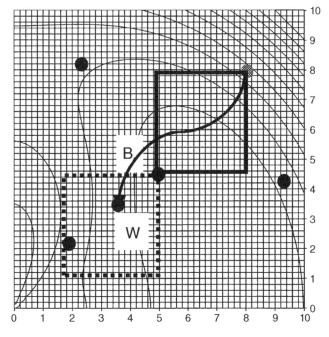

Figure 11.17 LF explanation: (3) The leap-over.

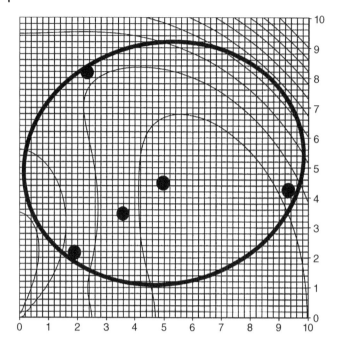

Figure 11.18 LF explanation: (4) Test for convergence.

DV space. However, the range of the OF deviation from the best has advantages as a convergence criterion.

Since the team has not converged, repeat the procedure. Identify B and W. The best is still the best. The former worst now has a mid-contour value, so a new worst is identified. Note: There is no need to search through all the player OF values to find the best. If the new leap-to player OF value is better than former best, it is the new best; else, the old best remains the lead. Also note, if the leap-to spot was either infeasible (for any reason, an explicit DV constraint, an execution error in the OF evaluation, etc.) or had a worse OF value than the leap-from point, that player remains as worst. There is no need to search for the best, and the algorithm only needs to search for the worst if the OF value of the leaping player improves. This aspect reduces computational burden.

In Figure 11.19, the new W leaps over B. One player is moved. Now the OF of the new location needs to be determined. In this illustration, the leap-to position is closer to the optimum, so the former W will become the new B.

Note, in Figure 11.20 that the team of players is both contracting and moving toward the optimum. The team is converging.

After repeating the steps, a few times, the players converge on the optimum, as Figure 11.21 reveals, and in a few more leap-overs, convergence will be declared. When converged, report the DV and OF of the best player in the cluster as DV* and OF*.

On average each leap-over moves the player with the worst OF ½ of the distance closer to (but on the other side of) the player with the best OF. If the best player remains the best, all players converge to it. If the best changes, the cluster leapfrogs downhill.

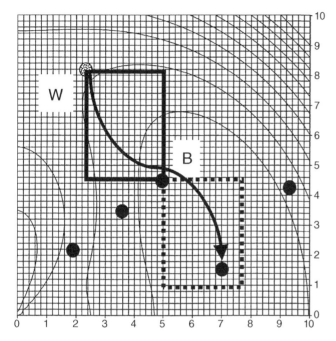

Figure 11.19 LF explanation: (5) Next leap-over.

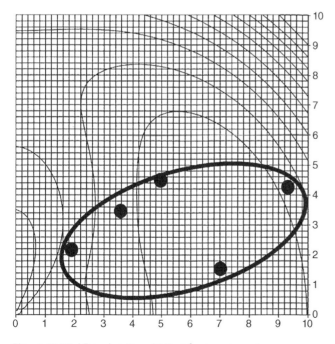

Figure 11.20 LF explanation: (6) Test for convergence.

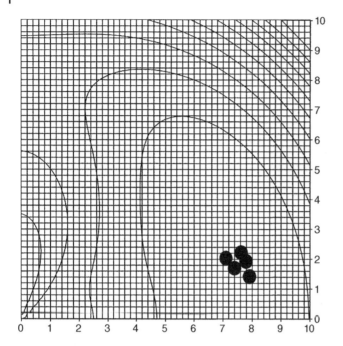

Figure 11.21 LF explanation: (7) Players are converging.

If there are M players (trial solutions), there are M initial OF evaluations. Then each leap-over only costs one additional OF evaluation. Nominally, I recommend that the number of players is 5 times the number of DVs, $M = 5N$, but with a minimum of 20.

$$M = \min\{20, 5N\} \tag{11.8}$$

However, this number is from experience, when seeking the number of players to maximize the probability of finding the global optimum while minimizing the NOFE to converge. It seems to be partly dependent on the leap-over amplification factor, which for simplicity and convenience, I recommend as unity (see Chapter 30 for an analysis).

Note: If there are N dimensions, each leap-over changes all N-DV values for the jumping player.

Certainly a user could consider one iteration to be one cycle through the algorithm—one search for W, one leap-over, and one OF evaluation. However, that would represent only one OF evaluation. In HJ, each iteration requires from $1 + N$ to $1 + 2N$ (averaging $1 + 1.5N$) function evaluations. In ISD each move of the trial solution requires $1 + N$ function evaluations. A simple Newton-type approaches $N^2/2$ function evaluation per iteration. In those optimizers, each iteration provides more information about the surface than just one point, and collection of information is used to direct the next trial solution. Accordingly, I chose to term an iteration in LF as N leap-overs, one leap-over per dimension, representing N OF evaluations. This is roughly equivalent to the work per iteration of the incremental steepest descent method. Not all players will necessarily move in an iteration—the worst could jump to a new spot and remain the worst.

Eliminating the worst is the driving philosophy of the leapfrogging and NM-SHH simplex searches. It is also the logic in searches called "Tabu." It is also the philosophy in Pareto-optimal selection. Also, to be shown, genetic algorithms and evolutionary algorithms remove the worst. By contrast, most optimizers seek to improve the best.

As a direct search, LF does not require surfaces to be continuum valued. It can handle cliffs and discontinuities, flat spots, and ridges that would result from discretization of the DV, time or space discretization in numerical methods to determine the OF value, conditionals that switch models and hence create discontinuities, or sampling discretization in time series models. Also, LF can handle constraints of any sort (explicit or implicit DV equations or as surprise non-executable encounters on variables within the OF).

Example 4   Figure 11.22 presents a 3-D view of an OF based on a model that seeks to predict player rank from player attributes. It is Function 84 in the 2-D Optimization Examples file. Rank is a discontinuous function. Like running a race, Players 1 and 2 remain first and second whether they are 1 cm or 20 m apart, as long a third does not pass second. As a result, the OF has flat spots, plateaus with zero slope, and separated by vertical cliffs.

Gradient and second-order (SQ and Newton-type) optimizers will encounter non-executable operations because of the flat spots. Single trial solution and direct search optimizers will converge wherever they are initialized because on each flat spot no direction is downhill.

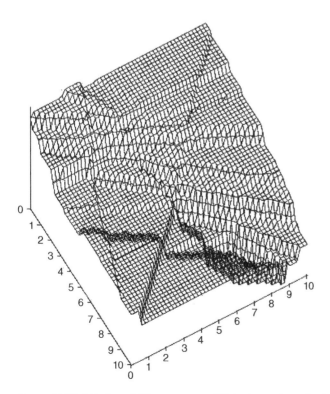

Figure 11.22 3-D view of OF response to DVs in the rank application of Function 84.

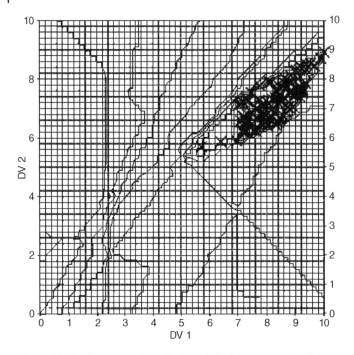

Figure 11.23 LF converged results from 100 trials on Function 84.

Figure 11.23 reveals the LF optimization from 100 trials, independently initialized. Note that 98 of the 100 trials found the global optimum. Although the model coefficient values were not exactly the same, the rank calculations placed the players in a ranking that exactly matched the data. The other two trials found a second-best choice for the rank model.

11.7.1 Convergence Criteria

Any of many criteria could be used to determine convergence of the LF algorithm. One of the simplest is the range on the DVs. For each DV calculate the range as the highest less the lowest. Test for each DV. Test after an iteration, after N leap-overs. When the range on each is small enough, then the team has converged. Report the best of the M players as the optimum. While simple to understand, this requires the user to identify a range for each DV (the implication of the DV on the OF might not be obvious) and the computer to search over all DVs for their ranges (computational time).

Another simple option is to claim convergence when the range of the OF values is small. It may be easier for a user to relate to the OF value and to determine what incremental change would be inconsequential than to do this for the DV values. Further, since the LF algorithm requires identification of the best and worst OF values, there is no additional computer burden (as there would be in determining the DV range). I recommend setting the OF threshold range for convergence to be an order of magnitude smaller than what the user decides is an inconsequential deviation from OF perfection, from a not quite optimal OF that is still fully acceptable in use. Test after an iteration, after N

leap-overs. Once converged, report the DV and OF values of the best of the M players. This is my preference.

There are many variations on such criteria. One could use rms (root-mean-square) deviations, for instance, on either OF or DV values. Or you could use the sensitivity of the OF to the DV and project the combined impact of the all of the DV ranges on the OF.

11.7.2 Stochastic Surfaces

A stochastic (noisy) application is one in which repeated samplings of the OF, at exactly the same DV trial solution, returns a distribution of OF values. Examples include experimental sampling or Monte Carlo simulations. If the surface is stochastic (noisy), then the best player might only appear to be the best because of the vagaries of sampling the surface (the OF value). Another sample at exactly the same location might reveal that the value at the same location is not so good. So to prevent an optimizer from finding the phantom best, a fortuitous artifact of all possible stochastic outcomes, reevaluate the OF of the best. Use the worse of the two (or more) replicate outcomes. Do this reevaluation prior to each leap-over. If a worse outcome is likely, it will eventually be expressed, and another player will become the best.

In this case, prior to moving the worst, reevaluate the best. Assign to the best, the worst of the replicate OF values. The extra OF evaluation(s) constantly checks "Are you still the best?" and permits searching for the "worst of the best" not the ultimate, if ever there was a "lucky best." If the best remains the best, leap over it. However, if the former second-best player becomes the best, leap over it.

In this case, at the end game of surrounding the optimum, the team does not converge on one spot. Instead, players leap around in the vicinity. Accordingly, an alternate criterion for convergence is needed. What I recommend is to observe the OF value of the worst player. As the team gathers in the vicinity of the optimum, the worst player OF value will fall to that of the local area, but will not asymptotically approach a single value; it will be subject to the vagaries of the stochastic surface, and its OF value will relax from the initialization worst to a noisy steady state representing the end-game vicinity. The player labeled the worst will likely change at each iteration. This is not observing the original worst player. Use steady-state identification on the worst OF value at each iteration and claim convergence when the worst (at each iteration) comes to steady state.

11.7.3 Summary

Balancing all measures of goodness, I think leapfrogging is the best of all optimizers. However, when simplicity is a priority, I think the cyclic heuristic direct search is best.

11.8 Particle Swarm Optimization

This optimization algorithm mimics a swarm of gnats, a flock of birds, or a school of fish, as they search for food, a safe place, or whatever. Such types of algorithms are called mimetic. They mimic the "intelligence" that we find in nature. Simulated annealing mimics the organization of molecules as they cool from a liquid to a structured solid. Ant colony algorithms are based on the human interpretation of the logic underlying the behavior of ants. The incremental steepest descent algorithm is based on a cube of ice sliding downhill in a gravitational field. In all, there seems to be some intelligence in the inanimate

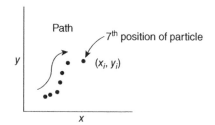

y

Path

7th position of particle

(x_i, y_i)

x

Figure 11.24 Path of a particle in 2-D space.

object or colony/swarm/association of not very smart indivi-duals. Other mimetic algorithms include genetic, evolution-ary, synthetic biology, bee colony, etc. Particle swarm optimization (PSO) is one of the earliest mimetic algorithm and remains as an archetypical example.

11.8.1 Individual Particle Behavior

Before considering the swarm, we'll look at the individual. A "particle," is a single player, an "individual," or a "trial solu-tion." Figure 11.24 shows a particle path through 2-D DV space.

The particle could move smoothly, deterministically, rationally, like a marble rolls downhill, but my observation of gnats in nature seems that they do it differently. The position of an individual within the swarm seems to have a random component. Further, similar to the manner that movie film or a strobe light discretizes the appearance of how things move in time, we will consider a par-ticle in discretized time intervals that has a random positional component. Using "i" as the iteration counter number, (x_i, y_i) is the x–y position of the particle at the ith iteration. Random motion can be modeled as

$$x_{i+1} = x_i + c_x(r_{ix} - 0.5)$$
$$y_{i+1} = y_i + c_y(r_{iy} - 0.5)$$

(11.9)

Here, r_i are independent random numbers, with a uniform distribution and range of $0 < r \le 1$, UID [0,1]. These random perturbations are independent on x and y as well as for each iteration. The coeffi-cients c_x, c_y are scale factors that determine how large the random position perturbations are. Nominal c values might be 0.01 * DV range.

Alternately, you could use Gaussian NID(0, σ_x) and NID(0, σ_y) for the random perturbations.

The Equation set (11.9) defines a random walk behavior. It is also a position model, as opposed to a velocity or momentum model. Alternately, you could model velocity changes as if there were attractive forces on the particle. Start this by using Newton's law of motion:

$$F = ma \to \frac{F}{m} = \frac{dv}{dt} \cong \frac{\Delta v}{\Delta t}$$

(11.10)

Expanding the x-dimension as a finite difference solution to the aforementioned:

$$v_{x_{i+1}} = v_{x_i} + \Delta t \cdot \frac{F_x}{m}$$

(11.11)

If the driving force is a uniformly distributed random phenomena and coefficients Δt and m are lumped into one factor, c,

$$v_{x_{i+1}} = v_{x_i} + c_x(r_{ix} - 0.5)$$
$$v_{y_{i+1}} = v_{y_i} + c_y(r_{iy} - 0.5)$$
$$x_{i+1} = x_i + \Delta t \cdot v_{x_i}$$
$$y_{i+1} = y_i + \Delta t \cdot v_{y_i}$$

(11.12)

Motion equation set (11.9) shows random walk behavior. It is a position model. Set (11.12) includes momentum and shows an autocorrelated random walk behavior. It is a velocity model. It is often termed an ARIMA (auto regressive integrating moving average). Δt is simply another parameter in the optimizer, the concept is force-related perturbations on velocity, but there is no time concept in the optimizer. Each iteration is one Δt time interval, one epoch, so use $\Delta t = 1$.

Both the position and velocity models work in PSO. I find the rationale a user provides for a particular choice is simply a personal preference, more so than either a mechanistic logic related to the application or to a comprehensive analysis related to maximizing optimizer desirables while minimizing undesirables.

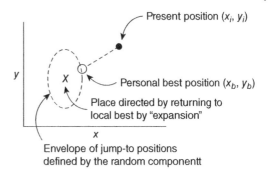

Figure 11.25 Illustration of influences on an individual particle.

If the particle can "recall" a previous best location, then it might be driven by both a random component and a "desire" to return toward the previous best. For the position-based model the motion equations become

$$x_{i+1} = x_i + c_x(r_{ix} - 0.5) + d_x(x_{\text{best}} - x_i)$$
$$y_{i+1} = y_i + c_y(r_{iy} - 0.5) + d_y(y_{\text{best}} - y_i) \tag{11.13}$$

The logic of Equation (11.13) is illustrated in Figure 11.25. The solid dot represents the present (x_i, y_i) position of the particle, and the open circle is the location of the personal historical best, (x_b, y_b). The historical best location and corresponding OF value need to be kept in computer memory and revised whenever a chance location finds a better spot. The spot indicated by the X indicates the deterministic leap-to location defined by the "d" terms in Equation (11.13), and the dashed circle reveals the local region within which the "c" terms would randomly place the particle.

Here the factor "d" means that the particle would desire to jump toward its personal best position or perhaps beyond. But you might only want it to jump halfway back to its personal best then use the random move. So the factor "d" might have reasonable values in the 0.2–1.5 range. The d value could be adjusted during the optimization based on patterns in the search that would indicate strong confidence in being near the global or not.

This desire to return to the personal best could also be modeled as a driving force in the velocity mode model:

$$v_{xi+1} = v_{xi} + c_{1x}(r_{ix} - 0.5) + c_{2x}(x_b - x_i)$$
$$v_{yi+1} = v_{yi} + c_{1y}(r_{iy} - 0.5) + c_{2y}(y_b - y_i)$$
$$x_{i+1} = x_i + \Delta t \cdot v_{xi}$$
$$y_{i+1} = y_i + \Delta t \cdot v_{yi} \tag{11.14}$$

You could add gradient, steepest descent, and action:

$$x_{i+1} = x_i + \cdots + c_{3x}\left[-\frac{\partial f}{\partial x}\right] \tag{11.15}$$

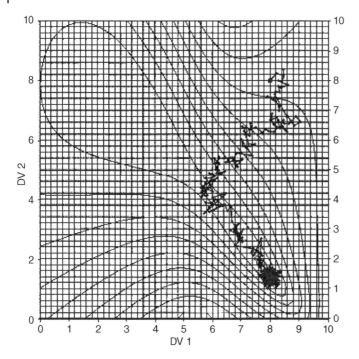

Figure 11.26 Illustration of an individual particle trend toward the minimum.

This will accelerate motion in the right direction but adds the problems associated with gradients (noise, discontinuities, choice of scale factor), and probably does not match the sensor ability of gnats and mosquitoes who seem to direct their motion by "smell" of CO_2 or H_2O intensity, not a gradient.

One particle with either Equation Set (11.13) or (11.14) can progress toward the optimum. Figure 11.26 illustrates one realization of the particle path from the initial trial solution at about (8, 8) toward the optimum at about (8, 1.5). It uses the simpler positional form of the equations. Note that the distance between each step in the particle path is not the same; it is influenced by the random driver. Recognize also that the particle generally moves downhill but often takes random diversions to an uphill point. That may appear to be not very smart, but in some instances it permits a particle to fall into a surprise minimum. With the simple logic of "explore locally with a small tendency to move back toward the personal best," the particle can find the minimum.

However, also observe in Figure 11.26 that the particle never converges to the minimum, but once in the vicinity it keeps randomly fluctuating about. The particle motion could include a perching rule, such as "if no better location has been found in 10 steps, and then begin reducing the c values" to diminish the random perturbations. Then it would converge to a point, and classic criteria could be used to claim convergence and report a single answer. However, the random movement could be continued, and convergence claimed when the OF w.r.t. iteration trend relaxes to a noisy steady state. There are many options.

Also realize that the random vagaries driving the particle could have as easily sent it downhill toward the left of the initial trial solution and led it on a path to the local optimum at about (2, 8).

11.8.2 Particle Swarm

In particle swarm, many particles are initialized in random locations in the DV space. Each particle follows the rules as previously. However, if we allow "wireless" communication between particles or permit them to "see" each other, such that this ability instantly transmits knowledge of the best overall particle position, and include a desire by each to also move toward the global best of all particles in the swarm, then the x-position model becomes

$$x_{i+1,j} = x_{i,j} + c_{1x}(r_i - 0.5) + c_{2x}(x_{bj} - x_i) + c_{3x}(x_g - x_{i,j}) \tag{11.16}$$

The variable $x_{i,j}$ is the x-position of the jth particle at the ith step. The variable x_g is the global best of all time for all particles in the swarm. The coefficient c_{3x} scales the desire for the individual particle to move toward the global best. There are similar equations for all particles and for each dimension of the DV. There are similar sets for either position or velocity models.

There are many implementation options:

1) You could model the global best information as diminishing with distance from global best such as illustrated in Figure 11.27:

$$c_3 = e^{-d^2/s^2} \tag{11.17}$$

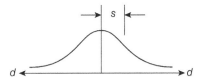

Figure 11.27 Radial basis function scaling of draw.

where "s" is a scale factor. Effectively there is hardly any magnitude for c_3 if $d > 3s$.

Additionally, one could have the global "draw" randomly occurring:

$$c_3 = r \cdot e^{-d^2/s^2} \tag{11.18}$$

2) You could have the c_1 factor diminish as the particle finds an optimum, and identify this agitation magnitude by moves that were not better. The rule might be *If $f_{j_{i+1}} > f_{j_i}$ then contract c_1, else expand c_1.*

3) If the particle wants to go to an x, y-position that violates a constraint, then return it to the previous position:

$$x = xold + c_1(r - 0.5) + c_2 C \dots$$

If   a constraint is violated

$$x = xold$$

else

$$xold = x$$

4) There are another trillion modifications, for example:
   - Claim swarm convergence when cluster size is within statistical randomness.
   - When the swarm is in a common vicinity, reduce the random perturbations to have them all converge on a common spot.
   - Use Gaussian perturbations instead of uniform.
   - Introduce an occasional swarm scattering event to enhance exploration.
   - Include gradient information gathered from the random explorations.
   - Accelerate motion of global best individual toward its global best.

- Include a repulsion from past vacated (not best) places.
- Reposition the worst in a random spot. The worst adds little value. Use that function evaluation to explore uncharted areas.
- At the end game, when the cluster seems to be in one common vicinity, eliminate particles to reduce computational burden.

What would you use as the criteria/criterion for convergence with such options?

Example 5   PSO is used to find the minimum of Function 11 with steady-state stopping criterion. I suggest that you observe this in the 2-D Optimization Examples program to enjoy the show. The path to convergence is revealed by Figure 11.28, a plot of the worst player OF value w.r.t. iteration. Steady state was claimed at iteration 140.

When convergence was claimed, the players were swarming about the global optimum, as revealed in Figure 11.29. The X within the open circle indicates the global best at convergence, and it is a true indicator of the optimum. (Note: The player that once found that spot will likely have left it to explore the local region. Likely no player is in the spot at convergence.) The solid dots indicate the player positions at convergence. Note that they were randomly initialized throughout DV space and gathered to the common best location. Note that they were permitted to continue random perturbations; there was no perching rule to permit end-game convergence to a single spot.

You should observe this and enjoy the show. Choose the "Y" option in the "Observe" field and enter a small value (like 0.01) in the "Pause" field to give the screen time to update.

Note: There are 20 particles in this swarm, and each has an OF evaluation at each iteration. With initialization and 140 iterations this search for the optimum took $20(1 + 140) = 2820$ function evaluations. That is much more than other optimizers take. The justification for particle swarm is that it has a high probability of finding an otherwise hard to find global optimum.

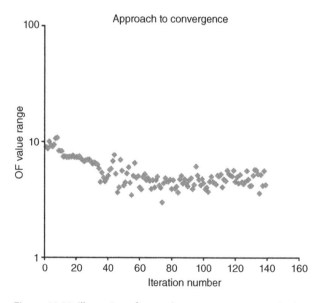

Figure 11.28 Illustration of a steady-state convergence criterion on the OF value of the worst particle.

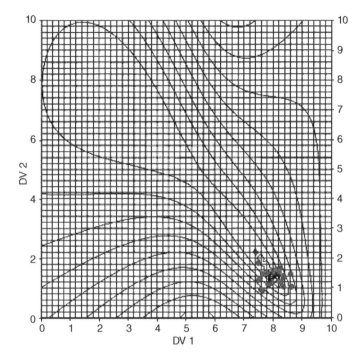

Figure 11.29 Particle positions at convergence.

Note: Even if the OF is deterministic, since this search is driven by random perturbations, the search will not be repeatable. The optimizer is in the category termed a stochastic algorithm. This is different from a OF that is stochastic, but either situation makes the OF value w.r.t. iteration be a random variable.

11.8.3 PSO Equation Analysis

The position model is equivalent to a first-order ordinary differential equation, a very familiar equation. To see this, start with the position model:

$$x_{i+1} = x_i + c(r-0.5) + e(x_b - x_i) + d(x_g - x_i)$$ (11.19)

Rearrange it to collect x_i terms, and you have

$$x_{i+1} - x_i + (e+d)x_i = ex_b + dx_g + w_i$$ (11.20)

Expressing the finite difference as a differential, dividing by $(e+d)$, and defining $\tau = \Delta t/(e+d)$, results in

$$\tau\frac{dx}{dt} + x = \frac{e}{e+d}x_b + \frac{d}{e+d}x_g + w_i$$ (11.21)

This is a first-order ODE with individual best, global best, and random perturbation as the forcing functions.

Similarly, the velocity mode model is simply a set of two first-order sequential ODEs or one second-order ODE. The models for v and x

$$v_{i+1} = v_i + c(r-0.5) + e(x_b - x_i) + d(x_g - x_i) \tag{11.22a}$$

$$x_{i+1} = x_i + \Delta t \cdot v_i \tag{11.22b}$$

can be combined into

$$\tau^2 \frac{d^2x}{dt^2} + 0\frac{dx}{dt} + x = \frac{e}{e+d}x_b + \frac{d}{e+d}x_g + w_i \tag{11.23}$$

Equation (11.23) is a second-order ODE with individual best, global best, and random perturbation as the forcing function

11.9 Complex Method (CM)

Contrasting the simple pattern of local points in the simplex, the complex method randomly places a good number of trial solutions (players) in feasible DV space, creating a multiplayer complex. The worst is reflected through the center of gravity of the other players. The move-to spot for each DV dimension is calculated from the same formula. Here is the concept for a nominal move in dimension, x:

$$x_{j,i+1} = \bar{x}'_i + \alpha\left(\bar{x}'_i - x_{j,i}\right) \tag{11.24}$$

The subscript j indicates the index value of the worst player, and $x_{j,i}$ is that player's x value at the ith iteration. The variable $x_{j,i+1}$ is the move-to x location and $\bar{x}'_i$ is the average of all players excluding the jth. The scale factor α indicates how far on the other side of the center of gravity to move the player.

This algorithm has elements that are very similar to LF. The worst one player moves at a time and vacates the prior position. However, instead of leaping over the best, in CM it leaps over the centroid of all other players. Also, instead of including a stochastic randomizing element in the leap-to position the complex method moves to the alpha point along a straight line.

If the move-to location is infeasible or results in a worse OF value, have the value of the distance coefficient α. Ravindran, Ragsdell, and Reklaitis report, from empirical experience, the recommended number of players is $M = 2N$ (twice the number of dimensions) and $\alpha = 1.3$. This value for alpha would progressively expand the cluster, if a move led to a better OF value, but if not, the value of $\alpha = 1.3/2 = 0.65$ would cause the cluster to contract.

It would seem that calculating the centroid of the position of the other players at each iteration is a computational burden. However, once the centroid of all players is calculated, $\bar{x}_i = (1/M)\sum_{j=1}^{M} x_{i,j}$, it can be incrementally updated after each successful move:

$$\bar{x}_{i+1} = \bar{x}_i + \frac{1}{M}\left(x_{i+1,j} - x_{i,j}\right) \tag{11.25}$$

Then the value of $\bar{x}'_i$ can be simply calculated by removing the value of $x_{j,i}$ from the overall centroid.

$$\bar{x}'_i = \frac{1}{M-1}\left(M\bar{x}_i - x_{i,j}\right) \tag{11.26}$$

11.10 A Brief Comparison

This section explores results from running several optimizers on Function 11 in the 2-D Optimization Examples file. The function is a linear regression application; however rather than using the common vertical least squares, a student, Chong Vu, was exploring a total least squares (or normal or perpendicular least squares technique). His first choice of (x, y) points to fit with a straight line seemed to give the model two options, one with a very steep negative slope and one with a modest positive slope. To keep the large negative slope solution in bounds on the 2-D contour, I added a penalty for off-graph values. Figure 11.30 is a 3-D view of the function. Note that there is a global optimum at about (8, 1.7) and a local optimum at about (3.3, 7.8).

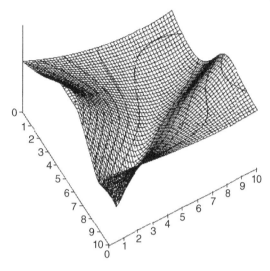

Figure 11.30 Surface of Function 11.

Table 11.1 reveals several optimizer outcomes. Column 1 lists the optimizers, and Columns 2–4 some relevant performance aspects. The convergence criterion for all optimizers except the last listed (PSO w SSID) is based on the DV change. Each must be <0.001. However, although a uniform criterion, it does have different meanings for the different optimizers. For gradient and second-order optimizers, it refers to the trial solution change in the DV values. For HJ it refers to the DV range on the pattern exploration. Similarly for CHD it is based on the incremental explorations. Finally, for the multiplayer algorithms, LF and PSO, it is applied to the DV range of all players. With 20 players, the best, when all are within convergence criterion, will likely be very close to the optimum. Note, although a seemingly identical convergence criteria is used, it has different implications for precision of the OF*.

The data reflects the outcomes of 100 trials from independently, randomly initialized trial solutions. ANOFE means average number of function evaluations to converge over the 100 trials. *P* global is the

Table 11.1 Optimizer results on Function 11.

Optimizer label	ANOFE	*P* global	Solution precision (OF* standard deviation)
CSLS (with heuristic search)	459	0.37	0.068
ISD with filtered ∇	145	0.37	0.038
NR	217	0	NA
LM	559	0.34	0.029
RLM	516	0.46	0.018
CHD	184	0.34	0.14
HJ	109	0.42	0.35
LF	283	0.33	0.012
PSO with contraction	3124	1.00	0.006
PSO with SSID	2474	0.99	0.15

probability that the optimizer found the global optimum—the fraction of successes in the 100 trials. The standard deviation is the precision associated with the trials that found OF*.

Note that each statistic in the table is the result of 100 trials, representing a sample of possible realizations, not the entire population. A new set of 100 trials will return similar values, but generally, each new average value is about ±5% of the true average.

The optimizer with the fewest number of function evaluations to converge would be preferred. In this case it is HJ. However, it has the worst precision. It does not have the worst precision because it is a bad optimizer; it is because the convergence criterion permitted it to stop prior to getting as close to the optimum as the other optimizers. When the convergence criterion threshold is changed to 0.00001, then the precision (standard deviation on the OF* values) was 0.032 that is of the same values as ISD or LM. However, the extra work to achieve that precision resulted in ANOFE = 269. So one should not look at a single performance metric. A comparison of optimizer work (ANOFE) needs to be on an equivalent precision basis.

Similarly, the optimizer with the highest probability of finding the global would be preferred. In this set of 100 trials, PSO (with end-game contraction of the swarm) found the global in 100% of the trials. However, that is the result of limited sampling, in many more trials the figure seems more like 99%. Regardless, of these optimizers searching on this function, PSO has the highest probability of finding the global. However, the benefit comes with a tremendous increase in the number of function evaluations needed to converge. So, again, in comparing optimizers there needs to be equivalence between all performance metrics.

Further, replicate trials (of 100 runs each) with RLM report that the probability of finding the global ranges between 34 and 46%. The one trial set that reveals 46% in the table is not the population truth, but the reveal of one particular sample. So the conclusion is if you want definitive results or to be able to differentiate optimizers on a fine scale, you'll need the results from many more than 100 trials.

Note that NR never found the global. Even when initialized in close proximity to the optimum and using central difference estimates of the derivative, surface shape sent it off on a wild goose chase, and it landed on either the local optimum, maximum, or the saddle point.

This particular function is continuum valued. There are no discontinuities from discretized DV values, numerical integration step sizes, conditional rules in the model, or category/rank considerations. Accordingly, optimizer analysis on this one function cannot be used as a universal statement of performance. Robustness to surface aberrations (a term referring to all of the non-continuum effects) needs to be considered when ranking one optimizer against another for general purpose. So perform comparison tests on functions with diverse characteristics.

11.11 Takeaway

Direct searches can handle discontinuities and constraints, and although their algorithms do not have the mathematical sophistication of gradient-based or second-order optimizers, direct search algorithms are more robust—they are generally more applicable. Multiplayer algorithms have a higher probability of finding the global optimum than do single TS optimizers. An iteration in one algorithm can take many more or fewer function evaluations than another. It is not iterations that define algorithm speed, but the number of function evaluations required to find the optimum and the internal work (such as that related to sorting or linear algebra procedures) to calculate the next TS. When comparing optimizers, choose conditions that result in equivalent performance in other important metrics.

My favorite optimizer is LF, and my second favorite is CHD. One could claim that I have a bias, but I will claim that the choice is grounded in balancing the number of function evaluations to find the global solution with precision, simplicity, and robustness of the algorithm.

11.12   Exercises

1   Start at a point near the middle left side on the contour in Figure 11.31, and show initialization and four iterations of the Nelder–Mead simplex search toward the optimum. Repeat from a point in the middle right side.

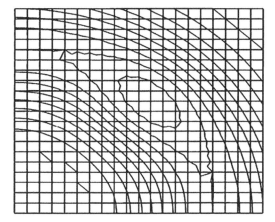

Figure 11.31 Illustration for Exercises 1 and 2.

2   Start at a point near the middle left side on the contour in Figure 11.31, and show three iterations of a Hooke–Jeeves search toward the optimum. Repeat from a point in the middle right side.

3   Figure 11.32 shows contours of a 2-D function. The minimum is near the (x, y) point $(5.5, 5)$. From position $(5, 8)$ illustrate the first three iterations of the cyclic direct search with the initial $\Delta x = +2$ and initial $\Delta y = +2$.

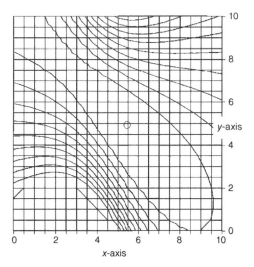

Figure 11.32 Illustration for Exercises 3 and 4.

4 From position (5, 8) in Figure 11.32, illustrate the first three iterations of the Hooke–Jeeves search with the initial exploration values of $\Delta x = +2$ and initial $\Delta y = +2$.

5 For a Hooke–Jeeves 2-D search, random surface structure would lead to an average of four function evaluations per pattern move and explorations in the identification of the new base point. Consider this rule to modify HJ: If the pattern was moving in the right direction, then it seems likely that the first explorations should be in the same x and y direction of the last successful x and y exploration moves. How might this be implemented? What would be the impact on number of function evaluations in the initial search, in the vicinity of the optimum?

6 Show the first three player moves of a leapfrogging search on Figure 11.33. Be sure to randomize your new position for the player in the jump-to area. Start the players at (horizontal, vertical) positions of (6, 6), (4, 3), (2, 5), and (7, 2). The minimum is indicated by the black dot at about (4, 4). Explain your jumps.

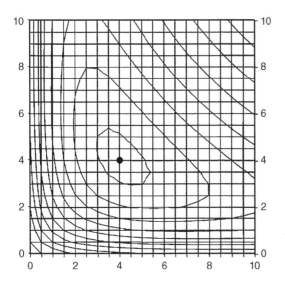

Figure 11.33 Illustration for Exercise 6.

7 In this chapter, the 2-D cyclic heuristic search used $x1$ and $x2$ (Cartesian coordinate system x and y) as the DVs. Could the cyclic search have been set up for a cylindrical coordinate system using r and θ? Defend your "Yes," "Maybe," or "No" answer.

8 Show the first two pattern moves on Figure 11.34 for a Hooke–Jeeves method (local search, pattern move, local search, pattern move). (i) Start at DV1 = 8 and DV2=8, and use the initial search pattern to be +1 DV units. Start at DV1 = 5 and DV2=9, and use the initial search pattern to be −1 DV units.

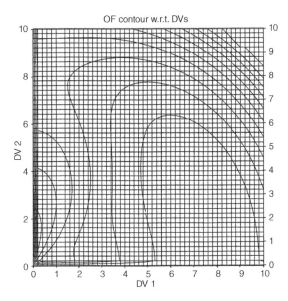

Figure 11.34 Illustration for Exercise 8.

9 One can define points in a plane in either Cartesian (x, y) or polar (r, θ) notation, and there is a unique one-to-one translation between (x, y) and (r, θ) values. (i) What aspect of an optimization application would make one notation more convenient than another? (ii) If you chose the leap-frogging method and polar coordinates, how many players are recommended?

10 On Figure 11.35, show the first three leap-overs, include the leap-to windows. Three LF players are located at DV values of $(4, 2)$, $(6, 6)$, and $(9, 5)$. The optimum is about $(6, 4.5)$. Clearly label the moves as (1), (2), and (3).

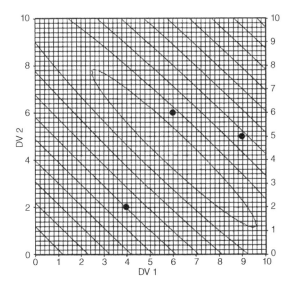

Figure 11.35 Illustration for Exercise 10.

11 Explore momentum in the optimization trial solution calculation. Observing the trial solution change with iteration number, we actually "see" it change in our observation time. It appears to be an object moving in the 2-D space, speeding up, curving, or slowing down as the optimizer rules direct the next trial solution. Momentum is a concept from the motion of objects. It would keep an object moving along the linearized path, unless some force causes velocity to change. Often people write the particle swarm equations in the velocity mode, using momentum, with the forces proportional to the deviations from particle local and swarm best. However, I wrote the PS equations in the position mode, with no momentum and with the jumps in position proportional to deviation from the local and swarm best. Determine whether momentum is helpful or not. Your evaluation should be somewhat comprehensive (tested on a selection of functions that exemplify a range of OF features and based on a diversity of measures of goodness).

12 Provide a flowchart and supporting text explanation for HJ.

13 Provide a flowchart and supporting text explanation for NM.

14 Provide a flowchart and supporting text explanation for LF.

15 Provide a flowchart and supporting text explanation for CHD.

16 Provide a flowchart and supporting text explanation for CM.

17 Provide a flowchart and supporting text explanation for PSO.

18 Use the 2-D Optimization Examples file to explore a variety of optimizers on a variety of functions. Show that the optimizer TS path affirms what is expected. Explain your choice of trials and conditions and assessment metrics. State why you would prefer one algorithm over another.

12

Linear Programming

12.1 Introduction

Linear programming (LP) is a technique for the special case of a linear OF and linear constraints. In a 2-DV case, both OF and constraints have the linear form of $a + bx_1 + cx_2$. While this may seem to be a trivial and restrictive model, it actually represents broad categories of applications in scheduling, blending, allocation, and logistics. LP applications are the mainstay of business optimization, and the applications go well beyond the 2-DV realm, with 100–1000 DVs not uncommon.

This text has a focus on nonlinear optimization. Although linear applications constitute an important category of optimization applications and linear programming (LP) is a mainstay for many, this chapter only provides an introduction. The reader should have an understanding of the application category and LP technique, but this book does not provide details to make one an expert in its implementation.

A few examples will introduce the linear aspects of the application.

Example 1 An optimization application that leads to an LP structure, and which may be familiar to the reader, is planning meals. The objective is to minimize the cost of food items purchased for the week of meals while preparing meals that meet minimum daily requirements of nutrition elements. Here, only two constraints, Vitamin A and salt, are explicitly indicated; but the many S.T. relations would include both minimum and maximum values and on all essential features such as water, fiber, protein, Vitamin C, calcium, etc. The optimization statement is

$$\begin{array}{c} \min \\ \{\text{quantity of each item}\} \end{array} J = \text{Cost of meals} = \sum_{i=1}^{N} c_i n_i$$

$$\text{S.T.}: \ \text{nutrition}: \sum \text{vitamin A} = \sum a_i n_i \geq \text{MDR}$$

$$\sum \text{Salt} = \sum b_i n_i \geq \text{MDR}$$

(12.1)

where n_i is the quantity of foot item i; c_i the price/unit of food item i; a_i, b_i, ... the amount of $a,b,...$ in food item i.

Note: The units on all variables are distinct: the quantity of rice may be gram; water, mL; the cost, $; and minimum daily requirements (MDR), USP units per day. Further, some variables may be continuum valued (32.4 oz water), but some may be integers (five pealed carrot sticks).

Note that both the OF and the constraints are linear.

Engineering Optimization: Applications, Methods, and Analysis, First Edition. R. Russell Rhinehart.
© 2018 R. Russell Rhinehart. Published 2018 by John Wiley & Sons Ltd.
Companion website: www.wiley.com/go/rhinehart/engineeringoptimization

Example 2 A woodworking crafts person wants to make foot stools and picture frames for an upcoming event. There is a fixed supply of time and wood, and the objective is to maximize the sale value of the items at the event. Using s as the number of stools, p as the number of picture frames, and a_k as the expected sale price of the items, the OF is $OF = a_s s + a_p p$. Using t as the time to make each item and w as the amount of wood needed for each item, the constraints are $t_s s + t_p p \leq T$ and $w_s s + w_p p \leq W$. Additionally, one cannot make a negative number of items, so $p \geq 0$, and $s \geq 0$. Additionally, s and p must be integers.

The optimization statement is

$$\begin{array}{l} \max\limits_{\{s,p\}} J = z = a_s s + a_p p \\ \text{S.T.:} \quad t_s s + t_p p \leq T \\ \qquad w_s s + w_p p \leq W \\ \qquad p \geq 0, \quad \text{integer} \\ \qquad s \geq 0, \quad \text{integer} \end{array} \tag{12.2}$$

Note: The units on each item are not identical. The DVs are integer numbers of items. The OF value is perhaps \$. The time constraint, T, has the units of days, and the material constraint has the units of board feet.

Note: Both the OF and constraints are linear functions of the DVs.

In general, the application can be stated in vector notation:

$$\begin{array}{l} \min\limits_{\{n_1,n_2,n_3,\ldots\}} J = \sum\limits_{i=1}^{N} c_i n_i = \mathbf{c}^T \bullet \mathbf{n} \\ \text{S.T.:} \quad \sum a_i n_i = \mathbf{a}^T \bullet \mathbf{n} \geq d_a \\ \qquad \sum b_i n_i = \mathbf{b}^T \bullet \mathbf{n} \geq d_b \end{array} \tag{12.3}$$

Other applications that lead to similar OF application structures include the following:

1) Blending sand (for Si), shale (for Al), and limestone (for Ca) to make cement
2) Purchasing: crude oil from suppliers A, B, C, etc., to refine to desired products
3) Allocating: personal time to meet all commitment and maximize wellness
4) Allocating: production of units to different manufacturing sites
5) Scheduling: when to make different products

LP applications are common to business management, organization management, logistics, and such.

The DVs could be continuum valued such as the time allocated to something or the mass in a package. Alternately, the DVs could be integers representing the number of items, deliveries, people, stages, etc.

LP has proven to be so useful, robust, and fast that it is not unusual for people to linearize a nonlinear application (both OF and constraints) about an initial TS and use LP to seek a solution. This would provide an approximation to a solution. Then successively repeat the local linearization about the approximate optimum and generate a new LP solution until the successive optima show no change.

12.2   Visual Representation and Concepts

Figure 12.1 illustrates a 2-DV situation with the OF calculated as OF $= a + bx_1 + cx_2$. The OF value is a plane in the three-dimensional view. The DVs are the axes that define the lower horizontal plane, and the OF is the vertical direction. With the OF plane so illustrated, the maximum is always farther to the right (higher x_1 values) and to the back (larger negative x_2 values). Similarly the minimum is to the front and left. In either case, however, farther along the x_1, x_2 axes gives a better OF value, which tends to be an unrealizable infinity at infinite x_1 and x_2 values. To prevent unrealizable solutions, constraints are indicated on the x_1–x_2 plane as lines that can be expressed in any number of manners. For example, $x_2 \leq d + ex_1$, or $d + ex_1 + fx_2 = g(x_1, x_2) \geq 0$.

Without constraints, the optimum is at x_1, $x_2 = \pm \infty$, depending on signs of OF coefficients b and c. With inequality algebraic constraints on the DVs, the shaded area defines the feasible DV region.

The inequality constraints could be dependent on either continuum or integer values, less than or less than or equal to conditions, or based on individual DVs or their combinations. Here are some examples placed in standard form:

$$dx_1 + ex_2 \geq f \tag{12.4a}$$

$$gx_1 + hx_2 \leq \text{integer} \tag{12.4b}$$

$$0 \leq x_1 \leq p \tag{12.4c}$$

$$0 \leq x_2 \leq \text{integer} \tag{12.4d}$$

If there is a greater than, $>$ or $\geq$, constraint, multiply the equation by -1 to convert it into a $\leq$ constraint to place it in the canonical form for LP.

Because the OF is a plane, the net lines on the OF surface are all parallel straight lines, as are contour lines.

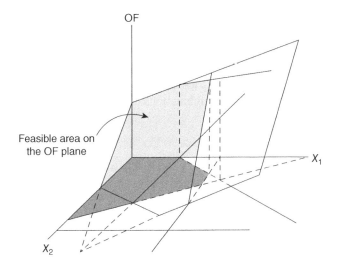

Figure 12.1 Illustration of LP concepts.

The optimum is on an intersection of constraints. To understand this, take any DV combination that is an unconstrained interior point. Moving a DV in the right direction improves the OF, but one can only move the DV to a constraint. Then moving along the constraint in the right direction can improve the OF. Note now that along the constraint, both DVs will have to change as one moves along a constraint. But one cannot move to infinity. One can only move to the point where the DV path hits another constraint. The optimum is at the intersection of constraints, and the optimum is likely unique. In the rare event that the constraint defining the best OF value is parallel to that best contour, then the maximum is any point along that constraint. But if the constraint is just a tiny bit not parallel to the contour, moving along the constraint to its intersection with another constraint finds the unique optimum. The best point is on an intersection of constraints. There may be equivalent points along a constraint, but a search only needs to evaluate the points at constraint intersections to find the best.

What would indicate that the constraint with the best OF value is parallel to a contour in the OF plane? First, consider the OF model $OF = a + bx_1 + cx_2$. On a contour the OF value is a constant OF_c, which defines the x_1, x_2 relation as $x_2 = (OF_c - a)/c - (b/c)x_1$. The slope of the contour in the DV plane, the relation between DVs is $-b/c$. A constraint could be represented as $dx_1 + ex_2 \leq f$, and on the constraint the relation is $dx_1 + ex_2 = f$, which leads to a equation for a line in the DV plane, $x_2 = f/e - (d/e)x_1$. In the special case where the slope of a constraint is equal to the slope of the contour projection, if $-b/c = -d/e$, then any point along that constraint will be an equivalent solution.

Even so, the intersection of that constraint with any other constraint is as good as any other place on the boundary.

Note: The feasible DV region illustrated in Figure 12.1 is convex. There is one contiguous feasible area. More restrictively, if you choose any two feasible DV sets, a straight line connecting them will lie wholly within the feasible region. This is condition of the introductory LP method, although enhanced algorithms can cope with non-convex constraints.

Also, note that there are more constraints than DVs in the illustration. If the number of constraints is equal to the number of DVs, then there is only one solution and one common intersection. If the number of constraints is fewer than the number of DVs, then the optimum will be at an infinity extreme, which indicates that some constraint was not identified.

Finally, note that in Figure 12.1 one of the constraint intersections is outside of the feasible area. Not all intersections are feasible.

This elementary presentation will be for the common LP application with a convex feasible region and the number of constraints greater than the number of DVs.

How many constraint vertices are there? Consider a system with m number of DVs and $n > m$ number of linearly independent constraints with a convex DV feasible space. (Linearly independent means that none of the constraint equations can be generated by linear combinations of any others.) The set might appear as

$$a_{1,1}x_1 + a_{1,2}x_2 + \cdots + a_{1,m}x_m \geq b_1$$

$$a_{2,1}x_1 + a_{2,2}x_2 + \cdots + a_{2,m}x_m \geq b_2$$

$$a_{3,1}x_1 + a_{3,2}x_2 + \cdots + a_{3,m}x_m \geq b_3$$

$$a_{4,1}x_1 + a_{4,2}x_2 + \cdots + a_{4,m}x_m \geq b_4 \tag{12.5}$$

$$\vdots$$

$$a_{n,1}x_1 + a_{n,2}x_2 + \cdots + a_{n,m}x_m \geq b_n$$

Or in compact vector matrix notation as

$$\mathbf{A}x \ge b \tag{12.6}$$

On any active constraint the inequality condition becomes its equality limit. With m number of variables, any m number of independent linear equations can be selected to define a constraint vertex. How many choices of m equations are there in a set of n? The answer is the number of combinations:

$$N = \binom{m}{n} = \frac{n!}{m!(n-m)!} \tag{12.7}$$

With $n = 5$ sets of constraints, and $m = 3$ DVs, there are only 10 intersections to explore. Here, an exhaustive search could be simply implemented. However, with 10 DVs and 20 constraints, there are about 185,000 vertices. We'd like to have an efficient search procedure that does not need to explore all possible vertices. LP is a clever way to structure the search for the best among all constraint intersections, greatly reducing the number of vertices that need to be evaluated.

LP has been around quite a while, and over time people have developed enhancements and procedures for special cases and ways to interpret variables. But since I am focusing this text on nonlinear optimization, LP is not a central method that I would suggest. I have included LP basics in this text, however, because it is important. If you desire greater depth, I recommend the text by Hillier and Lieberman.

In a 2-DV problem with constraints on the DVs only, the constraints are line boundaries in the DV1–DV2 plane as illustrated.

If there are also constraints included on the OF, then the constraint equations could appear to have three terms:

$$ax_1 + bx_2 + cz < j \tag{12.8}$$

However, since the OF is a linear response to x_1, x_2, this remains a linear constraint on the two DVs.

If there are 3 DVs, then the OF is a hyperplane in 4-D and the constraints are planes in 3-D space:

$$ax_1 + bx_2 + cx_3 < j \tag{12.9}$$

In these cases every pair of planes intersect in a line, and because this line will intersect another constraint plane at a point, three constraint planes will intersect at a common point. Again each intersection point represents the limit of the constrained boundary that might be the best point.

In high-order DV problems, the constraints are intersections of hyperplanes.

Could one use gradient or direct search techniques to solve an LP application? Yes, but they will likely get stuck on a constraint and not move along the constraint to the optimum. Consider a simple gradient-based logic. From a TS it moves in the steepest ascent direction, which is the direction of highest slope of the OF plane (or hyperplane). When it hits a constraint, it still wants to move across the constraint, not parallel to it in a direction that is not steepest ascent. Direct search methods have similar problems in wanting to jump across a constraint rather than be guided by them. A possible solution is to use a reduced gradient scheme, which will move along the constraint. However, such algorithms require many function (and gradient) evaluations relative to the constraint following LP logic.

12.3   Basic LP Procedure

The LP search logic goes like this:

Step 1: Start on an intersection of the constraint limits, which is a feasible vertex, it is a TS. Part of the LP procedure makes it easy to determine a feasible initial point. Some constraint intersections may be outside of the feasible region.

Step 2: If the initial TS is not the best point, then moving along one of the constraint lines (or one of the 3-D ridges, or one of the higher-D hyper-ridges) will move toward a better intersection. Move along a constraint. But, which one? It must be selected. The LP procedure determines it with this logic: At the TS, search among the active constraint boundaries to find the one that has the greatest potential in improving the OF.

Step 3: Move along that constraint in the beneficial direction, improving the OF, until the next constraint intersection blocks progress. How to know how far? The LP procedure determines it. This is the new TS.

Return to Step 2, and again, evaluate the constraint offering the best impact on the OF. Then return to Step 3, and move along it to the next to the next intersection. At Step 2, when no new direction offers an OF improvement, the procedure has found the optimum.

Note: The convergence criterion is that no move along a constraint leads to a better solution. This does not have a threshold value to choose. This does not look at the incremental changes in the DV value or the OF value. This does not require human choices, consequently, and in contrast to the nonlinear techniques, the LP solution precision and accuracy are not dependent on human choices (excluding the models, coefficient values, and other givens).

12.4   Canonical LP Statement

Start with a conventional linear statement, converting any greater than equalities to less than by multiplying by negative 1:

$$\min_{\{x\}} J = \sum_{i=1}^{i=m} c_i x_i$$

$$\text{S.T.:} \quad \sum a_{1,i} x_i \le b_1$$

$$\sum a_{2,i} x_i \le b_2 \tag{12.10}$$

$$\vdots$$

$$\sum a_{n,i} x_i \le b_n$$

$$x_i \ge 0 \quad \text{for all } i, 1 \le i \le m$$

Include a slack variable to each constraint to make each constraint an equality. For example, if there are m DVs, the slack variable added to the first constraint will be an $m + 1$st variable:

$$\sum_{i=1}^{m} a_{1,i} x_i + x_{m+1} = b_1 \tag{12.11}$$

The slack variable will have a nonnegative value $x_{m+1} \geq 0$

If there are n number of inequality constraints, there are n number of slack variables added to the statement. This converts all interacting constraints to equality constraints, leaving only the ≥ 0 condition on $n + m$ variables, the m DVs, and the n slack variables.

Note that the slack variables are not the same as the original DVs. The slack variables indicate how far from the constraint is the set of conditions, and they have the units of the constraint conditions, not necessarily the units of the DVs.

Also note that the slack variables have the same symbol as the DVs, x, because mathematically they will be treated as optimizer DVs.

The problem statement becomes

$$\min_{\{x^+\}} J = \sum_{i=1}^{i=m} c_i x_i$$

$$\text{S.T.}: \quad \sum_{i=1}^{m} a_{1,i} x_i + 1 x_{m+1} + 0 x_{m+2} + \cdots + 0 x_{m+n} = b_1$$

$$\sum_{i=1}^{m} a_{2,i} x_i + 0 x_{m+1} + 1 x_{m+2} + \cdots + 0 x_{m+n} = b_2 \qquad (12.12)$$

$$\vdots$$

$$\sum_{i=1}^{m} a_{n,i} x_i + 0 x_{m+1} + 0 x_{m+2} + \cdots + 1 x_{m+n} = b_n$$

$$x_i \geq 0 \quad \text{for all } i, \ 1 \leq i \leq m+n$$

This would appear to have complicated the problem, because now there are $m + n$ DVs, and the DV vector, x^+, has $m + n$ elements. However, since there are n linear constraints and each can be used to reduce the DV dimension, there is still only m number of DVs.

By not differentiating between slack and original DVs in the OF, the optimization statement is

$$\min_{\{x^+\}} J = \left[\sum_{i=1}^{i=m} c_i x_i \right] + 0 x_{m+1} + 0 x_{m+2} + \cdots + 0 x_{m+n} = c x_{\text{DV}} + 0 x_{\text{slack}} = c^+ x^+$$

$$\text{S.T.}: \quad A^+ x^+ (= A x + I x_{\text{slack}}) = b \qquad (12.13)$$

$$x^+ \geq 0$$

12.5 LP Algorithm

The LP procedure can be considered in the following manner:

Step 1: It is convenient to start a TS with the original DVs set to zero $x_{\text{DV}} = 0$. This do-nothing, buy-nothing, or make-nothing choice is a feasible solution, and the n linear equations of the constraints very easily compute the corresponding n slack variable values. For this initial special case, $x_{\text{slack}} = b$. For this solution the OF value is zero, which is not a very desirable value. The decision variables with a value of zero are termed nonbasic. There are m number of these, the same number as the DVs. The nonzero variables are termed the basic variables. There are n number of these, the same number as the inequality constraints.

Step 2: Change a DV value to get a better OF value. Which DV should be changed? Choose the one that has the largest impact on the OF for a unit change in the DV, the DV with the largest "c" coefficient value. Perhaps it is the ith, x_i.

Note this is a "greedy method" heuristic. It is like looking at one play in chess, not planning future moves. It takes the immediate best choice. However, it might be that the immediate best choice leads the path into a series of constraints, and each must be worked through to get to the optimum. It might be that the local second best path has fewer constraint intersections to negotiate on the alternate path to the optimum.

How much to change x_i? One can only change it to the point that a constraint is hit. This change can be determined by the coefficients in the constraint equations. In any constraint the maximum amount that x_i can be moved is the amount that makes the slack variable become zero, $\Delta x_i = b_j/a_{j,i}$. Here, subscript j represents the constraint equation index. Change x_i by the minimum over all j equations, $\Delta x_i = \min\{b_j/a_{j,i}\}$. This moves the trial solution to the jth constraint. It also sets the jth slack variable to zero and makes the ith DV nonzero. Consequently, the jth slack variable leaves the basic set and becomes one in the nonbasic set; and the ith DV leaves the nonbasic set and becomes one in the basic set.

Recognize that j is now a special constraint. It is the only one without slack capacity.

Now, any continued change in x_i will require a compensating change in another x-value, x_k, because there is no more slack in the jth constraint. The jth constraint will now require $\Delta x_k = -\left(a_{j,i}/a_{j,k}\right)\Delta x_i$. This means that a further change in x_i will have two impacts on the OF, due to both Δx_i and Δx_k. So, $\Delta\text{OF} = c_i\Delta x_i + c_j\Delta x_j = \left(c_i - c_k\left(a_{j,i}/a_{j,k}\right)\right)\Delta x_i$.

Step 3: Change the ith DV value along the jth constraint to get a better OF value. Which is the kth DV that should be chosen to make a compensating change to preserve the jth constraint? Choose the one that has the largest joint impact on the OF for a unit change in the DV with the largest $\left(c_i - c_k\left(a_{j,i}/a_{j,k}\right)\right)$ coefficient value.

How much to change x_i? One can only change it to the point that another constraint is hit. This change can be determined by the coefficients in the remaining constraint equations. In any constraint, the maximum amount that x_i can be moved is the amount that makes the slack variable become zero when both x_i and x_k are changed. For the lth constraint, $\Delta x_i = b_l/\left(a_{l,i} - \left(a_{j,i}/a_{j,i}\right)a_{l,i}\right)$. Here, subscript j represents the prior constraint equation index, the constraint we are moving along, and subscript l represents the constraint equation index that we are moving toward. Change x_i by the minimum over all l equations. This moves the trial solution to the intersection of the lth and jth constraints. It also sets the lth and jth slack variables to zero. Recognize that constraints l and j are both without slack capacity.

One cannot move further along the jth constraint, because the lth constraint has been encountered. Additionally, there is no sense in moving backward along the jth constraint, because the OF there is worse. So, search along the lth constraint.

12.6 Simplex Tableau

The procedure can be mechanized and presented in a simple row–column table, termed the simplex tableau. In it rows represent the OF function and constraints, and columns represent DVs and slack variables. In each stage of the LP simplex procedure, sequential row–column manipulation and renormalization of coefficients are similar to the $b_l/\left(a_{l,i} - \left(a_{j,i}/a_{j,i}\right)a_{l,i}\right)$ operation earlier. Each move changes the trial solution from the initial zero, successively along constraints, to the optimum.

The simplex tableau and procedure is a very clever device. Options can be implemented for diverse situations. And there are ways to interpret slack variables. It is a powerful technique but limited to the linear class of functions and constraints. Accordingly, because this book has a focus on nonlinear applications, this only presents a token introduction.

12.7 Takeaway

When the OF is nonlinear and the optimum is in an interior bound by constraints, LP does not find the optimum, because it only searches on the constraint boundary. However, if the OF is linear, the optimum will be on a constraint boundary. And if the constraints are also linear, the optimum will be on a constraint intersection. Then LP is a powerful optimization tool.

By contrast, in my experience, most engineering modeling and process or product design applications are nonlinear, with nonlinear constraints, interior optima, and often multiple optima, and sometimes stochastic. In those cases LP is not applicable.

On the other hand, many scheduling, logistics, and blending applications are linear with linear constraints. In such cases LP is very efficient.

LP finds an on-constraint solution. It will not be an interior optimum; but within uncertainty on model structure, model coefficients, and other "givens," the LP on-constraint solution may be indistinguishable for the true best. It may be close enough and defensible when considering work effort to determine the optimum, adequacy of a solution, and uncertainty in the givens. Accordingly, LP is often used industrially. It is the mainstay of the optimizer in model-predictive control to get a good enough control action. In successive LP, the original statement is linearized, LP provides a solution, and the statement is re-linearized about that point and re-optimized.

The LP simplex refers to a tabular procedure. It is not the SHH or NM simplex representing the simplest $N+1$ dimension shape in N-D space. Unfortunately they have the same name.

Often LP applications are confounded by DVs that are limited to discretized or integer values.

12.8 Exercises

1 Devise your own LP application.

2 Compare CSLS, NR, ISD, HJ, CHD, and LF on an LP application.

3 Could LP constraints be converted to penalty functions and solved with conventional optimizers?

13

Dynamic Programming

13.1 Introduction

Decision variables can be considered as the influence (the independent variable), while the dependent variable, as the process response (state). Most optimization techniques seek the influence values as the trial solution variables and calculate the process response using the model in the normal calculation procedure: given the input values, the output values are determined.

Contrasting these approaches, dynamic programming (DP) is a method of optimization by choosing a state path (the response) through stages or time. Here, the DV values are the outputs of the model. Then the inputs are calculated to achieve the state path. This can be considered as using the inverse of the process model.

DP is applied to processes that change in time, such as process control or scheduling of events as a process evolves. Although other optimization methods can be used on such applications, and often more efficiently, there are three paradigm shifts that are associated with DP that make it notable.

First, the DV for optimization, the TS, is the output of the process, not the input. Conventionally, as mentioned earlier, the TS is a DV choice of the influence or independent variable values, and the process states are the response. In most optimization procedures, the model is used to determine the values of the process states that respond to the TS. By contrast, DP seeks to choose the dependent variable values (the states) and to specify a best evolution of the state values over time. The TS is the state, and the input variable values are calculated from the inverse of the model.

Second, in DP the initial and final states are specified, and the objective is to seek the best path between them. This is like solving a boundary value problem in differential equations. By contrast, most optimization applications of dynamic processes specify an initial state and seek the best final state that can be achieved. Conventional optimizers can do this by using a final state as a constraint and determining the best path to it. DP does it differently, which is a notable innovation.

Third, once the optimization is completed for a particular start-to-finish path, the work has been done to determine the best path from any intermediate starting state. Once the nominal decision has been made, the model-required calculations are complete for all contingency situations.

Although other optimization techniques can solve the transient path optimization application, DP has a paradigm shift that the reader should understand. Further, there is a category of applications for which it has become the standard tool, and a reader should have that perspective.

The objective is to minimize or maximize some objective function, which is usually cost. Since the process changes in time (which can be considered with Δt sequential stages), the model needs to relate process state to action on the dynamic process.

Engineering Optimization: Applications, Methods, and Analysis, First Edition. R. Russell Rhinehart.
© 2018 R. Russell Rhinehart. Published 2018 by John Wiley & Sons Ltd.
Companion website: www.wiley.com/go/rhinehart/engineeringoptimization

Example descriptions can help reveal the application.

Example 1   What is the best airplane takeoff path? The graphs in Figure 13.1 illustrate airplane elevation as it changes with either distance or time. Elevation is the response to control action. Elevation is the state. The vertical axis is the state, and the horizontal axis is the stage. Note that if you are looking at elevation w.r.t. time, as in Figure 13.1a, time is the stage. But, in Figure 13.1b, a graph of elevation w.r.t. distance, distance is the stage. Note also that the control action, normally considered to be the DV, is not indicated. Typically, a dynamic model uses time as the increment for solving the ODEs, but even so, one can plot state (elevation) w.r.t. either time or distance away from the takeoff point. The plane starts at the elevation of the runway, and the flight plan indicates that it needs to fly at a target altitude.

Evaluating "best" as minimizing fuel consumption, the question is "What is the most fuel-efficient path to get to target altitude after 5 miles distance?" Alternately, the end stage could be defined as "after 20 min of flying":

$$\min_{\{h_1, h_2, h_3, ...\}} J = \text{fuel consumption} = \sum_{i=1}^{\#stages} fc_i$$

$$\text{S.T.}: \quad \left\{ \begin{array}{l} h_0 = 0 \\ h_N = h_{\text{target}} \end{array} \right\} \tag{13.1}$$

$$fc_i > 0$$

The DV listing $\{h_1, h_2, h_3, ...\}$ is the state path sequence—a list of desired h-values at each stage (in time or distance), not the position of the flaps and not the control action.

Stage is counted by index i. The OF is the sum of costs associated with the ith to $i + 1$st stage transitions. The S.T. constraints specify the start and end states or feasible values.

The inverse of the process model is used to determine the control action needed to implement stage-to-stage transitions and to assess their feasibility. Since the process model is likely to be a system of ODEs that define how the plane elevation responds to control action, the inverse will probably require iterative root-finding procedures.

(a)

Figure 13.1 (a) Time as the stage; (b) distance as the stage.

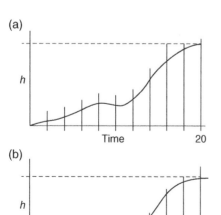

(b)

The economic model will determine the fuel consumption at each stage in the path, which is dependent on speed, elevation, and change. The determination of the OF value would likely require control action values (engine speed, flaps setting, etc.) as well as duration in the stage and air properties at elevation h.

Example 2   What is the best control action for heating a liquid that is flowing through a tank? The illustration in Figure 13.2 shows liquid flowing into a well-mixed tank, with control action either the rate of heat addition, $\dot{Q}_h$, or the rate of heat removal, $\dot{Q}_c$. The transient graph in Figure 13.3 is a possible sketch of how temperature should change in time. Initially the tank contents were at a desired T, and the set point changes to a new value.

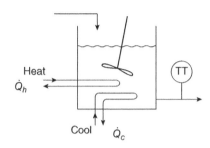

Figure 13.2 Process heater sketch.

The controller must move the process to the desired new T by a specified time, a grace period. How should the controller act? If the controller maximizes initial heating rate, then heat is wasted in the rise, and a greater amount is consumed during the at-temperature duration in the grace period. That would not be desirable; however the process gets to the set point quickly, which would be desirable. Alternately, if the controller incrementally increases the heat rate so that the temperature hits the target at the end of the grace period (not sooner), then energy consumption is minimized, but the deviation from the set point during the 0–N period is larger. The objective will be to balance accumulated deviation from set point (a quality metric) with accumulated cost (an economic metric).

The application statement is

$$\min_{\{T_1, T_2, T_3, \ldots\}} J = \sum_{i=1}^{N} \text{cost energy}_i + \sum_{i=1}^{N} \text{cost of } \left(T_i - T_{\text{sp}}\right)$$

$$\text{S.T.}: \quad T_0 = T_{\text{initial}} \tag{13.2}$$

$$T_N = T_{\text{sp}}$$

$$Q_h > 0, \ Q_c > 0$$

Figure 13.3 Possible trend of temperature w.r.t. time returning to the set point.

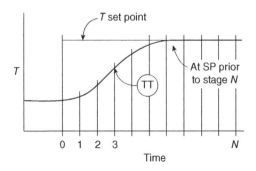

The development of the models is relatively easy. Start with a dynamic energy balance on the tank:

$$\text{Inflow} - \text{Outflow} + \text{Internal\_Generation} = \text{Accumulation} \tag{13.3}$$

This results in the ODE:

$$\frac{V}{F}\frac{dT}{dt} + T = T_{\text{inlet}} + \frac{\dot{Q}_h - \dot{Q}_c}{F\rho Cp}, \quad T(0) = T_{\text{initial}} \tag{13.4}$$

Using Euler's method to solve for how $T(t)$ responds to Q-dot,

$$T_{i+1} = T_i + \Delta t\left[T_{\text{inlet}} + \frac{\dot{Q}_{h,i} - \dot{Q}_{c,i}}{F\rho Cp} - T_i\right]\frac{F}{V} \tag{13.5}$$

where i is the counter for time increments.

The model converts to $T_{i+1} = f\left(T_i, \dot{Q}_{h_i}, \dot{Q}_{c_i}, F, V, \rho, C_p\right)$, where the manipulated variables are $\dot{Q}_{h_i}$ or $\dot{Q}_{c_i}$. Traditionally a dynamic model reveals how the state response changes with MV. In DP, however, the states are specified, so the inverse of the model is used to determine the MV value that will implement the desired state change:

$$\dot{Q}_i = F\rho Cp\left[\frac{V}{F}\frac{T_{i+1} - T_i}{\Delta t} + T_i - T_{\text{inlet}}'\right] \tag{13.6}$$

The trial solution is the state sequence. Choose the state sequence, use the model inverse to calculate the resulting $\dot{Q}_i$ sequence, and then use costs of $\dot{Q}_i$ and costs of deviation from target T to calculate the OF.

Using Δt_{stage} as the time increment for the stage duration, and the cost of heating as the heat rate times duration times cost factor, the energy cost of a particular path is $\sum c\Delta t_{\text{stage}}\dot{Q}_i$. The cumulative undesirability of T from the set point could be modeled as a traditional quadratic penalty, $\sum \Delta t_{\text{stage}}(T_{\text{SP}} - T_i)^2$. The units on these two OF elements are not the same, so a scaling factor is needed. Illustrated here is the equal concern approach, and the statement for DP is

$$\min_{\{T_1, T_2, T_3, \dots\}} J = \frac{1}{EC_\$}\sum_{i=1}^{N} c\Delta t_{\text{stage}}\dot{Q}_i + \frac{1}{EC_{\Delta T}^2}\sum_{i=1}^{N}\Delta t_{\text{stage}}(T_{\text{SP}} - T_i)^2 \tag{13.7}$$

13.2 Conditions

The stage variable needs to progress strictly monotonically in time. For the airplane, stage could be distance that progressively increases with time or fuel content that progressively drops with time. But if velocity falls during a climb portion, then velocity would not be a strictly monotonic function of time.

A second condition is that the path to get to a particular state is independent of the best path in forward states, and the remaining best path is independent of how it got to a particular state in the past. The states wholly define the process. In the heater example, it does not matter how the T_i obtained its value. It could have been slowly heated or overheated and then cooled in prior stages.

But once it has a temperature value in stage i, the best path forward is independent of the past. As another example, if you have $1,000,000 at age 30, the best investment choice for the future of the money is independent of whether you were given $1M, given $2M and gave half of it to charity, or started with $500k at age 25 invested in land and sold it for doubled the price 5 years later. This is forgetfulness, or no-inertia, no-recycle, or no-future-consequences property. This is termed Bellman's "Principle of Optimality," which might be more properly stated as a "condition for DP to work." Typically, if all the states are included, and if there is no recycle (of past states to influence future states), this condition is valid.

13.3 DP Concept

Normally, an optimizer will determine the stage-to-stage manipulated variables (those used to implement control action, the influences) and use the model to assess the consequences. However, any optimizer can use the stage-to-stage state values as the DVs and the inverse of the model to assess the costs and penalties. The advantage of the DP approach is that once one stage is analyzed, and an optimum decision determined for that stage transition, that decision is independent of the path through other stages.

In a one-state application, the state–stage graph is represented as a 2-D grid as shown in Figure 13.4. In this example the state is discretized into M intervals and the stage into N intervals, and the axis values on the gird are the state number, j, and stage number, i. State and stage values are directly related to the respective index:

$$s_i = \frac{s_0 + i(s_N - s_0)}{N} \tag{13.8}$$

$$S_j = \frac{S_0 + j(S_M - S_0)}{M} \tag{13.9}$$

In the heater example, state is temperature, while stage is time interval.

From any state in the second to last stage, the $N - 1$ stage, you must go to the final, desired state, in stage N. What does that transition cost? For the stirred tank heater, the energy balance determines the

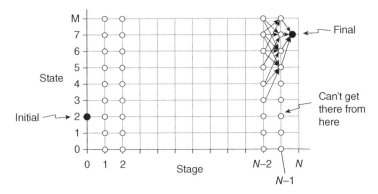

Figure 13.4 State and stage grid for DP.

energy required for the transition, and the energy required determines the cost. If the stage is a time interval of Δt, then the finite difference, explicit approximation can give the Q required to go from $T_{i=N-1}$ to $T_{i=N}$:

$$Q_{i=N} = F\rho C_p \left[\left(\frac{v}{F}\right) \frac{T_N - T_{N-1}}{\Delta t} + T_{N-1} - T_{in} \right] \tag{13.10}$$

If $\dot{Q}_i > 0$, it requires heating. Alternately, if $\dot{Q}_i < 0$, it requires cooling. If $\dot{Q}_i >$ max limit on capacity, then that state transition is not feasible.

The costs for heating or cooling would be different, and there is a cost of energy, for that stage (time interval):

$$\text{Cost of energy in the stage} = \left(C_h \cdot \dot{Q}_{h,i} + C_c \cdot \dot{Q}_{c,i}\right) \Delta t \tag{13.11}$$

There is also a cost associated for not being at the desired set point, T_{sp}. Possibly it could be related to either

$$\text{Cost of product deficit in the stage} = C\left(T_i - T_{sp}\right) \Delta t \tag{13.12}$$

$$\text{A penalty for a deviation in the stage} = \lambda\left(T_i - T_{sp}\right)^2 \Delta t \tag{13.13}$$

Neither of these costs may be linear functions of the MV (Q) of CV(T). These might be calculated as an average over the entry and exit states:

$$\text{Cost of product deficit in the stage} = C\left(\frac{T_i + T_{i-1}}{2} - T_{sp}\right) \Delta t \tag{13.14}$$

$$\text{A penalty for a deviation in the stage} = \lambda\left(\frac{T_i + T_{i-1}}{2} - T_{sp}{}^2\right) \Delta t \tag{13.15}$$

However you would choose to calculate the cost and record the transition costs from feasible paths. Figure 13.5 shows transitions that are feasible from four states at the $N-1$ stage to the final state in Stage N. The transition costs are 5 (representing required cooling) and 3, 6, and 15 representing heating. A transition from a lower state is not possible; there is inadequate heater power to raise the temperature by the transition amount within the one-time interval. This has several meanings. First, the path from the initial state in Stage 0 must go through one of the four states in the $N-1$st stage; otherwise it is not possible to get to the final state in the final stage. Second, regardless of the path that takes you from the initial state to State 5, in Stage $N-1$, the cost to subsequently move from State 5 to State 7 remains at 15.

The requirement is "no memory"; it does not matter how you got to a particular state. Once you are in it, the only thing that matters to the incremental change to the OF is what the next state is. Recycling of past products will violate this "no memory" basis.

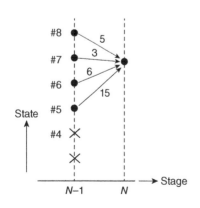

Figure 13.5 Illustration of transitions in the final stage.

Now consider the Stage $N-2$ and transitions indicated in Figure 13.6. If you are in State 5, you can transition to State 3,

Figure 13.6 Illustration of transitions in the next to last stage.

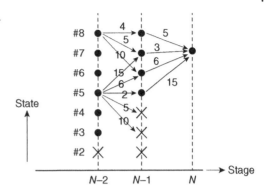

4, 5, 6, or 7 in the next stage. There is adequate heating or cooling power to permit those transitions, those are feasible transitions, and the transition costs are respectively 10, 5, 2, 6, and 15. But, at Stage N − 1, you cannot get from States 3 and 4 to the final state. Consequently, from State 5 in Stage N − 2, there are only three feasible paths to the ending state. One path to the end, from States 5 to 7, has an accumulative cost of 15 + 3 = 18. Another, from States 5 to 5 to 7, has a cost of 2 + 15 = 17, making that a better path from State 5 in Stage N − 2 to the final state. The best path, however, is from States 5 to 6 to 7, with an accumulated cost of 6 + 6 = 12.

It doesn't matter how you got to State 5 in Stage N − 2. If you are there, the best path to the end has now been determined, and it has a cost of 12.

This state transition of the best path to the ending state and accumulated cost will be determined for each state in Stage N − 2. The result is indicated in Figure 13.7.

Then the same state-to-state transitions are evaluated for the Stage N − 3 to N − 2 transition and choice of best path to the end determined. Then for prior stages, until Stage 0 is reached, only transitions from the initial state need be considered.

Figure 13.7 Illustration of the second to last stage transition.

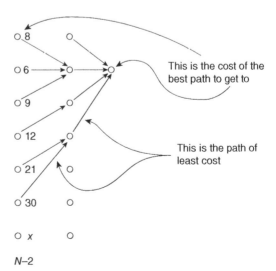

For the 1-D state example, with M states and N stages, and final and initial states specified, there are only M transitions for consideration in Stage $N-1$ (moving to Stage N) and in Stage 0 (moving to Stage 1). However, each internal stage-to-stage transition requires the evaluation of M^2 transitions. Many of those transitions are infeasible and do not engage the computer in detailed calculation of the model inverse or the associated cost. At most with N stages (where two states are fixed) and M states, there are $2M + M^2(N-2) = M[2 + M(N-2)]$ transitions that need to be analyzed.

Further, if the state-to-state transition cost is independent of stage, then only M^2 number of state transitions need be evaluated. Place the values in a lookup table, and refer to them for each stage analysis.

By contrast, looking at each possible path through an exhaustive search, there would be M^N number of transitions and many function evaluations. In a sense, the DP solution uses prior knowledge to minimize the number of function evaluations. The benefit in DP is the reduction in number of function evaluations relative to an exhaustive search, $M^2 < M[2 + M(N-2)] \ll M^N$.

In DP, once a solution has been found, if each optimal path through the stages is recorded, the NOFE work for all other solutions is also done. Regardless of how you might be in State Q, even if it is not on the original optimal path, all the necessary transitions have been analyzed to find the best path to the desired end point. There are no new function evaluations needed. You just need to calculate the new path through the table of already generated data.

DP discretizes states and only permits transitions to the discretized state value. This has an aspect that is similar to integer or discrete problems. If the best path is through an in-between state, then DP cannot find it. Often once a coarse DP solution is determined, the approach to fine-tune it is to use a finer discretization spacing above and below the optimal coarse path and to redo the DP on the finer state and/or stage mesh.

Dynamic programming (DP) is an optimization method capable of handling systems that are non-linear and in which the variables, objective function, and/or constraints can be of any form (e.g., linear or nonlinear, continuous or discrete, or deterministic or stochastic). DP is a standard tool for such applications as process control, inventory management, production scheduling, and design optimization.

I have not directly compared DP with the other optimizers in terms of NOFE, precision, robustness, PNOFE, etc.

13.4 Some Calculation Tips

DP can, for complicated applications, require a large computer memory and extensive computational time, instigating techniques to improve the efficiency of the DP method. Fortunately, there are many techniques that can do the same, and a few are summarized here.

For many applications the stage-to-stage transition costs are independent of the stage number. One way to reduce online computational requirements is to generate the stage-to-stage transition calculations once and store results for access as needed. Then only M^2 number of stage-to-stage state transitions needs to be evaluated.

Search only over heuristically determined rational states. Frequently, a variable is constrained by physical laws of nature or by the design of the system, so that exit states of any stage can be limited to those that satisfy these constraints. For instance, in the process controller example, realize that overheating and then cooling wastes energy, so only search through exit states that are in temperature higher than that in the entrance state. This type of heuristic limit on the search can reduce the computational work. One caution is that the optimal path may have an unexpected form, and if this path starts to approach the heuristically constrained states, the analysis constraint should be reduced or eliminated.

Reduce the size of the grid. The optimal policy can be approximated by initially using a coarse grid that restricts the DP calculations to a few states. This only roughly approximates the true solution, but once an approximate solution is known, a second DP optimization can be performed considering only a few more closely spaced states that bracket the rough solution. This process of reducing the grid dimension around the previous approximation is continued until an acceptably precise policy is identified. Although such an analysis requires additional complexities in the computer program to automate the analysis, these are minor. The net computational savings can be significant.

Employ a nonlinear grid. If the approximate form of the optimal state path is already known, then the initial grid can be divided nonlinearly into states and/or stages that are more densely packed in certain regions where fine-tuning is desired. There is no requirement in the DP method for either the state or the stage axis to be linearly incremented.

13.5 Takeaway

Dynamic programming is a versatile, but unusual, optimization method that also has broad applicability, including process control, inventory control, production scheduling, design optimization, and flight path (or any path) optimization. It can be applied to processes or systems with multiple states and constraints. If nonlinear, the inverse calculation could require iterative numerical procedures. I suspect, however, that contemporary direct search techniques would be easier to implement and faster to compute and provide greater accuracy in the solution. Those of us coming from a DP legacy should consider alternate optimization algorithms. The key advantage of DP is that once an optimal path has been determined, the results also reveal all contingency paths.

13.6 Exercises

1   You know that the shortest path between two points is a straight line. Use DP to find the path that minimizes distance from (x_1, y_1) to (x_2, y_2). Use x as the stage variable and y as the state variable. Use $d_i = \sqrt{(x_i - x_{i-1})^2 + (y_i - y_{i-1})^2}$. The objective is to find the state sequence that minimizes $\sum d_i$. Note: If ratio of number of states to stages (M/N or alternately N/M) is not an integer, then a straight line path through the states–stage intersection points is not possible. Choose either the ideal or nonideal number of states and stages.

2   Implement the stirred tank heater example.

3   Use DP to find a shortest time path between two points. Use your own application or that of Function 47.

4 Contrast the DP work (number of function evaluations) with other optimizers. In LF, for example, if there are M number of players, and one is near the optimum and each takes N leaps to converge to a Tolerance = Initial_Range$(1/2)^{N\_\text{leaps}}$, then at the end game, NOFE $\approx M(\ln(\varepsilon/\text{Range})/\ln(1/2))$. Perhaps it takes two times that many function evaluations to locate the vicinity of the global, so maybe NOFE $\approx 3M(\ln(\varepsilon/\text{Range})/\ln(1/2))$. Is this less than that expected for DP, $M[2 + M(N-2)]$, or possibly M^2? What number of DP states, M, or stages, N, is required for solution precision to be equivalent to that obtained with a conventional convergence threshold on LF?

14

Genetic Algorithms and Evolutionary Computation

14.1 Introduction

Genetic algorithms (GA) are mimetic approaches to the "intelligence" behind natural evolution embodied by random selection and survival of the fittest, which seems to direct evolution in biological species. These algorithms make progress toward an optimum in a logic that mimics our understanding of genetic evolution. Hence, the term evolutionary computation, or evolutionary optimization, is often used. However, in some disciplines evolutionary optimization means incremental process set point or controller coefficient adjustment in a manner similar to a CHD search. Accordingly, I prefer the term GA over "evolutionary."

The concept from genetics is that genes in the DNA of an individual define attributes or traits. Genes are functions within the DNA. Chromosomes are sequences of genes that define a trait such as height or eye color. To illustrate the concept, the cell entries in Table 14.1 represent genes, and the grouping of the first four comprises the chromosome that will relate to height. Similarly, the last two genes that are indicated comprise the chromosome for eye color. This is a concept and not the biological reality.

With two genes for tall and two for short this individual would be of medium height.

For brevity, the gene labels will just use the leading letters T, S, B, and G. The two rows in Table 14.2 could represent two individuals, both of medium height, one with brown eyes and the other with blue-green (hazel) eyes.

If these two were to "mate" by producing a descendant formed by random selection of sequential genes, then the result might be as shown in Table 14.3.

Children (progeny, descendants, offspring) have the same gene–chromosome structure as the parents, but the specific expression will be different. In this case the two medium height parents (TTSS and STST) generate a medium-short child (STSS) with a different eye color from either.

14.2 GA Procedures

In GA, there can be many procedures to compose descendants from parents. Here is one: The transfer of genes is randomly selected. Perhaps it is a 50/50 probability at each gene, but gene selection could be driven by alternate probabilities, conditional probabilities, or nonuniform distributions.

Engineering Optimization: Applications, Methods, and Analysis, First Edition. R. Russell Rhinehart.
© 2018 R. Russell Rhinehart. Published 2018 by John Wiley & Sons Ltd.
Companion website: www.wiley.com/go/rhinehart/engineeringoptimization

Table 14.1 Example of two chromosomes.

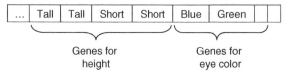

Table 14.2 Gene sequencing in two individuals.

...	T	T	S	S	B	G	
...	S	T	S	T	Br	Br	

Table 14.3 Random selection of parent genes makes the child chromosome.

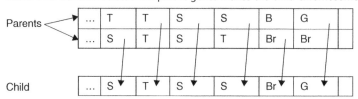

The parents can have more than one child, and each individual (either parent or child) is a "trial solution." Note that the initial population of two individuals in Table 14.2 has become three individuals with the addition of a child in Table 14.3. Parents could form any number of progeny, and alternate individuals could be selected to generate children. There are no social rules about mating, no feelings in the individuals, or no economic constraints on the number of children. So, in one iteration, an initial set of 10 trial solutions could spawn any number of children. If 90 children (arising from two for each possible parent pairing), in total there would be 100 trial solutions.

Some trial solutions are more fit for the environment than are others. For example, if a predator was spooked by blue eyes, then blue-eyed individuals will have a higher likelihood of surviving. Or, for example, if the air had a poison that was removed by low-growing plants, then short individuals will have a better chance to thrive than tall ones. If the criteria of survival were to cull vulnerable individuals, perhaps 90 of them, then the 10 that remain would be shorter and have bluer eyes. Since selection of the viable survivors will have these genes and these attributes, these functional advantages will become more prevalent in the successive populations. This creation of progeny, culling the least fit for the environment to leave the fittest, is termed a generation.

This random process of blending genes sometimes leads to an improvement from the parents. More often, it is not an improvement.

The 10 survivors become the parents of the next generation, and selection w.r.t. the criteria for fitness leaves the parents for the next generation.

Although this discussion considered 10 individuals, with each pairing to generate progeny, there is no such requirement on the initial number of trial solutions, the number or progeny, or selection pairing. Further, the fittest could be preferred for creating progeny.

Table 14.4 Illustration of several types of mutations.

Parents								
...	T	T	S	S	B	G	
...	S	T	S	T	Br	Br	

Child 1						
...	S	T	S	S	T	

Child 2					
...	S	A	S	S	

Child 3					
...	T	T	T	B	

In a GA, individuals do not die from old age. As long as they remain fit for the environment, they survive.

In GA, the child genes can also be subject to mutation. Table 14.4 illustrates several types of mutation for the height chromosome. In Child 1, the 4-gene sequence in the parents could become a 5-gene sequence in a child. The chromosome value could become something new, as in Child 2. And illustrated in Child 3, crossover could place a chromosome from one gene into another.

Although this random process of mutations sometimes leads to an improvement from the parents, more often, it is not an improvement.

14.3 Fitness of Selection

Continuing the concepts from evolution, where "survival of the fittest" determines which species prospers or becomes extinct, "fitness" is the term used to express the OF value. Often the environment has multiple measures of fitness. One individual who is good at finding food and water is more fit in that category than one not so good in doing so. But the one not good at finding food and water may have better predator-evading skills and be more fit for the environment in that sense. Accordingly, fitness can be considered as multiple separate objectives.

Figure 14.1 illustrates two criteria for fitness and two independent objective functions. Note that this reveal of an optimization is neither a graph of OF w.r.t. DV nor a contour response to two DVs. The

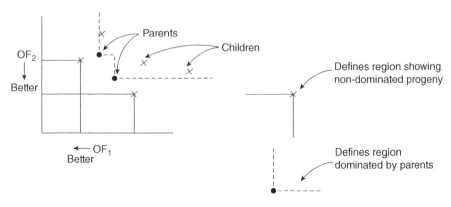

Figure 14.1 Illustration of selection of the fittest.

illustration is a state-plane representation. The individuals each have their own value of OF_1 and OF_2 and are shown in the figure. Parents are circles, and progeny are illustrated with x-symbols. The solid rectangles to the left and below an individual indicate the individual's dominance in the two categories. Although one individual has a better OF_1 value and the other a better OF_2 value, no other individual is better in both categories. There are no individuals within either rectangle.

By contrast, the dashed rectangles in Figure 14.1, which go to the right and vertical, indicate that the trial solutions (termed parents) dominate the ones to the upper right. The parents are better in both fitness categories than the children included in the dashed rectangles. Not illustrated, the upper parent is dominated by the far left child.

Dominated individuals are removed from the environment, as illustrated in Figure 14.2. This example shows a former parent and three children, dominated and removed. It also reveals that the population does not need to remain static. Two individuals in one generation have left three in the next.

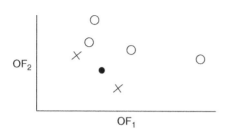

Figure 14.2 The fittest survive.

The non-dominated set of individuals again produce progeny by random gene donation and mutation.

There is no concept of "marriage," gender, or family to restrict the blending of genes. Each individual can mate with each other. There are a number of rules to pair parents to create new individuals. For instance, randomly select individuals from the non-dominated set and use them as parents to create children. Specify the number of progeny from each parent pair and randomly combine genes from the parents to create progeny. I imagine there could be more than two parents involved; three could donate genes to create children.

An epoch (iteration, era) generates a new set of progeny and removes dominated individuals.

You can have any number of rules to override purely random transfer of genes or mutations. Heuristics, human logical analysis of trends, can shape selection of parents or preference in gene selection to accelerate evolution toward the goal.

You can have any number of rules related to the removal of individuals. For example, mildly dominated individuals could be retained. You could retain those individuals who are only dominated by one other individual. Or you could retain Rank 1 individuals, those that are only dominated by individuals of Rank 0, who are not dominated by others, as illustrated in Figure 14.3. In Figure 14.3a the number associated with an individual represents "rank" of individual.

(a)

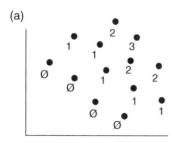

(b)

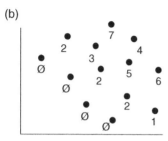

Figure 14.3 Showing (a) rank only dominated by individuals in the lower rank, (b) rank is based on the number of individuals with better fitness.

0 = non dominated

1 = dominated by 0's only

2 = dominated by 0, 1 only

3 = dominated by 0, 1, 2's

By contrast, in Figure 14.3b, the number associated with an individual represents the number of individuals with a higher fitness.

0 = dominated by none

1 = dominated by only 1 individual

2 = dominated by 2 individuals

3 = dominated by 3 individuals

The locations of the individuals in Figure 14.3a and b are identical. Note, however the difference in numerical value of the ranking.

An epoch (an evolutional term) is an iteration (an optimization term) in which new progeny are created, the OFs for each are evaluated, and the fittest are kept. A selection rule could be as follows: Keep all individuals with rank = 0 or 1 to survive each epoch or keep all with no more than three better individuals to survive each epoch.

The outcome is not a single DV*, but a set of Rank 0 individuals that are non-dominated. Note: The individuals shown are w.r.t. their OF values, not their DV values. However, each has the DV sets represented by chromosomes. Some individuals are better at OF_1, while others are better at OF_2. Some are intermediate in both OF_1 and OF_2, but none are better than any other in both categories. All are legitimate solutions.

Also note: The saved set does not have to be only the Rank 0 individuals. It could be Ranks 0 and 1 or Ranks 0, 1, and 2 representing both the non-dominated and nearly as fit solutions.

Accepting multiple solutions is a new concept at this point in the book. To convey understanding, consider the product line of cars. Some are big (good) but expensive (not good). Some are inexpensive (good) but small (not good). Some products are more suited as a second vehicle used for commuting, while others for family trips. There is a great diversity of successful, non-dominated products. All are legitimate offerings; all are DV* solutions.

These concepts of evolution, chromosome and gene, mutations, and generation of progeny take liberties with genetics. Also, it does not reveal how the optimization evaluates fitness from the genes, but hopefully this introduction relays the GA concept for a reader. An example with numerical values may fill in the gaps.

Example 1   A simple illustration of the evolution process and generation of trial solutions. Here $x1$, $x2$, and $x3$ are trial solution values. In Parent 1, $x1$ has the value of 3.142, $x2$ = 2.303, and $x3$ = 0.623. In Parent 2, $x1$ = 1.234, $x2$ = 5.678, and $x3$ = 9.012.

Parent 1	3	1	4	2	2	3	0	3	0	6	2	3
Parent 2	1	2	3	4	5	6	7	8	9	0	1	2
Child 1	3	2	3	2	2	6	0	8	0	0	1	2
Child 2	1	7	3	4	5	6	7	8	0	6	2	2

Arrows indicate which genes were randomly transferred to make the trial solution for $x1$ in Child 1 equal 3.232.

To determine which parent donates a gene, flip a coin. If an "*H*," then from Parent 1; if a "*T*," from Parent 2. Or call a random number. If <0.5 from Parent 1, otherwise from Parent 2. Or if <0.3 from Parent 1, if >0.8 from Parent 3, else from Parent 2.

The random numbers could also select parents, perhaps blending individuals from a Rank 0 category with one from a Rank 1 category. The probabilities do not have to be equal (50/50, or 33/33/33).

The trial solution for $x1 = 1.734$ in Child 2 contains a value of 7 that does not appear in either $x1$ for Parent 1 or Parent 2. The value of 7 could be a transfer from the $x2$ trial solution in Parent 2, or it could be because of a random mutation. The two new trial solutions are $(x1, x2, x3) = (3.232, 2.608, 0.012)$ and (1.734, 5.678, and 0.622).

If the criteria were to find individuals that have a number sequence that is closest to $(\pi, e, \gamma) = (3.14159..., 2.71828..., 0.61803...)$, then Parent 1 would be the best for the first two metrics of fitness, and Child 2 best in the third.

The prior example is a bit trivial. If one knew the value of π in order to check how good the optimization outcome was, then one would not need the optimizer. Hopefully, however, the examples help the reader understand the procedure.

Genes represent the decision variables, and there are many ways to structure the DVs, as the following example (more relevant to a GA application) reveals.

Example 2   Generate human linguistic rules (fuzzy logic) to describe how process response y depends on $x1$ and $x2$. Figure 14.4 shows possible influences and response.

A human linguistic rule is one that is expressed in text, not in mathematical equations. An example could be "IF $x1$ is high AND $x2$ is low, THEN in a short while y will be high." We use such rules all the time. For instance, in crossing the street we have a rule, "IF the incoming car is far away AND moving slowly, THEN cross the street now." In correcting our children, we might have several rules: "IF this is the second offense AND in the recent past, THEN remind the child of expectations" and "IF this is the third offense AND in the recent past, THEN place them in time-out now." In each case there is an antecedent (the IF condition) and a consequent (the action following the THEN). If the antecedent is true, then execute the consequent. In some cases there is a compound antecedent "AND" where two things have to be true for the antecedent to be true.

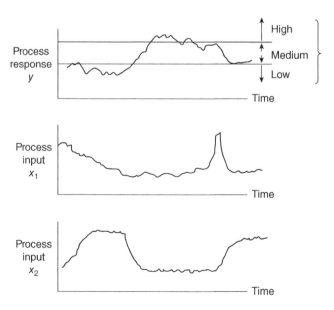

Figure 14.4 Illustration of process response and influences over time.

Certainly, however, the conjunction could be replaced with a disjunction, OR. Additionally, the true or false parts of the antecedent could be "is not" as opposed to "is."

Finally, in human linguistic reasoning we don't use exactly precise numerical values. Instead we use qualitative descriptions. Mom says, "It is cold and windy. Wear your heavy coat." She does not refer to the wind speed being 23.94 mph. Instead, she uses the label "windy." We cannot contradict the label windy, but we could challenge any number. "No, Mom. You're wrong. It is 23.942 mph."

Human linguistic rules guide our jobs. "When the boss comes in, act like you just finished talking with a customer." "When the solids are all dissolved, then add compound B." "Head east-by-northeast, you'll arrive in late afternoon."

I believe that there is great potential benefit in developing linguistic representations of process/system behavior. Once discovered, the rule can provide cause and effect understanding, affirm process intuition, and provide anticipation of outcomes for management preparedness.

In classic fuzzy logic (FL) representation, the user defines a linguistic category for key variables. In this example they are represented as high, medium, and low categories. But the membership in a category is not a simple 0 or 1. Consider an outside temperature of 100°F (about 38°C) is it hot? Definitely, yes, 100% hot. But what of 90°F (about 32°C)? Sort of hot, maybe 80% hot. The interpretation of 80% is not precise. It could mean that 80% of the people will say 90°F is hot. Or it could mean that in 80% of the time one particular person will think 90°F is hot. What of 80°F (about 27°C)? Well, that is between warm and hot, maybe 50% hot. What of 70°F (about 21°C)? Not hot, unless you are working in the sun, so maybe 10% hot.

In classic FL, a triangular membership function, a linear relation, is used to quantify the belongingness (membership) of a variable to a linguistic category. Figure 14.5 shows H, M, and L membership functions and the membership value, $\mu_c(x_1)$, of variable x_1 in a particular linguistic category, $c = \{H, M, L\}$.

In Figure 14.5, the x value indicated with a moderately high value has these membership function values:

$$\mu_L(z) = 0.0$$
$$\mu_M(z) = 0.2 \qquad\qquad (14.1)$$
$$\mu_H(z) = 0.8$$

A fuzzy logic rule for the data in Figure 14.4 might be

$$\text{IF}(x_1 \text{ is } H, \text{ AND } x_2 \text{ is } L, \text{ AND } x_2 \text{ has persisted } L) \quad \text{THEN } (\text{in a short while, } y \text{ will be } M) \qquad (14.2)$$

This rule can be parsed into genes, with the "IS" condition represented by a membership value greater than ½.

Figure 14.5 Illustration of membership functions for x_1 in the categories of low, medium, or high.

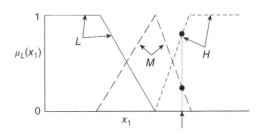

IF(| $\mu_H(x_1)$ | > | 0.5 | AND | $\mu_L(x_2)$ | > | 0.5 | etc.

And further parsed as

IF(| x_1 | H | > | 0.5 | AND | x_1 | L | > | 0.5 | etc.

The variable choices could be x_1, x_2, y. The linguistic categories could be *H, M,* and *L*. The comparison could be greater than (meaning is) or less than (meaning is not). The numerical values for the membership could be 0.5 (at least somewhat in the category), 0.9 (very confidently in the category), 0.3 (could be in the category), or other value. The conjunction could be AND, OR, NOT, etc.

An initial population would have several rules representing each possible chromosome. They could be initialized with human supposition of the rules or randomly structured. Maybe these would be the initial segments:

IF(| x_2 | M | < | 0.3 | AND | x_2 | H | < | 0.5

IF(| x_2 | L | > | 0.2 | AND | x_1 | L | > | 0.4

IF(| x_1 | H | > | 0.8 | AND | x_2 | M | > | 0.6

Once the initial population is created, use genetic algorithms (GA) or evolutionary optimization (EO) to guide the evolution (transfer and mutation) of genes to make best fuzzy logic rules. Evaluate best as minimizing

Complexity: OF_1 = the number of terms in the antecedent

$T - 1$ error: OF_2 = number of times rule predicts y is medium when it is not

$T - 2$ error: OF_3 = number of times rule does not predict y is medium when it was

14.4 Takeaway

GA is not the best choice when DVs are all numerical values but appears to have strong utility when DVs are attributes, descriptors, conjunctions, and other "class" variables.

Purely random selection of genes usually leads to non-fit, dominated individuals that do not move the population toward an optimum. Directed evolution, where the supervisor begins to recognize beneficial patterns in the genes and influences the probability of what genes are preserved and randomized, increases speed of the process. Much research is performed in that area.

14.5 Exercises

1 Consider the chemical structure of a new agent. The choices might be where the hydroxyl is added to a nominal structure, the length of a carbon–carbon backbone, or such. If this is a drug, the

desirables might be the binding to a virus particle and non-interaction with other bodily features. If a paint, desirables might be sunlight resistance and impermeability after curing. Envision your own application and define the genes and OFs that might make it a GA application.

2 Consider the design of an automobile. Design choices might be trunk size, shock absorber stiffness, etc. Envision your own application (such as a go-kart, grandmother car, or freight transporter) and define the genes and OFs that might make it a GA application.

3 Consider a game of chance with dice. The genes could relate to the game rules and the OFs to the variety of desirables. Define the genes and OFs that might make it a GA application.

4 Consider selecting players in a fantasy football team, or any team, such as a school project or national leadership. State what might be the attributes that would lead to player selection and what might be the interaction or fill-in-the-gap outcomes that would represent the OFs.

15

Intuitive Optimization

15.1   Introduction

As humans, we seem able to safely cross a country road. We can guide our muscles to avoid stepping into the mud in ruts or tripping on a rock. If a truck is coming, we can decide whether to cross or to wait, without quantifying the exact location of the vehicle position or speed and without a double-precision calculation of the time for truck arrival.

We plan weekly menus and shop for food without exact calculation of the diverse nutrients that we are supposed to heed in nourishment, or item cost (after-tax, after-coupon, after-special-offer, after transportation to and from) even though we are seeking to minimize cost, subject to meeting all of the nourishment constraints.

We choose topics of social conversation to avoid potential conflicts, to maximize happiness in interaction, or to initiate or terminate a visit. There is no model for what is often a stochastic situation.

Although this book guides you to develop models that quantify how DVs affect the OF, in nearly every daily human decision we are also attempting to optimize, but do not have models. We don't have definitive knowledge. We don't use optimization algorithms. We live life making intuitive decisions. And it works!

Regardless of whether the optimization is based on a "gut feel," design heuristics and "rules of thumb," or whether it is an approximate optimization with an idealized model or coarse convergence criterion or rigorous optimization, the user needs to understand what constitute permissible choices (what are the DVs and their constraint values), the measures of goodness or badness to be optimized (quantification of desirable and undesirable outcomes), the balance of importance of those issues (assembling the OF), and how the DVs impact the OF. Once the user has a good feel for those issues and relations, an intuitive decision can often be adequately close to optimum.

Optimization is the process of seeking the best. Often the intuitive solution of the optimization application is best when one considers uncertainty in the context, adequacy of an acceptable DV^*, and speed of getting the answer.

So, overriding the mathematical formality and computer code that this book represents, you have my permission to practice intuitive optimization, in which the human chooses the DV values. But first understand the OF and constraints and models for DV impact in light of the descriptions in this book.

Further, you have my permission to attempt to structure the human logic that you might use to make adjustments in a DV (such as using linguistic rules, or fuzzy logic), then have a computer use that set of rules as your surrogate in making decisions.

Engineering Optimization: Applications, Methods, and Analysis, First Edition. R. Russell Rinehart.
© 2018 R. Russell Rinehart. Published 2018 by John Wiley & Sons Ltd.
Companion website: www.wiley.com/go/rhinehart/engineeringoptimization

15.2 Levels

The intuitive procedures can be blended with mathematical procedures, generating a range of right solution methods.

Level 1—Totally intuitive.
Level 2—Interactive. Let the computer do some and the human choose some. For example, the human may choose DV values for DVs that represent categories, and the computer optimizes numerical values of the continuum variables. The human may choose initial trial solutions and let the computer take it from there. The human may observe iteration-to-iteration progress of a computer procedure and change optimizer parameters (making it more or less aggressive) as the situation seems to indicate.
Level 3—Automate the human logic. For instance, use fuzzy logic to express human rules to approach the optimum, and then let the computer execute that logic. Or use Bayes belief to automate change in logic or optimizer parameters as the evidence mounts.
Level 4—Totally mathematical/numerical.

15.3 Takeaway

Optimize optimization. Appropriate the material in this book for your context. The application statement is intuitively developed by human trial and error, making it imperfect. So, is mathematical perfection in the solution justified? Is a sophisticated algorithm to achieve perfection justified? Balance sufficiency with perfection.

15.4 Exercises

1   State your own examples in which intuitive decisions would be better than the use of formal optimization procedures (using mathematics or computer code). Explain the "why" behind your examples.

2   State your own examples in which formal optimization procedures (using mathematics or computer code) would be better than the use of intuitive decisions. Explain the "why" behind your examples.

3   In dynamic modeling, in making a first-order differential equation match data, the time-constant and gain are continuum-valued variables. Nearly all conventional algorithms can easily find those DV values to minimize the sum of squared deviations between model and data. However, if delay is to be included, it needs to be an integer number of sampling or numerical simulation intervals. It is both discrete and nonlinear. Worse, large delay means fewer data to be included in the model SSD, so, even if the quality of fit is unchanged, SSD will decrease with fewer data to be included. In fact, an extremely large delay value would leave zero data to fit, so the SSD would be zero, making the model appear to be perfect. Describe how you would integrate the human selection of the delay and the computer selection of time-constant and gain variables.

4 Use fuzzy logic to generate the rules that a human might follow to best fit a linear model to data, $y = a + bx$.

5 Consider a feed that is split into three portions and each fed to a catalytic reactor. The reactivity of the catalyst in each reactor changes individually in time. So, the completeness of the reaction in each reactor is different. The objective is to choose the three portions to make the output of each reactor the same. Since choosing two of the three feed rates fixes the third, there are just two DVs. The composition of the reactor outlets is not easily measured, but temperature is, and temperature is strongly related to composition. So, the objective is to choose the feed portions so that the three reactor outlet temperatures are the same. Assume the reactors are adiabatic and the reaction is exothermic. Essentially then, the reactor temperature will be high if the feed rate is very low, because there is enough time to let the reaction go to completion. Alternately, the temperature will be low if the feed rate is very high because there is not time to catalyze much reaction. As flow rate increases from zero to infinity, the temperature starts at a high value and then makes an S-shaped curve to lower temperatures, approaching the inlet temperature as flow rate goes to infinity. How would you balance the feeds? How would you get a computer to do it?

16

Surface Analysis II

16.1 Introduction

Modifying objective functions, in general, does not make the optimization solution faster, but it can have advantages in visualizing the surface, in seeing issues, in scaling it for presentation, or in simplifying a calculation. If you modify the OF, however, you need to ensure that your modification does not change the DV* solution.

Modifying the objective function is often useful for simplifying the optimization exercise. For example,

$$\text{maximize } J_1 = \prod_{i=1}^{N} \frac{1}{\sqrt{2\pi}\sigma} e^{-1/2((y-\tilde{y})/\sigma)^2} \tag{16.1}$$

finds the same model coefficients as

$$\text{minimize } J_2 = \sum_{i=1}^{N} \left(\frac{y-\tilde{y}}{\sigma}\right)^2 \tag{16.2}$$

And

$$\text{minimize } J_3 = \sqrt{\frac{a}{2 \quad \cos\theta \quad \sin\theta}} \tag{16.3}$$

finds the same θ value as

$$\text{maximize } J_4 = \sin 2\theta \tag{16.4}$$

Here are proofs for various OF transformations.

16.2 Maximize Is Equivalent to Minimize the Negative

maximize OF is the same as minimize $-$OF

Engineering Optimization: Applications, Methods, and Analysis, First Edition. R. Russell Rhinehart.
© 2018 R. Russell Rhinehart. Published 2018 by John Wiley & Sons Ltd.
Companion website: www.wiley.com/go/rhinehart/engineeringoptimization

Proof: Consider that $OF = f(x)$ and that a maximum exists at x^* with an OF^* value and then $OF(x^*) \geq OF(x)$ for any x-value. Multiplying by -1 changes the equality: $-OF(x^*) \leq -OF(x)$ for any x-value. So, $-OF(x^*)$ is the minimum of all possible values. Minimizing the negative of the OF returns the same DV values as maximizing the OF.

16.3   Scaling by a Positive Number Does Not Change DV*

If x^* minimizes OF(x) then it minimizes $a[OF(x)]$, where $a > 0$.

This will be an inductive logic proof.

Proof: If $OF(x^*)$ is a minimum, then $OF(x^*) \leq OF(x)$ for any x-value. Multiplying by the positive value a: $OF_2(x^*) = aOF(x^*) \leq aOF(x) = OF_2(x)$ for any x-value. And defining $OF_2(x)$ as $aOF(x)$, then $OF_2(x)$ is the scaled $OF(x)$, then $OF_2(x^*) \leq OF_2(x)$ for any x-value. So, scaling by a "+" value does not change the x^*, although it does change the value of the $OF(x^*)$.

16.4   Scaled and Translated OFs Do Not Change DV*

If x^* minimizes J(x) it also minimizes $J'(x) = \alpha + \beta J(x)$, where $\beta > 0$.

This will be a mathematical proof.

Proof: At the minimum two attributes are true: $dJ(x)/dx|_{x^*} = 0$ and $d^2J(x)/dx^2|_{x^*} > 0$.
Take the first derivative of $J'(x)$ evaluated at x^* at the minimum of $J(x)$:

$$\left.\frac{dJ'}{dx}\right|_{x^*} = \frac{d\alpha}{dx} + \beta \left.\frac{dJ}{dx}\right|_{x^*} = 0 + \beta \left.\frac{dJ}{dx}\right|_{x^*} = 0 + \beta 0 \tag{16.5}$$

If $\beta \neq \infty$ then $dJ'/dx|_{x^*} = 0$.
Take the second derivative of $J'(x)$ at x^* at the minimum of $J(x)$:

$$\left.\frac{d^2J'}{dx^2}\right|_{x^*} = \beta \left.\frac{d^2J}{dx^2}\right|_{x^*} \tag{16.6}$$

Since $\beta > 0$ and $d^2J/dx^2|_{x^*} > 0$, then $d^2J'/dx^2|_{x^*} > 0$.
Accordingly, since both $dJ'/dx|_{x^*} = 0$ and $d^2J'/dx^2|_{x^*} > 0$, the minimum of $J(x)$, x^*, also is the minimum of $J'(x)$.

16.5   Monotonic Function Transformation Does Not Change DV*

If f is a strictly monotonic increasing function over the range of the OF, then x^* that minimizes $OF'(x) = f[OF(x)]$ also minimizes OF(x).

This is more generic than just translating and scaling. Figure 16.1 illustrates a strictly increasing function. There are no discontinuities and no points of either negative or zero slope.

This means, $\partial f/\partial y > 0$ for all values of y.

The hypothesis is that if $f(OF(x^*)) \leq f(OF(x))$ for all x-values, then $OF(x^*) \leq OF(x)$.

At the minimum of $f(OF(x))$, $df/dx|_{x^*} = 0$ and $d^2f/dx^2|_{x^*} > 0$.

Using the chain rule, $df/dx = (\partial f/\partial OF)dOF/dx$.

Since at the minimum of f $df/dx = 0$, then $0 = \partial f/\partial OF|_{x^*}$ $dOF/dx|_{x^*}$.

Since the function is strictly monotonic $\partial f/\partial OF|_{x^*} \neq 0$, therefore $dOF/dx|_{x^*} = 0$.

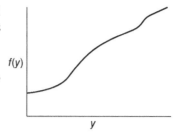

$f(y)$

y

Figure 16.1 A strictly positive monotonic function.

Could it be a minimum, saddle point, or a maximum for OF? Let's explore.

If $f(x^*)$ is a minimum, then $d^2f/dx^2|_{x^*} > 0$.

Using the chain rule,

$$\left.\frac{d^2f}{dx^2}\right|_{x^*} = \frac{d}{dx}\left(\frac{df}{dx}\right) = \frac{d}{dx}\left(\frac{\partial f}{\partial OF}\cdot\frac{dOF}{dx}\right) = \left.\frac{\partial^2 f}{\partial x\partial OF}\right|_{x^*}\left.\frac{dOF}{dx}\right|_{x^*} + \left.\frac{\partial f}{\partial OF}\right|_{x^*}\left.\frac{d^2OF}{dx^2}\right|_{x^*} \tag{16.7}$$

By expanding the mixed partial derivative, it becomes

$$\left.\frac{d^2f}{dx^2}\right|_{x^*} = \left.\frac{\partial^2 f}{\partial OF^2}\right|_{x^*}\left[\left.\frac{dOF}{dx}\right|_{x^*}\right]^2 + \left.\frac{\partial f}{\partial OF}\right|_{x^*}\left.\frac{d^2OF}{dx^2}\right|_{x^*} \tag{16.8}$$

As long as $\partial^2 f/\partial OF^2$ is bounded (not infinity) for all OF values, since $dOF/dx|_{x^*} = 0$, then

$$\left.\frac{d^2f}{dx^2}\right|_{x^*} = \left.\frac{\partial f}{\partial OF}\right|_{x^*}\left.\frac{d^2OF}{dx^2}\right|_{x^*} \tag{16.9}$$

Since $d^2f/dx^2|_{x^*} > 0$, then $\partial f/\partial OF|_{x^*}d^2OF/dx^2|_{x^*} > 0$.

Since $\partial f/\partial OF|_{x^*} > 0$ for all x-values, then $d^2OF/dx^2|_{x^*} > 0$.

Note the underscored caveat: "If f is a strictly monotonic increasing function *over the range of the OF*, then x^* that minimizes $(OF(x))$ minimizes $OF'(x) = f[OF(x)]$." Figure 16.2 illustrates an example that reveals the caveat violation. The relation $y = x^2$ is strictly positive if $x > 0$. If $x < 0$, then $y = x^2$ decreases with increasing x. If $x = 0$, then the derivative is zero. The top illustration in the figure reveals an OF that has a sign change, and the middle shows the square of that function. When squared, the minimum of OF is converted to a local maximum, and the two zeros of the OF become the minima.

Also note: Some strictly positive functions (such as log and sqrt) require the argument to be a positive value.

As a corollary, if the function is strictly monotonic decreasing over the OF range such as illustrated in any one of the elements in Figure 16.3, then the x^* value, which is the max $OF(x)$, is equivalent to the min of $f(OF(x))$, and the min $OF(x)$ is equivalent to the max of $f(OF(x))$.

Commonly used strictly monotonic increasing functions are illustrated in Figure 16.4.

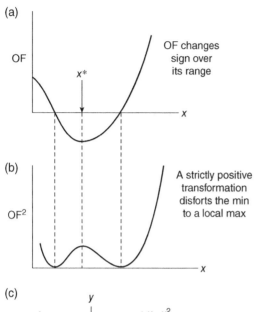

(a)

OF

x^*

OF changes sign over its range

(b)

OF^2

A strictly positive transformation disforts the min to a local max

(c)

y

$y=x^2$

Decreasing   Increasing

x

Figure 16.2 Illustration of a caution on using what seems to be a strictly positive monotonic function over a limited range: (a) the original OF, (b) the OF squared, (c) and a reveal of the limit of arguments to make squaring a variable a strictly increasing transformation.

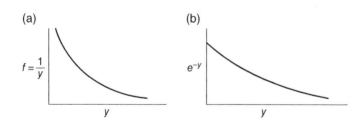

(a)

$f = \dfrac{1}{y}$

y

(b)

e^{-y}

y

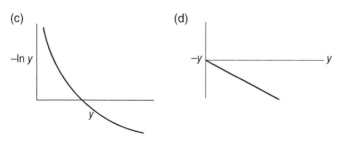

(c)

$-\ln y$

y

(d)

$-y$

y

Figure 16.3 Illustration of some strictly negative monotonic functions: (a) reciprocal, (b) negative exponential, (c) negative of the log or log of the reciprocal, and (d) negative.

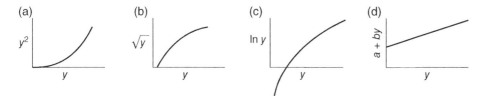

Figure 16.4 Common strictly increasing functions: (a) squared, (b) square root, (c) log, and (d) scaled and translated.

16.6 Impact on Search Path or NOFE

Do modified objective functions affect the search? In general, they have no impact on the search path or number of function evaluations. Consider a strictly monotonic increasing function as shown in Figure 16.1. This means, $\partial f/\partial y > 0$ for all values of y. There are no flat spots.

I'll use f' to represent the modified OF and f the original OF. For example, $f' = \ln(f)$.

For a direct search algorithm, if $f_{new} < f_{old}$, then $f'_{new} < f'_{old}$, because it is a strictly positive monotonic transformation, and the decision to proceed, expand, and contract is exactly the same. The trial solution path is exactly the same. If convergence is based on Δx, there is no difference in the steps (number of iterations or TS values) to converge. If convergence is based on ΔOF, the transformed convergence criteria needs to be based on the transformed OF. For example, claim convergence if $\Delta f < \varepsilon$ needs to be transformed to if $\Delta f' < \varepsilon^* \exp(-f')$.

For a gradient-based search, $\Delta x = -\alpha \nabla f' = -\alpha (df'/df) \nabla f = -\alpha' \nabla f$. Since df'/df is a positive value, this just scales the step size by a common value; the steepest descent direction is unchanged. A Cauchy sequential line search (CSLS) will follow the same path.

An incremental steepest descent (ISD) search will follow the same initial direction, and if the step size is based on DVs, $\alpha = \Delta s/\sqrt{\sum (\partial f/\partial x_i)^2}$, it too will follow the same exact path. Factoring the scaling term from the gradient

$$\Delta x = -\alpha \nabla f' = -\alpha \frac{df'}{df} \nabla f \tag{16.10}$$

Expanding α,

$$\alpha = \frac{\Delta s}{\sqrt{\sum (\partial f'/\partial x_i)^2}} = \frac{\Delta s}{\sqrt{\sum ((df'/df)(\partial f/\partial x_i))^2}} = \frac{\Delta s}{(df'/df)\sqrt{\sum (\partial f/\partial x_i)^2}} \tag{16.11}$$

And combining

$$\Delta x = -\frac{\Delta s}{(df'/df)\sqrt{\sum (\partial f/\partial x_i)^2}} \frac{df'}{df} \nabla f = -\frac{\Delta s}{\sqrt{\sum (\partial f/\partial x_i)^2}} \nabla f = -\alpha \nabla f \tag{16.12}$$

shows that it is the conventional ISD.

For Hessian-based searches, it is more complicated. Expand the second derivative in the Hessian:

$$\frac{\partial^2 f'}{\partial x_i \partial x_j} = \frac{\partial}{\partial x_i} \left(\frac{\partial f'}{\partial x_j} \right) = \frac{\partial}{\partial x_i} \left(\frac{df'}{df} \frac{\partial f}{\partial x_j} \right) = \frac{d^2 f'}{df^2} \frac{\partial f}{\partial x_i} \frac{\partial f}{\partial x_j} + \frac{df'}{df} \frac{\partial^2 f}{\partial x_i \partial x_j} \tag{16.13}$$

The locally common df'/df factor scales the conventional Hessian elements (on the second term on the right hand side of the previous equation) and the Jacobean elements. However, the Hessian has a new term, the first on the RHS, which changes the Newton–Raphson algorithm:

$$\mathbf{H}' \Delta x = -\nabla f' \tag{16.14}$$

becomes

$$\left[\left(\frac{d^2 f'/df^2}{df'/df} \right) \left| \frac{\partial f}{\partial x_i} \frac{\partial f}{\partial x_j} \right| + \mathbf{H} \right] \Delta x = -\nabla f \tag{16.15}$$

You should recognize the conventional Newton–Raphson, $\mathbf{H}\Delta x = -\nabla f$, pattern in the previous text, but the Hessian elements have the additional term on the left-hand side. The parenthesis represents a scalar multiplier for the matrix of the product of first derivatives. This new term presents an adjustment to the direction and magnitude of the Δx steps. If the new term is significant, it will affect the NR path to the optimum. If the OF transformation is linear, however, there is no impact.

For successive quadratic, the new model is $f' = a + bx_1 + cx_2 + dx_1^2 + ex_2^2 + fx_1 x_2$. If the OF transformation is linear, then the solution path is unchanged. Can the reader discover if the path remains unchanged if it is a nonlinear transformation?

For direct searches (Hooke–Jeeves, cyclic heuristic direct, leapfrogging) and for steepest descent searches (CSLS, ISD), OF transformations do not make a difficult surface easier to find the minimum. They do not remove surface aberrations or multiple minima. They do not reduce the number of function evaluations needed to find the minimum. They do not change the path to the minimum.

Alternately, for searches that use a second-order (quadratic) model of the OF response to DVs (Newton, SQ), a nonlinearly transformed OF does change the functional relation for the model and consequently the jump-to location. If the function is basically quadratic, a nonlinear OF transformation makes it not quadratic, which degrades the Newton or SQ action. Alternately, a transformation that could make a non-quadratic function becomes quadratic-ish, which will improve the speed of convergence with a Newton or SQ approach. For example, $f = (a - x_1)^4 + (b - x_2)^4$ becomes more quadratic-like if the square root is transformed by $f' = \sqrt{(a - x_1)^4 + (b - x_2)^4}$. The reader could use the 2-D example program to test this statement.

So it might be either an NOFE advantage or disadvantage for a Newton-type or successive quadratic algorithm.

The advantages of OF transformations are as follows:

1) Simplifying the OF calculation
2) Converting max to min or min to max
3) OF scaling for representation
4) Modifying for improved visualization of OF features in a particular range (such as the minimum) when their contours might get masked by the high values

The generic disadvantages of a transformed OF are as follows:

1) The functional transformation distorts human understanding of the modified OF, which is at least a barrier to assigning convergence criteria.
2) It could either increase or decrease the NOFE in second-order type searches.

16.7 Inequality Constraints

Inequality constraints can be converted to equality constraints by including a slack or a surplus variable. For example, an inequality constraint, $g(\boldsymbol{x}) > 0$, can be converted to an equality constraint by including variable s, $g(\boldsymbol{x}) - s^2 = 0$ with the condition $s \neq 0$.

Nominally there are N elements in the DV vector, $\boldsymbol{x}$. For each such inequality constraint transformation, the optimizer can search for the value of s, and $N-1$ of the DVs, and use the new equality constraint to determine the remaining DV value. If the original application was

$$\begin{array}{l} \min_{\{\boldsymbol{x}\}} J = f(\boldsymbol{x}) \\ \text{S.T.:} \quad g(\boldsymbol{x}) > 0 \end{array} \qquad (16.16)$$

With $\boldsymbol{x}^-$ representing the reduced set of original DVs, the new statement is

$$\begin{array}{l} \min_{\{s, \boldsymbol{x}^-\}} J = f(\boldsymbol{x}) \\ \text{S.T.:} \quad s \neq 0 \end{array} \qquad (16.17)$$

There are still N number of DVs. However, this may substantially change the shape of the OF response to the new DV set, making solution easier. It is also predicated on a mathematically convenient method to solve the inverse of the equality constraint for the remaining DV.

Note: If the constraint had been either $\leq$ or $\geq$, then there would be no restriction on the value of s, even $s = 0$ would be permissible.

16.8 Transforming DVs

Commonly, we linearly scale DVs to make units compatible or to make the values compatible with desired display or value ranges. However, DV scaling could be a nonlinear transformation.

16.9 Takeaway

Strictly positive functional transformations of the OF do not change the DV* value, but have some conveniences. They do not improve NOFE in direct search methods but may either increase or decrease NOFE in second-order methods. Converting inequality constraints to equality constraints substantially changes the shape of the response, but not the DV* solution values.

16.10 Exercises

1 Use a procedure similar to that of Section 16.5 to show that minimizing $f(x)$ is the same as maximizing $1/[1 + f(x)]$.

2 Use a sequence of OF transformations to show that maximizing $J = \ln\left(1/\left(a + b(x-c)^4\right)\right)$ (where $b > 0$) yields the same x^* value as minimizing $J = (x-c)^2$.

3 The equation $\Delta x = -\alpha \nabla f$ defines the DV increments in the direction of steepest descent. Use it to show that a strictly positive OF transformation does not change the direction of steepest descent in DV space.

4 Explain and defend the OF transformations that indicates solving $\displaystyle\max_{\{\theta\}} J_1 = \sqrt{5/\cos\theta}\ \ \sin\theta$ provides the same θ-value as $\displaystyle\min_{\{\theta\}} J_2 = \cos\theta\ \sin\theta$.

5 An OF transformation by a strictly positive monotonic function does not change the values for DV*, the optimum. Demonstrate this. (a) Take a function of your choice in the 2-D Optimization Examples code find the minimum. (b) Then change the VBA code to use a strictly positive monotonic function transformation of the OF. (c) Find the minimum. (d) The DV values from steps (a) and (c) should be identical, but the OF values will not. Explain why this is so. Use the two contours and the two 3-D OF versus DV plots to assist your explanation. (e) But what is the result if you transform an OF twice?

6 Minimize $J = x + 1/x$, S.T. $\sqrt{x} \geq a$, and to keep real numbers $x \geq 0$. Then convert the inequality constraint to an equality, $\sqrt{x} + s^2 = a$, and use s as the DV and solve for x from s. Show that the solutions are identical. Comment on advantages or disadvantages.

7 Use your own optimization application with an inequality constraint, and convert the constraint to an equality with a slack or surplus variable. Does the transformation simplify?

17

Convergence Criteria 2

N-D Applications

17.1   Introduction

This chapter builds on Chapter 6 and elaborates on convergence criteria that can be used in optimization. The concepts can be classified as either deterministic or stochastic and as either single player or multiplayer applications. Single trial solution searches (such as NR, LM, CHD, etc.) have one trial point and seek to move it. Here, convergence is based on the sequential trial solution, point-to-point changes. Multiplayer searches (such as LF and PSO) and pattern-type optimizers (such as HJ and NM) have a cluster of trial solutions. Here, convergence would be based on the range of the cluster or pattern.

  However, there is a lead player in the pattern-type and multiplayer optimizers, and ignoring the other active players in a multiplayer procedure, the convergence criteria could be based on the iteration-to-iteration changes in the best in the cluster. So, the single TS convergence criteria could be applied to a multiplayer algorithm.

17.2   Defining an Iteration

Convergence is tested after each iteration. In many optimizer procedures the definition of what constitutes an iteration seems obvious. An iteration is when one complete cycle of analysis and consequential TS move is complete. In CHD an iteration constitutes N surface explorations, one for each dimension, each DV. In HJ, an iteration has an average of $1.5N + 1$ surface explorations. In CSLS, there are $N + 1$ function evaluations to define the univariate search direction, then perhaps 15 (plus or minus a few) for the line search, for a total of about $N + 15$ NOFE per iteration. For SQ, with cross product terms, $(N^2 + 3N + 2)/2$ surface explorations are required to obtain the surrogate model. Newton-type procedures, also a second-order algorithm, have a similar number of function evaluations per iteration as SQ. If the DV dimension is 5, for instance, then CHD will have 5 function evaluations per iteration, and second-order methods will have 21 per iteration.

  However, in LF, if one player move is considered an iteration, there is only one surface exploration per iteration. This means that any one LF cycle of analysis does not provide nearly as much surface understanding or criteria for move direction as the others. Further, after one leap-over, the best player might remain unchanged. Accordingly, I define an LF iteration to have N leap-overs, which provides equivalent surface analysis per iteration as CHD, which is still much less than other methods. This choice seems to make sense for convergence testing, as I interpret observations of convergence assessment and optimizer progress with player moves.

Notably, in PSO an iteration constitutes the local, one-move exploration of each player. If there are M players (with M about 10 times the DV dimension N), then each iteration provides $M \cong 10N$ function evaluations. But since the best player only makes one randomized move in the local vicinity of the historical global best, there might not be any improvement in the DV* value with an iteration. So, PSO also should not use an iteration-to-iteration observation of the global best to define convergence. Further, in many PSO algorithms there is no end-game convergence of the player swarm; there is no reduction in the random perturbations influencing each player. As a result, after converging on the optimum, the end-game swarm size remains constant, but stochastically varying, with iteration. Accordingly, a deterministic criterion for convergence should not be used.

17.3 Criteria for Single TS Deterministic Procedures

Chapter 6 presented techniques for testing either the change in DV or the change in OF or the relative change in either to determine convergence. These are the most common approaches.

For visual presentation of the results, one could take the perspective that the ability to discriminate alternate DV solutions would define the convergence criterion. Perhaps the criterion should be to claim convergence when the change in the DV is less than one-tenth of what will be visually detectable. Alternately, if the DV is to be implemented, then one would select convergence to represent an order of magnitude smaller than the implementation discrimination ability. In either case such choices are dependent on the scale (value range) of presentation or implementation. If there are multiple DVs, then each should have its own convergence threshold, representing the discrimination ability for each DV. Some choose to use a common distance such as $\|\Delta x\| = \sqrt{\Delta x_1^2 + \Delta x_2^2 + \Delta x_3^2 + \cdots}$, but this presumes that all DVs have the same dimensional units and the same scale of DV impact on the OF. Usually the DVs are disparate. If a common threshold is desired, a solution is to use scaled variables.

Often, however, it is not the DV that is important, but the OF. For example, one can balance the checkbook to the penny, but when negotiating salary or the price of a house, when within $1000 claim convergence. So, consider the practicality of the OF, not what is possible to discriminate.

In a multivariable situation, the DVs may be making significant, but compensating changes as the trial solution converges on the DV* point. In this case the OF might have little net change, yet each individual DV influence on the OF value may be separately significant. In such a case, consider a convergence test on the OF, but test for the sum of all possible DV influences by propagating maximum error:

$$\Delta \mathrm{OF}_{\max} = \sum_{i=1}^{N} \left| \frac{\partial \mathrm{OF}}{\partial x_i} \Delta x_i \right| \tag{17.1}$$

This scaling by $\partial \mathrm{OF}/\partial x_i$ normalizes the Δx_i values on the OF value and permits the addition of DV impacts regardless of dimensional units. This approach however requires the partial derivative, which might not be analytically possible or even numerically representative near a discontinuity.

It could be that the OF model has uncertainty in its coefficients. For instance, if the OF is a function of DVs and coefficients, $J = f(c, x)$, then there is uncertainty in the OF due to the coefficients. Here, there is no sense in seeking perfection in the DV values. Perhaps claim convergence when the relative impact of the DV changes on the OF is an order of magnitude smaller than the OF uncertainty resulting from the coefficients:

$$\sum_{i=1}^{N}\left|\frac{\partial OF}{\partial x_i}\Delta x_i\right| = \Delta OF_{max,\, DVs} \leq 0.1\Delta OF_{max,\, c} = 0.1\sum_{i=1}^{n}\left|\frac{\partial OF}{\partial c_i}\varepsilon_{c_i}\right| \tag{17.2}$$

In regression, there is uncertainty in both the dependent and independent variables, which would cause residuals, the deviations between model and data. A convergence criterion could be to stop iterations when the reduction in residuals is two orders of magnitude less than the residuals. Here, the concept is that the model adequately (sufficiently, not perfectly) matches the cause–effect mechanisms in the data, and at the end of the optimization procedure, the residuals (the deviations between model and data) reflect experimental uncertainty. Then, end-game residuals cannot be removed, and they indicate a best possible outcome. If SSD is the objective function, then a measure of average residual is the rms value, $rms = \sqrt{SSD/N}$, where N is the number of data sets. This convergence criterion would be to claim convergence if $\Delta rms \leq 0.01\, rms$ (equivalently $\Delta SSD \leq 0.0001\, SSD$). This is a relative measure on the OF.

However, there are many variations on any theme, and an alternate convergence criterion for regression could be to stop iterations when the rms value from a random subset of the residuals approaches stochastic steady state w.r.t. iteration (see Chapter 29).

If the OF is stochastic, then the previously mentioned metrics are not applicable.

17.4 Criteria for Multiplayer Deterministic Procedures

For multiplayer or pattern approaches, the DV range could be either the difference between DV extremes or an rms deviation. If there are multiple dimensions, then there needs to be a range for each dimension. A conventional rms deviation is the standard deviation, the rms deviation from the average. However, one could choose it to be the rms value from the DV with the best OF value. Again, if multidimensioned, each DV needs to have an rms calculation. Also, if the units are not the same on all DV dimensions, then each DV either needs an individual threshold or uses scaled variables.

Similarly, the range of the OF values could be the difference between largest and smallest, but range could also be measured by any number of alternate statistics, for instance, the rms value from the average OF value or the rms deviation from the best OF value.

For either the DV or the OF case, the best player is likely closer to the DV* value than the extreme players, and the rms value from the best is probably more representative of the characteristic proximity of the players from the optimum.

If the rms value is used to represent the characteristic deviation in the DV, then convergence criteria could be applied to either the rms OF or DV values. However, the maximum possible impact of the variation in the DV values could be propagated to the OF and convergence based on

$$\Delta OF_{max} = \sum_{i=1}^{N}\left|\frac{\partial OF}{\partial x_i}rms_{x_i}\right| \tag{17.3}$$

It could be that the OF model has uncertainty in its coefficients. For instance, if the OF is a function of DVs and coefficients, $J = f(c,x)$, then there is uncertainty in the OF due to the coefficients; here there is no sense in seeking perfection in the DV values. Perhaps claim convergence when the relative impact of the DV changes (as indicated by the rms value) on the OF is an order of magnitude smaller than the uncertainty resulting from the coefficients:

$$\sum_{i=1}^{N}\left|\frac{\partial \text{OF}}{\partial x_i}\text{rms}_{x_i}\right| = \Delta \text{OF}_{\text{max, DVs}} \leq 0.1\Delta \text{OF}_{\text{max, c}} = 0.1\sum_{i=1}^{n}\left|\frac{\partial \text{OF}}{\partial c_i}\varepsilon_{c_i}\right| \tag{17.4}$$

17.5   Stochastic Applications

There are several sources of stochastic behavior. In one, the OF could be a Monte Carlo simulation or an experimentally obtained result; in either case it includes natural vagaries. With the same DV value, a resampling gives a similar but different OF value. In another, the algorithm may be a stochastic search that continually perturbs trial solutions with random perturbation. In a third, which I find useful in regression of models to data, the OF and optimizer algorithm may both be deterministic, but the random sampling of data for convergence testing will generate a noisy SSD or rms value that relaxes to a noisy steady state as iterations progress to the optimum. In any case, convergence can be claimed when there is no improvement in the observed value w.r.t. iteration.

In optimization of stochastic functions, use replicates on the best player, but if the replicate makes the second best become the best, then let the second best guide the search. Either in optimization of stochastic functions or when using a stochastic optimizer (such as PSO), then observe the worst player OF value w.r.t. iteration. (Not the OF value of the initial worst player, but the worst OF value of all players at each iteration.) Then declare convergence when the OF value of the worst player comes to steady state w.r.t. iteration. Alternately, observe the OF range (the deviation between best and worst) w.r.t. each iteration. This would be useful if the OF* is a large value. However, don't observe the best OF value w.r.t. iteration because if best starts near a local optima, then watching it might lead to an SS claim before the team has found the vicinity of the global. By contrast, even if the best fortuitously starts near the global optimum, the worst will make the largest change from initial to local final values.

17.6   Miscellaneous Observations

Internal root finding and nested internal optimization both have convergence criteria, which need to have a much smaller impact on the DV or OF value than the convergence criterion for the primary (outside) optimization. Similarly, numerical methods of solving differential equations and estimates of derivatives or integrals will cause numerical variation, and these also must have a much smaller impact than the convergence criterion might detect.

Convergence threshold can change during an optimization procedure as the optimum is approached. For example, in CSLS, the initial line search does not need to precisely find the optimum. Use an initial coarse convergence threshold to speed up the search for the vicinity of the optimum, then as the sequential searches do not make much change, tighten the convergence criteria to the desired value for the final answer.

17.7   Takeaway

Convergence criterion and corresponding threshold values are human choices. The choice of criterion needs to relate to the utility of the application context and be appropriate for the optimizer. Further, the threshold value needs to support desired precision, but not cause excessive NOFE.

17.8   Exercises

1   An ideal orifice flow rate calibration equation is $F = a(i - i_0)^{0.5}$. The value for coefficient "a" will be determined by a best fit of the equation to data. Define ΔDV stopping criterion so that uncertainty on the DV has a much lower impact on the calculated flow rate than measurement uncertainty on "i," the transmitted signal.

2   Explain the concept behind the convergence criterion based on propagation of maximum uncertainty, $\epsilon_J = \sum |(\delta J/\delta DV_i)\Delta DV_i|$, and state one advantage and one disadvantage relative to the conventional ΔDV criteria.

3   Explain the concept behind the convergence criterion based on detecting steady state of the SSD (sum of squared deviations) or rms (root-mean sum of squared deviations) in a randomly sampled subset of data, and state one advantage and one disadvantage relative to the conventional ΔDV criteria.

4   Here are three convergence criteria for stopping the optimization iterations in least-squares regression. State advantages and disadvantages of each.
   A $|\Delta DV_i| < \varepsilon_i$,   for each DV
   B $\sum |(\partial SSD/\partial DV_i)\Delta DV| < \varepsilon_{SSD}$
   C   A plot of rms of a random subset of data w.r.t. iteration is steady state

5   Explain why a convergence threshold of 0.01 on the LF cluster range leads to about the same solution precision as a convergence threshold of 0.001 on the ΔDV on a single TS direct search approach such as HJ or CHD.

6   Compare several (three or more) convergence criteria. You choose the DV and the OF, and the challenge is to $\underset{\{DV\}}{\max} J = OF$ Provide experimental data and analysis. (i) Define what concept you will use to decide whether one is better than another and the data you will need to assess/measure the concept. (ii) Decide on the DV(s). (iii) Use the 2-D Optimization Examples Excel-VBA file to generate data. (iv) Define the experiments you will use to generate results, and explain why they are a legitimate and critical test of your concept of better. (v) Generate data, and describe the results. Of course: sketch, erase, sketch, erase, sketch, ink. Your first attempt at understanding the application will be primitive, and as you develop the path for your primitive understanding, you'll become aware of a more comprehensive view. Would you get the same conclusion independent of optimization method, or function, or initialization?

18

Enhancements to Optimizers

18.1 Introduction

The objective is to improve the optimizers, to optimize the optimization algorithms. Specific metrics would be to maximize the probability of finding the global optimum and robustness to aberrations (constraints, nonlinearities, stochastic response, discontinuities, flat spots, etc.) while minimizing computational work (of both the algorithm and number of function evaluations) and minimizing complexity (for either programmer or user). In any one optimizer, there seem to be dozens of enhancements the lead to improvements.

This chapter summarizes a few techniques that I think are relevant, archetypical, and broadly functional.

18.2 Criteria for Replicate Trials

Consider a deterministic (not stochastic) response. One run of an optimizer may converge at an optimum, but it may be a local optimum, not the global. Another run from a different initialization may lead to the same local optimum. Or it might be the single global optimum has been found twice. How can one tell whether the global has been found? That is one question.

Even if there is only one optimum, the global, successive runs from independent initializations will each converge in the vicinity of the optimum, but not exactly on the true DV* spot. Consequently, each solution will likely have slightly different DV* and OF* values. How can one tell whether the different values indicate a common solution or different solutions? That is another question.

The true optimum will likely not be found, so the first question should be "How can one tell whether the vicinity of the global has been found?," and the second question should be "How can one tell whether the different values indicate expected precision at a common solution or different solutions?"

Unless there is an exhaustive analysis of the OF response over the entire DV space, perhaps an analysis of the functionalities in the models, I don't think that there can be a definitive, 100% confident, statement that the global has been found. However, I find two approaches useful in establishing confidence as to whether the global has been found. Both are grounded in probability analysis, one is *a priori* (before evidence) one is *posteriori* (after evidence).

Engineering Optimization: Applications, Methods, and Analysis, First Edition. R. Russell Rhinehart.
© 2018 R. Russell Rhinehart. Published 2018 by John Wiley & Sons Ltd.
Companion website: www.wiley.com/go/rhinehart/engineeringoptimization

The best-of-N technique indicates the number of optimizer trials from independently randomized initializations that are needed to have a user-desired confidence, c, and after N trials, the best of the N outcomes will represent one from the best possible fraction, f, of all outcomes:

$$N = \frac{\ln(1-c)}{\ln(1-f)} \tag{18.1}$$

See Chapter 23 for derivation details. The N calculated by Equation (18.1) needs to be rounded up. Further, this supposes that the initialization ranges are large enough so that the optimizer is able to find one solution in the desired fraction.

I find this useful. For example, if I suspect that there are 50 local optima, but don't know where the global best might be hiding, and want to be 99.9% ($c = 0.999$) confident that a solution is one of the best possible 2% ($f = 0.02 = 1/50$) that is possible, then $N = 342$ independently initialized trials.

Noteworthy, the arguments of the log functions are less than 1, so the numerator and denominator are both negative numbers. Then if the desired confidence is 100% ($c = 1$), the numerator term is the log of zero, negative infinity, and N is infinity. This model reveals that 100% surety is impossible to achieve with only empirical evidence. Also, if the true exact optimum were desired, for which none are better, $f = 0$, then the denominator is the log of 1, which is zero, again, making $N = \infty$ and revealing that it is not possible to know that the true DV* value has been found without analytical support.

Use of the best-of-N equation would be an *a priori* decision procedure. It may be, however, that after observing the DV* and OF* responses by the 100th trial, it is becoming obvious that there were not 50 local optima, but only 3. Of course, one cannot know whether continued trials will find a rarely encountered 4th optimum or whether there are just the 3. But if evidence keeps revealing only 3, then *a posteriori* belief that the global has been found could be used to stop future trials and accept the best found as the global.

Snyman and Fatti (1987) published a paper on the use of a Bayes conditional probability technique to indicate that the belief that the global has been found is adequate to stop successive optimizer trials. Implementation of the technique is relatively simple, and I have been pleased with results.

See Chapter 23 for derivation details. Here is my VBA code for the method.

```
Sub belief()    'belief that global has been found, Bayes conditional
probability
    'Snyman-Fatti (also Zielinski) method, Snyman page 145-150
    nbelief = k        'number of trials run by the optimizer
    If nbelief = 1 Then    'initialize values
        abelief = Cells(2, 37)   'a priori estimate of probability global will be
found on any one trial
        rbelief = 1       'number of trials in vicinity of best so far
        Pbelief = abelief  'initial belief that global has been found on one trial
        ebelief = Cells(3, 37)   'OF proximity that determines whether a new
global has been found
        fbelief = OF_Value_Old
    Else
        P11belief = abelief / (rbelief + 1)  'update conditional probabilities
        P12belief = 1 - P11belief
        P21belief = abelief + (1 - abelief ) / (nbelief + 1)
        P22belief = 1 - P21belief
```

```
    If OF_Value_Old < fbelief Then      'better, update belief
        Pbelief = Pbelief * P11belief / (Pbelief * P11belief + (1 - Pbelief) *
P21belief)
    Else                       'not better, worse or within vicinity
        Pbelief = Pbelief * P12belief / (Pbelief * P12belief + (1 - Pbelief ) *
P22belief )
    End If
    tbelief = Sheets("Converged_Data").Cells(3, 4) 'start count for r
    For rbelief = 1 To 100
      If Sheets("Converged_Data").Cells(2 + rbelief, 4) > tbelief +
ebelief Or_
          Sheets("Converged_Data").Cells(2 + rbelief, 4) = "" Then Exit For
    Next rbelief
    If rbelief > 1 Then rbelief = rbelief - 1
    If nbelief > 10 Then abelief = 0.9 * abelief + 0.1 * (rbelief / nbelief )
    fbelief = tbelief
  End If
```

In the Snyman–Fatti method, the user must define proximity criteria to specify whether a solution is in the vicinity of the best found so far and also an estimate of the probability that any particular run might find the global. In a sense this also requires *a priori* data.

The case of stochastic functions is different, because the function evaluation is not repeatable; each response to the same set of DV values returns a response that is randomly perturbed. Sometimes the response is termed noisy. Here, chance events would lead to a smaller (or larger) OF value than the average, and taking them at face value may lead the optimizer to a DV set for which the stochastic vagaries gave an unusual best. This would be a phantom best, because replicate samples of the OF at the same DV would yield not as good values. Often, the optimization objective is not to find the chance best, or to find the average, but to minimize a 95% worst case (or any other desired extreme value). There are several approaches to determine a 95% worst case.

One is to replicate perhaps $N = 10$ samples and then use normal statistics to estimate the 95% confidence limit:

$$OF_{0.95} = \overline{OF} \pm t_{N-1,1-0.95} \frac{\sigma}{\sqrt{N}} \qquad (18.2)$$

where $t_{N-1,1-0.95}$ is the *t*-statistic value for the degrees of freedom and confidence, $\overline{OF}$ is the average, and the $\pm$ operation is determined by whether the larger or smaller extreme is relevant.

Alternately, if one cannot presume that the noise on the OF is normally distributed (Gaussian), then sample N times and report the worst of the N as the representative limit. The formula to determine N based on a confidence, c, that one sample representing one from the worst fraction, f, is

$$N = \frac{\ln(1-c)}{\ln(1-f)} \qquad (18.3)$$

This version of the weakest-link-in-a-chain is easily derived. The probability of an event is the complement of the probability of not an event:

$$p(\text{at least 1 of } N \text{ samples in the worst fraction}, f) = 1 - p(\text{none of } N \text{ are in} f) \qquad (18.4)$$

For a composite event to be true, each event must be true:

$$p(\text{none of } N \text{ are in } f) = p(1 \text{ is not, AND } 2 \text{ is not, AND } 3 \text{ is not, AND...AND } N \text{ is not})$$

(18.5)

When events are independent, the AND conjunction is equivalent to multiplying individual probabilities:

$$p(\text{none of } N \text{ are in } f) = p(1 \text{ is not in } f)p(2 \text{ is not in } f)p(3 \text{ is not in } f)...$$

(18.6)

Using the complement probability,

$$p(\text{none of } N \text{ are in } f) = [1 - p(1 \text{ is in } f)][1 - p(2 \text{ is in } f)][1 - p(3 \text{ is in } f)]...$$

(18.7)

If events are independent,

$$p(1 \text{ is in } f) = p(2 \text{ is in } f) = ...$$

(18.8)

Then

$$p(\text{none of } N \text{ are in } f) = [1 - p(\text{any is in } f)]^N$$

(18.9)

The probability that any is in the worst fraction is:

$$p(\text{at least } 1 \text{ of } N \text{ samples in the worst fraction, } f) = c = 1 - [1 - f]^N$$

(18.10)

Rearranging to solve for N yields Equation (18.1).

18.3 Quasi-Newton

If the OF permits analytical second derivatives, then each element in the Hessian of a Newton-type optimization will require one function evaluation. If continuum-valued, the mixed derivatives have the same value whether one variable is first or the other is first:

$$\frac{\partial^2 f}{\partial x_i \partial x_j} = \frac{\partial^2 f}{\partial x_j \partial x_i}$$

(18.11)

So, rather than N^2 equations to be evaluated, there are only $N(N-1)/2$ function evaluations in the Hessian, if they are analytically tractable.

However, if analytic versions of the second derivatives are not possible (the OF is from experimental data not an equation, the OF is from a complicated procedure, the equations are too complicated, the function has discontinuities, etc.), which is usually the case in relevant applications, then numerical estimates of the derivatives are used. If a central difference is chosen, which is a best method for slope characterization at the end game, then the main diagonal elements can use the same function evaluations as the gradient uses, but the $N(N-1)$ may not be symmetric about the main diagonal and each requires four OF evaluations, increasing NOFE by $4N(N-1)$ at each iteration.

Quasi-Newton methods avoid that computational work by using past changes in the gradient as estimates of the Hessian second-derivative methods. This is akin to the secant version of estimating the slope in a Newton-type of root-finding algorithm. The approximation is not any more illegitimate than the basis for a Newton method that the truncated second-order Taylor series local surface approximation is globally valid or that a numerical estimate of the derivative is valid.

There are many quasi-Newton options that have been devised for incrementally updating the Hessian elements with change of past gradient values scaled by the corresponding past DV changes. Some are presented in Chapters 9 and 10.

18.4 Coarse–Fine Sequence

If, on average, it takes an optimizer N_s function evaluations in the initialization and initial search to find the vicinity of the optimum, then on average N_e function evaluations in the end-game fine-tuning to converge on the solution; then one run of an optimizer requires an average number of function evaluations of ANOFE $= N_s + N_e$. If N trials are being used to provide adequate confidence that the best of N will represent the global, then the expected number of function evaluations is $(N)(\text{ANOFE}) = N(N_s + N_e)$.

A method to reduce the amount of work is to not perform the end-game fine-tuning on each of the N trials. Only do the initialization and initial search for the coarse proximity of the optima. Then return to the best of the N coarse search results, accept the best as representing the proximity of the global, and do the end-game search there once. This reduces the expected number of function calculations to $N(N_s) + N_e$. The ratio in NOFE reduction is $1 - N_e / (N_s + N_e)$, which can be substantial.

It may be difficult to distinguish the initial search phase from the end-game convergence phase. However this duality can be easily implemented by using a coarse threshold in the convergence criterion for the initial N trials, then return to the best of N and continue the search with the desired convergence condition.

The loss of insight that is associated with this approach is the analysis of precision of the answer. With only one solution appropriately converged, there is no basis to reveal precision (uncertainty of the DV* or OF* values) as measured by standard deviation of replicate solutions.

18.5 Number of Players

In multiplayer algorithms (LF, PSO, GA), the more players that are involved, the better is the entire surface characterization, hence the ability to find the global. However, the more players there are, the greater is the associated NOFE. There are several tricks to preserve surface exploration yet reduce NOFE, and one is to initialize the surface with many players, enough so that one is likely in a location that will draw it (and all others) to the global optima. If one has an estimate of the fraction of DV space that will draw a player to the global, then the best-of-N formula indicates the number of players. Here, f_i is the fraction of the ith DV space that will definitely draw a player to the global, and c is the desired confidence that at least one player will be in that space; the number of players for initialization is

$$N_i = \frac{\ln(1-c)}{\ln\left(1 - \prod f_i\right)} \tag{18.12}$$

After the initial surface overpopulation, start the optimization procedure with the nominal number of players for that algorithm. This approach does require *a priori* knowledge of either the f_i values or a characteristic f-value and replacing $\prod f_i$ with f^N.

An alternate approach to reducing NOFE is akin to the coarse–fine mechanism of the previous section. Start with either the recommended number or an excess number of players for the algorithm, and then once it is obvious that the team is converging on one spot (perhaps by the range in DV values or OF values of all players), then reduce the number of players. If, normally, there is an initial search phase and an end-game convergence phase, then the average number of function evaluations can be represented as ANOFE = $N_s + N_e$. If culling half of the players when the end-game is identified, then ANOFE = $N_s + N_e/2$.

I have explored using Bayes belief methods to determine confidence that the vicinity of the global has been found, and when adequately confident, to initiate any number of decisions. These include stopping additional trials or reducing the exploration window in leap-overs. It does reduce NOFE, but I am not sure that the benefit is worth the complexity or risk of choices in the Bayes probabilities, leading other applications to false decisions.

18.6 Search Range Adjustment

Expansion–contraction logic can be applied to the size of the search mechanism. In LF, the leap-to window can be expanded or reduced, by the value of the factor α:

$$x_{w, \text{new}} = x_b - \alpha r(x_w - x_b) \tag{18.13}$$

A value $\alpha = 2$ would provide considerable wide area exploration, and $\alpha = 0.5$ relatively fast contraction. In PSO the factor c_{1x} scales the random search amplitude:

$$x_{i+1,j} = x_{i,j} + c_{1x}(r - 0.5) + c_{2x}(x_{bj} - x_i) + c_{3x}(x_g - x_{i,j}) \tag{18.14}$$

The algorithm could initially search with a large value for the leap-to range or random perturbation, and when the end game is identified, reduce these values to accelerate convergence. I have explored many approaches to automating the coefficient value to an end-game value. If the surface exploration is so great as to prevent players from gathering in the proximity of one spot, then observing the best OF as iterations progress and applying Bayes belief probability that the global has been found can indicate when to convert exploration into the end-game convergence. Although this works in reducing NOFE, I am not sure that the benefit is worth the complication.

18.7 Adjustment of Optimizer Coefficient Values or Options in Process

This idea has parallels to adjusting either the number of players or the magnitude of the search range in response to whether the optimization procedure is in the initial search phase or the end game of solution fine-tuning. There is a diverse classification of optimizer coefficients or choices. Included coefficients might be the tempering factor in a Newton method or the expansion or contraction factor in a heuristic search. Choices might represent a switch from forward difference to central difference numerical techniques for estimating derivatives or to use a conjugate gradient approach when there appears to be direction oscillation, or an LM-type switch between ISD and NR as confidence builds that NR is applicable. It could be an initial search with one optimizer type, like LF, followed with a second-order optimization approach, like NR, to the optimum.

My classroom students and I have all envisioned such enhancements. They have merit. The key issue in automating the switch is to robustly and universally identify when a search should switch to an end-game style from an initial search phase.

18.8   Initialization Range

If there is prior knowledge about what a right solution might be, then initialize the trial solution there. In a multiplayer algorithm, initialize at least one player there (at the hypothesized or former DV* solution). Often, once an optimization has thoroughly explored a situation and has arrived at the DV*, the answer will remain nearly right for a future period. For example, once you find the right workout for your body, it will be right tomorrow and the next day. But as your body develops or ages, in a year or so, the workout should be adjusted. The right workout would not be dramatically changed, just progressively adjusted. Similarly, once an extensive optimization has determined operating conditions for a plant, they will be valid as long as market prices and equipment efficiencies hold original values. However, when these values change, the optimum will shift. But it will not be a dramatic shift; it will be a slight shift from the prior optimum.

Alternately, for new applications, or when there is suspicion that the situation has completely changed, initialize trial solutions over the entire feasible DV range.

18.9   OF and DV Transformations

Chapter 16 provides an analysis that reveals that transforming the OF with a strictly positive monotonic function does not change the DV* value. A linear transformation with a positive slope, $b > 0$, is an example. $OF' = a + bOF$. The value of the intercept is irrelevant. If the OF is always positive valued, then the log transform is another example. $OF' = \ln(OF)$. In these cases $dOF'/dOF > 0$ for all OF values.

The transformed OF is useful for scaling the response and revealing detail in a region of greatest interest, like a square root or log transform might do for low OF values.

In direct search or steepest descent algorithms, OF transformation has insignificant benefit in reducing ANOFE. In second-order algorithms (SQ or Newton-based), the transformation may make the surface more quadratic-like, in which case it will reduce NOFE; but then, it might make it less quadratic-ish and increase NOFE.

Recognizing that linearly transforming an OF does not change the solution can be useful in avoiding "good ideas." For instance, an optimization might be seeking a model to characterize outcomes in proper categories, such as looking at pixel patterns from writing and interpreting which letter it represents. The optimizer may be seeking to maximize the number of right classifications, $OF = N_R$, out of a total of N_T. The user may wonder if including a penalty for wrong classifications would help the optimizer, $OF' = N_R - N_W$. This appears to be a new OF, a new assessment of desirability and undesirability, but since $N_W = N_T - N_R$, then $OF' = N_R - N_W = N_R - (N_T - N_R) = 2N_R - N_T = bOF + a$, which reveals it is just the linearly strictly positive monotonic scaling that will return the same DV* values. There is no sense exploring it.

The OF can be modified by including a penalty for constraint violations. This can be done in any number of ways, additive or multiplicative, or if additive with either quadratic or log-barrier functions.

If the penalty function is only active near to, or where a penalty is violated, then it will not affect the interior search results. When penalties are incorporated in the OF, then they can avoid the discontinuity associated with hard constraints, and it will facilitate finding a common DV* point.

In addition to OF transformations, the DVs can also be transformed. Scaled DVs each make units compatible for gradient-based methods and for applying a common convergence threshold on DV range. Scaled inputs make NN models useful.

Selecting alternate DVs are also useful. For instance, switch a search from the nominal DVs to an auxiliary variable (such as a distance factor) or a slack variable (to change the impact of constraints).

18.10 Takeaway

There are many ways to enhance the basic algorithms. Specific software providers include many.

18.11 Exercises

1 Don't believe me. Test concepts for yourself. Devise appropriate criteria to observe (Bloom's Analysis Level 4), trials to obtain performance data (Bloom's Synthesis Level 5), and how to interpret the data to determine if it works (Bloom's Evaluation Level 6). Do this for these techniques (or alternate options that people have claimed to be good).
    A Section 18.2
    B Section 18.3
    C Section 18.4
    D Section 18.5
    E Section 18.6
    F Section 18.7
    G Section 18.8
    H Section 18.9

Section 4

Developing Your Application Statements

The optimization algorithm is important, but more relevant to the optimizer answer are developing the appropriate and complete statements about the objective function and constraints and choosing fitting criteria for convergence. Further, understanding how to assess and improve the probability of finding the global is critical. And understanding how uncertainty in the givens affect the uncertainty of the optimum and how to optimize within uncertainty is essential. This section contains 11 chapters related to those topics.

19

Scaled Variables and Dimensional Consistency

19.1 Introduction

The term "primitive variables" would refer to the variables we normally use when we measure or model a system, whether the system be a process, a procedure, a device, a product, or a combination. Primitive variables have units, and primitive variables can be either relative or absolute, or continuum or integer. The units may be very different for the same variable, which can be confounding for human understanding, for example, distance in miles or microns or light-years, speed in rpm or kps, mass in kg or lbs, number of people or number of events, composition in mole fraction or ppm, energy in BTU or kJ, etc. Additionally, the mix of relative and absolute variables can also be confusing: temperature in °C or R, pressure in psig or Pa, and time in hours from the beginning of the local day or in years from the big bang.

By contrast, scaled variables represent a ratio of a primitive variable to its unit dimension, and common examples might be a fraction of a population as the number divided by total number, or distance as a portion of the total trip left. The scaling might be by a combination of variables, for instance, when activation energy in chemical kinetics is scaled by the product of temperature and the gas law constant or when fluid velocity is scaled by density, viscosity, and a characteristic dimension to form the Reynolds number. Scaled variables are dimensionless.

Although primitive variables directly relate to a natural variable, these are the proper variables to use in a phenomenological model and often have a convenient meaning to a user; many aspects of primitive variables make them undesired. Several of these undesired aspects are related to dimensional consistency:

1) When $Q = f(x,y,z)$ and the x, y, and z distances all have the same units, then the Pythagorean relation can be used to combine them into a common distance. In many functions, however, the DVs do not have either the same units or meaning. In simple linear regression, the two DVs are slope and intercept. In a viscoelastic model of a film, they are relaxation time-constant and pre-exponential stress. In a reaction optimization, the DVs might be reactor temperature and initial feed composition. In these examples the DVs have disparate units, which cannot be unified. However, many optimization algorithms simply treat the DVs as abstract numerical values, permitting combinations and violating engineering propriety.

2) Steepest descent (gradient based) methods use a common scalar to calculate the incremental step size for all dimensions, in the relations $\Delta x = -\alpha(\partial f / \partial x)$, and $\Delta y = -\alpha(\partial f / \partial y)$. Looking at the first

Engineering Optimization: Applications, Methods, and Analysis, First Edition. R. Russell Rhinehart.
© 2018 R. Russell Rhinehart. Published 2018 by John Wiley & Sons Ltd.
Companion website: www.wiley.com/go/rhinehart/engineeringoptimization

relation, the units on α are $[x^2/f]$. But if x and y have different units, then the second and subsequent equations are dimensionally inconsistent. Again, this violates engineering propriety.

Other objections to primitive variables relate to range consistency:

3) Considering value (range) consistency in variables, for instance, in the Earth's surface distances (miles) and elevation (feet), a unit change in one is 5280 times larger than a unit change in the other.

4) There may also be value (magnitude) consistency in derivatives. Similar to the issue earlier, on a large continent, temperature sensitivity with respect to elevation is much larger than with respect to surface distance.

5) In display of a variable that has a range of several orders of magnitude, important details in the low value behaviors may be obscured by the small range they occupy on a graph.

6) In neural networks, the output of a neuron with a bipolar sigmoidal transfer function is in the range from -1 to $+1$. It is a scaled variable. Accordingly, either it needs to be unscaled to match the data or the data needs to be scaled on an appropriate range to match the range of the neural network output variable.

7) Using primitive variables, the gradient and Hessian elements may have values that are orders of magnitude different, which can result in ill-conditioned matrices (numerical solution issues for linear algebra techniques).

8) If a model includes powers of a variable (e.g., $y = a + bx + cx^3 + dx^5$) and the independent variable has a range that is not on the order of unity (e.g., $1000 < x < 5000$), then the numerical value of the coefficient d must be 15 orders of magnitude smaller than the value of coefficient a. For consistency in the magnitude of the coefficients in an empirical power series model, use scaled variables.

Other objections to primitive variables are related to user convenience:

9) If one DV has the units of temperature and another of moles per liter, then a $1°C$ deviation in one is much smaller in impact than a 1 mol/L deviation in the other. Accordingly, a user must devise appropriate convergence thresholds for each DV. A uniform convergence threshold cannot be applied. As an example, in optimizing value-to-price for an automobile tire, DVs may be tire diameter, number of fabric plies, number of steel belts, tread thickness, rubber compound compositions, and curing temperature—apples and oranges.

10) Similarly, when functions are additively combined as penalties within the objective function, the user must devise equal concern scalars (weighting multipliers) that reflect the relative importance of the several issues. This might be easier if all variables are scaled on the same range.

11) When a user decides to change the dimensional units, then other thresholds and optimizer scale-dependent coefficients (e.g., initial step size) might have to change. For example, one user might prefer a choice between lbs and kg, psi and Pa, or miles and feet that is different from what another user prefers. If using scaled variables, the dimensional preference is irrelevant.

12) The solution to a particular application in scaled variables is identical regardless of the system size, density, or heat capacity. If the problem is to define a 2-D shape that maximizes area to perimeter, the solution is the same whether the perimeter is given as 50 cm, or 800 light-years, or the area is given as 30 ft^2. The solution is a circle. With scaled variables, one solution, one graph, one equation, and one set of coefficients are common to all situations.

13) When creating a program to display and optimize diverse relations, using scaled variables facilitates programming.

There are advantages to converting equations into dimensionless terms:

14) Reduce the number of terms in models.
15) Facilitate graphing.
16) Generate universal models.

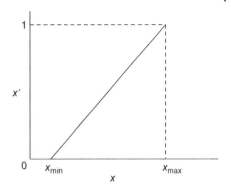

Figure 19.1 Illustration of a scaled variable.

19.2   A Scaled Variable Approach

One approach to scaling variables is to convert them to proportion between the maximum and minimum values as indicated in Figure 19.1.

The equations for scaling a variable in the range of 0–1 over the entire range are

$$x' = \frac{x - x_{min}}{x_{max} - x_{min}} \tag{19.1}$$

$$y' = \frac{y - y_{min}}{y_{max} - y_{min}} \tag{19.2}$$

Now, regardless of dimension or range, each variable is dimensionless. So, scaled distance has dimensional consistency, and it is also dimensionless:

$$\Delta S' = \sqrt{\Delta x_1'^2 + \Delta x_2'^2} \tag{19.3}$$

Ideally, each scaled variable is between zero and unity.

Maximum and minimum values used in scaling should be consistent with the expected range, but if the estimated range for scaling was too large, then a particular variable might only range between 0.2 and 0.9. That is OK. If the range estimate was too small, then a scaled variable might range between −0.1 and +1.3. That also is OK, as long as the models can generate values that cover the range.

For convenience, let the optimizer work with scaled variables. Unscale their values to let the procedure calculate the objective function value.

19.3   Sampling of Issues with Primitive Variables

DVs might have a broad and disparate range of values. For example, the Continental United States has width of about 3500 miles across east to west and a north to south height of about 1500 miles. However, the tallest mountain is about 3 miles high. So, the range on position variables is significantly different, as summarized in Table 19.1.

If one is looking at humidity changes from Pikes Peak (roughly located near the center of the country) to the Florida Gulf Coast (on the lower right), which on a typical day, the relative humidity

Table 19.1 Illustration of disparate ranges.

X	Ranges from 0 to 3500
Y	Ranges from 0 to 1500
Z	Ranges from 0 to 3

might go from 25% on Pikes Peak to 95% on the Florida coast, the overall (average, superficial, secant) derivative values might be

$$\nabla f = \begin{bmatrix} \dfrac{\partial \%RH}{\partial x} \\ \dfrac{\partial \%RH}{\partial y} \\ \dfrac{\partial \%RH}{\partial z} \end{bmatrix} \sim \begin{bmatrix} \dfrac{95-25}{3000-1500} \\ \dfrac{95-25}{500-800} \\ \dfrac{95-25}{0-3} \end{bmatrix} = \begin{bmatrix} 0.047 \\ -0.233 \\ -23 \end{bmatrix} \begin{bmatrix} \%RH \\ \overline{mile} \end{bmatrix} \tag{19.4}$$

The magnitude of the value of 23 is much different from 0.047. The magnitude of one is about 3 orders of magnitude different from the other. These are disparate values, which can lead to numerical problems with an optimizer.

Here all ∇f elements have the same units, but elevation might be reported in feet:

$$\nabla f = \begin{bmatrix} 0.047\%RH/mile \\ -0.233\%RH/mile \\ -0.005\%RH/ft \end{bmatrix} \tag{19.5}$$

With conventional scaling based on the variable range,

$$X' = \frac{X-0}{3500-0} \tag{19.6a}$$

$$Y' = \frac{Y-0}{1500-0} \tag{19.6b}$$

$$Z' = \frac{Z-0}{3-0} \tag{19.6c}$$

$$\nabla' f = \begin{bmatrix} 164.5\%RH \\ -349.5\%RH \\ -70\%RH \end{bmatrix} \neq \nabla f' \tag{19.7}$$

Now the values of the elements in the gradient are of similar order of magnitude. The ratio of largest to smallest, about $5:1$, is less than one order of magnitude. Note that the equation indicates that the gradient with respect to scaled variables, $\nabla' f$, is not the gradient of the scaled function, $\nabla f'$. The function has units of %RH, and when the gradient is based on scaled distances, the dimensions of the elements on the gradient also have the %RH units. However, the scaled function, f', would be dimensionless; so the elements of $\nabla f'$ would have the reciprocal units of each DV, perhaps remaining inconsistent.

Note also that one certainly could use the scaled function and the scaled decision variables. Then elements of $\nabla'f'$ are dimensionless, which makes them dimensionally consistent. Additionally, since all values are scaled on a zero-to-one basis, the secant derivatives all have the same value of about unity.

Unscaled, the choice of units on $x_1\ x_2\ x_3$ can have a dramatic effect on values or what a step size of 20 means for each. If the gradient is -35% RH/mile, it is equivalently -0.007% RH/ft. Then a change of $+20$ means -700% RH if in miles, but -0.014% RH if in ft.

For many reasons, it is often convenient to scale the DVs to a common basis.

19.4 Linear Scaling Options

In linear scaling, the scaled variable is linearly related to the primitive variable. There are several approaches in linearly scaling a variable:

Scale on 0-to-1 basis (this is most conventional) and was illustrated in Figure 19.1:

$$x' = \frac{x - x_{\min}}{x_{\max} - x_{\min}} \tag{19.8}$$

Alternately, scale on 0–10 basis,

$$x' = 10\frac{x - x_{\min}}{x_{\max} - x_{\min}} \tag{19.9}$$

This assigns a perfect 10 as the maximum. Alternately, you could scale on a 0-to-100 basis to reveal percent of full scale, or a 0-to-1000 basis if you wish to be able to differentiate parts per thousand. The choice is more grounded in user familiarity with the basis. I choose the 0–10 scaling for the 2-D Optimization Examples file.

Scale on a -1 to $+1$ basis as illustrated in Figure 19.2:

$$x' = \frac{x - x_{\mid}}{x_{\mid} - x_{\min}} \tag{19.10}$$

This is commonly applied in the input variables for neural network modeling.

Note: The midrange of x' is the value of zero, which could lead to a divide by zero when using relative convergence criteria on the scaled DV.

Scale on a -0.8 to $+0.8$ basis, as illustrated in Figure 19.3:

$$x' = 0.8\frac{x - x_{\mid}}{x_{\mid} - x_{\min}} \tag{19.11}$$

Again, here, the midrange value of $x' = 0$ could lead to a divide by zero.

Note also: The -0.8 to $+0.8$ range permits extreme values, which might arise if an optimizer wanders outside of an initially expected variable range (or actual data range) to remain satisfyingly between -1 and $+1$. This scaling practice is common in neural network scaling of both input and output variables. The advantage of the -0.8 to $+0.8$ scaling is that if a variable needs to exceed the hypothesized

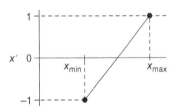

Figure 19.2 Scaling on a -1 to $+1$ basis.

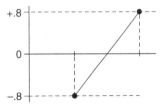

Figure 19.3 Scaling on a -0.8 to $+0.8$ basis.

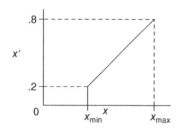

Figure 19.4 Scaling on a 0.2–0.8 basis.

x_L to x_H range by a small amount, it remains within the −1 to +1 limits of the neuron output value.

However, each of the earlier includes zero as a DV value, which might cause a problem if some part of a calculation has a divide by x' or $\ln(x')$. So, be sure to unscale in calculations that require the primitive variable or scale from a 0.2-to-0.8 basis, as illustrated in Figure 19.4:

$$x' = 0.2 + \frac{x - x_{mid}}{x_{mid} - x_{min}}(0.8 - 0.2) \tag{19.12}$$

Note: Some primitive variables are already scaled on a 0-to-1 basis, such as mole fraction, x, or weight fraction, w, or volume fraction, v, in composition. However, although they seem dimensionless (moles/moles and grams/grams), they have disparate units of mole fraction and weight fraction; and 0.5 mole fraction does not mean the same as either 0.5 weight fraction or 0.5 volume fraction. However, if the minimum value of the primitive variable is defined as zero, and the range is defined as unity, then for the 0-to-1 basis for scaling, the numerical value of the scaled variable is identical to the numerical value for the unscaled (primitive, original) variable. While the definition $x' = (x - x_{min})/(x_{max} - x_{min}) = (x - 0)/(1 - 0) = x$ may be confusing, it can be justified when desiring dimensional consistency.

19.5 Nonlinear Scaling

Here, two approaches are offered for nonlinear scaling. Many others are possible. First, consider nonlinear transformation of a single variable. You might use square root, reciprocal, or log functionality to substantially linearize a relation. For example,

$$x' = \sqrt{x} \tag{19.13}$$

This could make presentation simpler and may ease the optimization difficulty. Nonlinear scaling can also reveal detail in one region of the variable that is relatively insensitive on the entire scale but important to the application. For these reasons, we often log transform graph axes when plotting variables to either see detail in a region of interest or somewhat linearize the trend for convenience.

However, recognize that Equation (19.13) or similar transformations do not preserve the dimensions on the variable, change the dimensions, and are not dimensionless scaled variables. Perhaps the primitive variable can be divided by a reference value to make it dimensionless prior to the nonlinear transformation:

$$x' = \sqrt{\frac{x}{x_{reference}}} \tag{19.14}$$

In the second approach, scale by combining several variables into dimensionless groups such as Re, Pr, Nu, Sc, etc. This scales variables by a functional impact, not by the range of a variable. For instance, the Reynolds number is the ratio of momentum advection with fluid flow to the momentum diffused in the perpendicular direction by viscous effects. Dimensionless group scaling often leads to divisors and reciprocal relations, which can be a problem if linear relations are required by an optimizer. But, in addition to nondimensionalizing and to converting a class of applications to a common equation, it often reduces the number of variables that are considered in the relation. For instance, if heat transfer coefficient in a heat exchanger is considered to be related to diameter, density, viscosity, velocity, and thermal conductivity,

$$h = f(D, \rho, \mu, v, k) \tag{19.15}$$

Then a simple empirical model might have seven coefficients, yet only express linear and independent functionality:

$$h = a + bD + c\rho + d\mu + ev + fk \tag{19.16}$$

However a classic dimensionless model using the Nusselt and Reynolds numbers, $Nu = hD/k$, $Re = Dv\rho/\mu$, may have just two coefficients, express nonlinear relation, and capture the interaction between primitive variables:

$$Nu = a(Re)^b \tag{19.17}$$

Such dimensionless groups could be developed by "nondimensionalizing" the PDE (converting all state and independent variables to scaled variables, then rearranging coefficients to make each term dimensionless) or by an empirical method, such as the Buckingham Pi method. In either case, the power law relation of Equation (19.17) is not the mathematical solution. It is just a relation that has been found to often provide an adequate relation for data.

19.6 Takeaway

Why use scaled variables? There are many reasons, and 16 are given in the beginning of this chapter. However, as a caution, scaled variables might lead to a divide by zero in a model or in a convergence test. Further, although this chapter has promoted the use of scaled variables, be aware of your customer and the application. Humans and phenomenological models may prefer the use of primitive variables.

In optimization, it might be best to let the optimizer choose values for the scaled variables as the DVs; but since you probably understand the primitive variables and developed models using them, unscale the optimizer DVs to give primitive variables to the models to calculate the OF, to decide convergence, and for you to observe.

19.7 Exercises

1 Choose a section of a map that has temperature differences due to both elevation and latitude and longitude directions. Quantify, approximately, how the overall gradient component values depend on dimensions (your choice of temperature or distance dimensions). Show what happens to these average values if scaled variables are used for the temperature and distance dimensions. Choose the scaled variables to be on a zero-to-one basis.

2 In a 2-D situation, use the equation that determines Δx to show that the Newton–Raphson method is dimensionally correct. Show that a DV scaling approach makes the incremental steepest

descent calculation within the blended approach (such as Levenberg–Marquardt) also dimensionally correct.

3 Explore nonlinear scaling of the OF in Function 34 (zero gravity) of the 2-D Optimization Examples file. Unscaled, the contours of the net force on an object in the zero gravity points are totally obscured by the magnitude of forces near the planets. However, when square root or log transformed the contours in the proximity of the zero gravity, points can be seen.

20

Economic Optimization

20.1 Introduction

In optimizing something related to economics, the objective function, the statement of what is to be optimized, is often termed a profitability index. A viewpoint of the categories in economic optimization reveals two classes of considerations—the one-time initial capital investment (the purchase cost of an item) and the annual costs and income associated with operating the item. These yield three types of considerations—the one-time initial capital investment, the annual cash flow, and a combination of capital and annual cash flow. A combination of capital and annual cash flow is often preferred, and it should account for the time value of money and the schedule of income and outlay. Although there are several such metrics, and a particular enterprise might prefer one over another, they have similar aspects, and one example is adequate to understand economic optimization. I'll present a technique termed net present value (NPV). Often it is termed present value (PV) or present worth (PW).

As one confounding effect, annual expenses can include a probable penalty cost associated with an undesired event. This attribute of an investment is termed risk. As a second confounding effect, any profitability index is predicated on the value of the "givens" (the cost of electricity, the tax rate, the anticipated sales volume, etc.), which all have uncertainty over a typical future forecast of 20 years or so. We might know the cost of electricity today, but what should we use for the next 20 years, the period that we expect an investment to work for us?

This chapter provides a brief development of those issues and how they are combined to determine the value of the net present value profitability index. Diverse business types and economic scenarios have given rise to a multitude of profitability indices. But all of the indices have similar elements, so if you can understand the concepts of one metric, and how to use it in optimization, you can do it for the other particular profitability index that your customer would prefer. I have chosen to use the present value technique in this book, a common metric, keeping the focus on optimization, not on all of the nuances of economic analysis. Similarly, I have chosen to mention the cost factors in the analysis, but not provide the details of how to generate cost values. There are entire books related to estimating capital and expenses of processes and products and describing the diverse subtleties of analyzing depreciation, risk, and taxes. Since this book has a focus on optimization, with economic optimization as an application, this chapter merely provides an overview of how to formulate and optimize a profitability index.

Engineering Optimization: Applications, Methods, and Analysis, First Edition. R. Russell Rhinehart.
© 2018 R. Russell Rhinehart. Published 2018 by John Wiley & Sons Ltd.
Companion website: www.wiley.com/go/rhinehart/engineeringoptimization

In addition to introducing economic concepts, this section also reveals nonadditive combinations of elements in the objective function (OF). Often multiplicative or other combinations of elements are more phenomenologically valid than adding terms.

20.2 Annual Cash Flow

In one economic optimization category, the item to be optimized (a procedure, a device, a process, a system) has been designed, it exists, and its physical attributes, size, or rule features are a constraint. Here, economic optimization might be to choose operating conditions that minimize manufacturing costs, minimize use costs, maximize productivity, minimize time to finish, minimize variation, determine the best product distribution, etc. Although financial issues are key, the capital has been invested, the item exists, and the optimization is related to operating expenses and/or profitability, often termed annual cash flow. The expenses or profits, the cash flow, could be related to either $ per time unit (year, day, week) or per production unit (item, rail car load, lb of product, transaction). The cash flow would include the costs of labor, raw material, operating expenses, taxes, overhead, royalties, waste processing, sales price, marketing, etc. This does not consider the capital investment to create the item.

As an airplane example, after it is purchased, the optimization might seek the best choice of flying altitude to minimize fuel consumption per mile of travel. In a chemical plant, the example might be the choice of reaction temperature to minimize expenses while maximizing product value. A laptop example might be the time-out interval to deactivate the display to maximize battery life. An elevator example might be to decide whether a not-in-use elevator should move to a middle floor (to minimize time to meet a call) or remain on the last delivery floor (to save energy). A personal car example might be to acquire acceleration and breaking behaviors that minimize fuel and maintenance.

Simplistically the OF is after-tax annual profits, and one would wish to maximize J:

$$J = (1-T)(S-E) \tag{20.1}$$

where S is the annual sales income, E is the annual operating expenses (royalties, raw material, utilities, labor, etc.), and T is the tax rate, the sum of the portions of profit that goes to the federal, state, or local governments.

Alternately, if one is optimizing how to operate the process or item, then the sales income is not affected by the operating protocol. Then, effectively minimizing the after-tax impact of the expenses is an equivalent OF. Minimize J:

$$J = (1-T)E \tag{20.2}$$

Often such annual cash flow metrics include depreciation, an annual set-aside of $ to be able to refurbish worn-out equipment or reservoir value. If the life of the item is n years, and the capital is C, then a straight-line (linear) depreciation, D, is $D = C/n$. If tax regulations treat depreciation as a business expense, then the after-tax profits are

$$J = (1-T)\left(S-E-\frac{C}{n}\right) + \frac{C}{n} \tag{20.3}$$

20.3 Including Risk as an Annual Expense

Risk is the probability of an event times the penalty for the event:

$$\text{Risk} = c_{\text{of undesired event}}p(\text{undesired event}) \tag{20.4}$$

Although this could be applied to windfall fortuitous events, risk is usually only related to undesired events. Such undesired events could be faulty products that are returned by the customer, fines for violating environmental discharge limits from the manufacturing site, a labor strike that stops production, a rail transportation disaster that prevents raw material delivery, a toxic release that leads to personal injury and related medical and insurance expenses, etc.

The probability of an undesired event could be stated on a per time basis for a process that operates continuously. For instance, the probability of a fitting on a particular tank failing and consequently releasing flammable or toxic material to the environment might be 0.0001 events per 1000 h of operation. $P(\text{undesired event per tank per}1000\text{h}) = 0.0001$. This needs to be converted to the basis for the economic analysis, usually on a per year basis. Nominally, there is 8766 h of continuous operation in a year with 365.25 days. So, the tank operates for 8.766 of the 1000 h periods in each year.

We need the probability of the tank failing in a year. It is easier to calculate the probability of the complement, not failing:

$$p(\text{fail in a year}) = 1 - p(\text{not failling in a year}) \tag{20.5}$$

For tank to not fail in a year, it has to not-fail for each of the 8.766 periods:

$$p(\text{not failing in a year}) = p(\text{not failing in 1st period, AND, not in 2nd, AND not in 3rd,} ...) \tag{20.6}$$

Because of the AND conjunction, the probabilities are multiplied; and assuming the probability is independent of the 1st or 2nd or 3rd, ... interval, they are identical. Then

$$p(\text{not failing in a year}) = [p(\text{not in any period})]^{n\,\text{periods}} \tag{20.7}$$

Returning to the complements,

$$p(\text{fail in a year}) = 1 - [1 - p(\text{failing in any period})]^{n\,\text{periods}} \tag{20.8}$$

For the specific example, Equation (20.8) indicates $p(\text{fail in a year}) = 1 - [1 - 0.0001]^{8.766} = 0.00087626$.

If there are three tanks, then any one tank may have a failure, or two of the three may fail, or all three may fail. Assuming that the failure probabilities for any one tank are independent of the others, the binomial distribution can be used to determine the probability of zero failing, or 1 or 2 or 3, in a particular year:

$$p(x\,\text{failing|given}\,n\,\text{items}) = \binom{n}{x}p^x q^{n-x} = \frac{n!}{x!(n-x)!}p^x q^{n-x} \tag{20.9}$$

For the case of the 0.00087626 probability of any one tank failing in a year, the second column of Table 20.1 reveals the probability of zero through all three tanks failing in a year.

If the cost to fix a single event is $10,000, then the risk associated with one tank of the three failing is $26.24. It would seem unlikely that if a second tank fails in the same year, the penalty for the second

Table 20.1 Risk of failure.

N failing/year	p(N|3) failing/year	Cost of event	Risk
0	$9.97E{-}01$	$0	$0
1	$2.62E{-}03$	$10,000	$26.24/yr
2	$2.30E{-}06$	$30,000	$0.07/yr
3	$6.73E{-}10$	$70,000	$0.00/yr
—	—	Total =	$26.31/yr

failure would be the same as the first (e.g., the second speeding ticket on a driver's record costs more than the first). So, in Table 20.1, the penalty for the second tank failure in 1 year is twice the first with a total cost of the two-tank failure as $30,000. Since the probability of this event is very low, the risk is $0.07. The total risk of all such tank failures is the sum of the risks for each, $26.31/yr. When considering risk, this can be included in the annual cash flow, as are other annual expense items.

This used a per time basis for the probability. However, the undesired event may be based on a per-event-of-use or per item basis. For instance, a per item basis might be that 1 out of 500 items we manufacture has a defect that leads to customer return. Another per-event-of-use application might be that the probability of a pair of scissors becoming dysfunctional is 1/600 cuts. In a manner similar to that aforementioned, the per event of use or per item basis can be combined with the number of items or instances per year and translated to the per year probability, and then risk can be included in annual expenses.

There are many things that can go wrong. For a comprehensive analysis of an investment, all must be considered, and the cumulative risk added to the annual analysis.

The simple development mentioned earlier considered that the probability of events is independent, like rolling a fair die, where the value appearing on any roll is independent of the prior values. However, there is a concept of common causes. All tanks are likely given similar use, have similar age, and have identical maintenance protocols. So, if one tank fails, it is likely that others will fail for the same reason. Rather than modeling the risk with independent probabilities, a technique that is truer to reality might be to use conditional probabilities. For example, the probability of any one tank failing is 0.00087626, but if one fails then the probability of another failing may be 0.10.

How does one get the probability of undesired events and the cost of each? Many industry sectors have associations that track and compile such data from member experience. Often the data is also tracked by federal regulatory agencies. The events that need to be considered and the associated event probabilities and penalty values are specific to industry sectors, intensity of use, maintenance protocols, etc. Further, the impact of a failure depends on the proximity of people, containment methods, response speed, etc. The values for such factors need to be identified by a person knowledgeable about the situation. But once these are determined, risk can be calculated and included in the annual cash flow.

If straight-line pretax depreciation is considered as well as risk, the annual cash flow to be optimized is

$$J = (1 - T)\left(S - E - R - \frac{C}{n}\right) + \frac{C}{n} \tag{20.10}$$

Again, if the sales income is independent of the process design or operational choices, then maximizing the after-tax profit is equivalent to minimizing this portion:

$$J = (1 - T)\left(E + R + \frac{C}{n}\right) - \frac{C}{n} \tag{20.11}$$

Increasing the facility maintenance and operator training schedule to reduce risk will increase expenses. Part of the optimization seeks that balance.

Although the OF appears as a single one-line equation, values for E and R would be obtained from somewhat complicated procedures.

20.4 Capital

The second economic application category looks at capital investment. These are one-time initial costs to acquire the item; it does not consider recurring costs of using the item. These could be the initial purchase price of a device, or the construction cost of a system, and can also include the engineering design and prototyping costs.

A chemical plant example of initial capital would include the land purchase, the construction costs, the raw material inventory needed to fill the in-process pipes and vessels, and the cash on hand needed as working capital to be able to pay monthly bills. A software example for a user would be the initial purchase price of the software and the one-time costs associated with learning how to use it. A car example for a purchaser might be to choose the vehicle that meets the use objective (perhaps four passengers) at minimum purchase cost. If you are a purchasing agent, charged with saving purchase money, and have a myopic view that does not account for consequential expenses, the least cost item wins the bid.

If you want to get something running, but are short on capital, the least cost item would be preferred. The OF seems simple:

$$J = C \tag{20.12}$$

However, minimizing initial capital, giving it a priority over in-use costs or in-use convenience attributes, often leads to increased future expenses. A cheap car may be easy to buy today, but it may become a maintenance burden shortly.

Again, this discussion does not tell you how to determine the capital cost of an item. Each discipline has its own catalog of capital costs associated with design choices for equipment and procedures. However, once the value of the capital has been defined, the discussion here guides you to its inclusion in profitability indices.

20.5 Combining Capital and Nominal Annual Cash Flow

The third application category looks at both cash flow and capital investment. This would be appropriate if the item is not yet constructed, not yet designed, and not yet acquired; and the optimizer is selecting item features as well as the operating conditions. Here, both capital (the initial investment in the item) and the operating cash flow (expenses and profits) are necessary considerations. Usually, both capital and annual cash flow are considered in the design stage of a product, or process, or procedure, or renovation.

Realize, however, that capital is a one-time cost, with the units of $; but annual expenses and annual sales income are recurring cash flow events, with the units of $/yr. Capital, C, has different units from either expenses, E, or sales, S; and the three cannot be added. A simple solution that can be used to combine the economic issues is to consider an expected life of the item, n years, and allocate the initial capital on a per year basis, (C/n), which is how much you need to allocate each year to have enough to pay for the replacement cost. Then annualized capital can be added to annual expenses as the optimizer objective function:

$$J = S - \frac{C}{n} - E \tag{20.13}$$

As a personal car example, car choice A might have an expected lifetime of 10 years, a cost of $20,000, and annual operating expenses (fuel, maintenance, etc.) of $7000. Combined, the average annual cost is $9000/yr. Car B might have an expected lifetime of 15 years, a cost of $45,000, and annual operating expenses (fuel, maintenance, etc.) of $4000. Combined, the average annual cost is $7000/yr. Over a 30-year period, choice A costs $270k and choice B costs $210k. If you can afford the higher initial capital outlay, car B is the lower lifetime cost option.

Considering the after-tax profit, the OF is again

$$J = (1 - T)\left(S - E - R - \frac{C}{n}\right) + \frac{C}{n} \tag{20.14}$$

However, now, choices about design of the item affect C, E, and R.

Alternately, one could consider the initial capital and the expenses and income over the lifetime to compute a total cash outflow objective:

$$J = nS - C - nE \tag{20.15}$$

There are many profitability indices like these that consider a nominal year. Often these are expressed as a ratio. Payback time (PBT) is the time for nominal sales to pay back initial capital and nominal expenses. A simple formula is based on the years of nominal production for which after-tax profits pay back the initial capital investment $C = n(1 - T)(S - E)$. Here,

$$\text{PBT} = n = \frac{C}{[(1 - T)(S - E)]} \tag{20.16}$$

Long-term return on assets (LTROA) is the annual after-tax profit based on the average capital investment:

$$\text{LTROA} = \frac{[(1 - T)(S - E - D) + D]}{(C/2) + W} \tag{20.17}$$

There are many variations for these and other profitability indices.

Example 1   Here is a very simple idealized example of a single decision variable optimization. If you are a chemical engineer, you can derive the relations. Assuming you are not, I'll provide each relation.

A company is using an ideal plug-flow reactor to convert a waste stream containing chemical A to a low enough concentration value for stream reuse or discharge. The higher the reaction temperature, the smaller is the reactor that is needed. A smaller reactor means less capital, but the higher operating temperature requires greater utility expense to heat up the fluid. For simplicity, consider that the reaction is first order and has an Arrhenius temperature dependence, making this reaction rate model:

$r = -k_0 e^{-E/RT}[A] = -k[A]$. Here the symbol $[A]$ means concentration of component A, normally in moles per liter, but it could be any measure of quantity per volume. The ideal plug-flow model for the isothermal reactor is $F(d[A]/dz) + a\varepsilon k[A] = 0$, which when integrated gives $[A](z) = [A]_{\text{inlet}} e^{-za\ \varepsilon k/F}$. If the objective is to reduce the concentration of A to a value of $[A]_{\text{outlet}}$, then the reactor length is $L = \left(F/\left(a\varepsilon k_0 e^{-E/(RT)}\right)\right)\ \ln\left([A]_{\text{outlet}}/[A]_{\text{inlet}}\right)$. The reactor length depends on the choice of the reaction temperature, and the capital cost of the reactor depends on its length. Nominally, the price to purchase a unit, capital, scales as the six-tenths power of size:

$$C = C_n \left(\frac{L}{L_n}\right)^{0.6} = C_n \left(\frac{\left(F/\left(a\varepsilon k_0 e^{-E/(RT)}\right)\right)\ln\left([A]_{\text{outlet}}/[A]_{\text{inlet}}\right)}{L_n}\right)^{0.6}$$

This is the relation that reveals how the capital cost of the reactor, C, depends on the decision variable value, T.

The cost to heat up the reaction product is proportional to the heat required: $E = 8766cQ = 8766cF\rho Cp(T - T_{\text{inlet}})$.

If the expected life of the reactor is 5 years, then the simple annualized capital method of combining capital and expenses is

$$J = E + \frac{C}{n} = 8766cF\rho Cp(T - T_{\text{inlet}}) + \frac{C_n}{5}\left(\frac{\left(F/\left(a\varepsilon k_0 e^{-E/(RT)}\right)\right)\ln\left([A]_{\text{outlet}}/[A]_{\text{inlet}}\right)}{L_n}\right)^{0.6} \quad (20.18)$$

Values for $c, F, \rho, Cp, T_{\text{inlet}}, C_n, a, \varepsilon, k_0, E, R, [A]_{\text{outlet}}, [A]_{\text{inlet}}, L_n$, could be determined by a subject matter expert. With convenient but still reasonable values, the optimization statement becomes

$$\min_{\{T\}} J = 2(T - 300) + 100\left(\frac{1}{e^{-1000/T}}\right)^{0.6} \quad (20.19)$$

where T is in Kelvin.

The OF w.r.t. DV plot is shown in Figure 20.1, which has an optimum of annualized capital at a reaction T of about 380 K.

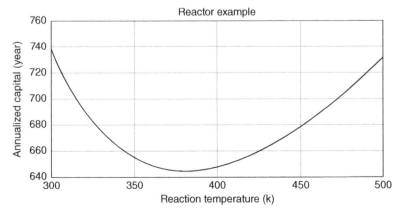

Figure 20.1 Reaction single DV example.

Note that this example considered very simple reactor relations and did not include risk, maintenance, or the design implications of a heat exchanger to heat the fluid or to cool the product. It did not show how to obtain the coefficient values. The example was meant to reveal the procedure of determining capital and annual expense values and combining them in the simple annualized capital economic metric.

20.6 Combining Time Value and Schedule of Capital and Annual Cash Flow

Those nominal year metrics, however, do not consider either the time value of money, or the start-up use rate, or the initial investment schedule. First, consider the cash flow schedule, with a chemical plant example: the capital outlay might be over a 3-year period of design, land purchase, land development, construction, initial inventory, people hiring and training, start-up testing, etc. All of this has to be completed prior to making a product. The entire capital is perhaps a series of monthly outlays during these several phases, not a one-time payment 3 years prior to start-up. Then, once constructed, the plant does not start up at full production. Likely, it starts at a reduced rate until marketing develops a sales base, and it may take 5 years to reach full production. So, the nominal S and E values might not occur for 8, or so, years after the first capital outlay. Further, as items wear out, there are additional periodic capital investments tempered by the scrap value of the items; then at the end life of the plant, it has a scrap value, and the land also can be sold. Looking only at the total capital and the nominal S and E values does not properly reveal the monthly cash flow situation, the schedule of income, and outlay over the item lifetime, which is essential for business management.

We need a better way to combine capital and annual cash flow. Further, a comprehensive profitability index needs to consider the time value of money.

There are two basic time-value concepts. One is that if you have capital, you'd invest it to have it grow and you would not place it in a safe (or under the mattress or bury it in a treasure chest) for keeping until it is needed. The second is that the price of things generally rises in the future due to inflation.

Let's consider the impact of investment, with this example situation: you need to pay out a capital investment of $100,000 at the beginning of each year for each of 5 years in order to construct a manufacturing facility. It appears that you need 5 × $100,000 = $500,000. But you don't. If you initially had $500,000 and used $100,000 for the first payment, what would you do with the remaining $400,000? A rational business would invest it. Let's say that the business is averaging 10%/yr on investments; then at the end of the first year, the $400,000 has grown to $440,000, at which time you make the second $100,000 capital outlay and are left with $340,000. After the second year, it has grown by 10% to become $374,000, at which time you make the third $100,000 payment, leaving you with $274,000. Eventually, after making the last $100,000 payment, at the beginning of the 5th year, you still have $121,540. So, you did not really need an initial $500,000. An initial capital of about $417,000 would be adequate to provide the $500,000 total capital.

The second issue in the time value of money is inflation. If the inflation rate is 2%/yr, that means that each year the price of goods and services rises 2% above the prior year. So, in the second year the needed $100,000 will have become $102,000, and by the 5th year, the annual capital investment would need to be $108,243. The sum of capital investments is not 5 × $100,000 = $500,000, but is $520,404.

Usually, the two factors (inflation and business investment return) are combined as one single discount factor.

For economic optimization we need a method to combine capital and annual income and expenses, to schedule the cash flow on a yearly (or perhaps monthly or weekly) basis, to include risk, and to account for the time value of money. There are several methods. Present value (PV), future value (FV), and discounted cash flow rate of return (DCFRR) are commonly applied. I believe that if you understand the principles and algorithm of any one, you can easily use one of the others. I'll reveal present value for this book. If it is what your customers use, great. If not, use what your customer wants.

20.7 Present Value

Present value is alternately termed net present value or present worth. First, define the schedule of cash outflow and inflow associated with the item, perhaps on an annual basis. Table 20.2 is a simple example of a project that has a 10-year life. In year zero, $400k is invested as capital to construct the item. The capital includes land purchase, facility construction, raw material to fill the unit, and working capital. There is no income that year. In year 1 production starts at a reduced rate, and expenses are $25k and income is $112k. The expenses include all costs, fees, taxes, etc. As years progress, production rises to its nominal level with sales income of $210k/yr and expenses of $50k/yr. However, it is expected that in year 5, capital refurbishing projects will require an additional outlay of $50k for a total cash outflow of $100k that year. By year 6 we expect operating experience to have developed processing improvements that reduce expenses from $50k/yr to $40k/yr. But the life of the product is anticipated to be 10 years only (perhaps it is an automobile model, or a drug, or a computer accessory, or a mining operation with a limited reservoir); so, starting in year 8 production begins to drop with associated drops in expenses and income. In year 9, the facility is shut down, and all raw material is converted to product. In year 10, all in process inventory is sold, the facility is sold for scrap value, and the land is sold for a total income of $250k.

Table 20.2 Present value example.

Year	Out	In	Discount	PV
0	400	—	1.000	−400.0
1	25	112	0.909	79.1
2	40	140	0.826	82.6
3	50	210	0.751	120.2
4	50	210	0.683	109.3
5	100	210	0.621	68.3
6	40	210	0.564	96.0
7	40	210	0.513	87.2
8	30	190	0.467	74.6
9	20	140	0.424	50.9
10	—	250	0.386	96.4
—	—	—	NPV =	464.6

All outgo and income data are expressed in today's value. They are not inflated for inflation.

The "Discount" column in Table 20.2 reveals the discount factor, which is the reciprocal of the compounded discount rate: $f = 1/(1 + i)^{year}$. Here the value of the discount (investment rate plus inflation rate) is 10%, $i = 0.10$.

The entries in the "Net" column of Table 20.2 are the present worth of the annual cash flow (income minus outflow) discounted by the discount factor. This scales each future year activity to the amount of money needed today to finance that future cash flow transaction. The entries in the PV column are the present values, today's equivalent values of the future transactions. For example, the scrap value of the facility is $250, but it is income in year 10. If we had $96.4k today, and invested it in something that has a 10%/yr return, it would grow to become $250k in 10 years. The future income of $250k is equivalent to a quantity of $96.4k today.

The net present value (NPV) is the sum of all of the PV for each year. The message is that if you give away $400k to finance this project, the net present value at the 10% discount rate is $464k. The $400k invested today is equivalent to having $464k. It is better investment than those that would generate the nominal 10%/yr return.

The reader may realize that an estimate of the future sales, expenses, product lifetime, etc. is needed for an NPV analysis and ask, "How can an investor know the future?" The answer is that you cannot predict the future, but you can anticipate what it might be from experience in similar projects. Usually, teams of experienced individuals put together such a plan.

However, this book is about optimization, which requires that the objective function be properly and comprehensively specified. What constitutes that is enterprise specific.

20.8 Including Uncertainty

You might know the current tax rate, but the future value may fluctuate as government boards decide to encourage business growth or decide to finance their municipality activities. You may know the current labor cost, but the future value may be different as diverse factors affect it. The marketing department has provided a projection of sales over the 20-year future, which may be a best good-faith estimate, but of course it is predicated on how they believe future economics, fashion, technology, competition, and such will influence the consumer demand for your product and your company's market share. Although we can forecast reasonable values and trends, every aspect of the "givens" that enter a profitability index has uncertainty.

Every value has uncertainty. Consider the calcium content of the limestone from a quarry to a cement mill. It will change with the location within the quarry as material is removed and new section is mined. Alternately, consider the heating content of fuel, it will change as suppliers and fuel origins and manufacturing processes change over a 20-year period. It is all uncertain, and different values for the "givens" lead to different optimum DV^* values.

Making 10% changes in the coefficients in the reaction Example 1 results in the dashed line relation between annualized capital and reactor T in Figure 20.2. The solid line is the original. You can see that the change in coefficient values shifts the T^* value from about 380 to 350 K. It changes the associated optimum reactor length as well as the value of the annualized capital. Reversing the 10% changes leads to an optimum of about 420 K, also changing the length of the reactor that should be purchased. If you accept that the 10% variations in combined coefficients is a reasonable possibility, then the dashed curve represents one possible realization of the model. There are many possible realizations.

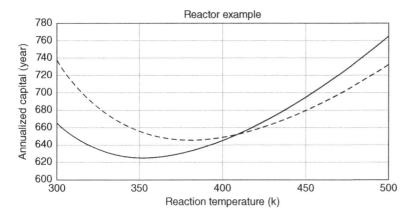

Figure 20.2 Reaction single DV example—impact of coefficient uncertainty.

I think a 10% uncertainty in production projections, F, in energy costs, c, in reactivity coefficients k, E, ε, or in unit duty, T_{inlet}, $[A]_{\text{outlet}}$, $[A]_{\text{inlet}}$, is probably an underestimate of the likely range of values. This would make the dashed line realization in Figure 20.2 a not very extreme possibility.

Now what to do? If the model with the nominal value of the "givens" indicates a reaction T^* of 380 K, then you could use low-pressure waste steam as the heating medium. But if another realization indicates that you need high-pressure steam, that requirement may affect other design choices.

Continuing thought about this impact of uncertainty is that if the nominal value of the "givens" indicates a reactor length of 1 m, but another realization indicates 0.25 m, and another indicates 4 m, which should you buy? If you wanted to be sure the process can reduce the $[A]$ adequately, then you might buy the 4-m reactor. If you realized that it will be several years prior to rising to full production, and wanted to minimize capital, you might buy the smaller reactor (which would be adequate for initial production) and leave space to install a larger one if the future indicates it is needed. You might want a greater certainty in the optimization outcome and determine what model coefficient values are the main culprits and acquire less uncertain, more precise estimates of those values.

In any case, you cannot understand the uncertainty without investigating the realizations that are possible. Whether the profitability index is just capital, or just expenses, or one of the many combinations (such as net present value), use a Monte Carlo technique to determine the range of outcomes of the realizations.

In a Monte Carlo analysis, set the base case for each "given" in the metric, and set a reasonable range for those values. For one realization, generate coefficient values by a random sample within the range. A simple approach is to use a uniformly distributed sampling:

$$c_k = c_{\text{nominal}} + (r_k - 0.5)(c_{\text{high}} - c_{\text{low}}) \tag{20.20}$$

With such a realization value, independent for each uncertain coefficient, perform the optimization to determine the DV* value(s) and associated OF* and other consequential design values. Repeat this for a large number of realizations (more than 100, probably less than 10,000). Plot the distribution of values as a pdf(x) or CDF(x) w.r.t. x graph. When the CDF of results does not change noticeably with realizations, with k-value, you've done enough.

If the *c*-values are held steady from optimizer iteration to iteration (over the case study life span), then the function to be optimized is deterministic. Including uncertainty, by changing the realization values of the givens at each function call, makes the application stochastic.

The most extreme values are improbable. Although possible, the extreme values indicate a conflu-ence of 10 or so variables, each pushing in the same direction and each being at an extreme value. It is more likely that uncertainty on the "givens" is independent and that some will be pushing high, while others are pushing low, and it is unlikely that all uncertainties will be at the extreme. Accordingly, you might want to consider the 90% extreme values of the optimization results, the *x*-values with 0.05 and 0.95 CDF values, which bracket 90% of all outcomes. Alternately, the quartile values (0.25 and 0.75 CDF values) are often considered. The decision is then made with knowledge of the impact of the uncertainty.

For instance, a nominal application might be to determine process design conditions to maximize NPV. But considering uncertainty, such a best design for nominal conditions might have a significant probability of a downside of poor economic performance. Alternately, you might want to design to maximize the 10% extreme worst case. This may not provide the best nominal value, but may provide an adequate value that has a higher likelihood of future success.

Figure 20.3 illustrates the result of an economic optimization based on NPV and on the 90% worst NPV value in a realization. The solid line reveals the result from Case Study 1 (Chapter 36) when pipe diameter is the decision variable used to minimize NPV at nominal conditions. The DV* is about 0.43 m. However, each of the givens has uncertainty, and with that nominal optimum pipe diameter, the uncertainty of the givens leads to a distribution of possible NPV values. Notice that the maximum (worst-case situation) NPV has a value of about $1575k, but also that the best-case NPV is $1040k. The dashed line represents the result of minimizing the 90 percentile NPV value. At each optimizer trial solution, the NPV distribution is computed by the Monte Carlo method, then the 90 percentile NPV value, the 90% worst value, is used as the optimizer OF. When the optimizer seeks to minimize the 90% worst NPV realization, the optimum diameter is about 0.46 m, and the cumulative distribution of NPV at that diameter is indicated by the dashed line. Notice that the best possible NPV outcome of this choice is just $1100k, but the worst case is now only about $1400k. Notice also that the 50 percentile

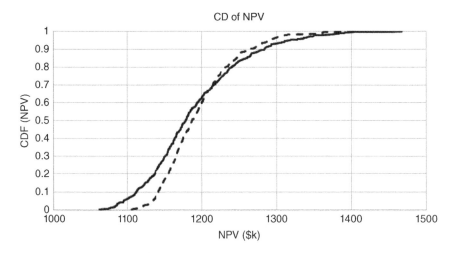

Figure 20.3 Optimization of nominal and 90% worst case.

NPV value for the nominal case, \$1185k, is a bit better than that of the case in which the worst NPV is minimized, \$1190.

Note the discontinuities (irregular non-smooth aspect) in the curves. This simulation used 500 realizations to generate the CDF. More realizations would provide a curve with less variation. Something on the order of several thousand is recommended; but for this illustration, 500 realizations provide a definitive curve and also a view of the impact of too few realizations.

This reveals that optimizing a worst case results in an alternate design choice, with a more secure decision, but nominal values of the profitability index are nearly the same.

So, how does one get the values for the uncertainty of the givens? Some come from engineering experience, and some come from business experience. All are estimates as to what the reality might be.

20.8.1 Uncertainty Models

The example of choosing a realization value for a coefficient in Equation (20.20) presumed that the likelihood of any value within the c_{high} to c_{low} range is equally probable, that the range was centered on the $c_{nominal}$ value, and that the value of any one coefficient will be independent of any other. The model then uses a uniformly distributed random variable, r_k, to generate the realization. This will often be a fully adequate technique.

However, a better model for your particular situation might be a Gaussian or a Poisson or a lognormal distribution. Further, in your application there may be cross correlation between variables. For instance, if inflation causes the price of electricity to increase, it likely also causes fuel and other raw material prices to increase. So, rather than independently perturbing each variable, a more representative analysis might have categories of variables perturbed by the same random variate.

If the Gaussian (normal or bell-shaped) distribution of perturbations is right for the application, then I like the Box–Muller method for generating a simulated value for the given. In Equation (20.21) the nominal value is perturbed with $NID(0, \sigma)$ fluctuation. The values r_1 and r_2 are uniformly distributed and independent random values $0 < r \le 1$:

$$c_{perturbed} = c_{nominal} + \sigma\sqrt{-2\ln(r_1)}\sin(2\pi r_2) \tag{20.21}$$

Finally, if your coefficient values are changing in simulated time of the project, then they might be autocorrelated. If fuel price is high 1 year, it is unlikely to randomly drop in the next year. There are many equations that can be used to generate an autocorrelated variable. I like an autoregressive moving average (ARMA) model. Based on a first-order ODE with w as a perturbation driver,

$$\tau\frac{dc}{dt} + c = c_{nominal} + w \tag{20.22}$$

In response to this ODE, c, will change in time. A better representation might be to use $c(t)$. In a digital simulator, $c(t)$ will not be a continuum variable, but a discretized variable that has values at the ith simulation interval. Accordingly, c_i will be a better representation than $c(t)$. As time progresses, the value of w will change, so perhaps the symbol $w(t)$, or w_i is more representative.

To keep the average value of c at the value of $c_{nominal}$, the noise driver, w, should have an average value of zero. The distribution of the w_i values could be a simple uniform, or Gaussian (normal), or other as the user decides is appropriate to represent the source of variation. A uniform distribution is simplest to program, $w_i = a(r_i - 0.5)$, where a is an amplitude factor. The Gaussian distribution would better represent "a driver" that is actually a confluence of several independent sources of variation.

Transformed by a finite difference approximation to the derivative, the next value in a time series of an autoregressive variable is

$$c_i = \left(1 - \frac{\Delta t}{\tau}\right) c_{i-1} + \left(\frac{\Delta t}{\tau}\right)(c_{\text{nominal}} + w_i) \tag{20.23}$$

Figure 20.4 illustrates the trend from two influence models. The markers represent independent, normally distributed deviations about an average value of 5, with a standard deviation of 1. They are attached to the left vertical axis. Notice that the band is locally horizontal. Take a window of 10 or 20 samples and consider the average. Notice that the range is from about 2.5 to 7.5 roughly a 5-sigma range. Notice that the markers are independent: if one has a high value, the next may be either high or low.

By contrast, the line represents an autoregressive moving average model with a time constant of 50 samplings and a driver sigma of 20 units. It has the same nominal average of 5 but is attached to the right-hand vertical axis to easily distinguish it from the markers. The driver for the ARMA trend is the same independent perturbation series represented by the markers. However, notice that in the 0–15 time interval, the average or the ARMA line persists around values that are a bit <5. In the 20–60 interval, it persists around a value of about 6.

In my experience of observing how both natural and economic influences change in time, the ARMA pattern is more representative. However, in batch processes, batch-to-batch influences can be independent. Similarly, when sampling random individuals, there is no autocorrelation. In those, what influenced the last sample is now long past, and new fluctuations influence the next. So, both models represented in Figure 20.4 are important.

It is certainly possible to use a higher-order ODE representation, and some simulations do. However, my experience is that the simple models of Equations (20.21) and (20.22) are fully adequate representative models.

Propagation of variance on Equation (20.23) reveals that

$$\sigma_c = \sigma_w \sqrt{\frac{\lambda}{2-\lambda}} \tag{20.24}$$

where $\lambda = \Delta t / \tau$.

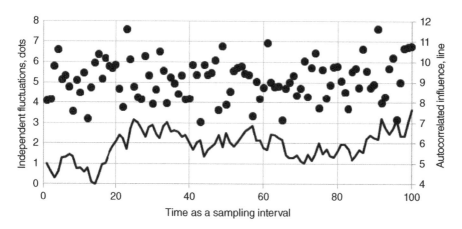

Figure 20.4 Comparison of independent and autocorrelated perturbations.

Choose the time constant τ to represent the duration of persistence that you think characterizes the ARMA influence. In a first-order process the duration of event persistence is about three time constants. Choose σ_c to represent the range of the ARMA variable, roughly $\sigma_c = \text{Range}/5$. Then use the inverse of Equation (20.24) to determine the σ_w value needed to drive the noise. Observe the resulting c_i values to ensure that you have a realistic simulation.

20.8.2 Methods to Include Uncertainty in an Optimization

Following are five methods (approaches) to include uncertainty in an optimization. For methods 3, 4, and 5, there are two options: (i) optimize the nominal statistic and (ii) optimize a 90% (or other extreme such as 75 percentile, or 99.9%, or 3-sigma) worst case. These methods are listed in order of increasing complexity but also increasing value. Methods 1 and 2 are of the bootstrapping type and create deterministic OFs to optimize. Methods 3, 4, and 5 are each of the Monte Carlo type, generating a stochastic OF to solve.

Method 1—Optimize the nominal deterministic case, and acquire DV^* values. Then explore possible consequences of variation in the givens with the nominal DV^* solution. This reveals the potential upside or downside to the DV^* choice, and it may provide adequate comfort that the deterministic DV^* solution is adequate. However, it does not adjust the DV^* values to minimize undesirable aspects of the downside consequences.

Method 2—Sample values for the givens from their distribution (nominal value plus its random perturbation). This would be one realization of the givens. Keep those values for the entire economic simulation and optimize. This would be a deterministic OF. Repeat the sampling (generating a realization) and optimization 1000 times. This would generate a distribution of DV^* and OF^* values, which might reveal the impact of uncertainty of the givens on possible solutions. But which DV^* set to choose?

Method 3—Sample values for the givens from their distribution (nominal value plus its random perturbation). This would be one realization of the givens. Keep those values for the simulation life. With a TS from the optimizer, obtain the OF value. For the same TS, sample and generate a realization and OF value for N (perhaps 100–100,000) times. From the N results calculate either (i) the average value or (ii) the statistical extreme worst OF value (lower quartile, 2-sigma extreme, 90% worst, etc.). Return that to the optimizer as the OF value. The optimizer will devise a new TS. Repeat. This will return a stochastic value to the optimizer, but it has the benefit of optimizing within uncertainty.

Method 4—It is more realistic that the givens will change values each year. So, for each year of the simulation, generate an independent realization of the givens. At the end of a 30-year simulation, each given will have 30 independent values. Certainly, this could be done on a monthly basis. At the end of the simulation, the OF will be a stochastic response to the TS. One could seek to optimize the nominal profitability index (that obtained from the 1 lifetime simulation) or an extreme undesired value (such as the 2-sigma worst case estimated from 10 lifetime simulations, or the 90% worst from 100,000 realizations). This is one step better than method 3, because it is more likely that givens will change from year to year, than holding fixed values for the project lifetime.

Method 5—It is progressively more realistic that any particular given value will not independently change from year to year. If energy cost was low (or high) 1 year, it will likely not independently jump to an alternate value in the next year. Likely, the situation will persist from some time, and although moving to higher or lower values, it will move in an autoregressive manner. So, instead of independently generating given realizations on each year, generate new realizations from an autoregressive moving average sampling, such as Equation (20.23). Again, one could seek to optimize the nominal profitability index (that obtained from the 1 lifetime simulation) or an extreme undesired value (such as the 2-sigma worst case estimated from 10 lifetime simulations, or the 90% worst from 100,000 realizations). This method also makes it relatively easy to have correlation between givens. If fuel costs rise, then it is likely that electricity and organic feedstocks prices will also rise.

20.9   Takeaway

In economic optimization the human choices related to the profitability index as the OF can have substantial impact on the DV*. Further, uncertainty in the estimation of future cost estimates, risk, and future production rates all have a substantial impact. Optimization should consider uncertainty and use Monte Carlo techniques. At least do method 1 from Section 20.8.2, and reveal the potential downside to a nominal DV*, by showing what might happen with that DV* choice. Preferentially, use method 3, 4, or 5, the stochastic approaches. This stochastic surface will require an optimizer that can handle that aspect.

20.10   Exercises

1   Formulate an economic optimization problem for a scenario of your choice. For example, how thick should the insulation be on a pipe to "eliminate" heat losses? What diameter pipe is best for transporting fluids? How many, and what size, smaller production units in parallel should replace one large unit? How large should be the fire extinguisher in a building hallway? Should one stop at a BS, MS, or PhD degree before getting a full-time job? You could use long-term return on investment, discounted cash flow rate of return, or any appropriate balance of cash-flow-to-capital investment as the objective function. Clearly describe OF, DV, and constraints with equations. Don't stress about the validity of the equations. Include reasonable assumptions. Focus on the optimization problem statement, not the discipline validity.

2   Determine the optimum number of months for a car loan. The more months, the lower the monthly payment, but the greater the total cost. Consider a penalty for excessively high payments and a penalty for not paying off the car within a useful life expectancy of the car. Seek to minimize NPV, not total cost.

3   As an undergraduate, you had engineering economics particular to your major and learned about an economic index that is used for evaluating a design or operation. Revisit that. Describe a particular situation and the resulting DVs, constraints, objective, OF, and models (or methods) that permit calculation of the OF value and constraint penalties. State who the stakeholders are, and include alternate issues that they might want to be integrated with the optimization search for the DVs that max or min the OF.

21

Multiple OF and Constraint Applications

21.1 Introduction

Often there are multiple and diverse objectives in an optimization. You might want to balance several competing objectives by combining them into one OF, and the question is "How to combine diverse and opposing objectives?" Here are some examples:

Find the best route for an automobile road trip: This seems like a simple statement, but how does one interpret best? There are diverse measures of desirability (goodness) and they have disparate units:

- Minimize cost (fuel, meals, nights, tolls)—$
- Minimize travel time—hours
- Maximize scenery and tranquility—what units?
- Maximize safety (slower, back roads, no traffic, proximity to assistance, not isolated, probability of an accident)—what units?

The turnpike might be the answer to minimizing travel time, but tolls and gas consumption and impact of an accident may prompt you to prefer another road.

Find the best design for a manufacturing plant: Again, this simple concept could have many meanings:

- Maximize profitability (DCFROI, NPV, LTROA).
- Minimize capital cost.
- Minimize risk.
- Minimize resource utilization and waste generation.
- Maximize security.
- Maximize flexibility.
- Maximize operability.

Minimizing capital might conflict with risk, waste, flexibility, and operability objectives.

Find the best automobile design: Again, there are many meanings:

- Maximize gas mileage.
- Maximize carrying capacity.
- Maximize useful life.
- Minimize purchase cost.
- Maximize safety.

Engineering Optimization: Applications, Methods, and Analysis, First Edition. R. Russell Rhinehart.
© 2018 R. Russell Rhinehart. Published 2018 by John Wiley & Sons Ltd.
Companion website: www.wiley.com/go/rhinehart/engineeringoptimization

And again, the objectives are contradictory. Purchase cost and gas consumption could be minimized by thinner metal, smaller size, and plastic materials; but such would undermine useful life, capacity, and safety desires.

Find the best control actions: Again, multiple desires:

- Minimize cost to make product.
- Maximize equipment life (low stress—pressure and temperature changes).
- Maximize quality (minimize variability).
- Minimize waste during product changeovers.

Here, a controller that is aggressive and rapidly returns a process to set point conditions will highly use utilities and place stress on the equipment. Consider a driver that aggressively accelerates or decelerates a car to the new speed limit. Such extreme action wastes fuel and stresses tires and joints.

This chapter will present three solutions for combining the disparate considerations. The classic method is to additively combine them in the single OF. But the understanding of deviation may lead to another functional combination such as multiplication. Another approach is to keep the functions separate in a multifunction Pareto analysis.

21.2   Solution 1: Additive Combinations of the Functions

A Solution: Often you can reduce the multiple and conflicting objectives to a common $ value, for example, cost to make a product. If control action, x (a vector of actions or set points or tuning values that leads to particular control aggressiveness), can be related to stress on equipment, which can be related to equipment life (capital per product unit) or maintenance (cost per product unit), then

$ process cost/unit $= f(x)$
$ maintenance cost/unit $= g(x)$
$ capital reinvestment allocated/unit $= h(x)$
Present value = $ value today of all future income and expenses
$ risk = (cost per event) $*$ (no. events expected in the time period of evaluation)/(no. units)
$ waste/unit $= q(x)$

Then they all can be included in the objective function (OF): if they all have the same importance, just add them:

$$\min_{\{x\}} J = f + g + h + q + cp \tag{21.1}$$

The cp element in the OF is a penalty for violating a constraint. c represents the cost of an event and p represents the probability of an event. It could be termed a penalty function. Alternately, if your business might decide that staying within regulations is an absolute must, a hard constraint, then

$$\min_{\{x\}} J = f + g + h + q$$
$$\text{S.T.}: \quad \text{pollution} < \text{regulations} \tag{21.2}$$

Difficulty 1: The problem with hard constraints is that they create nonanalytical functions that create problems for gradient-based methods. How can the optimization algorithm evaluate $f(x)$ if x is infeasible or if dJ/dx_i is at a discontinuity?

Difficulty 2: As an additional multi-objective issue, you might not be able to reduce all of the multiple objectives to a common basis. For the trip example, you may want to minimize cost, but if the value (scenery or landmarks) of a less direct path is extraordinary, you'd be willing to spend more on a less direct route, with only slightly more $, if merely an interesting sidelight; but a good bit more $, if it is an extraordinary experience. How does one equate tranquility or experience of the trip to $?

Difficulty 3: Further, you might not be able to definitively calculate a probability of a future event—like getting a fine for violating a regulation.

21.2.1 Solution 1a: Classic Weighting Factors

One approach is to add all individual objectives, but to weigh them by factors (similar to what is termed a Lagrange multiplier). This is implemented by the lambda coefficients in the following optimization statement:

$$\min_{\{x\}} J = \lambda_1 OF_1 + \lambda_2 OF_2 + \lambda_3 OF_3 + \dots \qquad (21.3)$$

Note: In this formulation each OF would need to be minimized. If an OF should be maximized, multiply it by -1 or use its inverse, OF^{-1}.

The more important an OF is, the larger the λ value. Some people say that $\sum \lambda_i$ should $= 1$. However there is no need. Further, if there are N OFs, only $N - 1$ need a weighting factor:

$$\min_{\{x\}} J = OF_1 + \lambda_2 OF_2 + \lambda_3 OF_3 + \dots \qquad (21.4)$$

What is important is that λ values represent and preserve the relative importance of the several elements in the OF.

An issue: The values of λ will be critical to the solution, and the question is how to determine them. The next section provides insight to the iterative development of an OF.

21.2.2 Solution 1b: Equal Concern Weighting

My favorite approach is to weight multiple objectives with the "Equal Concern" scaling method, which I first learned from the folks at Dynamic Matrix Control (DMC) Corp. who sold a linear, multivariable, constraint-handling, model-predictive controller. DMC Corp. was ultimately bought by AspenTech.

In the equal concern approach, list all OFs and constraints:

$f_1 = f(x)$ minimize travel time
$f_2 = g(x)$ minimize travel cost
$f_3 = h(x)$ never be more than 25 miles from a gas station (security)
$f_4 = p(x)$ keep the total cost, desirably, $< \$2000$

Which is the OF? Which is the constraint? It is actually difficult to say, and it depends on your perspective.

Choose one OF, say, travel time, and consider what might be an ultimate, best-case, minimum time, for example, 3 days. Now consider if $f(x) = 4$ days. How much concern does the $4 - 3 = 1$ day delay create? If your schedule is tight (perhaps you'll just make it to a wedding on time), then this may have a high concern. If, alternately, you don't need to be there for 6 days because your furniture will not arrive until then, but arriving a day or 2 early would be nice to provide time to scout the town and set up phone, water, mail, etc., then the 1-day delay may have very little concern. Feel the concern for being 1 day behind schedule. This may not be able to be equated to a $ value. It may be more influenced by how others perceive the situation and the implications of the event than by how you perceive it all. When setting equal concern factors, feel the collective concern of all stakeholders. Perhaps rate the level of concern on a 0–10 scale, for example, a hospital might ask a patient to rank pain on a 0–10 basis. Set this deviation from the nominal case as the equal concern value for $\Delta f_1 = 1\,\text{day}$, $EC_1 = 1\,\text{day}$.

Then consider f_2. What might be an absolute minimum, travel cost? Say, it is $1000. How much $ deviation would cause the same concern as the 1-day delay for f_1? If you have ample $ to do all that you need (up to the $2000 constraint) but would like to have some extra $ upon arrival, just in case, then a trip cost that is $100 over the ultimate minimum might have the same concern as being 1 day late. Then the equivalent deviation of f_2 from ideal, EC_2, is $100. However, $700 might cause an equal concern as 1 day, or $50 as 1 day. It all depends on your particular situation. Again, feel the concern of all stakeholders and choose a deviation for f_2 that creates the same collective level of concern as the basis deviation from f_1.

Figure 21.1 illustrates the procedure for assigning deviations that represent equal concerns. Step 1: Choose a deviation from ideal for one OF. Step 2: Feel the concern that deviation raises. Step 3: Choose a deviation from ideal for another OF that raises the same level of concern. The lines representing how concern rises with deviation are often unquantifiable. Accordingly, they are represented by dashed lines.

Consider f_3. If you're paranoid about running out of gas, on strange roads, at night, then 5 miles extra between gas stations could have a very high concern. However, if there are several people out for a relaxing drive on a pleasant day (each with cell phones and friends), then 15 extra miles between stations might be of little concern. Choose a value for EC_3 that creates the same concern as do deviations EC_1 and EC_2. Do the same for $EC_{4,5,6\ldots}$ and then state the OF as

$$\min_{\{x\}} J = \frac{OF_1 - OF_{1,\text{target min}}}{EC_1} + \frac{OF_2 - OF_{2,\text{target min}}}{EC_2} + \ldots + \frac{OF_n - OF_{n,\text{target min}}}{EC_n} \tag{21.5}$$

Some of the objective function elements, OF_i, may be the constraint functions.

Alternately, more simply, since the target minimum and EC value for each function will not change with the optimizer search for the DV* values, these are constants and irrelevant to the minimization. So the formulation reduces to

$$\min_{\{x\}} J = \sum \frac{OF_i - OF_{i,\text{target min}}}{EC_i} = \sum \frac{OF_i}{EC_i} - \sum \frac{OF_{i,\text{target min}}}{EC_i} = \sum \frac{OF_i}{EC_i} - \text{constant} \tag{21.6}$$

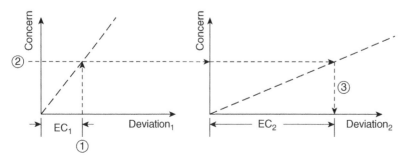

Figure 21.1 Illustration of how to assign equal concern values.

As Chapter 16 shows, the equivalent formulation of the optimization statement reduces to

$$\min_{\{x\}} J = \sum \frac{OF_i}{EC_i} \tag{21.7}$$

Note: Although the hypothesized, target or anticipated, minimum was involved in the concept, it is not needed in the optimization.

Note: I like this approach. EC values reflect what you feel and do not have to be analytically evaluated, and often the best approach to assign the EC values is the collective consensus (or customer's opinion). The concern can include institutional culture, political scenario, public image, ability to manage risk, confidence in forecasts, issues that are impossible to quantify explicitly but that can have consensus, and learned agreement, even if subjective.

Note: In this formulation, EC_i has the units of the OF_i. This makes each term in the optimization statement (Equation (21.7)) a dimensionless term.

21.2.3   Solution 1c: Nonlinear Weighting

Sometimes the concern is not linear, as the aforementioned formulation would reflect. In considering constraints, there is only a concern if the constraint is violated. Further, small deviations from ideal are often totally inconsequential, but larger deviations become important, and very large deviations are very very (more than just very) important. In least squares regression we square the deviation from ideal to discount small deviations but make the large ones very important. In calculating standard deviation in a set of numbers, we use squared deviation from the average. This squared functionality arises from the Gaussian distribution model of the impact of many independent perturbations that might normally affect experimental data. The squared penalty for a deviation also matches human penalty for legal violations. For example, the fine for speeding on a road does not linearly increase with speed excess. Similarly, the quadratic assessment of deviation is part of the Taguchi methods in design.

Accordingly, first determine if there is a constraint violation and then square the terms representing the magnitude of a constraint deviation. The constraint might be that total cost should be less than $2000. Or, in general, $f_4 < a$. As long as f_4 is less than a, there is no reason to consider the constraint. However, if $f_4 > a$, then the measure of badness, of undesirability, is $B_4 = f_4 - a$:

$$B_4 = \begin{cases} 0, & f_4 < a \\ f_4 - a, & f_4 \geq a \end{cases} \tag{21.8}$$

Using the squared deviation as the penalty for a constraint violation, the optimization statement becomes

$$\min_{\{x\}} J = \sum \frac{OF_i}{EC_i} + \sum \left(\frac{B_j}{EC_j}\right)^2 \tag{21.9}$$

This still preserves the nominal EC values for nominal excesses or violations.

This squared deviation from target better represents human perception of "badness," the impact of undesirable outcomes.

Example 1 Chlorine Dose Optimization for Water Treatment Plants

(Thanks to Professor Greg Wilber, OSU Civil and Environmental Engineering, September 13, 2010, for providing the basis of this example.)

An adequate dose of chlorine is needed to ensure sufficient kill of pathogens in drinking water. However, an excess dose of chlorine can lead to excess formation of trihalomethanes (THMs), which are carcinogenic and regulated under the Safe Drinking Water Act.

One simple model that can be used for pathogen destruction as a function of chlorine dose and residence time (time of exposure) is Chick's law:

$$N(t) = N_0 \exp[-kC^n t] \tag{21.10}$$

where $N(t)$ is the microbe population (CFU/mL or other measure) (It changes in time.); N_0 the microbe population at $t = 0$; C the chlorine dose (mg Cl_2/L); k the microbial decay coefficient (h^{-1}); n the empirical coefficient (usually taken to be 1.0 for chlorine); t the time (h).

Actual process designs are based on the required fraction removal of a given microbial pathogen (e.g., cryptosporidium). It is more complicated than just using the aforementioned equation to determine the C and t required to achieve N/No, but that is the basic idea. You set N/No equal to a desired removal level (e.g., 0.001 for 99.9% removal) and then determine t given a certain C. You would like to choose C to minimize residence time, because this minimizes equipment size. Obviously, the larger C is, the better is the residence time.

One equation that can be used for THM formation is given by the following:

$$\text{TTHM} = 0.00309[(\text{TOC})(\text{UV}_{254})]^{0.440}(\text{Cl}_2)^{0.409}(T)^{1.06}(\text{pH} - 2.6)^{0.715}(\text{Br} + 1.0)^{0.036}(t)^{0.265} \tag{21.11}$$

where TTHM is the total THMs (mg/L); TOC the total organic carbon (mg/L); UV_{254} the UV absorbance at 254 nm (cm^{-1}); Cl_2 the chlorine dose (mg Cl_2/L); Br the bromide concentration (mg/L); T the temperature (°C); t the time (min).

So, given an influent water quality (i.e., a known TOC, pH, temperature, and bromide concentration), there will be a maximum Cl_2 dose that does not exceed the maximum contaminant level (MCL) for TTHM (usually 0.80 mg/L).

To be in compliance with the Safe Drinking Water Act standards, there are rules regarding chlorine dose versus contact time, so design isn't as simple as finding the optimum dose that gives the lowest THMs while giving adequate pathogen disinfection. And best practices and regulations change in time. However, here, the objective is to determine the Cl dose that minimizes THMs with adequate disinfection.

Equations are rearranged and simplified so that t is in minutes for both.

$$t = \frac{a}{C}$$

where a is a combination of everything in Chick's law as well as the conversion of time from h to min. And $\text{TTHM} = bC^{0.409}t^{0.265}$, but since t is a function of C, $\text{TTHM} = dC^{0.144}$.

The objective is to find C to minimize t, but there is also the desire to minimize TTHM. Also, there is a constraint that TTHM must be lower than the permissible limit.

Using the "Equal Concern" approach to scale terms in the OF,

$$\min_{\{C\}} J = \frac{t}{\text{EC}_t} + \frac{\text{TTHM}}{\text{EC}_\text{TTHM}} = \frac{(a/C)}{\text{EC}_t} + \frac{dC^{0.144}}{\text{EC}_\text{TTHM}} \tag{21.12}$$

$$\text{S.T.:} \quad dC^{0.144} < \text{TTHM}_\text{limit}$$

Notable: Residence time is not the primary concern, but the process cost is. So, this simplification of the problem really needs to include process capital and operating expenses. Further, a high Cl_2 treatment means that not all of the Cl_2 is used, and it must be vented or recovered. If Cl_2 is vented, then there is another environmental concern to be included in OF and constraint. Explicit knowledge of the process and situation is required to construct a right OF and set of constraints, to derive appropriate model equations, to see all of the DVs, to determine EC values, and to determine coefficient values in the equations. However, regardless of how complete the description is, the concept of assigning EC values and constructing the additive combined OF is the same as the aforementioned statement.

21.3 Solution 2: Nonadditive OF Combinations

In the pipe diameter optimization of Case Study 1 in Chapter 36 and in the economic objectives on Chapter 20, initial capital and annual expenses are the two important aspects; but they were economically combined as net present value. They were not combined in an additive OF formulation. Other common business profitability indices are long-term return on assets (LTROA) and discount cash flow rate of return (DCFRR). Both combine capital and expenses with sales and taxes, but not in an additive formulation.

In Chapter 29, about choosing the complexity of a regression model, I'll recommend final prediction error (FPE), a multiplication of sum of squares (SSD) residuals and a complexity factor. $FPE = (N+m)/(N-m)SSD$, where N is the number of data, m is the number of model coefficients, and SSD is the resulting match to the data.

In Chapter 33, in evaluating optimizers, I'll offer that both number of function evaluations (NOFE) and the probability of finding the global are important. The probable NOFE to find the global is $PNOFE = ANOFE(\ln(1-c)/\ln(1-f))$. Again, the several criteria for desirability and undesirability are not added, but otherwise combined.

Don't jump to the classic mathematical, traditional, isolated-from-reality additive OF formulation. Reconsider. Do what is right for your customer, which is not necessarily the traditional or introductory optimization legacy. Sketch, evaluate, erase, sketch, s, e, e, see, see, ...

21.4 Solution 3: Pareto Optimal

Although combining all OF elements into one OF is convenient, many times we cannot combine all concerns into a single lumped OF. This happens when different users, customers, and stakeholders have different views of the balance that should be applied to the disparate desirables and undesirables (of the EC weighting factors). This also happens when certain applications have different behaviors.

For example, consider that we want to design a product, a car: Some people want a large car for long trips, many friends, lots of Christmas presents, capacity, comfort, and so on. Others want a smaller car to reduce annual expenses, but large enough. Others choose a motorcycle instead of a car. One product will not satisfy the utility balance view of all people. The best (annual expense) large car will be excessive for a single person. The best (gas mileage) car will not suit the needs of a family of 6.

However, a small car with high operating expenses may not satisfy anyone. If a small car is acceptable, then there are many designs that have lower expenses, and low expenses beat high expenses. Alternately, if large expenses are acceptable, then the functionality of a large car beats that of a small

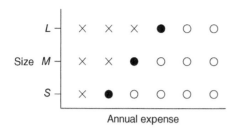

Figure 21.2 Initial concept for multi-objective treatment of two desirability metrics.

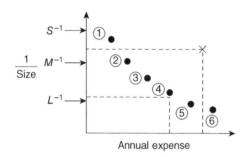

Figure 21.3 Proper structure of axes variables.

car. Of course there are more than two categories to judge a car; if personal image is a third category, then a small, hot antique roadster convertible with very high stature may be acceptable, even with high maintenance expenses. But a small car with the same high expenses, but only modest stature, would be dominated by (or preferred by a potential customer over) the one with higher stature.

To evaluate goodness when there are disparate OF components that you do not want to combine into one composite OF, the concept is to create a graph of OF2 w.r.t. OF1 and plot the (OF1, OF2) pair for each DV trial solution. Figure 21.2 illustrates the concept for two desirability aspects of a car. The vertical axis is size, and the horizontal axis is annual operating expense.

The x-ed-out possibilities are infeasible car designs. The open circles are feasible car designs. And the closed circles indicate the best designs—for any given size, the option with least operating expense or, for any given operating expense, the option with largest size.

Although this is the concept, it is more convenient to convert the axes so that worst is farther from the origin and best is nearer to the origin (see Figure 21.3). You can use reciprocal (shown) or negative of the function.

Note: This graph is not a plot of how the OF depends on the DV. The DV is not shown. This is a state graph of how the several OF values change with DV values.

Here, car design (X) is "dominated" by car designs (2), (3), (4), and (5). This means that each of (2), (3), (4), and (5) have a better combination of size and expense than design X. Folks who are willing to support an annual expense of car (5) or (6) will get a much larger value with design (5) or (6) than design (X). Graphically this is indicated by the dashed lines from X to the axes. Designs 2, 3, 4, and 5 are each within the rectangle. And folks who need a size of (3) will get much better annual expenses of (3) versus (X). So, reject design (X).

The set of options (1), (2),..., (6) are all non-dominated. There is no solution that is better in both categories. Solution (2) is better than (3) for annual operating expense, but (3) is better than (2) for size. As illustrated, there is no solution in the rectangle between solution (4) and the origin. No solution dominates design (4). Accordingly, all non-dominated solutions are kept as possible designs, possible products, and possible choices. One individual may choose solution (2) as the best that meets their needs, while another may choose (6) as best for their situation.

After exploring design options and alternate trial solutions in a product design, the optimization may not be able to find any new solution that dominates the (1)–(6) designs that are shown. Then the solutions represent the Pareto optimal set.

The Pareto optimal solution is not a single point item or DV value. The solution is a set of choices (labeled as 1, 2, 3, etc., in the aforementioned graph) for which no other design choice is better in two categories. Each is a Pareto optimal result of the design choices.

There are more than two dimensions, of course. Criteria for selecting a car include safety, customer's personal image, initial cost, annual cost, function, and many others.

Further, the number of decision variables does not have to be related to the number of OFs that characterize the product. There may be thousands of DVs, or just a few.

Regardless of DV dimension, the non-dominated solutions are those for which no other solution is better in all of the OF categories. Better is not the best in one category. The set of best solutions is the non-dominated outcomes of the trial solutions.

This multidimensional OF space that creates non-dominated solutions (best in at least one category—no other solution is best in all categories) leads to many viable designs.

A graph is not needed to identify the dominated and non-dominated solutions. A computer can test to see if a solution is dominated; if some other solution is better in all OF categories, then the first is dominated.

The axes of the Pareto optimal OF–OF plane can be log transformed, inversed, and scaled, and as long as the scaling is strictly monotonic (see Chapter 16), the transformation does not switch relative positioning, which is all that matters in identifying dominated and non-dominated outcomes. Log transformation may be very convenient for presentation of data if the OF value spans several orders of magnitude.

Additionally, the axis may be rank, such as first, second, third, and fourth place. Rank does not reveal ratio or relative proportion. First, for example, may be far ahead of second and third, which may be close to each other and far ahead of fourth, representing a total failure. Rank does not reveal relative desirability, but it does reveal order of desirability and can be used as an axis in a Pareto optimal set selection.

Conceptually, natural evolution of genes is similar to progressive changes in DVs, which leads to life-forms with altered attributes, hence survivability. Then survival of the fittest sets up a multi-objective space for selection of the best. Dominated species, not best in any category, do not survive. Non-dominated species survive. Pareto optimal selection parallels concepts of biological evolution. In this sense it is a mimetic technique. Optimization algorithms that seek to mimic the "intelligence" of nature (which guided evolution or is evidence in the behavior of schools of fish, flocks of birds, or genetic evolution) are termed mimetic optimization algorithms.

There are many non-dominated solutions.

Example 2   Which Optimizer Is Best?

Which univariate optimizer is best: successive quadratic, Newton's method, leapfrogging, heuristic direct, Golden Section, etc.? The application for the optimizer may be to find the value of a decision variable, x, that minimizes a function of x:

$$\min_{\{x\}} J = f(x) \tag{21.13}$$

*B*ut the question is not to determine the value of x^*, a numerical variable. Your objective is to choose an optimization algorithm that minimizes the bad aspects of the process of optimizing on the particular $f(x)$ function. Your DV is the optimizer algorithm, not the variable x:

$$\underset{\{\text{optimizer type}\}}{\min} J = \text{undesirable attributes of the optimization} \tag{21.14}$$

You have to define the undesirable aspects, the measures of badness. You must define a metric to quantify those undesirable aspects. You must define a method to combine the several metrics.

Suppose that you decide that algorithm complexity is a bad thing, because simplicity is a good thing. Complexity is an abstract concept, and you may not be able to assign a continuum value representing it; so, you might rank the optimizers from simplest to worst. The order might be HD, Newton–Secant, LF, GS, Newton's method, and SQ. The ranking first, second, third, and so on would be your personal assessment (the numerical values, metrics, that provide a measure of the abstract concept), but it should have a defensible basis (lines of code, education level required to understand the algorithm, number of user inputs, etc.). The problem with a rank, such as first, second, third, and so on, is that first and second may be very close and third may be a distant third. Rank does not provide relative position, but it is fully acceptable to use as a measure of goodness. There are diverse metrics to assess the concept of algorithm complexity. (In your mind, which is the right metric? How would you rank the optimizer complexity?)

Suppose you also decide that computer work is a bad thing. Work is an abstract concept of badness. The number of iterations to converge is a metric that you could use to assess work. However, some optimizers need more function evaluations per iteration than others, so perhaps number of function evaluations (NOFE) to converge could be a better metric to use as the assessment of work. However, successive quadratic, in the form of my algorithm in the univariate search examples, only uses one function evaluation at each iteration, but it does all of the computer work to solve for the a, b, and c coefficients in the linear equation set. So, perhaps execution time is a better measure of work than NOFE. Whatever you decide as a measure of "work," you do trials to get that metric for each DV. Again, there are diverse metrics to assess this concept. For instance, should one consider computer size requirements, as measured by RAM required to store variable values?

In this example the test objective is to find the point on the curve $y = x(x^2 - 3)$ that is closest to the (x, y) point $(1, 15)$. Note that y has a unique value for any given x, so that x can be the DV in the univariate search to find the point on the curve. Consider that the x and y dimensions have the same units. Then the distance between the point and the curve is $d = \sqrt{(x-1)^2 + (y-15)^2} = \sqrt{(x-1)^2 + (x(x^2 - 3) - 15)^2}$. Accordingly the objective is

$$\underset{\{x\}}{\min} J = \sqrt{(x-1)^2 + (x(x^2 - 3) - 15)^2} \tag{21.15}$$

This application has two minima and a maximum between them. You might want to graph the function and also the OF. Accordingly, I'll start the optimizers within the range of $2 \le x \le 5$ to be within the inflection points in the region of the global optimum. The convergence criterion will be on the DV (either its range (for LF and GS) or on the incremental changes for the other optimizers, with $\varepsilon = 0.001$.

Table 21.1 shows results. The NOFE column has two entries for HD, and the parenthesis indicates that this is the result of starting the search at either the $x = 2.5$ or $x = 4$ extreme. Since LF is a stochastic procedure, the random leap-over positions will not exactly reproduce the exact path to the optimum.

Table 21.1 A Pareto optimal comparison of optimizers.

Optimizer	Subjective complexity rank	NOFE on the application
HD	1	22(2.5), 22(4)
Secant (with numerical derivatives)	2	14
LF	3	~65
GS	4	20
N (with numerical derivatives)	5	—(2.5), —(4)
SQ	6	8

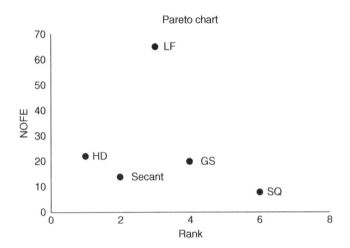

Figure 21.4 Multi-objective Pareto chart.

Accordingly, the table acknowledges that the NOFE was about 65 for many runs. For Newton unless the initial trial solution was very close to the optimum, the solution kept jumping to either the adjacent min or max. Accordingly, NOFE is listed as "---."

All of the optimizers found the same optimum $x^* = 2.8646$.... However, the objective was not to find x^*, but to identify the best optimizer that balances the complexity rank and NOFE on the exercise. Figure 21.4 is a Pareto graph of the OF2 (NOFE) w.r.t. OF1 (rank). Since Newton's method did not find the solution, it is not on the graph. Note that LF is dominated by both HD and secant and that GS is dominated by secant. HD, secant, and SQ are each non-dominated. The best between the non-dominated three algorithms for this particular univariate search application is the user's choice of how to balance complexity rank with NOFE.

Having done this exercise, however, raises several realizations:

1) Robustness should be considered. Newton's method did not work even within the limited range. SQ and secant approach don't work outside the ranges. GS will find the other minimum if initialized in a different range.

2) Precision should be considered. GS has four trial solutions and LF has 10 players, and they both bracket the optimum. As a result, when the high to low range is the convergence criterion, the middle points are closer to the optimum than the extremes. Alternately single-trial solution optimizers may incrementally move toward the optimum, and when the incremental step size meets convergence criterion, the solution still might be far from the optimum.

3) Many diverse applications should be considered. This one application is not necessarily representative of all that might be used, so the data from one application cannot be used to universally classify the NOFE results from all.

4) The impact of initialization range should be considered. The $2.5 \leq x \leq 4$ initialization range was selected to ensure that the optimizers commonly found the global. However, this requires user *a priori* knowledge. The probability of finding the global without such knowledge should be considered.

Suppose you also decide that robustness is a good thing. Robustness would be the likelihood that an algorithm could be successful, or not get confounded by surface aberrations. How would you measure robustness? Perhaps take 50 test applications that have such problems, start each optimizer 100 times in random places, and count the number of times in the 5000 trials that an optimizer had a problem. You'd desire to minimize the number of problems. What else might be an appropriate way to assess robustness?

You could combine the several OF aspects to create one overall OF: min NOFE + Rank. If you so choose, how would you scale complexity (rank) with NOFE? You could use the equal concern (EC) approach. You'll need to defend your choice of EC values and will realize that they are situation dependent. Sometimes we run (getting there in a hurry is more important than being out of breath when we arrive). Sometimes we walk.

Alternately, you could plot the result of each metric of desirability on a graph (e.g., NOFE w.r.t. rank for all optimizer choices) and then see if any optimizers dominate the others. See which comprise a Pareto optimal set.

21.5 Takeaway

Whatever you do to evaluate best, the outcome will depend on

1) Your best interpretation and your concepts of goodness or badness or desirability or undesirability
2) Your choice of metrics to assess that aspect
3) Your choice of experiments to gather assessment data
4) How you decide to combine and interpret the assessment data

Life is like that. Personal and business decisions are like that. To be appreciated, to be valued, and to have your recommendation accepted, you need to be sure that all stakeholders accept your choices in the four aforementioned steps. So, you need to reveal choices in steps 1–4.

21.6 Exercises

1   The more lights installed on the ceiling, the easier it is to see, but the greater the capital and operating expenses. The objective function is to determine the number of light fixtures to balance

function and costs. What are the equal concern values if the location is (i) a jewelry store, (ii) a bedroom, or (iii) great grandmother's kitchen?

2   A person goes to bed at midnight and has to be on the job at 8 : 00 a.m. the next day. Before going to bed, (s)he makes a decision about the time, T, for the alarm clock to ring. The longer (s)he sleeps, the more sleep benefit (s)he gets. Figure 21.5a shows a qualitative relation of benefit w.r.t. hours of sleep. However, the longer (s)he sleeps, the greater is the anxiety about, and potential cost associated with, being late. Figure 21.5b shows a qualitative cost of wake-up time. State these multiple and competing objectives as a single OF. Use these models Benefit $= 1 - e^{-T/3}$, Badness $= 1/(8 - T)$

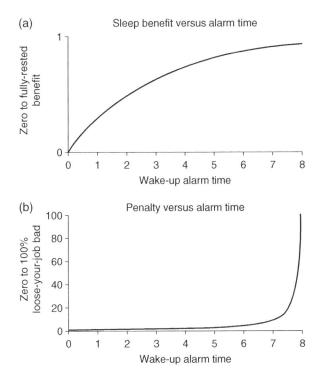

Figure 21.5 (a) Sleep benefit and (b) late penalty w.r.t. wake-up time.

3   Decide how to handle the multiple objectives associated with a trip. Consider these and any other aspects that you find important: Minimize cost (fuel, meals, nights, tolls). Minimize travel time. Maximize scenery and tranquility. Maximize safety.

4   Convert the hard penalty of the Cl dose Example 1 to a soft penalty and decide the EC factors.

5   Write the pseudocode to select non-dominated solutions.

6   When you choose what outer garments to wear, you have choices. What are the objectives? These might include match color, appropriate for the weather, appropriate for the anticipated activities,

or lowest cost to maintain. Add several of your own desirables and undesirables. When you choose the clothes, what is your OF, and how do you combine aspects? State this in optimization terms.

7 Should you take the stairs or use the elevator? State this in optimization terms.

8 What should you say in an outgoing voice mail message to a caller? State this in optimization terms.

9 Use the concepts from Chapter 16 to prove that the lambda values in Equation (21.3) or (21.4) do not have to sum to 1. Show that as long as the relative ratios of the weighting factors remain the same, the DV* is unchanged.

10 Prove that a strictly positive transformation of any axis in a Pareto analysis of a multi-objective optimization does not change the DV* solution.

22

Constraints

22.1 Introduction

Constraints are often as important to the optimization application as the objective functions. Pay attention to constraints.

They can be classified as "hard" or "soft," as "equality" or "inequality," and as "explicit" and "implicit."

Hard constraints are relations or consequences of the trial solution that cannot be violated. For example, keep the temperature above the freezing point, or else the pipes will burst. Or the material balance must be preserved: $\sum x_i = 1$. By contrast, soft constraints are ones that should not be violated, but a bit of a violation is not too bad. For example, only eat one piece of cake for dessert. Or obey the posted speed limit. Or only wear a white hat during the summer.

Explicit constraints are those that are explicitly revealed in the S.T. part of the optimization statement. Implicit constraints are not. If temperature is a DV, then an explicit constraint example might be to keep the temperature above the freezing point: $T > T_f$. If x and y are DVs, the explicit constraint may be that they need to be beyond a unit circle of the origin: $x^2 + y^2 > 1$. Each of these explicitly uses the DVs. However, some constraints may not be easily attributed to the DVs. For instance, find the equation of the projection of a path on a map (in x–y space) that does not exceed a slope as it goes through the mountains. The slope is the change in elevation, z, w.r.t. the distance along the path, S, and the constraint is $|dz/dS| < \varepsilon$ for all points on the path. Since the value of dz/dS cannot be calculated until after the DVs are placed in the model of how elevation depends on x–y position, the slope cannot be explicitly calculated only using the DVs, making dz/dS an implicit statement. Other implicit constraints might be applied to auxiliary variables or future variable in a model. For example, find the best valve position to get the composition to the set point, but don't let the tank go to empty, $M(t) > 0$ for all simulated time. Even though these conditions are stated in the S.T. section, they cannot be assessed until after the model is used to generate the OF value. They are not solely based on DV values. Additionally, implicit constraints are associated with non-executable statements in the OF calculation such as a divide by zero, conditions for which a formulation of DVs may not be possible.

Further, each constraint classification can have two types of relations: equality and inequality.

That $2 \times 2 \times 2$ classification (hard–soft, implicit–explicit, equality–inequality) means that there are eight combinations.

Engineering Optimization: Applications, Methods, and Analysis, First Edition. R. Russell Rhinehart.
© 2018 R. Russell Rhinehart. Published 2018 by John Wiley & Sons Ltd.
Companion website: www.wiley.com/go/rhinehart/engineeringoptimization

22.2   Equality Constraints

These are helpful!

22.2.1   Explicit Equality Constraints

Explicit equality constraints are actually helpful. Consider this optimization statement:

$$\begin{matrix} \min \\ (x,y,z) \end{matrix} \quad J = f(x,y,z)$$
$$\text{S.T.}: \ z = a + bx + cxy + dy + ey^2 \tag{22.1}$$

It appears that there are three DVs (x, y, and z). However, an equal sign in constraints means that there are fewer independent variables. Given values for x and y, z is constrained to one value. You cannot independently choose a value for z when x and y values are chosen. There are really only two DVs. You are not free to independently select values for three DVs; you are only free to independently select two values. Equality constraints reduce the degree of freedom (DoF), the number of independent values you can set. An equality constraint is a friend!

Each linear constraint (independently linear from the others) reduces a degree of freedom. Many types of nonlinear constraints (as illustrated earlier) also each reduce a DoF, but the implementation may not be so easy if they require root-finding methods or permit multiple solutions. Consider these several examples of constraints that permit several z values given x and y choices:

$$\text{A periodic function}: \sin(z) = x \tag{22.2a}$$

$$\text{A quadratic function with a} \pm \sqrt{\Box} \text{ solution}: x^2 + y^2 + z^2 = r^2 \tag{22.2b}$$

In some instances an equality constraint can complicate the solution. But often there are alternate formulations that may return to convenience. For instance, choose z to be the DV, and then x is deterministically set to a unique response in the $x = \sin(z)$ constraint. Or restrict the solution to one quadrant, and then there is only one acceptable sign in the quadratic relation.

In general, an optimization statement might be

$$\begin{matrix} \min \\ (x) \end{matrix} \quad J = f(x), \text{ where } x = \left\{ \begin{matrix} x_1 \\ \vdots \\ x_N \end{matrix} \right\}$$
$$\text{S.T.}: \ g_i(x) = 0 \ \text{for} \ i = 1 \text{ to } M \tag{22.3}$$

Here the M equality constraints reduce the x, DV, dimension from N to $N - M$, (if constraints are linearly independent and if the constraint does not permit multiple values).

If constraints are independent and provide unique values, the number of DVs is reduced. This is a benefit for optimization. Often, the work an optimizer does in finding a solution, the number of function evaluations required to guide it to DV*, is proportional to the square of the number of DVs. So, reducing the number of DVs from 8 to 7 reduces the work by about 49/64ths or roughly by about 25%.

In general, a revised optimization statement might be

$$\begin{matrix} \min \\ (x_{\text{subset}}) \end{matrix} \quad J = f(x) \tag{22.4}$$

where $\boldsymbol{x}_{\text{subset}} = \left\{ \begin{array}{c} x_1 \\ \vdots \\ x_{N-M} \end{array} \right\}$, and $\boldsymbol{x}_{\text{complement}} = \left\{ \begin{array}{c} x_{N-M+1} \\ \vdots \\ x_N \end{array} \right\}$ is calculated from $g_i(\boldsymbol{x}) = 0$ for $i = 1$ to M.

There is no need for either "f" or "g" to be a simple, one-line assignment statements. Either function could be a complicated calculation sequence, as long as the choice of $\boldsymbol{x}_{\text{subset}}$ can be used to calculate a unique $\boldsymbol{x}_{\text{complement}}$.

22.2.2 Implicit Equality Constraints

In the cases where the equality constraints do not admit a convenient solution, here are two options: The first is to apply a root-finding technique to solve for the value of the dependent variable. This, however, requires a nested procedure and a root-finding convergence criterion threshold that is small enough to have no detectable impact on the optimization OF.

The other approach is to convert the implicit equality constraints to penalty functions. Consider the optimization statement:

$$\begin{array}{c} \min \\ \{x,y\} \end{array} J = f(x,y,z) \tag{22.5}$$
$$\text{S.T.}: g(x,y,z) = 0$$

Conceptually, you can calculate $z = g^{-1}(x,y)$, with an iterative procedure. And this would reduce the number of DVs, but this might not be convenient or tractable.

In this case, consider converting the equality constraint to a penalty for its violation. In the equal concern method, the OF becomes

$$\begin{array}{c} \min \\ \{x,y,z\} \end{array} J = \frac{f(x,y,z)}{\text{EC}_f} + \left[\frac{g(x,y,z)}{\text{EC}_g} \right]^2 \tag{22.6}$$

One could also use the Lagrange-type weighting factor method:

$$\begin{array}{c} \min \\ \{x,y,z\} \end{array} J = f(x,y,z) + \lambda [g(x,y,z)]^2 \tag{22.7}$$

Either of these converts the hard equality constraint, from $g(x,y,z) = 0$, into a soft constraint based on the degree of deviation from the constraint. There is a penalty for a violation, which permits slight violations of the constraint as will be shown.

Whether the equality constraint is explicit or implicit, the penalty for its violation can be added to the OF.

22.3 Inequality Constraints

Inequality constraints permit any value until some limit has been reached. Examples include keeping the temperature above freezing, your bank account balance above zero, or the solvent concentration in air less than the lower explosive limit, or choosing quadratic equation coefficients to keep $b^2 - 4ac \geq 0$. Here, any value is acceptable until the constraint is met. If the constraint is not violated, if the condition is permitted, then the constraint is termed "inactive." Alternately, if the DV causes a constraint to be violated, the constraint is termed "active." As an optimizer seeks DV*, the inequality constraints could switch from inactive to active and back.

A feasible trial solution is one for which all inequality constraints are not active. The optimizer should be initialized with a feasible trial solution (or trial solutions as the case may be for the NM Simplex or the multiple players in LF).

Inequality constraints could be on the decision variable and explicitly stated, but as frequently they are on an auxiliary variable and are implicit (unknowable until the OF model is calculated). They can be $>$, $<$, $\leq$, or $\geq$. In the following statement there is a nonnegative condition on the z DV, and a less-than constraint on some formulation of the three DVs:

$$\begin{array}{ll} \underset{(x,y,z)}{\min} & J = f(x,y,z) \\ \text{S.T.:} & z \geq 0 \\ & g(x,y,z) < 0 \end{array} \tag{22.8}$$

Less-than constraints can be converted to greater-than by multiplying by -1. Further, all constraint expressions can be rearranged to the common reference to zero. For example, the constraint $x = y + z < 7$ can be rearranged to $x = y + z - 7 = g(x,y,z) < 0$.

The g-function could be algebraic, as well as differential. For example, the rule might be "Don't let the rate of elevation of the road exceed 1000 ft/mile." Or the constraint might be based on an integral (don't exceed 200 kg of sand in the cart). In any case the representation, $g(x,y,z) < 0$, seems like it might be an algebraic relation between DVs, which is a misleading suggestion of triviality. It may include derivatives and integrals.

To explore potential inequality constraints, Figure 22.1 roughly frames the limits of E–W and N–S choices in the Continental United States in an x–y rectangle. If the search is for a best place to live in the United States, then the x, y combination must be on land and in the United States.

Maybe "best" is defined as number of days with the high and low temperature between 85 and 55° F. However, the location must have rain and be livable. This might be the optimization statement:

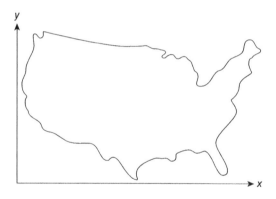

Figure 22.1 Sketch of the United States in an x–y frame.

$$\begin{array}{ll} \underset{(x,y)}{\max} & J = N_{\text{days between 55 and 85°F}} \\ \text{S.T.:} & (x,y) \text{ not in an ocean} \\ & (x,y) \text{ not in a lake or river} \\ & (x,y) \text{ not in Canada or Mexico} \\ & \text{Rainfall} \approx\, > 30 \text{inches/yr} \\ & x > 0 \text{ miles and } x < 3500 \text{ miles} \\ & (x,y) \text{ within 10 miles of a superstore} \\ & (x,y) \text{ in the lower 75 percentile for cost of living} \\ & N > 100/\text{yr} \end{array} \tag{22.9}$$

Constraints may be of many types. Illustrated here, constraints may be on

- A DV: $x > 0$ miles and $x < 3500$ miles
- A combination of DVs: (x, y) not in an ocean
- An auxiliary variable (neither DV nor OF): Rainfall approximately (not precisely) > 30 inches/yr
- On the OF, $N > 100$

Constraints are not necessarily on a DV value. In other applications, constraints may be on

- Rate of change: Δ accelerator pedal position $< 3°$ arc/s, slope must be less than 500 ft/mile.
- A sequence of events: "Home" must be last stop, or the "Mix" operation must follow adding A and B.
- A future variable value: When driving, don't build up so much momentum now that you slide out on an upcoming curve.
- Accumulation: Don't let tank go dry, don't let bank account go below $1000.

Here are some more diverse examples of constraints:

- Win the automobile race: Drive aggressively, but don't use up so much gas that the tank goes empty or that the tires go bald.
- Maximize airline revenue: Overbook passengers so that plane is full even after no-shows, but don't overbook to an extent that you can't take all ticket holders that show and cause a customer revolution.
- Load up on carbs before running the marathon race, but not to an excess that it makes you sick.
- Buy things on credit for joy today, but not so excessively as to constrain future cash flow (penalty for loan interest).

Inequality constraints may be active or inactive. Active means that the constraint is being violated. Inactive means that the trial solution is unconstrained, but is not violating an inequality.

22.3.1 Penalty Function: Discontinuous

If a constraint is violated, add a penalty value to OF:

$$
\min_{\{x_1 x_2 x_3\}} \quad J = \text{OF}(x_1 x_2 x_3) + \text{Penalty} \tag{22.10}
$$

$$
\text{Penalty} = \begin{cases} 100 & \text{if any constraint is violated} \\ 0 & \text{if none are violated} \end{cases}
$$

This is easy to implement but there are several issues.

One difficulty is in determining an appropriate value for the penalty. It requires knowledge of OF values. In Figure 22.2 is a penalty of 100 enough? In the illustrations, the constrained DV range is illustrated by the cross hatching. In Figure 22.2a an OF penalty of 100 is large enough to make all J values in the constrained region worse than the unconstrained best. But in Figure 22.2b the 100 is not a large enough penalty. Choosing an appropriate value requires knowledge about the OF.

Another difficulty is that near the constraint the discontinuity invalidates both df/dx and d^2f/dx^2 operations, which will confound some optimizers.

22.3.2 Penalty Function: Soft Constraint

Constraints are often inequality functions ($<, >, \geq, \leq$). For example,

$$
f(x_1, x_2, x_3, \ldots) \geq a \tag{22.11}
$$

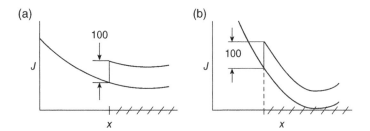

Figure 22.2 Illustration of a fixed penalty; (a) large enough, (b) not large enough.

Convert this to a zero on the RHS:

$$g(x) = f(x) - a \geq 0 \tag{22.12}$$

And consider the magnitude of the constraint violation (the deviation from the constrained value) as the measure of undesirability (badness):

$$B = \text{badness} = \begin{cases} f(x) - a & \text{if } f(x) < a \\ 0 & \text{if } f(x) \geq a \end{cases} \tag{22.13}$$

Either square the value, B, or take the absolute value to get a positive value. Scale it by either λ or an EC value to make magnitude of the penalty appropriate to the OF. Then add the penalty to the OF:

$$\min J = \text{OF} + \lambda B^2 \tag{22.14}$$

$$\min J = \text{OF} + \lambda |B| \tag{22.15}$$

$$\min J = \text{OF} + \frac{|B|}{\text{EC}} \tag{22.16}$$

$$\min J = \text{OF} + \left(\frac{B}{\text{EC}}\right)^2 \tag{22.17}$$

Although many users choose the absolute value of the violation as the penalty, using the squared value permits continuous second derivatives and provides a functionality that is consistent with human perception of badness.

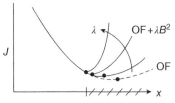

Figure 22.3 Illustration of a quadratic penalty added to the OF.

This method is termed "soft constraints" because the OF may include a little constraint violation at the minimum. This aspect is illustrated in Figure 22.3. The DV is constrained to be in the left portion of the x-axis. The OF is the left curve that becomes the dashed curve in the constrained region. The several solid curves in the constrained region represent the OF plus the quadratic penalty for the constraint violation. Increasing λ (reducing EC) makes the penalty larger. The optimum points for the several choices of λ are illustrated by the dots. When λ is small, the addition of the penalty is small, and the drop in the unconstrained OF in the constrained x region is larger than the penalty. As the λ-value increases, this optimum point is moved back toward the constraint, but it cannot prevent a bit of a constraint violation.

In general, the rate of change of OF w.r.t. DV is df/dx, and for small Δx deviations from the constrained x-value, a, the OF value is approximately $f(x) = f(a) + (df/dx|_{x=a})(x-a)$. The penalty for a value of $x > a$ is $\lambda(x-a)^2$. The optimum with the penalty is roughly at the point when the increase in penalty is equivalent to the decrease in f. So, with a soft penalty based on the x-deviation from the constrained x-value, $x^* = a - df/dx|_{x=a}/2\lambda$. Increase λ to make the permissible violation approach zero. However, large λ-values result in effective ∇f and $\mathbf{H}f$ discontinuities and require $\lambda = \infty$ to make the constraint violation become zero.

This is interpreted as a "soft constraint," meaning a small violation is accepted, even if not desired.

There are several adaptations to fixing the violation of a soft constraint formulation. In an application of this type,

$$\min J = OF$$
$$\text{S.T.}: f(x) \geq a \qquad (22.18)$$

Change the value of a by ϵ, a bias or correction or safety factor:

$$\text{S.T.}: f(x) \geq a - \epsilon \qquad (22.19)$$

Then determine the degree of violation from $(a - \epsilon)$:

$$B = \begin{cases} f(x) - (a-\epsilon) & \text{if } f(x) < (a-\epsilon) \\ 0 & \text{otherwise} \end{cases} \qquad (22.20)$$

Then, as illustrated in Figure 22.4 since the safe constrained value is shifted, the penalty begins to accumulate prior to hitting the constraint.

Still, appropriate λ and ϵ values require understanding of the system so that you can determine right values and graph relevant variables as iterations proceed.

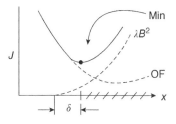

Figure 22.4 illustration of a quadratic soft penalty shifted by ϵ.

22.3.3 Inequality Constraints: Slack and Surplus Variables

Inequality constraints can be converted to equality constraints by including a slack or a surplus variable. If the constraint is $g(x) \geq 0$, then $g(x) = 17$ is a fully acceptable value, and it is 17 g-units away from the constraint. Here $g(x)$ has a surplus of 17 g-units. If the surplus, S, is removed from the g-value, then the inequality becomes an equality. $g(x) - S = 0$. There is now a condition on S, $S \geq 0$. Although this adds a variable to the optimization, the equality constraint removes a variable. Although the number of DVs and number of inequality constraints are unchanged, this provides some flexibility in the optimization statement. You could choose to use S as a DV and use the equality constraint to solve for one of the x elements, if an explicit or convenient-to-calculate arrangement of $g(x) - S = 0$ permits it.

Often the surplus variable is squared, using $g(x) - S^2 = 0$. This removes the constraint on the S-value, further simplifying the search if S is used as an alternate DV.

If the constraint is $g(x) > 0$, as opposed to $g(x) \geq 0$, then, when the surplus variable is removed to form the equality constraint, $g(x) - S^2 = 0$, all S-values are permissible except for the singular value of $S = 0$. This may be an issue for mathematicians, but I am not aware of engineering applications that would require the single-point exclusion.

The complement of this approach is applicable to less-than constraints. If the constraint is $g(x) < 0$, then $g(x) = -5$ is a fully acceptable value, and it is 5 g-units away from the constraint. Here $g(x)$ has a slack of 5 g-units. If the slack, s, is added to the g-value, then the inequality becomes an equality. $g(x) + s = 0$. There is now a condition on s, $s \geq 0$. Although the number of DVs and number of inequality constraints are unchanged, this, again, provides some flexibility in the optimization statement. You could choose to use s as a DV and use the equality constraint to solve for one of the x elements, if an explicit or convenient-to-calculate arrangement of $g(x) + s = 0$ permits it.

Again, often the slack variable is squared, with the equality stated as $g(x) + s^2 = 0$. This removes the constraint on the s-value, further simplifying the search.

Slack and surplus variables are often differentiated by the lower-case s or capital S. As often, however, authors use the lower case s for both. There is no need to explicitly differentiate them. The care must be in the addition or subtraction in using s (or s^2) when converting the inequality to an equality.

Note: The units on the slack or surplus variable are the same as the units on the g-function, or in the alternate formulation they are the square root of the units on the g-function. Although they could be chosen as DVs, their units may be distinct from the remaining DVs.

Note: To be an effective conversion of an optimization statement, the g-function must be invertible to solve for an x element. If the constraint is either a differential or an integral, it might not be invertible. If the constraint is a nonlinear function, it may require a numerical root-finding procedure.

Note: If the inequality constraint is implicit (based on internal variables to the model, not the DVs), then one may not be able to use this technique to convert the inequality to an equality and alter the DV set.

Note: You must decide which DV is to be replaced by the slack or surplus variable. As the examples will show, some choices simply shift the hard constraint to another form. Some choices remove a constraint. It could be that a particular application does not permit a convenient DV transformation.

Example 1

$$\begin{array}{c} \min \\ \{x\} \end{array} J = x + \frac{1}{x}$$
$$\text{S.T.}: \quad \sqrt{x} \geq 1.25 \tag{22.21}$$

Here the unconstrained minimum $x^* = 1$ violates the constraint. Rather than using a constraint handling optimizer, convert the inequality to an equality constraint using a surplus variable $\sqrt{x} - 1.25 - s^2 = 0$. Now there are no constraints on the value of s. Using s as the optimizer DV, solve for x from the s-value, $x = (1.25 + s^2)^2$. The optimization becomes

$$\begin{array}{c} \min \\ \{s\} \end{array} J = x(s) + \frac{1}{x(s)} = (1.25 + s^2)^2 + (1.25 + s^2)^{-2} \tag{22.22}$$

Example 2   Here is a simple example that was presented in Beveridge and Schechter:

$$\begin{array}{c} \max \\ \{x_1, x_2\} \end{array} J = x_2^2 - 2x_1 - x_1^2$$
$$\text{S.T.}: \quad x_1^2 + x_2^2 \leq 1 \tag{22.23}$$

The following analysis is mine. The feasible range of the DVs must be within the unit circle. As a hard constraint it causes optimizers to get stuck on the constraint, in the vicinity of the optimum. Figure 22.5 shows the results of 100 trials from randomized initializations with a Levenberg–Marquardt optimizer.

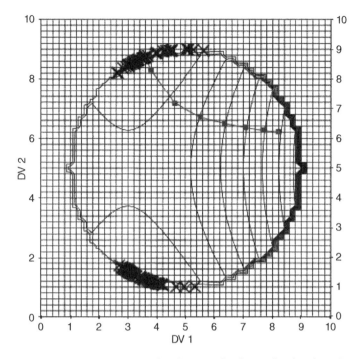

Figure 22.5 Optimization results of 100 randomly initialized trials on Equation (22.25).

A possible trick to alleviate the constraint discontinuity is to convert the inequality to the standard form then to an equality by adding a slack variable:

$$x_1^2 + x_2^2 - 1 + s^2 = 0 \tag{22.24}$$

Then use the equality to solve for $x_2 = \pm\sqrt{1 - x_1^2 - s^2}$. This adds a constraint that the relation between x_1 *and* s must keep the argument of the square root nonnegative. It also bifurcates the x_2-value into a positive and negative case. However, in this convenient example, when x_2 is substituted in the OF, both issues are eliminated. The converted optimization statement is

$$\max_{\{x_1,s\}} J = 1 - s^2 - 2x_1 - 2x_1^2 \tag{22.25}$$

When translated and converted to minimization,

$$\min_{\{x_1,s\}} J = s^2 + 2x_1 + 2x_1^2 \tag{22.26}$$

Now the solution is simple and reproducible. Again, after 100 LM trials from randomized initializations, Figure 22.6 reveals that the trials all converge on the same spot.

When it is feasible to invert the constraint to solve for the DV, adding a slack or surplus variable keeps the DV dimension the same, eliminates the inequality constraint, and improves solution precision.

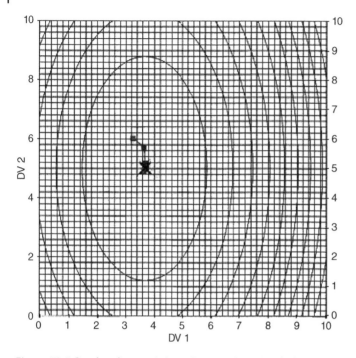

Figure 22.6 Results when optimizing Equation (22.28) a slack variable transformation.

The optimal solution is the same whether the DVs are retained as x_1, x_2 or converted to x_1, s. However, the generic nonlinear functions may not provide the convenience that lets this trick remove the active constraint. Often it just changes how the constraint is encountered.

Example 3 Function 67 in the 2-D Optimization Examples file places a point at (x_0, y_0, z_0) over (or below) the Peaks surface (Function 30), $z = f(x,y)$. The objective is to determine the point on the surface that minimizes distance to the point:

$$\begin{aligned} \min_{\{x,y\}} \ & J = \sqrt{(x_0 - x)^2 + (y_0 - y)^2 + (z_0 - f(x,y))^2} \\ \text{S.T.:} \ & \ln(x) - \sqrt{y} \le -0.5 \\ & y \ge 0 \\ & x > 0 \end{aligned} \tag{22.27}$$

To keep variables in the real domain (as opposed to imaginary or complex), the constraints on x and y are required by the logarithm and square-root functions. The primitive implementation with x and y as the DV, and a hard constraint, the contour and results of 100 randomized RLM trials are shown in Figure 22.7. The minimum is roughly in the middle of the figure, but the hard constraint blocks the optimizer from getting to the exact spot.

When the inequality constraint is converted to an equality constraint by adding a slack variable,

$$\ln(x) - \sqrt{y} + 0.5 - s^2 = 0 \tag{22.28}$$

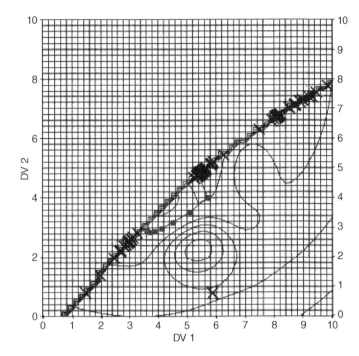

Figure 22.7 RLM solution to Equation (22.29).

And s and y are used as the DVs, and the equality constraint of Equation (22.28) is used to solve for x:

$$x = e^{\sqrt{y} - 0.5 + s^2} \tag{22.29}$$

Then the optimization is converted to

$$\begin{array}{l} \min_{\{s, y\}} \ J = \sqrt{(x_0 - x)^2 + (y_0 - y)^2 + (z_0 - f(x,y))^2} \\ \text{S.T.}: \ x = e^{\sqrt{y} - 0.5 + s^2} \\ \qquad y \geq 0 \end{array} \tag{22.30}$$

With this choice, the calculation of x necessarily satisfies the constraint required by the log function, and there is one unique x-value. Alternately, one could use x and s as the DVs and use Equation (22.28) to solve for y. However, that arrangement would result in two possible y-solutions and would have the constraint that $\ln(x) + 0.5 - s^2 > 0$.

Figure 22.8 shows the results of 100 randomized RLM optimization trials. The path shown leads to the global optimum, found with precision in about 50% of the trials.

22.4 Constraints: Pass/Fail Categories

For direct search methods, the value of J can be "fail." If a constraint is violated, the new trial solution value is "fail," not OF + penalty:

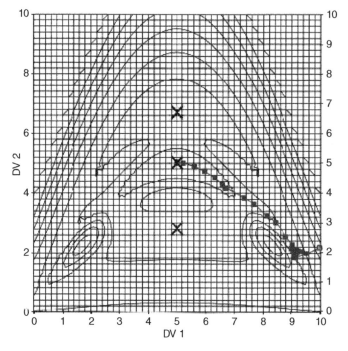

Figure 22.8 Results of 100 optimizer trials from randomized initializations for Equation (22.30).

$$\min J = \left\{ \begin{array}{ll} \text{OF} & \text{if no constraint is violated} \\ \text{"fail"} & \text{if any constraint is violated} \end{array} \right\} \quad (22.31)$$

The direct search logic can then be: calc x_{new}, calc J_{new}. If $J > J_{old}$ or if $J =$ "fail", then return to x_0, and use alternate search logic.

Use of soft constraints can permit a multi-DV search to go around the constraint "mountain" on the OF "surface" to seek an unconstrained minimum. Hard constraints can end the search at a constraint boundary. The word "can" does not mean "always"—it means under some conditions.

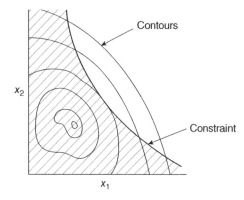

Figure 22.9 Hard constraints can block an optimizer.

22.5 Hard Constraints Can Block Progress

A hard constraint means you cannot use that trial solution. If an optimizer wants to go downhill, but is told, "You can't go there," perhaps it takes a smaller step in the downhill direction. When an inequality constraint is active, hard constraints force the solution to be on a constraint. Figure 22.9 is a 2-D illustration of the contours and the constraint w.r.t. the DVs.

Start somewhere in the upper right and follow the steepest descent downhill. When your path hits the constraint, steepest descent will want to cross over the constraint, and all moves will be "FAIL," leaving the trial solution somewhere on the constraint, but not at the true minimum.

Near or on the constraint, direct searches will have a low probability for finding the narrow region of better trial solutions.

Options/Solutions

1) Run many trials from independent initializations and take the best of N. But this adds work.
2) Switch to an alternate search approach: radial scan, line search along constraint, or $(N-1)$ DV search along the constraint (this is an active constraint strategy or a reduced gradient strategy). But this logic gets complicated for solving for the constrained solutions or when multiple constraints are encountered: how to recognize that the logic needs to be changed to $N-1$ or back to N or to $N-2$ and how to determine which of the $N-1$ DVs are to be chosen.
3) Convert hard constraints to soft constraints. But this permits some violation of the constraint.
4) Convert hard constraints to soft constraints and progressively increase the Lagrange-type multiplier (reduce the equal concern factor) as the optimizer approaches the end game. But what rule? When is the penalty adequately large to impose a negligible constraint violation?
5) Use only absolutely necessary constraints. This frees the optimizer to choose large DV relations. Let the optimizer run unconstrained to find the solution, and then check to see of any are violated. If none are violated, great! If there is a violation, then add that one as a constraint. In distillation modeling, the tray composition must progressively rise on each successive tray. One could think that the constraints must be $x1 > x2 > x3 > \ldots$. But without this constraint specification, minimizing the deviations on the material and energy balances on each tray leads to a feasible solution without imposing the hard constraints.
6) Use a GRG type of approach (see Chapter 10) that converts inequality constraints to equality constraints with slack or surplus variables, and then locally linearize the constraints. The local linearization permits a bit of constraint violation but probably inconsequential relative to uncertainty on the relations and coefficient values.
7) Keep the nonlinear constraints and, as illustrated in Section 22.3, use slack or surplus variables to convert them to equality constraints. Then use slack or surplus variables as the DVs and choose which of the original DVs are solved for using the constraints. This retains the nonlinear functionalities, but not all inversions of nonlinear equations will lead to unique or constraint free solutions. You need to make choices.

22.6 Advice

If a constraint is not necessary to specify, don't. For instance, (i) the composition should rise on each tray of a distillation column, so it may seem helpful to use $x(i) < x(i-1)$ as a constraint on an optimizer seeking to solve for a composition profile. But that will be the natural result, and imposing a hard constraint makes it difficult for the optimizer to find the solution. (ii) The placement of parking lot lamps to best light the lot must be within the property boundary of the lot. Optimizing to find light post locations that best light the lot will end with all within the boundary, because, if a trial solution is outside of the lot boundary, the light will not be as effective on the lot. But imposing the constraint on the location DV's just restricts, confounds, and complexifies the search.

If you are not sure if a constraint needs to be specified, don't. Run the optimizer. If the constraint is not violated in the solution, then it was not necessary to specify it. Alternately, state it as a soft constraint, an OF penalty for violation. This frees the DV TS values from restriction, facilitating the search. Again, if the constraint is not violated in the solution, then it was not necessary to specify it. However, if you find that a constraint is violated, then state it.

Active constraints block what can be implemented in the real world. If you have identified what the constraint is, take an alternate route to a solution than optimization of the original statement. See what can be done in the real world (with the physical device or process or with the legal contractual procedures) to eliminate the active constraint. Transfer out of the optimization technologist view of how to include the constraint in the exercise, and move into the business view of how can the constraint be eliminated from the situation, or at least diminished or extended. Don't seek mechanisms to remove all constraints; just consider the ones that are active and block optimization.

22.7 Constraint-Equivalent Features

One problem that constraints create for optimizers is the slope or level discontinuity in the OF. Another is deciding what new direction the optimizer should take when it hits a constraint and is told, "No, you can't go there."

Conditionals in either the OF or models can create sharp ridges or valleys or cliff discontinuities that similarly confound an optimizer. Here are two examples of conditionals: IF–THEN statements. In a fluid dynamics application, there might be a transition between laminar and turbulent flow. The model might have a statement if Re < 2100, then use the laminar model for differential pressure; otherwise use the turbulent model. In a separation tank, the required diameter could be determined by particle settling in the vapor phase, liquid degassing in the liquid phase, gas flow rate to prevent shear at the liquid surface from re-entraining droplets, or required liquid inventory holdup; and the tank diameter would be the largest of each of the four conditions.

Conditionals can be of many types, including conjunctions or disjunctions, either simple or compound. Either way, they create equivalent discontinuities in the OF. (Disjunctions are conditionals that are combined with an OR conjunction. For instance, IF(p or q) THEN. If either p or q is true, then the consequence is imposed.)

22.8 Takeaway

Be aware of the diversity of constraints, and recognize that they often are as important to the optimization solution as the OF. Be aware of the diversity of approaches in including or transforming constraints, and the advantages or drawbacks to those approaches. If a constraint becomes active, consider redesigning or re-specification of the physical system to remove the issue. Optimization applications are not about the intellectual challenge of the mathematics of solving the constrained statement, but are about bringing the physical realm into fruition.

22.9 Exercises

1   There are $2 \times 2 \times 2 = 8$ combinations of constraint categories. Provide an example of each of the 8.

2   A favorite childhood game of mine is to race "cars" on along a path drawn on quadrille lined paper. Search the web for "graph paper race track game." Draw an S-shaped race track of about 3–5 unit

width, and add a start and finish line. Place dots to represent the cars on quadrille line intersections on the start line. Cars can accelerate or decelerate in either the vertical or horizontal direction by one space each move, in each player's turn. Cars can only occupy points of intersection of the quadrille lines. For example, if the last move of a car was +3 horizontal and +1 vertical, the next move of the car can be to any of the nine points in the +2 to +4 horizontal and 0 to +2 vertical intersections. Players take turns. A car may not occupy a spot that has another car, nor can it go through another car on its move. A car cannot leave the course. One would like to accelerate at each move to go as fast as possible, but doing so will build momentum and cause the car to crash out of bounds on a curve. How does this relate to optimization? Find a strategy to minimize time (moves) from start to finish, but don't violate a constraint. Decisions now affect a future constraint. Get a friend, and enjoy the game. Relate your experience to constrained optimization.

3   If the objective is to minimize $f(x)$, the constraint is $g(x) < 0$. Using a quadratic penalty, and c is the value of x that makes $g(x) = 0$, show that x^* occurs when $dJ/dx = df(x)/dx + 2\lambda g(x)(dg(x)/dx) = 0$. If $x = c$, then the optimizer will realize that moving away from the constraint limit into the constraint will improve the OF value, unless $\lambda = \infty$.

4   How many DVs need to be included in the following optimization search? Explain. $\underset{\{x,y,z\}}{\text{Min}} J = f(x,y,z)$. S.T.: $g(x) < a$, $h(y,z) = b$, $c < x < z < d$, $e < y < g$

5   Explain the differences between a hard constraint and a soft constraint. Feel free to use sketches or equations. Be brief and clear! Think about your complete answer, and then write the essence.

6   Here is an optimization application to determine a path from point A to point B that maximizes rewards gathered along the way. It could be the path from home to classroom that encounters the most smiles. It could be the broom path that gets most of the dirt in one sweep. The path in the x–y plane is described by the equation $y = a + bx + cx^2$. The optimization question is what are the best values for the a, b, and c coefficients in the path model. Since it should start at point A and end at point B, the path is constrained. Along the way $\rho(x, y)$ represents the reward density (points, number of smiles, quantity of dirt) at an x–y location. If the ending point B does not have to be exactly at the (x_A, x_B) location, but there is a desire to be in the proximity of (x_A, x_B), then the second hard constraint can be converted to a penalty related to the y deviation from point B. Reformulate the optimization statement to reflect this. $\underset{\{c\}}{\text{max}} J = \int_A^B \rho dS$, S.T.: $y_A = a + bx_A + cx_A^2$, $y_B = a + bx_B + cx_B^2$

7   Add a constraint to any application of your choice, and express "badness" (violation of the constraint) as a penalty function in the OF. Choose an equal concern factor (or Lagrange-type multiplier) to scale the constraint w.r.t. the unconstrained OF. The constraint could be on the OF value, on the DV value, on the rate of change of DV/iteration or OF/DV, etc. You choose, but make the constraint active at the unconstrained optimum. This is the soft constraint approach. Explore the impact of EC (or λ) on the optimized DV and OF values. A graph comparing the unconstrained OF w.r.t. DV and penalized OF versus DV for several EC (or λ) values will be revealing.

8   A soft constraint on a DV violation of a DV boundary permits some deviation. (i) Choose a function and an optimizer and set a DV constraint value that would block the optimizer from finding

the unconstrained minimum. (ii) Add a code to implement the constraint violation as a penalty. (iii) Demonstrate the impact of the value of the penalty factor. (iv) Comment on how you would determine a right value of the penalty factor.

9   In the soft penalty method for handling constraints, a penalty is added to the primary objective function. Explain how a constraint penalty could be handled as a multi-objective Pareto optimization.

23

Multiple Optima

23.1   Introduction

One of the several attributes of optimization applications that can lead an optimization astray is the existence of multiple optima, multiple local minima. If an optimizer trial solution is in the vicinity of local optima, then it will seek the bottom of that local hole, and not climb out to see the deeper hole next door. The concept is illustrated in Figure 23.1.

Each figure illustrates three local minima. Figure 23.1a could illustrate delivery costs for a shipping company as it changes with business size. With small-sized service, buy a small truck. The more you ship, the less the truck cost/delivery. However, when shipping volume exceeds the small truck capacity, you need to either trade it for a larger truck or buy a second small truck. This extra capital increases cost/delivery of the first increment over the small truck capacity but permits you to move toward larger volumes. The OF has local optima with discontinuities.

In contrast, Figure 23.1b shows an OF as a continuous function with multiple optima. It also has three optima, one at the feasible DV left boundary.

In each case, one optimum is clearly the global best. This is called the global optima. The others are local optima.

Also, in each case the OF is deterministic, not stochastic. There is a repeatable, single OF value for repeated trials with identical DV values. This chapter provides guidance for deterministic functions with multiple optima. If stochastic, then techniques of this chapter will lead to selecting a fortuitous best, not a representative best.

We want our optimizer to find the global overall best. What would a Golden Section search algorithm do in the figures? What would the steepest descent search do?

The concept extends to higher dimension applications. In a 2-DV application the OF can be graphed as a contour plot looking at the DV map. In Figure 23.2, the 2D contour map shows three local optima at 0, +1, and +4 values, with some of the contour values labeled. Starting in the upper right, a downhill search will end up at the local optimum near (7, 8) with a value of about 1. Starting in the lower right, it will end in the local optimum near (8, 2) with a value of about 4. However, the global optimum is at about (3, 4) with a value of about 0.

To clarify the phenomenon, consider seeking the best location to live in the central-southwest United States. If best location means within a large population center and you start near Tulsa, Oklahoma, the direction of best improvement takes you to Tulsa (about 600,000 people). However near Oklahoma City, the best local direction leads you to OKC (about 800,000 people). At either of these,

Engineering Optimization: Applications, Methods, and Analysis, First Edition. R. Russell Rhinehart.
© 2018 R. Russell Rhinehart. Published 2018 by John Wiley & Sons Ltd.
Companion website: www.wiley.com/go/rhinehart/engineeringoptimization

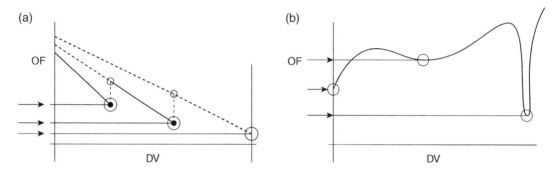

Figure 23.1 Multiple optima (a) discontinuous (b) continuum values. *Source:* Rhinehart (2016b). Reproduced with permission of Wiley.

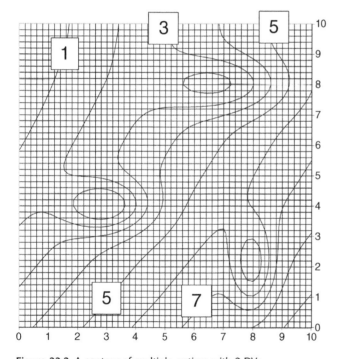

Figure 23.2 A contour of multiple optima with 2 DVs.

moving away from the city center moves you toward open country, low population density, toward a "worse" local OF value (if a large population is "best"). If large population is "best," then Dallas, Texas, wins, and Tulsa, OKC, Kansas City, Abilene, …, are all local optima. Of course, truly, "best" includes many other considerations.

Many optimization applications lead to multiple optima. The challenge is to find the global best and not become misdirected to local optima. In a familiar application, prior knowledge may be able to guide the optimizer initialization with a trial solution in the proximity of the global, so that it will move to the global, not a local. However, for new applications the surface might not be understood.

The challenge is to find the global minimum with minimal effort and high probability. This chapter will reveal several methods:

- Multiple starts from randomized initializations, take the best-of-N results.
- Multiple starts from directed initializations that would appear to not find already discovered optima.
- Initial surface exploration, then optimization from the best of the test sites.
- Coarse convergence to scope the vicinity of the global, then one final fine search for the best.
- Adjust the optimizer for early phase exploration, then have it switch to an end-game convergence.

23.2 Solution: Multiple Starts

One solution to increasing the chance of finding the global is to have multiple optimization runs from randomized initial values and to use enough runs so that there is adequate confidence that at least one of them will have found the global. The number N can be derived from probability analysis.

Here is a diversion, but the familiar exercise of analyzing coin flipping may help the reader with probability concepts. If you wanted to flip a coin and get a head (H), is one flip enough? The probability of an H is 0.5. In N number of randomized flips (independent trials), the probability of getting at least one H is p(at least $1H|N$ trials) $= 1 - 0.5^N$. If you want to be 99.9% confident in getting at least one H from independent flips, then you are desiring that p(at least 1 $H|N$ trials) $= 0.999 = 1 - 0.5^N$, which can be solved for $N = 10$ trials. We'll use a similar analysis to determine the number of randomly initialized optimizer runs that are needed to be confident in finding global optima.

First, we need to analyze the probability of locating the global optima. Figure 23.3 illustrates what a steepest descent search would do in the 1-D application. Starting with a trial solution anywhere in Region (1) leads it downhill to the OF_1 optimum. Starting in Region (2), it finds OF_2, and in Region (3) it will end up at OF_3.

If Region (1) has 25% of the DV range (or area if 2 DVs, or volume if 3 DVs, …), and Region (2) 55%, and Region (3) 20%, and if the starting point $(x_{10}, x_{20}, x_{30}, …)$ was randomly selected from the permissible DV range, then there is a 25% chance of finding OF_1, a 55% chance of finding OF_2, and a 20% chance of finding OF_3.

$$p_1 = \text{probability of finding } OF_1 \, (= 25\% \text{ in the example})$$
$$p_2 = \text{probability of finding } OF_2 \, (= 0.55) \tag{23.1}$$
$$p_3 = \text{probability of finding } OF_3 \, (= 0.20)$$

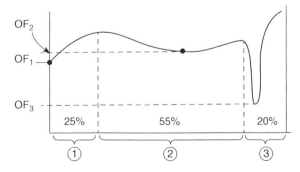

Figure 23.3 A univariate function with three minima and region of attraction labeled. *Source:* Rhinehart (2016b). Reproduced with permission of Wiley.

The question is "What is the probability of finding OF_3 (the global) in N random starts?" There are two ways to evaluate this: The very complicated way sums the probability of finding it in any one or two or three, …, of the N trials:

$$p(\text{finding } OF_3 \text{ in at least 1 of } N) = p(OF_3 \text{ in 1 but not 2, 3, 4,…,}N) + p(2 \text{ but not 1, 3, 4,…,}N)$$
$$+ p(1 \text{ and 2, but not 3, 4,…,}N) + p(1 \text{ and 3, not 2,4,…,}N) + \cdots$$
$$+ p(1 \text{ and 2 and 3, but not 4,…,}N) + \cdots$$

$$(23.2)$$

Alternately, the much simpler way is to consider the complementary, NOT, event: $p(\text{not finding } OF_3 \text{ in any of } N)$

$$p(OF_3 \text{ in at least 1 of } N) = 1 - p(\text{not in any of } N) \quad (23.3)$$

Expanding the NOT in any,

$$p(\text{not in any of } N) = p(\text{not in 1st, and not in 2nd, and,…, and not in } N\text{th}) \quad (23.4)$$

Using the AND conjunction to signal multiplication of probabilities and converting each NOT to an IS:

$$p(OF_3 \text{ in at least 1 of } N)$$
$$= 1 - \{[1 - p(\text{find } OF_3, \text{ in 1st})][1 - p(\text{finding in 2nd})]…[1 - p(\text{finding in } N\text{th})]\} \quad (23.5)$$

Since $p(\text{finding OF})$ is purely random and independent of trial number (when the algorithm trial solution or other initializations are random and independent), then

$$p(\text{find } OF_3, \text{ in } i\text{th}) = p(\text{find } OF_3, \text{ in } j\text{th}) \quad (23.6)$$

Then

$$p(OF_3 \text{ in at least 1 of } N) = 1 - [1 - p(OF_3 \text{ in any trial})]^N \quad (23.7)$$

For example, with 20% representing the probability of finding the global in any one trial with a randomized trial solution initialization, $p(OF_3 \text{ in any one random trial}) = 0.20$, then $p(OF_3 \text{ in at least 1 of } N)$ is given in Table 23.1.

If you have no knowledge of where the optimum is, but the probability of finding it in any one trial is 20%, then with 20 random starts the probability of finding it increases to 99%. Is a 99% probability of finding the global optimum acceptable? It may be. But then it may not. If you are considering personal safety, you'd want a higher confidence. If you want to be 99.999% sure that the optimizer finds the global solution when there is a 20% chance of finding it on any particular trial, then $N = 52$ trials.

Optimizers do not actually stop exactly at the local or global optimum. They stop in the vicinity of it when the convergence criterion indicates that it is close enough to claim convergence. Consider a histogram of OF values from a thousand, 1000, random starts in the OF represented in Figure 23.3. Ideally, it will appear as three bars representing the three optima, as indicated in Figure 23.4.

However, because the stopping criteria stops short of perfection (which would require ∞ iterations of progressive improvement), the optimizer converged solutions at each OF will not exactly stop at the same place. Looking finer at the histogram, in Figure 23.5, one sees the "spread" in OF values.

Table 23.1 Probability of at least one success in N trials.

N	1	2	3	4	5	10	20
$p(OF_3 \text{ in } N)$	0.20	0.36	0.49	0.59	0.67	0.89	0.99

Note: Although nominally there are just three optima, there are many converged OF* solutions.

Note: In each grouping of the three nominal solutions, the far-left value is closest to the true local optimum, and the locally higher OF values represent solutions that converged in its proximity.

Note: The histogram of OF* results depends on each of several things—the objective function, the optimizer, the convergence criterion, and the threshold for convergence. The objective function determines the number and location of the local optima. The optimizer algorithm rules may give it a higher probability of jumping over a local minimum. For example, with a very large initial step size, the heuristic direct search will initially jump to and fro, increasing its likelihood of finding the global. If the convergence criterion is on the incremental changes in the OF value, then the width of each peak in the pdf(OF) graph will be similar. But if the convergence criterion is on the changes in the DV value, then the width of the steep OF₃ peak will be broad. Finally, if the convergence threshold is smaller, the width of each peak will be smaller. The character of the histogram depends on many aspects, including the optimizer type and its initialization choices.

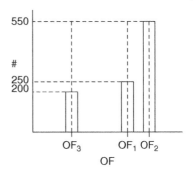

Figure 23.4 Histogram of optima with randomized initializations in the OF of Figure 23.3. *Source:* Rhinehart (2016b). Reproduced with permission of Wiley.

Normalizing the histogram (divide # by 1,000,000, or however many optimizer trials from independent random initializations) and smoothing the curve displays the probability density function of the objective function value, pdf(OF), as in Figure 23.6.

We don't know what the pdf(OF) graph looks like until after we get the solution to many (this may require on the order of 10^6) trials with random initializations. However, we do know that whatever

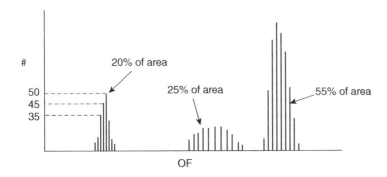

Figure 23.5 Detail of histogram. *Source:* Rhinehart (2016b). Reproduced with permission of Wiley.

Figure 23.6 The pdf(OF*) solutions associated with Figure 23.3. *Source:* Rhinehart (2016b). Reproduced with permission of Wiley.

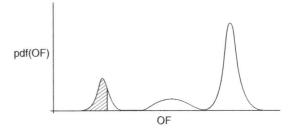

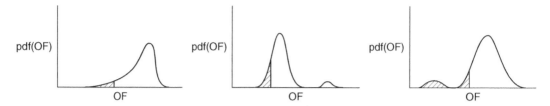

Figure 23.7 Diverse pdf(OF*) functions with the best 10% area shaded. *Source:* Rhinehart (2016b). Reproduced with permission of Wiley.

shape the pdf(OF) has, there is a best 10% of all possible OF values, as shown in the shaded area in the several illustrations in Figure 23.7.

23.2.1 *A Priori* Method

From the probability analysis:

$$p\left(\begin{array}{l}\text{of finding a solution that is one of the best 10\% of}\\\text{all possible solutions in at least 1 of } N \text{ random starts}\end{array}\right) = 1 - [1 - 0.1]^N \tag{23.8}$$

If you want p(best f fraction) in at least 1-of-N trials to have a certain value, say, 99%, you can calculate N. The relation is

$$p_{\text{success}} = 1 - [1 - \text{Best fraction}]^N \tag{23.9}$$

which, when solved for N

$$N = \frac{\ln(1 - \text{desired } p_{\text{success}})}{\ln(1 - \text{desired best fraction})} = \frac{\ln(1 - c)}{\ln(1 - f)} \tag{23.10}$$

Here $c = p_{\text{success}}$ is the desired probability, the confidence, of finding one of the solutions from the best fraction of all possible outcomes, f. You choose values for the desired confidence, c, and desired best fraction, f.

For example, if you want to be 99% sure that the procedure (OF application and the optimizer and convergence methodologies) will find at least one result in the best 5% of all possible results that procedure would find, then $c = 0.99$ and $f = 0.05$. Then the number of independently randomized trials is

$$N = \frac{\ln(1 - 0.99)}{\ln(1 - 0.05)} = 89.78113496... \tag{23.11}$$

Round to an integer (or alternately, round to the next higher integer).

$$N = \text{INT}\left(\frac{\ln(1 - 0.99)}{\ln(1 - 0.05)} + 0.5\right) = 90 \tag{23.12}$$

Ninety trials are required, 90 random initializations, to be able to find one of the best 5% of all possible OF values at least once, with a 99% confidence.

If you know that the optimizer has a 30% chance of finding the global and you want to be 99.9% sure of it being found, then the minimum number of optimizer trials from randomized locations is

$$N = \text{INT}\left(\frac{\ln(1-0.999)}{\ln(1-0.3)} + 0.5\right) = 20 \tag{23.13}$$

If you want the optimizer to find a best 1% of all possible solutions and you want to be 99.9% sure of it being found, then

$$N = \text{INT}\left(\frac{\ln(1-0.999)}{\ln(1-0.01)} + 0.5\right) = 688 \tag{23.14}$$

Note: Diverse probability applications result in the same type of analysis and consequently the same relation for N. These include weakest link in a chain and system reliability.

Here is how to interpret this equation: Consider M trial sets. In each trial set, randomly initialize and optimize N times using $N = \ln(1-c)/\ln(1-f)$.

Trial	
1	N random independent initializations and save best of N
2	N random independent initializations and save best of N
3	...save best of N
4	...save best of N
M	...

This experiment will mean $N \times M$ number of optimization searches. For the $c = 0.99, f = 0.05$ choices, and $N = 90$. If $M = 10{,}000$, this is 900,000 optimizer runs. The expectation is that in 99% of the M trial sets (which is 9900 of the 10,000 sets), the best-of-N independent optimizations will find one of the best possible 5% OF values.

To critically test this best-of-N relation, you need a large value for M. If $M = 1$, one set of $N = 90$ optimizer trials, then that set might not contain one of the best 5%. Then $p = 0.00$, when you expect 0.99. Alternately, that one set might contain one of the best 5%. Then $p = 1.00$, when you expect 0.99. If $M = 2$, the possible outcomes are $p = \{0.00, 0.50, 1.00\}$. Elementary binomial probability can reveal the number of sets, M, needed to see if the outcome is the expected value.

It is the same analysis that is used in analyzing probability in flipping coins. Again, the coin flipping diversion may be helpful. Flip a coin; the $p(\text{Head})$ is 50%. Flip a coin three times and count the number of heads. It will not be the expected $3 * 0.50 = 1.5$ number of heads. One cannot experimentally obtain half a head on a flip. If you flip the coin 100 times ($M = 100$ trials) you might get 47 heads, or an experimental $p(\text{Head}) = 0.47$ when the expected is $p(\text{Head}) = 0.50$. Experimentally, this is a binomial distributed outcome (H and T represent two possible outcomes). Similarly, the best-of-N success or failure is also a binominal category (the best of N is either within the best 5% of all possible solutions, or not within the best 5%). For a binomial process, the standard deviation of the experimental count of successes is $\sigma_S = \sqrt{Mpq}$ where M is the number of trials, p is the probability of success, and $q = 1-p$ is the probability of not success.

Experimentally, the probability of success, $p = $ number of successes/number of trials $= s/M$, and the standard deviation for the probability is $\sigma_p = \sigma_S/M = \sqrt{pq/M}$.

With sigma so calculated, the expected 2-sigma range for an experimental probability is $\pm 2\sqrt{pq/M} = \pm 2\sqrt{s(M-s)/M^3}$ where s is the count of the number of successes in the M trials.

If you expect p to be 5% (0.05), and you want the experimental range on p to be relatively small (± 0.005) (just one order of magnitude smaller), then you are desiring to have a value of M such that

$$2\sqrt{\frac{0.05(1-0.05)}{M}} < 0.005 \tag{23.15}$$

Solving, the value of M needs to be 7600.

That is a large number of simulations, but it is good to know that a large number is needed for testing probability outcomes. It usually requires a large number of trials to see differences with statistical confidence. A 100,000 or 1,000,000 is not unusual.

Regardless of what you specify for the best fraction of all possible OF values, in a large number of randomized optimization trials, there is always a number of results that will fall in that best fraction. If there is a single optimum, for instance, randomized trials will not end in exactly the same place, so there will be a distribution of OF values and a best fraction of that distribution. But if there is a single optimum, then all solutions are in the vicinity of the global optimum, not just the best fraction.

A problem with the previous formulation of the required number of trials to find the global in an application with multiple optima is that the formulation presumes knowledge of the probability that the vicinity of the global will be found on any random optimization trial. The probability of finding the vicinity of the global depends on the optimizer type, as well as the surface, and without that knowledge the best-of-N equation is just a reasonable estimate for the number of trials to be confident in finding the vicinity of the global. It is an *a priori*, before the evidence, estimate.

For calculating N to find the vicinity of the global, the previous analysis is predicated on the initialization range permitting the optimizer to find the global. If, for example, the optimizer is always started at the vicinity of a local optimum, it may always find the latter, not the global.

23.2.2 *A Posteriori* Method

One could observe the sequence of results of independent trials and stop when there is ample confidence that the global optimum has been found. This would be an *a posteriori* method, after the evidence.

Inspired by work of Jan Snyman, here is a Bayesian belief approach. Consider the population of OF values that would arise from millions of independent trials. If there are multiple optima, the CDF(OF) graph might look like Figure 23.8. In it, α represents the fraction of trials, the probability of finding the global optima, and the complement $(1-\alpha)$, the probability of finding the local optima. One would not see this CDF(OF) curve perfectly with only 100 trials, and one would see a limited sampling representation, but it exists. Let n represent the number of trials from randomized conditions and r represent the number of those trials that found the global (that ended in the vicinity of the global, within a desired precision).

In Figure 23.8, the first step in the CDF seems to rise vertically, but the reality is that not all OF values at the global will have exactly the same value. Figure 23.9 reveals a close-up of the CDF(OF) in the global region, showing that the seeming repeated global OF* values are not identical.

If $r = 1$, meaning that only one trial has found the global but also that the vicinity of the global has been found, where on the CDF(OF) curve might the OF value be? Assuming that on average half of future global OF values will be above and half below this one, which means that if $r = 1$, a reasonable estimate for the OF value is that it represents CDF(OF) $= \alpha/2$. This would mean that the probability of finding a better OF value is $\alpha/2$. Similarly, with r trials finding the global and assuming they are

Figure 23.8 Illustration of a CDF(OF*).

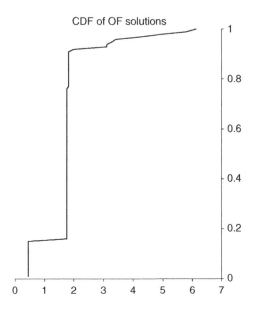

Figure 23.9 Detail of the initial portion of Figure 23.8.

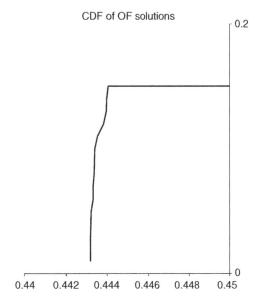

uniformly distributed along the CDF curve, then the probability of finding a better OF value is $\alpha/(r+1)$. Using OF* to represent the true global OF value and OF_n^* to represent the best of n trials, the probability that a new trial will find a better OF value than OF_n^* is $\alpha/(r+1)$. Alternately, the probability that a new trial will find an equal or worse OF value than OF_n^* is $1-\alpha/(r+1)$.

Alternately, if the vicinity of the global has not been found (if $r = 0$) after n trials, OF values from the n trials are in the $1-\alpha$ CDF region. Again, assuming that they are equiprobable at any CDF value

Table 23.2 Possible conditional probabilities.

	Probability of a new trial *finding* a better value than OF_n^*	Probability of a new trial *not finding* a better value than OF_n^*
If the truth is the vicinity of the global *has been* found	$p_{11} = \alpha/(r+1)$	$p_{12} = 1 - \alpha/(r+1)$
If the truth is the vicinity of the global *has not been* found	$p_{21} = \alpha + (1-\alpha)/(n+1)$	$p_{22} = 1 - \alpha - (1-\alpha)/(n+1)$

means that the probability of finding a better OF value is $\alpha + (1-\alpha)/(n+1)$. Alternately, the probability that a new trial will find an equal or worse OF value than OF_n^* is $1 - \alpha - (1-\alpha)/(n+1)$.

Table 23.2 represents the conditional probabilities of a new trial finding or not finding a better OF value, given that either the vicinity of the global has been found or not.

After $n = 1$ trials, the probability that the vicinity of the global will have been found is α. The belief that the global has been found is $B = \alpha$. Set $r = 1$. Then calculate the conditional probability values with those α, n, r values. On each subsequent trial, compare OF_n to OF_{n-1}^*.

If the new OF_n is better than OF_{n-1}^*, then a new best has been found. Update the OF_n^* value, also update the belief that the global has been found with a Bayesian approach:

$$B = \frac{Bp_{11}}{Bp_{11} + (1-B)p_{21}} \tag{23.16}$$

If the new OF_n is not better than OF_{n-1}^*, then

$$B = \frac{Bp_{21}}{Bp_{21} + (1-B)p_{22}} \tag{23.17}$$

If the new value is equivalent to (in the vicinity of) the prior best, then it appears that the global has been rediscovered, so, increment the r count, $r = r + 1$. Alternately, if the OF_n^* is not in the proximity of the OF_{n-1}^*, it is either a worse solution (a local optima that is not the global) or a better solution (indicating a new global has been discovered). If it indicates that a new global has been discovered, then reset $r = 1$.

As trials progress and subsequent OF values are better or worse than the former OF*, the belief that the global has been found will rise and fall. Eventually, subsequent trials will suggest that the global has been found, and B will rise toward 1. When $B = 0.99$ (99% confidence that the global has been found), stop trials. Of course, you could specify $B = 0.9999$ if you wanted to be 99.99% confident.

I find this approach to provide reasonable stopping criteria for the sequence of independent optimization trials to find the global, except for the following:

The method is predicated on the *a priori* estimate of two values; one is α, the probability of finding the global. If this were known, then it could be used in the best-of-N relation to set the number of trials. The other is the condition that is used to determine if an OF solution is in the vicinity of the best so far. Should it be based on the OF value or the ΔDV values? What threshold? Perhaps this can be related to a convergence criterion.

Additionally, this approach is predicated on the initialization range including the global, and not always misdirecting the search to a local. So, be sure that the initialization range you choose permits the global to be found.

Further, some applications have identical OF values at independent DV values. For instance, consider the objective to find points in space that minimize gravitational pull. These zero-gravity points lie between bodies. If there are three bodies, there are up to four zero-gravity points between the three bodies, all with the identical minimum gravitational pull of zero net force. If the user wishes to distinguish between equivalent solutions, then both OF and DV values need to be considered in determining whether a solution is in the vicinity of OF*.

Further, when either hard constraints or shallow valleys or coarse convergence leads to solutions that are in a wide vicinity of the global, it may be that some past OF* value that was counted in the r set has now evolved to be a large distance from the current OF* value. So rather than incrementally updating r, I sort past solutions and count the number within the vicinity of the current best.

Although I find this approach to provide reasonable stopping criteria for the sequence of independent optimization trials to find the global and not overly sensitive to a reasonable estimate of α, there are some impracticalities associated with the *a priori* estimates of both α and the criterion to determine whether an OF solution is in the proximity of OF$_n^*$. If the value of α could be known, then the best of N provides the number of trials needed, which is a simpler procedure.

23.2.3 Snyman and Fatti Criterion *A Posteriori* Method

Snyman and Fatti (1987) propose an alternate *a posteriori* method to estimate the confidence that the proximity of the global has been found, which appears to require less *a priori* information than the method previously. It is also based on a Bayesian analysis. The derivation is based on the belief that the vicinity of the global has been found and is based on n, the number of randomized trials, and r, the number of trials that resulted in an OF value in the vicinity of the best OF found so far. The value of r will be at least 1. Even if $n = 1$, the one solution is the best in the set of n. If r is a large number, this indicates that the best OF value has been repeatedly discovered. If both n and r are large, then one can be confident that the global has been found.

Their formulation is

$$p\left(f_{\text{best in }n\text{ trials}} \overset{m}{=} f^*\right) = 1 - \frac{(n+1)!(2n-r)!}{(2n+1)!(n-r)!} \tag{23.18}$$

where p is the probability that the best of the n trials represents the global, which is in the desired proximity of the global.

Although factorials are difficult to compute, a fortunate aspect of the formulation is that there are the same number of terms in the numerator and denominator, and the value can be easily calculated by multiplying the sequential ratio of terms in a FOR–NEXT loop.

```
T = 1
For i = 1 To n
   T = T * (n - r + i) / (n + 1 + i)
Next i
P = 1 - T
```

For selected n and r values, probability values from the formula are presented in Table 23.3.

Snyman recommends claiming that the global has been found and stopping trials when the probability is greater than 0.99. This would accept that when 5 trials out of 10 are in the proximity of the best of the n, the vicinity of the global has been found. One could alternately specify that the threshold for stopping trials is $P > 0.999$.

Table 23.3 Snyman and Fatti *p* values w.r.t. *r* and *n*.

	n = 1	2	5	10	20	50	100
r = 1	0.666667	0.7	0.727273	0.738095	0.743902	0.747525	0.748756
2		0.9	0.878788	0.87594	0.875235	0.875038	0.875009
3			0.954545	0.944862	0.940901	0.938794	0.938136
4			0.987013	0.977296	0.972846	0.970343	0.969539
5			0.997835	0.991486	0.987932	0.98579	0.98508
10				0.999997	0.999888	0.9997	0.999613
20					1	1	1

This approach is predicated on the initialization range including the global, and not always misdirecting the search to a local. So, be sure that the initialization range you choose permits the global to be found.

This approach is predicated on the user's definition of when two solutions are in the proximity of each other. If the user sets too large a range for the proximity, then all solutions will be counted in the *r* set. Alternately, if too small a range, then only one will be counted. I would suggest that to determine if a solution should be counted in the *r* set, (i) the proximity value should be the same as that based on the convergence criterion. However, if there are two or more DV solutions that result in the same global optimum value, then any of the several DV sets would find the optimum, but if the criterion to determine whether solutions are in the vicinity of the best of *N* is DV range, then equivalent solutions will not be counted. So I also suggest that (ii) the criterion be based on OF proximity, not DV proximity.

This approach is also predicated on there being a reasonable probability of finding the global. If, for instance, the global is a pinhole with a very small region of attraction in the DV range, then it may not have been discovered in 100 trials, but the proximity of the best of the other solutions in the set of *n* = 100 may be very high, and the formula will falsely indicate that a high probability of finding the vicinity of the global has been found. In fact, a basis of the Snyman–Fatti derivation is that α, the probability of finding the global on any one randomized trial, is larger than the probability of finding any other local optima.

In my experience, the three methods give similar results. Using the *n* and *r* values of Table 23.3, to determine the probability of finding the global, Table 23.4 represents the $p = r/n$ values.

The probability of finding the global is *f* in Equation (23.12).

Recognize, however, that the statistics of small numbers can permit a significant uncertainty on the *p* calculated. Since $\sigma_p = \sqrt{pq/n}$, the 2σ uncertainty on probabilities in the *n* = 5 column are about ± 0.35, which is very uncertain. In the *n* = 100 column, the 2σ uncertainty on probabilities is about ± 0.03.

Note that if *r* = 1 and *n* = 1, the estimate of the probability of finding the global is 100%, which is not meaningful. The ratio of *n* to *r* representing the *p*(global) is only valid for large *n*.

Table 23.5 presents the confidence of finding the global, using the inverse of the best-of-*N* formula to estimate the confidence that the global would be found in *n* trials with the probability of finding it from Table 23.4.

Table 23.4 Probability as a ratio of r to n.

	$n = 1$	2	5	10	20	50	100
$r = 1$	1	0.5	0.2	0.1	0.05	0.02	0.01
2		1	0.4	0.2	0.1	0.04	0.02
3			0.6	0.3	0.15	0.06	0.03
4			0.8	0.4	0.2	0.08	0.04
5			1	0.5	0.25	0.1	0.05
10				1	0.5	0.2	0.1
20					1	0.4	0.2

Table 23.5 p global using n and f and best-of-N formula.

	1	2	5	10	20	50	100
1	1	0.75	0.67232	0.651322	0.641514	0.63583	0.633968
2		1	0.92224	0.892626	0.878423	0.870114	0.86738
3			0.98976	0.971752	0.96124	0.954669	0.952447
4			0.99968	0.993953	0.988471	0.984534	0.98313
5			1	0.999023	0.996829	0.994846	0.994079
10				1	0.999999	0.999986	0.999973
20					1	1	1

Note: For large n, where the $p = r/n$ estimate has greater certainty, there is strong similarity in confidence values of the Snyman–Fatti *a posteriori* criterion (Table 23.3) and the best-of-N *a priori* prediction (Table 23.5).

Note: Both techniques indicate that $n \approx 100$ trials are needed to be confident that the global has been found.

So specify that the initialization range will permit the optimizer to find the global. Specify a desired confidence that you wish for the optimizer to find the global. Use prior knowledge to estimate the probability of finding the global. Alternately, specify a desired best fraction of all possible converged OF values that you desire at least one trial to find. Use the best-of-N relation to estimate the required number of trials. Once trials start, use the Snyman–Fatti formula to update the confidence that the global has been found. When it reveals the desired confidence, terminate trials.

If the global is a pinhole in a vast region that leads nearly all optimizer trials to a local optimum, then the Snyman–Fatti formula will be used to conclude that the global has been found when the local is repeatedly found. If you believe that the probability of finding the global is reasonable, then use the S–F formula. If you believe there is a small probability of finding the global, then use the *a priori* best-of-N number of trials.

23.3 Other Options

There are alternate methods to increase the probability of finding the global minimum, yet minimizing computational work.

1) Guide the DV initialization. Start in places where there were no prior starts, and don't start in places that seem to draw the search to the same, repeated optimum. If an operator is observing the resulting OF from a random start, and then another, the operator might come to realize that all starts in some particular region lead to the same optimum and use heuristic rules to explore other starting places—not random numbers. There are many approaches one could use to direct new DV choices. I would consider using Fuzzy logic (an embodiment of human linguistic rules) to permit a computer to automate a human logic. However, success with this is predicated on having a logical method to "see" patterns in past search outcomes to direct initializations in places likely to discover new features. In a high-dimension application, what would you use as a metric to quantify the optimum that an initial TS might lead to?

2) Choose 100 random starting places, as surface probes, and evaluate the OF at each TS. Then optimize, but only starting from the best of the 100. This supposes that the best of 100 surface probes will be in the vicinity of the global, and the optimizer will go to the global. Or start 1000 initialization surface probes, or 10,000, etc. How many? If f represents the fraction of the DV range that leads to the global and it is a single DV application, then to be c confident that at least one of the initial probes will be in the fraction that will lead to the global, the number of probes is $N = \ln(1-c)/\ln(1-f)$. If, however, it is a multiple DV application, then the probe that leads to the optimum must be in the combined fraction of best of all DVs. Using $f_{\text{characteristic}}$ as a representative (characteristic) fraction of the portion of each DV range that is needed to find the global, then $N = \ln(1-c)/\ln\left(1-f_{\text{ch}}^{N_{\text{DV}}}\right)$, where N_{DV} is the number of DVs, the application dimension. But that supposes that the hypervolume of DV space that represents the minimum is a hyperrectangle. In my experience it is more like a hypersphere. The volume of a hypersphere is $V = \pi^{n/2} r^n / \Gamma(1 + n/2)$. If f_{ch} represents the range fraction, then, here, the effective hypersphere radius is half the diameter, half the range, $r = f_{\text{ch}}/2$. Roughly, the coefficient $\pi^{n/2} r^n / \Gamma(1 + n/2)$ has a value that is around 4, meaning that $N = \ln(1-c)/\ln\left(1-4(f_{\text{ch}}/2)^{N_{\text{DV}}}\right)$. Using the exact value for the gamma function is not necessary because the estimate of f_{ch} would be imprecise. With reasonable values for the range fraction, this approach provides a number of initial surface probes, in which optimizing from the best fulfills the expectation of the confidence in locating the global. Unfortunately, the number is sensitive to the estimate of f_{ch}, and for a 6-DV application in which $f_{\text{ch}} = 0.2$ and $c = 0.95$, $N = 750,000$, which may be an excessive number.

3) For multiplayer algorithms start with a coefficient value that leads to higher exploration, and then as the belief increases that the vicinity of the global has been found, switch to coefficient values that lead to convergence. In leapfrogging this could be the leap-to window amplification factor and/or the window origin. I have been pleased with starting LF with a window amplitude factor of 2 (the leap-to window is twice the length of the leap-from window) and then returning it to an amplification of 1 and shifting the window origin corner ¼ from the best to the worst when there is confidence that the global has been found. In particle swarm this could be the draw to the global best and/or the magnitude of the random driver. In any case, the user must choose a logic that triggers the belief that the vicinity of the global has been found, and although such methods work (they

increase the probability of finding the global without excessive work), I am not aware of either a universal metric or basis for a decision to switch modes. I have used Bayesian logic to estimate a belief that there is no better vicinity and to switch to end-game convergence, but this is predicated on my choice of probabilities. I have used a simple counter of the number of leap-overs with no improvement and switched to end-game convergence when this is a high number. But I need to choose the number. I have used the calculation in method 2 to determine the number of leap-overs, and switched to the end-game coefficient values when n is achieved, but, again, this requires my estimate of a value for the critical of f_{ch}.

4) Swarm intelligence: Start N places simultaneously. (Actually, unless parallel computation, the computer has to place each particle, one by one, individually.) These N trial solutions are called "particles" or "individuals" or "players." Evaluate f for each individual, and do one search iteration for each. As the OF for each, use the individual's local value weighted by a penalty for being far from the best. For example, for the ith individual instead of the OF being OF_i, it would be $OF_i + \lambda[(x_i - x_b)^2 + (y_i - y_b)^2]$. This adds a movement toward the swarm's best along with the local move downhill. One iteration is a move of each individual. There are many versions of how to evaluate the OF and how to choose the next trial solution for the particles.

5) Start searching with one initial trial solution, when near a minimum. Start adding random perturbations to next guess which are intended to have the solution individual and jump out of the local hole or away from a constraint. Again, the human must provide the metrics to assess the situations that are near convergence, the search window to find some other locale, and the number of explorations.

6) Start in the DV region of historical, traditional, expected successes.

23.4 Takeaway

In all there is no guarantee of finding the global optimum. In each approach, increasing N, the number of trials from independent initializations, increases likelihood of finding one of the best few.

If you have an experience base, and know from past solutions about what the solution will be, then start near the former DV^* as the initial guess. Then, perhaps, only that one run of the optimizer is needed. But if it is a new application and you are not certain of the location of the optimum or the surface features, then I think that you should start with about $N = 100$ trials, and as you see the character of the surface and $CDF(OF^*)$ being revealed, use that information to adjust choices of f and c, hence N.

Specify that the initialization range will permit the optimizer to find the global. Specify a desired confidence that you wish for the optimizer to find the global. Use prior knowledge to estimate the probability of finding the global. Alternately, specify a desired best fraction of all possible converged OF values that you desire at least one trial to find. Use the best-of-N relation to estimate the required number of trials. Once trials start, use the Snyman–Fatti formula to update the confidence that the global has been found. When it reveals the desired confidence, terminate trials.

If the global is a pinhole in a vast region that leads nearly all optimizer trials to a local optimum, then the Snyman–Fatti formula will believe that the global has been found when the local is repeatedly found. If you believe that the probability of finding the global is reasonable, then use the S–F formula. However, if you believe there is a small probability of finding the global, then use the best-of-N number of trials.

23.5 Exercises

1 How many times must an optimization be started to find the global optimum?

2 Consider that you are automating an optimization, have no knowledge of the surface, but suspect that there are local optima, so you are going to have the computer randomize the initial trial solution. Show how to calculate the number of optimization runs from randomized initial DVs that is required to be 99% confident that the computer will find one of the best possible 1% of all possible solutions.

3 State the meaning of N, f, and c in the formula $N = \text{int}\left(\ln(1-c)/\ln(1-f) + 0.5\right)$. You may use sketches to illustrate concepts.

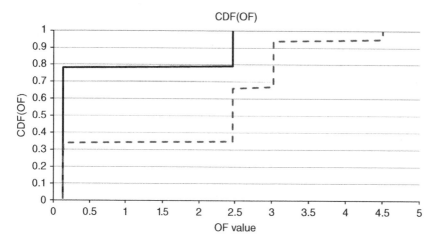

Figure 23.10 Illustration for Exercise 4.

4 Figure 23.10 is a graph of the CDF(OF) for two separate optimizers on the same problem.

A In total, how many minima were found?

B The optimizer represented by the solid line has an ANOFE of 153. Show how to calculate the PNOFE to be 99% confident that you did enough independent starts to find the global minimum OF value of about 0.15.

C Repeat B, but for the optimizer represented by the dashed line, which has ANOFE = 76.

D Which optimizer is preferred? Why?

E What is the probability that the optimizer represented by the dashed line will find the optimum at an OF value of about 3.

5 Demonstrate that best-of-N works as claimed. Find or create a function that has more than one convergence point. Choose an optimizer. Run the optimizer many times (on the order of 10,000) from random initializations. This will give you 10,000 or so results, some of which may be "No

Convergence." Choose a value of f and c. (i) Use the best-of-N equation to determine N. (ii) Determine the OF value, OF_f, which represents the value for which only f (fraction) have equal or lower values. (iii) Take your optimizer results, sequentially, in sets of N. If you used 10,000 trials and N is 25, then there will be $10,000/25 = 400$ sets of N. Count the number of sets that found at least one OF value equal or lower than OF_f. The fraction of sets with at least one OF value below OF_f should be c. But this is a stochastic process, so the experimental number will not be exactly c. For example, flip a coin 100 times. You expect 50% will be heads, but if you get 53% heads, you will accept the ideal 50% calculation is accurate. (iv) Do the OF-less-than-OF_f count for several f and c values to show that your result was not a happenstance of a fortuitous choice. You only need to generate the 10,000 optimization trials one time, on one function, for one optimizer. Use your programming skills to automate the generation of OF values from random starts and the assessment of results to validate the equation. But once the experiment is automated, it is easy to change function or optimizer and obtain results on other combinations.

6 Compare the several *a priori* and *a posteriori* methods to determine the adequate number of trials needed to be confident in finding the global. Consider utility, functionality, user ease, computational burden, robustness to surface aberrations (equivalent OF, underspecified, discontinuities, etc.), sensitivity to human choices, etc. Design experiments to gather data to support your conclusions.

24

Stochastic Objective Functions

24.1 Introduction

A deterministic function returns the exact same value each time it is given the same inputs. What is the value of 3 times 4 (in base 10 arithmetic)? Whether last year or next year, whether in Barbados or on the Moon, whether by human or by computer, the answer is 12. By contrast a stochastic function does not return the same value when given supposedly identical conditions. If I roll this cubical die, what value will appear? Even if I do it, in one setting in the same location, trying my best to be repeatable, the answer is a 1 or a 2 or a ... or a 6. This stochastic function returns one of 6 values with an equal probability. It does not consistently return the same value.

How does this relate to optimization? Following are several example situations:

Example 1   Consider that a fast-food business wants to determine the optimum amount of salt on fried potato strings. This is a single DV application, and the OF will be customer rating. Each batch of fries is hand-salted by counting the shakes of the salt canister. Since each batch of fries does not have exactly the same number or orientation of fries and each shake of salt does not deliver exactly the same amount of salt, there is batch-to-batch variation in salt per fry. There is even greater variation within each sample of fries from the batch given to any customer. The data will be obtained experimentally by asking customers to rate the fries—too salty, just right, and not salty enough. In addition to the variation in salt content from the preparation, there are differences in customer preferences. Accordingly, even if the preparation recipe is unchanged, the results will have variation. The DV is the number of shakes per batch, and the objective is to maximize the OF, the portion of customers that say "just right."

Here is some data from a thought experiment. One day, 1294 customers bought fries, which required 20 batches, each of which are salted with 4 shakes, and the customer response was 9% (reporting too salty), 32% (just right), and 53% (needed more salt). In general the data would indicate more salt should be added. The next day the business experimented with 5 shakes per batch, and the survey of 985 customers was 16, 45, and 39%. Since opinion is moving in the right direction, the next day they used 6 shakes per batch, and the 1533 customer response was 23, 44, and 34%. This has dropped the "just right" response a bit; but is the response the absolute truth or just an uncertain value, a perturbation about the truth due to experimental vagaries associated with the demographics of the customers, what else they ordered, and the random influences in the food preparation?

Engineering Optimization: Applications, Methods, and Analysis, First Edition. R. Russell Rinehart.
© 2018 R. Russell Rinehart. Published 2018 by John Wiley & Sons Ltd.
Companion website: www.wiley.com/go/rhinehart/engineeringoptimization

A deterministic view of the response would conclude that since the "just right" portion has decreased, one should return to 5 shakes per batch. So, they did, and the response of 1157 customers was 18, 39, and 43%, which is worse than 6 shakes, not better. A human observer might conclude that 5 shakes per batch is about right. But what would an algorithm do? The last DV = 5 was worse than the DV = 6, so try DV = 7 shakes?

Stochastic responses confuse an optimizer by masking the true response with random perturbations.

Example 2   Consider a computer-calculated simulation, which, in contrast to experimental outcomes, is usually deterministic; but in assessing a business venture, the future is quite uncertain, and an investigator might use a Monte Carlo approach to select a possible realization. Uncertainty rises with projecting costs (e.g., electricity, labor, raw material), sales (dependent on competition and demand factors), and government actions (taxes, regulations, constraints, embargos). There is much uncertainty in these givens, and they change over time in the forecast life of the venture. One way to assess profitability is to sample from possible values of each uncontrolled impact for each year in the business forecast and determine the business metric for that particular realization. It is not the truth, but just one possibility. The simulation might be rerun for 1000 realizations, giving 1000 possible values of the profitability metric. The optimization objective may be to find the DV values that minimize a 95% worst-case outcome over 1000 realizations. This OF value is not deterministic. Another set of 1000 realizations from nominally identical conditions will return a different 95-percentile value. Following a fortuitous best set of OF values, the DV trend can lead the optimization into a generally undesired region. Perhaps it is possibly a good decision, but only under a rare confluence of external influences.

Example 3   A similar technique in system reliability simulates the life of a system when subject to realizations of possible events of its use and component failure thresholds. In designing a reservoir, bridge, or building, as well as traffic light sequence, number of elevators needed, and many similar situations, the designers often test the outcomes with Monte Carlo simulations of event realizations over an extended period. Applications for optimizing control of stochastic processes also include systems with varying time delays, defensive trajectory policy for the interception of incoming missiles, subway timetable optimization, triggers for financial action to regulate an economy, controller tuning for processes with uncertain parameter values, and robust fault detection.

Example 4   Consider the design of a road. A DV could be the number of parallel lanes in a road. A deterministic OF could be the cost of the road. The more lanes, the more paving; the more land, the more cost. If you repeat the calculation tomorrow, with the same road plan, and landowners, you get the same OF value, the same cost of the road.

A stochastic OF might include the cost of accidents on a road, and the number and severity of accidents will depend on external variables such as traffic intensity, driver fitness, and weather. A 10-lane road would provide plenty of space for fast cars to pass slow cars. By contrast, a two-lane road (one lane in each direction) will have some people passing others by crossing over into the oncoming lane. You could simulate traffic patterns (slow cars and fast cars), human patience (willing to go slow, wanting to pass immediately), and weather conditions (wet, icy, or dry roads) over a year and count the number of cars that will have near misses with oncoming traffic. Depending on the random number sequence, the number might be 80 near misses. But if you repeat the simulation, it might return 90 or 60 near misses. Each run of the simulator, each replicate trial, provides a different OF value.

Example 5 Here is a simple dice game: Pay $100 to play, and get $400 if you win. Winning is getting a 1 on the dice in N rolls. What does N have to be for a player and the house to break even? Here the chance of winning, p(at least 1 in N), times the prize for winning, $400, would equal the cost of playing, $100. The probabilities are as follows:

$$p(\text{not a 1 in } N \text{ rolls of a die}) = \left(1 - \frac{1}{6}\right)^N \tag{24.1}$$

$$p(\text{at least 1 in } N \text{ rolls}) = 1 - \left(1 - \frac{1}{6}\right)^N \tag{24.2}$$

The challenge is to determine the value of N such that $400\left[1 - (1 - 1/6)^N\right] = 100$. This root-finding exercise could be solved analytically. However, here, use optimization to find the "root." Square the difference and seek a minimum:

$$\min_{\{N\}} J = \left(100 - 400\left[1 - \left(1 - \frac{1}{6}\right)^N\right]\right)^2 \tag{24.3}$$

Although the game is stochastic, this probability of an average outcome creates a deterministic OF. A particular value of N always returns the same J value.

However, you could do a Monte Carlo simulation as illustrated by this pseudocode:

```
Set COUNT = 0 (number of wins)
Choose MTRIALS (you choose the number, M, of trials of N dice rolls each)
For each of MTRIALS
Set GAME = 'LOSE'
For each of NROLLS (N is the DV)
            Get a random number, r (uniformly distributed between 0 and 1)
            If r<1/6 then set GAME='WIN'
      Next NROLL
If GAME='WIN' then COUNT=COUNT + 1
Next MTRIAL
J = (100-400*COUNT/MTRIALS)^2
```

Increasing the value of M leads to less variability in J, but does not eliminate variability.

For such a simple situation, the analytical approach is easier to implement, and more precise. However, for many probabilistic problems, the ideal analytical math is impossible, intractable, etc. For instance, one could add complexity to the game with a rule that if successive roles of a 1 happen, then both 1s are negated. Or the added rule could be if you lose but decide to play again, the prize value raises by $100. Due to conditional probabilities or multiple stochastic variables, Monte Carlo simulation of outcomes is often preferred. A noisy surface creates problems for both the optimization search algorithms and in deciding convergence criteria.

Returning to the introductory discussion, stochastic OFs confound an optimizer search for the best DV value. There are several issues: (i) Near the minimum the stochastic variability might be larger than the DV impact, and the superficial trend might make the optimizer move the DV in the wrong direction. Additionally, (ii) optimizers that seek to use gradient information to direct a search may be wholly confounded by the noise impact on the values. Further, (iii) optimizers that seek a minimum (or maximum) may follow a chance path of fortuitous best points into a DV region that is generally undesired. Finally, (iv) the stopping criteria must not require a tolerance or precision that is greater than the replicate variability of the OF.

To include uncertainty in the OF, first develop the nominal deterministic model. Next, identify all of the "givens" that might have uncertain values. Then, estimate the uncertain range and the distribution pdf for each of those values (see Chapters 25 and 26 for some guidance). Finally, at each call of the function, randomly sample each of the givens to obtain a realization, a particular possible value; and use this value to calculate the OF value. This is method 3 in Section 20.8.2. Also consider methods 4 and 5 to autocorrelate or cross-correlate perturbations.

In general, stochastic responses confound an optimizer, and the questions are "How can an optimizer take rational decisions on stochastic inputs?" and "What should be used as the criterion for convergence?" One answer is to "Resample the best, and use the worst of the samplings to characterize the OF" and "Claim convergence when there is no statistical improvement of the OF w.r.t. iteration."

Now that computers are fast enough and have become a common personal tool for engineers and analysts, it appears that stochastic simulations (Monte Carlo analysis) are gaining traction in the engineering and business community. However, classic optimization techniques were developed for deterministic objective functions, and they get confounded by stochastic response. Accordingly, the human is often left to intuitively guide the DV evolution to optimize the stochastic function. So, procedures that can handle stochastic responses will permit automation of the optimization. This chapter presents one approach that I have found effective and practicable.

24.2 Method Summary for Optimizing Stochastic Functions

This section is based on the publication by Rhinehart, R. R., "Convergence Criterion in Optimization of Stochastic Processes," Computers & Chemical Engineering, Vol. 68, 2014, pp. 1–6:

1) Use a multiplayer, direct-search optimizer to search for the optimum such as LF.
2) Identify the best player, and replicate the trial solution values. Take the worst of replicate evaluations as the representative value.
3) If the replicated values reveal it is no longer the best player, use the remaining best player (formerly second best) to guide the search.
4) Use a steady-state identifier on the iteration sequence of the OF value of the worst player to determine convergence.

24.2.1 Step 1: Replicate the Apparent Best Player

Leapfrogging (LF) is a multiplayer, direct-search approach. Trial solutions (players) are randomly initialized through decision variable (DV) space. One player will have the worst objective function (OF) value, while another, the best. The worst player leaps over the best and "lands" at a random spot in the reflected DV space on the other side of the best. If the new leap-over position is infeasible, then the same player leaps from the infeasible spot back over the best player and into a new random location within the new reflected DV space. Normally, for deterministic functions, whether the leap-to location is better or worse, when the new location is feasible, the move is complete. Normally, an iteration is defined as the number of leap-overs the same as the DV dimension. This means that if leap-to positions are feasible, there are N function evaluations per iteration.

However, that definition of an iteration is arbitrary. An iteration in the CHD method has one function evaluation per dimension, but HJ has $1 + 1.5N$ function evaluations per iteration, ISD (with central difference) has $1 + 2N$, and CSLS and SQ and Newton types have many more.

If a leap-to position is up on a cliff, or otherwise worse than the leap from position, then that player will be the next to leap. As will be seen, if it jumps on and off of a cliff, this will make the local variation seem greater than the stochastic properties at the local optimum and confound the steady-state identification. Accordingly, for stochastic functions if the leap-to position is either worse or infeasible, then immediately leap it back. If it jumped up on a cliff, then the leap back will be an improvement. Call an iteration when there are N leap-overs with improvement.

In LF, the player with the worst OF value leaps over the player with the best, but the best player might have had a fortuitous realization. Rather than truly being best, it might be a phantom appearance. One would not want that to lead the optimization to move more players into an undesired DV region. To accommodate the stochastic OF attribute, here, the best player is reevaluated at each leap-over, and the worst of replicate values is taken as its representative OF value.

How many replicates? If you want to be c confident that the number of replicates will have found one of the worst fraction, f, of OF realization values, then

$$N = \frac{\ln(1-c)}{\ln(1-f)} \tag{24.4}$$

For example, if you desire to be 95% confident that there are enough replicates to find 1 of the 75 percentile worst values (1 in the 25% worst fraction), then

$$N = \frac{\ln(1-0.95)}{\ln(1-0.25)} = 10 \tag{24.5}$$

This would indicate the current OF value and 9 additional samples. However, if the best remains the best, 9 additional samples at the next leap-over would represent a total of 19, and 9 more on the next leap-over would represent 28. The worst of 28 is roughly the 90 percentile worst. I find that 1–3 replicates per leap-over is adequate. If the player is a phantom best, then after a few leap-overs, it will have enough replicates to acquire a value representing a local worst outcome.

The greater the number of replicates, the lower the chance that a converged DV* value will represent a bad solution masquerading as a good one. Choose c and f values to represent the situation. If it is a normal business economic analysis, then $c = 0.95$ could be an appropriate number; however, where there is less tolerance for risk, perhaps $c = 0.999$ is appropriate.

If reevaluation of what appears to have been the best player reveals that it is not the best, then the second best becomes the lead player and basis for the leap-over, redirecting the search. This prevents end-game convergence of the cluster in DV space and motivates the novel approach demonstrated here—stop when there is no improvement in the stochastic response with respect to iteration. Stop at steady state.

24.2.2 Step 2: Steady-State Detection

Normally, in optimization of deterministic functions, the best trial solution will progressively improve with each iteration, asymptotically approaching its optimum. However, with replicating stochastic functions, at the optimum, the value of the best player will generate what appears to be a noisy value at a steady-state average. As the optimizer moves toward the optimum on a stochastic surface, this noisy value will approach its noisy steady state (SS).

If one player is initialized in the vicinity of the global, then it may remain in the best spot, while others are gathering, and observing the OF response of the best player might identify SS prior to all players converging. However, if the worst OF is observed, then when it settles to SS, all players will be in the vicinity of the optimum. Accordingly the method is to observe the OF of the worst player and claim SS when it settles to a noisy SS w.r.t. iteration.

Appendix D provides a method to detect steady state in a signal. There are many methods to determine steady state (SS) in a signal. The method demonstrated here is based on a probable steady-state detection algorithm, chosen for simplicity and effectiveness. The method first calculates a filtered value, X_f, of the process measurements, and then the variance in the data is measured by two methods. Deviation d_2 is the difference between measurement and the filtered trend, while deviation d_1 is the difference between sequential data measurements.

If the process is at SS, X_f is almost the middle of the data. Then a process variance, σ^2, estimated by d_2 will ideally be equal to σ^2 estimated by d_1, and the ratio of the variances, $r = \sigma^2_{d_2}/\sigma^2_{d_1}$, will be approximately equal to unity, $r \cong 1$. Alternately, if the process is in a transient state, then X_f lags behind the data trend, the variance as measured by d_2 will be much larger than the variance as estimated by d_1, and ratio will be much greater than unity, $r \gg 1$. Being a ratio of variances, the statistic is scaled by the inherent noise level in the data and is dimensionless. Critical values for the ratio to confidently trigger the process being at steady state with independent and identically distributed variation (white noise) were developed. Filter coefficient values of 0.1 and an r-critical value of 0.85 provide agreement with human perception to claim SS.

The method presumes no autocorrelation in the "time series" (iteration sequence) of the data. This is ensured by observing the worst player position after each iteration, not the best. The best player makes sequential moves toward the optimum or remains stationary for several iterations, creating autocorrelation in the sequential OF values. By contrast, the worst player is from an iteration-to-iteration, new randomized surface sampling. As the players converge on the optimum, the worst player moves from the high OF values to the low ones. With iteration, the worst OF value progressively drops until, at the minimum, its stationary value represents independent stochastic OF values from the proximity of the optimum.

24.3 What Value to Report?

Because the surface is stochastic, the converged solution is not unique. Multiple trials that find the vicinity of the global optimum will stop at random locations in the vicinity. The best of those results is not truly the best, but just the fortuitous outcome of the confluence of stochastic events realized at the last sampling. Accordingly, the best of N trials is not the DV* "truth," but simply a representative value. Reporting only the best of N trials as the optimum will convey the appearance of a single point and high precision, which misrepresents the situation. An alternate could be to report the average of all DV* values that ended in the vicinity of the global. This provides a better location of DV*, but again a single-point value does not convey the uncertainty of range of possible DV* values. The average reported with some measure of its uncertainty (e.g., DV* range or DV* standard deviation) would better represent the vicinity of equivalent solutions.

However, the average might not be a feasible solution. If the DV must be an integer, an average of 3.14 does not make sense. If the DV* solutions surround a convex infeasible solution, then the average may be within the constraint.

Further, if one reports the average outcome of the trials that find the vicinity of the global, one first has to segregate trials with some indication of which subset qualifies to represent the global.

Accordingly, be sure to convey the criteria you use to segregate trials, the range of DV* values that represent the uncertainty associated with identifying the global, the number of replicates, and your understanding of how constraints might affect the interpretation of the representative value.

24.4 Application Examples

Several relatively simple examples illustrate the issues to reveal the applicability of the method for a variety of surface features. In each the DVs and the OF are scaled to a common 0–10 range. These are all 2-DV applications; however, in my testing the method is effective in 1-D or N-D applications. First, there is a brief description of each application and then the results.

24.4.1 GMC Control of Hot and Cold Mixing

Function 36 in the 2-D examples code presents this application. Hot and cold fluids are mixed in line, and the objective is to determine the controller outputs (ideally these are the control valve positions), o_1 and o_2, to produce the desired mixed temperature and total flow rate.

With no uncertainty on model values, the contour of the 2-D search for o_1 and o_2 appears as Figure 24.1a. However, with a relative uncertainty on each model coefficient (Gaussian with a 5% standard deviation), in Figure 24.1b the contours of one realization are obviously irregular, and each realization will generate a different irregular set of contours. The corresponding 3-D plot of OF w.r.t. DVs (Figure 24.1c) reveals one realization of the irregular surface, and each realization will distribute the spikes and wrinkles differently. In general, Figure 24.1c shows that the minimum is somewhere along a curved path from the front right to the back left.

24.4.2 MBC of Hot and Cold Mixing

This is similar to the aforementioned physical process; however, it is a single controlled variable application. The objective is to determine the optimum sequence of future control actions to make the temperature best match a linear reference trajectory to the set point. Control is subject to rate of change constraints.

24.4.3 Batch Reaction Management

This simulates an algae farm and offered as Function 63 in the 2-D Optimization Examples program. Algae is seeded in a wastewater pond, permitted to grow in response to sunlight and nutrients, and after a time the water is processed to harvest the algae. Profit depends on the mass of algae harvested. The longer the growth period, the more algae in the pond; but the mass asymptotically approaches a maximum as limits of nutrients make the algal death rate approach growth rate. Waiting a long time to

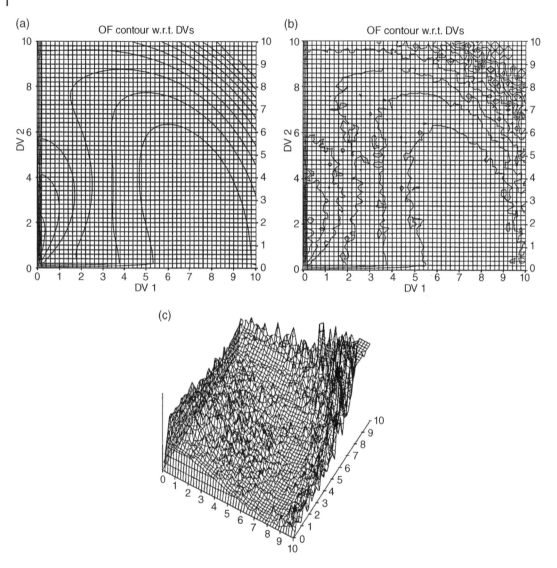

Figure 24.1 (a) Contour plot of the deterministic background. (b) One realization of the stochastic contours. (c) The surface in 3-D. *Source:* Rhinehart (2014). Reproduced with permission of Elsevier.

harvest 100% of what is possible reduces the number of batches that can be generated per year. It might be better to wait until the batch is only 80% developed and to be able to harvest twice as many.

Additionally, the deeper the pond, the more volume there is to grow algae and, consequently, mass that can be harvested. But as the pond gets deeper, less sunlight and CO_2 will diffuse to the lower levels. Too deep a pond leads to the cost of processing excess water with little extra product to be harvested.

The objective is to determine the batch growth time and pond depth to maximize annual profit. Figure 24.2 illustrates one realization of the surface. Contrasting Figure 24.1c, in which the smooth

Figure 24.2 Algae farm OF surface. One realization. *Source:* Rhinehart (2014). Reproduced with permission of Elsevier.

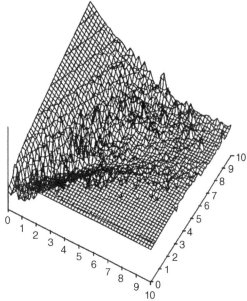

area is in the vicinity of the optimum, in Figure 24.2 the fluctuations are most severe in the region of the optimum.

24.4.4 Reservoir and Stochastic Boot Print

These two simulators represent a water reservoir design problem and are Functions 18 and 20 in the 2-D Optimization Examples program. The objectives of a reservoir are to trap excessive rain water to prevent downstream floods, to release water downstream to compensate for upstream droughts, and to provide water for human recreational and security needs. The design objectives also include minimizing the cost of the dam and land. The taller it is, the more it costs to build; but with its greater capacity, the probability of flood or drought impact will be lower, and the recreational and security features will be better. The fuller it is kept, the less it can absorb floods, but the drought, security, or recreational performance will be better. On the other hand, if kept nearly empty, it can mitigate any flood, but cannot provide recreation or drought protection. The OF represents the reservoir cost plus the economic risk (probability of an event times the cost of the event). The lower left axis in Figure 24.3 (one realization of Function 20) represents the dam height, while the lower right axis is the set point portion of full.

The OF value in Figure 24.3 depends on a probability of the flood or drought event, and each realization of the contour will yield a slightly different appearance. Note that starting in the middle of the DV space on the planar portion, a downhill optimizer will progressively move toward the far side of the illustration to a less costly reservoir, but into the region with the spikes. In that region for a small reservoir, or one that is kept too full or too empty, there is a probability of encountering a costly flood or drought event that the reservoir cannot mitigate. Moving in the downhill direction the optimizer may, or may not, encounter a costly event in the Monte Carlo simulation. If it does not, it continues to place

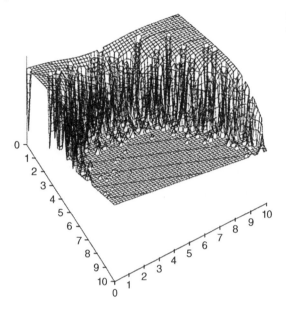

Figure 24.3 Reservoir. One realization. *Source:* Rhinehart (2014). Reproduced with permission of Elsevier.

trial solutions into a region with a high probability of a disastrous event and continues into the bad region as the vagaries of probability generate fortuitous appearing OF values.

Stochastic boot print, Function 18, is a surrogate for the reservoir, a simplified set of equations that computes faster. It's response is similar.

24.4.5 Optimization Results

Figure 24.4a shows the results of a Levenberg–Marquardt search on the GMC control problem with convergence based on sequential changes in the DV, a classical measure. Trial solutions were initialized at random locations within the DV space, 100 times, and converged at 100 wrong places. The pattern is independent of the threshold on the change in DV or whether convergence is based on a change in OF value. The results are similar when other single-point algorithms are used, even direct-search optimizers such as Hooke–Jeeves. These algorithms get stuck on a local fortuitous good spot and cannot find a way out of the apparent, phantom minimum. In contrast, Figure 24.4b reveals the results of a particle swarm optimizer search starting with 20 particles, with an end-game perching attribute to converge. The broad surface exploration and particle attraction toward the best leads the swarm to the vicinity of the global minimum. But, in that vicinity, the swarm converges on a fortuitous best, resulting in a range of phantom solutions.

The range of converged solutions is relatively insensitive to several orders of magnitude on the convergence criterion precision. In all figures the threshold value is 0.01 as an rms change on the DVs; however, a four-order-of-magnitude range of 0.001–1 was explored. By contrast, the number of function evaluations to converge is strongly dependent on the convergence threshold. This is an issue for a user, who must specify the threshold values for the convergence criterion, but cannot have rational, *a priori* criteria for relating criterion thresholds to results. In the 100 trials, PSO required 2180 ANOFE (average number of function evaluations per trial) to converge.

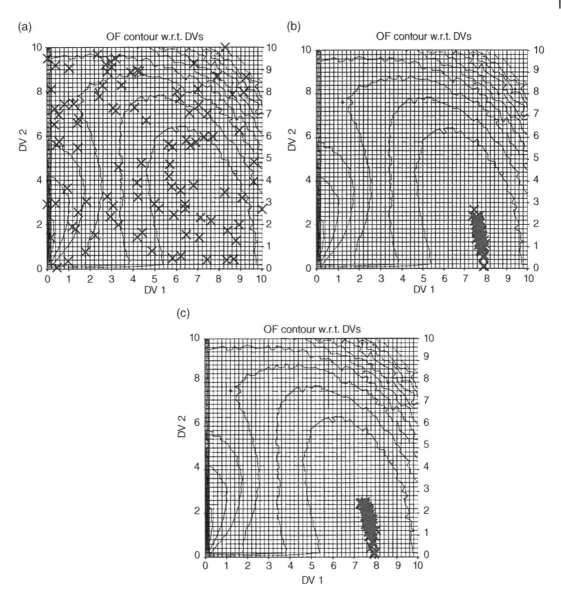

Figure 24.4 Conventional optimization results on the GMC problem: (a) Levenberg–Marquardt, (b) particle swarm, and (c) leapfrogging. *Source:* Rhinehart (2014). Reproduced with permission of Elsevier.

Figure 24.4c reveals similar results for LF. For the same convergence criterion, the ending range of solutions is nearly identical to PSO, but LF averaged 1150 ANOFE, therefore the computational work is significantly lower.

In contrast, Figure 24.5 illustrates the impact of the method presented here. It is the same problem as that in Figure 24.4c; however, the OF value of the best player is based on the worst of the prior and one replicate value, and convergence is based on the automated, scale-independent recognition of steady

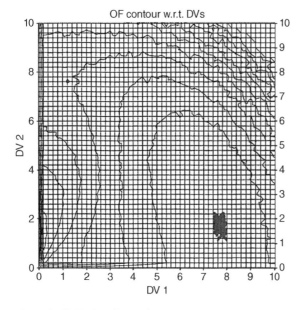

Figure 24.5 Method for stochastic optimization. *Source:* Rhinehart (2014). Reproduced with permission of Elsevier.

state. The replicates reduce the impact of fortuitous good spots from continuing to act as a draw, which significantly reduces the range of solutions nearer to the global precision. One might expect that replicates double the optimizer effort, but because of the SS convergence criterion, evaluations stop when continued work is substantially ineffective. The ANOFE for this set of 100 trials is 1090 (about 10% fewer), and the converged solutions are visibly more precise.

These illustrations motivate a choice of dual criteria for evaluating the method: (i) the range of converged solutions indicating precision and (ii) ANOFE indicating data generation cost.

Table 24.1 summarizes findings for each of the stochastic functions described previously, comparing LF with a conventional convergence criterion of DV rms less than 0.01 and LF with a replicate

Table 24.1 Experimental comparison—rms values of cluster scaled DV range/average number of function evaluations to converge.

	Function				
Optimizer	Hot and cold mixing GMC	Hot and cold mixing MPC	Stochastic boot print	Reservoir	Algae farm
LF with rms DV convergence based on 0.1% of DV range	0.8/1450	0.8/1300	1.8/930	2.2/610	1.2/1200
LF with replicate and SSID method	0.4/910	0.4/700	1.2/520	0.8/515	1.2/430

Source: Rhinehart (2014). Reproduced with permission of Elsevier.

evaluation and the SS detection on the worst player value. The table entries are the precision of the cluster of converged solutions (as indicated by rms value of the scaled DV range) and the work required to obtain the solution (as measured by the average number of function evaluations to converge). The values are based on 100 trials, rounded to reflect run-to-run variability. For similar converged cluster size (DV rms), the proposed convergence criteria finds the optimum with nearly half the ANOFE. Or for similar ANOFE the proposed method provides greater precision on the DV optimum.

24.5 Takeaway

Whether experimental responses or Monte Carlo simulations, stochastic objective functions are broadly applicable and include uncertainty in an optimization application. They are important in forecasting economics, reliability, etc.; but they confound classic optimization algorithms and convergence criteria. Using LF with replicates and SSID on the worst player, OF value locates the vicinity of the global with precision and/or fewer function evaluations.

24.6 Exercises

1 Explain what you would use as convergence criterion when you are optimizing an active process as it "runs." Describe such a process (economic process, chemical process, natural process, human process). There are usually several DVs and several OFs that are combined into one weighted-sum OF. For example, if it is the national economy, DVs may be the rates for prime interest and federal spending and OFs may be unemployment and gross national product. A natural process example is wildlife, in which population is controlled by zoning for land use and hunting and fishing licenses. The processes have a stochastic, not deterministic, response.

2 If OF values come from either physical experiments or a stochastic simulator, then experimental variability adds noise (uncertainty, perturbation, error, variation) to the OF value. Consider both direct and gradient-based search algorithms, and discuss how noise on the OF value affects the determination of the next trial solution in (i) the search stages that move toward the optimum and (ii) the "end-game" stage very close to the optimum.

3 The problem with a stochastic function (such as Function 18, Reservoir) is that optimizers seek the best. If the best represents a low-cost small reservoir, in a fortuitous simulation year with neither flood nor drought, the fortuitously good trial solution remains the best, the basis and an optimizer seeks a better solution from there. When it finds a smaller reservoir size in another fortuitous simulation year, the trial solution moves to a smaller lower-cost reservoir. But repeated simulations could reveal that fortuitous, one-time best trial solution was actually frequently disastrously expensive. (i) Run several trials of the Leapfrogging optimizer on Function 18, with a convergence criterion of 0.2 on the DVs. (ii) Report on your results. You should see the converged solution penetrate into the region with a high probability of a flood or drought disaster. (iii) Run several trials of the Leapfrogging with replicate optimizer, with the same convergence criterion, and five replicates. (iv) Report results. You should see that not only does the retesting of the best player and accepting the worst value of the replicates keep the converged solution in the disaster-free region, but also the NOFE is increased.

4 I have claimed that the steady-state identification method is a good approach to detecting convergence in optimization of a stochastic function. "Good" means that it works reasonably well, is scale independent, is user convenient, and is grounded in fundamentals. What do you think? Provide evidence to support your conclusion. What makes one method better than another? If the SSID approach is universal (if it is scale independent, does not require a person to choose threshold values, and does not require adding apples and oranges), then that makes it better than the other approach. If the NOFE is lower for one method than another while giving the same optimum DV value, then the one method is better. How will you define better? What will you measure to quantify better and to defend your conclusion?

5 Function 52 is a single-decision variable optimization. The DV is the number of rolls of a six-sided die. You pay for each roll, so there is a penalty directly related to the DV, N. If you get two or more 5s, you win. But if a 6 follows a 5, then the 5 does not count, and regardless of the number of 5s, if there are three 3s, you lose. Add a rule to the game and implement it in the 2-D Examples function. Perhaps, if there are two or more 1s, then the 3 rule is discounted. Or if there are four or more 5s, the reward is doubled. (i) Explain your new rule. (ii) Show the code. (iii) Show the surface. (iv) Compare several optimizers, and include LF with replicates (Optimizer #16) and the steady-state convergence criterion (Convergence #13).

6 The file 2-D Optimization Examples includes several stochastic functions. Choose one and explore the results from several optimizers and several convergence criteria.

26

Optimization of Probable Outcomes and Distribution Characteristics

26.1 Introduction

Normally, optimization seeks to minimize or maximize a deterministic single value. However, many situations deal with probability of events, which are the result of uncertainty or environmental vagaries. The objective is to minimize some statistic related to the process, procedure, or product. The statistic could be variance of a quality metric, quantity of off-spec material that results from manufacturing variability, an undesired event probability, system reliability, economic risk, etc. These objectives could be to minimize the 95% worst outcome, maximize the 99% minimum outcome, minimize variance, etc.

Sometimes the probability distribution associated with the application can be analytically obtained, providing a deterministic value for the objective, such as variance, or the 99% extreme case. However, more often, the outcome is a stochastic value.

Whether deterministic or stochastic, the topic of this chapter is about optimizing aspects of a probability distribution associated with an objective, which is in contrast to optimizing the nominal value.

I think that examples can best reveal the concept.

Example 1 How many dice rolls are needed to maximize the chance of getting at least 2 fives? (If you rolled 10 times and got 3 fives, then it is a success.) The more rolls, the greater the chance; so the answer is to maximize the chance of getting at least 2 fives take infinite rolls. But we'll penalize excessive rolls. Since minimizing the negative is the same as maximizing, the unweighted primitive statement is

$$\min_{\{N\}} J = -p(k \geq 2 \text{ fives in } N \text{ rolls}) + N \tag{26.1}$$

Of course, that is dimensionally incorrect and does not acknowledge the relative importance of the number of rolls with the probability. With EC factors of $N = 2$ (or more) being excessive and $p = 0.1$ (or less) having the same level of concern:

$$\min_{\{N\}} J = \frac{-p(k \geq 2 \text{ fives in } N \text{ rolls})}{0.1} + \frac{N}{2} \tag{26.2}$$

Engineering Optimization: Applications, Methods, and Analysis, First Edition. R. Russell Rhinehart.
© 2018 R. Russell Rhinehart. Published 2018 by John Wiley & Sons Ltd.
Companion website: www.wiley.com/go/rhinehart/engineeringoptimization

Multiplying the OF by a positive coefficient (a strictly positive monotonic transformation) does not change the solution. So EC factors are equivalent to Lagrange-type weighting factors:

$$\min_{\{N\}} J = -p(k \geq 2 \text{ fives in } N \text{ rolls}) + 0.05N \tag{26.3}$$

We need a model for how N affects the probability of getting at least 2 fives. The probability can be segmented into several terms, and then the binomial distribution can be used to determine how p changes with N:

$$p(k \geq 2|N) = p(k = 2|N, \text{OR} k = 3|N, \text{OR} k = 4|N, ..., \text{OR} k = N-1|N, \text{OR} k = N|N) \tag{26.4}$$

Using the OR conjunction in probability,

$$p(k \geq 2|N) = p(k = 2|N) + p(k = 3|N) + p(k = 4|N) + ... + p(k = N-1|N) + p(k = N|N) \tag{26.5}$$

From the binomial model, $p(k|N) = (N!/k!(N-k)!)p^k(1-p)^{N-k}$.

The probability of a success (getting a five on any one roll) $p = 1/6$, and of not a success $= q = 5/6$. So,

$$p(k \geq 2|N) = \sum_{k=2}^{N} \frac{N!}{k!(N-k)!} \left(\frac{1}{6}\right)^k \left(1 - \frac{1}{6}\right)^{N-k} \tag{26.6}$$

Then the executable optimization statement is

$$\min_{\{N\}} J = 0.05N - \sum_{k=2}^{N} \frac{N!}{k!(N-k)!} \left(\frac{1}{6}\right)^k \left(1 - \frac{1}{6}\right)^{N-k} \tag{26.7}$$

Alternately, one can use the complement to the probability of getting at least 2 fives, which is the probability of either getting 0 fives or 1 five:

$$p(k = 0|N) = \frac{N!}{0!(N-0)!} \left(\frac{1}{6}\right)^0 \left(1 - \frac{1}{6}\right)^{N-0} = \left(1 - \frac{1}{6}\right)^N = \left(\frac{5}{6}\right)^N \tag{26.8}$$

$$p(k = 1|N) = \frac{N!}{1!(N-1)!} \left(\frac{1}{6}\right)^1 \left(1 - \frac{1}{6}\right)^{N-1} = N\left(\frac{1}{6}\right)\left(\frac{5}{6}\right)^{N-1} \tag{26.9}$$

Then the optimization statement about a concern-tempered probability becomes a relatively simple deterministic function:

$$\min_{\{N\}} J = 0.05N - \left[1 - \left(\frac{5}{6}\right)^N \left(1 + \frac{N}{5}\right)\right] \tag{26.10}$$

Note: N, the DV, can only have integer values and is constrained $N \geq 2$.

Note: This is a deterministic OF, even though it was about a probabilistic process. Any person, any day, given a same DV value, N, the OF value will be exactly the same.

Suppose, however, the situation is more complicated. For example, suppose a six immediately after a five cancels the five. And suppose that exactly 3 threes means a failure regardless of the other roll values. How can one compose the OF with such conditions? I imagine that an analytical approach would become complicated and prone to multiple sources of derivation errors. I suggest that the answer is to use a Monte Carlo approach and simulate the game a large enough number of times to see the result with adequate precision. This, however, converts the deterministic OF to a stochastic OF.

See Function 52 (dice probability) in the 2-D Optimization Examples file. It reveals two categories of surface aberrations for optimizers: the discrete/integer possibilities generate discontinuities in the surface, and the stochastic nature does not make the local flat surfaces have a definitive value.

That example sought to maximize a probability, but the statistic might be to minimize the probability that a response value will be below an unacceptable threshold, as the next example illustrates.

Example 2 A company makes seatbelts, and there is a threshold requirement, a specification, for breaking strength of the product. They design the number of threads, N, and tensile strength, T, of each thread so that the belt meets the specification, but with minimum cost. In a very simple analysis, consider that the strength-bearing strands are parallel and each with the same strength, and then the collective strength of all is $S = NT$. One can increase collective strength by either increasing N or T.

Variation in manufacturing and in materials, however, means that there will be variation in the product strength. Accordingly, if the design target was the product specification, then some products would exceed, and others fall short of the specification. As a result, the product design target strength must exceed the product strength specification, but not by too much. The more the design exceeds the threshold, the more the belt costs. But there is still variability from belt to belt, and if the target is not much above the threshold, there is a probability that some belts will have a strength below the threshold.

The strength variability of the N-thread composite can be analyzed by propagation of variance. First a model of the composite strength,

$$S = T_1 + T_2 + T_3 + \dots = \sum_{i=1}^{N} T_i \tag{26.11}$$

Assuming variability in each strand is independent, then applying propagation of variance,

$$\sigma_S^2 = \sigma_{T_1}^2 + \sigma_{T_2}^2 + \sigma_{T_3}^2 + \dots = \sum_{i=1}^{N} \sigma_{T_i}^2 \tag{26.12}$$

If the variance on each strand is equal, then

$$\sigma_S^2 = N\sigma_T^2 \tag{26.13}$$

Or, taking the square root of each side relates the belt strength variability to the thread variability:

$$\sigma_S = \sigma_T \sqrt{N} \tag{26.14}$$

If the variation is normal (Gaussian), then the Gaussian distribution can be used to determine the fraction of belts with a strength below a threshold value:

$$\text{pdf}(S) = \frac{1}{2\pi\sigma_T\sqrt{N}} e^{-1/2\left((S-NT)/\sigma_T\sqrt{N}\right)^2} \tag{26.15}$$

$$p\left(S_{N,T} < S_{\text{specification}}\right) = \text{cdf}\left(S_{\text{specification}}\right) = \int_{-\infty}^{S_{\text{specification}}} \text{pdf}(S)\,dS \tag{26.16}$$

Choose N and T values, and then calculate the fraction of belts with strength below $S_{\text{specification}}$ from the aforementioned. Mathematically, there is always a value for the probability that $S_{N,T} < S_{\text{specification}}$, even if vanishingly small. Designing a belt such that the ideal of $p\left(S_{N,T} < S_{\text{specification}}\right) = 0$ is impossible. Designing to minimize $p\left(S_{N,T} < S_{\text{specification}}\right)$ leads to the absurd limits of $N = \infty$ or $T = \infty$.

So, choose N and T to minimize the probability of a product failing and to minimize costs to manufacture the belt:

$$\min_{\{N,T\}} J = p\left(S_{N,T} < S_{\text{specification}}\right) + C(N,T) \tag{26.17}$$

This statement of course is not dimensionally correct but could be made so by the concept of risk. Risk is the probability of an event times the cost of the consequence of the event:

$$\min_{\{N,T\}} J = C_{\text{failure}}\,p\left(\text{event where } S = S_{\text{specification}}\right)p\left(S_{N,T} < S_{\text{specification}}\right) + C_{\text{manufacturing}}(N,T) \tag{26.18}$$

Note: This is a deterministic OF, and the DV, N, is restricted to integer values.

Note: The Gaussian model for the individual strand tensile strengths and the independence of variation accepted in this example description are not necessary. One could use numerical methods to analyze both common cause events and alternate distributions. Monte Carlo techniques would make the surface stochastic.

Note: Calculating $P\left(S_{N,T} < S_{\text{specification}}\right)$ using the Gaussian distribution is not easy and requires a root-finding convergence. Often analysis uses the logistic distribution as an approximation. The simplification may be questioned, as well as the Gaussian and independent suppositions.

Note: In reality, the designer has much more to choose than just N and T, and many more specifications to meet. This example reveals the basic concept of including a probability distribution in the OF.

Example 3 In economic optimization the designer may be seeking to optimize a profitability index (P.I.) such as net present value or long-term return on assets. However, the actual value of the P.I. is subject to uncertainty on the givens, and the investigator may have used propagation of variance to estimate the impact on the 95% probable error on the index, $\varepsilon_{\text{P.I.}} = 1.96\sigma_{\text{P.I.}}$. If all of the uncertainties are independent, smallish, many, and equivalent in magnitude, then the distribution of the P.I. will essentially be normal (Gaussian).

Figure 26.1 shows the pdf of the P.I. for two design options, A and B.

Option B has the better nominal, average, or expected value, μ_B, but there is high variability in the P.I. of Option B, and there is much uncertainty. In fact, the illustration reveals that there is about a 20% probability (the shaded fraction of area under the curve B) that the Option B choice will lose money rather than make money. By contrast, sensitivity analysis of Option A only indicates a 2% chance of total failure, but A has a lower expected profitability. Which should the investors choose? If this is their last bit of investment capital, then they might have a lot of concern about the probability of failure, and chose A to minimize the probability of loss. The application statement would be

$$\min_{\{A,B\}} p(\text{P.I.} < 0) = \text{cdf}(\text{P.I.} = 0) = \int_{-\infty}^{0} \text{pdf}(\text{P.I.})d\text{P.I.} \tag{26.19}$$

Figure 26.1 The profitability index distribution of two designs.

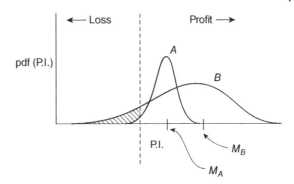

Alternately, they might revisit Option *B*, which offers the higher potential, and ask what changes in design choices, x, would minimize the probability of failure:

$$\min_{\{x\}} p_B(\text{P.I.} < 0) = \int_{-\infty}^{0} \text{pdf}_B(\text{P.I.}) d\text{P.I.} \tag{26.20}$$

On the other hand, if they can make many investments and can absorb an occasional loss, they might choose the higher potential Option *B* to maximize the profitability index:

$$\max_{\{A,B\}} \text{P.I.} \tag{26.21}$$

Note: The right choice is not a consequence of optimization. It is a consequence of the situation.

Note: This was developed as a deterministic OF, using analytical propagation of variance, but it certainly could be cast as a stochastic application using Monte Carlo techniques to reveal the distribution of P.I. values.

Note: In some cases the apparently continuum-valued DV (such as pump capacity and outlet pressure) may include design options (such as either a centrifugal or positive displacement pump). So, in such cases the DV includes a class variable, and the optimizer must be appropriate to such.

26.2 The Concept of Modeling Uncertainty

There is a general concept for the model and sources of uncertainty illustrated in Figure 26.2.

The variable y indicates the outcome of a process, procedure, or product. If it is desirable, then seek to maximize y, and if undesirable, seek to minimize y. Examples might include the following:

- Number of happy customers
- Sales income
- Profitability index
- Quality
- Safety

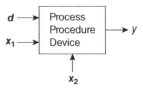

Likely, there are many outcomes of interest and the overall objective will be some joint metric of *y* items.

Figure 26.2 Illustration of the concept of modeling.

The variable d represents environmental, or wild, or uncontrolled influences on outcomes y. It also represents the "givens," which are never known with certainty. Again, there are many as indicated by the vector notation, $\boldsymbol{d}$. These might include the following:

- Raw material—quality, density, composition, molecular weight
- Environment—T, P, %RH
- Human foibles—dropped product, wet it
- Actual operating conditions—cook time, speed, T
- Actual dimensions—length, circularity

The vector $\boldsymbol{x}_1$ indicates operating choices. For example:

- Process—T, P, F, set points
- Sequence—in operating manual, assembly

And the vector $\boldsymbol{x}_2$ indicates design choices, such as:

- Size, weight, H, W, L
- Surface roughness
- Number of trays
- Shape of an airplane wing
- Valve C_v and characteristic
- Power frequency (50 or 60 Hz)

Each $\boldsymbol{d}$, $\boldsymbol{x}_1$, and $\boldsymbol{x}_2$ element has uncertainty or variation, which causes y variation. The variation can be in either individual samples or time of continuous operation.

Propagation of variance on y can be used to determine the uncertainty in the response:

$$\sigma_y = \sqrt{\sum \left(\frac{\partial y}{\partial d_i}\right)^2 \sigma_{d_i}^2 + \sum \left(\frac{\partial y}{\partial x_{1,i}}\right)^2 \sigma_{x_{1,i}}^2 + \sum \left(\frac{\partial y}{\partial x_{2,i}}\right)^2 \sigma_{x_{2,i}}^2} \tag{26.22}$$

It is likely that the derivatives would have to be evaluated with finite difference techniques. Component sigma values can be estimated in any number of ways (see Chapter 25).

Once σ_y is determined, the optimization objective could be any number of options:

- Minimize the probability of y exceeding a threshold:

$$p(y > y_{\text{threshold}}) = \int_{y_{\text{threshold}}}^{\infty} \frac{1}{\sqrt{2\pi}\sigma_y} e^{-1/2\left((y - y_{\text{nominal}})/\sigma_y\right)^2} dy \tag{26.23}$$

- Minimize the 95% worst (largest) y-value:

$$y_{0.95} = y_{\text{nominal}} + 1.65\sigma_y \tag{26.24}$$

- Maximize a nominal y-value:

$$y_{\text{nominal}} = f(\boldsymbol{x}) \tag{26.25}$$

- Minimize variance about the y-nominal value (or simply the uncertainty, since the $\sqrt{\Box}$ is a positive definite OF transformation):

$$\sigma_y = f(\boldsymbol{x}) \tag{26.26}$$

26.3 Stochastic Approach

The examples earlier led to deterministic objective functions. However, they were each predicated on assumptions that permitted analytical expressions. By contrast, a Monte Carlo approach to get the pdf (y), a numerical approach, has some advantages:

- Permits the use of a dynamic model with time evolution of response
- Permits use of conditional probabilities
- Permits nonlinear responses to be included
- Permits any non-Gaussian or nonindependent probability model
- Has less chance of calculus and algebra errors

However there are some disadvantages to a Monte Carlo approach:

- There is no explicit algebraic equation to reveal the sensitivity of y-variance to x- or d-variance that could guide a designer toward understanding how to improve the design.
- Computational burden is high, it is not uncommon to need 1000 realizations (1000 independent calculations of the OF) to assess key features of the distribution, or a million realizations when seeking to determine extreme values of a distribution.
- The stochastic aspect of the application creates confusion in directing the optimizer search and confounds detecting convergence (see Chapter 24 for a method to handle these issues).

To implement a Monte Carlo simulation, follow these steps:

1) Specify the variability of all x and d influences. This may be from experience and educated guesses as much as (i) understanding true pdf models (such as binomial, Poisson, Gaussian, beta, exponential, etc.) or (ii) having exact knowledge of the magnitude of the variation (such as range or standard deviation). Some how-to hints follow.
2) Select a DV trial solution X_t
3) Run enough "realizations" to get PDF(Y) so that OF can be calculated. This may mean 10^3–10^6 realizations

For realization = 1 to N

$$d_r = d_{\text{base}} + \varepsilon_d \qquad (d_r \text{ is a realization, a possible value. How to obtain } \varepsilon \text{ follows})$$

$$X_{1\,r} = X_{1\,\text{base}} + \varepsilon_{x1} \qquad (X_1 \text{ could be the trial solution})$$

$$X_{2\,r} = X_{2\,t} + \varepsilon_{x2} \qquad (X_2 \text{ represents uncontrolled influence factors})$$

$$\text{Calc } Y_r = f(X_{1\,r} X_{2\,r} d_r)$$

Next realization

4) Compute the statistic for y, for example, the probability of y being the unacceptable range. The objective function is f(statistic).
5) Check convergence and stopping criteria.
6) End, or let optimizer choose a better trial solution, and return to Step 2.

If Y is a time response, then the realizations would represent sequential samplings in time.

In Step 3, the question is "How to get values for ε?" Note that this ε represents the environmental perturbation to a "given," not the convergence criterion threshold.

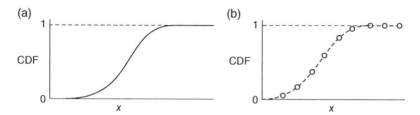

Figure 26.3 (a) Continuous or analytic and (b) experimental or discrete.

First, have a relation between X (an input variable, a "given") and CDF(X). This could be from data or from a theoretical distribution model. The continuous curve in Figure 26.3a indicates a theoretical model, and the dots connected by the dashed line in Figure 26.3b illustrate linear interpolation between data.

Recognize that in either case the CDF values are estimates. Although Figure 26.3a seems precise and exact, the model coefficients that constitute the theoretical model may have been obtained from a best fit of an ideal model to empirical data, or they may be from an ideal fundamental concept. So, the seemingly less sophisticated and precise linear interpolation in Figure 26.3b might actually better represent the truth of the situation.

The CDF represents the probability of an equal or lower x-value. The CDF has a 0–1 range, and each equal interval (e.g., 10% between 0.20 and 0.30 or between 0.86 and 0.96) has an equal chance of producing the associated x-value. So, as a random variable CDF is a uniformly distributed variable from 0 to 1.

Second, generate a random value, uniformly distributed on 0–1 interval. In your software these might be functions RAN( ), RAND( ), or RND( ), U().

Third, use the inverse of CDF(X) to determine the x-value that would generate that randomly distributed uniform, independent 0–1 value:

$$X_i = \text{CDF}^{-1}(U_i) \tag{26.27}$$

To get the ith realization value of x, you may need interpolation between data points if it is an empirical CDF. However, a theoretical model is usually a series or integral with a nonanalytic inverse, requiring a root-finding technique to solve for the inverse. In this case a piecewise linear approximation of the CDF will be useful. Prior to the Monte Carlo analysis, select a few x-values, calculate the CDF of each, and store in a table to be used as if it were empirical data.

If you accept the normal Gaussian distribution, you can easily generate a normally (Gaussian) and independently distributed variable with zero mean and sigma, NID($0,\sigma$), with the Box–Muller formula (1958):

$$\varepsilon = \sigma\sqrt{-2\ln(u_1)}\sin(2\pi u_2) \tag{26.28}$$

Here u_1 and u_2 are independent random numbers, uniformly distributed on a 0–1 interval, $0 < u \le 1$. Use the natural (Naperian) log, not the base 10 log. The mean (average) of many epsilon values calculated by Equation (26.28) is zero, and the variance is sigma squared. Either the cosine or sine function can be used with equivalent results. (This formula remains a magical thing to me. I don't see that it has a fundamental derivation based on the Gaussian distribution, but it generates a set of epsilon variables that is exceedingly true to NID($0,\sigma$).) I am impressed with its simplicity and fidelity.

26.4   Takeaway

Optimization of probable outcomes may be more appropriate than optimization of nominal values. However, to consider probable outcomes, estimates and models of the underlying uncertainty are needed.

26.5   Exercises

1   The Taguchi method directs design of a product or a process so that performance has minimum sensitivity to environmental vagaries. How would you state this optimization OF and DV?

2   Show that Equations (26.7) and (26.10) are equivalent in the values for $p(k \geq 2 | N)$.

3   Solve $\displaystyle\min_{\{N\}} J = 0.05N - \left[1 - (5/6)^N (1 + N/5)\right]$. Graph J w.r.t. N for a range of your own EC values.

4   Simplify the seatbelt example with a logistic function to approximate the Gaussian distribution of individual strengths. $\mathrm{CDF}(S) = \int \mathrm{pdf}(S)dS \cong 1/\left(1 + e^{-a(S-b)}\right)$. Choose your own values for coefficients a and b to reveal a reasonable CDF(S) shape. Obtain a 2-D contour graph of the optimization. Solve it.

5   Choose your own function of $y(x,d)$, but don't make it too complicated. Use propagation of variance to determine σ_y. Compare the x^* values for a variety of cases, for instance, to minimize σ_y, to maximize y, to minimize the 95% worst case, etc.

6   Add new rules to the dice game of Example 1, and optimize.

27

Discrete and Integer Variables

27.1 Introduction

Integers, the counting numbers, are a subset of discrete numbers. Even if a variable is continuum valued conceptually, in practice it might be discretized. An example is time in a digital simulator. Whether the time in a simulation is discretized into 0.1 s increments, millisecond increments, or whole-second increments, the simulated time can only have discrete values. Whenever some variable is placed in cells or array elements, the interval is discretized. Other examples include spatial increments in solving a differential equation (even though distance may be continuum valued) or bin increments in creating histograms.

Each category of variable in an optimization (DVs, constraints, intermediate variables, or OF values) may be limited to discrete values. These values may be point values, for which existence is only at one particular point, and no in-between values exist. Or the discretization may create flat spots, such as is often the result of a "dead-band" logic wherein a variable retains its past value until it exceeds some discretization value. The step-and-hold logic in a digital clock is just such an example. The time shows $11:23$ until 1 min later when it jumps to $11:24$.

Flat spots arise from several mechanisms. One type could be that of constraints in a physical process. For example, the level in an open bucket rises with the addition of liquid until the bucket overflows, and then adding more liquid does not change the level, but removing liquid can. The level response to flow rate will have a flat spot when the level is full and liquid is added. As another example, the consequences of controller tuning coefficients (the optimizer DVs) may have a controller exceed the limits of 0 or 100% during a simulation. If so, an override will reset the intermediate variable to a feasible value. There will be a range of DV values for which the signal will be capped at the limit, resulting in an OF flat spot, an OF value that is insensitive to local changes in the DV. Discretized OF values may be encountered when the economic value is rounded to the smallest monetary unit. As a final example, consider that the objective is to rank order players. If the person in third place improves, if the DV makes the fitness better, but not enough to change ranks, the OF will have a local flat spot.

Flat spots also happen in dynamic modeling due to a time delay that must have integer multiples of the sampling interval, such as in SOPDT modeling for controller design. Here, the gain and time constants would be continuum-valued variables, and the delay, an integer multiple of sampling interval, so this is a mixed integer programming application. Flat spots also happen when all of the DVs are continuum when the outcome is a rank or sequence, such as modeling attributes to determine fitness or when using the random keys method to solve a traveling salesman problem type.

Engineering Optimization: Applications, Methods, and Analysis, First Edition. R. Russell Rinehart.
© 2018 R. Russell Rinehart. Published 2018 by John Wiley & Sons Ltd.
Companion website: www.wiley.com/go/rhinehart/engineeringoptimization

Flat spots in the OF cause problems for gradient-type searches, and for classic single TS direct searches, because they make it seem to the algorithm that any change in the DV has no impact; so it must have converged. This happens in design for reliability situations. There are integer programming algorithms, but one usually prefers simpler to use algorithms.

Discretized point values cause a different set of problems. Variables that can only have point values include the number of people to be assigned to a project (3.83 people?) or the diameter of a pipe that could be purchased or the volume of milk to be added to a cookie dough mixture (would a recipe call for you to add 7.84193 mL?) or the length of a beam to fit a rafter (15 ft 5.72264/16ths of an inch?). Point values can only have that isolated value. Discretized variables in simulations typically include the time or distance increment in an initial value problem, or the integration step, or the values that are incrementally calculated. For example, remaining fuel in a rocket simulation might be calculated by the time increment (either simulation or sampled data) and fuel consumption. If the prior fuel content was 100,000 and the time increment and fuel rate are 0.1 and 400, then the incremental fuel values are the discretized values of 99,960, 99,920, 99,880, etc. Some or all the DVs are integers, or discretized intervals where in-between numbers are invalid.

Additional examples arise when the variables in an optimization (either DVs, constraints, OF, or intermediate variables) have attributes such as

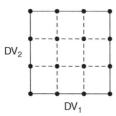

Figure 27.1 Illustration of point values that are feasible.

- Binary bit values 0001, 0010, 0011, 0100, etc.
- Delay as a number of sampling intervals
- Number of items, batches, parallel units, seats, people, check-out lines, police cars, etc.
- Round to convenient numbers for size (10 lb, 20 lb, 30 lb sacks) (6 pack, 12 pack, 24 pack)
- Axis graphing intervals of 1, 2, 5, 10, 20, 25, 50, etc.
- Prices where 1¢ is the smallest unit
- OF values that represent rank or category

Figure 27.2 Illustration of rounding the DV values to a feasible discretization.

Optimization algorithms presented in this book treat decision variables as a continuum, and do not jump from one point value to another. However, optimizer TS values could be rounded or discretized for the OF calculation. Invalid non-integer or non-sequence values and discontinuities cause problems for optimizers. They can arise in response to many forms of discretization.

Figure 27.1 shows a 2-DV situation for which the DVs may only have point values. The crossover points on the grid represent feasible solutions; but they have zero area, so the chance of randomly finding a feasible pair in real space is zero.

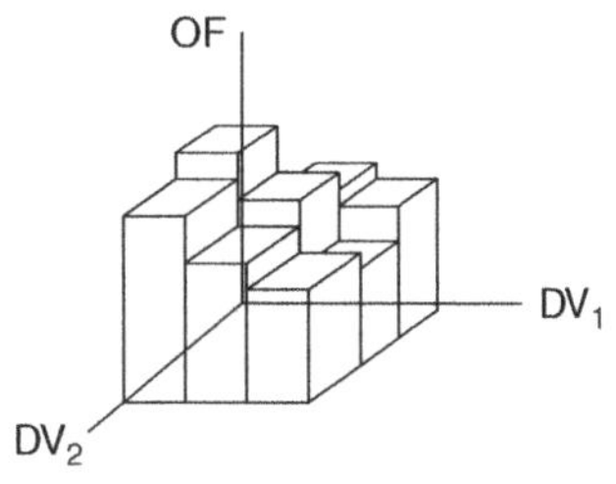

As one solution, the algorithm could round (or truncate) real DV values (continuous valued) to the integer, or discretized value. In Figure 27.2, any DV pair in the shaded space will be represented by the feasible point (1, 1).

This is an improvement in OF response surface continuity; but now the surface has flat spots with jumps between levels, as illustrated in Figure 27.3. This means that gradient-based and second-order

OF

DV₁

DV₂

Figure 27.3 An OF response to discretized DVs.

optimizers can't make sense of the surface because on the flat spot, gradient and Hessian elements will be zero. Successive quadratic is a second-order model-based optimizer. If all of its model exploration points are on a flat spot, then the model is $OF = OF_{base} + 0 * DV_1 + 0 * DV_2 + \cdots$.

Example 1   Integer DVs

How many items to use? Figure 27.4 shows a possibility of seven headlights on a motorcycle. You want to be sure to have enough light when one or some fail, but you also want to minimize the cost of the unit. Figure 27.5 shows a possibility of four exhaust fans for a building, where there is a requirement to move a certain volume of air per minute. If each of the fans is sized for one-third of the required capacity, then when one fails, the remaining three provide adequate capacity. More fans are better for reliability, but you also wish to minimize the capital cost.

Figure 27.4 A motorcycle headlight configuration.

In either case, any one item might fail. You could buy five full-sized items, operate one, and have four spares to be very confident that you'll have one working. But five full-sized items cost five times more than one. Suppose you buy seven half-sized items. Then using the 6/10th power law to scale unit cost to size, the total cost will be $7(1/2)^{0.6} = 4.6$ times one full sized. This is less expensive, but you need to use two in parallel. But you have five spares, provided in high confidence that you'll have access to two when needed.

The optimization question is "How many to buy and what size to buy to minimize cost, while meeting specification and a desired probability of success?"

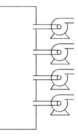

Figure 27.5 An exhaust fan configuration for a building.

n = no. items = DV_1

m = no. needed to operate at full capacity = DV_2

Q_{total}/m = capacity per item

$C_{full\ sized}((Q_{total}/m)/Q_{total})^{0.6} = C_{full}(1/m)^{0.6}$ = cost per item

$n \cdot C_{full}(m)^{-0.6}$ = cost of n items

p = probability that any one item fails

$q = 1 - p$ = probability that any one item works

If m items are required, then there are $n - m$ spares. The system is fully functional if 0 fail, 1 fails, 2 fails, ..., $n - m$ fail:

$$p_{success} = p(0 \text{ fail or } 1 \text{ fails or } 2 \text{ fails or...or } n - m \text{ fails})$$
$$= p(0 \text{ fail}) + p(1 \text{ fails}) + \cdots + p(n - m \text{ fail})$$
(27.1)

$$P(l|n) = \left(\frac{n!}{l!\,(n-l)!}\right)p^l q^{n-l}$$
(27.2)

$$P_{success} = \sum_{l=0}^{n-m} P(l|n)$$
(27.3)

$$\min_{\{n,m\}} J = nC_{full}m^{-0.6}$$

$$S.T.:\ n,\ m > 0$$
(27.4)

$$n,\ m \text{ integer}$$

$$p_{success} > \text{threshold}$$

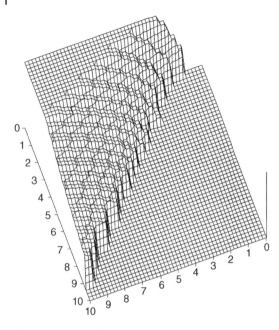

Figure 27.6 The OF response to Function 56.

See Function 56 in the 2-D Optimization Examples file and Appendix E. Figure 27.6 illustrates the surface. The code treats DV_1 and DV_2 as continuum values, but the function rounds them to the nearest integer for calculation in both the reliability constraint and product cost functions. Note: There are flat spots in the 3-D graph, with vertical cliffs between them.

27.2 Optimization Solutions

Diverse methods have been developed in the following text to solve optimization applications with discretized variables.

27.2.1 Exhaustive Search

Do an exhaustive search over all integer value combinations. This could be excessive. The number of possible combinations is $(N_1)(N_2)(N_3)...$ $= \prod N_i$. where N_i = no. possible discretized values for DV_i. This is a huge amount of work.

27.2.2 Branch and Bound

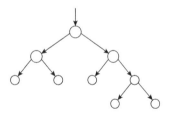

Figure 27.7 Representation of a branch-and-bound strategy.

Treat the integer DVs as continuum numbers and use an optimizer to find the trial solutions. Then round the trial solutions to a nearby possible feasible solution. Then fix one DV and search for the others over feasible point values in the neighborhood. Repeat until the neighborhood has been searched. This multiple strategy, illustrated in Figure 27.7, requires managing multiple paths with an optimization at each node. Also, this will get trapped in a local optimum. So, you need enough trials to be confident that the global has been discovered.

27.2.3 Cyclic Heuristic

When the DVs are discretized, use a cyclic heuristic direct search with expansion = 1 and contraction = –1. But here the increments are discrete values of the variable, not necessarily its units. For instance, if the DV values are roughly triplets on a log scale (1, 2, 5, 10, 20, 50, 100, etc.), then at 20 a "+1" incremental move of the DV goes to 50 and a "–1" moves to 10. Stop when the search keeps returning to the same base. But, again, this single TS approach will lead to local optima, so you need enough trials to be confident in finding the global.

27.2.4 Leapfrogging or Other Multiplayer Search

Let the algorithm keep continuous values for the DVs, but round them to the nearest valid discretization for the OF calculation and feasibility tests.

27.3 Convergence

If surfaces have flat spots, when all players have gathered to a common surface, leap-overs just contract the cluster in the continuum, but don't change the discretized solution. So, instead of convergence based on DV space, claim convergence when all OF values are the same. This is easy in LF because the worst and best OFs are known. If the worst OF value is the same as the best OF value, if the OF range is zero, then all players have the same OF value.

   Since all DV values are discretized, the test could claim convergence when all rounded DVs have the same value.

27.4 Takeaway

For any number of reasons, discretized variables are common. Use an optimizer and a convergence criterion that can cope with the discontinuity.

27.5 Exercises

1 Many of the functions in the 2-D Optimization Applications file have discretized, rank, or category variables. Explore optimization and convergence determination techniques on them. How will you rate optimizer or convergence criterion performance? Design experiments to acquire appropriate performance data and use it to reveal why one choice is better than others.

28

Class Variables

28.1 Introduction

Class variables are alternately called category, text, nominal, or string variables. They do not have numerical values. They are items or choices. For instance, "Which is better: a car, a motorcycle, or an airplane?" Here the DV is a method of transportation, and the assessment of best could be related to fuel and insurance cost, joy of travel, number of friends you can bring, operator training cost, speed of transportation, parking upon arrival, etc. The decision variable, the list of items, does not have a common, quantifiable metric (either continuum-valued or discretized) that defines the OF. In size, the airplane is largest and motorcycle smallest, but this aspect is not what defines the OF. The motorcycle might accelerate the fastest, and the car carries the most luggage per passenger, but, again, these are consequences of the DV.

Here are some more examples of choices that are an either–or situation. Which is better: a chair or a stool? Which is better: a ladder, stairs, rope-and-pulley, or an elevator? A turbine, piston, or centrifugal pump? Distillation, crystallization, or pressure-swing absorption? Jim, Ted, or Brad? An electrical engineer or a computer scientist?

Additionally, the list of items may constitute all of the things that need to be completed, and the question is "What is the best sequence of implementing the operations?" For example, in preparing a batch reaction, components A, B, and C need to be added, the material mixed, heated to react, then cooled. But what sequence? Add A, heat, start stirrer, add B, …? Add A and B, stir, add C, heat, …? The right sequence has no relation to common aspects that might relate the operations, such as the alphabetical order of the names, the number of letters with curved parts in the name, the date it was first invented, etc.

Another example of a sequence optimization is termed the traveling salesman problem (TSP). Nominally the person must visit N cities; home is the start and must be the last, but what is the sequence of the in-between visits? The DV is the sequence and must include all cities. Simplistically, this OF is often, "Find the sequence to minimize the total distance traveled," but it certainly can have constraints, such as City J must come before City B to pick up something at J and deliver it to B. Or time to travel between any two cities must be less than 6 h, or the date of visiting City D must be compliant to contract schedules or availability of the customer there. And the shortest distance objective might be tempered by cost of that path (travel, meals, overnight).

My wife is a realtor, and each Thursday morning the local realtors group-tour about a dozen new listings in town. They want to minimize travel cost and travel time, avoid school zones or road repair

Engineering Optimization: Applications, Methods, and Analysis, First Edition. R. Russell Rhinehart.
© 2018 R. Russell Rhinehart. Published 2018 by John Wiley & Sons Ltd.
Companion website: www.wiley.com/go/rhinehart/engineeringoptimization

events, maximize convenience and safety, comply with schedule conflicts related to home owner and listing realtor, etc.

Although called TSP, this optimization of a sequence of events, operations, and activities is common to logistics, scheduling, and planning. It could be applicable to working sections of a puzzle, writing chapters in a thesis, sequencing stages in building a business, getting dressed in the morning, delivering bad news, cooking a meal, etc. It would relate to optimizing trash pick-up sequence, power line repair sequence, customer service sequence, and personal getting-ready sequence (brush teeth, comb hair, put on shirt, put on shoes, put on jacket, eat breakfast, cook breakfast). Possible objective functions could be time to complete all tasks, risk of encountering an undesired situation, spoilage or time-out exceeded, distance, cost of the sequence, work time of a crew between states, convenience of arrival/departure times, etc. Some transitions may be inexpensive, but the reverse very expensive (detours on one side of a road will affect travel in that direction only). Constraints could be related to sequence (socks must be put on before shoes, must pick-up prior to delivery, etc.).

In sequencing class variables, the OF could include many alternate considerations. Similarly, constraints could be on any number of aspects related to a sequence.

Optimization with class variables is very common. Again the attributes of a list of class variables does not have a common quantifiable property that can be described with a common numeric value—either a continuum, discrete, or rank value.

There are two categories of class variable optimization. In one the DV is dichotomous; it is an either–or choice, a selection of one item from a list. In the other the DV is a sequence that must include all items in the list.

In either case there will likely be other continuum or discrete variables associated with the application. For example, in the batch reaction, continuum DVs might include the temperature that is best for cooking, mixer size, rate of addition of B that is best, etc. For the TSP continuum DVs might include what time to start a trip, the time prior to arrival to call a customer, and how much to spend on dinner with the customer.

If dichotomous, and there are n choices in the DV, then an exhaustive search could optimize the continuum variables for each of the n choices. If, however, there are n_i choices in the ith DV and no constraints of one DV choice on the others, then there are $N = \prod n_i$ separate optimizations. Also simplistically, if there are n cities (other than home) that need to be visited, then there are $N = n!$ number of sequences that would need to be considered in an exhaustive search. In either case, optimizing each possibility in an exhaustive search through either all dichotomous combinations, or all sequence possibilities, may be impractical.

Optimization procedures that are guided by either heuristics (go to the nearest neighboring city next) or human logic (fuzzy logic or expert systems) are useful. This chapter presents the random keys method. My thanks to Scott Essner, who introduced it to me.

28.2 The Random Keys Method: Sequence

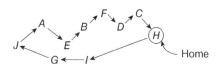

Figure 28.1 Illustration of the TSP.

For simplicity, consider a traveling salesman problem (TSP)—find the sequence of cities to minimize the total distance path to visit all cities. The distance between the ith and jth city is $d_{i,j}$. The trip must start and end at home. In the illustration of Figure 28.1 there are nine other cities to visit: A, B, C, D, E, F, G, I, and J.

The simple optimization statement is

$$\underset{\{sequence\}}{min} J = \sum d_{i,j} \qquad (28.1)$$

Assign each city a random number value (the key). For example,

A	0.832
B	0.106
C	0.218
D	0.507
E	0.034

And so on

Then sort the cities by their key, as though it were a continuum DV value, and calculate the from–to transition cost; in the path when sorted by the random key "*H*" starts, "*E*" with a value of 0.034 is next followed by "*B*" with a value of 0.106:

$$J = \Sigma d_{i,j} = d_{E,B} + d_{B,C} + d_{C,D} + ... \qquad (28.2)$$

If there are N cities to visit and another is home, there are $N + 1$ cities. There are N transitions. Except for the last two transitions (the last is go home, the next to last is go to the unvisited city), each transition must choose a go-to city from the unvisited cities. An exhaustive search would have $N!$ possible combinations.

By contrast, in the random keys method, each of the N cities has a random key, which directs the search. The optimizer's job is to determine random key values that optimize the OF. For a problem of sequencing N tasks (N cities to visit plus home for a total of $N + 1$ cities), there are N DVs.

Note: If the random key value for City C was 0.217 or 0.109, or 0.505, the sequence would be unchanged, so the OF value would be unchanged. This reveals that there are flat spots on the OF versus DV sequence.

If using leapfrogging as the optimizer, each player has N DV values. For example, one player, Player1, might have the key values from the table earlier.

Player1 (0.832,0.106,0.218,0.507,0.034,...)

Another player would be initialized with an alternate set of randomized key values. For example,

Player2 (0.117,0.588,0.941,0.001,0.773,...)

The random numbers in each player element define that trial solution sequence. Calculate the OF for each player.

Have the worst leap over the best. Each DV dimension does an independent random leap over. For a leap-over illustration, consider these key values for elements in the worst and best players:

Worst: 0.117, 0.588, 0.941, 0.001, 0.773, ...
Best: 0.832, 0.106, 0.218, 0.507, 0.034, ...

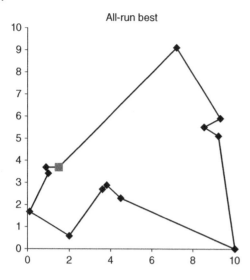

All-run best

Figure 28.2 A best solution to the Realtor Tour TSP example.

Leap to: $0.832 + r_1(0.832 - 0.117)$,
$\qquad 0.106 + r_2(0.106 - 0.588)$,
$\qquad\qquad 0.218 + r_3(0.218 - 0.941)$,
$\qquad\qquad\qquad 0.507 + r_4(0.507 - 0.001)$,
$\qquad\qquad\qquad\qquad 0.034 + r_5(0.034 - 0.773)$,

Calculate the OF for the new player position.

Example 1 The Realtor's Tour example (visit www.r3eda.com for a copy) shows the location of 12 houses for a Thursday morning realtor's tour of new listings. Each Wednesday afternoon when the listings are set, the board executive plans a sequence to visit each house. With the simple objective to minimize cumulative distance as the crow flies, the optimal path is shown in Figure 28.2. The vertical axis is the north–south direction scaled to a 0–10 value, and the horizontal axis is the scaled E–W direction.

For this example there are many near optimal paths with similar distance. Certainly, the OF could be related to travel time and have penalties for high traffic, congested streets, and school zones.

28.3 The Random Keys Method: Dichotomous Variables

Other examples of class or category variables include dichotomous (mutually exclusive choices). For instance, in design, the choice might be between a batch and a continuous reactor, tray and packed column separation, and to build in Location A and Location B. Again, assign random numbers to the class variables and segregate DV choices by a range of values. For example, if the value for $DV_1 > 0.5$, then it is a batch, else a continuous reactor.

28.4   Comments

The OF "surface" response to the class variable DV will have flat spots. Consider that a sequence begins with these three choices E (with a key value of 0.034) to B (with a key value of 0.106) to C (with a key value of 0.218). The value of DV_2 is 0.106, and the sequence (consequently the OF) remains unchanged for a DV_2 range of $0.034 < DV_2 < 0.218$. The OF surface remains the same value for the DV range. Here the flat area will expand and contract as the next lower and next higher DVs change value, and the location of the flat surface will shift as the DVs evolve. Accordingly, an optimizer must be able to cope with flat spots.

Because of the flat spots, because there is a range of DV values that lead to the same solution, a reasonable convergence criterion would be to stop searching when all OF values are the same, when $0 = \sum (OF^* - OF_i)^2$, or when $OF_{high} - OF_{low} = 0$.

Likely, the application will trap the optimizer in local optima or even flat spots that are adjacent to a better choice. Accordingly, use multiple starts to increase the possibility of finding the global.

It might seem to be a good idea to use a heuristic solution as either an initial trial solution or as the initial basis for one player in a multiplayer optimizer. However, that near best condition could draw all random players toward it and cause a better solution to be bypassed. Accordingly, randomize the player initialization. However, using the heuristic solution might be a good benchmark.

28.5   Takeaway

Optimization with class variables can be handled with optimization algorithms using the random keys method to assign numerical values to otherwise dichotomous or sequential variables. The OF response flat spots that result are similar to those encountered when truncating a continuum-valued DV to its integer or discrete values, or using ranks, or classification variables.

28.6   Exercises

1   Does it matter within what range the random keys are initialized (such as 0 to 1, or −1000 to +1000)?

2   Can the procedure work if the optimizer assigns a new key value that is outside of the initialization range for the key values?

3   Create your own multi location TSP application. Explain it. Explain how the OF would be calculated. Explain how random key values would be initialized. Explain how the key values would be used to define the sequence.

4   Explain how you would adjust a standard TSP application to include alternate criteria to define the OF (such as travel time, or tolls) and to include constraints (such as Location C must come within 2 days of Location E).

5   The Realtor Tour Excel-VBA file on the www.r3eda.com site has 12 cities. Explore its operation, and then change the location and/or number of the cities.

29

Regression

29.1 Introduction

This chapter is a summary of key issues and solutions related to nonlinear regression–fitting nonlinear models to data. The details of diverse issues are revealed in the book Nonlinear Regression Modeling for Engineering Applications: Modeling, Model Validation, and Enabling Design of Experiments by Rhinehart, R. R., John Wiley & Sons, Inc., Hoboken, NJ, 2016b. Here, they are summarized and presented with an optimization application perspective.

Regression is the procedure of fitting a model to data and gives rise to the following optimization questions:

1) Classic nonlinear regression procedures direct the user to transform the model so that linear regression can be applied. Should one?
2) In regression with dynamic models, it is not uncommon to have a delay, an integer multiple of time-sampling intervals. If delay is a DV, then it is a discretized variable. Compounding the associated difficulties, in regression of nonlinear models, the continuum variables can be constrained, and the models have discontinuities. What optimization method is best?
3) Classically, regression minimizes the squared deviation between model and experimental response, the vertical deviation. This presumes that there is no uncertainty on the input variables and that the y-variance is uniform over all data. Often these are not reasonable assumptions, and maximum likelihood is a better concept to use. But maximum likelihood is computationally complex. Is there a useful compromise technique?
4) Classic convergence in regression is based on incremental changes in the model coefficients. But how should the thresholds be selected? A better technique is to claim convergence when there is statistically no improvement in the model w.r.t. natural data variability. How should the variability be determined?
5) Often in empirical modeling we progressively add flexibility to a model (perhaps by adding power series terms, or by adding hidden neurons, or by adding logical rules), but improved model flexibility comes at the cost of increasing complexity. We wish goodness of fit to be balanced by model simplicity. What is the OF for determining model order?

Engineering Optimization: Applications, Methods, and Analysis, First Edition. R. Russell Rhinehart.
© 2018 R. Russell Rhinehart. Published 2018 by John Wiley & Sons Ltd.
Companion website: www.wiley.com/go/rhinehart/engineeringoptimization

29.2 Perspective

Some ask why use sum of squared deviations as opposed to the sum of absolute values of deviations as the objective function in regression. Both objective functions (both measures of undesirability) will work. Both provide reasonable fits of model to data. Minimizing absolute value of the deviations is intuitive, and it seems reasonable. Theoretically, however, as Section 29.5 shows, the sum of squared residuals is grounded in the normal (Gaussian) distribution associated with data variation. If the model is true to the phenomena, and the data are perturbed by many small, equivalent, random, independent influences, then the residuals (data to model deviation) should be Gaussian distributed. Maximizing the likelihood that the model represents the process that could have generated the data, can be functionally transformed, and is equivalent to minimizing the sum of squared residuals.

Additionally, the sum of squares method is continuous with continuous derivatives. The absolute value of the deviation has a slope discontinuity, which can confound classic optimizers (such as Newton types and successive quadratic).

For those two reasons, the classic method, the method used by most modeling software providers, is to minimize sum of squared residuals.

Neither, however, gives the true model. Even if the model were functionally perfect, the perturbations in the experimental data would cause errors. Remove one data and replace it with another data point supposedly at the same conditions. The new one will not be in the same exact place, and the new model will be a bit different.

Further, the truth about nature is elusive. While the model may provide adequate, close enough functionality for engineering utility, and regression may find the best coefficients for the model, the model remains wrong. All we can hope for is a surrogate model that reasonably matches the data. Either minimizing the sum of squared residuals or the sum of absolute values of the deviations provides reasonable models.

29.3 Least Squares Regression: Traditional View on Linear Model Parameters

The objective is to obtain a model that best fits data from an experimental response. First, generate data experimentally. Run an experiment. "Dial in" input conditions, x, measure the response, y. Then find the best model, $\tilde{y} = \tilde{f}(c, x)$. The tilde over the f and y explicitly indicates that it is the model, a mathematical approximation, not the true description of Nature. The vector c represents the model coefficients.

Models can have many functional forms. Here are four of an infinite number of common empirical choices for a y-response to a single x-variable. The coefficients are a, b, c, d, e. The user must decide on the functional relations appropriate to the application:

$$\tilde{y} = a + bx \quad \text{linear } y - \text{response to } x \tag{29.1a}$$

$$\tilde{y} = a + bx + cx^2 \quad \text{quadratic } y - \text{response to } x \tag{29.1b}$$

$$\tilde{y} = a + bx + cx^2 + dx^3 + ex^4 \quad \text{higher order } y - \text{response to } x \tag{29.1c}$$

$$\tilde{y} = a + bx + cx^2 + dx^{-1} \quad \text{mixed functionality} \tag{29.1d}$$

The optimization statement is

$$\min_{\{c\}} J = \sum (y_i - \tilde{y}_i)^2 = \sum \left(y_i - \tilde{f}(c, x_i) \right)^2 \tag{29.2}$$

where the index i represents data set number. Note that coefficient values, c, and model functionality, $\tilde{f}$, are common to all of the i-data points.

Classically, we formulate the model to be linear in the coefficients, linear in the DVs for optimization. Then we can use linear regression techniques and have the convenience of knowing that there is one unique solution and that the DV values can be explicitly computed.

A model is linear in coefficients if the modeled response is independently linear to each coefficient— or if the second derivative w.r.t. any combination of coefficients is zero:

$$\frac{\partial^2 \tilde{y}}{\partial c_j \partial c_k} = 0 \tag{29.3}$$

The four models are all linear in coefficients a, b, c, d, e. The advantage of this is that the analytical method can be used to determine model coefficient values. Set the derivative of the OF w.r.t. each DV to zero:

$$\frac{\partial J}{\partial c_j} = 0 \tag{29.4}$$

This results in N linear equations, each linear in the N DVs. For example, in the quadratic model, $\tilde{y} = a + bx + cx^2$, the three derivatives can be rearranged to this set of linear equations in the three coefficients:

$$\frac{\partial J}{\partial a} = 0 \rightarrow (N)a + \left(\sum x_i\right)b + \left(\sum x_i^2\right)c = \sum y_i \tag{29.5a}$$

$$\frac{\partial J}{\partial b} = 0 \rightarrow \left(\sum x_i\right)a + \left(\sum x_i^2\right)b + \left(\sum x_i^3\right)c = \sum x_i y_i \tag{29.5b}$$

$$\frac{\partial J}{\partial c} = 0 \rightarrow \left(\sum x_i^2\right)a + \left(\sum x_i^3\right)b + \left(\sum x_i^4\right)c = \sum x_i^2 y_i \tag{29.5c}$$

This set of equations is known as the "normal equations," and regardless of the number of DVs, it can be represented as $\boldsymbol{Ma} = \boldsymbol{RHS}$, for conventional linear algebra solution.

29.4 Models Nonlinear in DV

However, often, models are nonlinear in the coefficients. Here are three examples:

$$\tilde{y} = ax^p \quad \text{power law with zero intercept,} \tag{29.6a}$$

$$\tilde{y} = a + bx^p \quad \text{power law with } y-\text{intercept} \tag{29.6b}$$

$$\tilde{y} = ae^{-b/x} \quad \text{reaction rate} \tag{29.6c}$$

The optimization statement is still the same:

$$\min_{\{c\}} J = \sum (y_i - \tilde{y}_i)^2 = \sum \left(y_i - \tilde{f}(c, x_i)\right)^2 \tag{29.7}$$

where the index i represents data set number. Again, the coefficient values, c, and model functionality, $\tilde{f}$, are common to all of the i-data points.

Classically, we formulate the model to be linear in the coefficients; however, the power law and exponential models are not.

A common linearizing transform for a power law model, $\tilde{y} = ax^p$, is to log-transform it:

$$\tilde{y}' = \ln(\tilde{y}) = \ln(a) + p\ln(x) = a' + px' \tag{29.8}$$

The relation between $\tilde{y}'$ and x' is now linear in coefficients a' and p. The optimization application becomes

$$\min_{\{c'\}} J = \sum (y_i' - \tilde{y}_i')^2 \tag{29.9}$$

Once linear regression is performed on the transformed data, inverting the transform will return the desired coefficient values. For example, once the value for a' has been determined, $a = e^{a'}$.

There are a wide variety of linearizing transforms for the variety of model functionalities, and conventional practice has offered such as the right way to solve the regression problem. But what is accepted as best in human history has changed from the precomputer era to today. Once, the convenience of linear calculations overrode the desire for model fidelity. Now, I think with the convenience of nonlinear regression model fidelity can be preserved.

Chapter 16 showed that log-transforming an OF is a strictly positive transformation that does not change the DV values. The log-transform of the OF would mean

$$\min_{\{c\}} J' = \ln(J) = \ln\left[\sum (y_i - \tilde{y}_i)^2\right] = \ln\left[(y_1 - \tilde{y}_1)^2 + (y_2 - \tilde{y}_2)^2 + \ldots\right] \tag{29.10}$$

However, log-transforming the individual data changes the OF:

$$\min_{\{c'\}} J = \sum (y_i' - \tilde{y}_i')^2 = \sum (\ln(y_i) - \ln(\tilde{y}_i))^2 = \sum \left(\ln\left(\frac{y_i}{\tilde{y}_i}\right)\right)^2 = 2\left[\ln\left(\frac{y_1}{\tilde{y}_1}\right) + \ln\left(\frac{y_2}{\tilde{y}_2}\right) + \ldots\right] \tag{29.11}$$

You can represent this as a weighted least squares OF, where the original OF terms are weighted by a factor, λ_i:

$$J = \sum_{i=1}^{N} \lambda_i (y_i - \tilde{y}_i)^2 \tag{29.12}$$

Using the linearized model $J' = \sum_{i=1}^{N} (\ln(y_i) - \ln(\tilde{y}_i))^2 = \sum_{i=1}^{N} (\ln(y_i/\tilde{y}_i))^2$, then the weighting factor is

$$\lambda_i = \left[\frac{\ln(y_i/\tilde{y}_i)}{y_i - \tilde{y}_i}\right]^2 \tag{29.13}$$

The weighting factor depends on both the value of y and the deviation between model and data. It is not a uniform value over all data sets. The DV solutions from the original nonlinear statement of Equation (29.10) and the optimization using the linearizing transforms of Equation (29.11) return different DV values (model coefficient values).

If, however, the model truly represents the data, and the experimental error is small and/or the range is large, then the deviation between the nonlinear and linearized DV values is small. Linearizing transforms have served mankind well, but it should not be considered as best with today's tools. To prevent the distortion of data importance that results from linearization, do not linearize. Use a numerical optimization method on the nonlinearized data.

29.4.1 Models with a Delay

The model might represent how a process evolves over time. If first order and linear, the ordinary differential equation would be

$$\tau \frac{dy(t)}{dt} + y(t) = ku(t) \tag{29.14}$$

Alternately, the explicit finite difference representation (Euler's method) would be

$$y_{i+1} = y_i + \frac{\Delta t[ku_i - y_i]}{\tau} \tag{29.15}$$

Or the equivalent autoregressive moving average (ARMA) representation is

$$y_{i+1} = ay_i + bu_i \tag{29.16}$$

Here the subscript i indicates the sampling interval, and time is the product of the sample interval number and the time increment, $t = i\Delta t$. If the solution to the differential equation is numerical, then Equations (29.15) and (29.16) represent the model. Or if the process data were to be periodically sampled, then values would only exist at discrete time values. In either case, the variable, i, has integer values, and time is discretized.

If not first order or linear, the discretized models are more complicated, but the discretization aspect remains.

There may be a delay between the model input and the impact on its output. For instance, charging on a credit card today has no immediate impact on your bank account, but after the time delay, you have to pay the bill. Then it has impact. Similarly, eating does not nourish the body until after the food is digested and moved to where sugars can be absorbed. There is a delay. Delays can also be due to information or material transfer in a process or communication system.

If first order, linear, and with a delay, the ordinary differential equation would be

$$\tau \frac{dy(t)}{dt} + y(t) = ku(t - \theta) \tag{29.17}$$

Alternately, the explicit finite difference representation (Euler's method) would be

$$y_{i+1} = y_i + \frac{\Delta t[ku_{i-n} - y_i]}{\tau} \tag{29.18}$$

Or the equivalent autoregressive moving average (ARMA) representation is

$$y_{i+1} = ay_i + bu_{i-n} \tag{29.19}$$

Here n represents the number of samplings that equal the delay $n = \theta/\Delta t$, but n must be an integer either for either numerical solution or for matching model to sampled data. In the model, θ, or equivalently n, is a coefficient with unknown value to be determined by regression:

$$\min_{\{a,b,n\}} J = \sum_{i=n+1}^{i=N} (y_i - ay_{i-1} - bu_{i-n-1})^2 \tag{29.20}$$

Even if the model was classified as linear in ODE terminology (meaning that the principle of superposition can be applied to devise a complete analytical solution as the sum of elements), the regression exercise is nonlinear in how θ, or equivalently n, appears in the model. Not only is the DV discretized to integer values, but it also has a nonlinear impact on the OF. Further complicating things, the data prior to n samplings cannot be used to calculate the OF, because the model does not have u-data prior to $i = 0$. So, the trial solution for n affects how many elements are in the OF sum.

That was for a dynamic model, but the issues are similar even if the model represents a process for which there is no dynamic evolution, just a delay. This could represent a pipeline transport with plug flow:

$$\tilde{y}_i = \tilde{f}(c, x_{i-n}) \tag{29.21}$$

Again, even if the model might be linear in the other coefficients, it is nonlinear in the impact of n on the model. Accordingly the optimization is nonlinear, and the DV is discretized.

For such, I recommend a direct search technique with multiple players.

29.5 Maximum Likelihood

We usually pretend (we often say "assume," "accept," "given," or "imply") that the value of x, wherein the influence on an experiment was perfectly known, was exactly implemented and has no uncertainty. Usually, experimental uncertainty is wholly allocated to the response variable, as the concept is illustrated in Figure 29.1.

Further, we often pretend that the total uncertainty in the location of the data pair is due to random, independent, symmetrically distributed, often Gaussian measurement noise on y.

My use of the term pretend is a bit stronger than most would prefer to use. It does imply a purposeful misrepresentation. But is not the alternate statement "assume the basis for this analysis" also a purposeful misrepresentation? Anyway, don't use the word pretend when you describe your work. Consider the impact on your audience, and use a term that seems to provide a legitimate underpinning for the assumptions and simplification choices that you made to permit the analysis.

I think that Figures 29.2 and 29.3 provide a truer representation for an experimental process.

Both figures illustrate uncertainty on the measured response, y_{meas}, as a perturbation being added to the true y-value. The uncertainty reflects fluctuation, noise, variation, experimental error, etc.; and it may be independent of the y-value or scaled

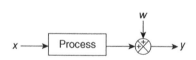

Figure 29.1 Noise perturbing an experimental measurement.

Figure 29.2 Uncertainty on the experimental inputs—concept 1.

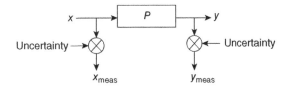

Figure 29.3 Uncertainty on the experimental inputs—concept 2.

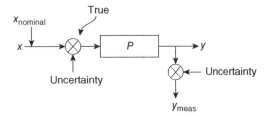

with the y-value. It may be symmetric and Gaussian or from an alternate distribution. When independent for each measurement, it is often called random error. If it is a constant value, uniform over many samplings, then it is termed bias or systematic error.

Both figures also reveal uncertainty impacting the input to the experiment. In Figure 29.2 the experimenter sets an input and reads its value from a scale. The reading has uncertainty, so x-meas is not the same as the x-true that influences the process. In Figure 29.3 the experimenter chooses the value for x, x-nominal and attempts to implement it by adjusting the final control element to make the x-value match the desired value, but experimental uncertainty makes the true influence to the experiment a bit different. The fluctuation could be a bias or systematic error with a fairly constant value over many samplings; alternately it could be independent for each sampling representing random noise effects.

Returning to uncertainty on the y-measurement, normally, it is considered to represent the uncertainty surrounding the measurement, or lab procedure, or analytical device. But partly, fluctuation on y-meas is due to uncontrolled experimental aspects that make y-true vary. Uncontrolled disturbances and environmental effects include raw material variability, relative humidity, electrical charge, experimenter fatigue, air pressure, etc. Certainly, one can model fluctuation to other inputs to the process in a manner of modeling the vagaries on the x-inputs. But this analysis will pretend that uncertainty on those uncontrolled fluctuations is included in the uncertainty on y-meas.

With this concept of variation on both the independent and dependent variables, consider the illustration of one data pair in Figure 29.4. The point indicates the experimental values, but the possible true values for either x or y could come from a distribution about the point. Frequently, the variation in either value is the result of many independent, random, and equivalent perturbations, making the distribution normal (Gaussian), as illustrated. The farther away the possible x-value is from the x-measurement, the less likely it could have actually generated the x-measurement. A similar argument goes for the y-value. Accepting that the x-variation and y-variation are independent, the

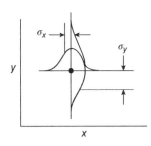

Figure 29.4 Pdf of x and y overlaid on the experimentally obtained data point.

probability that the measured point could have been produced for alternate x- and y-values is the joint probability:

$$P(x \text{ and } y) = P(x)P(y) \qquad (29.22)$$

which is illustrated in Figure 29.5.

If Gaussian, the joint probability distribution for the possible location of the point about the measurement (x_0, y_0) is

$$
\begin{aligned}
\text{pdf}(x, y | x_0, y_0) &= \frac{1}{\sqrt{2\pi}\sigma_y} e^{-(1/2)\left((y-y_0)/\sigma_y\right)^2} \frac{1}{\sqrt{2\pi}\sigma_x} e^{-(1/2)((x-x_0)/\sigma_x)^2} \\
&= \frac{1}{2\pi\sigma_x\sigma_y} e^{-(1/2)\left[((x-x_0)/\sigma_x)^2 + \left((y-y_0)/\sigma_y\right)^2\right]}
\end{aligned}
\qquad (29.23)
$$

In maximum likelihood, the objective is to find the model coefficients that maximize the possibility that if the model represented the process, data point (x_0, y_0) could have been the result. The concept is illustrated in Figure 29.6. Note that the points labeled A, B, and C are each on the curve and that point B has the highest likelihood, the highest joint pdf value. There is a possibility that the data point could have been generated from points A and C on the curve, but the point on the curve with the highest likelihood of generating the data point is B. Accordingly, the optimization is a two-stage procedure. First, choose the model coefficient values, then search along the curve (a univariate search) to find the point of maximum likelihood of generating (x_0, y_0), and label this point on the curve $\left(x_0^*, y_0^*\right)$, recognizing that if y is the modeled response to x, the point is $\left(x_0^*, \tilde{y}\left(x_0^*\right)\right)$.

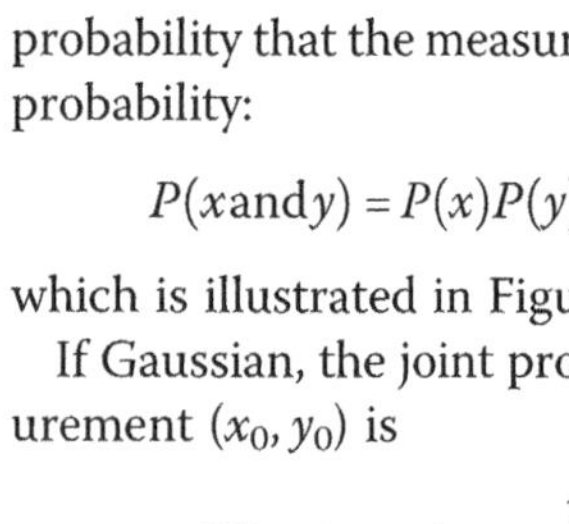

However, this must be done for each data point. If the variation is independent for all x- and y-values, then

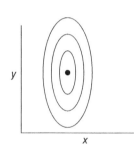

Figure 29.5 Likelihood contours surrounding the experimentally obtained data point.

$$\underset{\{c\}}{\max} J = \prod \left\{ \frac{1}{2\pi\sigma_x\sigma_y} e^{-1/2\left[\left((x_i^*-x_i)/\sigma_x\right)^2 + \left((\tilde{y}(x_i^*)-y_i)/\sigma_y\right)^2\right]} \right\} \qquad (29.24)$$

This can be simplified. Since $2\pi\sigma_x\sigma_y$ is a positive constant, scaling the OF by it does not change the DV* values. Then, since the log-transformation is a strictly positive transformation, maximizing the product of exponential terms is the same as maximizing the sum of the arguments. Then, since minimizing the negative is the same as maximizing, and since multiplying by 2 does not change the DV* values, the objective becomes

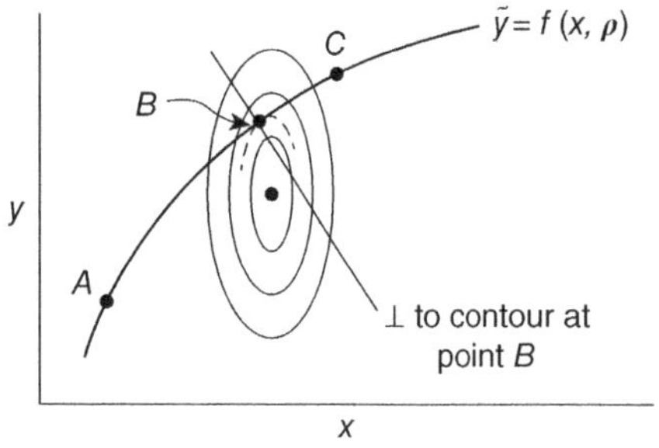

$$\underset{\{c\}}{\min} J = \sum \left[\left(\frac{x_i^* - x_i}{\sigma_x} \right)^2 + \left(\frac{\tilde{y}\left(x_i^*\right) - y_i}{\sigma_y} \right)^2 \right] = \sum J_i$$

Figure 29.6 The point on the curve with maximum likelihood of generating the experimental data.

$$(29.25)$$

In Figure 29.6 caption area:
$\tilde{y} = f(x, \rho)$

$\perp$ to contour at point B

Although simplified, this is still a two-stage procedure. The outer optimizer assigns values for model coefficients, c. Then for each experimental data set, the inner optimizer performs the univariate search to find the $[x_i^*, \tilde{y}(x_i^*)]$ values from which the J_i values can be calculated. The sum of the J_i values is the OF for the outer optimizer.

Note: If the uncertainty on the independent variable is ignored, then $x_i^* = x_i$, and if σ_y is independent of the y-values, the OF can be scaled by the positive multiplier, σ_y^2, which reduces the OF to the common vertical least squares minimization.

Note: This can be extended for data and model with multiple x-values.

Note: Values for σ_y and σ_x need to be known. The values for σ_y and σ_x represent expected variability if all influence values are kept unchanged for replicate measurements. Do not use the standard deviation of the entire list of all the experimental data. Replicate a few conditions at a low medium and high value and evaluate σ_y and σ_x within the replicate sets. Recognize, however, that even with 10 replicated values, there is much uncertainty on the experimentally calculated σ. The 95% confidence limits are roughly between 0.5σ and 2σ.

Note further: It may not be possible to experimentally evaluate the σ_x values, which may have to be estimated using techniques from Chapter 25.

Note: Replicate trials to generate y-values from which to calculate σ_y, and include the impact of uncertainty on x on the y-data. However, the concept in Equations (29.23)–(29.25) is that the σ_y value excludes the σ_x effects. Propagation of uncertainty could be used to back out the x-variability effects on the y-data.

Fortunately, exact values are not necessary. Classic vertical least squares regression implies that $\sigma_x = 0$; and even an approximation of σ_y and σ_x values will allow the maximum likelihood approach to return model coefficient values that are closer to true than does vertical least squares.

However, this two-stage optimization is somewhat complex, and it is further complicated by removing idealizations used in its development (constant variance, independence of variation, and the Gaussian pdf model). As developed, this has a bunch of pretends and is not perfectly aligned to the maximum likelihood concept. However, it does reveal a method for accommodating independent variable uncertainty; and in my investigations it provides model coefficient values that are closer to true than the simpler vertical least squares. (Such studies must be performed with simulated data, permitting the true coefficient values to be known, so that they can be compared with the regression values.)

Accepting the validity of the simplifications earlier is an optimization exercise. The idealizations simplify the procedure yet retain much of the benefit of the maximum likelihood concept in accommodating the dual uncertainty of both x- and y-values. It is an intuitive conclusion. The traditional least squares OF is opinion based on a diversity of assumptions and only valid if they are true.

29.5.1 Akaho's Method

Continuing the intuitive evaluation of simplifications that reduce work, yet adequately retain the functionality, I like Akaho's approximation. It reduces maximum likelihood to a one-level optimization. Again, the simplification takes the procedure one step further from the ideal concepts; but to me, the benefit (reduction in complexity and computational work) exceeds the loss in veracity of the model coefficient values (as revealed in simulations).

First, scale the x- and y-variables by their estimated standard deviation:

$$y' = \frac{y}{\sigma_y} \tag{29.26a}$$

$$x' = \frac{x}{\sigma_x} \tag{29.26b}$$

$$\sigma'_y = \frac{\sigma_y}{\sigma_y} = 1 \tag{29.26c}$$

$$\sigma'_x = \frac{\sigma_x}{\sigma_x} = 1 \tag{29.26d}$$

This makes the likelihood contours become circles with a unit variance, as illustrated in Figure 29.7.

Recall that along a univariate search, the maximum (or minimum) is when the contour is tangent to the line and that steepest descent (or ascent) is $\perp$ to the line. For x- and y-variables scaled by σ_x and σ_y, the contours are circular, and the radius is the $\perp$ to the circle. That radius is also the shortest distance from the data point to the line.

So, if $\sigma'_x = \sigma'_y$, then maximum likelihood regression is equivalent to finding the shortest or $\perp$ distance to the line. Suppose the x–y relation is a curve. Still, the maximum likelihood for any data point is on a circle that is tangent to the $f(x, \boldsymbol{c})$ functions (see Figure 29.8a and b). If it was not tangent, it either crossed the curve or did not touch the curve. Either case means another contour was the one that will be tangent to the function. Then, the best curve, the one maximizing likelihood for the point,

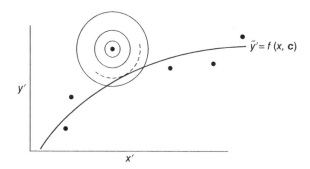

Figure 29.7 Maximum likelihood with scaled variables.

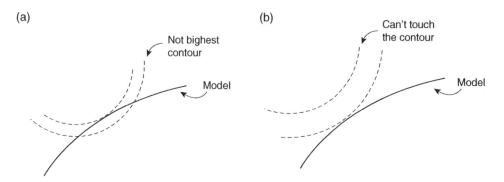

Figure 29.8 The point of maximum likelihood is tangent to the highest likelihood contour: (a) the curve crosses one contour to a better contour, and (b) the curve does not touch the better contour.

is tangent to the likelihood contour, which is a circle. This means that the distance to the (x'_i, y'_i) point is the radius of the circle, the minimum distance between center and circumference, and that the vector from curve to $(x_i\, y_i)$ is $\perp$ to the curve.

With scaled variables, the maximum likelihood method seeks to find the model coefficient values that minimize the perpendicular distance from the model to the data points, as illustrated in Figure 29.9.

Note: In scaled space the x- and y-variables are dimensionless, and they can be combined into a common dimensionless distance:

$$S'_i = \sqrt{\left(x'_i - \tilde{x}'_i\right)^2 + \left(y' - \tilde{y}'_i\right)^2} = \sqrt{\left(x'_i - \tilde{x}'_i\right)^2 + \left(y' - \tilde{y}\left(\tilde{x}'_i\right)\right)^2} \tag{29.27}$$

The objective is

$$\min_{\{c\}} J = \sum S'_i = \sum \sqrt{\left(x'_i - \tilde{x}'_i\right)^2 + \left(y' - \tilde{y}\left(\tilde{x}'_i\right)\right)^2} \tag{29.28}$$

This is still a two-stage procedure. The outer optimizer chooses c-values, then the inner univariate search finds the $\tilde{x}'_i$-values that minimize each S'_i, and the sum of the S'_i values becomes the OF for the search for c-values.

This approach is variously termed normal least squares, or perpendicular least squares, or total least squares. Be sure to scale the x- and y-variables by their standard deviations so that (i) the likelihood contours are circles and (ii) x and y can be dimensionally combined into a joint distance.

Figure 29.10 illustrates multiple data points and a nonlinear $y = f(x)$ relation graphed on scaled space. Note that the shortest distance between data and model is not the vertical distance. Also note

Figure 29.9 Points of maximum likelihood are normal to the curve when variables are scaled by their standard deviation.

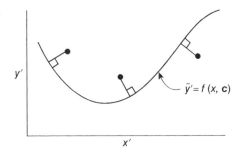

Figure 29.10 Model coefficient values to maximize the product of each likelihood in scaled variables.

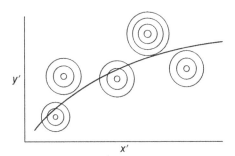

that the direction from the line to data is not the same for each data point and the direction of the normal to the curve changes with position. Finally, note that the curve is not equidistant from the data. Some are on a high contour, and others are on a low contour.

For the special case of a linear relation between y and x, the normal (or $\perp$) deviations all have the same slope (the negative reciprocal of the model line). Accordingly, in this special case, the $\perp$ deviations are a fixed proportion of the vertical deviations.

Now comes Akaho's technique that approximates the normal distance and reduces the optimization to a one-stage procedure. The technique is presented in Figure 29.11.

All are in scaled variables $x' = x/\sigma_x$, $y' = y/\sigma_y$. This makes the likelihood contours around the appearance of the data point (where measurements or settings make the point appear to be) be concentric circles. Point A is the ith experimental data pair (x_i, y_i), point B is the ith modeled value $(x'_i, \tilde{y}'_i)$, and point C is the closest point on the model curve to the (x'_i, y'_i) experimental data. Because the likelihood contours are circles, line AC is the radius of a circle, and it is perpendicular to every circle perimeter, one of which is the likelihood contour tangent to the curve. So, line AC is also perpendicular to the curve, a normal to the curve. The data pair (x'_i, y'_i) is a point on the normal or perpendicular line.

The tangent to the model at the modeled $(x'_i, \tilde{y}'_i)$, at the ith (x') value, is illustrated; and point D is the closest point on the tangent line to the data location (x'_i, y'_i). Line AD is perpendicular to the tangent line.

Any point on the tangent line can be modeled as a linear deviation from the tangent point $(x'_i, \tilde{y}'_i)$ along the line of slope S:

$$\left(y'_T - y'_i\right) = \left(\frac{d\tilde{y}'}{dx'}\right)\left(x'_T - x'_i\right) = S\left(x'_T - x'_i\right) \tag{29.29}$$

(x', y') is a data pair on the perpendicular to the tangent line from the data apparent location, (x', y'). It is a radius of the circle centered at the apparent data point location. The slope of a perpendicular line is the negative reciprocal of the tangent line. The relation of any (x,y) pair on the radius line is

$$\left(y_\perp - y_i\right) = -\frac{1}{S}\left(x_\perp - x_i\right) \tag{29.30}$$

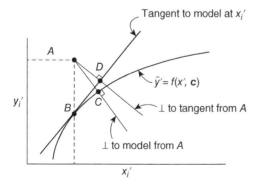

Tangent to model at x_i'

Figure 29.11 Akaho's method to approximate likelihood. *Source:* Rhinehart (2016b). Reproduced with permission of Wiley.

$\tilde{y}' = f(x', \mathbf{c})$

$\perp$ to tangent from A

$\perp$ to model from A

At the intersection of the tangent and perpendicular lines, point D, x on the tangent is equal to x on the radius, and y on the tangent is equal to y on the radius. $x'_{Ti} = x'_{\perp i}$ and $y'_{Ti} = y'_{\perp i}$ call them (x'_{Di}, y'_{Di}). Using the two line models, solve for (x'_{Di}, y'_{Di}):

$$x'_{Di} = x'_i + \frac{y'_i - \tilde{y}'_i}{s + (1/s)} \tag{29.31a}$$

$$y'_{Di} = y'_i - \frac{1}{s}\left(\frac{y'_i - \tilde{y}'_i}{s + (1/s)}\right) \tag{29.31b}$$

The distance from data point to tangent, from point A to D, is

$$d'^2_{Ti} = \left(x'_i - x'_{Di}\right)^2 + \left(y'_i - y'_{Di}\right)^2 \tag{29.32}$$

which can be reduced to

$$d'^2_{Ti} = \frac{\left(y'_i - f'(x'_i)\right)^2}{1 + s^2} = \frac{\left(y'_i - f'(x'_i)\right)^2}{1 + (df'/dx')^2} \tag{29.33}$$

Here the normal distance from data to model is approximated by the vertical distance scaled by the slope of the model. Once the optimizer chooses c-values, $f'(x'_i)$ and df'/dx' are directly calculated. The inside search for each of the ith points on the model is avoided. Accordingly, Akaho's approximation to the maximum likelihood optimization is a one-stage procedure:

$$\min_{\{c\}} J = \sum_{i=1}^{n}\left(d'_{Ti}\right)^2 = \sum_{i=1}^{n}\frac{\left(y'_i - f'(x'_i)\right)^2}{1 + (df'/dx')^2_i} \tag{29.34}$$

For a model with multiple inputs, the method is

$$\min_{\{c\}} J = \sum_{i=1}^{n}\left(d'_{Ti}\right)^2 = \sum_{i=1}^{n\,\text{data sets}}\frac{\left(y'_i - f'(x'_i)\right)^2}{1 + \sum_{j=1}^{N\,\text{inputs}}\left(df'/dx'_j\right)^2_i} \tag{29.35}$$

Example 1 Develop the regression optimization statement using Akaho's technique and the model $y = ax^b$. (This does not mean that the process that generated the data has that relation.)
  The scaled variables are

$$y' = \frac{y}{\sigma_y}, \quad x' = \frac{x}{\sigma_x} \tag{29.36}$$

Then the model in scaled variables is

$$y' = \frac{a}{\sigma_y}x^b = \frac{a}{\sigma_y}(x'\sigma_x)^b = \frac{a\,\sigma_x^b}{\sigma_y}x'^b = a'x'^b \tag{29.37}$$

The scaled model coefficients are a' and b:

$$\frac{df'}{dx'} = a'bx'^{b-1} \tag{29.38}$$

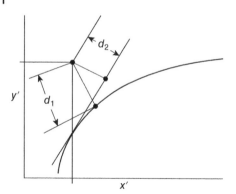

Figure 29.12 If variability is large, Akaho's approximation loses validity.

In scaled space find a' and b from this optimization:

$$\min_{\{a',b\}} J = \sum_{i=1}^{n} \frac{\left(y_i' - a' x_i'^b\right)^2}{1 + \left(b x_i'^{b-1}\right)^2} \tag{29.39}$$

Then revert to the model $a = a' \sigma_y / \sigma_x^b$, and the unscaled the optimization is

$$\min_{\{a,b\}} J = \sum_{i=1}^{n} \frac{\left(y_i - a x_i^b\right)^2}{\sigma_y^2 + \sigma_x^2 \left(ab x_i^{b-1}\right)^2} \tag{29.40}$$

The advantage is that Akaho's method accounts for x-variability but does not require the nested optimization needed to maximize likelihood or to minimize total least squares. It is a single-level optimization.

The disadvantage is that this simplification of maximum likelihood assumes constant σ_y and σ_x throughout the range and known values for σ_y and σ_x. Further Akaho's approximation is not good if curvature makes tangent not represent the distance. See Figure 29.12, where the tangent is a considerable departure from the curve. So, y- and x-uncertainty, σ_y and σ_x, should be small relative to how curvature changes the model.

However, in my opinion, Akaho's method is fully adequate as a replacement for maximum likelihood, especially when considering the myriad of presumptions related to known sigma values, pdf models, linearity, fidelity of the model to the true natural phenomena, etc.

29.6 Convergence Criterion

Whether using conventional vertical least squares or a maximum likelihood version in nonlinear optimization, traditional convergence criteria for the optimization would be on $J, \Delta J, \Delta c_1, \Delta c_2, \Delta J / J$ etc. As ever in optimization, choosing right values for the thresholds on any of those requires understanding of the sensitivity of J to each c and/or to the relative improvement in J or c. These thresholds are scale dependent. They would also be dependent on the sensitivity of the model to the coefficient value. Such aspects confound the user in making a right choice of the threshold.

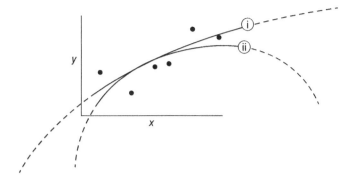

Figure 29.13 Alternate models with equivalent validity to the data.

Further, if the data are very noisy, there should be no pretense about being able to know the truth of the function parameter values, and convergence criteria might as well be "loose." However, if the noise amplitude is small, then it is reasonable to "see" the model with greater precision, and threshold convergence criteria could be justifiably small. Reasonable convergence criteria should be relative to the noise on the data, which masks the truth. This is illustrated in Figure 29.13, where model (i) is as good as (ii) within the range. The choice of either model is irrelevant considering the data scatter. As a thought exercise, change the location of a data point, reflecting an alternate realization, and consider the impact on the model location.

However, if the variation in the data was much lower, then the data might reveal one model is superior to another. Less variation would also more precisely place the model.

The question is, "How does one set thresholds appropriate to achieving best precision (as justified by variation in the data—noise, uncertainty, fluctuation, or random error) with minimum optimizer work?" An answer is "When model improvement is inconsequential to inherent variability in residuals (data from model), then claim convergence."

Consider the objective function, for example, the sum of squared deviations (SSD) or its root mean value (rms), as a measure of improvement with iteration:

$$J = \mathrm{OF} = \mathrm{SSD} = \sum_{i=1}^{N} (y_i - \tilde{y}_i)^2 \tag{29.41}$$

$$J = \mathrm{rms} = \sqrt{\frac{1}{N}\sum_{i=1}^{N}(y_i - \tilde{y}_i)^2} \cong \sigma \tag{29.42}$$

Figure 29.14 reveals how J might improve with optimization stages. As iterations progress, the SSD or rms progressively drops to its best possible value for the variability in the data and the imperfection in the model.

Since $\mathrm{rms} \cong \sigma$ when $\Delta\sigma \leq 0.01\sigma$, alternately when $\Delta\mathrm{rms} \leq 0.01\mathrm{rms}$, claim convergence and stop. If a single TS algorithm, use the iteration-to-iteration change in the rms. However, if a multiplayer algorithm, use the

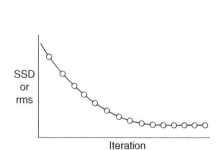

Figure 29.14 How SSD or rms might relax to a minimum value with progressive iterations.

rms range of players after an iteration. This is a convergence criterion based on the relative change of the OF.

By contrast, as a convergence criterion, at each iteration, measure SSD or rms on a random sampling of about one-third of the data. (Here, I use random to mean without replacement. Don't sample a point twice, but I'm not sure that the computational overhead of managing the "no resampling" has any real benefit. In bootstrapping, the data is sampled with replacement.)

$$\text{rms}_{\text{random}} = \sqrt{\frac{3}{N} \sum_{i=1,\text{random}}^{N/3} (y_i - \tilde{y}_i)^2} \tag{29.43}$$

Figure 29.15 shows how the $\text{rms}_{\text{random}}$ value changes with iteration. In general there is a reduction with iteration as the optimizer improves the model, but the random sampling makes it a noisy trend. At some point, the progressive improvement is inconsequential to the noise on the $\text{rms}_{\text{random}}$ signal; at some point the $\text{rms}_{\text{random}}$ value comes to steady state w.r.t. iteration. This would indicate that there is no meaningful improvement in the model relative to data variability, and convergence should be claimed.

Use any technique you prefer to identify steady state to the $\text{rms}_{\text{random}}$ w.r.t. iteration trend. My choice is the ratio of filtered variances of Cao and Rhinehart (1995). Since the sampling is random, at steady state the "noise" on the rms subset is independent from iteration to iteration, which satisfies the basis of their technique. This is a scale-free approach. If you change the units, the ratio remains dimensionless with the same threshold value. The threshold value required to detect SS is independent of the number of DVs or whether the stochastic nature is due to Monte Carlo simulation, experimental outcomes, or random data subset sampling. It is relatively independent of the distribution of the deviations. It is a computationally simple approach. Details are described in Appendix D. Here is a summary. Measure the variation from the average in about the past 10 data iterations, and measure the variation along the trend, as shown in Figures 29.16 (Not at SS) and 29.17 (At SS).

When not at SS, the deviation from average is much greater than deviation from a trend line:

$$\frac{\sigma_{\text{avg}}}{\sigma_{\text{trend}}} \gg 1 \tag{29.44}$$

By contrast, when at SS, the deviation from average is indistinguishable from deviation from the trend:

$$\frac{\sigma_{\text{avg}}}{\sigma_{\text{trend}}} \sim 1 \tag{29.45}$$

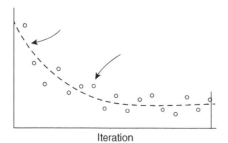

Iteration

Figure 29.15 How SSD or rms of a random data sample relaxes to a noisy steady state with iteration. *Source:* Rhinehart (2016b). Reproduced with permission of Wiley.

Figure 29.16 A not-at-steady-state trend. *Source:* Rhinehart (2016b). Reproduced with permission of Wiley.

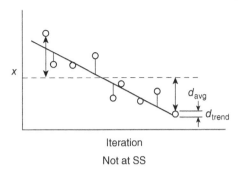

Iteration

Not at SS

Traditional methods use the sum of all data to calculate the signal values, which is a computational burden. It requires storage of N past data and updating sum of squared deviations at each iteration. So revisit the concept of calculating the variance, and use a first-order filter approach:

$$x_{f_i} = \lambda(x_i) + (1-\lambda)x_{f_{i-1}} \tag{29.46}$$

There is a relation between λ, the number of data, and a time constant for the filter:

$$\frac{1}{N} = \lambda = 1 - e^{-\Delta t/\tau_f} \tag{29.47}$$

Equation (29.46) is also called an exponentially weighted moving average (EWMA).

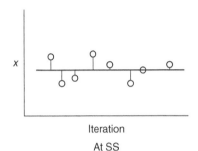

Iteration

At SS

Figure 29.17 A steady-state trend. *Source:* Rhinehart (2016b). Reproduced with permission of Wiley.

The expectation (the mean or many data average value) on x_f is the expectation on average value, which is the average:

$$E(x_f) = E(\bar{x}) \tag{29.48}$$

It doesn't matter what quantity x represents. Normally, it represents an individual measurement. But it could represent a scaled variable, or a deviation, or a deviation squared. If defined as the first-order filter of the squared deviation from the average, it is termed an exponentially weighted moving variance (EWMV):

$$s_{f_i}^2 = \lambda(x_i - \bar{x})^2 + (1-\lambda)s_{f_{i-1}}^2 \tag{29.49}$$

Instead of using $\bar{x}$, use the computationally simpler x_f. In the following equations, $n_{f_i}^2$ is the measure of variance between data and average used in the numerator, and $d_{f_i}^2$ is the measure of the variance inherent in the data (between successive data) and used in the denominator:

$$\sigma_1^2 \stackrel{m}{=} n_{f_i}^2 = \lambda_2(x_i - x_{f_{i-1}})^2 + (1-\lambda_2)n_{f_{i-1}}^2 \tag{29.50a}$$

$$\sigma_2^2 \stackrel{m}{=} d_{f_i}^2 = \lambda_3(x_i - x_{i-1})^2 + (1-\lambda_3)d_{f_{i-1}}^2 \tag{29.50b}$$

$$\bar{X} \stackrel{m}{=} x_{f_i} = \lambda_1 x_i + (1-\lambda_1)x_{f_{i-1}} \tag{29.50c}$$

$$r_i = \left(2 - \lambda_1\right) \frac{n_{f_i}^2}{d_{f_i}^2} \qquad\qquad (29.50d)$$

The filter on x is performed after calculating the numerator measure of variance to eliminate correlation of $x_{f_{i-1}}$ to x_i. Otherwise, the r-statistic is more complicated. To return the ratio $n_{f_i}^2 / d_{f_i}^2$ to represent the variance ratio, it needs to be scaled by $(2 - \lambda_1)$. If $r_i \gg 1$, then the trend is in a transient. If $r_i \cong 1$, then the trend is in a steady state. Since data vagaries cause the statistic to vary when the signal is at SS, the values for r_i range between about 0.8 and 1.5. Accordingly, if $r_i < 0.85$, then accept that the trend is at SS. I initialize all values with zero.

Figure 29.18 illustrates the distribution of the r-statistic when a signal is not at SS (right-hand curve) and all r-statistic values are greater than 3. The left-hand curve shows the distribution of the r-statistic when the process is at SS. Note that half of the values are above the ideal value of 1 and half below. The middle curve shows the distribution for a signal that is nearly, but not quite at SS. Note that the almost at SS signal has occasional r-statistic values of less than 1, but it does not place values in the 0.85 region. Accordingly, 0.85 is a useful trigger to determine SS. I use 0.80 when I want to be very sure and 0.9 if less strict.

Since $\lambda = 1/N$, effectively in filtering $1/\lambda = $ number of data the window that is observed. Ten to 20 data points is a reasonable quantity.

I have been fairly satisfied with $\lambda = 0.1$ and $r_{\mathrm{crit}} = 0.9$. When the process is at SS, these values will claim steady state faster (fewer iterations). However, they may accept a near to steady state, but not quite there process as at steady state. You might want to use those values when exploring or screening to make a sequence of trials end in less time. However, $\lambda = 0.05$ and $r_{\mathrm{crit}} = 0.80$ values will provide greater assurance that the procedure has really achieved steady state. Although they will require more iterations, they provide greater confidence that a type II error is not being made.

If optimization starts either far from optimum on a very flat surface, or if the optimizer is just getting adjusted to the right step sizes to make progress, then the initial progress in OF w.r.t. iteration could be slow. As Figure 29.19 illustrates, the initial iterations may not reveal a substantial improvement.

Exploring options for using SS to claim convergence, one approach is to not start testing for SS until after there is confidence that a transient has happened. However, if by chance the initial trial solution is very near the optimum, it will not detect a transient; then never begin testing for SS, and run the maximum iterations. I find that initializing all variables with zero is a great solution. First, it is simple. There is no need to collect N data to initialize values for $n_{f_i}^2$, $d_{f_i}^2$, or x_{f_i}. Second, even if the optimizer

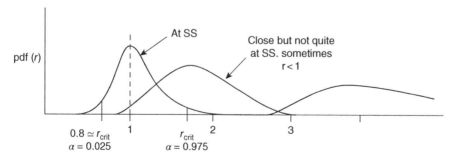

Figure 29.18 The distributions of the r-statistic for three events. *Source:* Rhinehart (2016b). Reproduced with permission of Wiley.

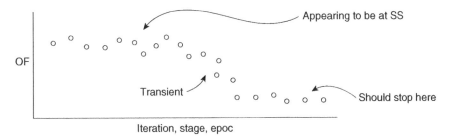

Figure 29.19 Observation of rms from a random data sampling as optimizer iterations progress.

initial trial solution fortuitously defined the correct model, the SSID initialization makes it appear that there is an initial transient as $n_{f_i}^2$, $d_{f_i}^2$, and x_{f_i} approach their steady values.

29.7 Model Order or Complexity

Regardless of the optimizer choice, objective function choice, or method of convergence, the user makes choices with regard to the model. If the model is phenomenological (indicating that the user chose a mechanistic, or first principles approach), then the model structure is fixed. However, if the user chooses empirical modeling, the question is, "What order to use?" Nominally this would mean "What powers of x are to be used." But also this could mean how many wavelets, or how many neurons in the hidden layer, or how many categories in a fuzzy logic linguistic partition, or how many sampling intervals in an ARMA model, or how many orthogonal polynomials.

A Taylor series analysis indicates a power series is "right" for any relation. For a y-response to a single x-independent variable, the series is

$$y(x) = y_0 + \frac{\partial y}{\partial x}\bigg|_{x_0}(x - x_0) + \frac{1}{2!}\frac{\partial^2 y}{\partial x^2}\bigg|_{x_0}(x - x_0)^2 + \frac{1}{3!}\frac{\partial^3 y}{\partial x^3}\bigg|_{x_0}(x - x_0)^3 + \dots \tag{29.51}$$

Although the derivatives seem like a function, they are evaluated at the single base point, x_0. Accordingly, they are coefficients in the model with constant values. By grouping coefficients of like powers of x, one obtains the classic power series model:

$$y = a + bx + cx^2 + dx^3 + ex^4 + \dots \tag{29.52}$$

Again, in empirical modeling, the question is, "What model order is right?" In empirical regression, if there are too many model coefficients, the model can fit the noise and provide a ridiculous representation of the process. In Figure 29.20, the true $y(x)$ relation is represented by the dashed line. It is the unknowable truth about nature. The dots represent the results of five trials to generate y-values from x-values. The solid line is a five-coefficient (fourth-order) polynomial model fit to the data. It exactly goes through each data point, but it is a ridiculous model. It fits the noise and does not reveal the underlying trend. Depending on the discipline, this is termed "memorization" or "overfitting."

A classic regression modeling heuristic to prevent overfitting is to have three (or more) data pairs per model coefficient. Then, to balance the work and cost associated with generating data, if three experimental sets per model coefficient are adequate, then do not generate more than three per

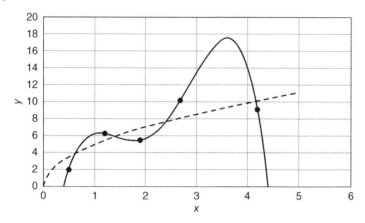

Figure 29.20 Overfitting.

coefficient. So, for six data points you should only consider a two-parameter (linear) model. If there are 15 data points, you could use a 5-parameter (quartic) model. This heuristic does not say that a quartic model is right. Perhaps a linear model would be better.

What does better mean?

- More representative of an intuitive understanding of behavior (trends and extrapolation)?
- More functional in use (invertible with a unique solution)?
- More sufficient (meets customer specifications for variance)?
- Lower cost?

You should consider what issues constitute complexity/simplicity and desirability/undesirability in your application and add your perspectives to this list.

Many people have investigated this issue and devised a variety of metrics to assess complexity of a regression model. What follows is one approach that looks at one metric for model goodness of fit and relative complexity.

Let's look at variance reduction as a measure of goodness of fit. The commonly used r^2 is a measure of SSD reduction, but does not account for the degrees of freedom in the variance:

$$\text{SSD}_{\text{data-to-model}} = \text{SSD}_{\text{residuals}} = \sum_{i=1}^{N} (y_i - \tilde{y}_i)^2 \tag{29.53}$$

$$r^2 = \frac{\text{SSD}_{\text{original data}} - \text{SSD}_{\text{data-to-model}}}{\text{SSD}_{\text{original data}}} = \frac{\sum_{i=1}^{N}(y_i - \bar{y})^2 - \sum_{i=1}^{N}(y_i - \tilde{y})^2}{\sum_{i=1}^{N}(y_i - \bar{y})^2} \tag{29.54}$$

A total reduction of residual variability, a perfect fit to the data, would result in $r^2 = 1$. However, a perfect value of unity does not mean that the model is right. The basic model architecture could be wrong. For example, if the unknowable truth is $y = a - b/(x + c)$, and noisy data is generated within the $x = 1$ to $x = 4$ region, the graph may appear to be the left half of a parabola (see Figure 29.21), and a quadratic model $y = a + bx + cx^2$ may seem to provide a reasonable fit.

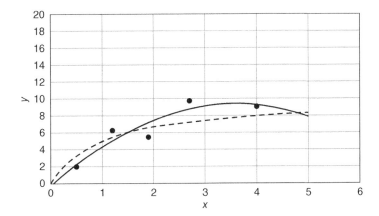

Figure 29.21 A wrong model may appear to provide a reasonable fit to data.

Model goodness of fit will be evaluated by $SSD_{residuals}$. How will complexity be assessed? Using N as the number of data sets and m as the number of model coefficients, the quantity $N + m$ represents the total work related to number of experiments and modeling effort. The quantity $N - m$ represents the degrees of freedom. The ratio $(N + m)/(N - m)$ represents the effort-to-freedom aspect. Consider that the model has m coefficients. For a quadratic model (order 2 in the independent variable), there are $m = 2 + 1 = 3$ coefficients. In a sixth-order polynomial, there are $m = 7$ coefficients.

There are several methods to balance improvement of fit relative to model complexity, as measured by number of coefficients, m. A simple one is the final prediction error (FPE) criterion, originally introduced by Akiake, but promoted and brought to wider acceptance by Ljung (in the 1980–2000 period, about 15 years later):

$$\text{FPE} = \frac{N + m}{N - m}\text{SSD}_{\text{residuals}} = \frac{N + m}{N - m}\sum_{i=1}^{N}(y_i - \tilde{y}_i)^2 \tag{29.55}$$

$$\text{FPE} \overset{m}{=} \text{reduction in error penalized by complexity per DoF} \tag{29.56}$$

The procedure is to start with a low order model, small m, and progressively add coefficients (terms or functionalities) until FPE rises. As m increases from an initial small value, the model will get better, and SSD will drop. Initially, usually, the reduction in SSD will be greater than the increase in the complexity factor, and FPE will drop with increasing m. Incrementally increase m until FPE rises, which indicates that the previous m was best. This is a univariate optimization with an integer DV:

$$\underset{\{m\}}{\min} J = \text{FPE} \tag{29.57}$$

Note: It does not have to be a power series model as I've been illustrating. It could be a neural network model (adding hidden neurons) or a first principles model (adding nuances).

This FPE optimization does not provide the values for the model coefficients, just the model order. Choose an m, model order. Use regression to determine coefficient values, and calculate FPE. Choose another m, and repeat. This is a nested optimization.

Given a model architecture and number of coefficients, m, any optimization technique could be used to find the model coefficient values:

$$\min_{\{a, b, \ldots\}} J = \sum_{i=1}^{N} (y_i - \tilde{y}_i)^2 \tag{29.58}$$
$$\text{S.T.}: \tilde{y}_i = a + bx_i + cx_i^2 + dx_i^3 + ex_i^4 + \ldots$$

Regardless of the choice of optimization technique (successive quadratic, Cauchy, LM, NM Simplex, Hooke–Jeeves, leapfrogging, … or classic analytical), they should find the same best coefficient set values. Which optimization technique is used to determine the $\{a, b, c, d, \ldots\}$ values is irrelevant to the decisions about model architecture or the use of FPE to determine the correct model order.

If you have a mechanistic model (first principles, phenomenological, physical), use it. The Taylor series expansion is mathematically defensible and indicates that your phenomenological model can be converted to a power series. If you have no concept of the mechanistic model, then an empirical model is permissible. But if you have a phenomenological understanding, use it. A phenomenological model has fewer coefficients. A phenomenological model extrapolates. A phenomenological model represents the best of human understanding. Don't use $y = a + bx + cx^2 + dx^3 + \ldots \ldots$ to override phenomenological models.

If your modeling software does not provide SSD, but provides the r-square value, you can use the r^2 value instead of SSD in FPE. From the definition of r^2,

$$r^2 = \frac{\text{SSD}_{\text{original data}} - \text{SSD}_{\text{model residuals}}}{\text{SSD}_{\text{original data}}} \tag{29.59}$$

This can be solved for SSD:

$$\text{SSD}_{\text{model residuals}} = (1 - r^2) \text{SSD}_{\text{original data}} \tag{29.60}$$

Substituting in the FPE equation,

$$\text{FPE} = \frac{N+m}{N-m} \text{SSD}_{\text{residuals}} = \frac{N+m}{N-m} (1 - r^2) \text{SSD}_{\text{original data}} \tag{29.61}$$

The objective is to find the number of model coefficients to minimize FPE:

$$\min_{\{m\}} J = \text{FPE} = \frac{N+m}{N-m} (1 - r^2) \text{SSD}_{\text{original data}} \tag{29.62}$$

Since $\text{SSD}_{\text{original data}}$ is a positive-valued constant, scaling the OF by it does not change the m^* value. So,

$$\min_{\{m\}} J = \text{FPE}' = \frac{N+m}{N-m} (1 - r^2) \tag{29.63}$$

The $\text{SSD}_{\text{original data}}$ value is not needed.

FPE is one approach to determining a balance between model complexity and goodness of fit. Many others have been developed and promoted. There are alternate measures of the desirables and undesirables and alternate ways to combine the balance. My personal experience is that FPE (and similar methods that only look at the SSD and number of model coefficients relative to number of data) permit greater empirical model complexity than I often prefer. Accordingly, I like an equal concern approach, because I can adjust the weighting based on my feel about the application:

$$\min_{\{m\}} J = \left(\frac{m}{\text{EC}_m}\right)^2 + \left(\frac{\text{rms}}{\text{EC}_{\text{rms}}}\right)^2 \tag{29.64}$$

Regardless, I like FPE for its simplicity and general relevance if the user does not have specific and alternate views about the balance of desirables.

Also, I like FPE because it reveals a nonadditive approach to combining the several OF considerations.

29.8 Bootstrapping to Reveal Model Uncertainty

Uncertainty in experimental data leads to uncertainty on the regression model coefficient values, which leads to uncertainty in the model-calculated outcomes. We need a procedure to propagate the uncertainty in experimental data to that on the modeled values, so these aspects of model uncertainty can be appropriately reported.

In linear regression, this is relatively straightforward, and mathematical analysis leads to methods for calculating the standard error of the estimate and the 95% confidence limits on the model. However, the techniques are valid only if the following conditions are true: (i) the model functional form matches the experimental phenomena; (ii) the residuals are normally distributed because the experimental vagaries are the confluence of many, small, independent, equivalent sources of variation; (iii) model coefficients are linearly expressed in the model; and (iv) experimental variance is uniform over all of the range (homoscedastic), and then analytical statistical techniques have been developed to propagate experimental uncertainty to provide estimates of uncertainty on model coefficient values and on the model.

However, if the variation is not normally distributed, if the model is nonlinear in coefficients, if variance is not homoscedastic, or if the model does not exactly match the underlying phenomena, then the analytical techniques are not applicable. In this case numerical techniques are needed to estimate model uncertainty. Bootstrapping is the one I prefer. It seems to be understandable, legitimate, and simple and is widely accepted.

Bootstrapping is a numerical Monte Carlo approach that can be used to estimate the confidence limits on a model prediction.

One assumption in bootstrapping is that the experimental data that you have represents the entire population of all data realizations, including all nuances in relative proportion. It is not the entire possible population of infinite experimental runs, but it is a surrogate of the population. A sampling of that data then represents what might be found in an experiment. Another assumption is that the model cannot be rejected by the data, but the model sufficiently expresses the underlying phenomena. In bootstrapping,

1) Sample the experimental data with replacement (retaining all data in the draw-from original set) to create a new set of data. The new set should have the same number of items in the original, but some items in the new set will likely be duplicates, and some of the original data will be missing. This represents an experimental realization from the surrogate population.

2) Using your preferred nonlinear optimization technique, determine the model coefficient values that best fit the data set realization from step 1. This represents the model that could have been realized.

3) Record the model coefficient values.

4) For independent variable values of interest, determine the modeled response. You might determine the y-value for each experimental input x-set. If the model is needed for a range of independent variable values, you might choose 10 x-values within the range and calculate the model y for each.

5) Record the modeled y-values for each of the desired x-values.

6) Repeat steps 1–5 many times (perhaps over 1000).

7) For each x-value, create a CDF of the 1000 (or so) model predictions recorded in step 5. This will reflect the distribution of model prediction values due to the vagaries in the data sample realizations. The variability of the prediction will indicate the model uncertainty due to the vagaries within the data. For some applications 100 samples may be adequate, and others may require 100,000 samples. When the CDF shape is relatively unchanging with progressively added data (steps 1–5), you've done enough.

8) Choose a desired confidence interval value. The 95% range is commonly used.

9) Use the cumulative distribution of model predictions to estimate the confidence interval on the model prediction. If the 95% interval is desired, then the confidence interval will include 95% of the models; or 5% of the modeled y-values will be outside of the confidence interval. As with common practice, split the too high and too low values into equal probabilities of 2.5% each, and use the 0.025 and 0.975 cdf values to determine the y-values for the confidence interval.

10) Repeat steps 7–9 for each model coefficient value recorded in step 3 to reveal the uncertainty range on each.

This bootstrapping approach presumes that the original data has enough samples covering all situations so that it represents the entire possible population of data. Then the new sets (sampled with replacement) represent legitimate realizations of sample populations. Accordingly, the distribution of model prediction values from each resampled set represents the distribution that would arise if the true population were independently sampled.

Bootstrapping assumes that the limited data represents the entire population of possible data, that the experimental errors are naturally distributed (there are no outliers or mistakes, not necessarily Gaussian distributed, but the distribution represents random natural influences), and that the functional form of the model matches the process mechanism. Then a random sample from your data would represent a sampling from the population; and for each realization, the model would be right.

If there are N number of original data, then sample N times with replacement. Since the central limit theorem indicates that variability reduces with the square root of N, using the same number keeps the variability between the bootstrapping samples consistent with the original data. In step 1, the assumption is that the sample still represents a possible realization of a legitimate experimental test of the same N. If you use a lower number of data in the sample, M, for instance, then you increase the variability on the model coefficient values. You could accept the central limit theorem and rescale the resulting variability by square root of M/N. But the practice is to use the same sample size as the "population" to reflect the population uncertainty on the model.

In step 6, if only a few resamplings, then there are too few results to be able to claim what the variability is with certainty. As the number of step 6 resamplings increases, the step 9 results will asymptotically approach the representative 95% values. But the exact value after infinite resamplings is not the truth, because it simply reflects the features captured in the surrogate population of the original N data, which is not actually the entire population. So, balance effort with precision. Perhaps 100 resamplings will provide consistency in the results. On the other hand, it is not unusual to have to run 100,000 trials to have Monte Carlo results converge.

One can estimate the number of resamplings, n, needed in step 6 for the results in step 9 to converge from the statistics of proportions. From a binomial distribution, the standard deviation on the proportion, p, is based on the proportion value and the number of data:

$$\sigma_p = \sqrt{\frac{p(1-p)}{n}} \tag{29.65}$$

Desirably, the uncertainty on the proportion will be a fraction of the proportion:

$$\sigma_p = fp \tag{29.66}$$

where the desired value of f might be 0.1.

Solving Equation (29.65) for the number of data required to satisfy Equation (29.66),

$$n = \frac{((1/p)-1)}{f^2} \tag{29.67}$$

If $p = 0.025$ and $f = 0.1$, then $n \approx 4000$.

Although $n = 10,000$ trials is not uncommon, and $n = 4000$ was just determined, I think for most engineering applications 100 resamplings will provide an appropriate balance between computational time and precision. Alternately, you might calculate the 95% confidence limits on the y-values after each resampling, and stop computing new realizations when there is no meaningful progression in its value, when the confidence limits seem to be approaching a noisy steady-state value.

In step 9, if you assume that the distribution of the $\tilde{y}$-predictions are normally distributed, then you could calculate the standard deviation of the $\tilde{y}$-values and use 1.96 times the standard deviation on each model prediction to estimate the 95% probable error on the model at that point due to errors in the data. Here, the term error does not mean mistake; it means random experimental normal fluctuation. The upper and lower 95% limits for the model would be the model value plus/minus the probable error. This is a parametric approach.

By contrast, searching through the $n = 4000$, or $n = 10,000$ results to determine the upper and lower 97.5 and 2.5% values is a nonparametric approach. The parametric approach has the advantage that it uses values of all results to compute the standard deviation of the $\tilde{y}$-prediction realizations and can get relatively accurate numbers with much fewer number of samples. Perhaps, $n = 20$. However, the parametric approach presumes that the variability in $\tilde{y}$-predictions is Gaussian. It might not be. The nonparametric approach does not make assumptions about the underlying distribution, but only uses two samples to interpolate each of the $\tilde{y}_{0.025}$ and $\tilde{y}_{0.975}$ values. So, it requires many trials to generate truly representative confidence interval values.

Unfortunately, the model coefficient values are likely to be correlated. This means that if one value needs to be higher to best fit a data sample, then the other will have to be lower. If you plot one coefficient w.r.t. another for the 100 resamplings recorded in step 3 and see a trend, then they are correlated. When the variability on input data values is correlated, the classical methods for propagation of uncertainty are not valid. They assume no correlation in the independent variables in the propagation of uncertainty. Step 10 would have questionable propriety.

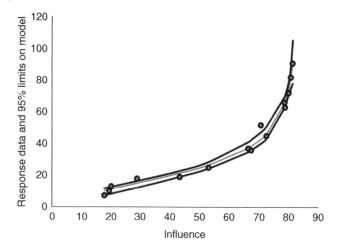

Figure 29.22 Use of bootstrapping to reveal model uncertainty due to data variation. *Source:* Rhinehart (2016b). Reproduced with permission of Wiley.

Also, step 9 has the implicit assumption that the model matches the data, that the model cannot be rejected by the data, and that the model expresses the underlying phenomena. If the model does not match the data, then bootstrapping still will provide a 95% confidence interval on the model; but you cannot expect that interval to include the 95% of the data. As a caution, if the model does not match the data (if the data rejects the model), then bootstrapping does not indicate the range about your bad model that encompasses the data, the uncertainty of your model predicting the true values.

Figure 29.22 reveals the results of a bootstrapping analysis on a model representing pilot-scale fluid flow data. The circles represent data, and the inside thin line is the normal least squares modeled value. The darker lines indicate the 95% limits of the model based on 100 bootstrapping realizations evaluated at the x-values of the data set of 15 elements. (The y-axis is the valve position and the x-axis is the desired flow rate. Regression was seeking the model inverse.)

29.8.1 Interpretation of Bootstrapping Analysis

Here is a bootstrapping analysis of the response of electrical conductivity to salt concentration in water. The data in Figure 29.23 are generated with concentration of salt as the independent variable and conductivity as the response; but as an instrument to use conductivity to report salt concentration, the x- and y-axes are switched. The figure represents a calibration graph. The model is a polynomial of order 3 (a cubic power series with four coefficients).

The conductivity measurement results in a composition error of about ±2 mg/dL in the intermediate values and higher at the extremes. The insight is that if ±2 mg/dL is an acceptable uncertainty, then the calibration is good in the intermediate ranges. If not, the experimenters need to take more data to use more points to average out variation. Notice that the uncertainty in concentration has about a ±5 mg/dL value in the extreme low or high values. So, perhaps the experiments need to be controlled so that concentrations are not in the extreme low or high values where there is high uncertainty.

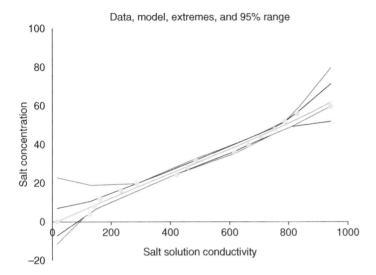

Figure 29.23 Instrument calibration of salt concentration w.r.t. conductivity.

There are five curves in the figure. The middle curve is the best least squares model. It is a continuous function. The extreme curves represent the extreme values of the models, at selected *x*-points, from the 100 bootstrapping trials. The extreme curves are a connect-the-dots pattern, not a continuous model representation. The close-to-extreme curves are connect the dots of the 95% upper and lower model results.

Figure 29.24 is another example of data from calibrating index of refraction with respect to mole fraction of methanol in water. Because the response gives two mole fraction values for one I.R.

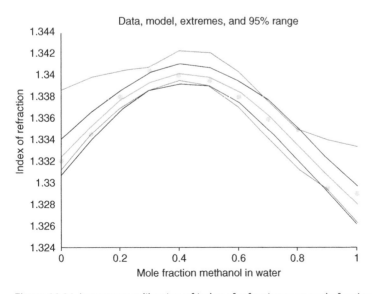

Figure 29.24 Instrument calibration of index of refraction w.r.t. mole fraction methanol in water.

value, the data represents the original x–y order, not the inverse that is normally used to solve for x given the measurement. This model has a sine functionality, because it gives a better fit than a cubic polynomial.

How to interpret? If the distillation technician takes a sample, measures index of refraction, and gets a value of 1.336, the model (middle of the five curves) indicates that the high methanol composition is about 0.72 mol fraction. But the 95% limits indicate that it might range from about 0.65 to 0.80 mol fraction. This large range indicates that the device calibration produces ±0.07 mole fraction uncertainty or ±10% uncertainty in the data. However, if the index of refraction measurement is 1.339, the model indicates about 0.4 mol fraction, but the uncertainty is from 0.3 to 0.6. Probably, such uncertainty in the mole fraction would be considered too large to be of use in fitting a distillation model to experimental composition data.

Bootstrapping analysis reveals such uncertainty, when it might not be recognized from the data or the nominal model. The model can be substantially improved with a greater number of data in the calibration.

The data in Figures 29.23 and 29.24 was generated by ChE students in the undergraduate unit operations lab at OSU. I greatly appreciate Reed Bastie, Andrea Fenton, and Thomas Lick for sharing their index of refraction data for my algorithm testing and to Carol Abraham, John Hiett, and Emma Orth for sharing their conductivity calibration data. The fact that this analysis suggests that they need more data is not a criticism of what they did. One cannot see how much data is needed until after performing an uncertainty analysis.

Hopefully, these examples help you understand how to use and interpret model uncertainty.

29.8.2 Appropriating Bootstrapping

For steady-state models with multivariable inputs (MISO), or for time-dependent models, the bootstrapping concept is the same. But the implementation is more complex.

For convenience, I would suggest to appropriate the bootstrapping technique. Hand select a portion of data with "+" residual values and replace them with sets that have "–" residual values. Find the model. Then replace some of the sets that have "–" residuals with data that has "+" residuals. Find the model. Accept that the two models represent the reasonable range that might have happened if you repeated the trials many times. I think this is as reasonable a compromise to bootstrapping as it is to what would happen if truly generating many independent sets of trial data.

A first impression might be to take all of the data with "+" residuals out, replacing them with all of the data with "–" residuals. But don't. Consider how many heads you expect when flipping a fair coin 10 times. You expect 5 heads and 5 tails. You would not claim that something is inconsistent if there were 6 and 4 or 4 and 6. Even 7 and 3 or 3 and 7 seem very probable. But 10 and 0 or 0 and 10 would be improbable, an event that is inconsistent with reality. So, don't replace all of the "+" data with "–" data or vice versa. Take a number that seems to be at the 95% probability limits. The binomial distribution can be a good guide. If the probability of an event is 50%, then with $N = 10$ data, expecting 5 to have "+" residuals, the 98% limits are about 8 and 2 or 2 and 8. So, move 3 from one category to the other. For $N = 15$ data, the 95% limit is about 11 and 4 or 4 and 11; so, move 4. For $N = 20$ data, it is about 14 and 6 or 6 and 14; so, move 4. Use this as a guide to choose the number of data to shift.

29.9   Perspective

In nonlinear regression you could use any optimizer and any convergence criterion. You could use Levenberg–Marquardt with convergence based on the cumulative sensitivity of the DV changes on the OF. You could use Hooke–Jeeves with convergence based on the pattern exploration size. I use leapfrogging and, for convergence, the steady-state condition on the random subset rms value. Any combination can be valid.

However, regardless of the optimizer or the convergence criterion, nonlinear regression beats regression on the linear transformed model, because the linear transformations change the weighting on some data points. I strongly encourage you to use nonlinear regression. However, if there is a copious amount of data so that deviations are balanced, or if the variation is small, then linearizing transforms could represent best balance of convenience and sufficiency over perfection.

Regardless of the optimizer there could be local optima traps in nonlinear optimization, so you should use a best-of-N-independent initializations approach. Some people use $N = 10$. I suggest $N = \ln(1 - c)/\ln(1 - f)$.

I like nonlinear regression over linear regression on linear transformed models because it is as easy for me and gives results that are truer to reality. I like $N = \ln(1 - c)/\ln(1 - f)$ because it provides a basis for the value of N. I like leapfrogging because it provides a balance of simplicity and performance (robustness, PNOFE) that wins in most test cases I've explored. For regression, I like the convergence criterion based on random subset rms steady-state condition because it is scale independent, is data based, and does not require a user to have to input any thresholds. I like the r-statistic filter method for steady-state identification because it is robust and computationally simple.

Although I think my preferences for nonlinear regression are defensible, they will not be shared by everyone.

The program "Generic LF Static Model Regression" combines what I think are best practices: best of N for repeat runs, steady-state stopping, and leapfrogging.

29.10   Takeaway

Your model is wrong. First, nature is not bound by your human cause-and-effect modeling concept. Second, even if your model could be functionally true, because experimental data are noisy, not exactly reproducible, then a best model through the data will be corrupted by the data variability. For non-linear models, use a bootstrapping technique to reveal the uncertainty of the model resulting from the data variability.

Vertical least squares minimization is grounded in normal (Gaussian) variability but pretends that all input values are perfectly known. Akaho's method is a nice simplification of maximum likelihood to account for input uncertainty; but it presumes uniform variance. The classic maximum likelihood technique uses the Gaussian distribution, but alternate models could be used. However, it is a two-stage optimization procedure, and the conceptual perfection might not be justified by the data-induced uncertainty in the model, or the necessarily specified estimates for the σ-values of the input and response variables. For me, Akaho's method provides the best balance of accounting for input variability with minimum complexity.

Many nonlinear regression applications have a single optimum with a smooth, quadratic-like response to the model coefficient values. Gradient or second-order optimization techniques are often excellent. However, as often in my applications, nonlinearities or striations in the response confound gradient and second-order procedures. In any case the user must choose a convergence criterion and appropriate threshold, which requires *a priori* knowledge.

For generality (robustness, simplicity, effectiveness), I like LF with best-of-N randomized initializations and convergence on the rms of a randomly sampled subset of the data. I like bootstrapping as a method to assess uncertainty of the model due to data variability.

29.11 Exercises

1 Show that you know what the "normal equations" are. You can explain, derive, describe, or whatever; but keep your answer direct to the point and simple.

2 State which is better, and explain why: the use of $\Sigma|d|$ or Σd^2 in regression.

3 Figure 29.25 shows a curve fit to three data points. Variables x and y have the same units and scale. Since values for σ_x and σ_y change with x–y position, the likelihood contours are shown with varying aspect ratio. Values 0.7, 0.8, and 0.9 represent the contours of constant likelihood. In a maximum likelihood optimization to best fit the curve to the data, determine the value of the objective function.

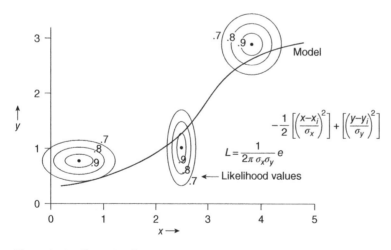

Figure 29.25 Illustration for Exercise 3.

4 Linearization of a model by log-transforming the data distorts the data and causes the optimizer to find a set of optimal parameter values that is less true to the underlying trend than when the optimizer uses the untransformed data. What conditions would make such linearization acceptable? Defend your answer.

5   Linearization of a model by log-transforming the data distorts the data and causes the optimizer to find a set of optimal parameter values that is less true to the underlying trend than when the optimizer uses the untransformed data. But log-transforming the objective function (a feature taken advantage of in the transformation of maximizing likelihood to minimizing squared deviations) does not change the optimal solution. Explain why one log-transformation changes the optimal solution but the other does not. You may use equations and sketches to help reveal your explanation.

6   If you did not know the phenomenological functional form of a relation, you could use a power series expression, $\rho = a + bT + cT^2 + dT^3 + \ldots$ to generate a $\rho(T)$ relation. How would you determine the number of terms in the power series expansion? Explain.

7   Wikipedia reports that the phenomenological model for temperature dependence of resistivity in noncrystalline semiconductors is $\rho = A\exp\left(T^{-1/n}\right)$ where $n = 2$, 3, or 4 depending on the dimensionality of the system. If you knew that functional relation, you would have one coefficient, A, to fit the model to data. Is that linear or nonlinear regression? Explain.

8   Figure 29.26a is a plot of three data pairs of y versus x. Sketch the linear relation, $y = a + bx$, that is the best line in a traditional least squares approach (vertical deviations). I suggest you use your pencil as the trial line, move it around to "eyeball" the best line, and then sketch the line. Figure 29.26b indicates likelihood contours about the same three data points. Sketch the linear relation, $y = a + bx$, that is the best line in a maximum likelihood approach. Again, I suggest you use your pencil as the trial line and move it around to "eyeball" the best line. Discuss the similarity and contrast of the two solutions.

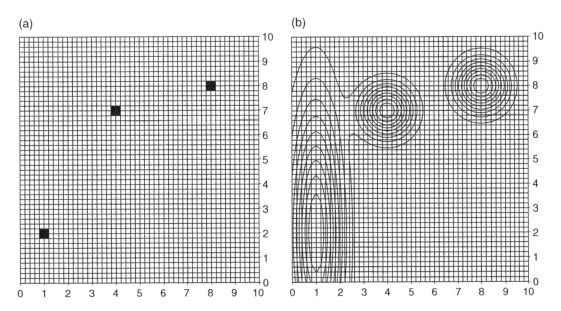

Figure 29.26 (a) Data, (b) data with uncertainty indicated by likelihood contours.

9 What limiting conditions would make vertical least squares return just as good a model as maximum likelihood? Defend your answer.

10 FPE is a method to determine when to stop adding functionalities to a model. Can FPE be used if a functionality is a reciprocal or exponential? Consider this question: "The current $m = 3$ model is $y = a + bx + cx^2$, can I use FPE to determine if I should use an $m = 4$ model of the form $y = a + bx + cx^2 + d/x$?" Choose, and explain your answer: Yes, No, Possibly.

11 If the logarithm is a strictly positive transformation that converts the maximum likelihood concept to the least squares concept for regression, why should data not be log-transformed to linearize the model in regression?

12 Use objective function transformations to show how maximizing likelihood becomes minimizing vertical least squares. Start with this optimization statement (a Gaussian likelihood model that assumes no x-uncertainty and constant y-variability). The model is a power law model. N is the number of data, and i is the data index:

$$\max_{\{a,p\}} J = \prod_{i=1}^{N} \frac{1}{\sqrt{2\pi}\sigma} e^{-(1/2)[(y_i - \tilde{y}(x_i))/\sigma]^2}$$

S.T.: $\tilde{y} = ax^p$.

13 If the model is $\tilde{y} = ax^p$, and σ_y and σ_x have different values, but are constant, Akaho's technique to approximate maximum likelihood scales the variables by the standard deviations, then seeks a solution by the scaled notation below. Unscale the variables to convert this to the user-familiar variables of $a, p, x,$ and $\tilde{y}$.:

$$\min_{\{a',p'\}} J = \sum_{i=1}^{N} \left[\frac{y'_i - \tilde{y}'(x'_i)}{1 + (d\tilde{y}'/dx')^2} \right]^2$$

14 In regression modeling, we often log-transform a power law model to obtain an equation that is linear in the model coefficients. Then we can apply the analytical least squares of vertical differences approach to generate the "normal equations," from which linear algebra gives the exact solution for model coefficient values. Quantify the weighting distortion on the squared deviation values due to the log-transformation.

15 Derive equations that show, for regression to a linear model, minimizing the sum of squared vertical differences between data point and line provides different results from minimizing the sum of squared normal (perpendicular or orthogonal to the regression line) differences.

16 You do not need to understand chemical reactions to do this exercise. A chemical reaction model of a packed-bed reactor is a follows:

$$k_1(C_i - C_o) + \ln\left(\frac{C_i}{C_o}\right) = \frac{Vk_o}{F} e^{-E/RT}$$

where

C_i is the inlet concentration of reactant, gmol/L
C_o is the outlet concentration of reactant, gmol/L
V is the reactor volume, mL, $V = 1000$ mL
F is the reactant inlet flow rate, mL/s
T is the reaction absolute temperature, K
R is the gas law constant 8.314, kJ/kmol-K
k_1 is the mass transport resistance coefficient, $(\text{gmol/L})^{-1}$—nominal range 5–100
k_0 is the "pre-exponential" reaction rate coefficient, s^{-1}—nominal range 0.0001–1
E is the reaction activation energy, kJ/kmol—nominal range 5000–40,000

The unknowns are the values of reaction properties k_1, k_0, and E. Use optimization of the least squares C_o deviations from data to model (vertical deviations) to determine values for the reaction properties k_1, k_0, and E. The known values belong in several categories. The gas law constant, R, is a universal constant of nature. The reactor volume, V, is a physical property of the experimental equipment. Design of the experimental conditions leads to user choices of the three variables: C_i, F, and T. Finally, the response of the reaction system is the measured value for C_o.

This will be root finding nested with the optimization. Either Newton's method or successive substitution works well for the root finding. The optimizer will "guess" at values for the three DVs (k_1, k_0, and E). Then root finding is needed to solve for the modeled C_o value. Then the sum of squared deviations in outlet concentration (model minus measured) can be evaluated.

Use steady state based on random sum squares sampling as the convergence criterion. Use N independent trials to be 99% confident that you'll have found at least one of the best 1% possible solutions (Table 29.1).

17 In regression modeling, we often log-transform a power law model $y = ax^p$ to obtain an equation that is linear in the model coefficients. Then we can apply the analytical least-squares-of-vertical-differences approach to generate the "normal equations," from which linear algebra gives an exact solution for model coefficient values. Alternately, we can use nonlinear numerical optimization techniques to approximate the coefficient values. Nonlinear optimization is more work, but does not cause the distortion of the log-transformation. Choose coefficient values for your power law. That model represents the unknowable truth about nature. Simulate experimental data. Perform both nonlinear regression and linear regression on the log-transformed model, with the data, to get model coefficients. Compare results. On a y versus x plot, there is a steep portion of the power law model and a not-so-steep portion. Your data points should include both slope regions; otherwise linear regression will be right. If your data has zero noise, then both regression approaches should give identical coefficient values, equaling the true (but unknowable) values; however, the convergence criteria on the numerical algorithm will limit precision. If the noise is too high, neither method will give good values. If there are many data points, then in any one region, "+" and "−" deviations will balance the "pull," and the linearized approach should not be biased by the weighting distortion. However, if too few (just 2) data points, the two approaches should return identical, but bad wrong, coefficient values. You choose the number, location, and noise aspects of the data so that your comparison of the two regression techniques reflects the method, not the number, placement, and noise on the data. In generating data, be true to reality. Choose a nominal x-value, but perturb it a bit before calculating the y-value. Then add noise to the modeled y-value to represent the measurement. Your data representing your

Table 29.1 Data for Exercise 16.

Run	Nominal C_i (gmol/L)	Nominal F (ml/s)	Nominal T (K)	Measurement C_o (gmol/L)
1	0.5	1	300	0.387994
2	0.5	1	325	0.266296
3	0.5	1	370	0.194875
4	0.5	2	300	0.585542
5	0.5	2	325	0.486686
6	0.5	2	370	0.283753
7	0.5	3	300	0.395264
8	0.5	3	325	0.345275
9	0.5	3	370	0.394145
10	1	1	300	0.885848
11	1	1	325	1.0322
12	1	1	370	0.595486
13	1	2	300	0.920307
14	1	2	325	0.696697
15	1	2	370	0.948037
16	1	3	300	0.788529
17	1	3	325	1.129603
18	1	3	370	0.881381
19	2	1	300	1.988986
20	2	1	325	2.039343
21	2	1	370	1.48991
22	2	2	300	2.250937
23	2	2	325	1.778625
24	2	2	370	1.646426
25	2	3	300	2.110882
26	2	3	325	1.777817
27	2	3	370	1.99707

experimental knowledge of the process should be the nominal x- and y-measurement pairs. You cannot know either the actual (perturbed) x-value or the true y-value.

18 I claim that FPE is a good method for selecting model order. The method finds the number of model coefficients that minimizes $FPE = [(N + m)/(N - m)] * SSD_m$. What do you think? Provide evidence to support your conclusion. Here is one approach: use a linear y versus x relation to generate data, and add noise to the y-data. Then let Excel (or any software) find the best linear model ($m = 2$), then the best quadratic model ($m = 3$), then cubic ($m = 4$), etc. At each m, use the r-square as a measure of SSD. You can show that $SSD_m = SSD_1 * (1 - r\text{-square}_m)$. Since FPE′, FPE scaled by a positive multiplier, does not change the DV optimum, you can use $FPE′ = [(N + m)/(N - m)] * (1 - r\text{-square}_m)$ versus m to determine the best value of m. Because you know that the generating model was linear, if the method is good, then the results should show that the minimum FPE occurs at $m = 2$. If the FPE method is good, then the value of m should be consistent with the generating model (e.g. $m = 4$, if the generating model was cubic). And the approach should be independent of either the number of data, or the amplitude of the noise, or the slope of the generating line, or values of the other model coefficients. I suspect that the results will not be deterministic. You will not get a clear "Yes, it works perfectly, there was no error in the optimum m" result. The statistical vagaries of noise perturbations on the true y-values in one particular test may make it appear that a cubic model is best when the truth is quadratic. So, your assessment and conclusion as to whether FPE works or not cannot be based on an expectation that the method is universally perfect.

19 I have claimed that the steady-state identification method is a good approach to detecting convergence in optimization of model coefficient values in nonlinear regression. "Good" means that it works reasonably well, is scale independent, is user convenient, and is grounded in fundamentals. What do you think? Provide evidence to support your conclusion. What makes one method better than another? If the SSID approach is universal (if it is scale independent, and does not require a person to choose threshold values, and does not require adding apples and oranges), then that makes it better than the other approach. If the NOFE is lower for one method than another, while giving the same optimum DV value, then the one method is better. How will you define better? What will you measure to quantify better and to defend your conclusion?

20 Develop an experiment to compare Akaho's method to approximate total least squares to classic vertical least squares. Here is a guide: generate N (x, y) pairs from the truth about nature (a reference model). Add independent variation to the x- and y-values to obtain the simulated experimental data pairs. Assume that the simulated x-values are the truth, and use nonlinear regression on the conventional vertical y-deviations to determine model coefficient values. Also, scale the data with the respective variances. Use nonlinear regression on the Akaho distance to determine scaled model coefficient values. Unscale the coefficient values. Do this for several noise realizations (perhaps 10), and comment on the accuracy and precision of the coefficients from the two methods. Accuracy is the average deviation from the true value. You could compare the average coefficient values of the 10 runs with the values used to generate the noiseless, true, data to have a measure of accuracy. Precision is a measure of repeatability. You could use the standard deviation of coefficient values from the 10 realizations (or range of the values) as a measure of repeatability.

21 (i) Choose a common nonlinear constitutive relation from your discipline. For example, the ideal gas law, $PV = nRT$, would indicate that pressure is inversely related to volume, $P = nRT/V$. Such models might be the temperature dependence on resistivity, the displacement response to load, viscosity response to shear rate, etc. Nature is not as ideal as the simple model, and each of such relations has a progression of one-better models. The van der Waals relation is one-better than the ideal gas law. $P = nRT/(V - nb) - a(n/V)^2$. (ii) Use your one-better relation to generate a nonlinear response (such as P vs. V), and simulate an experiment in which you would change the

input variable and measure the noisy response (nature masks the true response with NID(0,σ) noise). Simulate taking 10 measurements over a range of conditions that might be practicable. For example, a volume range of 0 to ∞ is possible but not practical. (iii) Use FPE to determine the best order of a power series model that represents your output (response, result, dependent variable, y) to input (influence, cause, independent variable, x) response. A power series model is $y = a + bx + cx^2 + dx^3 + ex^4 + \ldots$.

22 Refraction is the bending of a ray of light as it passes through a change in media. It is based on electron density and can be used to analyze composition of a mixture. In calibrating an index of refraction (y) device to measure methanol composition in a sample (x), we generated the following data (Table 29.2):

Table 29.2 Data for Exercise 22.

x	y
0	1.332
0.1	1.3345
0.2	1.338
0.3	1.3405
0.4	1.34
0.5	1.3395
0.6	1.338
0.7	1.336
0.8	1.335
0.9	1.3295
1	1.329

We use a conventional power series model $y = a + bx + cx^2 + \cdots$ to get the calibration relation. What order equation is "right?" Use FPE (or some alternate measure of complexity and performance) to determine the right order of the calibration relation.

23 Generate a simulation experiment that demonstrates that a linearizing transform of a model to permit linear regression does not provide as good a result (accuracy and precision) as nonlinear regression. For example, a power law nonlinear model is $y = ax^p$, and linearizing by log-transformation, it is $\ln(y) = \ln(a) + p\ln(x)$. (i) You choose a model and coefficient values. Perhaps use an equation from your discipline that is nonlinear in coefficients and can be transformed with a clever rearrangement to become linear in coefficients. (ii) Generate (x,y) data that represents the unknowable truth. (iii) Add NID noise (white noise, Gaussian distributed; I like the Box–Muller technique) to the y-data. You choose sigma. This pretends that x can be perfectly known and that all measurement uncertainty is on the y-value. (iv) Use a least squares optimization to solve for the model coefficient values that make the nonlinear model best fit the data and the

transformed model best fit the transformed data. Record the regression values for the model coefficients. (v) Repeat steps (iii) and (iv) a bunch of times. (vi) Evaluate accuracy and precision on the regression coefficient values. Accuracy is the closeness of the average coefficient value to the truth. Precision is the variability of the coefficient values. The better technique will have lower values for both accuracy and precision. Note: If your noise amplitude is extremely large, then neither procedure will work well. Or if noise is very small, then both techniques will be equivalent on the "noiseless" data. If the number of data is excessively large, then "+" and "–" deviations will balance the impact of each other, and the two techniques will seem equivalent. If the x-range is small relative to the nonlinear distortion of the functional transformation, then the two techniques will seem equivalent. For example, use a log-transformation and the x-range is from 200 to 210. So, choose conditions that seem reasonable from your experience. Finally, how many is a bunch? How many independent data realizations and optimizations are needed to confidently claim that the accuracy of one technique is better than the accuracy of the other? Reverse the t-statistic equation to determine what value of N is required to claim that one is better than the other. (If you need more guidance on finding a model, use this orifice calibration application, in which the truth is $F = 3.1415926535(i-4.012)^{0.456789}$, where $4.012 < i < 18.9$ and $\sigma_F = 0.76543$. Here F represents the flow rate, and i represents the milliamp electrical signal from the orifice device to the control computer. And use this power law model $F = a(i\text{-}4.012)^p$ to determine the coefficient values from the simulation-generated data.)

24 Choose a model and model coefficient values, consider that to be the truth about nature, and test several aspects of this chapter. Generate data by the method of Figure 29.3, adding Gaussian variation to both model input(s) and output. Choose the functional form of your regression model to exactly match the simulator model. Use regression to determine the model coefficients that best fit the data. As a test, if there is zero variability on the inputs and output, then the regression model coefficient values should exactly match those of the simulator. With input and output variability, compare vertical least squares to Akaho's method. Compare choice of convergence criterion and threshold. Use bootstrapping to reveal the uncertainty of the regression model and compare with the unknowable truth about nature.

Section 5

Perspective on Many Topics

Now that optimizers and applications have been developed, a perspective on how to optimize optimization is in order. The last of five chapters in this section provides a troubleshooting guide, which may help you find clues to improving your applications.

Engineering Optimization: Applications, Methods, and Analysis, First Edition. R. Russell Rhinehart.
© 2018 R. Russell Rhinehart. Published 2018 by John Wiley & Sons Ltd.
Companion website: www.wiley.com/go/rhinehart/engineeringoptimization

30

Perspective

30.1 Introduction

This chapter contains diverse points of information, which can be clarifying or of interest. It is a step back to gain perspective of the forest. I often sense that the sequence of details on one topic can obscure its relation to other topics. Many have been presented. This chapter seeks to connect diverse topics so that the reader can understand the choices and interactions.

30.2 Classifications

Optimizers can be classified as to their algorithms. Additionally, applications can be classified as to the issues that they present. And convergence criteria can be classified as to their basis. Often the name of the category of one is misapplied to describe the other.

1) Optimizers can be classified by their methodology:
    a) Analytical—Use calculus to exactly solve for the solution. This may provide an explicit solution, or it may result in iterative root finding on the equations.
    b) Direct search—Only use function values. Guide the next trial solution with human heuristic rules.
    c) Region elimination—A methodology applied to univariate direct searches to eliminate a DV range that probably does not contain the optimum.
    d) Gradient based—Use local surface slopes (sensitivity of OF to DV) to direct the search.
    e) Second-order techniques—Use a quadratic model of the OF response to DVs to direct the next TS.
    f) Surrogate model based—Generate an approximating model of the OF response to DVs, and then use its optimum to direct the next TS. SQ uses a quadratic model as a surrogate model and the best few TS results to generate the next model. But the surrogate model could be any preferred functionality (such as a neural network), and the past points do not need to be discarded. If the OF calculation is very expensive (or time consuming), then surrogate models can be helpful.
    g) Constraint following—An unconstrained search is not blocked by constraints. When a constraint is active, the optimizer can follow the constraint toward an optimum or a point when

Engineering Optimization: Applications, Methods, and Analysis, First Edition. R. Russell Rhinehart.
© 2018 R. Russell Rhinehart. Published 2018 by John Wiley & Sons Ltd.
Companion website: www.wiley.com/go/rhinehart/engineeringoptimization

the search direction becomes unconstrained. There are many approaches (LP, GRG, ridge following, reduced gradient, etc.).

h) Combinations—No optimizer is perfect for all applications, and diverse combinations have been formulated to make the best of several approaches.

i) Inverse—Most optimizers use the DVs as the TS and use these influences on the model to calculate the states and the OF from the states. By contrast, DP uses the states as the DVs and calculates the influences required to make the state-to-state transition.

j) Single TS or multiplayer—A single TS algorithm uses one trial solution. This point might actually be a point (as in CHD), but it might also be a local surface exploration (as in HJ, NM, or NR). So, perhaps "local TS" is a better term than "single TS." By contrast, a multiplayer algorithm scatters "simultaneous" trial solutions throughout the feasible DV space and uses the collective understanding to guide the next search.

k) Stochastic or deterministic—Here, this dichotomy in nomenclature is applied to the optimizer, not the OF. Deterministic means that if the iteration were repeated, the TS would be exactly the same. Stochastic means that the optimizer uses a random number generator in making the next TS. As a direct search, CHD and HJ are deterministic, and LF and PSO are stochastic.

2) Applications can be classified by their character:

a) Deterministic means that the OF is a deterministic response to the DV. Any time a TS is replicated, exactly the same OF results:

i) Linear OF—The solution is at a boundary defined by the constraints.

ii) Nonlinear with continuous value and derivatives—This may have multiple optima, and nonlinearity may confound algorithm.

iii) Nonlinear with discontinuous values, derivatives, or flat spots—These difficulties may be the result of conditionals, discretization, category, rank, integer variables, and min/max operations in either the DV, the OF, or the constraints. They may be revealed as sharp valleys or ridges, flat spots, or cliff discontinuities. They confound gradient-based or second-order searches.

b) Stochastic—This means that the OF is responding to random events and replicate TS values will not return exactly the same OF value. This is characteristic of experimentally obtained OF values or Monte Carlo simulations. Use replicates and steady-state convergence criteria, and report proximity and range of solutions, not a single value.

c) Constraints block the path to the optimum. There are many types. These may be on the DV, the OF, or auxiliary state variables. They may be explicit (as a function of the DVs) or implicit (can't know that a constraint is violated until after the model reveals the consequences of the TS). They may be on a future state, or on a rate of change, or on an accumulation. Constraints can be handled in several ways:

i) Hard—No violation is permissible. This creates a discontinuity that confounds many optimizers.

ii) Soft—The magnitude of a constraint violation is added to the OF as a penalty. This alleviates the discontinuity issues, facilitates the optimizer in finding the unique solution, and permits a small constraint violation.

iii) Constraint following—Here the optimizer seeks to ride the constraint until it is no longer active. Some locally linearize the constraint to facilitate solution. In general the constraint cannot be exactly followed, and often these result in either a bit of constraint violation or suboptimal solution.

d) Multiple OF—There is not a single solution, but a set of solutions that are nondominant. Reveal the set. Let a user decide what balance best suits their application context.

e) Difficulties—I have labeled these as surface aberrations:

   i) Integer or discretized variables, which may be either DV, OF, or auxiliary states.

   ii) Slope or level discontinuities in the OF due to conditional (IF–THEN) statements associated with equations or procedures.

   iii) Striations, microridges on the otherwise apparently smooth surface that guides optimizers down the local valley to a local optimum. These could be due to numerical methods to solve ODEs or integrals.

   iv) Flat spots in the OF response to DV arise in rank or categorization OFs or with class-type DVs.

   v) Multiple optima, in which the TS may be led to a local optima not the global best.

   vi) Uncertainty in the givens (whether these are models, conditions, situations, forecasts, environmental variables, constants, material properties, etc.) creates uncertainty in the DV*.

   vii) Nonlinearity or non-quadratic relations in the OF response to DVs can confound many optimizers.

3) Convergence

   a) In optimization of a deterministic OF. Convergence can be based on either of the following:

      i) DV—Such as incremental iteration-to-iteration steps or end-of-iteration player-to-player distance measures.

      ii) OF—Again, this could be as incremental iteration-to-iteration steps or end-of-iteration player-to-player distance measures. Alternately, it could be related to DV impact on OF or relative impact w.r.t. uncertainty due to the givens.

   b) In optimization of a stochastic OF—For either stochastic algorithms or stochastic OFs, claim convergence when there is no improvement in OF w.r.t. iteration.

   c) In regression—Claim convergence when the rms of residuals from a random subset of data comes to steady-state w.r.t. iteration.

   d) In accepting that the global has been found—This repeats optimizer trials from randomized initializations with the aim to increase the probability of finding the global best. The number of trials can be based on confidence that global has been identified (best-of-N, Snyman–Fatti, Bayesian belief).

30.3 Elements Associated with Optimization

There are many elements, stages, and choices associated with optimization. The user needs to understand and specify each. The following elements are separate considerations; but they are interacting and often vaguely differentiated, which confounds analysis, reporting/communicating, troubleshooting, implementation, and outcomes. The following elements are situation specific. One set of choices is not universally applicable. Finally, the following elements are not mathematical analysis based. They are not one-choice-is-right, all-others-are-wrong. Each choice involves understanding the limits of the techniques and the computational implications of the situation or choice:

1) Understand the OF response to DV, the OF character, and the constraints to be able to choose appropriate procedures.

2) Choose the number of independent searches (to be confident in finding the global minimum).

3) Specify stopping criteria to end a futile search.
4) Select initial trial solution or range and incremental Δx, Δs explorations.
5) Choose the optimizer procedure (successive quadratic, heuristic direct, Newton's, Golden section, Cauchy, incremental steepest descent, LM, RLM, LF, PS, GA, GRG, etc.).
6) Choose a constraint accommodation.
7) Choose how to evaluate the OF for each trial solution. It may be a simple explicit function or a complicated procedure with nested root finding. It may be obtaining experimental results, or hand calculations, or surveys, etc.
8) Choose how to evaluate OF derivatives at each iteration. Should it be analytical or numerical? If numerical, should it be central difference, backward, or forward; and with what increment?
9) Choose what to use as the OF and how to blend multi-objective issues (multiplicative or additive, what equal concern factors, worst of best possible, 95% extreme, Pareto optimal set, etc.).
10) Choose models to relate constraints, penalties, and OF to DV.
11) Choose convergence criteria and thresholds (OF is close enough to expected minimum, $\Delta X < 0.1$ or 0.0001, on ΔX or ΔOF or something else).

30.4 Root Finding Is Not Optimization

There are similarities between root finding and optimization. Both have Newton's method, both use degree of freedom language, and both have convergence criteria; and this often causes confusion. Root finding seeks a value of x that makes $f(x) = 0$. By contrast, optimization seeks a value of x that minimizes $f(x)$.

Optimization can be used for root finding. If the objective is to find a value of x that makes $f(x) = 0$, then square the function and use optimization to find the value of x that minimizes $[f(x)]^2$. The advantage is that optimization provides more robust and efficient options than root finding.

However, root finding can be used for optimization. At the optimum, the derivative is zero. So root finding on the derivative of $f(x)$, $df/dx = g(x)$, will find the optimum of $f(x)$.

Although, generally, optimization is more robust, optimization is not infallible. The algorithm might be misdirected by the surface feature, the technique might find a local optima, convergence choices might stop it too early, and so on.

30.5 Desired Engineering Attributes

Engineering is not just about technical competence. State-of-the-art commercial software beats novice humans in speed and completeness with technical calculations. By contrast, engineering is a decision-making process about technology within human enterprises, human value systems, and human aspirations.

In an exercise with our faculty and Industrial Advisory Committee members to understand our constituent's desires, it became apparent that engineering is a process of balancing opposing ideals; and we developed the following list in an effort to capture the timeless balance of values that guide the activity of engineering. Originally developed in about 2002, we periodically reviewed this with our IAC members and periodically fine-tune the wording.

A straight line is very long. Maybe the line goes toward those who do rigorous theoretical science on one end, and toward those who unquestionably push the Run button and accept the computer output on the other end. No matter where one stands, the line disappears into the horizons. No matter where one stands, it feels like the middle, the point of right balance between the extremes, because the other extremes are out of sight, or because the individual is personally comfortable that far from the two extremes. But the person over there, too far, thinks they are in the middle, also. The right place on the line is not based on proximity to an ideal desired extreme, or distance from an undesired extreme. The right place is based on the utility of the choice within application context. Finding the right place is an optimization exercise.

To be prepared for industrial careers, graduates need to understand the issues surrounding what we have termed desired engineering attributes, a description of the right balance between opposing ideals.

Engineering is an activity that delivers solutions that work for all stakeholders. Desirably, engineering

- Seeks **simplicity** in analysis and solutions while being **comprehensive** in scope
- Is **careful**, correct, self-critical, and defensible; yet is performed with a **sense of urgency**
- Analyzes **individual mechanisms** and integrates stages to **understand the whole**
- Uses state-of-the-art **science** and **heuristics**
- Balances **sufficiency and practicality** with **perfection**
- Develops **sustainable solutions**—profitable and accepted **today**—without burdening **future stakeholders**
- Tempers **personal gain** with **benefit to others**
- Is **creative**; yet **follows codes**, regulations, and standard practices
- Balances probable **loss** with probable **gain** but not at the expense of EHS&LP—**managed risk**
- Is a collaborative **partnership activity** that is energized by **individuals**
- Is an **intellectual analysis** that leads to **implementation and fruition**
- Is **scientifically valid**, yet **effectively communicated** for all stakeholders
- Generates **concrete** recommendations that honestly reveal **uncertainty**
- Is grounded in **technical fundamentals** and the **human context** (societal, economic, and political)
- Is grounded in **allegiance to the bottom line of the company** and to **ethical**, **legal**, **and social standards of technical and personal conduct**
- Supports **enterprise harmony** while seeking to **cause beneficent change**

30.6 Overview of Optimizers and Attributes

30.6.1 Gradient Based: Cauchy Sequential Line Search, Incremental Steepest Descent, GRG, Etc.

- Follows the local steepest descent direction. Perhaps not the direct path toward the minimum
- Zigzags wall to wall across a long descending valley
- Derivative can be confounded by nonanalytic or stochastic surfaces
- Gets stuck on hard constraints or steep valleys (except for GRG)
- Goes to the local minima
- Stops if gradient = 0

30.6.2 Local Surface Characterization Based: Newton–Raphson, Levenberg–Marquardt, Successive Quadratic, RLM, Quasi-Newton, Etc.

- Uses a second-order local characterization (quadratic surrogate model) of the surface to jump to a supposed minimum. Predicated on a quadratic OF function of each DV. If true, first jump-to is the optimum. If nearly true, jump-to is near the optimum.
- Gets confounded if surface has discontinuous level or derivatives, flat spots, stochastic response, or inflections.
- Newton's portion or SQ will seek min, max, or saddle points.
- Newton's portion or SQ will encounter a divide by zero in a flat or linear section.
- Zigzags wall to wall across a long descending valley.
- Gets stuck on hard constraints or steep valleys.
- Goes to a local minima.
- NR (and Quasi-Newton) and SQ jump to absurd out of bounds places if outside of the inflection point. LM, RLM, blend ISD, and NR to preserve a downhill progress.
- OF transformations can make a surface more or less quadratic-ish and change the ease that the optimizer finds the minimum. But the OF transformations that make the local minima more quadratic-ish might not provide user convenience (for viewing results, for relating results to the situation, etc.).

30.6.3 Direct Search with Single Trial Solution: Cyclic Heuristic, Hooke–Jeeves, and Nelder–Mead

- Sequentially moves the single trial solution
- Uses function values only (no derivatives) within human logic to direct the next trial solution
- Robust to many surface features that cause problems for gradient and model-based optimizers
- Gets stuck on hard constraints or steep valleys
- Goes to a local minima
- Effectiveness is independent of OF transformations that are strictly positive monotonic

30.6.4 Multiplayer Direct Search Optimizers: Leapfrogging, Particle Swarm, and Genetic Algorithms

- Same advantages as direct searches earlier
- Less impacted by hard constraints or steep valleys
- Increases probability of finding the global minima
- Effectiveness is independent of OF transformations that are strictly positive monotonic

30.7 Choices

The choices a user makes regarding the application are more important to the outcome than the optimization algorithm or what executes it. Don't think that because you choose BFGS over SQ, or PSO over HJ, or cloud computing over a calculator, there is some enhancement to the solution validity. Other user choices dominate the validity of the solution. These include:

1) Defining, and redefining, the OF (criteria to consider, models)
2) Defining how opposing aspects are combined
3) Choosing values for weighting coefficients
4) Defining the DVs (separating givens from those that could be changed as DVs)
5) Defining constraints (identifying them and choosing hard or soft)
6) Choosing an algorithm appropriate to the surface and attributes for this application context
7) Choosing algorithm methodology—numerical or analytical derivatives (what delta to use, forward or central), quasi-Newton, etc.
8) Choosing convergence criterion and threshold
9) Choosing a methodology to solve the model—analytical or numerical? If numerical what step size and method (for ODE and PDE Euler's, Runge–Kutta, collocation) (for integration rectangle rule, Simpson's, Gaussian quadrature, etc.)
10) Choosing number of starts (to provide adequate confidence that the global has been found)
11) Choosing algorithm coefficients—expansion and contraction, tempering, and switch criteria to end-game values
12) Choosing TS initialization—randomize, over what DV range, how many players?
13) Evaluating confidence in the DV* OF* result w.r.t. uncertainty in the models, "givens" basis
14) Choosing number of realizations in stochastic situations—number of replicates, what confidence limits to use

Be sure to clearly reveal your choices, and make choices that are consistent within the application context.

30.8 Variable Classifications

DV is the decision variable—it is the value of the condition that you choose. For example, lifestyle is dependent on your salary. You can choose work time, work effort, and job type, but not salary. Salary is a response to what you can choose. Lifestyle is a function of salary, but salary is not the DV. Lifestyle = f (salary) = $f(g$(work hours, training investment, etc.)). Similarly, if you want to increase productivity, you cannot use time/unit as the DV. You can change work processes, change time/unit, and change productivity.

There are a variety of naming conventions for numbers, representing the nomenclature preferences of mathematicians, philosophers, psychologists, statisticians, and computer language. But the concepts are aligned. Here is my explanation from an engineering view. (It substantially follows the S. S. Stevens 1946 categorization, which many number purists argue is inadequate.)

30.8.1 Nominal

Nominal (name) has no quantitative relation. It is a category or classification. The OF may be the count of items that are classified correctly. In a rock collection, there is the first rock you got, then the second, third, and so on. But the numbers are just names and do not represent any quantity related to properties of the rock. The names could have well been "A," "B," "C," or "alpha," "beta," "gamma." We name trial runs in randomized sequence. The "1" in the name "Trial 1" has no relevance to the flow rate, the impact, the difficulty, the benefit, and so on. Some numerical labels may have a meaning. The first digit on an American football player's jersey indicates the player's position and relative importance of that position to scoring. The second digit indicates the player. But the player number is just an alternate name, not a value to insert into a formula.

DVs may represent nominal labels, for example, Product x or Product y, distillation or crystallization.

30.8.2 Ordinal

Ordinal (order) means sequence or rank. A runner comes in first place, or second, or third, and so on in a race. Here the number is a measure of placement, or goodness, or desirability and relates to some property of the event or property of an item (largest, second largest, etc.; brightest, second brightest, etc.). Ordinal numbers reveal order. But they do not indicate a relative or proportional quantification of the property. First and second may be nearly tied and far ahead of third. Alternately, first may be far ahead of second and third. Ranking indicates relative relationship, but does not quantify it. In spite of this lack of relative positioning, you can average or find the median of rank numbers. I usually come in about fifth in a nuclear family mini-mini-marathon. Perhaps, never came in fifth, but came in first once (in a two-person race) and ninth another time (in a nine-person race). You can average them, but there is still little relative meaning.

A Likert scale is a ranking scale that people use in surveys, to get a collective opinion. These are often called interval, because the user choices would be 1, 2, 3, 4, etc. We use it in the bubble form for Student Survey of Instruction to get student opinion on "Rate the instructor's preparation." One student's opinion may be the instructor "seemed OK" and rate a 3 out of kindness. Another opinion may be the instructor "seemed OK" and rate a 2. Another may think the ordinary job was "as good as possible" and rate it a 4. The average of all students in the class could be a 3.141, which is not a number on the interval scale, but can be interpreted as not much better than "OK."

Like ranking values, interval scales may or may not be linearly related to the aspect they quantify. Here is an example: Although student end-of-course grade might have a linear correspondence to technical perfection to course material, when we curve a course based on the normal distribution, there may be 15% "A" grades, 25% "B"s, 50% "C"s, and so on. This would make the student grade in one course, or the student overall GPA (comprised of interval values of $A = 4$, $B = 3$, $C = 2$, etc.) not be linearly related to academic perfection.

Either DVs or OFs could be ordinal labels.

30.8.3 Cardinal

Cardinal numbers are linearly related to the quantity they represent. They could be continuum or integers, or of another discretization interval. They could be scaled to start at zero (such as zero degrees Kelvin means absolutely zero heat) or relative to a reference value (such as centigrade where zero is the freezing point of water). They are variously classified as integer, rational, real, or complex. Within optimization applications, I think the key concepts are discriminating between continuum and discretized values.

Integers are the counting numbers, the whole numbers. They indicate the number of whole items. They can only have integer values. They can only have values that exist at intervals of unity. DVs could be integers add one Tsp, add two Tsp. Try a linear model, second-order model, third-order model, and so on.

Integers are a type of interval number. But the value has a linear correspondence to the measured quantity, magnitude, or attribute. The interval could be half, or quarters, or 16ths. The Westminster

Clock chimes on the quarter hour. Its interval is 1/4 h. Or is it 15 min? Either way the time interval is not unity, and it has intervals of 0.25 or 15. A digital thermometer may read in interval increments of 0.5°C. Even though the temperature might be 24.3681°C, the display will reveal 24°C, because the temperature did not cross the 24.5°C threshold to report 24.5°C. In bit representation of computer storage, an 8-bit storage location might be 00101101. The discrete interval is 1 bit. But representing a display range of 0–100%, the 2^8 possible numbers in an 8-bit storage have a 1-bit discretization interval that represents a $(100–0\%)/2^8 = 0.390625\%$ interval. Observing such a number display, you would see all reported values as multiples of the 0.390625% interval. There would be no in-between numbers. I call this discretization error or uncertainty. Look at a table of viscosity or table of *t*-statistic critical values, and you will find that the table does not report infinite numbers for each entry. It may report one decimal digit, in which case the discretization interval is 0.1 or 4 decimal digits, in which case the discretization interval is 0.0001.

We often imagine that properties of space, time, mass, and temperature are continuous and that the property can be divided into infinitesimal intervals. Real numbers or continuum numbers permit having infinite decimal places. The ratio of 1 to 3 is 0.3333333…. But on an atomic view, mass is not continuous. If you want to increase the amount of water in a glass, the smallest increment you can add is one molecule of water. Effectively, on a large engineering scale, the continuum view seems valid, and we consider the measurements of properties to be real numbers or continuum, with the interval effectively zero.

From a mathematics view, continuum numbers are either rational (ratio) and can be represented by a ratio of integers or irrational and cannot be represented by a ratio of integers. When the decimal part has a repeated pattern, the number is rational. If there is no repeated pattern (sqrt of 2, Pi, base of natural logarithm, golden ratio, etc.), the real number with infinite decimal digits is irrational.

Real numbers (rational or irrational) and integers (which include interval numbers that are linearly related to the attribute) are directly related to the property of the event or item (time, mass, weight, intensity, value, etc.). As a result real and discretized numbers have a ratio or proportional property. If the numerical representation of one event is twice in magnitude as the numerical representation of the other, then the one event has twice the magnitude, impact, and value of the other. Real, integer, and discrete numbers preserve this ratio or proportional relation, making them useful for mathematical and engineering analysis. Most DVs are of the real number category and include coefficient values, dimensions, timing, quantity, and so on.

When DV choices have neither ordinal nor cardinal value, when they cannot be ranked or sequentially ordered according to the degree of some property (whether continuous or discrete, quantifiable or subjective), then progressive logic related to better DV values cannot be used. For example, consider these DVs: What car to buy: Chevrolet-Mini Cooper-Ferrari? How to travel: Air-bus-walk? In these cases, you need to do an exhaustive search and evaluate the OF for each choice. This book covers logic and algorithms for progressive-better-choice, primarily on cardinal (continuous numbers that provide relative values). Ordinal values permit ranking, but cannot be used in a ratio to establish relative value. "Flavor" is ordinal. You can rank foods in a taste preference. But you cannot say that French fried potatoes taste 3.15 times better than rhubarb.

30.9 Constraints

A) Constraints are conditions that cannot be violated.
B) May be on the DVs, OFs, or auxiliary variables.
C) May be on rate of change or value.

D) May happen now, or some place along the future path or future time.

E) Hard—must not be violated (or may not, should not, could not?). These are often related to human laws or dangerous catastrophic situations.

F) Soft (needs EC or λ value if added to the single OF, could define Pareto optimal solutions if the penalty becomes a separate OF).

G) Whether a fixed penalty value or a progressive one that increases with the extent of deviation, these are "soft" in the sense that they could be slightly violated by the optimum DV values.

H) With constraints added as penalty functions in the OF, the search algorithm is permitted to explore DV values that violate the constraint. By contrast, constraints are termed "hard" when they are not included as penalty functions; but when they define non-accessible, not permissible regions in the search. An algorithm might want the next trial DV solution values in a combination that violates a constraint, but the constraint override does not permit a search in that area. In a "hard" constraint implementation, when a constraint is active, the OF function value is "FAIL."

I) Constraints are not secondary OFs. We minimize OFs. Constraints have no influence as long as they are not violated. One method of handling constraints is to add a penalty function to the OF, and even in this constraint-handling approach, the penalty value is zero as long as the constraint is not violated.

J) The minimum of the OF is not a constraint. You could say that the minimum value of an OF represents the best and that you are constrained to no better values. But, in optimization, that is not termed a constraint, but the minimum value. There may be a constraint on an OF; for instance, "We do not invest in a project unless LTROI is greater than 40%." But in such a case, the best design might result in a 56.2% LTROI, so the optimum is not the constraint.

K) If the OF is a linear function of the DV, then the minimum is negative infinity. However, this occurs at an infinity value of the DV, which, although mathematical possible, is impossible to implement. So, the minimum value of the OF lies on the DV constraint.

L) Risk-based soft constraint. Often constraints are a bit arbitrary. Keep the tire air pressure below 40 psig, but above 30. Too high an air pressure should explode the tire, too low and it flexes too much and will wear out. But are 30 and 40 definitive catastrophic values? No. Realizing uncertainty on tire pressure gages and how people might overfill and then bleed back, "they" probably tell you don't go over 40 psig knowing that the real limit for danger is 50. And you can run a tire at 25 psig, but it does not instantly ruin, and it just wears faster. The event is not exceeding the 40 or 30 psig limits. The event is the modest consequence of violating a constraint, not a catastrophic response to doing so.

Whatever the constraints are, they are really just guides, not absolutes. What about composition? The sum must be 100%, it appears. But when composition is analyzed, it has uncertainty. When it is calculated, it has digital truncation error. So, a violation of 100% within expected uncertainty is fully acceptable. The posted speed limit is not a constraint. On good-weather low-traffic days, going somewhat above the limit is permissible. During dangerous conditions, you should not go "permissibly" fast.

Risk is the probability of an undesired event times the cost of that event. We make life decisions based on risk. The undesired event might be getting a speeding ticket, and the "cost" would include the sum of all impacts (arrest time, embarrassment, fine, insurance premium increase, etc.). The undesired event might be getting sick from eating tainted food:

$$r = p(\text{Event}) \cdot c(\text{Event}) \tag{30.1}$$

In optimization with constraints, the event might be the violation of the constraint—either equality $f(x) = a$, or inequality $f(x) > a$. The extent of the violation would be $\varepsilon = |f(x) - a|$ if equality, or if inequality, $\varepsilon = f(x) - a$ if $f(x) < a$, otherwise $\varepsilon = 0$.

One could use a logistic model to indicate the probability of the penalty event (ticket, tire failure) based on the extent of constraint violation:

$$p(\text{Event}) = \frac{1}{1 + e^{s(f(x) - c)}} \tag{30.2}$$

where s is a scale factor and c is the center (the value of $f(x)$ that has a 50% probability of leading to the undesired event). You could determine values for s and c that make sense. What value of $f(x)$ would have a 10% chance of leading to the event? What value of $f(x)$ would have a 90% chance of leading to the event? From these two relations (or similar relations), it is easy to solve the two logistic model coefficients.

The cost of the event might be modeled in any of a number of ways. The penalty may be a constant like the cost of a tire, the cost might scale linearly with the extent of violation, or with the square of the violation extent $c(\text{Event}) = b\varepsilon^2$.

In any case, choose a functionality that makes sense with the situation and uncertainty in the constraint "givens."

30.10 Takeaway

Any one step in the optimization procedure is fairly simple, but the assembled whole is somewhat complicated. There is interaction in the elements. There are many desirables to satisfy. Be sure to see the impact of the elements within the optimization context.

30.11 Exercises

1 State an advantage and a disadvantage for using multiple optimization approaches in one application. Use a specific example to illustrate your point. Feel free to use a sketch to support your explanation.

2 In each list, two items have similar attributes (uses, properties, methods, etc.). Encircle the item that is most unlike the other two, and state why.
   A Newton–Raphson, successive quadratic, Cauchy's sequential line search method
   B Constraint, OF value, Cauchy's sequential line search method
   C Equal concern, multi-objective, best-of-N
   D Levenberg–Marquardt, incremental steepest descent, CSLS
   E Cyclic direct, root finding, interval halving

3 Categorize each of these searches as either D for direct, G for gradient-based, S for surface characterization based, or some combination of D, G, and S.
   A Cyclic heuristic
   B Particle swarm

> **C** Cauchy
> **D** Successive quadratic
> **E** Golden section
> **F** Newton–Raphson
> **G** Levenberg–Marquardt
> **H** Incremental steepest descent
> **I** Hooke–Jeeves
> **J** Dynamic programming

4 For each pair listed, describe the advantages and disadvantages that specifically contrast the two techniques:

> **A** LP versus HJ
> **B** Threshold on ΔDV versus steady-state stopping criteria
> **C** Best-of-N number of starts versus more particles in the swarm
> **D** Methods that can deal with hard constraints versus converting hard to soft constraints
> **E** Scaled variables versus primitive (unscaled) variables
> **F** Multi-OF Pareto optimal versus weighted multiple objectives in a single OF

5 How can an underspecified optimization problem be (i) recognized and (ii) fixed?

6 Match each surface feature to the most appropriate optimizer. Explain your choice(s). (i) Nearly quadratic, (ii) sharp valley or cliff, (iii) multi-minima, and (iv) flat section.

7 Is an optimization iteration comprised of one function evaluation? Defend your "Yes," "No," or other answer.

8 With a brief discussion (and possibly supporting illustration), explain why numerical estimates of the derivative are preferred over analytical true values.

9 With a brief discussion (and possibly supporting illustration), explain whether the numerical estimate of the gradient elements should be central difference or forward difference.

10 Here is a list of features of optimization applications. Briefly describe the difficulty each creates. (i) Integer DVs, (ii) probabilistic/stochastic models, (iii) parameter correlation, (iv) IF-THEN conditionals (make jumps or changes in the model), (v) experimental OF data (as opposed to a computer calculation), and (vi) hard constraints.

11 If scaling DVs to a 0–1 range, such as $DV' = (DV - DV_{min})/DV_{range}$, does not change the DV^* value, why do it?

12 With a brief discussion (and possibly supporting illustration), explain, in assessing precision of an optimizer, whether you should look at the range (or some similar measure such as rms) on the DV or on the OF?

13 Match the items on the left column with the letter identification from one item on the right with the strongest connection (not the two or three letters that it could be).

Scaled variable	A	Convergence
Soft	B	Search algorithm
Cost	C	OF
Precision	D	DV
Stochastic	E	Constraint
Mimetic	F	Propagation of variance
NOFE	G	Aberration
p(global)	H	Initialization
Line search	I	Optimizer goodness metric
Levenberg–Marquardt	J	Model
$N = \ln(1 - c)/\ln(1 - f)$		
Flat		
Penalty for undesirability		
Reduces number of DVs		
Ridge discontinuity		
Strictly positive transformation		
Class		
Steady state with iteration		
Simplicity		
State and stage viewpoint		

14 Briefly explain the relation of each of the following fundamental topics to optimization. (To formulate a brief but complete answer, think through the entire explanation before you start writing.)

A $-\nabla f$
B Likelihood
C CDF
D $\epsilon_{OF} = \sum |(\partial OF/\partial DV)\epsilon_{DV}|$
E Local quadratic model

15 Explain why convergence threshold on the ΔDV is not the same as the precision of the DV solution.

16 Compare two convergence criterion: one based on the search range, Search_range = Sqr((dx1^2 + dx2^2)/2), and the other based on the maximum influence that the DV changes could have had on the OF. Stop if $\Delta OF_{max} = |(\partial OF/\partial x_1)\Delta x_1| + |(\partial OF/\partial x_2)\Delta x_2| \leq$ search range threshold. Which approach is better for a deterministic function (one that returns an exactly repeatable value each time it is called with the same DV values—not a stochastic function)? You need to explain (i) what attributes (plural) would make one method better, (ii) how would you quantify the attributes, (iii) what data is required for the analyses, and (iv) how would you structure evaluation tests to generate data that provides a comprehensive evaluation. Then you can (v) run the trials, (vi) organize the data, (vii) describe your results, and (viii) defend your choice of better.

17 Investigate something with the course material, not already done in the course, and which personally interests you. The effort that your project defines should be about the same as your effort in completing past course assignments. Don't make this a major project. For any mission that you choose, define a way to quantify goodness/effectiveness for the comparison, and devise trials to gather data that is sufficient to make claims. Your report should clearly state (i) your objective (perhaps as a hypothesis for example, "I think that faster change from incremental gradient to Newton will be better"), (ii) your approach to defining and assessing better/goodness/effectiveness, (iii) your design/explanation/derivation of the new rules/functions, (iv) your choice of tests and how they provide a comprehensive situations, (v) your test results and a discussion about how you know the data appropriately reveals features and permits comparison claims, and (vi) conclusions. Here are possible ideas:

A Explore an optimization application related to your work/research, present or past.

B Convert one of the 2D optimization algorithms that I provided into an *N*-D general method, and validate it (Visit the companion site.).

C Compare particle swarm with something else you decided was the best so far. Use multiple criteria and test situations.

D Explore changing the parameter values in an optimizer. For example, change the rate that the RLM shifts from incremental steepest descent to Newton's, or change the number of individuals in the leapfrogging approach. Use multiple criteria and test situations in the comparison.

E Explore new functions representing difficult optimization situations. Search the Internet or literature for test problems that others have created. Create your own functions that include an optimization difficulty. Show how the specific difficulty confounds an optimizer.

F Explore linear programming, dynamic programming, or genetic algorithm.

G Explore a Pareto optimal approach to a multi-objective problem. Return to something done in the course, but make it a multi-objective optimization (instead of combining multi-objectives or soft constraints into a single OF).

H Explore a method that permits an optimizer to retain the historical information it "saw" about the surface, and use that information (not just the local gradient or exploratory knowledge) to decide where to place the next trial solution.

18 Here are two ideas on improving an optimizer when the deterministic function has multiple minima. The objective is to reduce PNOFE = $N*$ ANOFE, the probable number of function evaluations to be confident in finding the vicinity of the global. Choose one idea and explore it: (i) implement it, (ii) reveal results, (iii) analyze outcomes, and (iv) comment on whether it is good. If you have an equivalent interesting idea, describe it.

A Idea 1: Coarse then fine convergence criterion in a single trial solution procedure. If there are several minima, some searches will lead to a local optimum. There is no need to have tight convergence and waste time and function evaluations at that local minimum. So, initially explore the $N = \ln(1-c)/\ln(1-f)$ trials with a coarse (loose, tolerant, wide) convergence criterion threshold. After the N coarse search trials, and the best-of-N optimum is coarsely located, initialize the final search at that DV value and continue the search with a tight convergence criterion. Since it is the same optimizer, f should be the same whether convergence is coarse or fine. On the one trial that is continued with the fine convergence, NOFE should be the same as if it were started with the fine convergence. But for the $N-1$ other trials, NOFE should be smaller. This approach should reduce NOFE, keeping f the same.

B Idea 2: Large then small leap-to window in leapfrogging. If there are several minima in a function, then a larger leap-to window in LF will permit better surface exploration and increase the probability that the global optimum is found. $DV_{Worst,\ New} = DV_{Best} - \alpha\,r(DV_{Best} - DV_{Worst,\ Old})$. Here, α is the leap-to window expansion factor. A window amplification factor of larger than 2.71828 leads to unstable cluster expansion. So, keep $\alpha < 2.71828\ldots$ However, a large leap-to window will make convergence very slow. A small window size, $\alpha = 0.1$ perhaps, will accelerate convergence, but will rapidly gather players very near to the local best, and can make the cluster converge on the side of a hill. So, perhaps $\alpha = 0.25$ is a good value to accelerate convergence without causing premature convergence. But such a value will prevent surface exploration. So, the idea is to use $\alpha = 2$ for the initial leap-overs, and then once the vicinity of the global has been found, switch to $\alpha = 0.25$. This initial exploration approach might increase the probability of finding the global optimum and increase f. And the end-game alpha-value might accelerate convergence reducing NOFE. What criterion will you use to switch alpha-values? One might consider that if the lead player has not switched in M number of leap-overs, then the vicinity of the global has been discovered. Another might progressively change α from 2 to 0.25 with increasing number of leap-overs. Another might use a Bayes' conditional probability logic to trigger the α-value change when the belief is adequately high that the global best has been found.

31

Response Surface Aberrations

31.1 Introduction

The term surface aberrations refers to features of the OF response to DVs, features of the topology, that make it difficult for optimization algorithms. Finding the optimum of an unconstrained quadratic surface is a trivial exercise. However, for many applications several types of OF features cause difficulty for the algorithm. These are easily visualized in the three dimension of a 2-D application (OF response to DV1 and DV2) but translate to 1-D and N-D applications. The following illustrations show the 3-D net and/or 2-D contour map patterns.

Unlike many classic test functions, most of the illustrations here are not simple functions that are contrived by the mind of man to reveal an aberration to challenge an optimizer. Most of these examples naturally appeared in my applications and are described in the following sections.

31.2 Cliffs (Vertical Walls)

Function 41, "parallel pumps," has level discontinuities (cliffs, or vertical walls) in the 3-D graph of the OF response to the DVs. This is illustrated in Figure 31.1.

The contour plot of Figure 31.2 also reveals the level discontinuities as multiple collinear contours.

Functions 33, 41, 46, and 79, and several others in the companion site have these sort of level discontinuities. These could be the result of a phase change (liquid to vapor; solid to liquid), equipment change (diesel to gas; batch to continuous), constraint, or penalty. Although Function 41 has the discontinuities aligned with the axes, and as straight lines, this is not universal. The cliff discontinuities can be off-axis and curved.

Gradient-based optimizers cannot evaluate the derivative at the level discontinuity.

31.3 Sharp Valleys (or Ridges)

Figure 31.3 is a 3-D illustration of an economic optimization of the dimensions of a tank for disengaging liquid and vapor (Function 27). In it, liquid and vapor flow together into one end of a horizontal cylindrical tank. As they flow to the other end, the liquid droplets fall to the bottom. The vapor and liquid then exit separately at the top and bottom of the other end of the tank. The DVs are tank

Engineering Optimization: Applications, Methods, and Analysis, First Edition. R. Russell Rhinehart.
© 2018 R. Russell Rhinehart. Published 2018 by John Wiley & Sons Ltd.
Companion website: www.wiley.com/go/rhinehart/engineeringoptimization

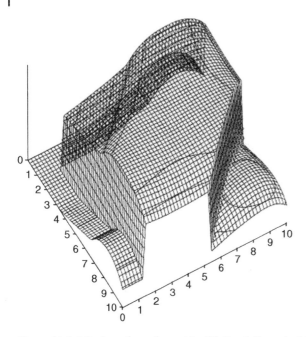

Figure 31.1 3-D view of a surface with cliffs (level discontinuities).

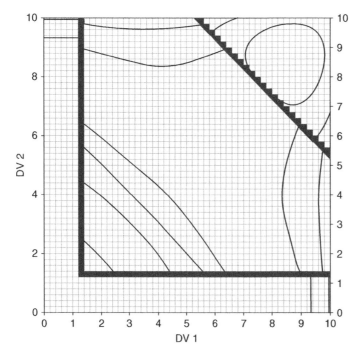

Figure 31.2 Contour map of a surface with level discontinuities (cliffs).

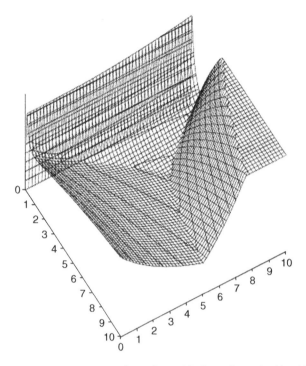

Figure 31.3 3-D view of a surface with slope discontinuities (sharp valleys).

diameter and tank length. There are IF–THEN conditionals that require the dimensions to be large enough to meet droplet settling time and liquid holdup. The sharp valley discontinuities are due to using the larger of the dimensions resulting from constraints on either the liquid or vapor holdup.

As Figure 31.4 shows, the contours have kinks that reveal the surface slope discontinuity, and the result is that, at the ridge (the bottom of a valley going across the valley), the optimizer thinks it has found the optimum. The sharp valley makes it improbable for the optimizer to move down the sharp valley. The locus of X-markers indicates where 20 independent optimizer trials stopped. The connect-the-line dots reveal the path of the last optimizer trial.

Such sharp valleys could be the result of a change in equipment size or law/regulation change (discontinuity). These can be due to the discretization of time or space of numerical methods used in integrating within the OF, or from convergence criterion in iterative root-finding subprocedures in the OF. Perhaps it is just an aspect of the DV impact on the OF.

That illustration was for sharp valleys or slope discontinuities. However, effectively the same situation can occur with curved bottoms to long valleys between relatively steep walls. Figure 31.5 illustrates this. Functions 54, 62, and several others in the companion site have similar issues.

Gradient-based optimizers cannot evaluate the derivative when there is a slope discontinuity. The steep slope to the valley walls relative to the valley bottom will make the search bounce from side to side. It does not have to be a discontinuity. Any valley between steep walls relative to the slope of the valley floor will create the problem. If encountered, consider changing the scale of the DVs so that the slope of the valley floor is similar to its walls.

Conjugate gradient techniques can help, if the function is continuous. I like multiplayer algorithms.

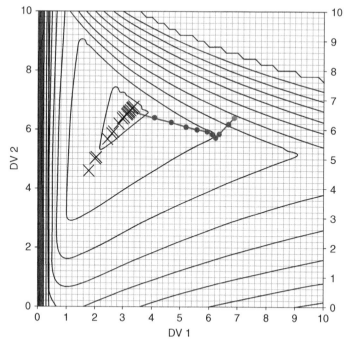

Figure 31.4 Contour map of a surface with slope discontinuities (sharp valleys).

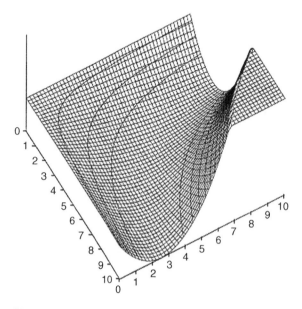

Figure 31.5 A surface with a valley with steep side walls and a gentle bottom slope.

31.4   Striations

These are small-scale corrugations or grooves or parallel ridges in the surface. Often, these are the result of numerical methods to solve differential equations or to integrate functions, and the issue they create is that the local valleys guide an optimizer to follow the local channel to its bottom. They prevent an optimizer from "seeing" the overall trend in the surface.

Figure 31.6 shows a 3-D close-up of such in Function 15, a study of investment for retirement, where the numerical time discretization in solving the model causes the striations. Figure 31.7 shows the contour of the local area.

On a large scale the striations are not visible, but as Figure 31.8 shows, they guide an optimizer along the vertical channels to a local minimum in the channel and prevent it from moving sideways toward the global optimum. Shown are the results of 25 runs of LM optimizer from randomly initialized locations. The Xs reveal the converged points. The path is that of the last trial. Note that the path is vertical, along a ridge, not diagonal toward the optimum.

Whether LM, RLM, ISD, CSLS, CHD, or HJ, the solution outcomes are very similar. The issue can be eased with smaller time discretization, but that adds burden to evaluating the OF. Multiplayer algorithms are better at finding the global optimum.

31.5   Level Spots (Functions 1, 27, 73, 84)

Figure 31.9 is a 3-D view of Function 1, the cork board. The objective is to determine the dimension of a picture frame to enclose a pattern of wine corks, but the dimension is limited to integer numbers of corks in rows and columns. The discretization creates flat spots in the surface.

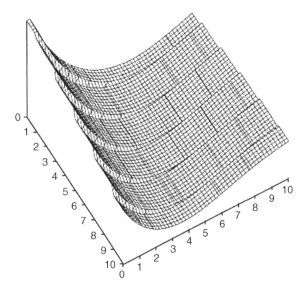

Figure 31.6 A closeup 3-D view of a surface with striations due to numerical discretization.

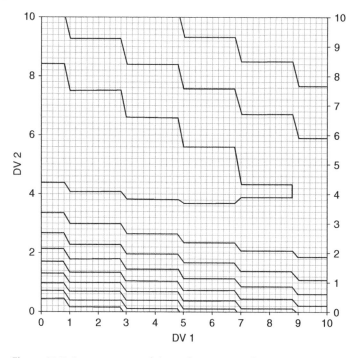

Figure 31.7 A contour map of the surface section of Figure 31.6.

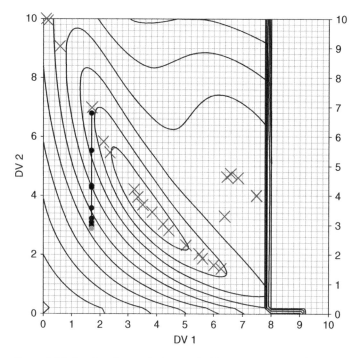

Figure 31.8 Results of an optimizer guided by striations, even though not visible in the large overview.

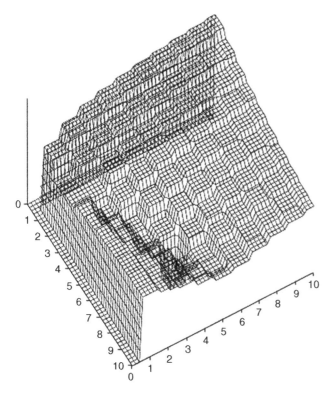

Figure 31.9 A 3-D view of a surface with level spots.

The 2-D contour is shown in Figure 31.10 along with 25 randomly initialized solutions using HJ as an optimizer. Unless the initial trial solution is very near a ridge, the exploration phase does not see a better direction, contracts the pattern, and converges in the same place that it was initialized. ISD and CSLS will have the same issues. Worse yet, second-order optimizers (SQ or Newton-types) will encounter a divide by zero in executing their algorithm.

Flat spots could be the result of insensitivity of OF to DV, which may occur in one particular region or due to discretization of either the DV or OF values. Discrete or integer values create flat OF response to small DV changes that are less than the interval. Flat spots also arise when using rank, classification, or category variables (nominal or ordinal).

Should your algorithm stop when OF_new = OF_old or should it continue across the level? If it stops, then there is no chance of crossing the level to find a new downhill. If it keeps going, it may go to a rise on the other side, or it may never leave the infinite level.

If the OF is insensitive to the DV, reconsider the DV. If insensitive in a particular range, then maybe it should not be a DV.

Multiplayer optimizers are a good solution. Although they will not always find the global best, they have a high probability of doing so.

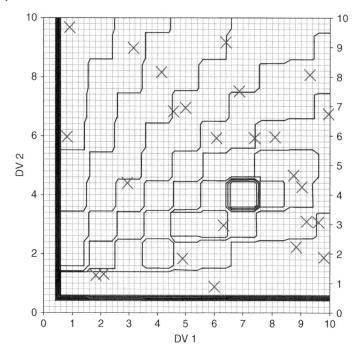

Figure 31.10 A contour map of a surface with level spots.

31.6 Hard-to-Find Optimum

If there are multiple optima and the global is a thin hole or other small feature relative to the DV range that is hard to find, or happens to be situated in a DV region that has a small "area of attraction," then it will be improbable to find that point. Figure 31.11 illustrates such for Function 29, "Mountain Path." The objective is to determine the equation of a path from one point to another. Most trial solutions, initialized over the DV space illustrated, will find the local optima. The global is near the point (9, 1) in the front-center of the illustration; but many optimizers will become misdirected to local optima near the center of the surface. For example, in the contour view of Figure 31.12, LM has only a 45% chance of finding the global best, and if a larger DV range were to have been selected, the probability of finding the global would be smaller.

It may take many random starts to find the area of attraction. Surprisingly, on this application, tests reveal that CSLS has about a 98% chance of finding the global best. The relative success of one optimizer over another suggests that one should have an arsenal of optimizers and use all and select the one that returns the best solution. Alternately, multiplayer algorithms also have a very high probability, nearly 100% in this case, of finding the hard-to-find optimum.

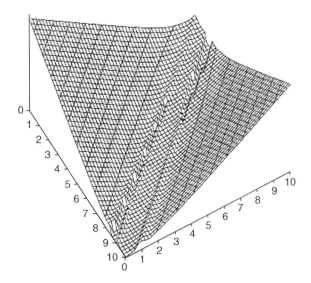

Figure 31.11 3-D view of the search for a mountain path, revealing a hard to find minimum.

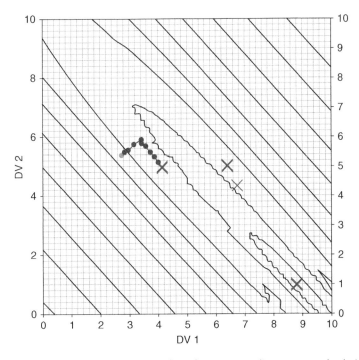

Figure 31.12 2-D contour map of the function revealing converged solutions.

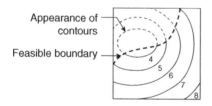

Appearance of contours

Feasible boundary

Figure 31.13 Illustration of the impact of an infeasible operation encountered within the OF calculation.

31.7 Infeasible Calculations

Infeasible calculations include non-executable operations such as $\sqrt{(-)}$, $\ln(-)$, $1/\emptyset$, $[f_1(X_1) - f_2(X_2)]^{(\text{non integer power})}$. Alternately, it could be due to a violation of a material or energy balance, $\sum X_i \neq 1$, an undesired future value of a variable value, or an aspect of a path. In any case, these would not be recognized by looking at the TS values; they would only be discovered during an attempt to evaluate function. As Figure 31.13 illustrates, an optimizer may want to cross over a feasible boundary toward the unconstrained minimum.

Hard constraints could be considered an equivalent to infeasible calculations. If a DV is not permitted to violate a hard constraint, then DV values that would cause a violation are infeasible. In this case, however, the constraint would be explicitly stated in the ST list, and the constrained CV boundary would be explicitly known. Given other DV values, the value of one that limits the constraint could be calculated from the ST relation. Whether the ST relation could be solved for the critical DV value explicitly or implicitly, the boundary can be known before the OF is calculated. By contrast, Figure 31.13 illustrates the feasible boundary with a dashed line, indicating that the constrained relation cannot be known until the OF calculation is explored. The infeasibility is implicit to the procedure.

31.8 Uniform Minimum

If the OF is underspecified, then there might be multiple solutions that provide exactly the same OF value. This is illustrated in Figure 31.14, in the optimization results of Function 43. The locus of converged DV* values is a rational line, indicating a definitive correlation between DV1 and DV2, and notably all have effectively the same OF* value.

Several of the functions express this, or a close kin to it, the slightly sloping valley (Functions 54 and others in the companion site).

If the OF is underspecified, the solution will result in multiple equivalent choices. This could suggest that there is an unidentified DV, which could be used to discriminate choices. Consider this simple optimization statement:

$$\min_{\{x,y\}} J = \left(y - x^2\right)^2 \tag{31.1}$$

Here, many x, y pairs provide the same minimum. The choice of (9, 3) cannot be distinguished from (0, 0) of (4, −2).

In such cases, reconsider the choice of DVs and the OF. Or if, for instance, there is a preference that the y-value be in the proximity of 7, add such a feature to the OF.

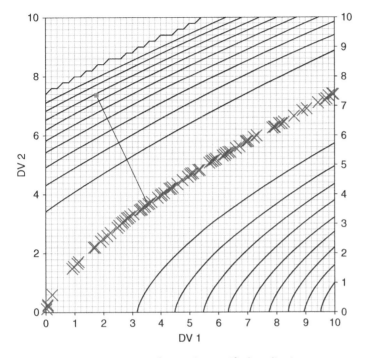

Figure 31.14 Consequences of an under-specified application.

31.9 Noise: Stochastic Response

A stochastic function does not return the same answer when asked the same question. An example question is "If I roll these five dice, will one of them show a 3?" A deterministic function does return the same answer. As an example, "What is 3 times 5?" We are mostly familiar with deterministic situations because they comprise nearly all of what we were taught in school, and by such programming we often think that deterministic responses represent reality. But experimental results are not exactly reproducible. And when the model is forecasting what might be a realizable future, the uncertainty in the givens is reflected in the variation of the OF calculation. Functions 18, 20, and 35 all have uncertainty. Many others in the companion site do or could, if you choose to include uncertainty in the givens.

Figure 31.15 is the 3-D illustration of Function 18, "Reservoir." The application seeks the height of a dam (lower left axis) and the fraction full to maintain the reservoir (lower right axis) to minimize costs. There are costs associated with size of dam (the larger it is, the more the dam, and the land to be flooded by the reservoir will cost). However, the larger capacity it has, the better it can save downstream life from floods or draughts. Keeping the reservoir nearly full or nearly empty provides a one-way safety feature for flood or draught, but not for the other. The chance of flood or draught conditions depends on upstream rain fall, which cannot be deterministically forecast for a 100-year future. Even if the reservoir is deterministically optimized for the 100-year worst case, this does not guarantee that either (i) an event will not exceed the given or (ii) the given will be nearly attained. So, the illustration reveals what might happen over a several-year period.

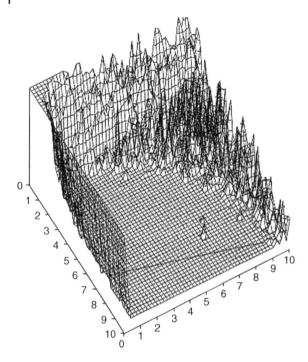

Figure 31.15 3-D illustration of a stochastic response.

In the middle region, the cost of the dam and reservoir linearly increases with size and has a slight linear trend associated with the recreational and water supply security benefit of keeping the water level high. In the middle DV region, there is ample capacity to absorb excessive rainfall or supply water downstream during draught conditions. However, near the not-large-enough upper left, or not-enough-spare capacity of the upper right, there is a reasonable chance that a confluence of conditions will cause a flood that cannot be managed where the fortuitous best chance confluence of conditions led to a phantom good OF value.

If the new trial solution is better or worse, can the algorithm use that information to direct the search? An apparent improvement might just be the statistical vagaries, and one should not let that misdirect the search. As Figure 31.16 reveals, an optimization algorithm grounded in a deterministic function places converged solutions well into the too-small or too-full region.

One could alleviate this by running many more simulations for each OF evaluation then taking the 99% worst case. But this increases the computational work and does not eliminate the stochastic effect. Using the central limit theorem, averaging reduces variation by $\sqrt{N}$, but cannot eliminate it. Additionally, the nature of stochastic functions invalidates many of the classical criteria for optimizer convergence.

My preferred solution is to use an optimizer that can handle a stochastic function.

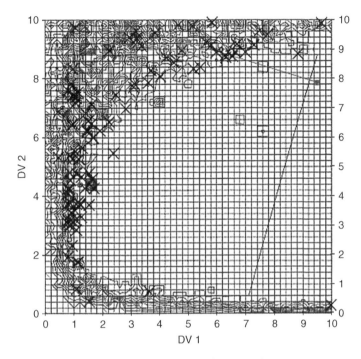

Figure 31.16 Optimization results on the function of Figure 31.15.

31.10   Multiple Optima

Function 30, in Figure 31.17, is the popular "Peaks" function, attributed to MathWorks. Many of the functions in the 2-D examples file have multiple minima; however the equations in "peaks" are relatively simple to implement, understand, and adjust to explore special cases of interest.

In Figure 31.17 there are three mountain peaks rising out of the nominal surface and two depressions or valleys. However, there is also a local high valley between the three peaks. The best of these three minima (the lowest) is the one in the near front. However, the "ground" also gently slopes from the peaks toward the boundaries. So a trial solution to the right of the peaks will go downhill to the back or right boundary. Figure 31.18 shows results from 100 second-order optimizer runs from randomized initializations. The 100th run, the last in the set of trials, found the second best valley, and the connect-the-dots path revealed its approach to the local optimum. Note that all three valleys were discovered and that many solutions wandered off toward a boundary. In this set, the best solution, roughly at (5.5, 2.5), was discovered in 20% of the trials.

Local minima can draw the search to a local minimum, not the global. There is no guarantee that your search will find the global (overall best) optima. Although starting many times increases the probability of finding the global, it is never 100% certain. Multiplayer algorithms have a higher probability of finding the global. Figure 31.19 shows LF results on the peaks challenge problem. LF only converged on the best two valleys, finding the best in 71% of the runs.

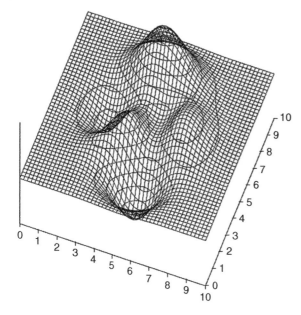

Figure 31.17 The popular peaks function for testing optimizers.

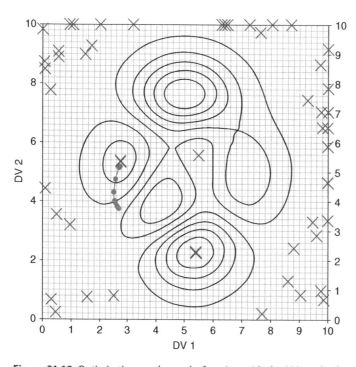

Figure 31.18 Optimization on the peaks function with the LM method.

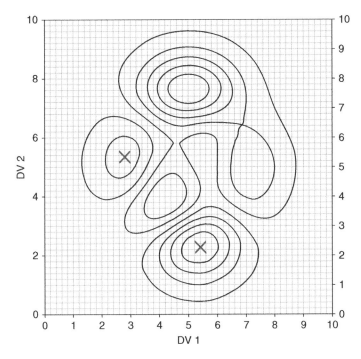

Figure 31.19 Optimization on the peaks function with leapfrogging.

31.11 Takeaway

Recognize the challenges that your particular application might provide. These issues termed surface aberrations are not limited to 2-DV, 3-D applications. They scale to higher dimension. Be able to use diverse optimization algorithms.

31.12 Exercises

1 "Surface aberrations" is my term for surface features that challenge an optimizer. Choose two. For each, (i) sketch the surface feature in a 2-DV situation (3-D or contour illustration), (ii) explain the challenge in text, and (iii) state an optimizer that would be appropriate to the situation.

2 For each of five of the nine surface aberrations listed in the chapter: (i) Explain the problem feature. (ii) State why it creates a problem for some particular optimizer. (iii) From a printout of either the contour or 3-D plot, explicitly point out that problem feature. You could show what an optimizer does on the contour plot to support your explanation. You'll need to look at either the contours or 3-D plots to find functions that have the problem features. Do that first. Select functions that express five different response behaviors that lead to difficulties for optimizers. Some functions will express more than one problem. But clearly reveal answers (i), (ii), and (iii) for five separate problems.

3 Include the preference that the *y*-value be in the proximity of 7 in the uniform minimum example of Equation (31.1).

32

Identifying the Models, OF, DV, Convergence Criteria, and Constraints

32.1 Introduction

In routine applications, the legacy of solving the optimization and using the results will have given customers and implementers a good understanding of the context of the application and the appropriate choices for the models OF, DV, convergence criteria, thresholds, and constraints. However, for new applications, the implementer must design these procedures and must make the appropriate choices. The question is "How?" The answer is "trial and error." Try something, and when you have the result, consider all aspects of the outcome. Then evaluate whether all aspects are satisfactory. If not satisfactory, improve the procedure.

That answer, the solution, and the method of getting optimization right is qualitative. Accordingly, this section has no mathematical relations. It is also fuzzy, providing a general direction, but not a specific rule to execute.

Although not rigorous, or sophisticated, or quantitative, the following material is important to application success. Perhaps this section is more important than being an expert in the optimization algorithm.

The user of a new application will want to know how to define the best way to create the model and to choose the DVs and the convergence criterion. I still revert to trial and error. It is a design problem. Sketch a possible definition of the application, get the solution, and explore it. Decide what are the desirables and undesirables associated with all aspects of the exercise, and seek improvements in the design of the application to improve the balance. I am pretty sure that first tries are usually incomplete or inadequate, and I do not know of any recipe that leads to the right choices on the first try. Here is an example of one of my trial-and-error sketch–evaluate–erase procedures.

I was seeking to solve the Goddard problem, which is how to schedule rocket thrust to maximize height of a rocket. It is Case Study 4, Chapter 39, in this book. I first decided to schedule thrust with elevation, divided an elevation range into 20 elements, and then had the optimizer seek the 20 thrusts as 20 DVs, as the trial solution. After each optimizer iteration the list of 20 thrusts was used by the model to see how high the rocket would get, which was the OF. I quickly realized that to schedule the thrust by elevation, I needed to know the elevation range, but I could not know this until after I saw how high the rocket would get. However, I could schedule thrust with remaining fuel; it goes from 100 to 0%, and there is no question about the limits. So, I changed the basis to 20 increments of fuel content—100–95%, 95–90%, 90–85%, etc. The first thrust would be held until fuel dropped to 95%, then the second until 90%, etc. Then, looking at the optimizer best DV set, it seemed that all of the initial thrusts were nearly 100%. This makes sense (see discussion in Case Study 4). So, I set the initial thrust as 100% and used the optimizer to find the fuel content threshold to start backing off to lower thrust values, and then divided the remaining fuel into 10ths to schedule. If the threshold content was 40%, then the schedule would be 40–36%, 36–32%, 32–28%, etc. This reduced the DVs to 11, which is a big improvement. Then, I observed that the 10 after-threshold thrusts seemed to represent a trend that might be modeled as quadratic or cubic. So, I decided to model the after-threshold thrust as a cubic: $T = a + b(M - Mo) + c(M - Mo)^2 + d(M - Mo)^3$. Now, there are only 5 DVs: Mo, a, b, c, and d. In doing this, it appeared that the after-threshold trend of T w.r.t. M is nearly linear, so I reduced the model to $T = a + b(M - Mo)$, with only 3 DVs. Yea!

Is the reduction defensible? 3 DVs is much preferred over 20, or 11, or 5. The reduction of the model from 20 scheduled thrusts to the 3 DV model actually permitted better height, because the best transition of 100% thrust to the lower value happens in between the mass increments that defined the 20 DVs. Not restricting the change to 5% increments permitted better timing of the thrust reduction from the initial 100% to the schedule. I also discovered that the improvement in height in using the cubic model is insignificant relative to the impact that uncertainty on air density and humidity and BTU content of the fuel, and drag coefficient of the rocket have on elevation. So, the linear schedule is all that is rationally justified.

The evolution of the application statement is an optimization. The objective of defining the application is to make the answer best match the ideal perfection with minimal optimizer work, within the uncertainty from external vagaries. I don't know how to automate such an evolution in modeling choices or to know, before trying, what will be the best basis for the design of the optimization problem. Certainly, the design of the optimization application is an optimization: Maximize desirables (ability to improve the OF, fidelity to the real-world context, customer desires, etc.) and minimize undesirables (complexity, computational time, impact of uncertainty, etc.) by redesigning the application (models, DVs, convergence criterion, optimizer choice, etc.).

32.2   Evaluate the Results

After a first pass at formulating the problem and getting the solution, do the following to reconsider the OF and constraints:

1) Look at the extremes of the DV and consider who (which person in the supply chain or product stage, what organization observing the product or procedure, what customer or client) would have a concern about some aspect (what aspect of use, impact, influence, cost, weight, size, association, etc.). This will help you understand your customers and what is important about the context of the application. This may lead you to correct misperceptions embodied in the original OF statements.
2) State the concerns as precisely and quantitatively as possible. Be able to communicate and understand the concern with precision. Do not use imprecise words such as "too large," because it does not guide how much different a right solution might be. Clearly state why "it is too large to fit under the kitchen sink." Such precise statements will permit defining an appropriate constraint on, for instance, size.
3) Reconsider how those customers will use the result. State other concerns they might have (that relate to your application). The item may have optimized cost, but be too heavy for a delivery person to lift. The procedure may be fast, but too complicated for a user to understand. The solution may soundproof a car from road noise, but not permit a driver to hear emergency vehicle sirens.

4) Choose the important concerns.

5) State which are parts of the OF (the primary concerns or desirability you want to minimize or maximize) and which are constraints (conditions that must be met).

6) Determine how to quantify the OF concern(s). It must become a numerical value that is proportional to the magnitude of the concern.

Consider the OF* and DV* values and the values of other state or auxiliary variables in the model:

1) Look at the OF* value. It may be the best, but it may not be implementable. Consider how your customers will react to such an OF* value, and use that insight to redefine DV, constraint, OF, and convergence choices. For example, you may have minimized cost within the "givens" (initially directed constraints and values), but the design might still not have a value-to-price ratio that customers need. Consider what else you can change. Often the "givens" could become alternate DVs.

2) Look at all of the model values. You might find that at the optimum conditions, some flow-pressure relation would lead to cavitation or that a tank becomes too near empty during an optimal transition, or that in 1 month cash flow is negative. Use the insight to redevelop constraints.

3) Adjust the OF or constraints or convergence criterion as appropriate.

Reconsider the DVs:

1) Now that the OF concerns are identified, reconsider all of the things (items, constraints, inputs, conditions, parameters, values, shoulds, etc.) that you are free to change (or that someone else might have the authority to change) to improve the OF. Often there is much greater flexibility in the real implementation and in the real world than in the initial statement of the situation. What barriers could be removed?

2) Consider the "givens," from the original formulation, as possible DVs that could be changed. Don't be constrained by prior statements of the givens or methodologies.

Reconsider the models:

1) With this progressive experience, understanding, and evaluation, revisit the models that the constitutive relations used in the analysis. Are they complete and adequate models for the application? You may have used the ideal gas law or a turbulent flow relation, but now realize that the optimum conditions would require an equation of state that better models near-condensing phenomena or a fluid dynamics model in a laminar flow regime. If a tube length is short, then you may need an entry length model, not a fully developed flow model.

Consider uncertainty:

1) It is associated with coefficient values, forecasts, and expectations. It is associated with models and procedures. If the impact of the uncertainty is significant (on the OF* or DV* values), then reconsider models with greater fidelity or seek more definitive values of coefficients.

Consider convergence criteria and threshold:

1) Make the criterion and threshold appropriate to the application context. Choose values for adequate (not excessive) precision of the DV* and OF* values.

The bottom line is to consider all you do as tentative, until you are certain that it is not. Sketch the application, then evaluate the solution, and then erase and repeat. An artist does not start a new picture concept with ink, but with pencil, and progressively erases or discards, and re-sketches until satisfied with the whole. Similarly, in developing an optimization application, sketch, evaluate, erase, s, e, e, s, e, e, see, see, ..., and, finally, ink.

A few simple examples can reveal the methodology:

Example 1 Choose an insulation thickness on a house to minimize heating cost over a 15 year period:

$$\text{Cost of insulation} = aAt \tag{32.1}$$

where a = \$/volume of the insulation, A = area to be covered, and t = insulation thickness.
   Using a simple model for heat losses,

$$\text{Heat loss rate} = \dot{Q} = \frac{k}{t}A(T_h - T_a) \tag{32.2}$$

where T_h = house T and T_a = external, ambient air T.
   Now the heating cost for the 15 year period can be calculated as

$$\text{Heating cost for 15 years} = 15b\dot{Q} \tag{32.3}$$

where b = \$/Btu.
   The primitive optimization statement is

$$\min_{\{t\}} J = aAt + \frac{15bkA}{t}(T_h - T_a) \tag{32.4}$$

Consider the models: The heat cost, b, affects the OF, but you are not free to change it. You can't change T_a. Perhaps you could change insulation type and change "a." If yes, add "a" as a DV. Could you change T_h to lower the cost? Does this model account for the heat losses to the ground or to the attic? Does this model account for the enhanced heat transfer to the air due to wind or rain? Perhaps the model needs to account for convection, not just conduction. This model does not account for the conduction through the interior drywall or the exterior brick. Are these resistances important to consider?

Example 2   Noise Filter Design Example

The desire is to filter noise in a process signal. Consider the analog resistor–capacitor (RC) circuit approach to tempering direct current (dc) voltage fluctuations (Figure 32.1):
   The ODE that describes how V_f responds to V_s is

$$\frac{1}{RC}\frac{dV_f}{dt} + V_f = V_s \tag{32.5}$$

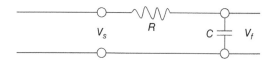

Figure 32.1 Illustration of an RC dc circuit.

where the time constant $\tau = 1/RC$.

The digital equivalent to the ODE solution is

$$V_{f_i} = \left(1-\lambda\right) V_{s_i} + \lambda V_{f_{i-1}} \tag{32.6}$$

where $\lambda = 1 - e^{-\Delta t/\tau}$, and $0 \le \lambda \le 1$.

In one extreme, if $\lambda = 0$, there is no filtering, and $V_{f_i} = V_{s_i}$. In the other extreme, *if* $\lambda = 1$, the filtering is so strong that V_{f_i} does not respond.

What is a right value for λ? Use propagation of variance to analyze the filter:

$$\sigma_{V_{f_i}}^2 = \left(1-\lambda\right)^2 \sigma_{V_{s_i}}^2 + \lambda^2 \sigma_{V_{f_{i-1}}}^2 \tag{32.7}$$

If stationary and if the signal noise amplitude is unchanging, then $\sigma_{V_{f_i}}^2 = \sigma_{V_{f_{i-1}}}^2$. Then variance reduction is

$$\frac{\sigma_{V_f}^2}{\sigma_{V_s}^2} = \frac{\left(1-\lambda\right)^2}{1-\lambda^2} = \frac{\left(1-\lambda\right)\left(1-\lambda\right)}{\left(1-\lambda\right)\left(1+\lambda\right)} = \frac{\left(1-\lambda\right)}{\left(1+\lambda\right)} \tag{32.8}$$

After converting the relation by the square root, variation reduction (noise amplitude reduction) becomes

$$\frac{\sigma_{V_f}}{\sigma_{V_s}} = \sqrt{\frac{\left(1-\lambda\right)}{\left(1+\lambda\right)}} \tag{32.9}$$

If the objective is to minimize the noise transmitted by the filter, then

$$\min_{\{\lambda\}} J = \frac{\sigma_{V_f}}{\sigma_{V_s}} = \sqrt{\frac{\left(1-\lambda\right)}{\left(1+\lambda\right)}} \tag{32.10}$$

$$\text{S.T.}: \ 0 \le \lambda \le 1$$

We could use an optimizer to determine the λ-value to minimize the noise amplitude, but, in this simple case, we don't need an optimizer. The DV* solution is obvious from the relation or graph, as illustrated in Figure 32.2.

The answer is to minimize noise, make $\lambda = 1$. Recognize that this is an extreme value, which is a clue that something is missing. Consider the implications of an extreme DV value. With $\lambda = 1$, the filtered signal never tracks the process signal. That is an undesired aspect that was not considered in the primitive optimization concept for the OF. So, consider what other aspects are missing from the formulation.

Filtering introduces a lag response when the signal really changes. Figure 32.3 illustrates a step change in the signal and the filtered response for several λ values.

The lag tempers information transfer, which degrades control and decision ability. So you also want to minimize τ.

Superficially, again, but this time seeking to minimize both lag and noise:

$$\min_{\{\lambda\}} J = \frac{\sigma_{V_f}}{\sigma_{V_L}} + \tau = \sqrt{\frac{1-\lambda}{1+\lambda}} - \frac{\Delta t}{\ln\lambda} \tag{32.11}$$

$$\text{S.T.}: \ 0 \le \lambda \le 1$$

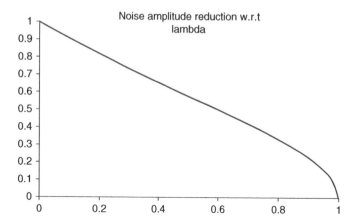

Figure 32.2 The trend of Equation (32.10) leads to an extreme DV value.

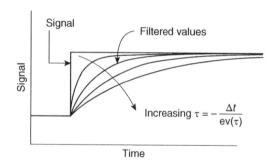

Figure 32.3 The impact of filter coefficient on time lag of the filtered response.

However, one cannot add the noise reduction ratio (dimensionless) and the time constant (units of time), variables with disparate units. So, scale each term between best and worst. The first term is already scaled between 0 and 1. Define $\tau' = \tau/\tau_{\text{maximum permissible}}$, with $\tau_{\text{maximum permissible}}$ as a user choice. Then the optimization statement is

$$\min_{\{\lambda\}} J = \sqrt{\frac{1-\lambda}{1+\lambda}} - \frac{\Delta t/\ln\lambda}{\tau_{\text{maximum permissible}}} \tag{32.12}$$
$$\text{S.T.} : \ 0 \le \lambda \le 1$$

This is now a balance of opposing ideals—total noise removal and lag.

The DV* value depends on the user's choice of $\tau_{\text{maximum permissible}}$, which assigns the balance of importance between the two parts of the OF. How would one decide what is the right value? Recognize that the right value is dependent on the application context. One could use any approach to optimization (analytical, Golden Section, heuristic direct, Newton's method, etc.) to find the same optimum value for λ^*. The optimization algorithm does not make the λ^* value right; the user's choices for the models, for the OF considerations, and for the weighting factor are the critical issues. Continually reassess the formulation for its fidelity to the application situation.

Example 3   Best Convergence Threshold

What is the best ε for convergence based on Δx? Is that OF statement grounded in a desire to minimize the deviation from x^*? Or, is the statement to minimize Δx? If so, here is one interpretation of the application to shape the optimization statement:

$$\underset{\{\Delta x\}}{\min} J = \Delta x \tag{32.13}$$

Obviously, the answer is $\Delta x = -\infty$, which is not realistic. Negative 1000 is below negative 1 on the number line. $\Delta x = -\infty$ is an extreme value, which implies that something is not quite right with the optimization statement.

Accordingly, and cleverly, let's consider the absolute value, which will make −1 smaller than −1000:

$$\underset{\{\Delta x\}}{\min} J = |\Delta x| \tag{32.14}$$

Now the answer is $\Delta x = 0$, another extreme value, which should also be your trigger to reconsider the formulation of the application. The implication for requiring $\Delta x = 0$ is that the optimizer will never achieve convergence.

What is good or bad about Δx being too large or too small? Let the answer to that question guide the development of the OF. Too small a Δx-value leads to excessive iterations. Too large and the optimizer will not stop close enough to the true optimum. Maybe add the two undesirables:

$$\underset{\{|\Delta x|\}}{\min} J = \text{number of iterations} + \text{deviation from } x^* \tag{32.15}$$

This has a balance of opposing ideals, but the units on the two terms don't match. Perhaps this can be fixed using equal concern factors. Maybe,

$$\underset{\{|\Delta x|\}}{\min} J - \text{number of iterations}/\text{EC} + \text{deviation from } x^*/\text{EC} \tag{32.16}$$

Since number of function evaluations, not iterations, is a better indication of computational work, and since the DV is ε not the consequential Δx, maybe use

$$\underset{\{\varepsilon\}}{\min} J = \frac{\text{NOFE}}{\text{EC}_1} + \frac{|x - x^*|}{\text{EC}_2} \tag{32.17}$$

But, how to know x^*? If x^* was known, why run the optimizer? Perhaps express proximity to the true optimum by closeness of the derivative to being zero. Maybe,

$$\underset{\{\varepsilon\}}{\min} J = \frac{\text{NOFE}}{\text{EC}_1} + \frac{|dy/dx|}{\text{EC}_2} \tag{32.18}$$

I am not claiming that this is the right answer nor that this is the comprehensive set of issues that should be considered. This example revealed the sketch, evaluate, erase, s, e, e, see, see, ... procedure.

32.3 Takeaway

Setting up an optimization application is an iterative procedure. The first statements of the OF, or the DVs, or the models, or the constraints, or the convergence criterion are usually incomplete or inappropriate and need correction. Here are some aspects to consider:

- Considering extremes in the DVs helps reveal OF and constraints.
- Quantifying the OF and constraint rules helps reveal DVs.
- Contemplation about constraints may change your opinion as to whether they are constraints or OF components to minimize.
- Considering how all stakeholders (clients, customers, participants, etc.) will respond to the DV*, OF*, values, and any values of variables in the models will help you to identify appropriate aspects.

32.4 Exercises

1   What aspects of a classroom homework assignment would make you want to keep it? (It was a complex victory. It clarified concepts. It has utility for future. It resulted in a good grade. It was fun. It revealed interesting new concepts.) What aspects did you not like? (It was repetitive. You thoroughly understood the concept and did not need the exercise. You could not get a solution. The instructions were unclear.) How can those concepts be quantified? Create an appropriate optimization statement that could guide the instructor to create good assignments, to maximize the "+" aspects, and to minimize the "−" aspects. Fill in the specifics of this generic statement:

$$\max_{\{\text{attributes of assignment exercises}\}} J = \sum \frac{\text{desired things}}{\text{EC}} - \sum \frac{\text{undesired things}}{\text{EC}} \quad (32.19)$$

2   Set up the optimization statement to determine the equation of a path in x (= East–West direction) y (= North–South direction) through mountains. The elevation is a function of E–W and N–S position, $z = f(x,y)$. The objective is to minimize the distance, S, along the mountain path from point A (x_A, y_A) to point B (x_B, y_B) on the condition that rate of climb or descent is always less than ts (too steep). The path is to be described by a fourth power relation $y = a + bx + cx^2 + dx^3 + ex^4$. Using this information derive the equations (in terms of $a, b, c, d, e, f, x_A, y_A, x_B, y_B, x, y, z$) for the objective function and the constraints. State the decision variables. When needed, use dy/dx instead of $b + 2cx + 3dx^2 + 4ex^3$.

3   As a batch process ages over several days, it makes a higher yield of desired product, but there is a diminishing returns approach of product made to an asymptotic limit. $M(t)$ is the mass of material produced after a batch time t, in days, and a simple diminishing returns equation is $M(t) = k(1 - e^{-t/\tau})$. When the batch process is terminated, the vessel is emptied to recover the

product, filled with fresh feed, and a new batch is restarted. The turnaround time for emptying and recharging is θ. The process owner wants to maximize annual production of the desired product. Letting the batch run longer increases the yield per batch, but it means fewer batches per year. How long should each batch run to maximize annual productivity of the vessel? Set up the optimization statement. Is annual productivity the right OF? Clearly indicate OF, DV(s), constraint(s) (if applicable), and the equations (models) that relate DV(s) to other elements.

4   When changing the speed of a car, you wish to follow an accelerator position pattern that not only moves the car to the new speed in a small time but also does not shock the system with large accelerations. Typically, to increase the speed, you press moderately hard on the pedal, and then back off as the car approaches the target speed. There is not one pedal position but a sequence. Simplify the analysis: use Newton's law with no drag forces, $F = ma/g_c = (m/g_c)dv/dt$, and consider that the force the motor places on the car is linearly related to the pedal position, $F = kp$. This results in the model $dv/dt = (g_c k/m)p$. (i) Formulate the optimization application in the canonical (standard) statement. (ii) Could the DV be either the car speed, v, or the pedal position, p?

5   In the optimum pipe diameter application of Case Study 1, Chapter 36, initial cost (capital), C, and annual expenses, E, are concerns. We would like to minimize both, but they are not additively combined as $J = \lambda_C C + \lambda_E E$ and they were combined as NPV. In a ranking exercise overall fitness F is a function of individual attributes f_1, f_2, etc.; but, instead of adding them as $J = F = f_1/EC_1 + f_2/EC_2$, I would suggest multiplying the individual attributes so that $J = F = f_1 {}^* f_2$. Explain what application issues would make the additive OF right or the other function right.

6   All of the approaches presented in Exercise 5 seek to combine several issues into a single OF, and define a single right DV value. In a Pareto analysis, the individual OF components are kept separate, and the optimization seeks a set of non-dominated solutions. Explain what application issues would make the combined, single OF right or the Pareto set of non-dominated solutions right.

7   Ideally $\dot{Q} = ca\sqrt{\Delta P/(1-\beta^4)}$ for an orifice flow rate measurement. Variable "a" represents the orifice area, β is the ratio of orifice to pipe diameter, and "c" is a coefficient combining several variables. Fluid passing through an orifice restriction creates a ΔP (pressure drop), ideally described by the inverse of the orifice equation. Instruments measure the ΔP and use the orifice equation to calculate flow rate, which is used for process control. But noise on an orifice measurement is due to turbulence-induced pressure fluctuations on the differential pressure sensor. Noise amplitude is proportional to Reynolds number in the orifice. So, a smaller orifice diameter generates higher in-orifice velocity, which increases dP fluctuation, which is unwanted, because it increases $\dot{Q}$ (flow rate) measurement uncertainty. However, a smaller orifice diameter increases the dP response to flow rate, which increases sensitivity of measurement. Sensitivity could be related to $\Delta P_{\text{maximum}}/\dot{Q}_{\text{maximum}}$. So, smaller orifice diameter is better for sensitivity. Ideally, the orifice ΔP is recovered after the fluid flow profile re-expands, but there is a permanent pressure loss from an orifice, roughly proportional to the same ΔP needed for measurement. Pumping power loss is related to the permanent ΔP times flow rate. Develop the optimization statement, so that it could be solved, but don't solve it. Show the equations that develop and state the OF.

8 Explore the impact of changing the *r*-statistic steady-state stopping threshold value. You need to define a way to quantify goodness and devise a sufficient number of trials to gather data that is sufficient to make claims. Associated topics are propagation of uncertainty that relates uncertainty on DVs to uncertainty on modeled values. Measures, like FPE, relating to how many model coefficients are justified. The scale-free steady-state stopping criterion is claimed to be right for any model or data variance. Your job is to figure out how to measure appropriate attributes, select test situations that provide legitimate conditions to measure those attributes, do the tests, and evaluate the results. Report what aspect you chose to explore, how you decided to evaluate goodness, how you decided to structure tests and test results, and a discussion about how you know the data appropriately reveals features and permits comparison claims and conclusions.

9 Consider optimizing the size of a gas tank on a car. You probably know that a spherical tank will contain the largest volume for the least surface area, hence tank material and weight. But, they don't use spherical tanks. (i) List the DVs and briefly describe about 5 OF issues. (ii) Describe a few constraints. (iii) Devise a few of the models that would relate your DVs to OFs. (iv) Create the optimization statement. Do not try for perfection in the optimization statement. What I want is that you experience the process: As you choose what a DV might be, considering its extreme values will increase your understanding of the application and reveal additional OF/constraint issues. As you consider an OF issue and how to make it better, you will see additional DVs.

10 Consider a competition of five players. It could be a cross-country race, American Ninja obstacle course, spelling bee, etc. One wins, one comes in second, etc. This ranks them as first, second, third, fourth, and fifth. Each player has attributes that would logically relate to fitness of that player for the competition. For example, height would be an attribute of fitness for basketball success. Speed would be a fitness attribute for many sports. Endurance, another. Mental focus would be an attribute for chess or golf. Balance would be an attribute for ballet, wrestling, or gymnastics. For this exercise consider that there are three attributes ($f1$, $f2$, and $f3$) that are related to success in the competition and that each player has unique values ($f1A$, $f2A$, and $f3A$ are player A values; $f1B$, $f2B$, and $f3B$ for player B; etc.). The objective is to devise a formula using the three attributes that would predict the outcome of a contest of new players. You choose a formula for relating the separate attributes to overall fitness. Consider that you have a database of the ranks when players A, B, C, D, and E played, and players F, G, H, I, and J, played, such as in Table 32.1 that shows the attributes of the A, B, C, D, and E contest. Consider that you have similar data from the F, G, H, I, and J contest, and the K, L, M, N, and O contest. Show how you would use that information to adjust coefficients in your model so that it can predict the ranks of any set of players.

Table 32.1 Data for Exercise 10.

Player	A	B	C	D	E
$f1$-value	0.7	0.4	0.7	0.4	0.3
$f2$-value	0.5	0.8	0.7	0.6	0.5
$f3$-value	0.8	0.6	0.6	0.9	0.7
Rank	4	3	1	2	5

Define the model and objective function, and state the DVs and constraints. (i) Express the optimization statement in canonical form. Clearly indicate what represents (ii) model, (iii) data, (iv) OF, (v) DV, and (vi) constraints to reveal that you can differentiate between the separate elements.

11 This exercise is seemingly simple: Just find the value of a single variable that minimizes a function. The background makes the application somewhat complicated. The objective is to determine the threshold value for a classifier. A classifier is some form of computation (neural network, fuzzy, vector machine, etc.) that takes input data (sensor readings on attributes) and determines the value of an output variable that predicts whether the sensed thing is or is not. For instance, is it a rabbit or a dog? The rabbits eat my vegetable garden plants, but the dogs don't. Size would be one input to the classifier, but a Yorkshire terrier is about the same size as a rabbit. So, let's have hair length as another input, but a Chihuahua is both small and short haired and would have the same sensor values as a rabbit. There might be five inputs related to object shape features. Ideally, the classifier will return a 1 if it is a dog and a 0 if it is a rabbit. But the objects might not be in a conventional pose. They might be standing, facing directly, lying, and so on. As a result, the classifier might not be perfect and might return a 0.8 (or a 0.4, for example) if the object is a dog or a 0.1 (or 0.5, for example) if the object is a rabbit. The objective is to determine the threshold value for the output that maximizes the classifier performance. For this particular classifier, the probability of identifying "dog" when the object is a dog is p_{DD}. Then, the probability of claiming "rabbit" when the object is a dog, returning the wrong answer, is $p_{DR} = 1 - p_{DD}$. Similarly, returning "rabbit," if rabbit, is p_{RR}, and the probability of a wrong answer is $p_{RD} = 1 - p_{RR}$. Data analysis from hundreds of trials reveals the following relations of how the probability of a correct classification depends on the threshold, T, value of the classifier output:

$$p_{DD} = 1.2 \left[\frac{1}{1 + e^{8(T-0.8)}} - 0.168 \right] \tag{32.20}$$

$$p_{RR} = 1.06 \left[\frac{1}{1 + e^{-7(T-0.5)}} - 0.029 \right] \tag{32.21}$$

A graph of p_{RR} and p_{DD} w.r.t. T is illustrated in Figure 32.4.

The solid line is p_{DD} and the dashed line is p_{RR}. If T is set at 0.6, then approximately 85% of the instances of dog it will correctly indicate "dog," but then 15% of the instances it will falsely report "rabbit." Eighty-five percent correct is a solid "*B*" grade. That's good. Also, however, if $T = 0.6$ and the object is a rabbit, the classifier will report "rabbit" about 65% of the instances, but report "dog" 35%. Now, the 65% is a grade of "*D*," not so good. A "confusion matrix," a new term for the classic statistical term "contingency table," presents the number or probability of the true and false responses (Table 32.2).

So, how to assess goodness of a classifier? It will be situation specific. For example, If you live on a lettuce farm, dogs are not a problem, but rabbits eat your crop and they need to be chased away. In this case you might not care at all about dogs, but want to detect any rabbit. If you choose $T = 0$, every rabbit will be reported, but so will every dog. If it is just a dog, you don't want the machine alerting you with a false alarm, "RABBIT IN GARDEN. WAKE UP. GO CHASE RABBIT."

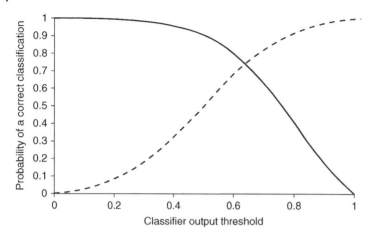

Figure 32.4 Illustration for Exercise 11.

Table 32.2 A confusion matrix for Exercise 11.

	Classifier reports dog	**Classifier reports rabbit**
Object is a dog	Right answer	Wrong answer
	Probability = p_{DD}	Probability = $1 - p_{DD}$
Object is a rabbit	Wrong answer	Right answer
	Probability = $1 - p_{RR}$	Probability = p_{RR}

So, there is a situation-based benefit of true and penalty for false classifications. (i) Explain your choice of a situation and consequential logic for the choice for the objective function that balances benefit and penalties. (ii) Use the p_{DD} and p_{RR} relations earlier to formulate the OF equation for your situation. (iii) Code it. (iv) Use any of the univariate search optimizers to find the optimal value for T. (v) Report your results.

12 The answer is 1, or 3 min if at a high altitude. You can find the answer by searching the Internet with the question "How long do I have to boil campsite stream-water to purify it for drinking?" I am not interested in the answer nor in the true details of this particular example. What I do want is for you to frame an optimization application. This application seems both rich in considerations and familiar enough for all to be able to do it. (i) Identify and explain five issues (there are many more issues) associated with the length of time to boil water in a container over a campfire. (ii) For each issue, sketch a graph of how the feature would change w.r.t. (with respect to) the boiling duration. Your sketch should clearly represent your concept, not misrepresent it. If something decays exponentially in time, don't show a linear sketch. If it decays to zero, don't make a crude sketch that shows it not reaching zero or going below. But, your sketch does not need to represent the truth about nature. Don't spend time researching, for example, the death rate of different microbes, or the exact evaporation rate of water at different altitudes. This is about the concepts of optimization, not the details of the campsite water purification. (iii) Explain

constraints. (iv) Use your sketches to show that too short a boiling time is bad and that too long a boiling time is bad. (v) Present the optimization application in the canonical form, clearly revealing OF, DV, and constraints.

13   Revisit Example 1 and take it to a next level of completeness. Solve.

14   Revisit Example 2 and take it to a next level of completeness. Solve.

15   Revisit Example 3 and take it to a next level of completeness. Solve.

16   Consider optimizing an optimization book chapter. What are the OF, DV, and constraints?

33

Evaluating Optimizers

33.1 Introduction

An optimizer is a procedure for determining the optimum. A classic statement is to determine the value of the decision variable to maximize desirables and minimize undesirables:

$$\max_{\{DV\}} J = \sum \text{desirables} - \sum \text{undesirables} \tag{33.1}$$

The optimization algorithms presented in prior chapters included the analytical method, Newton's, Golden Section, heuristic direct, Hooke–Jeeves, leapfrogging, etc. The question here is not "What is the result when an optimizer is used?" but "Which is the best optimizer?" Here the DV is not a variable in an engineering model, but the choice of the optimizer, a class variable. The statement is

$$\max_{\{\text{analytic, Newton, Golden, Heuristic, LF, ...}\}} J = \sum \text{desirables} - \sum \text{undesirables} \tag{33.2}$$

In order to evaluate an optimizer, we must define the desirables and undesirables, define metrics to assess them, and define a way to either combine them in a single OF or to alternately use a Pareto analysis.

As well as class variables for the DVs, each optimizer has coefficient values (such as the number of players, an acceleration/temper factor, initial step sizes, or thresholds for logic switching or convergence). These may be continuum valued or discrete. Further, there are other diverse category choices and other class variables, such as forward or central difference, surrogate model type, and local exploration pattern. Accordingly, the DV list should include the other optimizer choices. A more comprehensive statement is

$$\max_{\{\text{Method}, c, \text{Rules}\}} J = \sum \text{desirables} - \sum \text{undesirables} \tag{33.3}$$

As with any optimization application, begin by qualitatively considering the issues.

Engineering Optimization: Applications, Methods, and Analysis, First Edition. R. Russell Rhinehart.
© 2018 R. Russell Rhinehart. Published 2018 by John Wiley & Sons Ltd.
Companion website: www.wiley.com/go/rhinehart/engineeringoptimization

33.2 Challenges to Optimizers

Classic challenges to optimizers are objective functions that have the following:

1) Non-quadratic behavior
2) Multiple optima
3) Stochastic responses
4) Asymptotic approach to optima at infinity
5) Hard inequality constraints or infeasible regions
6) Slope discontinuities (sharp valleys)
7) A gently sagging channel (effectively slope discontinuities)
8) Level discontinuities (cliffs)
9) Flat spots
10) Nearly flat spots
11) Very thin global optimum in a large surface, pin-hole optima, improbable to find
12) Discrete, integer, or class DVs mixed with continuous variables
13) Underspecified problems with infinite number of equal solutions
14) Discontinuous response to seemingly continuous DVs because of discretization in a numerical integration
15) Sensitivity of DV or OF solution to givens

33.3 Stakeholders

Stakeholders, the people who might claim to have a stake in the outcome of the optimization or choice of the optimizer, include the following:

1) Software user—wants speed of solutions, no execution errors, no decisions/choices related to parameters, robust to surface features
2) Software programmer—efficient computation, low storage, scalability, few conditionals, and simple logic
3) Business manager—low number of function evaluations, utility of solution not compromised by complexity or propagation of uncertainty

33.4 Metrics of Optimizer Performance

There are many metrics that can be used to assess optimizer (and coefficient choice) performance. Start out by listing what you think are the OF and DV and constraint considerations, then let this trigger who might be stakeholders, and then look at it from their perspective. We aim to have the following:

1) Efficiency 1—A minimum number of function evaluations (NOFE). Each function evaluation represents either computer work or experimental cost. Be sure to include either numerical or analytical evaluations of gradients and Hessian elements in the NOFE. These derivative evaluations might be a part of the optimizer, or they could be within a convergence criterion. Count them all.

2) Efficiency 2—Minimum computer time. Nominally, the OF evaluation is the most time-consuming aspect, but where the computer must sort through multiple players, invert matrices, or apply convoluted logic, the optimizer logic may take more time than the OF evaluation. When you accumulate optimizer time, be sure to discriminate between the several-time consumers such as display I/O (especially the choices for your tracking convenience), the function evaluation duration, convergence testing, and the optimizer. Also, be aware that other processes that a computer might be running (such as a virus scan) may reduce space or time that the computer might use for the optimization. In any comparison use equivalent languages (don't compare an interpreter to a compiled .exe program) and equivalent computing devices (don't compare a laptop to a main frame).

3) Efficiency 3—User required level of understanding. How easy is it for the user to understand the optimizer logic and computations? How easy to set up the optimizer and choose optimizer parameter values? How easy is it for the user to establish confidence that the optimizer result is in desired proximity of the global? User ability to adapt the code? Algorithm complexity? This could be measured by the number of choices a user must make or number of lines of code related to the issues.

4) Efficiency 4—Consider storage requirements needed to compute or invert the algorithm.

5) Efficiency 5—Consider the complexity of the code, the number of conditionals, the number of alternate treatments or methodologies, and the number of switches in logic.

6) Interpretability 1—How well will the human understand and interpret the results? Considering the CDF of the OF* or the range of the DV* values, how will a user interpret the probability of finding the global?

7) Interpretability 2—How will a user make an actionable decision based on the optimization results with respect to uncertainty in the "givens," the model equations, and the probability of the optimizer results?

8) Probability of finding the global optima—When there are multiple optima, traps, diversions to a constraint, etc., the likelihood of any particular run of an optimizer in finding the global is important. You many need 1000, or more, runs from random initializations to be able to assess the probability of an optimizer to find the global. Probability could be measured by the fraction of trials that the optimizer identifies the global (converges in the vicinity of the global).

9) Accuracy—This is the closeness of the optimizer to the true optimum. It is often termed systematic error or bias. The optimizer does not stop exactly at the optimum, but stops in the proximity of it based on the convergence criterion. If the best of N trials is taken as the answer, what is the average deviation from true? If the distribution of converged solutions about the optimum is not symmetric, then an average DV* value will have a bias. Does convergence choice, initialization range, constraint accommodation penalty methods, single/double precision, discretization, etc., affect accuracy?

10) Precision—This is repeatability. The optimizers are numerical procedures, which have finite convergence stopping criteria. They will not stop exactly at the true optimum. Closeness to the optimum can be assessed either by the OF value or by the DV value deviations from the true optimum. We talk about finding the global optimum, but the reality is that the optimizer finds the proximity of an optimum, not the exact point. Precision could be measured by the rms (root-mean-square) deviation (either DV from DV*, or OF from OF*) from those trials that located the global. Precision also is dependent on convergence choice, initialization range, constraint accommodation penalty methods, single/double precision, discretization, etc. Accordingly, to compare optimizers, the impact of these other choices needs to be removed.

11) Robustness—This is a measure of the optimizer ability to cope with surface aberrations (cliffs, flat spots, slope discontinuities, hard constraints, stochastic OF values, a trial solution that is infeasible or cannot return an OF value). It also includes the optimizer ability to generate a next feasible trial solution regardless of the surface features without execution faults. In general, you want an optimizer to perform well (not necessarily perfectly) on a wide variety of applications with all reasonable DV initializations and optimizer parameter choices (number of players, convergence threshold, acceleration factor, etc.). You would like the procedure to be fully autonomous, not requiring human intervention to solve problems. You want the results to be relatively insensitive to human choices of rules or coefficient values. Perhaps a measure of robustness could be the fraction of times the optimizer can generate a feasible next trial solution. Or the fraction of applications that it can solve. Perhaps robustness should be categorized to application features (discrete, class, discontinuity).

12) Scalability—As the number of DVs increases, how does the computational time or storage increase? How does the NOFE increase? How do the requirements on the user (such as the number of user-chosen coefficient values that need to be specified for either the optimizer or the convergence criteria) increase? The burden might be acceptable for low-order applications but excessive for higher dimension ones. Do the diverse aspects rise linearly with DV dimension, as a quadratic, or exponentially? Is the result independent of user-selected variable dimensions or scaling?

33.5 Designing an Experimental Test

Since any of these metrics will depend on the initial trial solution(s) and the surface features of a specific objective function, you will need to run many trials and calculate an average value representing individual functions. However, replicate trials from random initializations will not produce exact duplicate results. For simple problems, perhaps 20 trials are fully adequate to have relatively certain values of the statistics. However, you may need 100–10,000 trials to determine representative values for probability statistics associated with other attributes. You should keep running trials until the standard deviation of the statistic of interest is small enough to confidently differentiate between statistic values representing different experimental conditions (optimizers, coefficients, convergence criteria).

Statistical comparisons need to be made on replicate results, such as a t-test of differences in NOFE. The central limit theorem reveals that the standard error of the average is inversely proportional to the square root of the number of replicate trials. The uncertainty of the average reduces with $\sqrt{N}$. The uncertainty on experimental probability is dependent on both N and the probability.

Further, one optimizer (Newton's, successive quadratic) might jump to the solution in one step if the function is a simple quadratic response, but it might become hopelessly lost on a different function. How do you compare optimizers? You need to choose a set of test functions that represent a diversity of features that your applications will present.

If one is assessing number of function evaluations (or iterations) and the precision of a solution, then it becomes understood that the choice of convergence criterion and the threshold for convergence also affect the results. "The optimizer" involves more choices than just the optimization algorithm choice.

With this understanding, the DV section of the optimization statement might be understood as a 3-D choice. For each optimizer, the first column in the DV list of Equation (33.4), there are a variety of convergence criteria that can be selected (second column); and for each criterion, there is a numerical threshold (third column):

$$\left.\begin{cases} \text{Newton} & \text{OF} & 0.1 \\ \text{Golden} & \Delta\text{DV} & 0.01 \\ \cdots & \cdots & \cdots \end{cases}\right\}^{\max} J = \sum \text{desirables} - \sum \text{undesirables} \qquad (33.4)$$

But there are also the initialization factors such as number of trials of randomized initializations and coefficient values (or initial step sizes, or initial range, or number of players):

$$\left.\begin{cases} \text{Newton} & \text{OF} & 0.1 & \text{initialization} \\ \text{Golden} & \Delta\text{DV} & 0.01 & \text{initial}-2 \\ \cdots & \cdots & \cdots & \text{initial}-3 \end{cases}\right\}^{\max} J = \sum \text{desirables} - \sum \text{undesirables} \qquad (33.5)$$

And this needs to be considered on a full spectrum of applications:

$$\left.\begin{cases} \text{Newton} & \text{OF} & 0.1 & \text{initialization} \\ \text{Golden} & \Delta\text{DV} & 0.01 & \text{initial}-2 \\ \cdots & \cdots & \cdots & \text{initial}-3 \end{cases}\right\}^{\max} J = \sum_{\text{application types}} \left(\sum_{\text{optimization}} \text{desirables} - \sum_{\text{optimization}} \text{undesirables} \right)$$

$$(33.6)$$

It can become complicated and confusing.

Now one must devise the test basis:

- The choice of test functions needs to span the features that are relevant. One function is not a comprehensive means of testing. Choose a number of functions that include all of the challenges that might be represented by the application.
- The number of trials from randomized starts needs to be high enough to eliminate statistical vagaries from misdirecting the answer. Use the t-statistic to determine the number of trials, N, needed to have adequate precision or discrimination between values.
- The initialization choices need to be representative of the application—initialize in the area of the global optimum or in local regions off optimum, or throughout a global DV range.
- The criterion for convergence and the threshold values need to be fair. In cases that cannot use the same criterion (such as a multiplayer in comparison to a single-trial solution procedure), the testing needs to choose values that provide an equivalence. For example, choose the criterion thresholds to generate equivalent precision of the DV^* values, and then evaluate NOFE.
- Use a t-test to compare averages.
- Use a t-test to compare probabilities.
- Combine metrics for efficiency. For instance, use PNOFE instead of separately p(Global) and ANOFE. PNOFE = ANOFE($\ln(1-c)/$ $\ln(1-f)$). Use a t-test to compare PNOFE values.

- Assess precision by the rms deviation of all solutions at the global best, from the best of N. This could be based on the DV* values or the OF* values. Use an F-test to compare rms values.
- Design the test for equivalence in some performance aspect so that the others can be easily compared. For instance, adjust the convergence threshold so that two optimizers have the same precision, and then compare PNOFE values at equivalent precision.
- Use EC factors when combining metrics, additively. Defend the EC values within the application context.
- Create an equivalent basis w.r.t. DV* or OF* precision. Don't choose an excessive precision, but make it compatible with the application.
- Use a Pareto optimal selection when not combining competing objectives into one expression.
- Rank optimizer by number of functions (and/or performance criteria) on which it is best (3 points) or second (2 points) or third (1 point).

Example 1 As a simple comparison example, I have compared a variety of optimizers on one function (Function 89 in the companion site—enzyme reaction) on a limited number of assessment criteria (ANOFE and p-Global) with the convergence criterion threshold chosen so that precision was equivalent (a standard deviation of global OF values of about $1E - 6$). The labels on the data in Figure 33.1 are the optimizer names, and the average number of function evaluations is plotted w.r.t. the probability of the optimizer finding the global. This is a Pareto-type analysis. In each axis, better is toward the origin.

In this analysis PS (particle swarm) and LM (Levenberg Marquardt) are dominated by LF, HJ, CHD, and CSLS. Each of those four is better in both of the assessment criteria. ISD (Incremental Steepest Descent) is dominated by HJ and CHD, while CSLS (Cauchy Sequential Line Search) is dominated by LF and HJ. The non-dominated algorithms in this application are LF (Leapfrogging), HJ (Hooke–Jeeves), and CHD (Cyclic Heuristic Direct). Although LF was worse than HJ for ANOFE, it was better in p(global). This, of course, only represents two of many criteria for assessing optimizer desirables, and only tests optimization on one function representing one set of issues (nonlinear with slight discontinuities due to time discretization), under one convergence precision outcome.

Example 2 Compare several optimizers with a variety of 2-DV test functions using PNOFE (the probable number of function evaluations needed to be 99% confident that the global will be found in at least 1 of N trials) at equivalent precision on OF* values. In each of the applications, convergence will be based on the DV.

In LF and PS (with an end-game perching, contraction feature), the convergence is on the 20-player cluster size. With HJ, the convergence criterion is based on the size of the exploration range. In LM the convergence is based on iteration-to-iteration DV changes. As a result, even if the convergence threshold is identical, the precision will be different. So, each test is based on an OF* precision based on rms deviation from the best of all solutions of $rms(OF^*) \cong 10^{-6}$.

Table 33.1 reveals the results of $N = 100$ trials on each of 11 different functions. The PNOFE values are calculated from the ANOFE of the 100 trials and the probability of that optimizer finding the global minimum. The PNOFE data is a stochastic response. Another set of 100 optimizer trials from randomized initializations will not exactly replicate the same patterns. The probability of finding the global is stochastic. (Flip a coin 100 times, and you expect 50 trials to be heads. But, obtaining only 47 of the 100 is not unexpected.) Additionally, the number of function evaluations to converge will be a stochastic response to the vagaries of initialization. Accordingly, the PNOFE values in the table have

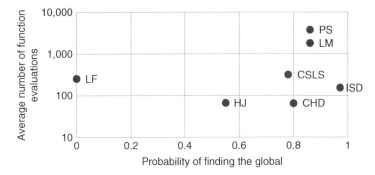

Figure 33.1 Pareto comparison of two contrasting performance criteria on diverse optimizers.

Table 33.1 PNOFE data for the optimizer comparison.

Function	LF	HJ	LM	PS
Three Springs (#12)	460	**410**	3.7k	2.1k
Fluid Bed Dryer (#90 in the companion site)	400	**260**	1.9k	2.5k
Peaks (#30)	**1.0k**	3.3k	5.8k	2.5k
Mountain Path (#29)	**1.6k**	13.7k	38.5k	10.4k
Parallel Pumps (#41)	**1.2k**	11.1k	45.0k	8.1k
Shortest Time (#47)	**160**	230	220	2.3k
Hose Winder (#46)	**1.1k**	2.6k	48.0k	2.3k
Classroom Lecture Pattern (#73)	**550**	3.1k	—x—	1.7k
Rank Model (#84)	**490**	14.7k	—x—	2.4k
Goddard Problem (#79)	**890**	4.2k	—x—	2.2k
Reliability (#56)	**2.6k**	—x—	—x—	8.4k

The numbers in bold indicate statistical winners, based on a t-test and 95% confidence. The —x— entries indicate optimizers that either encountered execution errors or converged at any initialization and failed to find an optimum.

about a 10% coefficient of variation (the rms value, or standard deviation, is about 10% of the PNOFE value).

Using points to score the first place (3 points), second place (2 points), and third place (1 point) optimizers, the scores are LF = 31, PSO = 19, HJ = 16, and LM = 3.

33.6 Takeaway

Design tests to compare optimizers to be a comprehensive and complete assessment of issues that matter to the application situation and its stakeholders.

33.7 Exercises

1   Figure 33.2 reveals the CDF(OF) of 50 trials of two different optimizers on the same function. Which optimizer is best? Explain your choice.

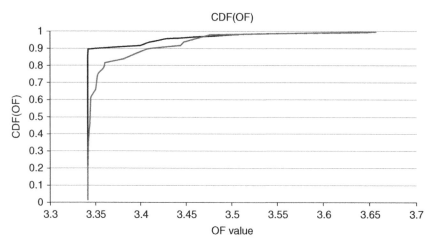

Figure 33.2 Illustration for Exercise 1.

2   State the number of function evaluations per iteration for each of these three optimizers. Explain your answer. Define "iteration" as one complete pass through all stages of the optimization algorithm on a 2-DV problem. (i) Cauchy Successive Line Search. (ii) Newton (Newton–Raphson), (iii) Hooke–Jeeves.

3   If the model is linear in coefficients, for example, the temperature dependence of specific heat is traditionally modeled as $c_p = a + bT + cT^2 + dT^{-1}$, and then linear regression can be used to determine model coefficient values. First, measure c_p at a variety of temperatures to generate the data. Then set the derivative of the SSD with respect to each model coefficient to zero, and you have the four normal equations. Linear algebra will provide the exact values of the coefficients for that particular experimental data set. Alternately, if you choose a numerical optimization procedure (Newton–Raphson, ISD, LM, HJ, LF, etc.), it will approach the optimum DV values and stop in the proximity of the true DV* values when the convergence criterion is satisfied. Which method is better? Briefly state three (or more) issues.

4   Here are four algorithms: linear programming, dynamic programming, leapfrogging, and Levenberg–Marquardt. For each, state what aspect the other three have in common that differentiates the one from the others.

5   With a brief discussion (and possibly supporting illustration), explain PNOFE.

6   If you do not know either DV* or OF*, explain how could you calculate precision (closeness to the DV* value) for an optimizer.

7 Explore the desirability of several 2-D optimizers. Use the file "2-D Optimization Examples." Explain how you will determine values of metrics that would assess such desirable attributes and also how you will design optimization experiments to reveal and obtain those values. Select functions with diverse attributes. Explore optimization algorithms that are most commonly used or have widest applicability. You should end up with an opinion about what optimization approach is best. To defend your opinion, you should present a table that summarizes desirability of the optimizers, within the uncertainty that your experimental metrics. Prior to the table, you should explain your experimental results. Prior to revealing experimental results, you need to explain your experimental plan and method of interpreting data. Finally, but this is the first part of your presentation, you need to explain how you will be assessing the attributes of desirability. I think all of this can be presented in five, or so, pages.

8 Compare your favorite of the gradient-based searches with your favorite of the direct search procedures. Run them a sufficient number of times on several of the simple and several of the difficult surfaces. Organize the assessment data in a table. Pick a best and defend your choice. The performance differences are not due to unknowable mystery, they are directly the result of the algorithm. Use your knowledge of the algorithms to explain why the results are as you obtained.

9 With your own selection of functions, optimizers, precision, etc., acquire assessment data to evaluate the optimizers. Justify your choices in the experimental design and evaluation procedures. Explain and defend your conclusions.

34

Troubleshooting Optimizers

34.1 Introduction

This chapter provides a collection of perspectives and insights that may help with identifying issues and solving them when applying optimization.

34.2 DV Values Do Not Change

If a DV value does not change as iterations progress, and if the DV keeps the initial value, then this is a signal of any of a number of undesired aspects. It could be that the coding for the optimizer does not include the DV that is not changed. Perhaps the variable name is misspelled. It could be that the OF calculation is independent of that DV (either because coding erroneously omits it from the OF calculation or because it is irrelevant to the OF or because the OF is effectively insensitive to the DV). It could be that the initial value of the DV is on a constraint or in a flat spot. Don't accept DV^* results if any DV does not change with iteration. Something is wrong. Investigate the surface topology. Check the coding, logic, and constraints.

34.3 Multiple DV^* Values for the Same OF^* Value

Again, there could be several reasons that distinctly different DV^* sets give the same OF^* value:

Discretization: The objective function may have flat spots w.r.t. DV value. This could be the result of an aspect that discretizes the continuum DV to integer or other finite values. Here, small changes in the continuum DV will lead to the same discretized value for the OF calculation, and at the optimum the continuum DV will have a range of values. You can let the optimizer use the continuum values, but report the discretized value.

Local insensitivity: The OF may be dependent on the DV in general, but the optimum may be in a region of flat response, no impact of the DV on the OF. This may be the result of conditionals (IF–THEN statements in the OF calculation) or a functionality that has zero local sensitivity. Here, a DV value will change until it enters the local insensitivity region. You should report the range of equivalent DV^* values, not one particular set. Further, consider that the local flexibility in the DV value indicates that the application is underspecified, and include an additional criterion in the optimization statement.

Multiple equivalent solutions: consider a quadratic equation solution. Two values return the same solution. The OF may have similar features.

Underspecified: If there are more DVs than needed, or if DV structure makes some redundant, then many DV combinations will result in the same OF* value. In such cases, consider an additional measure of desirability and include it into the OF, or remove redundant DVs.

34.4   EXE Error

It is not uncommon to encounter a non-executable computer operation. These include divide by zero, numerical overflow (or underflow, in integers or reals), negative argument values in powers or roots of log calculations, subscript out of range, etc. Overflow errors might be alleviated by switching to long or double precision variables. Similarly subscript range might be alleviated by increasing the dimension on an array. If not, this probably indicates an error in the program coding or the optimizer assigning an excessive DV value. There are billions of situations that could lead to .exe errors. Mostly, fixing them is just classic computer code debugging. Check the coding. Check the optimization concept and its embodiment. Add error trapping lines that look at argument values prior to a calculation and return a "constraint hit" situation to the optimizer.

When using second-order methods (NR or SQ) on a surface that is linear (planar), the matrix inversion will result in a divide by zero. If the surface is nearly linear, it will result in an ill-conditioned matrix. If this is the case, change optimizer type.

34.5   Extreme Values

The optimization may send a DV to a limit, perhaps to zero, or to unity, or to ±infinity. Observe the iteration-to-iteration DV trend. If the DV seems to be moving to a limit, then reconsider the choices in the OF, place constraints on the DV range, or initialize in the local region of importance (not in a range that sends the optimizer out of bounds).

34.6   DV* Is Dependent on Convergence Threshold

It is, of course. The DV* value is dependent on both the convergence criterion and its threshold. If the convergence threshold is made smaller, the DV* and OF* values will be more precise, more reproducible, and have a smaller variability. If the precision is not adequate, tighten the convergence threshold to get smaller variability. However, don't expect perfect precision, zero variability in replicate DV* or OF* values from randomized initializations. The optimizations will not stop exactly at the same DV* values. Convergence tests will stop the iterations when the solution is in the proximity of the optimum. So, don't seek perfect precision. Choose convergence threshold values to balance context-required precision with computational burden. Understand your application context.

34.7   OF* Is Irreproducible

If replicate trials do not provide the same OF* value, it could be because the surface is stochastic or experimental, or that the convergence criteria is too coarse, or that the region surrounding the OF* value has multi optima (such as due to numerical discretization), or that the function has many local optima.

Noisy: If the OF response is noisy, because it is either calculated from a stochastic function or experimentally obtained, use an optimizer algorithm and a convergence criterion that can handle the stochastic aspect. Don't report a single value, but acknowledge the uncertainty in the DV* and OF* ranges.

Coarse convergence: Consider reducing variation by tightening the convergence criterion threshold (see discussion in Sections 34.6 and 34.9).

Coarse discretization: If the numerical discretization is large (perhaps as a convenience to reduce computational work), then the discretization striations may be trapping the optimizer. Consider smaller discretization sizes or an alternate optimizer type.

Multi optima: Some test function applications use high frequency trigonometric functions that create a deterministic undulating surface response to the DV. Each local optima is near to the global but traps the optimizer. Multiplayer optimizers seem to be a good solution.

34.8   Concern over Results

S.E.E.: If the OF statement is not comprehensive, or if the mathematical implementation is not true to the concept, then the solution will raise concern; it won't feel right, and it will contrast intuitive expectations. If so, reconsider the models, the issues, the context, the choices, the optimization statement, the optimizer, and the convergence criterion. Sketch, erase, evaluate, s., e., e., …

This almost seems like my message: "If you don't get the answer that you want, then change the equations until you get the answer that you want."

Every stage of trying to define the application and getting the optimization leads to progressive understanding of the application and its context. Let this experience guide both your expectations and the evolution of the details in the application.

Critically test: Seek to create concern. Don't try to avoid it. Test from randomized initializations and coefficient values in the optimizers—the optimizer should find the same solution. Test over several cycles to ensure all internal variables are appropriately re-initialized—the first and last solutions should be the same. Test on a variety of options in the function (e.g., ideal approximations, alternate given values, etc.)—trends should be as expected. Test for the impact of uncertainty in the givens—if uncertainty in the results is not acceptable, refine uncertainty issues on the givens.

34.9   CDF Features

The CDF(OF*) graph can provide troubleshooting clues.

Probability of the global: Ideally, the DV values from many optimizer trials lead to the true DV*, with precision, and then the CDF(OF*) graph will make a sharp rise from 0 to 1 at OF*. However, if there are multiple optima, and each is found with precision, then the CDF(OF*) graph will make sharp steps at each local OF* value. Intervals between the CDF values at the steps indicate the probability that each local optima has been found. This knowledge would be useful in assessing confidence that the global best has been found.

Precision: The CDF(OF*) graph, however, will not make sharp steps. Precision of the optimized DV* values, and of the OF* values, depends on the convergence criterion and threshold. If convergence is "loose" (if the threshold value is large), then DV* and OF* variation will be visible. Precision will be poor. In this case the CDF(OF*) graph will not make a sharp step at each transition, but it will make an S-shaped transition between steps. If the width of the S-transition is undesirably large, if the OF* values are not precise enough, then make the threshold for convergence a smaller value, or perhaps change the convergence criterion from one based on DVs to one based on the OF.

Note: If there is one OF*, and the scale of the OF* axis on the CDF(OF*) graph is based on the range of the OF* values, then regardless of the threshold in the convergence, the graph will be S shaped. So, consider the range of the resulting OF* values when changing the threshold.

If there is an approximation operation within the OF calculation (such as numerical root finding or a function approximation by a series) that has a convergence test, or if there is time or space discretization in a numerical procedure, then trial-to-trial differences in the internal function convergence would cause OF* variation. At each optimum, instead of sharp or S-shaped steps in the CDF(OF*) graph, the result might appear as a ramp or series of many steps from the vicinity of one OF* value to the next. Consider tightening internal convergence thresholds or making the time or space discretization smaller.

Leading ramp or tail: If the first step on the CDF(OF*) graph has a ramp or tail to the left, then it may be that convergence is being claimed prior to the optimizer reaching the optimum. If the optimizer is hitting a hard constraint or the DV* solution is in a valley with steep sides and gentle bottom slope, or discretization, or loose convergence, then it could be stopping prior to reaching DV*. Consider changing the hard constraint to a soft constraint, replacing the DV with a slack variable, or using an optimizer with a conjugate gradient logic to better accommodate the valley. Alternately, the long valley may be a clue to redundant, or effectively redundant, parameters. See Section 34.10, and consider reducing the number of DVs in the application.

34.10 Parameter Correlation

Ideally, the DV* values end at exactly the true DV* spot, but because convergence criteria ends iterations when the search is close enough, the DV* values end in the close proximity, not exactly on, the true DV* value. Further, if using randomized initializations, the small deviations in DV* from the true value will be random and independent perturbations. If this is so, a plot of the values of one DV* w.r.t. the associated set of another DV* values will reveal a scatterplot. If the axes are scaled to match the convergence criterion on the DVs, then the scatterplot will be a circular "shotgun" pattern. This is what you would hope to find.

Note: This cannot be an observation of DV* values when they end in local or equivalent, but distinctly separate, optima. The reveal of DV* variation must be at the same DV* solution.

Note: If the convergence criterion threshold on one DV is larger than that on another, then the variation in one will be larger than the other, and the shotgun pattern will not be circular, it will be an oval with axes aligned with the DVs. Although there may be a trend in the pattern w.r.t. one variable, there is no correlation, and the ellipse will not be skewed in a joint direction, in an off-axis direction.

Note: It may take 20–100 independent trials, each ending at the same OF* to have enough data to confidently see the pattern.

Classic linear correlation analysis can reveal whether the parameter perturbations are independent or related. Hopefully, desirably, the DV* values are independent. Hopefully, there is no correlation between DVs.

If multiple DVs, correlation between each pair should be explored. If there are N DVs, then there are $N(N-1)/2$ cross correlations to consider. This is easy, since correlation analysis is a common software offering.

If there is parameter correlation in the DV* values, it reveals the presence of any of a number of undesired aspects.

Redundant coefficients: If there are more coefficients than conditions, if the application is under-specified, then the optimizer has an extra degree of freedom. It can assign a value to one DV then find DV* values that end at the same OF* value. In this case there will be a definite relation between DV* values, a locus of points that clearly define a curve. The view of the OF w.r.t. a correlated DV pair will have a valley with a common minimum. This extra degree of freedom means that another issue can be added to the optimization. Reconsider the desirables and undesirables associated with the extremes of the DV* values to help identify the missing condition in the OF.

Effectively redundant coefficients: It may be that the application is in a range in which there is a strong interrelation between DV values. An example is the hyper-elastic spring model between tension and elongation, (stress, σ, and strain, ε) $\sigma = A(e^{B\varepsilon} - 1)$. Here the model coefficients A, B seem independent; however, if either the experiments all use small ε values, or if the material is such that the value of B is small, then the product $B\varepsilon$ will be small, and the exponential $e^{B\varepsilon}$ will be approximately the same as $1 + B\varepsilon$, making the model effectively $\sigma \cong AB\varepsilon$. Here it is seen that coefficients A, B are redundant. If the combined value for the $\sigma(\varepsilon)$ relation is $AB = C$, then any A, B relation satisfying $A = C/B$ is equivalent to any other relation. Such effectively redundant coefficients arise in situation where there is a weak relation of one DV to the data or where the model choice is one step more complicated than the data justifies. In this case the correlation trend between redundant variables will not be a crisply defined line, but a fuzzy trend. The locus of DV* values on a graph of one DV w.r.t. another will reveal a trend, broadened by random variation. The view of the OF surface w.r.t. the two correlated DVs will reveal a valley with steep walls and a relatively gentle slope of the valley bottom. Use this to lead you to reconsider complexity in the model. Probably some simplification, reducing a DV, is appropriate.

Root finding: If calculation of the OF requires a root-finding procedure and the convergence criterion is too large, then it will find an approximate value. If the resulting value changes the OF response to the DVs, then DV* values will be correlated. Consider tightening the convergence criteria used in root finding, number of terms in a series or function approximation.

Hard constraints: Hard constraints often block an optimizer, and it converges with DV* values that form a pattern on the constraint. Here the locus of points of converged DV* values will be a crisp line, but they will not have the same OF*, and they will be on a constraint. If this is the case, consider converting the hard constraint to a penalty function, or changing the optimizer to one that better handles the hard constraint, or just report the best OF* from the constrained set (if all OF* values are acceptable). In some cases, converting the search DVs to a slack variable can remove the hard constraint.

Ridges, valleys, and striations: Discontinuities in the OF could arise from discretization of time or space, from conditionals that switch from one model or one variable selection to another, or in truncation or discretization of other variables in the OF (such as rounding to the nearest cent). These trap optimizer solutions in local striation features, or when on a ridge make them think that there is no better direction downhill. Consider smoothing the transition between conditionals in a fuzzy logic manner. Consider making numerical discretization smaller. Consider a multiplayer optimizer.

34.11 Multiple Equivalent Solutions

An application may have multiple solutions, each valid, and each with the same OF* value. Consider the solution to a quadratic equation. There are two independent solutions, and each is valid. Consider the desire to find the zero-gravity points between three large bodies in space. There are two to four locations depending on the size and location of the masses. If such is the case, report that there are two (or more) equivalent solutions. Alternately, consider how stakeholders would view one or the other, and see if there is an additional criteria for desirability that would select one over the other. For instance, the negative root of the quadratic relation may be mathematically valid, but not physically possible or one zero-gravity point may be less sensitive to spatial location than another, hence more desirable.

34.12 Takeaway

Debugging the initial attempt at defining the application statement and the mathematical formulation is often a long process. Don't expect to get a new application right the first time. Critically look at many aspects of diverse variables to guide evolution of the application.

34.13 Exercises

1   Revisit any of the optimization applications that you have explored, and look at the results in light of topics in this chapter.

Section 6

Analysis of Leapfrogging Optimization

Leapfrogging is a relatively new multiplayer direct search approach. Earlier chapters compared LF to other diverse optimization algorithms and provided experimental data that generally indicated LF was good w.r.t. several metrics (high probability of finding the global, low number of function evaluations, high precision, robustness to surface aberrations, simplicity, etc.). This chapter provides some fundamental mathematical analysis of the technique. Those interested in the theoretical aspects of algorithms (Myers-Briggs N types) may enjoy the chapter in this section. By contrast, if you are an S type, don't read this chapter. The empirical findings of the prior chapters will be more than adequate.

35

Analysis of Leapfrogging

35.1 Introduction

Leapfrogging is a relatively new procedure and can benefit from the depth of analysis that has been established for other optimizers. This chapter provides some analysis to reveal fundamentals about the technique and statements about critical aspects for an application user and reveals the analysis paths and opportunities for others to explore.

Leapfrogging (LF) is a multiplayer (multi-particle) optimization algorithm. In it, trial solutions (players) are randomly initialized throughout feasible decision variable (DV) space, and the objective function (OF) value for each is determined. There is a player at the best position (the trial solution with the best OF value; the default here is a minimum) and a player with the worst position. The worst leaps over the best, like the children's leapfrogging play, into a random spot in the reflected DV space, and the OF of the new trial solution (TS) is evaluated.

The basic algorithm determines the leaping player new position from this equation:

$$x_{new,j} = x_{best,j} + r_j\left(x_{best,j} - x_{old,j}\right) \tag{35.1}$$

in which x represents a DV value, for dimension j, and r is a uniformly distributed random number. The subscript "best" represents the x-position of the best player, the subscript "old" represents the leap-from x-position of the leaping player, and the subscript "new" represents the leap-to x-position of the leaping player. The subscript j indicates the DV dimension. The leap-over move is randomized for each dimension and for each leap-over.

LF does not require the computation of gradients or Hessian and therefore can handle hard constraints, discontinuities, flat OF response regions, and stochastic functions.

Like particle swarm and other population-based optimizers, the multiplayer aspect permits nearly global exploration; but unlike those algorithms, there is no local exploration by the intermediate players. This greatly reduces the computational or experimental work in function evaluations. LF has been demonstrated on a variety of applications and is either tied or better for outcomes (probability of finding the global and computational burden) and is much simpler to understand and code than a variety of other algorithms. Being a direct search algorithm, surface discontinuities are not an issue, as well as DV scale or dimensional incompatibility.

Illustrations and test functions in this analysis will consider the optimum as the minimum.

Engineering Optimization: Applications, Methods, and Analysis, First Edition. R. Russell Rhinehart.
© 2018 R. Russell Rhinehart. Published 2018 by John Wiley & Sons Ltd.
Companion website: www.wiley.com/go/rhinehart/engineeringoptimization

There are a few notable attributes of the LF optimization procedure:

- When the cluster is on the side of a hill, the worst player (based on OF) leaps a long way (in DV space), across the entire cluster, over the lead player on to a further downhill spot. This expands the size of the cluster of players.
- However, near the optimum, the cluster surrounds the best player in all dimensions. Since random leap-to positions cut worst-to-best distance in half (on average), the cluster converges on the best player. As the cluster converges, the best player changes when its leap-over places it closer to the true optimum.
- If the leap-to position is infeasible (for any reason based on DV, or OF, or constraints, or computational solution), the leaping player is still the worst and leaps from the infeasible DV values back over the best.
- If there is a tie for best (such as a flat optimum), use the first player in the list to consistently draw all players to one spot. If there is a tie for the worst, randomly choose the one to leap, giving all a chance to converge.
- After initialization, there is no need to search for the best: If the leap-over finds a better TS, then it is the new best. Further, the algorithm only needs to search for the worst in the cluster if the leap-to solution became better than the leap-from position.

Options: Nominally the leap-over lands the player in a random (uniformly distributed) spot on the other side of the best as defined by the DV axis. However, the leap-to position could be into an expanded or contracted location, or a shifted location (e.g., to center the leap into area on the best) or into a non-rectangular window (e.g., either aligned with the direction between best and worst, or a circular target), or determined by an alternate probability distribution. These all work, but for many enhancements, there doesn't seem to be adequate benefit for the added complexity. For the simpler enhancements, expanding or contracting the leap-to window is achieved by multiplying the movement rule by a factor, α:

$$x_{\text{new}} = x_{\text{best}} + \alpha r \left(x_{\text{best}} - x_{\text{old}} \right) \tag{35.2}$$

Large α-values, perhaps $1.5 < \alpha < 2$, encourage surface exploration. By contrast, small α-values, perhaps $0.5 < \alpha < 1$, accelerate convergence. The standard, nominal, value is $\alpha = 1$.

Shifting the leap-to window corner from the best, B, back toward the worst, W, by a fraction, β, of the B to W distance can also be easily implemented:

$$x_{\text{new}} = x_{\text{best}} + \alpha(r - \beta)\left(x_{\text{best}} - x_{\text{old}} \right) \tag{35.3}$$

With the standard implementation, $\beta = 0$, the leap-to window corner is started on the best player and W leaps to the other side. This would encourage exploration of new area. If $\beta = 0.5$ the leap-to window is centered on B, which would accelerate convergence, assuming B is in the vicinity of an optimum. If $\beta = 1$, and with $\alpha = 1$, the leap-to window corners are bounded between B and W.

35.2 Balance in an Optimizer

We want a balance of opposing ideals in an optimizer. This would include minimizing the number of function evaluations (NOFE) and maximizing the probability of finding the global (p-global). In leapfrogging,

1) We want the player cluster to expand to accelerate DV movement when on a downhill search. The larger the value of α (leap-to range factor) and the higher the number of players, M, the better this is impacted. There are several aspects of the downhill migration of the team of players:
   a) Avoid premature convergence of all players to one spot on the side of the hill. If either M (number of players) or α (leap-to window amplification) is too low, this could happen. This would desire large M and α values.
   b) Provide good surface exploration to improve the chance of a player jumping into a new portion of the DV space to fortuitously discover a better OF and become a new lead to progress the cluster to a global minimum. This would desire large M and α.
   c) Accelerate toward the minimum when moving in the right direction, as opposed to moving at a slow rate. Again, this desires large M and α.
2) By contrast, we want the player cluster to contract quickly when in the vicinity of a minimum. Opposing Desire no. 1, the smaller the value of α and the fewer the number of players, the better this is achieved.
   a) Contract quickly when at the optimum to reduce number of function evaluations (NOFE) to convergence. This would desire small M and α values.

We want to find values of M and α to maximize p-global and minimize NOFE. If p-global is used to define the number of trials, $N = \ln(1-c)/\ln(1-f)$, and average NOFE is combined with N, then PNOFE is the probable number of function evaluations to be c confident in finding one of the best f fraction of solutions. Then the optimization statement is

$$\min_{\{M,\alpha\}} J = \text{ANOFE} \frac{\ln(1-c)}{\ln(1-f)} \qquad (35.4)$$

But there are many alternate schemes toward improving LF optimizer performance:

1) Initialize with many excess players, M_i number, and then take the few best of them for the LF procedure. This increases the chance that one of the excess players will be in the vicinity of the global and draw all players to it. One could run N LF trials, each using randomly initialized M number of players, and take the best-of-N for the solution. Here, the NOFE is $N *$ ANOFE per trial. But with the excess initialization, the NOFE is M_i + ANOFE. Since players start in closer proximity, ANOFE is smaller. M_i can be estimated.
2) Adjust M, α, β during a trial. Perhaps start with high M, α values and $\beta = 0$. Then when it is obvious that exploration is over and the cluster is contracting at an optimum for the end-game search, reduce M, α values and make $\beta = 0.5$, or so, so that the window is centered on the B. Perhaps the end game could be identified by the number of leap-overs with no change in lead, or when the cluster has contracted to perhaps 10 times the convergence criterion. I've also used Bayesian belief mechanisms and a number of leap-overs that would be equivalent to the M_i number of excess explorations in Idea 1. All seem to have some benefit, but none are obviously functionally superior. Simple is best.
3) Use a coarse convergence in the first N trials to scope out the approximate location of the global. Then repeat one last trial with players initialized in the vicinity of the best-of-N, with fine convergence to find the global with precision.

Using Equation (35.4) to guide outcomes, experience on diverse optimization applications of 2–100 DVs has illustrated that $M = 5$–10 players per dimension (with a minimum of about $M = 20$) and a leap-to window of the same size of the leap-from window ($\alpha = 1$) have been a good numbers. However, that was an empirical finding. What is needed is an analysis of how M (number of players) and α (leap-to expansion factor) affect the search. Metrics for these would be (i) the probability of finding the global, (ii) the rate (number of function evaluations—NOFE) at which the cluster gathers in the vicinity of the optimum, and (iii) the NOFE for end-game convergence at the optimum. Also, what is needed is a data-based assessment that indicates whether the vicinity of the global has been found.

The following sections reveal analysis of selected LF attributes.

35.3   Number of Initializations to be Confident That the Best Will Draw All Others to the Global Optimum

This section is based on the publication, Manimegalai-Sridhar, U., A. Govindarajan, and R. R. Rhinehart, "Improved Initialization of Players in Leapfrogging Optimization," Computers & Chemical Engineering, Vol. 60, 2014, pp. 426–429.

In functions of multiple local optima (where the global minimum is not located at the bottom of a single continuous concave space), starting with a greater number of players for surface exploration increases the probability of at least one player being in the vicinity of the global optimum. With one trial solution in the vicinity of the global optimum, it would be the best player. Other players will be drawn toward its vicinity, and the team cluster will converge at the global optimum. Therefore, adding players increases the chance of finding the global optimum. On the other hand, a greater number of players require higher computational burden. Accordingly, there needs to be a trade-off between effort of initial surface exploration and the benefit. This trade-off uses number of function evaluations as the measure of both aspects.

Here M is the number of players, and D is the optimization dimension, the number of decision variables (DV). However, M initializations may not be sufficient to locate the vicinity of the global optimum, due to surface aberrations and/or complexity of the problem. In such cases, multiple independent random starts might be required to increase the confidence of finding the global minimum. If a large number of independent trials, N, are initiated, some trials will find the global, others local optima. With large N, there is increasing probability that the global will be discovered. In one approach to find the global optimum, run many optimization trials from random initializations, and select the best-of-N as the answer. A method to determine the number of independent starts required for the optimizer to find the vicinity of the global optimum (a result within the f fraction of best possible solutions) with a confidence of c is defined by (see Chapter 23)

$$N = \frac{\ln(1-c)}{\ln(1-f)} \tag{35.5}$$

However, the downside to a large number of independent trials is the computational burden. For this analysis, computational burden will be measured by the number of function evaluations (NOFE). If ANOFE is the average NOFE per trial, over N independent starts, the total NOFE will be $N * \text{ANOFE}$. If N is determined by Equation (35.5) to be c confident that at least one of the N trials will have discovered one of the best fraction of possible solutions, then $N * \text{ANOFE}$ is the probable NOFE (PNOFE) to find the global.

The value of f can be visualized by running the optimizer from random initializations many times (thousands), then plotting the cumulative distribution function (CDF) of the optimizer attaining an objective function (OF) value (or lesser). Figure 35.1 shows an example. The global minimum is

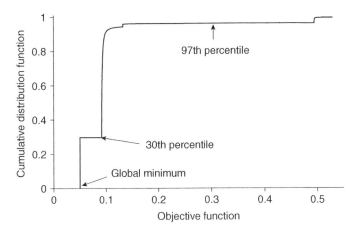

Figure 35.1 Cumulative distribution function. *Source:* Manimegalai-Sridhar *et al.* (2014). Reproduced with permission of Elsevier.

located at an OF value of 0.05, and this global is found in about 30% of the trials. This means that the probability that the optimizer finds a value in the vicinity of the global is $f^* = 0.3$. This value for f^* would be used in Equation (35.5), the user would choose a value for the desired confidence of finding one of the best 30%. For instance, if the user wants to be 99.99% confident that in N trials, at least one will return an OF in the best 30%, then $N = 26$ independent trials.

This analysis presents an improved multiplayer initialization approach using PNOFE as the measure of work required.

35.3.1 Methodology

The concept is not to run only one optimization trial, but to initialize with many more players than needed (to increase the chance of one player being in the vicinity of the global) and to select the best $M = 10 * D$ number of players from the initial excess of players.

In LF, the best player remains the best until the worst player leaps over and finds an OF value that is better than the current best. This implies that if one player is initially in the proximity of the global optimum, it will attract all other players toward it. Figure 35.2 illustrates the concept. Here, proximity to the global means that one player should have an OF value that is lower than the OF value at the local (second best) optima. Region A represents the global attractor DV region, in which the OF value is lower than the OF value at any local optima. Therefore, to guarantee finding the global optimum with LF, at least one player should be placed in

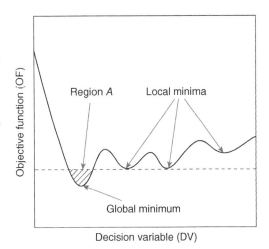

Figure 35.2 Global attractor area. *Source:* Manimegalai-Sridhar *et al.* (2014). Reproduced with permission of Elsevier.

Region A. To have a high confidence of finding the global, there needs to be a high confidence that initialization placed at least one player in Region A.

The desire is that at least one player of M_i initial players should be in Region A, and the question is "How many random initializations are required to be confident that at least one player is in Region A?" Simple probability provides the answer:

$$p(\text{at least 1 of } M_i \text{ Players in } A) = 1 - p(\text{none of } M_i \text{ Players in } A)$$

$$= 1 - p(\text{1st not in } A, \textit{and } \text{2nd not in } A, \textit{and } \ldots)$$

$$= 1 - p(\text{any one player not in } A)^{M_i}$$

$$= 1 - [1 - p(\text{any one player in } A)]^{M_i}$$

(35.6)

From Equation (35.6), the number of initial players required, M_i, in order to have a desired confidence that at least one player is randomly located in Region A is calculated as

$$M_i = \frac{\ln[1 - p(\text{at least 1 of } M_i \text{ Players in } A)]}{\ln[1 - p(\text{any one player in } A)]}$$

(35.7)

Equation (35.7) is simplified to

$$M_i = \frac{\ln(1-c)}{\ln(1-f_A)}$$

(35.8)

where c represents the confidence of achieving the global optimum, and f_A represents the probability that any one player will be randomly initialized in Region A, the region of attraction to the global.

The initialization only specifies players being in Region A, but with LF, leaping players may fortuitously land in the global attractor area even if they are not initially in Region A. So, the optimization technique could reach the global optimum with a probability that is higher than what is nominally expected from the choice of c. Accordingly, Equation (35.8) represents an upper bound on M_i.

If multiple DVs, the region of attraction would be the joint region. If independent, $f_A = \prod f_{A_i}$. Here, PNOFE is the ANOFE of one trial plus the extra function evaluations for initialization:

$$\text{PNOFE} = \text{ANOFE} + \left[\frac{\ln(1-c)}{\ln\left(1 - \prod f_{A_i}\right)} - M \right]$$

(35.9)

The improved initialization will be analyzed for two expectations:

1) Probability of finding the global: The value of c is a lower bound on the probability of finding the global optimum.
2) Reduction in PNOFE: The PNOFE for both the improved initialization and the original LF are calculated and compared with show the improvement.

35.3.2 Experimental

The initialization approach was applied to several test functions. Here, four are two-dimensional and one min–max application of up to 8 DVs was explored. Function 38 (in the companion site) is the popular Goldstein–Price function, which has an irregular-shaped somewhat flat valley between steep walls.

There are four local minima in the valley. Function 30 is the popular Peaks function that has three well-shaped minima. Functions 37 and 21 have more severe features. Function 37, "Sharp Troughs," has three minima, one conventional and two within steep valley walls with a gentle bottom slope. The global is at a slope discontinuity in the valley, leading nearly all search approaches to premature convergence, close to, but not exactly at, the optimum. Function 21, "Boot Print with Pinhole," resembles a boot print in snow attracting nearly all trials to the toe of the boot print; however, up on the snow surface there is a pinhole minimum. All test functions have local minima that can misguide a search. These two-dimension functions are scaled for DV and OF ranges of 1–10 and tested with $c = 99.99\%$.

This investigation also includes a high-dimension min–max application:

$$\begin{array}{c}\min\\ \{x_1, x_2, ..., x_D\}\end{array} J = \max(g_1, g_2, ...g_D)$$

$$\text{where } g_i = \alpha + \beta x_i + \gamma x_i^2 - e^{-\delta(x_i-\varepsilon)^2} \tag{35.10}$$

$$\text{S.T.}: \quad 0 \le x_i \le 10$$

The coefficient values can be used to select a variety of cases. The data presented here use $\alpha = 5$, $\beta = 1$, $\gamma = 0.1$, $\delta = 1$, and $\varepsilon = 9.25$, which provides a generally quadratic surface with a broad minimum value of 2.5 located at $x = 5$ and a global minimum near $x = 9.25$. The Region A, global attractor interval of each DV, is $f_i = 0.1555238...$ DV fraction of full range. A min–max example has broad applications, can be extended to high dimension, and provides an analytical tractable solution for this study. Here, the value of f_A to be used in Equation (35.9) is f_i^D, and the desired confidence of $c = 99\%$, kept the number of initialization players to a reasonable value at higher dimensions. The numbers appended to the min–max label in the tables indicates the number of DVs.

To use Equation (35.8) for determining M_i, values for c and f_A need to be chosen. Here, a 99.99% confidence is used for c in the 2-D experiments, and the f_A value is based on *a priori* (and just approximate) knowledge of the function/surface. In this improved initialization method, an optimization trial has two stages. First, populate the surface with the M_i excessive number of players. Second, run the optimization search with the best $M = 10D$ players selected from the initial M_i players. By contrast, in a conventional best-of-N method, the surface is populated with $M = 10D$ players, optimization goes to completion, and this is repeated N times. These procedures were repeated N times from random starts, so that results represent broad expectations, not an individual fortuitous trial. On the 2-D applications the optimizer used a traditional DV-based convergence criteria, stop if $\Delta DV \le 0.00001$. On the min–max application, the threshold was 0.001.

35.3.3 Results

Table 35.1 shows the number of initial players for test functions. The smaller the global attractor area, f_A, the higher the number of players required for initial surface exploration.

For Functions 30, 37, and 38, the f_A values were approximate. It is unlikely that an implementer would know the exact value for f_A *a priori*. However, the implementer might have a reasonable feel for an estimate. Accordingly, for example, a value of 0.08 was used on Function 37, when a closer value is 0.049.

Table 35.2 summarizes the results of the PNOFE obtained from the original best-of-N (middle two columns) and the improved initialization algorithm (rightmost two columns). The columns headed "CDF(%)" list the proportion of times the optimizer found the global optimum. (In Figure 35.1, e.g., the value is 30%.)

Table 35.1 Number of initial players for test functions—from Equation (35.9).

Function	c (%)	f_A (%)	M_i (players)
#38 (in the companion site)	99.99	1.0	916
#30	99.99	4.0	226
#37	99.99	8.0	110
#21	99.99	0.3	3,066
Min–max-1	99.00	15.55	27
Min–max-2	99.00	2.42	188
Min–max-3	99.00	0.376	1,222
Min–max-4	99.00	$5.85E-2$	7,869
Min–max-5	99.00	$9.10E-3$	50,611
Min–max-6	99.00	$1.42E-3$	325,433
Min–max-7	99.00	$2.20E-4$	2,092,508
Min–max-8	99.00	$3.42E-5$	13,454,598

Source: Manimegalai-Sridhar *et al.* (2014). Reproduced with permission of Elsevier.

Table 35.2 Comparison of PNOFE of original and improved initialization algorithms.

Function	Global OF value	Original LF		LF with improved initialization	
		CDF (%)	PNOFE	CDF (%)	PNOFE
#38 (in the companion site)	0.0447	79.27	1,369	100.00	1,073
#30	0.1326	79.23	1,283	100.00	399
#37	0.0800	97.78	865	99.79	633
#21	−8.7304	11.97	20,869	100.00	3,225
Min–max-1	1.2474	90.20	149	99.99	89
Min–max-2	1.2474	52.40	1,011	99.50	350
Min–max-3	1.2474	18.10	5,535	99.20	1,506
Min–max-4	1.2474	4.10	32,781	99.00	8,394
Min–max-5	1.2474	1.00	211,693	99.00	51,191
Min–max-6	1.2474	0.20	1,412,373	99.00	326,180
Min–max-7	1.2474	0.05	7,292,766	99.00	2,093,300
Min–max-8	1.2474	0.02	22,609,124	99.00	13,455,580

Source: Manimegalai-Sridhar *et al.* (2014). Reproduced with permission of Elsevier.

The improvement in the confidence from 99.99%, originally while determining the number of players required for the improved initialization to 100%, is due to fortuitous leap-overs into the global attractor area. In the case of Function 37, however, the experimental CDF, 99.79% was less than the original confidence of 99.99%. This can be attributed to the subjective nature of determining f_A.

The value used in Table 35.1 was 8%, which set the initial number of players to 110. The actual value is 4.88%. Using 4.88% and 110 players, the confidence in finding the global is 99.59% as calculated by the inverse of Equation (35.8). The actual probability of finding the global is higher.

The columns headed "PNOFE" represent the number of function evaluations that would be required to start enough times (N) so that the confidence in finding the global is 99.99%. Where the experimental CDF% value is 100.00, the PNOFE represents the ANOFE for one trial.

The 2-D tests report the results of 10,000 trials. The min–max investigations report the results based on the number of trials ranging from 100 to 5000, limited by practicalities of computational time.

In all test cases, the improved initialization provided a reduction in PNOFE, demonstrating its utility. The amount of the reduction was different for each function, based on the surface irregularities associated with the function.

The second to the last column of Table 35.2 indicates that the probability of this technique in finding the global is $\geq c$, as expected. The PNOFE columns of Table 35.2 indicate that the improved initialization approach reduces NOFE in each case (from about 15 to 70%).

As a perspective, first, the improved initialization approach is only useful if there are multiple optima, in which one is the global. If there is only one optimum, or if all local optima are equivalent, then one optimization trial is all that is needed when using a standard player initialization.

Second, there must be *a priori* approximate knowledge of the hiddenness of the global attractor area as a fraction of the total DV space. Perfect knowledge is not needed. The actual fraction of the DV area that will be the global attractor for the four 2-D functions are, respectively, 1.27, 3.34, 4.88, and 0.46%. The numbers used to estimate the initial number of players were 1, 4, 8, and 0.3%.

Third, finally, it is assumed that the initialization range encompasses the global. By contrast, if the initialization range is in the vicinity of a local optimum, none of the M_i players will be located in the vicinity of the global.

35.4 Leap-To Window Amplification Analysis

The uniform and independently distributed (UID) random number, r, over a range of 0–1, has an average value of ½. The expectation, or average value, can be calculated from the pdf:

$$E(r) = \bar{r} = \frac{\int_{r=0}^{r=1} r\,\mathrm{pdf}(r)dr}{\int_{r=0}^{r=1} \mathrm{pdf}(r)dr} = \frac{\int_{r=0}^{r=1} r/(1-0)\,dr}{\int_{r=0}^{r=1} 1/(1-0)dr} = \frac{1}{2}r^2\Big|_0^1 - \frac{1}{2} \tag{35.11}$$

A sum of N UID r-values then is expected to have a value of $N/2$.

So, it might be intuitively accepted that the average value of a product of N such random numbers will have an average value of $(1/2)^N$. A thought experiment will reveal this is not so. Consider each random number as its average of 0.5 with a deviation of $\pm\varepsilon$. Here ε is a uniformly distributed deviation ranging from $-\frac{1}{2}$ to $+\frac{1}{2}$. In a product of a large number of r-values, there will be $r_j = 1/2 + \varepsilon$, which will be balanced by an equally probable $r_k = 1/2 - \varepsilon$, and the product of the two will be $r_j r_k = (1/2 + \varepsilon)(1/2 - \varepsilon) = (1/2)^2 - \varepsilon^2$. The average value of the product is not $(1/2)^2 = 0.25$, but

$$\overline{r_j r_k} = \int_{-1/2}^{+1/2} \left[(1/2)^2 - \varepsilon^2\right]d\varepsilon = 0.16666.$$ The deviation toward $r = 0$ has a larger impact on the product than the deviation toward $r = 1$.

Although the average leap-over distance for a single leap is based on $\bar{r} = 1/2$, as it turns out for a large number, the effective leap-over distance is $r_{eff} = 1/e = 1/2.818...$, not ½.

That value can be derived: The product of a series of UID random numbers is

$$\Pi = r_1 r_2 r_3 r_4 ... r_N = \prod_{i=1}^{N} r_1 \tag{35.12}$$

Log transform the product:

$$\ln(\Pi) = \ln(r_1) + \ln(r_2) + \ln(r_3) + \cdots + \ln(r_N) = \sum_{i=1}^{N} \ln(r_1) = N\overline{\ln(r)} \tag{35.13}$$

where $\overline{\ln(r)}$ represents the average, expectation, of the $\ln(r)$ values.

What is the average value of the $\ln(r)$ when r is UID on the interval 0–1?

$$\overline{\ln(r)} = \int_0^1 \ln(x)\text{pdf}(x)dx = \int_0^1 \ln(x)\frac{dx}{1-0} = [x\ln(x) - x]\Big|_0^1 \tag{35.14}$$

$$= 1\ln(1) - 1 - 0\ln(0) + 0 = 0 - 1 - 0(-\infty) + 0$$

To evaluate $x\ln(x)$ at $x = 0$, use L'Hôpital's rule:

$$\lim_{x \to 0} x\ln(x) = \frac{\ln(x)}{1/x} = \lim_{x \to 0} \frac{1/x}{-1/x^2} = -x = 0 \tag{35.15}$$

resulting in

$$\overline{\ln(r)} = -1 \tag{35.16}$$

So, for large N,

$$\ln(\Pi) = N\overline{\ln(r)} = -N \tag{35.17}$$

exponentiating

$$e^{\ln(\Pi)} = \Pi = e^{-N} = \left(\frac{1}{e}\right)^N \tag{35.18}$$

This reveals that, for a large number of terms in the product, the basic LF algorithm (on average) reduces the distance between two active players at a rate of $1/e$, not ½.

With the exploration amplitude factor included, for large N, the product is

$$\Pi = \alpha r_1 \alpha r_2 \alpha r_3 \alpha r_4 ... \alpha r_N = \left(\frac{\alpha}{e}\right)^N \tag{35.19}$$

Consider two players separated by an initial distance of d_0. After a large number of leap-overs, the distance will average

$$d_N = d_0 \left(\frac{\alpha}{e}\right)^N \tag{35.20}$$

For large N on average, if $\alpha < e$, the product will become very small, and the leaping player will converge to the best. If $\alpha = e$, then the product tends to stay about unity, and the distance between players remains. However, if $\alpha > e$, the product will rise unbounded, and the pair will diverge.

Convergence can be based on distance between players compared with a threshold. If $d = (\alpha/e)^N d_0 \le \varepsilon$, then converged. Assuming two players, and either that the best is at the global optimum and the other player jumps back and forth until convergence, or that they leap over each other, then solving for N to exactly make $d = \varepsilon$, and for the case of $\alpha \le 1$, adding one more leap will result in convergence:

$$N = 1 + \frac{\ln(\varepsilon/d_0)}{\ln(\alpha/e)} \qquad (35.21)$$

If there are M number of players, each initially d_0 from the best, and they take turns leaping, then when one is at the global optimum, n is large enough to make $r_{\text{eff}} = 1/e$, and $\alpha \le 1$

$$N = (M-1)\left(1 + \frac{\ln(\varepsilon/d_0)}{\ln(\alpha/e)}\right) \qquad (35.22)$$

But they don't take turns, the worst player leaps, and depending on the topography, the worst might not be the farthest distance from the best. Further, the best will probably not have magically appeared at the global optimum and never relinquish position to a leaping player. The former best will likely yield the lead as the local surface is explored. Also, this equation is based on a large N and $\alpha \le 1$. Finally, using $r_{\text{eff}} = 1/e$ means that the analysis represents an average, not a particular leap-over series, in which there is a possibility that an early leap-over with a very small r-value would converge. Considering all of those issues, the relation should be accepted as an estimate of how M, α affect the trend in N, not an absolute count.

This analysis, however, indicates that N should be linear with $\ln(\varepsilon/d_0)$ and that all values of alpha should have a common value for r-effective. Here is a plot of average N to converge in 100,000 runs w.r. t. $\ln(\varepsilon/d_0)$. It is linear for a variety of α values. The markers are the data; the light lines are trend lines through the markers. The heavy lines represent Equation (35.22) (Figure 35.3).

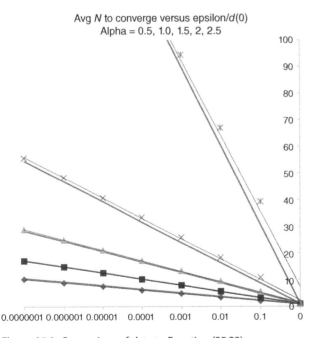

Figure 35.3 Comparison of data to Equation (35.22).

The common value for r-effective that best fits the actual n and the model

$$n = 1 + \frac{\ln(\varepsilon/d_0)}{\ln(\alpha r_{\text{eff}})} \tag{35.23}$$

has the value 0.37018, which is suspiciously close to the reciprocal of the base of the natural logarithm. (I appreciate Dr. Josh Habiger, Statistics, at OSU, for recognizing this, which led to the derivation of equations earlier.) Accepting $r_{\text{eff}} = 1/e$, then

$$N = 1 + \frac{\ln(\varepsilon/d_0)}{\ln(\alpha) - 1} \tag{35.24}$$

Simulation data reveals that if $\alpha > 1$, the intercept rises. This makes sense because if it is on the verge of convergence with $d_i = \varepsilon$, with $\alpha > 1$ the next leap-over may place the players at a farther distance. Empirically, I find that the intercept rises as $1/(\ln(a) - 1)$, but as of this writing has not been able to derive the fundamental reason. So the model is

$$N = 1 + \frac{\ln(\varepsilon/d_0)}{\ln(\alpha) - 1}, \quad \alpha \le 1$$

$$N = \frac{-1}{\ln(\alpha) - 1} + \frac{\ln(\varepsilon/d_0)}{\ln(\alpha) - 1} = \frac{\ln(\varepsilon/d_0 e)}{\ln(\alpha/e)}, \quad \alpha > 1 \tag{35.25}$$

Both are for the large N condition.

Conventionally, convergence is tested after the optimization algorithm changes the trial solution; therefore, there must be at least one leap-over before convergence testing. Then the N to convergence relation reduces to

$$N = \max\left\{ 1, \ 1 + \frac{\ln(\varepsilon/d_0)}{\ln(\alpha/e)} \right\}, \quad \alpha \le 1, \ \text{large } N \tag{35.26a}$$

$$N = \max\left\{ 1, \ \frac{\ln(\varepsilon/d_0 e)}{\ln(\alpha/e)} \right\}, \quad \alpha > 1 \tag{35.26b}$$

Figure 35.4 presents a comparison of the model (35.26) and data. It is an excellent fit.

Two questions remain: (i) What is the mechanistic reason for the intercept relation to α? And (ii) unfortunately, for small α values, there is a bias, perhaps due to the large N condition not being met. Defined as $N_{\text{model}} - N_{\text{data}}$, the bias seems independent of either N or ε/d_0, and only dependent on α. Why? For either, there is room for continued analysis.

Empirically, the bias with $\alpha \le 1$ has a quadratic-ish trend and is closely approximated by

$$b = \begin{cases} -0.857039(1 - \alpha) + 0.3959193(1 - \alpha)^2, & \alpha \le 1 \\ 0, & \alpha > 1 \end{cases} \tag{35.27}$$

Further, since N must be at least one leap-over,

$$N = \max\left\{ 1, \ 1 + b + \frac{\ln(\varepsilon/d_0)}{\ln(\alpha/e)} \right\} \tag{35.28}$$

The objective would be to select M and α that would minimize the number of function evaluations to convergence, N in the previous equations. They indicate that both small M and α minimize N. However, small M and α lead to premature convergence, an undesirable outcome, as the Section 35.5 reveals.

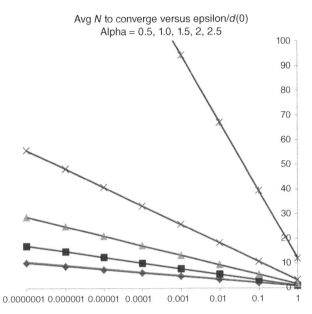

Figure 35.4 Comparison of data and Equation (35.26).

35.5 Analysis of α and M to Prevent Convergence on the Side of a Hill

To start, consider a one-dimensional optimization. If M players are placed on a surface (a line) that has a long downhill slope, then the back one leaps over the front one and becomes the new team lead. If the player positions in one dimension are x_1, x_2, x_3, ..., x_i, ..., x_M, sorted in OF order with x_M in the lead (best), then on average, x_1 jumps to the new location $x = x_M + (\alpha/e)(x_M - x_1)$. On average the value for the UID[0,1] random number is ½, but in a product of large number the effective value is $1/e$. Instead of preserving the label x_1 for the original worst player, call the $(i+1)$th trial solution as x_{i+1}. So, the subscript represents the trial solution sequence. Subscript i is the trial solution number in order from $i = 1$ (worst) to best. M is the number of players.

Using x as the position and $r_{\text{eff}} = 1/e$,

$$x_{i+1} = x_i + \frac{\alpha}{e}(x_i - x_{i-M+1}) \tag{35.29}$$

This is just a recursive relation for sequential terms in a series.
The new cluster range is

$$R_{i+1} = x_{i+1} - x_{i-M+2} \tag{35.30}$$

Substituting the new player position from Equation (35.29),

$$R_{i+1} = x_i + \frac{\alpha}{e}(x_i - x_{i-M+1}) - x_{i-M+2} \tag{35.31}$$

adding and subtracting x_{i-M+1} and rearranging

$$R_{i+1} = x_i\left(1 + \frac{\alpha}{e}\right) - x_{i-M+1}\left(1 + \frac{\alpha}{e}\right) - (x_{i-M+2} - x_{i-M+1}) \tag{35.32}$$

Using the old range

$$R_i = x_i - x_{i-M+1} \tag{35.33}$$

and substituting Equation (35.32),

$$R_{i+1} = R_i\left(1 + \frac{\alpha}{e}\right) - \left(x_{i-M+2} - x_{i-M+1}\right) \tag{35.34}$$

If we accept another primitive choice, that the players were uniformly spaced before the leap-over, then the initial distance between the last two players is the range divided by $n - 1$. If they continue to be uniformly spaced, then

$$R_{i+1} = R_i\left(1 + \frac{\alpha}{e}\right) - \frac{R_i}{M-1} = R_i\left(1 + \frac{\alpha}{e} - \frac{1}{M-1}\right) \tag{35.35}$$

Dividing by the prior range, the ratio of term values in the series is

$$\frac{R_{i+1}}{R_i} = 1 + \frac{\alpha}{e} - \frac{1}{M-1} \tag{35.36}$$

If $\alpha = 1$ and $M = 2$, then

$$\frac{R_{i+1}}{R_i} = 0.367879\ldots \tag{35.37}$$

With only two players, each leap-over cuts the distance between the pair of players by about 37%, and the players eventually converge on the side of the hill, as they progress downhill. They don't get to the bottom. That is undesirable and is termed premature convergence.

However, I find that when on a long downhill slope the players do not preserve a uniform spacing. Empirically, the distance between players seems to approach:

$$d_{i+1} = d_i\left(1 + \frac{\alpha}{e}\right) \tag{35.38}$$

where $d_{i+1} = x_{i+1} - x_i$

This is empirical. I suspect there is a fundamental relation, providing another opportunity for analysts.

And since spacing between players becomes related,

$$R_i = d_{i-M+2}\left[1 + \left(1 + \frac{\alpha}{e}\right)^1 + \left(1 + \frac{\alpha}{e}\right)^2 + \cdots + \left(1 + \frac{\alpha}{e}\right)^{M-2}\right] = d_{i-M+2}\sum_{k=0}^{k=M-2}\left(1 + \frac{\alpha}{e}\right)^k \tag{35.39}$$

rearranging the previous to solve for the trailing player difference as a function of the range and substituting

$$\frac{R_{i+1}}{R_i} = 1 + \frac{\alpha}{e} - \frac{1}{\sum_{k=0}^{k=M-2}(1 + \alpha/e)^k} \tag{35.40}$$

Desiring that $R_{i+1}/R_i > 1$, or perhaps $R_{i+1}/R_i > 1.25$ to provide a safety margin to ensure there is not premature convergence, the relation becomes

$$1 + \frac{\alpha}{e} - \frac{1}{\sum_{k=0}^{k=M-2}(1 + \alpha/e)^k} > 1.25 \tag{35.41}$$

This relation is a constraint on the values for M and α to ensure that the team does not converge on the side of a hill. But this is still ideal (1-D, and large N).

35.6 Analysis of α and M to Minimize NOFE

We would like to determine values for M and α that minimize NOFE, yet guarantee that we won't get premature convergence on a downhill trek. The OF is Equation (35.42), representing a 1-D application. One constraint is Equation (35.41). Further constraints are that $\alpha > 0$, $\alpha < e$, $M > 1$ and that M is an integer. Each player must converge, for the team to converge, and the number of function evaluations, after one player is at the optimum, N, is the sum of Nj values for each other player to converge to the optimum. Using optimization to find the number of players and alpha value that minimizes the number of function evaluations to convergence (once the global has been located), subject to not converging on the side of a hill becomes

$$\min_{\{\alpha, M\}} J = \sum_{j=1}^{M-1} b + \frac{\ln(\varepsilon/d_{0,j})}{\ln(\alpha) - 1} \tag{35.42}$$

S.T.: $\quad 0 < \alpha < e = 2.71828\ldots$

$$b = 1 - 0.857039(1 - \alpha) + 0.3959193(1 - \alpha)^2 \quad \alpha \le 1$$

$$b = \frac{-1}{\ln(\alpha) - 1} \qquad\qquad\qquad\qquad \alpha > 1$$

$M > 1$, integer

$$1 + \frac{\alpha}{e} - \frac{1}{\sum_{k=0}^{M-2}(1 + \alpha/e)^k} > 1.25$$

For this analysis, ε/d_0 is chosen as $0.001/10^{5r_i}$ for each player, where r_i is a uniformly distributed independent random number, $0 < r_i < 1$. Since each trial randomizes each initial player as $d_0 = 10^{5r_i}$, the surface is stochastic. Accordingly, I used leapfrogging optimization with replicates and steady-state convergence criterion on the worst player for the optimization. The contour is shown in Figure 35.5, where the vertical axis is scaled to be 4α, and the horizontal axis is scaled to be $M/2$.

The results are also shown in Figure 35.5. The x-markers indicate each of 100 trials, ending at the constraint. The inconsistency in the contours in the upper right is due to the stochastic nature of the function.

The results indicate that the global optimum is $M = 4$, $\alpha \cong 1.27$, with multiple local optima and an OF range of about $2:1$ between options. Fortunately, the optimum M and α values are fairly insensitive to several orders of magnitude changes to the ε/d_0 value.

The best M and α values seem harder to remember, as a generic rule, than the second best ($M = 5$, $\alpha \cong 1$), and since there is insignificant difference in their NOFE values, the easier-to-remember $M = 5$ and $\alpha \cong 1$ is my preference.

That analysis was for best convergence, but does not address the M and α needed for surface exploration to find the vicinity of the global. Accordingly, I like an empirical rule, which is supported by the ideal analysis: use $M = \max\{5N, 20\}$ and $\alpha = 1$.

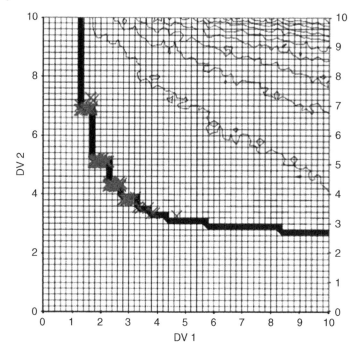

Figure 35.5 Optimization results from Equation (35.42).

The analysis is idealized, representing the most distant player leaping over the best, which is located at the optimum. By contrast, the location of the best will more likely progressively change as the cluster converges, and the leaping player, the worst, might not be the farthest in distance. Further, the convergence criteria may not be the distance between players. In spite of these issues, the analysis provides results, which are consistent with my experience, and I believe that the analysis provides a reasonable guide for LF design.

35.7 Probability Distribution of Leap-Overs

The aim in this section continues to seek a mathematical relation for the average number of leap-overs to convergence. But rather than seeking a large N approximation to the series, this seeks the evolving probability distribution for successive leap-overs. If the probability distribution for the player DV distance, d, can be described for each ith leap-over, given initial choices for $M, d_0, \alpha, \varepsilon$, $\text{pdf}(d|M, d_0, \alpha, \varepsilon, i)$, then the probability of convergence can be estimated for each ith leap and the average calculated.

The question continues to be "Given choices for $M, d_0, \alpha, \varepsilon$, how many leap-overs are expected for convergence?" After about three leap-overs, it appears to be too complicated to get a closed form mathematical solution for how the pdf evolves with leap-over. Numerical integration is possible, and by several tests it seems to provide right solutions. I've tested only on first 15 leap-overs.

This is an analysis of just two players, best, B, and worst, W. W leaps over B. There are two possible scenarios: (i) B remains the best and does not change DV location or (ii) they switch the lead position. Convergence is based on the DV distance between the two players.

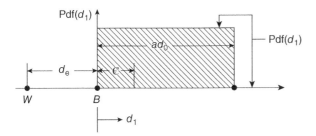

Figure 35.6 Illustration of the pdf of leap-to position for the first leap.

Again this has the same idealizations as the series analysis in Section 35.5. In an LF application, the best will likely change with every few leap-overs, and, as a new best is found, the DV location of B changes. Also, there are multiple players, and the DV convergence criterion will likely require that each player is within the convergence distance of the best. Further, players will not be located at uniform initial distances from the best, so some will require more leap-overs than others to be within the convergence criterion. Also, the leaping player is the one with the worst OF value, not the farthest in DV distance. So, if on a valley wall, across from the best, that player may be continually selected to leap over, even if it is much closer by DV distance to the best than one up the valley floor from the best. Accordingly, this is an elementary analysis. But the probability analysis clarifies the leap-over mechanism and supports the mathematical series analysis earlier.

Figure 35.6 illustrates the first leap. The worst player, W, leaps over the best, B. The initial DV distance between them is d_0. Since the leap-over location is uniformly distributed and random, r, times the window amplified distance, ad_0, the farthest a leap-over can be is to the DV location $d_1 = ad_0$, and the smallest is to $d_1 = 0$. The window amplitude factor is a, and the leap-to distance is d_1.

For the first leap-over from a point, d_0, the pdf(d_1) is uniform:

$$\text{pdf}(d_1) = \begin{cases} 0, & d_1 \leq 0 \\ \dfrac{1}{ad_0}, & 0 < d_1 \leq ad_0 \\ 0, & d_1 > ad_0 \end{cases} \tag{35.43}$$

Note that the d-distances are always positive, in the direction from B to W (on W's side of B). Even though d_1 is in the opposite side of d_0, both are positive measures.

Satisfyingly, it is easily confirmed that $\displaystyle\int_{-\infty}^{\infty} \text{pdf}(d_1)dd_1 = 1$.

If $d_1 \leq \epsilon$, then the convergence criterion is met. The probability that $d_1 \leq \epsilon$ is $\displaystyle\int_0^\epsilon \text{pdf}(d_1)dd_1 = \epsilon/ad_0$.

Figure 35.7 illustrates the second leap-over. If $d_1 < \epsilon$, there is no second leap-over—the process converged. W can only leap to a d_2 position if $\epsilon < d_1 \leq ad_0$. The farthest that W can leap to is $d_2 = a^2d_0$, and the smallest is $d_2 = 0$.

Consider the player at the position indicated by $d_{1,1}$. It can jump to any d_2 position between 0 and $ad_{1,1}$, and the leap-to pdf from $d_{1,1}$ will have a uniform distribution with amplitude of $1/ad_{1,1}$. The player at position indicated by $d_{1,2}$ can jump to any d_2 position between 0 and $ad_{1,2}$, and the

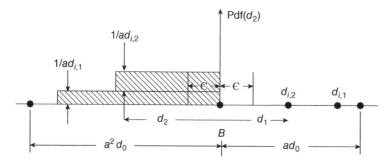

Figure 35.7 Illustration of the pdf of leap-to position for the second leap from two locations.

leap-to pdf will have a uniform distribution with amplitude of $1/ad_{1,2}$. Because the leap-to range of possible values for $d_{1,2}$ is smaller, the pdf amplitude of $d_{1,2}$ is larger than that of $d_{1,1}$, which is illustrated in Figure 35.7.

To get to any position d_2, the player must leap from a far enough away position, $d_1 \geq d_2/a$. To get to d_2, the player must be within the range of $d_2/a \geq d_1 \geq ad_0$. But also $d_1 \geq \epsilon$. So the leap-from position could be expressed as $\max\{\epsilon, d_2/a\} \geq d_1 \geq ad_0$.

The probability that a player is exactly at some $d_{1,i}$ value is zero, but the probability that a player is within a small Δd interval of $d_{1,i}$ is $\Delta d/(ad_0 - \epsilon)$. More generally, the fraction of leap-from area under the pdf(d_1) curve in the Δd interval is $(\Delta d(1/ad_0))/\int_\epsilon^{ad_0} (1/ad_0)dd_1$. Because the argument of the integral is a constant, the probability simplifies to $\Delta d/(ad_0 - \epsilon)$.

Then the pdf of d_2 can be calculated as the sum of all contributions to it, the pdf of leaping from the neighborhood of $d_{1,i}$ times the probability that a player is in that locale. Any d_2 value can be reached from the d_1 extreme of ad_0. However, to reach d_2, d_1 must be larger than d_2/a. But also, d_1 must be larger than ϵ. Accordingly the limit on the terms in the sum must be bounded by the maximum of either d_2/a or ϵ as the minimum and ad_0 as the maximum. In the limit of infinitesimal Δd, the sum becomes the integral:

$$\text{pdf}(d_2) = \sum_{d_{1,i}=\max(\epsilon,d_2/a)}^{d_{1,i}=ad_0} \frac{1}{ad_{1,i}(ad_0-\epsilon)}\frac{\Delta d}{} = \frac{1}{(ad_0-\epsilon)}\int_{d_1=\max(\epsilon,d_2/a)}^{d_1=ad} \frac{1}{ad_1}dd_1 \tag{35.44}$$

This permits a not-too-inconvenient analytical solution:

$$\text{pdf}(d_2) = \begin{cases} 0, & d_2 \leq 0, \ d_2 > a^2d_0 \\ \dfrac{1}{a(ad_0-\epsilon)}\ln\left(\dfrac{ad_0}{\epsilon}\right), & 0 < d_2 \leq a\epsilon \\ \dfrac{1}{a(ad_0-\epsilon)}\ln\left(\dfrac{a^2d_0}{d_2}\right), & a\epsilon < d_2 \leq a^2d_0 \end{cases} \tag{35.45}$$

Figure 35.8 illustrates the pdf of d_2 for the case of $a > 1$.

Satisfyingly, $\int_{-\infty}^{\infty} \text{pdf}(d_2)dd_2 = 1$. To confirm this, you'll need the identity: $\int \ln(u)du = u\ln(u) - u$.

Figure 35.8 Illustration of the pdf of leap-to position for the second leap from all leap-from locations.

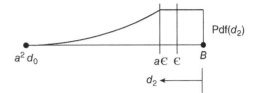

If $d_2 \leq \epsilon$, then the convergence criterion is met. The probability that $d_2 \leq \epsilon$ is $\int_0^\epsilon \mathrm{pdf}(d_2)dd_2 = (\epsilon/a(ad_0-\epsilon))\ln(ad_0/\epsilon)$.

Without considering that the pdf of the leap-from region is uniform, the generic version of Equation (35.44) is

$$\mathrm{pdf}(d_2) = \frac{\displaystyle\int_{d_1=\max(\epsilon,d_2/a)}^{d_1=ad_0} (\mathrm{pdf}(d_1)/ad_1)dd_1}{\displaystyle\int_{d_1=\epsilon}^{d_1=ad_0} \mathrm{pdf}(d_1)dd_1} \qquad (35.46)$$

The pdf of the third leap-over is still analytically tractable:

$$\mathrm{pdf}(d_3) = \frac{\displaystyle\int_{d_2=\max(\epsilon,d_2/a)}^{d_2=a^2 d_0} (\mathrm{pdf}(d_2)/ad_2)dd_2}{\displaystyle\int_{d_2=\epsilon}^{d_2=a^2 d_0} \mathrm{pdf}(d_2)dd_2} \qquad (35.47)$$

Also, when I solve it, satisfyingly, $\int_{-\infty}^{\infty} \mathrm{pdf}(d_3)dd_3 = 1$. If the reader enjoys tedious calculus, confirming this might be pleasant. But the exercise reveals a progression in the complexity of an analytical solution, and I have not pursued analytical solutions beyond pdf(d_3).

The recursion relation is

$$\mathrm{pdf}(d_k) = \frac{\displaystyle\int_{\max(\epsilon,d_{k-1}/a)}^{a^{k-1}d_0} (\mathrm{pdf}(d_{k-1})/ad_{k-1})dd_{k-1}}{\displaystyle\int_{\epsilon}^{a^{k-1}d_0} \mathrm{pdf}(d_{k-1})dd_{k-1}} \qquad (35.48)$$

Numerically it is relatively easy to sequentially calculate the pdf for $d_1, d_2, d_3, ..., d_{15}$ for a variety of d_0, a, ϵ values. My version uses the trapezoid rule of integration, and within numerical error, it affirms the analytical solution. It also determines the probability that the process leads to convergence by each leap-over. For $d_0 = 10, a = 1, \epsilon = 0.01$, the probability of convergence by the first 15 leap-overs is nearly 100% (99.3%) as Figure 35.9 reveals.

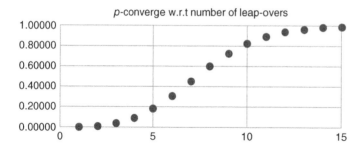

Figure 35.9 Numerical results of using Equation (35.48) to determine the cumulative probability of converging after N leap-overs.

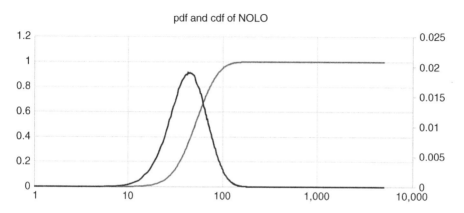

Figure 35.10 Probability distributions of convergence after N leap-overs. Note the semi-log scale.

35.7.1 Data

I have created a Monte Carlo simulation of leap-over series. In it the distance between two players is progressively multiplied by ar_k, where k is the leap-over counter, until $d < \varepsilon$, where the pair converges. The number of leap-overs to converge is used to increase the count in a histogram of 5000 bins. After each million trials the pdf and cdf of N to stop are revealed. See Figure 35.10 for the case of $\alpha = 2$ and $\varepsilon/d_0 = 10^{-7}$. NOLO represents the number of leap-overs to converge. The abscissa is on a log scale for presentation. The distribution is not symmetric, as the log transformation makes it appear, so the average N to converge is about 55.776 in this case. It is not at the peak of the pdf, but at the 50 percentile of the CDF curve.

The average N to converge is calculated as $N = \sum_{k=1}^{5000} k(\mathrm{bin}(k)/\text{total trials})$. For a large variety of choices of d_0, α, ε, the Monte Carlo average matches the model of Equation (35.28). Of course, the experimental average is a stochastic statistic with a distribution and never an exact match to the ideally expected value. So, the claim that data matches the equation is based on the 99% confidence interval of the average of 50 replicates of the experimental average. In each case, the ideal equation predicted value is within the 99% confidence limit of the simulations.

The VBA program uses double-precision variables. It does not use the built-in single precision RND () function, but a conventional UID pseudo-random number generator in double precision. The number of trials, 10^6, was chosen to make the pdf curve relatively smooth and repeatable. The number of

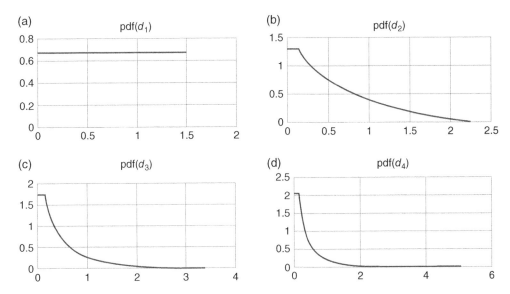

Figure 35.11 Pdf of leap-to position after (a) one leap-over, (b) after two, (c) after three, and (d) after four.

bins in the histogram, 5000, was chosen to ensure that extreme N situations would be identified. If in any of the trials a count exceeds 5000, the program provides a warning of insufficient number of bins.

Figure 35.11a–d reveals the pdf of N-to-converge, given choices of d_0, α, ε, pdf $(N|d_0,\alpha,\varepsilon)$. The several figures indicate the results of the first several leap-overs with $d_0 = 1, \alpha = 1.5, \varepsilon = 0.1$.

Note that at each leap-over the possible leap-to range increases by a factor of $\alpha = 1.5$. Note that the initial $d1$ distribution is uniformly distributed. Note the shoulder on subsequent pdf curves is the result of leap-from locations being limited to those larger than ε.

This program divides the d value into 1000 intervals and computes up to 15 leaps. If the probability of convergence by the 15th leap is greater than 99.9% the program uses the data to calculate the expected number of leaps to converge. Again, data from this numerical calculation of the pdf curves matches the Monte Carlo simulation very well.

The compounding complexity of analytical solutions suggests that there is little utility in seeking the general dependence of the number of leap-overs to achieve convergence given values for d_0, α, ε, although doing so may provide an analytical triumph. The numerical solutions are possible, but the result is time consuming, not invertible (can't answer the Q "What combination of a, ϵ leads to 99% probable convergence in 20 leap-overs?"), and predicated on multiple idealizations.

I think the value in this exercise is a confirmation of the mechanisms and understanding of the leap-over process. With my perspective on sufficiency I decided to stop here. Others may enjoy and value continued analysis.

35.8   Takeaway

The rest of the book has been written as a how-to guide to the practitioner, to the person who intends to use optimization. By contrast this chapter provides an introduction to analysis of the leapfrogging optimization technique and should be seen as is providing opportunities for researchers. Exercises that follow summarize next step challenges in analysis.

35.9 Exercises

Note to reader: These exercises are more aligned with those seeking to analyze LF. These are intended to be small research investigations. By contrast, exercises in other chapters and in the case studies have been devised to affirm the reader's understanding for optimization application.

1 How to extend the analysis to multiple dimensions? If convergence is based on the same ε/d_0 for each DV, convergence will be after all DVs have converged. But the $M - 1$ players each start with a unique d_0 value. Perhaps the analysis could be based on a characteristic d_0 as previously mentioned. But more legitimately, it should include the variation in d_0.

2 The analysis in this chapter considers that the farther is the player from the optimum, the worse is the player OF value, and always the farthest player is the leaping player. However, the surface may have steep valleys, and a close player in DV space may be high on a steep wall and may need to leap many times across the valley, from wall to wall, until it is very close to the optimum in DV space before the other players begin end-game convergence leaping.

3 The analysis of this chapter considered that the $M - 1$ players were converging on the one player already at the optimum; however, as the team converges, the lead player will change position, and the location of the best will change. This means that some player that had leaped to within ε of the lead and had converged may now be outside of the converged distance.

4 Suppose the convergence was based on rms DV distance $\mathrm{rms}\,DV_j = \sqrt{\sum \left(d_i^* - d_{i,j}\right)^2}$ where j is the player index and i is the DV index. If rms DVj is less than epsilon, the player has converged. Convergence of the cluster is claimed when all have converged. Develop the appropriate models for the analysis.

5 Suppose convergence was based on the OF value or rms of the OF values. How does the analysis change?

6 For large N, the number of leap-overs to make $d = \varepsilon$ is $\ln(\varepsilon/d_0)/\ln(\alpha/e)$, then with $\alpha < 1$, one more leap-over would guarantee convergence, making $N = 1 + \ln(\varepsilon/d_0)/\ln(\alpha/e)$. However, I find this to not exactly match the data and to deviate greater with smaller alpha values. Is it because the value of N is not large, and the smaller is alpha the smaller also is N? Does the few number of leap-overs make r_{eff} closer to the ideal one value of ½?

7 Along the same issue, for large N, the number of leap-overs to make $d = \varepsilon$ is $\ln(\varepsilon/d_0)/\ln(\alpha/e)$, then with $\alpha < 1$, one more leap-over would guarantee convergence, making $N = 1 + \ln(\varepsilon/d_0)/\ln(\alpha/e)$. However, with $\alpha > 1$ there is a probability that the next leap will expand the range, placing $d > \varepsilon$. For any particular leap and an UID r-value, the probability that it converges on the next leap is $\varepsilon/\alpha\varepsilon = 1/\alpha$, and the probability that it does not is $1 - 1/\alpha$. I find that an empirical correction to the N equation, instead of one more leap to converge it, is $1/-\ln(\alpha/e)$ leaps, making $N = \ln(\varepsilon/d_{0e})/\ln(\alpha/e)$. Can this be analytically derived?

8 On the side of a hill, with progressive leap-overs of the trailing player to become the lead, does the player spacing on average become $d_{i+1} = d_i(1 + \alpha/e)$?

Section 7

Case Studies

Ten case studies are provided. My view is that the simple end-of-chapter exercises are appropriate to experiencing, and thereby understanding, isolated concepts. However, the real world of applications is not the doing of isolated exercises. These case studies are simple enough to be learning tools, but complicated enough to help a reader see how to structure and perform optimization within a situational context.

Engineering Optimization: Applications, Methods, and Analysis, First Edition. R. Russell Rhinehart.
© 2018 R. Russell Rhinehart. Published 2018 by John Wiley & Sons Ltd.
Companion website: www.wiley.com/go/rhinehart/engineeringoptimization

36

Case Study 1

Economic Optimization of a Pipe System

36.1 Process and Analysis

Figure 36.1 illustrates a fluid transport example. Fluid starts in a tank on the left and is pumped through a long pipe to a receiver on the right. The question is "What is the right pipe diameter?" Here the pipe diameter is the single decision variable (DV), and we will use net present value (NPV) as the objective function (OF).

The smaller the pipe diameter, the lower the cost of the pipe. Also it lowers the cost of the product inventory to fill the pipe. Both are good. However, too small D makes the pressure drop (ΔP) large, requiring an excessively sized pump and drive (high capital), also with excessive energy expense to operate (annual cash flow). By contrast, a larger D leads to less expensive pump and reduced energy expense; however, it increases both pipe cost and the investment in in-process inventory. Basically, energy expense is proportional to power consumption, and capital cost is proportional to capacity to the six-tenths power.

As a helpful artifice, I like qualitatively sketching the mechanisms to be modeled. Figure 36.2 reveals the impact of the decision variable, diameter, on the key economic factors.

Figure 36.2a shows the cost of the pipe rising as the six-tenth power response to capacity. The in-pipe product inventory (Figure 36.2b) rises as the in-pipe volume, which scales with the square of the diameter. Figure 36.2c and d reveal how capital cost and operating expenses of the pump scale with pipe diameter.

36.1.1 Deterministic Continuum Model

Capital is the cost of the initial investment in equipment, inventory, and cash on hand to pay associated bills. Equipment capital, installed, usually follows a six-tenths power law. You can find base prices in texts on engineering economics or in your company data base:

$C_{\text{pipe}} = C_1 L (D)^{0.6}$ Cost is based on diameter and per unit length

$C_{\text{pump}} = C_2 \dot{W}^{0.6}$ Cost is based on capacity

$C_{\text{inventory}} = C_3 V$ Cost is based on volume in pipe

Engineering Optimization: Applications, Methods, and Analysis, First Edition. R. Russell Rhinehart.
© 2018 R. Russell Rhinehart. Published 2018 by John Wiley & Sons Ltd.
Companion website: www.wiley.com/go/rhinehart/engineeringoptimization

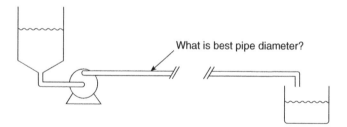

Figure 36.1 Process illustration. Determine the optimum pipe diameter.

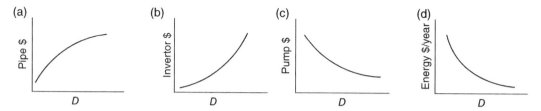

Figure 36.2 Impact of diameter on (a) pipe cost, (b) in-pipe inventory, (c) pump cost, and (d) pump energy expenses.

Classic fluid mechanics relations are needed to determine the power and volume dependency on pipe D:

$$\Delta \text{Pressure to move fluid} = f\frac{L_{\text{equ}}}{D}\frac{1}{2}\rho v^2 = f\frac{8\rho L_{\text{equ}}}{\pi^2 D^5}\dot{Q}^2$$

$$\text{Pump power} = \dot{Q}\cdot\Delta P = f\frac{8\rho L_{\text{equ}}}{\pi^2 D^5}\dot{Q}^3 = \dot{W}$$

$$\text{Pipe volume} = V = L\frac{1}{4}\pi D^2$$

Moreover, for electrical circuits, $E = iR = Ri$. In electrical current, the motivating force is proportional to the flow of electrons. For turbulent fluid flow, however, the motivating force is proportional to the square of the flow of fluid. For electrical circuits, power is iE; here it is the analog $\dot{Q}\cdot\Delta P$.

Note: Each of the capital costs depends on the decision variable, pipe diameter, D. The units are $:

$$C_{\text{pipe}} = C_1 L D^{0.6}$$

$$C_{\text{pump}} = C_2\left[f\frac{8\rho L_{\text{equ}}}{\pi^2}\dot{Q}^3 D^{-5}\right]^{0.6}$$

$$C_{\text{inventory}} = C_3\frac{1}{4}\pi D^2$$

Values for C_1 and C_2 depend on the operating pressure and will jump to higher levels with high pressure service.

Expenses are the annual costs associated with maintaining and running the process—the monthly bills, but the units are in $/yr. They are related to energy not only to run the pump but also for maintenance and depreciation. I am using a nominal industry characteristic value for maintenance, 4% of capital per year, and a 10-year straight-line depreciation model to regenerate capital:

Energy to run pump	$E = a\dot{w}$	Units are \$/day
Maintenance	$M = (0.04)(\text{installed fixed capital})$	Units are \$/yr
	$= 0.04\left(C_{\text{pipe}} + C_{\text{pump}}\right)$	
Depreciation	$D = \text{annual put back or savings to replace worn-out equipment}$	Units are \$/yr
	$= 0.1\left(C_{\text{pipe}} + C_{\text{pump}}\right)$	

I did not include the expenses for operations labor and supervision or associated overhead, because these will be independent of the decision variable in the process design. They are a constant in the OF.

Working capital is the cash on hand needed to pay monthly bills for energy, maintenance, raw material, etc.:

Cash on hand	$$\text{WC} = \frac{1}{12}(E + M + D)$$	Units are \$

After tax expenses are essentially (1 – tax rate) times profit and profit is sales minus expenses. But since sales is independent of the DV choice, it is a constant in the OF, and only the expenses need be considered. Here is the analysis:

$$\text{Pretax profit} = \text{Income} - \text{Expenses} = \text{Gross Profit}$$
$$\text{Tax} = r(I - E)$$
$$\text{After tax profit} = I - E - \text{Tax} = I - E - r(I - E)$$
$$= (1 - r)(I - E)$$
$$= (1 - r)I - (1 - r)E$$
$$\text{After tax expenses} = (1 - r)E$$

Present value takes all future income and expenses and adjusts them to a present value: If you invested x dollars today at a 10%/yr compounded rate of return, it would be worth y dollars in some future year. Table 36.1 illustrates the method. The first column indicates the year into the future. The second

Table 36.1 Cash flow table for PV calculation.

Year	Category	Value	Discount factor	Value today of future \$
0	Capital invest operating expenses	\$600k	$(1 + 0.10)^0 = 1.00$	\$600k
1	Operating expenses	\$100k	$(1 + 0.10)^1 = 1.10$	\$90.9k
2	Operating expenses	\$100k	$(1 + 0.10)^2 = 1.21$	\$82.6k
3	Operating expenses	\$100k	$(1 + 0.10)^3 = 1.331$	\$75.3k
⋮	⋮	⋮	⋮	⋮
			$(1 + i)^{\text{yr}}$	
				$\sum = P.V.$

column describes the cost category and the value column the amount of the cost. The discount factor column represents the inflation rate, at 10%/yr, and the last column is the value divided by the discount factor, indicating the value today of the cash flow in the future. For example, if one had $90.0k today, invested and earning 10%/yr, in a year it would have grown to $100k, which would meet the cash flow need in year 1 into the future.

Values for the "givens" are as follows:

$Q = 0.5 \text{m}^3/\text{s}$

$\rho = 850 \text{kg/m}^3$

$L = 500 \text{ m}$

$L_{\text{equ}} = 875 \text{ m}$

$f = 0.005$

$C_1 = 2 \text{k\$/m}^{1.6}$

$C_2 = 7 \text{k\$/kW}^{0.6}$

$C_3 = 5 \text{\$/m}^3$

$\alpha = 0.1 \text{\$/kWh}$

Use a 25-year life of the process with a 10%/yr discount value to calculate the NPV of the design.

36.1.2 Deterministic Discontinuous Model

The pump (and its drive) and the pipe cost jumps by 50% if high pressure service is needed: if $\Delta P > 150$ kPa, then $C_1 = 3 \text{k\$/m}^{1.6}$ and $C_2 = 10.5 \text{k\$/kW}^{0.6}$.

The diameter could be discretized to 1 mm increments, representing what is possible to purchase. Or you can choose to let the optimizer find the "perfect" value, such as 33.2568312 cm (if such a concept of 8-place precision is meaningful in light of the idealizations and simplicity of the models and the values of the coefficients and givens).

Risk can be added as an annual forecast penalty for the possibility of a pipeline break and loss of contents. The larger the pipe diameter, the greater the environmental impact. Risk is specifically calculated as the probability of an event times the cost of the event. I have modeled the probability of an undesirable event as independent of the pipe diameter. By contrast, the cost of the event will be related to the volume that might be released. I used a discontinuous relation for penalty increasing with pipe volume.

One could find the diameter to minimize capital, expenses, or NPV. Table 36.2 shows values that I get. You could consider these to be answers to the case study and use them to benchmark your findings. If you set up the economic analysis exactly like I did, the numbers should be the same. But if you used alternate, but also valid methods (like a monthly accounting or a different cost model for high pressure choices), then your results will not exactly match mine. However, the trends should be similar. The first column, "OF choice," indicates which index is used as the OF for minimization. The second column reveals the optimum pipe diameter for the OF choice. The remaining columns indicate the values for each index, when the pipe diameter is chosen to minimize the one.

Note the large differences in best diameter due to objective function selection: If strapped for capital, choose 33.3 cm to minimize capital. But it'll cost. Alternately, if flush with capital, choose 48.9 cm to minimize expenses, to generate best profit in the future. Alternately, use NPV as OF, and then 43.1 cm is best overall (at 10% interest).

Table 36.2 How various economic indices trend with choice of which to minimize.

OF choice	DV pipe diameter (M)	Nominal capital ($k)	Nominal expenses ($k/yr)	Nominal NPV ($k)	NPV at 90 percentile ($k)
Min capital	0.333	616	79	1334	2079
Min expenses	0.489	688	57	1205	1243
Min NPV	0.431	653	58	1184	1249
Min 90% worst NPV	0.45	664	58	1187	1241

In my implementation of this example, the investment was in year zero and operation immediately followed. However, a plant construction and design might have 3–5 years of up-front investment, which needs to be scheduled. Further, I did not include reinvestment of capital to replace the worn-out pump or a scrap-scale income from the terminated process. There are many options that a reader might want to explore.

36.1.3 Stochastic Discontinuous Model

The "givens" are never absolutely known. Who would definitively state that a, the cost of electricity, will be some particular value over the 25-year life? Accordingly, I have estimated the uncertainty on each of several key variables as follows:

The uncertainty on $\dot{Q}$ is $\pm 30\%$.
The uncertainty on f is $\pm 50\%$.
The uncertainty on L_{equ} is $\pm 50\%$.
The price of product inventory is $\pm 40\%$.
The price of electricity is $\pm 50\%$ uncertainty.

There are several approaches that can be used to account for uncertainty:

- Monte Carlo method—Randomly select each uncertain value from within its range. This is one "realization" of a possible future. Use the optimizer to find best DV value from each realization. Do this for 1000 or so realizations. Generate the probability distribution of best DV and OF. If range is unacceptable, investigate reducing variability. This is method 3 of Section 20.8.2. You certainly could also use methods 4 or 5.
- Analytical propagation of uncertainty method—Perform a sensitivity analysis to determine $\Delta\text{OF}/\Delta\text{"Given"}$. Then collectively propagate the uncertainty on all. If 95% probable uncertainty, $\varepsilon_{0.95,\,\text{OF}} \cong 0.8\sqrt{\sum (\Delta\text{OF}/\Delta G_i)^2 \varepsilon_i^2}$. If range is unacceptable, investigate reducing variability.
- Perform a sensitivity analysis to determine $\Delta\text{OF}/\Delta\text{"Given."}$ This implies that the range on $\text{OF} \sim (\Delta\text{OF}/\Delta\text{"Given"})\text{Range}_{\text{Given}}$. See which parameter range has the largest impact on OF and seek to improve certainty in that factor.
- Find the DV value that minimizes variability (uncertainty) in OF. Choose a DV, and use the Monte Carlo method to look at many realizations, enough to estimate uncertainty in the OF. Here the OF is not the profitability index but the range or standard deviation of the NPV realizations.

- Find the DV that minimizes the potential for the OF to be above a threshold, min p(OF> bad number). Or minimize risk, the probability of a bad event times the economic cost of the bad event. These OFs are very different from minimizing or maximizing the profitability index. In my example, I chose to minimize the 90% worst-case NPV. The bottom row of Table 36.2 shows the results.

36.2   Exercises

1   Implement the relations described and the NPV analysis. This is easily done in a spreadsheet but could equivalently be done in any structured computer code.

2   Show how elements in the several economic indices change with pipe diameter.

3   Explain when expenses, or capital, or NPV might be the basis for an economic optimization. This depends on company cash on hand, cost to borrow, etc. You might relate this to a large personal purchase, such as a car.

4   Implement a single DV search algorithm. I've used Golden Section, heuristic direct search, and Leapfrogging methods. Although Newton-type and successive quadratic techniques work well on the deterministic continuum model, the discontinuities of risk, pump cost with high pressure, and discretization of pipe diameter are problems for them. I like the direct search approaches. Further, for the stochastic NPV90, LF with replicates and the steady-state convergence is my choice.

5   Show optimizer results to minimize expenses, capital, and NPV. Describe the differences in diameter resulting from minimizing expenses, capital, or NPV.

6   Explain that pipe diameter can only have discrete values, and redo the optimization with discrete pipe sizes. Reveal differences.

7   Explain and illustrate sensitivity of the economic indicators to the "givens" F, f, electricity price, and inventory price. Perhaps change a value and see the impact it has on the optimum diameter and the profitability index. Show how a sensitivity analysis should focus engineering attention to elements that need fine-tuning, more investigation, high certainty, and acknowledgment in any recommendations.

8   Explain the infinite possibilities of combined deviations for the "givens" related to uncertainty of several variables, and each within a ±range of possible values. Explain a Monte Carlo experiment in selecting many realizations one by one, calculating the outcomes for each, and creating a histogram and CDF.

9   Use the Monte Carlo results to create a CDF of the OF due to the uncertainty of the "givens" for a particular pipe diameter choice. Explore the impact of too few realizations (perhaps 50) and too many (perhaps 10,000). With too few the CDF will be discontinuous and irreproducible—a new 50 realizations will give similar but clearly different results. As the number of realizations increases, the CDF will become more uniform and reproducible. With an excessive number of

realizations, there is no meaningful improvement in the CDF precision, but it consumes excessive computation time.

10 Explain making a decision on the 90% worst case, not the nominal case. Perhaps use coin flips as an illustration of the concept. You don't want a 51% chance of getting a good outcome. You want a 90% chance. (Or 95%, or 99.9%, or 99.999%, depending on the situation.)

11 Use your optimizer to minimize NPV90, and compare the results with minimizing NPV. Do this with too few number of realizations (perhaps 10), and you'll find that the optimizer results are not repeatable. Do this with about 1000 realizations, and you'll have adequately repeatable results w.r. t. the diameter discretization.

12 Show the one-by-one relation of uncertainty to standard deviation (width of CDF) and the use of this sensitivity to focus engineering attention on reducing uncertainty on the OF resulting from uncertainty on the "given."

13 Explain risk as probability of an event times the cost of the event, and add risk of a pipeline break to the expenses. The larger the pipe diameter, the larger the event, and the more severe the penalty. Show how adding risk changes the optimal pipe diameter.

37

Case Study 2

Queuing Study

37.1 The Process and Analysis

Queuing is a useful and widely applicable study. Although presented here as a classic number of check-out lanes in a store, it is fundamental to logistics as in elevator management, traffic light control, supply chain design, student–teacher specification, and many similar work-process situations. Since the concepts of queuing and standing in checkout lines at a store are familiar to most, it represents a classic archetypical exercise.

The objective here is to design the number of checkout lanes in a store to maximize desirables and minimize undesirables. From an investment view, more lanes means more floor space, counter/register initial cost, and clerk annual expenses. However, more lanes means greater reliability (an extra lane exits to replace one or more that are disabled, which can happen for any number of reasons). From a customer view, more lanes means less wait time. Additionally, a customer view could be that excessive number of lanes and idle clerks waiting for a customer, imply that the business inefficiency costs are compounded and they are charging me higher than needed product prices or, in contrast, that I deserve and can afford this privileged treatment. What might be other desirables and undesirables? What are constraints? How can the impact of objectives and constraints be assessed, quantified, or measured? How can diverse objectives be combined?

A classic OF would be to minimize some combination of cost (directly related to the number of lines) and customer aggravation (due to wait time). If just those two aspects and an additive OF model, you would need to define appropriate equal concern factors for the two competing terms in the OF. Alternately, you may choose to use a Pareto optimal procedure and retain all of the non-dominated solutions for a business manager to choose. Alternately, you may choose an economic model for the cost aspect that combines capital (investment in the lanes) and annual expense (of clerks and maintenance).

In design, the DV is the number of register lines to handle customers, which is a single DV problem with integer values. However, once a design is complete, in the operation mode, the DVs are the number of clerks to have on duty staffing the lanes, the number of trained personnel doing other in-store jobs, and the number of those who can shift tasks to open a new register when there is a backup of customers or no-show checkout clerks. Again these are integer DVs.

You can use leapfrogging with continuum DVs, let the jump-overs land in non-integer values, but round the value when given to the function to generate the OF value.

Classically, queuing simulations run the process for a time interval, let customers arrive in a Poisson distribution, and use a Gaussian distribution for the processing time. The probability that a clerk is not

Engineering Optimization: Applications, Methods, and Analysis, First Edition. R. Russell Rhinehart.
© 2018 R. Russell Rhinehart. Published 2018 by John Wiley & Sons Ltd.
Companion website: www.wiley.com/go/rhinehart/engineeringoptimization

able to work (out sick, injured, called away for either a personal or business emergency, etc.) could be modeled as a binomial distribution. Then from the Monte Carlo simulation over a period of time, relevant measures of desirability or undesirability are determined. For instance, if customer aggravation is related to wait time, then what should be the measure of undesirability: an average wait time, the number of customers waiting for more than 2 min (a threshold for aggravation) to be served, or the cumulative wait time of all customers waiting over 1 min as the OF? You choose. Since any statistic has a distribution (one Monte Carlo simulation will generate a different realization from another), should the metric be the average from one run, or the 95% worst in a run (an extreme worst) or the 25% worst (even a medium bad outcome is undesired)? You choose. How many customers, or what simulation time period, is needed to be included in a simulated situation to reveal a true representation of years of business? You choose.

For a Poisson process, λ is the expected (average) number of events in a time interval, and then the probability of k events in that time interval is

$$p(k) = \frac{\lambda^k e^{-\lambda}}{k!} \tag{37.1}$$

According to Law and Kelton (1991), the time interval between events that are Poisson distributed is

$$\Delta t = -\left(\frac{1}{\lambda}\right) ln(r) \tag{37.2}$$

where r is a random number uniformly distributed on an interval of 0–1.

The Box–Mueller method can be used to assign the NID(μ, σ) processing time for a particular customer:

$$Process\_time = \mu + \sigma\sqrt{(-2ln(r_1))}\sin(2\pi r_2) \tag{37.3}$$

where r_1 and r_2 are uniformly distributed random numbers $0 < r \le 1$.

The probability that m clerks are available when n are scheduled to be available is

$$p(m|n) = \binom{m}{n} p^m q^{n-m} = \frac{m!}{n!(n-m)!} p^m q^{n-m} \tag{37.4}$$

where p is the probability that a clerk is available when scheduled and $q = 1 - p$ is the probability that a scheduled clerk is not available. You might evaluate this every half-hour of simulated time.

The probability that a lane is operable will have the same binomial equation. You might evaluate this every several hours of simulated time.

This is a stochastic optimization application. The optimizer will choose a DV and run the simulator to determine the OF. Replicate runs (same DV value) will result in different customer loads, process times, and clerical availability, which will generate different wait time and other desirability/undesirability values. Stochastic processes confuse optimizers that are based on a deterministic response model. So, the stopping criteria need to be appropriate for a stochastic problem. And the choice of optimizer needs to be one that does not progressively lead the solution to DVs with fortuitously good values. Since the DVs are integers, the optimizer needs to be able to cope with level changes in the OF as well. I'd suggest leapfrogging with replicates and stopping based on the OF of the worst player achieving a steady state w.r.t. iteration count.

37.2 Exercises

1 Choose and defend reasonable values for $\lambda, \mu, \sigma, p_{\text{clerk}}, p_{\text{lane}}$ and the costs of a checkout line and clerk salary. (In a superstore, at peak Holiday shopping time, λ might be 3000 customers/hour. But, on a weekday morning at the same store, it might only be 50 customers/hour. But don't use my numbers. Relate this exercise to your personal experience.)

2 Create a simulator to simulate the store process for an extended time period. Probably this must be an executable code routine, neither an analytic set of equations nor a spreadsheet calculation.

3 Choose and defend an OF and its weighting values.

4 Set up an optimizer to solve for DV* in the design case—the number of lines to install.

5 Compare the results from a deterministic model (average λ, μ with no variability) with the results when variation is included.

6 Observe the time evolution of events, relate them to your personal experience, and revise the OF metrics and weighting criteria.

7 Consider that not all customers view wait time with the same concerns. Children may enjoy being distracted by the products that line up the queuing line. Adults may enjoy seeing the headlines and celebrity photos on the magazines or visiting with others in line. However, some customers may be late for their next activity and have an exaggerated undesirability impact of the wait. Include this random weighting of the penalty for wait time as a stochastic element in the model.

8 Use propagation of uncertainty on the deterministic model to estimate the uncertainty in the OF. Compare this to the range encountered in the simulation. This is a numerical technique, such as Method 1 in Section 20.8.2.

9 Calculate the duration needed for the simulator (perhaps number of customers or length of simulated time) under the nominal conditions to have a desired certainty in the result. What is a reasonable value for the desired certainty?

10 Consider that N lanes are optimal for the high customer load, but there is a 10% chance that a checkout lane will not be functional in any month and out of service for 2 days for repairs. How many spare lanes should be installed?

11 Optimize within uncertainty using either Method 3, 4, or 5 from Section 20.8.2.

38

Case Study 3

Retirement Study

38.1   The Process and Analysis

This is an optimization application, which has had strong practical relevance to me. I think it will have strong relevance to any professional. The question is simple: "While working, how much of my income should I invest in retirement plans; and at what age should I retire?"

This case study reveals several aspects of the process of a human defining the optimization and also several problems that the surface presents to optimizers. Hopefully, this case study is also of personal interest to you as well as it brings a practical reveal of some issues related to optimization.

Whether the models I use match those you would use, or not, the procedure represents a classic optimization application. Don't read this as my teaching the "way" to live a life. Look at this as a guide to creating an optimization application. Feel free to change the basis and equations to better suit you.

This analysis is for a person who works as an employee (such as a scientist, engineer, manager, or professor) for a career, invests in retirement accounts (such as stocks, mutual funds, or IRAs), then retires, and does not continue to earn money after retirement. Not everyone follows that path. Some start their own business, sell it for millions, and then "retire" to pursue the joys of starting another business. Some inherit a fortune or play the lottery so that they can have that fortune (but more probably lose a fortune). Some invest in rental properties, then "retire" with the income of landlords. Some decide that government support in retirement (Social Security and Medicare) or having your children take care of you is the "way." But those were not my choices. I substantially followed the standard path (invest while being an employee to fund a work-free retirement) and faced two key questions. (i) When should I retire? (ii) While earning income, what portion should I invest for retirement? I look at these two variables as the DVs.

You might have an alternate plan for how to "make it." But whatever the plan there are choices (DVs), and you'd like to choose the DV values that maximize the desired outcome.

Back to my analysis, the DVs are when to retire and how much to save each year. Let's look at the range of DV values to help understand the objective.

When? If I retire early, I have more time to pursue personal interests in retirement but less accumulated funding to support a healthy and ample post-retirement lifestyle. If I retire too late, I die or have diminished functionality and don't have time or ability to enjoy the remaining years in retirement.

What portion? If I invest a large portion of my career salary, then I'll have a lot of funding to support retirement joys. But this means a meager life while young and with family. Alternately, investing a

Engineering Optimization: Applications, Methods, and Analysis, First Edition. R. Russell Rhinehart.
© 2018 R. Russell Rhinehart. Published 2018 by John Wiley & Sons Ltd.
Companion website: www.wiley.com/go/rhinehart/engineeringoptimization

small portion means more funding to enjoy life while working but little investment to support retirement.

The next question is "What do I want to maximize?" What is the OF? I decided it was lifetime joy. You could decide to maximize the inheritance you leave for your children. But that would mean focused pursuit of higher income jobs and scrimping in life to invest all earnings in the savings for the inheritors to enjoy. Although I love my children and grandchildren, that choice did not appeal to me. Another person could choose to maximize the number of children he leaves. It could be to maximize life span. It could be to become at one with the spirit of the universe and live in natural harmony with nature. There are many options that a person could use to decide the objective of living. I decided it was lifetime joy.

One must define a method to quantify the abstract concept. Joy, today, is the personal pleasure in what I do, the relations I have, the experiences I have, personal health and wellness, the worry-free security about the future, the guilt-free worry about the past, and the pride in how the children and grandchildren are developing. Joy from yesterday is still felt today and is related to what I have preserved in pictures (parties, vacations, cars, houses, family, etc.), in my files (inventions, publications, methods to calculate the value of Pi, etc.), and in my house (paintings, mementos, things I've made, hand-me-downs, etc.). Joy, tomorrow, will be related to the time I have to pursue personal interests, the funding I have to support it, and the physical/mental health I have to be able to enjoy. It will also be related to the legacy I leave for my family, which will be either some inheritance for them or debts related to my poor end-game financial planning.

Figure 38.1 shows the OF response to the DVs. This is the result of my model for lifetime joy as a function of year to retire (the left-front axis) and portion of salary to invest while working (right-front axis). Both DVs are scaled on a 0–10 basis. It is the nominal response, the response to a particular set of given values.

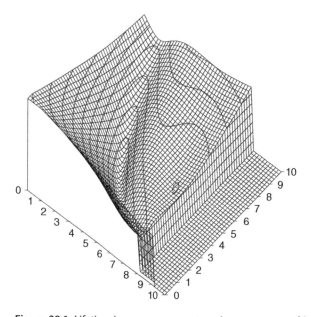

Figure 38.1 Lifetime joy as a response to retirement year and investment portion.

I choose to maximize accumulated joy. Accumulated joy is a path function—the sum of joys from age 25 (starting a career) to death (about age 85 is the forecast 50-percentile for my situation). During any particular time interval, accumulated joy is dependent on my physical/mental ability to participate in life and my income that enables me to have and pursue joy. My OF is accumulated joy from starting a career to death. My DVs are when (R = retirement age) and what portion (p) of salary to invest while employed.

$$\max_{\{R,p\}} J = \sum_{i=\text{age } 25}^{\text{death}} (\text{joy}_i) \tag{38.1}$$

In any ith year, joy will be related to the time available to pursue things that bring joy and income that facilitates pursuit and physical/mental functionality. I think this model makes sense: joy in any particular year is a product of *Disposable* income, *Hours* to pursue personal joys, and personal *functionality*.

$$j_i = D_i H_i f_i \tag{38.2}$$

Disposable income is the after-tax salary, also after the portion set aside for retirement is removed. However, because of inflation, salary needs to be discounted by the inflation rate, r.

$$D_i = s_i[1 - T(S_i)](1 - p_i) \tag{38.3}$$

$$s_i = \frac{S_i}{(1-r)^i} \tag{38.4}$$

where S is salary and s is discounted salary. In my lifetime inflation has averaged about 3.75%/yr, so I've used $r = 0.0375$.

The portion set aside is only operable while working; it is zero in retirement.

$$p = \begin{cases} p, & i \leq R \\ 0, & i > R \end{cases} \tag{38.5}$$

You should recognize that Equation (38.5) is a source of discontinuity.

The tax rate depends on after-savings salary. Equation (38.6) seems to be a reasonable (but unvalidated) model of the US personal tax structure from my data in 2015.

$$T = \frac{0.85}{1 + e^{0.1(35 - s(1-p))}} \tag{38.6}$$

Finally, salary rises with work experience and promotions, but it does not appear to me to be unlimited. Personal choices related to how high in the corporate world you wish to rise make salary asymptotically approach a limit. My initial salary (as a process engineer with an MS in 1969), S_b, was \$12.5k/yr. In 2016 it was about \$150k/yr, S_e. Because things are different today, your initial salary will be substantially different. My final salary reflects my career choice to become a professor. I am pretty sure that my classmates who stayed in industry are making a higher salary. When I model my annual salary progression, this seems a reasonable model for $S(i)$ while working.

$$S_i = S_{i-1}\left[1 + \frac{0.15(S_e - S_i)}{(S_e - S_b)}\right], \quad i \leq R \tag{38.7}$$

But what about income after retirement? Investment Capital compounds in time by the investment growth rate, q, and also the yearly accumulation of the contributions while working.

$$C_i = C_{i-1}(1+q) + pS_i, \quad i \le R \tag{38.8}$$

This simple model (and my VBA function code) considers that the investment is a one-time annual investment. Actually, monthly accumulation might make a more realistic model. In my lifetime, the growth rate of investments has averaged about 7%/yr, $q = 0.07$.

Upon retirement, I start taking funds out of the capital for my post-retirement salary. Then the capital will decrease due to each withdrawal, but annually it will continue to increase due to interest.

$$C_i = C_{i-1}(1+q) - S_i, \quad i > R \tag{38.9}$$

How much to withdraw annually? It might seem that I should divide capital by the years left to live and take out that portion each year. But I might live longer than the 50-percentile age of 85 to the 90-percentile age of 90. So, my plan is to take out a portion that would make the investment last until 90.

$$S_i = \frac{C_{i=R}}{90-R}, \quad i > R \tag{38.10}$$

Hours to pursue personal joy while working include vacation and weekends, but not all of that off-time is personal enjoyment time, because one has to do cooking, taxes, laundry, repairs, and such chores. Hours to pursue joy while working also include the joys of the job that are related to self-actualization, pride in accomplishment, etc. Hours to pursue joy in retirement might be 6 days/week, leaving 1 day/week for chores. Here is a model for the fraction of annual time to pursue joy.

$$H = \begin{cases} \dfrac{1*52 + 3*5 + 0.5(5*45)}{364}, & R \le i \\ \dfrac{6*52}{364}, & R > i \end{cases} \tag{38.11}$$

Again, the several conditions on the prior set of equations are a source of discontinuity in the OF response to DVs.

The functionality factor starts with a value of 1, meaning that I am fully functional both physically and mentally, but it drops toward 0 as age leads to disability. I think it can be modeled as a modified logistic function (a sigmoidal curve) and think this is reasonable.

$$f = 0.1 + \frac{0.9}{1 + e^{0.2(i-80)}} \tag{38.12}$$

The value of 80 in Equation (38.12) is the age at which overall functionality is cut to 50%, the 0.2 is a scale factor that determines when the drop-off begins and the 0.1 is an asymptotic value of 10%. This seems reasonable to me on average, based on people I observe.

I don't think that "they" will let me work if my function factor is below 50%. So, there is a constraint on the R value. One must retire if the age factor for functionality is less than 0.5.

$$0.5 \le 0.1 + \frac{0.9}{1 + e^{0.2(R-80)}} \tag{38.13}$$

Continuing, I don't think joy is directly related to disposable income. Below the poverty level, one might have income, implying positive joy; but the daily stress of not being able to support an enjoyable, healthy, plentiful, and beneficent life might actually create negative joy. Further, I don't think that a person who is 100 times as rich as me has 100 times the joy. There are only so many hours in the day to have joy and only so much impact in things we do. So, I model a joy factor, g, as a function of discounted salary. This model seems to make sense to me.

$$g = 1 - 2.5e^{-0.1s(1-p)(1-T)} \tag{38.14}$$

Finally, if I die destitute, with debts leftover from my end-life expenses and a burden to my children, my joy will be greatly diminished. If I die and leave $ to my children, I'll be happy. But I think in neither case is the joy penalty or enhancement related to the quantity of debt or inheritance. So, I made a modest addition to joy if leaving inheritance and larger penalty if not.

$$P = \begin{Bmatrix} +g, & C_{i=\text{death}} > 0 \\ -3g, & C_{i=\text{death}} \leq 0 \end{Bmatrix} \tag{38.15}$$

Putting it all together,

$$\min_{\{R,p\}} J = -\left[P + \sum_{i=\text{age }25}^{\text{death}} (g_i D_i H_i f_i) \right] \tag{38.16}$$

$$\text{S.T.}: \quad 0.5 \leq 0.1 + \frac{0.9}{1 + e^{0.2(R-80)}}$$

Note:

1) This OF is a sum of a product of factors.
2) This is about my fourth revision to this analysis. The first set of statements always seem a reasonable sketch, but they are usually incomplete. As I've explored this over several years, I've ended with this model. However, at this point I realize that the overall investment model did not include Social Security or Medicare benefits in retirement or employer contributions to health/medical benefits and to retirement plans. It also modeled the investment with annual contributions and withdrawals, not monthly ones. Finally, personal life joy in the early years leaves mementos such as photos, trip souvenirs, purchases, and baby shoes; and career joys from skill development have recurring benefit. So, I think the value of joy in the early years should be compounded, and I still have a bit of fine-tuning to consider for this analysis.
3) This is a deterministic model.
4) It is an interesting OF (R,p) surface. The lower left axis in Figure 38.1 is the scaled retirement age, and the right-side axis is the scaled investment proportion. The cliff to the left represents the infeasible R. There are two minima and one global. The global minimum is in a steep valley, with a nearly level floor.
5) Although it seems to be a continuously differentiable function that a gradient, or Hessian based, or surrogate quadratic model could handle, it is actually a discontinuous surface. The 100 solutions and final trace from start to end in Figure 38.2 reveal the LM optimizer results. Note that LM did not follow the apparent steepest descent path, but stayed next to the discontinuity ridge (striation) as it went downhill. Figure 38.3 is a close up 3-D view, which reveals the discontinuity due to the year-to-year discretization of retirement age. If I did it on a month-to-month basis, the surface would have the same problem but on a smaller scale, as it would with week-to-week or day-to-day time discretization.

This application naturally produces four surface aberrations (deviations from ideal)—multiple optimum, discontinuities due to the constraint, striations due to the time discretization, and steep valley with a shallow bottom.

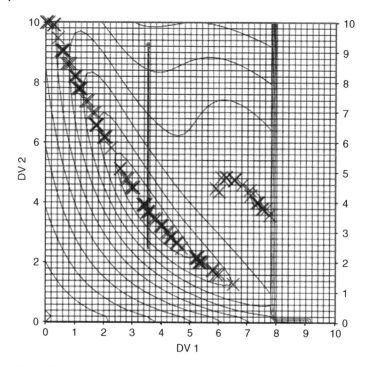

Figure 38.2 Optimizer results using LM.

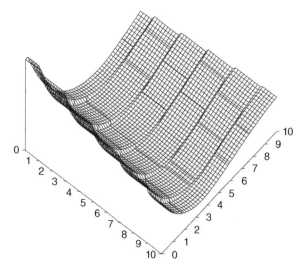

Figure 38.3 Close-up view of the surface revealing striations.

The optimum for this model of my life indicates that I should have been investing about 35% of my salary so I could retire at age 60. I did not. I retired at age 70, which is what the minimum of the surface indicates is the best age, if only investing about 20%/yr (which is what I did).

However, I'm going to accept this model as fairly reasonable but take a new direction in the analysis.

What was described previously is a deterministic optimization application. The OF value is a deterministic response to the DV values. Any time you replicate DV values, the OF response will be exactly the same. My modeled age of death, 90, is unchanged. The interest rate and inflation rates are unchanged. My forecast age at half function, 80, is unchanged. These are all projections, not certainties. And the actual realization of my life will likely be different from the nominal.

If I model the actual realization of my life coefficients as a Gaussian probability perturbation about the nominal value using a Box–Muller relation, then

$$x_{\text{realization}} = x_{\text{nominal}} + \sigma_x \sqrt{-2\ln(r_1)} \sin(2\pi r_2) \tag{38.17}$$

where x represents a "given" that might vary. Here, those are death age, half age, interest rate, and inflation rate. With this, I get a stochastic surface, in which each DV trial uses an independent realization for those four "givens." Figure 38.4 presents one realization of the surface. Now, the optimization problem is confounded in identifying the downhill direction. Each re-sampling gives a new point elevation and new elevation of the points in close proximity.

Let's reconsider the OF. I don't want the best choice of retirement age and savings portion for some single nominal case, but I'd like the best overall for what might be my life outcome. My life outcome could have any of a range of factors. I'd like the R and p values to consider the worst likely outcomes and find values that make the best of a worst situation. I'd like to maximize the minimum joy that might happen. Since I cannot run every possible realization, I'll desire R and p values to maximize the minimum joy that the realizations will find in 100 realizations.

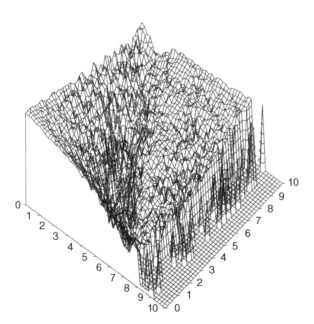

Figure 38.4 One realization of the surface when considering uncertainty in the givens.

Now the objective function is

$$\min_{\{R,p\}} J = -\min_{k=1,\,100} \left\{ \left[P + \sum_{i=\text{age }25}^{\text{death}} (g_i D_i H_i f_i) \right]_k \right\}$$

(38.18)

$$\text{S.T.}: \quad 0.5 \leq 0.1 + \frac{0.9}{1 + e^{0.2(R-80)}}$$

Function 15 in the 2-D optimization example program executes these OFs (See companion site.).

Feel free to adapt the function to better represent what you anticipate in life. Don't accept my model and method for assessing the OF. Part of optimization is defining the OF. Don't let this be an abstract intellectual exercise. Relate this exercise to your personal situation. If some entity is going to pay you to do optimization, they will want the models and assessments to be relevant to their situation, not a generic off-context result.

38.2 Exercises

Use Function 15 in the 2-D optimization example program to implement these exercises.

1 With the stochastic feature commented out as a deterministic function (any repeat of the same DV values will return exactly the same OF value), it appears to be a continuous function but actually has surface discontinuities based on the year-to-year age discretization. Reveal these.

2 A single TS (trial solution) optimizer will get stuck on the local valleys. A multiplayer optimizer will more likely find the global optimum. Compare the results of a single TS optimizer to a multiplayer optimizer. Define what criteria you will use for a comparison. Define how to structure experiments to provide legitimate results.

3 Convert the function to a stochastic OF, so that it accounts for possible uncertainty in key variables such as interest and discount rates and half age and death age and reports the worst of multiple possible outcomes with the same DV set. Contrast the deterministic and stochastic function attributes. Show that a different realization of the stochastic function leads to different results.

4 Compare any optimizer and convergence criterion to LF with replicates and the steady state of the worst OF convergence criterion.

5 Determine the number of trials from random initializations that are appropriate to adequately define the application and to define actionable DV* values.

6 Compare the DV* values from the deterministic and the stochastic models.

7 Apply propagation of uncertainty on the deterministic model to estimate the impact that uncertainty has on the OF* and DV* values. How does this compare to the range of DV* values from multiple optimizer trials or the range of LF trial solutions at convergence?

8 Change the models (and equations) to better suit your perspectives and your situation on what should be optimized.

39

Case Study 4

A Goddard Rocket Study

39.1   The Process and Analysis

Yes, it is rocket science; although simplified here, it is adequately rich. The objective is to schedule thrust (e.g., change thrust with time) to maximize the altitude a rocket can achieve. You could have full thrust until the fuel burns out, and then let the rocket coast up until gravity pulls it back. But this plan makes the rocket go fast in the lower altitude where the air resistance is high and wastes energy. A better plan to minimize the impact of air resistance is to move slowly in the low altitudes and then full throttle to fast speed in the high altitudes. But moving slow in the low altitudes, in the high gravitational field, means wasting fuel to fight gravity. In a limit of not enough thrust to overcome gravity, the rocket never rises; it burns all its fuel on the launch pad. This optimization is known as the "Goddard problem" in honor of rocket scientist Robert Goddard.

This will consider a simple situation of a single-stage vertical rocket (no changes in drag coefficient), moving vertically from a stationary flat earth (no Coriolis effects) through an atmosphere with nominal density dependence on height. Even with such idealizations, it will be sufficiently complex!

One generally accepted solution is to accelerate the rocket at full thrust to terminal velocity (the speed at which it would fall if pulled down by gravity when air drag resistance exactly balanced gravity) and then reduce thrust to maintain $v_{terminal}$. As the rocket rises both air density and gravitational pull reduce, so thrust would be scheduled with the local $v_{terminal}$. Then coast after burnout until gravity and air drag stop upward motion. Although the optimum velocity trajectory seems to be a bit lower than terminal velocity, this rule reveals the diversity of variables that could be used to schedule thrust.

There are many versions for this exercise. Considering options on the DV, thrust (rate of fuel consumption) could be scheduled w.r.t. any number of parameters that change in time: time, elevation, terminal velocity, speed, remaining fuel, or air density. The scheduling could be a continuum relation (e.g., $T = f(h) = a + bh + ch^2$) or a rule-based type of model IF $t < a$ THEN $T = 100\%$, ELSE $T = b\%$ or IF $v < v_{terminal}$ THEN $T = 100\%$, ELSE $T = a + bh$.

Further, the OF could be any of several: Perhaps maximize elevation at the top of its fuel-depleted upward drift or perhaps reach a given elevation with minimum fuel consumption.

Do you schedule the throttle with height, or time, or remaining fuel, or terminal velocity? This will be your choice. Let's consider height, and several options. (i) Should you section height into 10,000 ft intervals and find an optimal throttle position for each interval? This would step the throttle to the new position at each 10k ft interval. (ii) Why should the transition heights be in 10k ft intervals, and why should the intervals be uniform? Could the optimizer find both the transition heights and the thrust

Engineering Optimization: Applications, Methods, and Analysis, First Edition. R. Russell Rhinehart.
© 2018 R. Russell Rhinehart. Published 2018 by John Wiley & Sons Ltd.
Companion website: www.wiley.com/go/rhinehart/engineeringoptimization

values? (iii) Or should you have an equation that defines throttle position with elevation and use the optimizer to determine the coefficients in the equation? This would provide for smooth throttle transitions as elevation increases. What equation complexity would you use? What would you use for an equation form to ensure that throttle is bounded between 0 and 100%? (iv) Should you choose a combination of scheduling techniques, for instance, set the initial thrust as 100%, and then trigger an equation after some height? Let the optimizer determine the transition height and the post-transition equation coefficients.

This set of questions about the DV schedule will be applicable to any schedule parameter, of either elevation, or time, or fuel content. In any case, what would you do when fuel runs out, but the equation or schedule says go to 83% thrust? And how can you know how to define stages with elevation until you know how high the rocket will rise?

The model of the rocket can be found at http://bocop.saclay.inria.fr/?page_id=465. Thanks to Thomas Hays for pointing me to it while he was a PhD candidate in aerospace engineering. The situation can be illustrated as follows: Figure 39.1 shows the Earth with radius r_0 and the rocket rising vertically from the surface with an elevation (from the Earth center) as r. The rocket path is solid during the time when it had fuel and was thrusting and dashed when the fuel is spent and it is coasting away due to momentum. The elevation hits a peak when the combined gravitational pull and atmosphere drag reduces the upward v to zero and it begins to free fall back to Earth. The figure is not to scale, represents a concept, and illustrates a rocket that will reach more than an Earth diameter from the surface. However, in this case study, the rocket only reaches about 50 miles up, which is only about six-thousandths of a diameter.

Figure 39.2 illustrates how thrust might be scheduled with elevation. There are many ways to do this; for example, thrust could be scheduled with time, velocity, fuel content, etc. This schedule starts the thrust at 100% and then steps the thrust to certain values at certain heights (waypoints) and holds that thrust value until the next waypoint is reached. After the third waypoint, the u_4 thrust is held until fuel is depleted. When the fuel is spent, the thrust is zero. Here, the optimizer could choose the three waypoint elevations and the four thrust values, a total of 7 DVs. But it could be divided into 10 sections, giving the optimizer 19 DVs. Or the parameter for the schedule could be any other variable. Here the schedule is discontinuous. Alternately, the schedule could be defined by an equation and based on rocket fuel inventory, such as $T = a + bm + cm^2 + dm^3$, or a ratio of polynomials such as $T = a(1 + bm + cm^2)/(1 + dm^3)$. In either polynomial case, the optimizer would have four DVs, the schedule coefficients a, b, c, d.

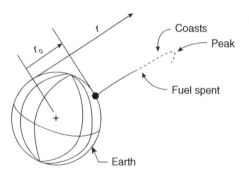

Figure 39.1 The concept.

Figure 39.2 An example of scheduling thrust with elevation way points.

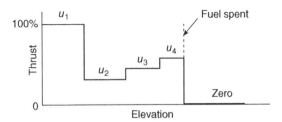

The model can be developed as follows: Start with Newton's law of motion for the rocket $F = ma/g_c$. Note that both F and m will change in time.

The three forces (thrust, gravity, and drag) will change in time. Thrust, $T = T_{max}u$, will be a fraction of maximum, full thrust, defined by the control signal u, which changes in time. Gravitational pull will be dependent on the rocket mass and distance from the Earth center, $F_{gravity} = GMm/r^2 = g_0mr_0^2/r^2$. Air drag will use the conventional (and elementary) turbulent drag on a body immersed in a large fluid field, $F_{drag} = AC_D(1/2)\rho v^2$, using a conventional (elementary) model for how air density depends on elevation $\rho(h) = \rho_0 e^{-kh}$, where elevation $h = r - r_0$ and ρ_0 is the air density at r_0 sea level. Note that h and v both change in time.

The rocket mass changes with fuel consumption, ideally modeled as a linear response to the control signal $dm/dt = \alpha u$, $m(t = 0) = m_0$.

The velocity and acceleration are related, and the initial velocity is zero. $dv/dt = a$, $v(t = 0) = 0$. Similarly, position and velocity are related. $dh/dt = dr/dt = v$, $h(t = 0) = 0$, or $r(t = 0) = r_0$.

Combining relations in Newton's law of motion and restating the position and mass models,

$$\frac{dv}{dt} = -g_0\left(\frac{r_0}{r}\right)^2 + T_{max}g_c\frac{u}{m} - \frac{AC_Dg_c}{2}\frac{v|v|}{m}\rho_0 e^{-k(r-r_0)}, \quad v(t = 0) = 0 \tag{39.1}$$

$$\frac{dr}{dt} = v, \quad r(t = 0) = r_0 \tag{39.2}$$

$$\frac{dm}{dt} = \alpha u, \quad m(t = 0) = m_0 \tag{39.3}$$

The $v|v|$ term could be expressed as v^2. As $v|v|$ the sign of the drag force works whether the velocity is upward or downward. Since this study only models the rocket on its upward journey, I will use v^2.

For convenience and generality, convert to scaled variables. Some are logically scaled by their initial values $r' = r/r_0$ and $m' = m/m_0$. One is logically scaled by the maximum value $u' = u/100\%$. Velocity and time can be similarly scaled, $v' = v/v_0$ and $t' = t/t_0$, but here the scaling values are not the initial values of zero. Convenient scaling values for v_0 and t_0 can be revealed when the three ODEs are converted to scaled variables. Choosing $v_0 = \sqrt{r_0g_0}$ and $t_0 = \sqrt{r_0/g_0}$, the model becomes

$$\frac{dm'}{dt'} = -\alpha u', \quad m'(t' = 0) = 1 \tag{39.4}$$

$$\frac{dv'}{dt'} = -\left(\frac{1}{r'}\right)^2 + b\frac{u'}{m'} - c\frac{v'^2}{m'}e^{-d(r'-1)}, \quad v'(t'=0) = 0 \tag{39.5}$$

$$\frac{dr'}{dt'} = v', \quad r'(t'=0) = 1 \tag{39.6}$$

According to the Bocop site, appropriate coefficient values for a rocket on Earth are $a = 24.5$, $b = 3.5$, $c = 310$, $d = 500$. Here are some un-scaling values: $r_0 = 3957$ miles, $g_0 = 32.174$ ft/s^2, $t_0 = 806$ s, and $v_0 = 25{,}950$ ft/s. Use the initial fuel mass as 90% of the full rocket. This means that when $m' = 0.1$, the fuel is spent.

There are many ways to solve the three ODEs. In my solution, I use the Euler method, a forward finite difference explicit approximation to the derivative with $\Delta t' = 0.00001$. A more proper value is $\Delta t' = 0.0000001 = 10^{-7}$. The small integration time step is needed to make the solution independent of it; however, it takes a long time to run, and given the idealizations in the model and precision on the model coefficients, the coarse consequence of the numerical method is probably inconsequential. I have also investigated primitive predictor–corrector and Runge–Kutta techniques to solve the ODEs, but again, I don't think that the model and coefficient values justify numerical methods that seek analytically exact solutions. So, I defaulted to the simple Euler method. The numerical solution is

$$m'_{i+1} = m'_i - dt'\,au'_i \tag{39.7}$$

$$v'_{i+1} = v'_i + dt'\left[-\left(\frac{1}{r'_i}\right)^2 + b\frac{u'_i}{m'_i} - c\frac{v'^2_i}{m'_i}e^{-d(r'_i-1)}\right] \tag{39.8}$$

$$r'_{i+1} = r'_i + dt'\,v'_i \tag{39.9}$$

where i is the time step counter.

Start with the initial conditions, and do until (solve the ODE set until) $v' < 0$, which is the indication that the rocket has reached its peak and started to fall back. At that point use $r = r'r_0$ as the OF value.

Function 79 (in Appendix E) and Functions 77 and 78 (in the 2-D Optimization Examples file on the companion site) explore three variations on the problem. They also show the finite difference-approach to modeling the solution to the three time-dependent differential equations (which are coupled and nonlinear). These functions have only 2 DVs. In Function 77 the DVs are initial and second-stage thrust. The second-stage transitions at 40k ft and lasts until fuel runs out. In Function 78 the initial thrust is 100% and the decision variables are the elevation to switch to the second thrust level and that level. Function 79 has an alternate objective—reach a desired elevation (about 100k ft), minimizing fuel content. The thrusts are 100% and then 20%, and the DVs are the elevations to transition from first stage to second and then second to thrust off (coast).

The OF has flat spots, cliff discontinuities, ridge discontinuities due to numerical discretization, and hard constraints.

39.2 Pre-Assignment Note

I explored several approaches to the DVs and would offer this guide, which minimizes the number of variables and finds nearly the max height with the minimum number of function evaluations:

DV1—the initial thrust, which should be 1.

DV2—the rocket mass when switching to a new thrust, the trigger to change from the initial thrust to the schedule. My optimizations place this in the $m' = 0.6$ region.

DV3 and DV4—the coefficients in a linear relation for thrust as a function of rocket mass after the initial thrust: $u' = DV3 + DV4 * (DV2 - m')$. Here DV3 would be the jump-to thrust at the point of change.

I chose to schedule the thrust with respect to rocket mass, because I know that it starts at $m' = 1$ and ends at $m' = 0.1$. Although the thrust could be scheduled w.r.t. either height or time, you don't know what the range of r' or t' is. Perhaps the thrust could be scheduled with terminal velocity. As I have explored higher-order models for thrust as a function of mass, I find that they describe nearly a linear relation.

The choice of DVs progressed with my investigation. My first attempts were to have 3 DVs representing thrust step and holds with height. When it was working, I extended it to 10 and then to 20 thrusts. This made me realize that I could not know the height range for the thrusts when the fuel runs out. So, I changed to schedule thrust with fuel content. The thrust patterns showed equivalently high thrust for the first several stages and then a sharp drop to lower thrusts that seemed all about the same. Looking at this pattern, I decided to use the DVs as aforementioned. I then decided to increase the thrust equation to a cubic, which resulted in 6 DVs. Then it seemed that the optimizer indicated that the initial thrust should be 100%. So, I set it to 100% and eliminated a DV. Very pleased with progress, I explored various models and began to realize that a linear schedule seems about as effective as the cubic, which reduces the number of DVs to 3. Sketch, evaluate, erase, sketch, evaluate, erase, …, ink.

The best strategy is to use full throttle to accelerate to a "high" velocity and then back off power; and as elevation increases, then increase throttle as air resistance diminishes.

My optimum schedule gets the rocket to an altitude of about 50 miles, which is very small relative to the 3960-miles radius of the Earth, making the Cartesian viewpoint seems reasonable. It peaks at 3.5 min. A characteristic velocity is about 1500 ft/s, which means that the rocket moves about 12 ft in one simulation discretized time step.

Doing this in a simulation, I have not had to make plans for where the rocket will land when it freefalls back.

39.3 Exercises

Use at least 4 DVs. Perhaps start with just 2 DVs and use the 2-D program to practice on. Then use the Generic LF optimization software to determine a solution for 4 or more DVs.

1 Explain the situation and basic equations in your own words.

2 Explore the surface in a 2-D simulation and comment on the several aberrations.

3 Explain your choice for the scheduling of the DVs: Discrete values staged by regular intervals of height, fuel, velocity, mass, or time; or variable stage interval defined by the optimizer; or an equation to schedule the DV continuously w.r.t. some parameter; or some combination.
   A Choose a 2-DV simple situation.
   B Choose an N-DV situation.

4 Explain your finite difference model and the triggers that you will use and actions that the code will take to recognize when the rocket hits the peak altitude and when the fuel is spent.

5   Implement your *N*-DV optimization in my base case version "Generic LF Optimization Goddard Problem." Clearly reveal your unique code and your unique DV schedule. Discuss the results.

6   Make the simulation go to the next step of rigor by using a better model for air density and/or air drag.

7   Explore the impact of uncertainty on model coefficient values and air density (which changes day to day with barometric pressure, humidity, and temperature) on the optimum height.

8   Investigate the OF response to DV values for any pair of DVs in the vicinity of the optimum. Explore a DV range of ± 0.000001 of the DV* value. You will see that the OF surface has flat spots. Explain why.

40

Case Study 5

Reservoir

40.1 The Process and Analysis

There is a Monte Carlo function in the 2-D Optimization Examples VBA program that simulates the stochastic outcome of a reservoir—Function 18. Reservoir capacity and nominal level are the decision variables. The models are described in Appendix E.

The larger the reservoir and associated dam, the greater the initial cost. Cost is the objective function. So, superficially, the solution is to build a small dam to reduce cost.

However, if the reservoir is too small, and/or it is maintained nearly full, it does not have enough capacity to absorb a flood due to exceptionally heavy upstream rainfall and it will transmit the flood downstream. Downstream flooding incurs a cost of damaged property. But the chance of a flood and the magnitude of the flood depend on the upstream rainfall. So, the simulator models a day-to-day status with a lognormal rainfall distribution for a time period. I set it to 20 years, but you can change it.

Contrasting flooding conditions are drought conditions. If the reservoir is too small, and/or maintained with little reserve of water, an upstream drought will require stopping the water release, which stops downstream river flow. Downstream dwellers, recreationists, or water users will not like this. There is also a cost related to zero downstream flow.

There is a fixed cost of the structure and a probabilistic or stochastic penalty cost of extreme events.

Since one 20-year simulation will not reveal the confluence of "100-year events," the simulator runs 50 reservoirs for 20 years each—50 realizations of the 20-year period. You can change the number of realizations.

The function is set up to return either the maximum cost for the 50 realizations or the estimated 99% probable upper limit on the cost. You could choose another performance indicator.

An excessively large reservoir kept half full will have ample reserve (to keep water flowing in a drought) and open capacity (to absorb excess rainfall and prevent downstream flooding), but it will cost a lot. A smaller reservoir will have less cost. But too small a reservoir will not prevent problems with drought or flood. So, there is an in-between optimum size.

If the nominal reservoir volume is near the full mark, then the reservoir will not be able to absorb floods, but it will have plenty of capacity for a drought. If the nominal level is too low, it will be able to absorb water and prevent a flood, but not keep water flowing for a drought event. So there is an in-between set point capacity that is best. The optimum set point for the level might not be at 50%. It depends on whether the vagaries of rainfall make floods a bigger event than droughts.

Engineering Optimization: Applications, Methods, and Analysis, First Edition. R. Russell Rhinehart.
© 2018 R. Russell Rhinehart. Published 2018 by John Wiley & Sons Ltd.
Companion website: www.wiley.com/go/rhinehart/engineeringoptimization

For any given sized reservoir, the fuller it is kept, the greater is the fresh water reserve (for supply to cities) and recreational area. So, other benefits are added as a negative penalty to the cost.

There is a fixed cost of the structure, a probabilistic or stochastic cost of extreme events, and a negative penalty for reserve and recreation benefits.

Figure 40.1 is a plot of the OF versus DV1 and DV2. The left-front axis represents reservoir size. The right-front axis represents level set point. Both are scaled 0–10 and the cost (vertical axis) is also scaled 0–10.

Figure 40.2 reveals a contour plot with size as the horizontal and level set point as the vertical axis.

Notice that for very large reservoir sizes, the set point level is irrelevant to upstream floods or droughts. The diagonal contours in the central region reflect a preference for lower cost tempered by the benefits of larger reservoir capacity and water levels. The optimum in this simulation is around $x1 = 2$ and $x2 = 3$.

Notice that the large central region of the function is planar. Cost is not affected by flood or drought events and scaled linearly with size and set point. Second-order optimizers will encounter a divide by zero EXE error when a trial solution is on the central plane.

Notice that the optimum and disaster-free region of diagonal contours is surrounded by a noisy boundary, and every realization gives different values, depending on the outcome of the realizations. How will you define stopping criteria? What optimizer should you choose?

Because this "realistic" simulation takes computational time, I added Function 20 (Stochastic Boot Print) to provide a computationally rapid surrogate with the same linear and stochastic features.

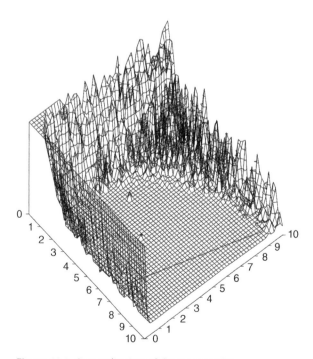

Figure 40.1 One realization of the reservoir OF.

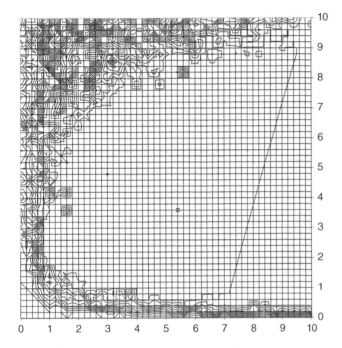

Figure 40.2 Contour plot of one realization of the reservoir.

40.2 Exercises

1 Explore each optimizer and appropriate convergence criteria on either Function 18 or its surrogate 20. You should see that most optimizers end with a trial solution deep into the high risk area. My choice would be LF with replicates and the SS convergence based on the worst player.

2 Adjust the probability models to your liking.

3 Change the convergence criterion from SSID on the worst player OF to something else that makes sense to you.

4 Convert the OF to any number of responses, such as the total cost over a 100-year simulation, or the worst over five 20-year simulations, or the 95% worst case based on the average and standard deviation of ten 1-year simulations.

5 Enhance the exercise by including equations that represent using the dam to generate hydroelectric power.

6 Explore the impact of uncertainty of the givens.

41

Case Study 6

Area Coverage

41.1 Description and Analysis

Consider a square parking lot. Where should one place overhead street lamps to provide light to best cover the area? Or overhead cameras to best monitor the area. If there are five devices, the location pattern could be a pentagon, or that of the five patterns on a die.

The objective could be to cover the area for customer visibility. But it could alternately be to observe the perimeter for security.

The intensity of the lamplight to the surface drops as the square of distance to the surface. From a pole h distance high, with a light directed vertically down in a right circular cone, the distance from the lamp to any point on the ground at radius r from the pole is $d = \sqrt{h^2 + r^2}$, and a simple model of light intensity at any point on the ground from the ith pole is $I(r_i) = I_0/(h^2 + r_i^2)$. Here, the radial distance from the ith pole to any position (x, y) is $r_i = (x_i - x)^2 + (y_i - y)^2$, making $I_i(x, y) = I_0/[h^2 + (x_i - x)^2 + (y_i - y)^2]$. This supposes that all poles are of equal height and all lamps of equal intensity, shining the same amount of light in downward directions. To calculate coverage at any particular location on the ground, one could sum the light intensity from each pole: $I(x, y) = \sum_{i=1}^{n} I_0/[h^2 + (x_i - x)^2 + (y_i - y)^2]$. Then to calculate the entire coverage of the area, one could integrate intensities over all (x, y) points on the ground. $I_{\text{total}} = \sum_{i=1}^{n} I_0 \int_{y=0}^{y=s} \int_{x=0}^{x=s} 1/[h^2 + (x_i - x)^2 + (y_i - y)^2] dxdy$

It would seem that the optimization about where to locate five poles to maximize the total light on the area would be

$$\max_{\{x_1, y_1, x_2, y_2, x_3, y_3, x_4, y_4, x_5, y_5\}} J = I_0 \sum_{i=1}^{n} \int_{x=0}^{x=s} \int_{y=0}^{y=s} \frac{1}{h^2 + (x_i - x)^2 + (y_i - y)^2} dxdy \qquad (41.1)$$

where n is the number of lights.

I suspect, however, that the answer would be to place all five poles at the same center of the square. The reason is that light falling outside of the square does not count. For each pole maximizing its light delivery to the parking lot or playground means placing the pole in the center.

So, this OF formulation misses some essential aspects. A better formulation might be to maximize the light to the one spot receiving minimum light.

Engineering Optimization: Applications, Methods, and Analysis, First Edition. R. Russell Rhinehart.
© 2018 R. Russell Rhinehart. Published 2018 by John Wiley & Sons Ltd.
Companion website: www.wiley.com/go/rhinehart/engineeringoptimization

Further, the calculus integration means a continuum representation, and although there may be an analytical solution to the integral, in more realistic formulations, I don't think that analytical solutions exist. Accordingly, to implement a meaningful objective function evaluation, a numerical discretization is needed. I would suggest a simple rectangle rule of integration:

$$I_{total} = I_0 \left(\frac{s}{m}\right)^2 \sum_{i=1}^{n} \sum_{j=1}^{m} \sum_{k=1}^{m} \frac{1}{\left[h^2 + (x_i - x_k)^2 + (y_i - y_j)^2\right]} \qquad (41.2)$$

where s is the length of the side of the square area and m is the number of x and y discretizations.

41.2 Exercises

1   Consider that the objective is to maximize the light delivered to the least lit spot on the area. Develop the OF formulation, choose a number of lights that would lead to nontrivial solutions (n = 5, 6, 7, 8, 10), and use an optimizer to determine the optimum placement of the n lights.

2   Consider that any ground light intensity over a minimum value is superfluous; then effectively the value of light at a particular point is $I_{threshold}$ if $I_{k,j} = I_0 \sum_{i=1}^{n} 1 / \left[h^2 + (x_i - x_k)^2 + (y_i - y_j)^2\right] \geq I_{threshold}$. Develop the OF formulation, choose a number of lights that would lead to nontrivial solutions (n = 5, 6, 7, 8, 10), and use an optimizer to determine the optimum placement of the n lights.

3   Consider a logistic function that describes ground light effectiveness as a function of intensity. $E = 1 / \left(\left[1 + e^{-d(I-c)}\right]\right)$, where c is a center value (the intensity that provides a 50% effectiveness) and d is a scale factor. Then as intensity becomes very large, the effectiveness approaches unity, and as I approaches zero, the effectiveness approaches $1 / \left[1 + e^{dc}\right]$. Develop the OF formulation, choose reasonable values for d and c, choose a number of lights that would lead to nontrivial solutions (n = 5, 6, 7, 8, 10), and use an optimizer to determine the optimum placement of the n lights.

4   Consider a likelihood function that describes ground light effectiveness as a function of intensity, $E = e^{-a\left[h^2 + (x_i - x_k)^2 + (y_i - y_j)^2\right]}$, and an effectiveness threshold. Develop the OF formulation, choose reasonable values for d and c, choose a number of lights that would lead to nontrivial solutions (n = 5, 6, 7, 8, 10), and use an optimizer to determine the optimum placement of the n lights.

5   Human visual effectiveness is not linearly related to light intensity. Find an appropriate model and then use it to redevelop the OF based on the human ability to see.

6   Consider that the objective is to monitor the perimeter. Here the center illumination is irrelevant. Develop the OF formulation, choose a number of lights that would lead to nontrivial solutions (n = 5, 6, 7, 8, 10), and use an optimizer to determine the optimum placement of the n lights.

7   Explore the impact of coarse or fine discretization of the x- and y-dimensions. Coarse will create ridges on the OF response and result in multiple nearly equivalent solutions. Fine will provide better accuracy but will increase the function run time by the square of m.

8 Any of the number of lights (=5, 6, 7, 8, 9, 10) should lead to multiple placement of the n lights—a global best and a local best. Show this.

9 Explore the impact of convergence criterion and threshold on p(global), NOFE, and precision.

10 Explore the impact of number of players on p(global), NOFE, and precision.

11 Explore the impact of leap-to window management on p(global), NOFE, and precision.

12 Explore the impact of number of initializations on p(global), NOFE, and precision.

13 Reshape the light from a right circular cone to an ellipsoid; now each lamp will have 3 DVs, x-position, y-position, and orientation of the ellipsoid axis.

14 Change the shape of the area from square to rectangular, or circular, or triangular, or to an alternate shape that you may have seen.

15 Consider the additional objective, which is also to minimize the loss of coverage if the device with the greatest impact were to fail.

42

Case Study 7

Approximating Series Solution to an ODE

42.1 Concepts and Analysis

A linear second-order, initial value problem (IVP), ordinary differential equation (ODE) is $A(d^2y/dx^2) + B(dy/dx) + Cy = u(x)$. The initial conditions are $y(x = x_0) = y_0$ and $dy/dx|_{x = x_0} = \dot{y}_0$. In an ideal case the forcing function is a constant, held steady, for all x after the beginning at x_0, $u(x \geq x_0) = u_{SS}$. If $A = 0$ it is a first-order ODE, not second-order. For $A \neq 0$, the analytical solution for this ODE is relatively simple (if one is practiced in solving ODEs) and results in a model of the form $y(x \geq x_0) = \alpha + \beta e^{-(x-x_0)/\tau_1} + \gamma e^{-(x-x_0)/\tau_2}$. There are three cases. In the case in which $B^2 > 4AC$, the process is monotonic and asymptotically stable with τ-values $\tau_1 = 2A/\left(B + \sqrt{B^2 - 4AC}\right)$ and $\tau_2 = 2A/\left(B - \sqrt{B^2 - 4AC}\right)$. Then $\alpha = u_{SS}/C$, $\beta = [\tau_1\tau_2\dot{y}_0 + \tau_1(y_0 - \alpha)]/(\tau_1 - \tau_2)$, and $\gamma = -[\tau_1\tau_2\dot{y}_0 + \tau_2(y_0 - \alpha)]/(\tau_1 - \tau_2)$.

If either $B^2 = 4AC$ or $B^2 < 4AC$, the functional form of the solution is not the two exponentials shown previously. But the convenient case of the "overdamped" situation is a commonly occurring dynamic model, which results from two first-order stable processes in series. The output of the first process is the influence to the second, and the output of the second does not feedback to influence the first. The resulting model is that given previously. For testing, you can first specify the time constant values of the individual processes and then calculate the ODE coefficient values from $A = \tau_1\tau_2$, $B = \tau_1 + \tau_2$, and $C = 1$.

However, when related to fluid dynamics, it is not uncommon for the state variable to have a power of 2 (when turbulent drag force is related to fluid velocity) or a power of 0.5 (when flow rate is related to pressure drop in a valve). Then the nonlinear ODE could be represented as $A(d^2y/dx^2) + B(dy/dx) + Cy^p = u(x)$. If the process is asymptotically stable (if $y(x)$ approaches a steady value, with the first and second derivatives approaching zero, with large x), then the final steady-state value of y is $(x = \infty) = \sqrt[p]{u_{SS}/C}$. Even with such a minor nonlinear change to the ODE, the solution becomes an infinite series.

Further, there are infinite other ways to make the ODE nonlinear. Again, for example, consider fluid flow processes. If one of the first-order processes represents a tank that is filling or emptying and the cross-sectional area is not uniform (perhaps it is a cylindrical tank on its side, perhaps a conical or spherical shape, or perhaps it is an elastic shape that stretches with inventory of contents or a visco-elastic shape that also changes in time), then the A, B, and C "constants" in the ODE are not constants.

Engineering Optimization: Applications, Methods, and Analysis, First Edition. R. Russell Rhinehart.
© 2018 R. Russell Rhinehart. Published 2018 by John Wiley & Sons Ltd.
Companion website: www.wiley.com/go/rhinehart/engineeringoptimization

Alternately, if one of the first-order processes is outflow through a valve, but the resistance to flow is not a constant (perhaps it transitions from the laminar to turbulent regime). then, again, the A, B, and C "constants" in the ODE are not constants. In many cases, the A, B, and C coefficients are functions of the state variable and/or the independent variable. A general representation of the nonlinear IVP ODE could be $f(x,y)(d^2y/dx^2) + g(x,y)(dy/dx) + h(x,y) = u(x)$, with initial conditions $y(x = x_0) = y_0$ and $dy/dx|_{x=0} = \dot{y}_0$.

Except for ideal cases, such nonlinear ODEs are mathematically intractable and are usually solved numerically. One method is to approximate the first and second differentials with a finite difference representation and sequentially use prior modeled y-values to predict the next. This widely accepted approach has methodologies termed Euler's, predictor–corrector, Runge–Kutta, Gear, etc. And the implementation choices can be implicit or explicit or a combination or forward or backward or central difference differential approximations. The problem with such methods is that they do not provide a mathematical function for the answer. They provide a data series. Often a mathematical function is desired for interpolation, extrapolation, and optimization. Other numerical methods to solve ODEs include collocation.

Here is an alternate method using an approximating surrogate model of the solution.

Recognize that if the solution $\tilde{y}(x')$ is a continuous function, it can be approximated as a Taylor series. Here $x' = x - x_0$, the deviation from the initial independent variable point. We often hope that the coefficient terms approach zero as the power of the term rises, so that we can use a truncated series of a few convenient number of terms. Alternately, there are many rearrangements of power series terms that lead to alternate embodiments of the truncated series (Newton's interpolating polynomial, Chebyshev functions, wavelets, etc.), which represent the solution as a truncated series of a few terms:

$$\tilde{y}(x') = a\tilde{y}_1(x') + b\tilde{y}_2(x') + c\tilde{y}_3(x') + \cdots = \sum_{i=1}^{n} c_i \tilde{y}_i(x').$$ Here, $\tilde{y}_i(x')$ represents individual wavelets, functionals, or trial solutions. The functionalities should be chosen to admit convenient analytical first- and second-order derivatives.

I like simple exponentials as the functionalities because they represent the ideal linear solution functionalities, which can be used to test a procedure. Additionally, they make it easy to visualize the resulting $\tilde{y}(x)$ shapes, and their derivatives are very simple. If $\tilde{y}_i(x') = e^{-x'/\tau_i}$, then $d\tilde{y}/dx = -(1/\tau_i)e^{-x'/\tau_i}$ and $d^2\tilde{y}/dx^2 = (1/\tau_i^2)e^{-x'/\tau_i}$.

Note that the τ-values, τ_i, and coefficient values, c_i, in the trial solution are as yet undetermined.

If the truncated series model is a valid representation of the solution to the ODE, then substituting it into the ODE will make the equation balance for any feasible x-value:

$$f(x,\tilde{y})\sum_{i=1}^{n} c_i \frac{d^2\tilde{y}_i}{dx_i^2} + g(x,\tilde{y})\sum_{i=1}^{n} c_i \frac{d\tilde{y}_i}{dx} + h(x,\tilde{y}) = u(x) \tag{42.1}$$

Alternately, if the coefficients or functionalities in the trial solution are not good, then the RHS will not match the LHS and the magnitude of imbalance can be assessed as the deviation between RHS and LHS:

$$\varepsilon(x) = u(x) - f(x,\tilde{y})\sum_{i=1}^{n} c_i \frac{d^2\tilde{y}_i}{dx_i^2} + g(x,\tilde{y})\sum_{i=1}^{n} c_i \frac{d\tilde{y}}{dx} + h(x,\tilde{y}) \tag{42.2}$$

where $\tilde{y} = \sum_{i=1}^{n} c_i e^{-x'/\tau_i}$

For this optimization application, make the penalty for the imbalance be the square of the deviation.

Rather than evaluating this for all possible x-values, we will just evaluate it for a limited number within a user-defined range $x_0 \leq x \leq x_{final}$. The optimization statement is

$$\begin{array}{c} \min \\ \{c, \tau\} \end{array} J = \sum_{j=0}^{N} \varepsilon(x)^2$$

$$\text{S.T.:} \quad x = x_0 + j\Delta x, \quad \Delta x = \frac{(x_{final} - x_0)}{N}$$

$$y(x = x_0) = y_0$$

$$\left.\frac{dy}{dx}\right|_{x = x_0} = \dot{y}_0$$

(42.3)

Note: The two initial conditions are equality constraints. Accordingly, once $n - 2$ coefficients in the set c are chosen by the optimizer, the initial conditions can be used to solve for the other two. So, the optimizer only needs to seek values for the n time constants, the reduced set of $n - 2$ coefficients. There are $2n - 2$ number of DVs.

Additionally, if the process is stable, then the asymptotic values of the trial solution elements must sum to the final value. For instance, if the ODE is represented as $A(d^2y/dx^2) + B(dy/dx) + Cy^p = u_{SS}$, then the final steady-state value of y is $y(x = \infty) = \sqrt[p]{u_{SS}/C}$. This could be a third equality constraint. Representing the reduced set of coefficients for the optimizer as c^-, the optimization statement becomes

$$\begin{array}{c} \min \\ \{c^-, \tau\} \end{array} J = \sum_{j=0}^{N} \varepsilon(x)^2$$

$$\text{S.T.:} \quad x = x_0 + j\Delta x, \quad \Delta x = \frac{(x_{final} - x_0)}{N}$$

$$y(x = x_0) = y_0$$

$$\left.\frac{dy}{dx}\right|_{x = x_0} = \dot{y}_0$$

$$y(x = \infty) = \sqrt[p]{\frac{u_{SS}}{C}}$$

(42.4)

But it gets easier!

With a power series model for the trial solution, $y(x) = c_0 x^0 + c_1 x^1 + c_2 x^2 + \cdots$, the functionality of each term is specified, and the only unknowns are the coefficient values and the user choice for the highest power to be used in the series. (Note that the $c_0 x^0$ term is simply the constant c_0.) Similarly, for polynomial functionalities (wavelets, orthogonal polynomials, etc.), the functionalities of each term are specified, and the unknowns are coefficients and the user's choice of the highest-order term to be included in the series. (Note also that the first term in such series is a constant.)

Following this tradition, the aforementioned optimization should only determine the coefficients to the exponential terms. But how to choose the tau-values? I find that permitting the optimizer to choose the largest tau-value and setting other tau-values as half of the prior tau-value works. So does setting each as one-third of the prior, or linearly spaced between zero and tau-max:

$$\min_{\{c, \tau_{\max}\}} J = \sum_{j=0}^{N} \varepsilon(x)^2$$

$$\text{S.T.:} \quad x = x_0 + j\Delta x, \quad \Delta x = \frac{(x_{\text{final}} - x_0)}{N}$$

$$\tau_i = \frac{\tau_{\max}}{2^{i-1}}$$

User specified n and N \hfill (42.5)

$$y(x = x_0) = y_0$$

$$\left.\frac{dy}{dx}\right|_{x = x_0} = \dot{y}_0$$

$$y(x = \infty) = \sqrt[p]{\frac{u_{SS}}{C}}$$

Now there are $n - 3$ number of c-values and one τ-value for a total of $n - 2$ DVs.

If one starts with a nonlinear ODE and gets an optimization solution, how does one know that it is the correct solution? I think that validating new procedures is an essential part of engineering and science. This sequence of exercises provides progressive benchmarks that will reveal if you have implemented the solution correctly. Once benchmarked (tested and found right) on a range of conditions with knowable or accepted results, then the procedure can be accepted as credible for applications with unknowable answers.

42.2 Exercises

1 Benchmark the optimization application with a simple case with an easy-to-obtain analytical solution. For example, consider that the ODE is the linear case with constant coefficients and RHS value: $8(d^2y/dx^2) + 6(dy/dx) + 1y = 5$ with initial conditions $y(x = 0) = 0$ and $dy/dx|_{x=0} = 0$. Here the time constant values are 2 and 4, conveniently chosen so that one is half the other so that when the optimizer chooses a correct value for $\tau_{\max}$, the rule $\tau_i = \tau_{\max}/2^{i-1}$ will determine the right value for the lower time constant. One DV is $\tau_{\max}$. Coefficient values for three terms in the approximating trial solution are set by the equality constraints. So, a 1-DV optimization application will have three terms in the series solution. A 2-DV application will seek the $\tau_{\max}$ value and the coefficient value for one of the four terms in the series. A 3-DV application will permit five terms in the series. Five terms in the series is two more than are actually needed. You could specify 10 DVs, which would permit a nine-term model, but this is excessive, and coefficient correlation actually makes it difficult for the optimizer. Any optimization should find the exact analytical solution. If you permit 3 DVs and it chooses $\tau_{\max} = 16$, then the τ_i values are 16, 8, 4, and 2. In this case the coefficients on the $\tau_i = 16$ and $\tau_i = 8$ terms should be zero. Alternately, if it chooses $\tau_{\max} = 8$, then the τ_i values are 8, 4, 2, and 1. In this case the coefficients on the $\tau_i = 8$ and $\tau_i = 1$ terms should be zero. In any case, the coefficient values for the $\tau = 2$ and $\tau = 4$ terms should be 5 and 10. If the optimization solution does not return results that match the analytical solution (within convergence criterion precision), then something is wrong. Fix it.

2 Repeat Exercise 1, but choose alternate values for the RHS and initial conditions. Again, the optimization solutions should exactly match the analytical solutions.

3 Repeat Exercise 1, but have $u(x)$ not a constant. The solution to an ODE should have the functional form of forcing function terms. For convenience have $u(x) = de^{-x/\tau_x}$, but choose τ_x to be independent of the ODE ideal solution values.

4 Repeat Exercise 1, but choose a set of A, B, and C coefficients that do not result in the $2:1$ ratio for the time constants. For example, $(A,B,C) = (12,8,1)$ lead to time constant values of $(3, 4)$. Now the optimization trial solution functionalities cannot exactly match the analytical ODE solution. Now the optimizer cannot find the analytical solution. However, with one or two extra terms in the series, it should be able to provide a model that comes very close to the analytical solution. Demonstrate that it is reasonably close by comparing a graph of the true analytical and approximating solutions.

5 A relatively simple nonlinear second-order ODE is $A(d^2y/dx^2) + B(dy/dx) + Cy^p = u_{SS}$. This can be solved numerically. With a central difference for the first derivative

$$A\frac{y_{k+1}-2y_k+y_{k-1}}{\Delta x^2} + B\frac{y_{k+1}-y_{k-1}}{2\Delta x} + Cy_k^p = u_{SS}$$

The user must choose an appropriate Δx value. Rearranged to solve for y_{k+1} results in the recursion formula,

$$y_{k+1} = \frac{u_{SS} - Cy_k^p + (2A/\Delta x^2)y_k + (B/2\Delta x - A/\Delta x^2)y_{k-1}}{(B/2\Delta x + A/\Delta x^2)}$$

Here $y_1 = y(0)$, and $y_2 = y(0) + \dot{y}(0)\Delta x$. These first two y-values are set by the initial value and initial rate of change. Then subsequent y-values are calculated by the recursion formula. Choose A, B, C, and p values. Compare the numerical solution to that found by the optimizer seeking to best fit the ODE with a series of approximating functions. The validity of the numerical solution is predicated on a small enough Δx, and a rule to determine the right value is to rerun the numerical method each time, halving the value of Δx. When the solution becomes relatively identical, that is the right Δx value.

6 In any application, the values of A, B, and C are not selected for convenience of the user. They are the result of the parameter values in any particular situation (tank diameter, valve coefficient, frontal area, etc.). Probably they are not integer values, and probably there is some uncertainty in these givens. Friction factors of new pipes are typically $\pm 15\%$ of nominal and can change significantly after in-use corrosion or scaling. Cross-sectional area of a tank could vary several percent as local dents, ridges, and baffles change with height or as thermal expansion changes with temperature. What is a best model if there is uncertainty in the model givens? In the limit of zero uncertainty, the stochastic solution should match the deterministic solution.

7 Once you have decided that the procedure is providing good models, choose functions $f(x, y)$, $g(x, y)$, and $h(x, y)$, and use the optimizer to determine an approximating model. As values for those functions approach the values of A, B, and C, the optimizer solution should approach the prior validated solutions.

43

Case Study 8

Horizontal Tank Vapor–Liquid Separator

43.1   Description and Analysis

The objective is to design a horizontal cylindrical tank to disengage liquid and vapor (see Figure 43.1). Two-phase flow enters the top of the tank on one side, the liquid drops out to the lower portion of the tank, and the vapor to the upper portion. They exit at the other end of the tank, separately. Liquid at the bottom. Vapor at the top. As the vapor flows from entrance to exit, mist droplets gravity-fall from the vapor to the liquid. The higher the vertical dimension of the head space, or the higher the vapor flow rate, the longer the tank must be for droplets to have time to fall out of the vapor. This places a constraint relating tank length to diameter. There are two additional constraints: the vapor flow rate must be small enough so that it does not make waves on the liquid surface and re-entrain droplets. And the liquid holdup must provide a reservoir to provide continuity of downstream flow rate in spite of inflow pulses. The process tank designer can choose tank aspect ratio (l/d) and liquid level in the tank (f, as a fraction of diameter), and then lower-cost tank diameter will be a consequence of compliance to the limiting constraint.

The objective is to minimize the capital cost of the tank:

$$\min_{\{l/d, f\}} J = \text{cost}(l/d, f, \max\{d_1, d_2, d_3\}) \tag{43.1}$$

I modified this problem from one given to me by Josh Ramsey, Chemical Engineering design instructor at Oklahoma State University, who modified it from an example that Jan Wagner had used when he taught the class, who created it based on the publication by W. Y. Svrcek and W. D. Monnery, "Design Two-Phase Separators within the Right Limits," Chemical Engineering Progress, 1993, vol. 89, pp. 53–60. There are actually three stages of separation in the tank: the initial impingement plate to deflect incoming liquid, a settling zone to let large droplets gravity-fall from vapor to liquid, and then demisters to coalesce small droplets. For simplicity, this exercise only considers the second-stage mechanisms of droplet settling and idealizes some aspects to keep the equations easily recognized and understood by the general reader.

What are the undesirables that need to be balanced? If liquid level is high in the tank, then the cross-sectional area for vapor flow is low, vapor velocity is high, and mist is re-entrained as the high velocity wind shears the liquid surface. To prevent this tank d must be large, which is costly. By contrast, if level is low, then liquid residence time is too low to provide surge capacity. To prevent this, either tank d or l/d must be large. There is an intermediate level that minimizes cost. Similarly, if l/d is low, then d

Engineering Optimization: Applications, Methods, and Analysis, First Edition. R. Russell Rinehart.
© 2018 R. Russell Rinehart. Published 2018 by John Wiley & Sons Ltd.
Companion website: www.wiley.com/go/rhinehart/engineeringoptimization

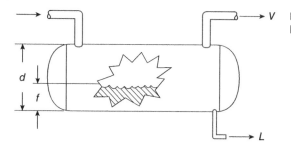

Figure 43.1 Illustration of a vapor–liquid separator, a horizontal right circular tank.

needs to be high to provide low gas velocity and high enough droplet settling and degas times and liquid holdup. This increases tank cost. Alternately, if l/d is large, the diameter is limited to ensure gas velocity constraints, and the long tank is expensive.

Tanks come in standard sizes. For this exercise, consider that l/d ratios will be restricted to half-integer values (1, 1.5, 2, 2.5, 3, 3.5, etc.). This means that one of the DVs, the l/d ratio, is discretized, which creates ridges in the OF response. Further, the tank diameter is limited to standard sizes; here, consider that they are based on 3-inch (quarter-ft) increments. This means that although the DV representing the choice for liquid level in the tank is continuum valued, the OF will have flat spots. Further, the three conditions on tank diameter (ensure settling time, ensure liquid holdup time, and ensure vapor velocity is low enough to not re-entrain droplets) create a discontinuity in the surface, intersecting sharp valleys, even if the DV and OF were continuum valued.

For simplicity, I consider that the cost of the vessel is directly related to the mass of the metal needed to make it (which will be the surface area times thickness times density), and I also modeled the vessel as a right circular cylinder with flat ends. Since density, thickness, and cost/mass are each constant multipliers of the surface area, the OF can be stated as $J = ((0.5 + l)/d)d^2$. Certainly, one could modify this to reflect alternate cost models or tank shapes.

To reveal detail of the lower values of the OF surface, I find it convenient to take the square root of J. A strictly monotonic function transformation does not change the DV* values: $J = d\sqrt{((0.5 + l)/d)}$.

The liquid level can be stated as a fraction of the tank diameter, and the geometry/trigonometry of a segment of a circle can be used to calculate the cross-sectional area faction for liquid and vapor flow (see Figure 43.2). Beta is the angle between the liquid surface and the segment line, which is the radius to the intersection of the segment and circumference. Alpha is the angle between the two radii. The

Figure 43.2 Geometry of a segment of a circle.

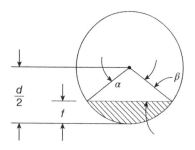

sine of beta is the distance between the segment and the center divided by the radius, $\sin(\beta) = y/r = 1 - 2f$. Then, with angles in radians, $\beta = \text{Arcsin}(1 - 2f)$, and $\alpha = 2((\pi/2) - \beta)$. The area of the segment, the cross-sectional area for the liquid flow rate, is $A_L = (\pi d^2/4)(\alpha - \sin(\alpha))$, which leaves the complementary area for vapor flow, $A_V = (\pi d^2/4) - A_L$. If the liquid level is above center, the calculations are the same, but the segment is for the vapor not the liquid.

Again, for simplicity, assume plug flow of both liquid and vapor, which is immediately developed at the entrance. Then the liquid residence time is $t = A_L l / \dot{Q}_L$, which must be greater than the holdup time desired. This can be rearranged to calculate a required tank d. Further, the vapor flow rate Reynolds number must be low enough to prevent re-entrainment: $\text{Re} = d_{\text{ch}} u \rho / \mu \leq \text{Re}_{\text{threshold}}$. This can be rearranged to determine the required tank d. For simplicity, use the maximum vapor space height as the characteristic diameter, $d_{\text{ch}} = d(1 - f)$. Finally, the vapor residence time must be longer than the time for droplets to fall the largest vapor distance, $A_V l / \dot{Q}_V \leq d((1 - f)/v_{\text{terminal}})$, which also can be rearranged to determine the required tank d. The required tank d is the maximum of the d determined by each of the three constraints and then rounded to the next higher 3-inch increment.

Without the increments on either l/d or d, the surface has discontinuities due to the choice of diameter-controlling mechanism. This can be seen as discontinuities in either the net lines or contour lines at the bottom of the three valleys in Figure 43.3. With discretized l/d and d values, the surface has additional ridges and many flat spots as Figure 43.4 reveals.

Even if there is no discretization (Figure 43.3), the discontinuity in the valleys is a problem for most optimizers. Gradient-based optimizers stop anywhere along the discontinuity, as shown by the X's in Figure 43.5.

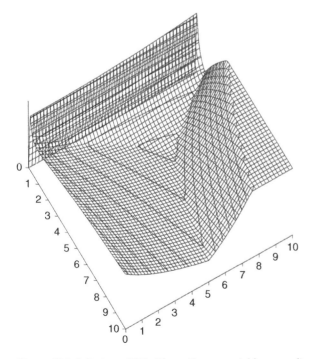

Figure 43.3 3-D view of OF with continuum variables, revealing the sharp valleys.

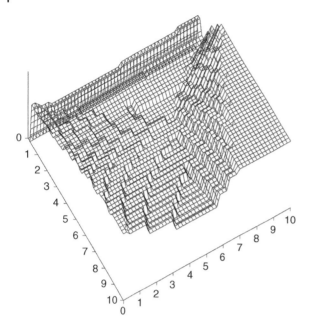

Figure 43.4 3-D view of the OF with discretized *D* and *L/D* values.

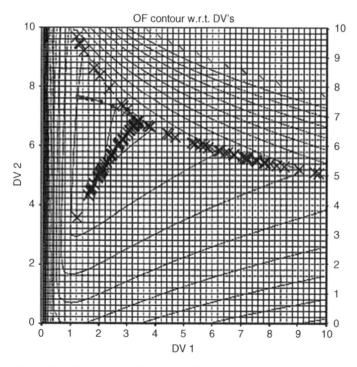

Figure 43.5 The impact of the sharp valleys on a gradient-based optimizer.

With the flat spots from discretization, most single TS optimizers stop where are they are initialized (on a flat spot, because no direction is downhill), as shown in Figure 43.6.

The optimum from the continuum version of the application can be rounded to the nearest l/d and tank d values; but this is not the optimum when the application considers the discretization.

This application is provided as Function 27 in the file "2-D Optimization Appilications.xlsm." It uses the values in Table 43.1.

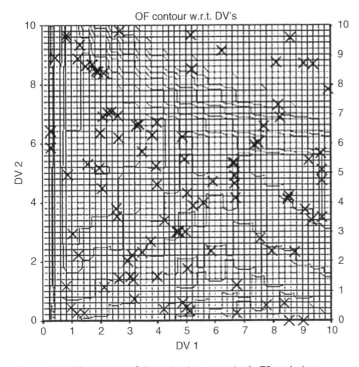

Figure 43.6 The impact of discretization on a single TS optimizer.

Table 43.1 Value of givens for the liquid–vapor tank separator application.

Property/condition	Value (units)
Vapor flow rate	2 cuft/s
Liquid flow rate	0.05 cuft/s
Vapor density	0.6 lbm/cuft
Vapor viscosity	0.0001 lbm/ft-s
Droplet terminal velocity	0.1 ft/s
Liquid holdup time	500 s
Maximum vapor Re	20,000 (dimensionless)

43.2 Exercises

1 Explore Function 27 in the 2-D Optimization Examples file and show how the code is developed from the description above.

2 Comment out the lines that discretize the l/d and d values, and explore the performance of several optimizers on the function. Use a single trial solution gradient-based optimizer, a single trial solution direct search algorithm, and a multiplayer direct search optimizer. Use data to assess key performance indicators such as the number of function evaluations to find the optimum and proximity of converged solution to the global. Since the results will depend on the randomized initialization, use data from many trials, and apply statistical tests (perhaps a t-test) to determine whether one optimizer is significantly better than another.

3 Activate the l/d and d discretization, and describe the optimization-relevant similarities and differences between the contour with and without discretization.

4 Repeat the comparison in Exercise 2, but with the l/d and d discretization.

5 Compare the $\mathbf{DV}^*$ and OF^* values from the continuum and discretized applications. Are the rounded values of the continuum $\mathbf{DV}^*$ equivalent to the $\mathbf{DV}^*$ values from the discretized optimization?

6 Add a consideration that the liquid capacity of the tank must also provide inflow surge capacity, a large enough volume to accumulate liquid if the inflow liquid rate temporarily increases. Note: If the liquid level rises, then the diameter needs to be increased to preserve the two required vapor flow rate conditions.

7 The ends of a tank will not be flat. A process tank is not a right circular cylinder. The ends will be domed or hemispheres. Add this geometrical aspect to the analysis of liquid holdup and cost.

8 The initial vapor–liquid disengagement zone and the impingement plate zone that deflects liquid to the bottom, followed by flow alignment planes to minimize vapor turbulence, might require a length of about 2 ft at the entrance portion of the tank. This extends the liquid volume, but not the length for mist settling. Similarly, a structured-packing droplet-coalescing demister at the exit will add another 3 ft to the liquid volume. Add these aspects to the application and investigate the impact on the $\mathbf{DV}^*$ values.

9 If you are a process person with fluid mechanics experience, you should be able to choose fluid properties and flow rates and calculate droplet terminal velocity for a particular system. If you have a strong enough fluid dynamics background, you may enjoy determining (i) the maximum permissible Re number to keep surface shear below the value that would re-entrain the droplets that should be falling out by gravity, (ii) an appropriate estimate of the characteristic length (the hydraulic diameter), or (iii) an appropriate non-plug-glow velocity profile.

10 Change the givens in the table above and explore that impact on the $\mathbf{DV}^*$ values. This is a Method 2 analysis of Section 20.8.2.

11 Of course, the givens (in any situation) are nominal values. Consider that they might be subject to the vagaries of process pressure and temperature and composition, which will change fluid properties and droplet settling velocity by $\pm 10\%$ from nominal and flow rates by $\pm 20\%$ from nominal. When flow rate is the factor that selects the diameter, higher flow rates would require a larger tank. When droplet terminal velocity selects the diameter, lower values lead to larger tanks. Any of several combinations of variable values could become the limiting diameter factor.... You could design for the worst case for all variables; but not all will simultaneously limit the diameter, and the confluence of worst cases (each of seven variables at its extreme, and pushing the size in the same direction) is improbable. So, instead of deterministic nominal or extreme values, sample each "given" from its realizable range and convert the exercise from a deterministic to a stochastic application. Use Method 3 of Section 20.8.2. Optimize.

44

Case Study 9

In Vitro Fertilization

44.1 Description and Analysis

I am grateful to Babatunde Ogunnaike (presently Dean of Engineering at the University of Delaware) for revealing this procedure, which I have appropriated as a case study. It could be a relatively simple exercise, a medium-complicated exercise as this case study outlines, or a considerably complex one as the side notes and exercises disclose.

Some couples cannot get pregnant the natural way, but are fully capable of carrying a child to birth. For them, *in vitro* fertilization is a possible solution. It is the process of harvesting an egg, fertilizing it *in vitro* (Latin for in glass), observing the cellular development, and implanting the zygote (newly fertilized egg) in the uterus. If there is only a 50% chance of one zygote leading to a successful pregnancy, would you implant 1, 2, 3, ... zygotes? The binomial distribution describes the probability of x-number of successes in n-number of trials, when an individual probability of success is p. Of course, $x \le n$, and the events must be independent for the binomial distribution to be valid:

$$P(x|n) = \binom{n}{x} p^x (1-p)^{n-x} = \frac{n!}{x!(n-x)!} p^x (1-p)^{n-x} \tag{44.1}$$

This is generally accepted to model data on IVF. However, it is an idealization. The value of p is not a universal constant. The value of p would depend on the skill of the procedure and the properties of the zygote and uterus. In general, p-values range from about 0.2 to 0.6. Further, for any given set of participants, the probability of a zygote successfully attaching and being carried full term depends on the number of zygotes implanted. The probability of success drops with increasing n. Perhaps a reasonable model is $p(n) = p_1 e^{-(n-1)/m}$.

The question is, "How many zygotes should be implanted?" It is a surgical procedure, with costs and risks. So, you would want to maximize success. If one zygote is implanted, then the probability of a success is p, but the probability of no success is $(1-p)$. If $p = 0.4$, this means that there is a 60% chance of not a success. You would like a surgical procedure to have a higher outcome. If two zygotes are implanted, then for a nominal $p = 0.4$ the probabilities for 0, 1, or 2 successes are

$$P(0|2) = (1-p)^n = (1-0.4)^2 = 0.36 \tag{44.2}$$
$$P(1|2) = np(1-p)^{n-x} = (2)(0.4)(1-0.4)^1 = 0.48 \tag{44.3}$$
$$P(2|2) = 0.16 \tag{44.4}$$

Engineering Optimization: Applications, Methods, and Analysis, First Edition. R. Russell Rinehart.
© 2018 R. Russell Rinehart. Published 2018 by John Wiley & Sons Ltd.
Companion website: www.wiley.com/go/rhinehart/engineeringoptimization

Now the probability of no success is reduced to 36%, but there is a 16% chance of having twins. But that may not be such a bad thing.

Supposing three zygotes are implanted, then

$$P(0|3) = 0.216 \tag{44.5}$$
$$P(1|3) = 0.432 \tag{44.6}$$
$$P(2|3) = 0.288 \tag{44.7}$$
$$P(3|3) = 0.064 \tag{44.8}$$

The probability of not a success is about 22%, which is an improvement. But there are about a 35% chance of multiple births and a 6% chance of the procedure leading to triplets. If four zygotes are implanted, then the probability of no success is only about 13%, but the probability of multiple births is 52%, which might be quadruplets. This multiple birth possibility may lead to a reevaluation of how success is evaluated!

Success is measured by not only minimizing $P(x = 0|n)$, or maximizing $P(x = 1|n)$, but also minimizing excessive births, perhaps $P(x \geq 3|n)$.

If the desirability of a singleton birth is equivalent to the undesirability of no birth and also equivalent to the weighted cost of the procedures, and the undesirability of triplets (or greater) has double the equal concern impact, and the outcome of twins is a neutral event (equivalently desirable or undesirable), then the optimization application is to find the number of zygotes to be implanted to minimize the sum of the separate probabilities:

$$\min_{\{n\}} J = 2\text{cost} + P(0|n) - P(1|n) + 2P(\geq 3|n) = 2\text{cost} + 2 - P(0|n) - 3P(1|n) - 2P(2|n) \tag{44.9}$$
$$\text{S.T.:} \quad n \geq 3 \text{ (If } n = 1,2 \text{ the } P(2|n) \text{ term does not exist)}$$

The reader should be able to show that if $n = 1,2$, the OF is $2\text{cost} + P(0|n) - P(1|n)$.

Certainly, however, some couples may not be adverse to triplets, but they may be very adverse to quadruplets or greater. Some couples may embrace the possibility of twins (some of my favorite people are twins, and twins form the basis of many romantic comedies) and weight twins as a better outcome than a singleton. Or another couple may weight the several probabilities completely differently. Feel free to customize the OF elements and their relative weighting to your customers' values.

Perhaps the optimization solution with $p = 0.5$ is $n = 3.4567$ zygotes. You can only implant an integer number. Do you implant 3 or 4? The DV must be discretized to integer values. Will you round up or truncate DV values to integer values?

Further, on a particular set of players, the individual probability of success may be $p = 0.7$. In this case

$$P(0|3) = 0.027 \tag{44.10}$$
$$P(1|3) = 0.189 \tag{44.11}$$
$$P(2|3) = 0.441 \tag{44.12}$$
$$P(3|3) = 0.343 \tag{44.13}$$

which leads to an 80% chance of twins or triplets, which may be perceived as not a success.

Alternately, the individual probability of success may be $p = 0.2$, and if so, implanting three zygotes has a 51% chance of no success. The value for p cannot be known until after the outcome of the procedure is discovered.

To account for the uncertainty in the p-value, one option is sequential IVF procedures. Harvest and fertilize many eggs. Implant n_1, and freeze the others for later. If there is no success (no birth), then implant n_2 on a second procedure. Because the p-value is unknown, be conservative, and use a nominally high p-value to determine n_1 for the first procedure. If no success, then use a lower p-value for the n_2. Now the OF is as above if a success on the first procedure: $OF_1 = 2\text{cost} - P(0|n_1,p_1) - 3P(1|n_1,p_1) - 2P(2|n_1,p_1)$. But if not, then the OF of the second implant procedure needs to include the cost of three procedures, the disappointment of the first, and the outcome of the second. $OF_2 = 3\text{cost} + P(0|n_1,p_1) + 2 - P(0|n_2,p_2) - 3P(1|n_2,p_2) - 2P(2|n_2,p_2)$. This can be modeled by comparing $P(0|n_1,p_1)$ with a random number (uniformly distributed on the interval 0–1). If $u \leq P(0|n_1,p_1)$, then it was not a success:

$$\min_{\{n_1,n_2\}} J = \left\{ \begin{array}{c} OF_1, \quad \text{if } u > P(0|n_1,p_1), \text{ else} \\ OF_2 \end{array} \right\} \tag{44.14}$$

$$\text{S.T.:} \quad n_1 \geq 3, n_2 \geq 3 \text{ (if either is 1, or 2, the last terms do not exist)}$$

This makes the OF a stochastic function. The n_2 term is only added if the random chance indicates it should be considered. So, now it is stochastic with discretized DVs.

And we can further confound this ideal by including uncertainty in the p_1 and p_2 values.

The VBA code is relatively simple and is provided by Function 31 in the 2-D Optimization Examples file.

With the discretization and stochastic effects removed, the OF response to the single DV, n_1 (Figure 44.1), reveals a nonlinear surface with both slope and value discontinuities associated with the change in the OF equation due to the possibility of triplets or greater. The optimum is just prior

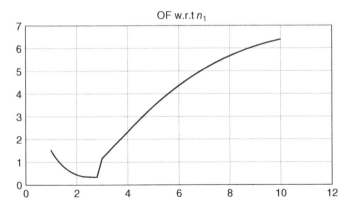

Figure 44.1 One aspect of the optimization if n could be a continuum.

to 3, perhaps about $n_1 = 2.9$. The optimizer must be able to cope with such surface discontinuities in both n_1 and n_2.

But one cannot implant a fraction of a zygote. With the discretization to integer values, the surface has flat spots and the optimum is at $n_1 = 2$ (see Figure 44.2).

With the lower p-value for n_2, the optimum is at $n_2 = 4$, but a value of $n_2 = 3$ is nearly equivalent (see Figure 44.3). The optimizer must cope with flat spots in both DVs and the broad valley, as well as the nonlinear and discontinuous trend changes.

Finally, with the stochastic features, and both n_1 and n_2 as DVs, the surface (shown in Figure 44.4) is perturbed with transient fluctuations, and the optimum is $n_1 \cong 3$ and $n_2 \cong 4$, but with nearly equivalent results in the n_1 choices of 3 or 4 in the n_2 range of 3–5.

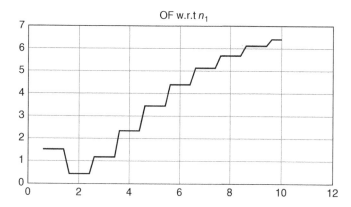

Figure 44.2 The OF response to n_1.

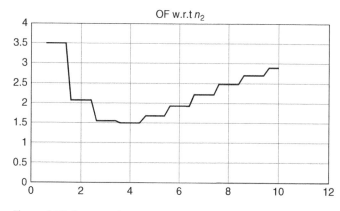

Figure 44.3 Response to n_2.

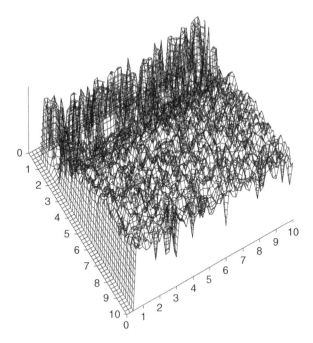

Figure 44.4 The stochastic response w.r.t. n_1 and n_2.

44.2 Exercises

1 Explore Function 31, which implements the application above. Indicate the stochastic and discretization portions of the code and their expression in the contours. You might want to remove or reduce the stochastic nature of the p_1 and p_2 values to see the impact of discretization and the stochastic contribution of the second OF term.

2 Investigate the impact of model coefficient values on the OF* and DV*. For example, suppose the probabilities for a zygote taking hold were different. Suppose the competition factor were different. Suppose the range of values on the two probabilities were different. Suppose the distribution of the probability uncertainty were different. Suppose the competition factor was not uniformly applied to each zygote (the probability of others becoming a success is only reduced when one is a success).

3 Change the OF to acknowledge how another couple might perceive the desirables and undesirables associated with $x = 0$ or $x > 1$ successes. Perhaps twins are also a desirable outcome and triplets are not bad either. Explore the impact it has on the surface. For instance, triplets may be moderately undesirable, but quadruplets or more may be highly undesirable.

4 Add features to the OF, and demonstrate that your model appropriately reflects the concept.
   A For instance, include a factor for the client's perception of the cost/risk/inconvenience of the separate surgical procedures of harvesting and implanting. What equal concern values would you choose to make these issues additive to the probabilities?
   B Or include the possibility that one of the zygotes could split to become identical twins.

C Or include a penalty for the second procedure to acknowledge the additional risk and anguish.

D Or consider the implications if $n_1 + n_2$ eggs are harvested and n_1 fertilized and the first implantation is a success. What is the fate of the n_2 eggs? This is partly a moral issue. At the least, it is biological material that must be properly eliminated to prevent unauthorized misuse. How would you include this undesirable in the OF? Further, not all eggs become zygotes *in vitro*, so the procedure would harvest and attempt to fertilize extra eggs to ensure that there is enough. Suppose there are an excess number of zygotes.

E Or change the procedure so that the n_2 eggs are only harvested if the initial implant procedure is not a success, and add an undesirable to reflect the second harvesting procedure on the OF.

F Or consider that the couple needs a surrogate to carry the baby. How might the surrogate perceive the desirables and undesirables? How does this affect the OF?

5 With the several stochastic features removed, but keeping the DV discretization, the OF surface response to the 2 DVs will have flat spots, but a single optimum. Test several optimizers on it. Use a statistical test of ANOFE as the criterion for comparison.

6 With the stochastic features included, compare several optimizers. Be sure to include LF with replicates and the SSID convergence criterion. Show that conventional optimizers seek a fortuitous phantom best, which is not the best for the population norm.

7 The procedure could be a 3-stage or 4-stage procedure with n_1, n_2, n_3, n_4. How would you adjust the OF to include subsequent stages? How would you adjust the optimizer to handle more than 2 DVs?

8 This analysis considered that the p-values for each zygote in a common procedure are independent. However, if the p-value for the nominal procedure is 0.6 for a particular confluence of patients and providers, it could be that an event in the procedure or temporal condition of the patient may make $p \cong 0$ for the first procedure for all zygotes. This is termed a common cause or a conditional probability. If something happens, it affects all probabilities—probabilities of each event are not independent. But the p-value may remain at the nominal 0.6 for a second procedure that does not have that common cause. How would you adjust the OF to account for a common cause event?

9 This analysis considered that if the first procedure was not a success, then the p-value for the second drops to 0.2. An alternate model might use Bayes' inference, a Bayes belief to change the p-value. Implement it.

45

Case Study 10

Data Reconciliation

45.1 Description and Analysis

Values of data from online sensors are corrupted by sensor drift, which could be due to ambient conditions (temperature, humidity, etc.), dust accumulation, element wear, corrosion, cumulative vibration damage, etc. It could also be that the sensor was not perfectly calibrated, which leads to incorrect values. These aspects mean that the value is biased, is in error, and has a systematic error.

Further, measurements don't occur just once; instead, process sensors are sampled once each minute, or second, or whatever the sampling interval. This produces a sequence of biased values.

Further, the sensors report a noisy value. The value with the systematic error is also confounded with random perturbations on each sampling, which can arise from turbulence, mechanical vibrations, electromagnetic interference, imperfect mixing, etc. This is termed random error. (These temporal perturbations about the nominal value might not be independent or Gaussian. There are often autocorrelation and non-Gaussian drivers. This exercise will not consider those cases.)

Random error can be reduced by averaging sequential data. However, even if the random error could be eliminated, the systematic bias remains. Data reconciliation is a procedure to use online data to estimate the systematic errors of the same online data. In data reconciliation the correction is based on redundant measurements (but not at the same location) and a model.

For example, per Figure 45.1, if liquid flows into a tank, and liquid flows out, then the inflow minus the outflow must equal the accumulation in the tank. However, if the flow meters are of the orifice type, then a 5% error is not unexpected. If the inflow meter reads 103 gpm, it could actually be 97 gpm. If the outflow meter reports 91 gpm, it could actually be 97 gpm. The readings would indicate an accumulation of 103 − 91 = 12 gpm. However, the level indicator might properly reflect the actual flow rates and report no change in level. Alternately, the in-tank inventory is not level, but volume, which may be calculated from level and tank cross-sectional area. If either the level sensor is not exactly calibrated or the tank area does not exactly account for ribbing or dents or the volume immersed devices, then the reported inventory accumulation is also wrong.

One solution is to adjust each process measurement by a bias. The corrected value, y', is the measurement, y, corrected with the bias, b:

$$y' = y + b \tag{45.1}$$

Engineering Optimization: Applications, Methods, and Analysis, First Edition. R. Russell Rhinehart.
© 2018 R. Russell Rhinehart. Published 2018 by John Wiley & Sons Ltd.
Companion website: www.wiley.com/go/rhinehart/engineeringoptimization

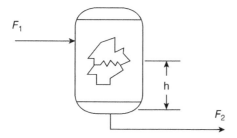

Figure 45.1 Illustration of a case of redundant measurements.

There are many measurements and each measurement is reported at each sampling. Here the subscript i indicates the ith measurement, and t represents the sample time. With the condition that b_i does not change in time, or changes very slowly over a time window,

$$y'_{i,t} = y_{i,t} + b_i \tag{45.2}$$

There are uncertainties on each measurement, which can be estimated as the standard deviation of the noisy value during a "steady" period, σ_i.

Desirably, the bias to the sensor would be determined by a calibration. However, it is often impractical to remove sensors from a process for offline laboratory calibration; and even if you did, in just a bit of time later, they would become out of calibration.

In data reconciliation, the b_i values are chosen to best reconcile all data and to best seek internal consistency in the data, with respect to expectations, or models. The models are typically material and energy balances. For example, assuming constant density, the material balance model about the tank is

$$A\frac{d\widetilde{h}}{dt} = \widetilde{F}_{\text{in}} - \widetilde{F}_{\text{out}} \tag{45.3}$$

From this, any two corrected measurements and the model can be used to predict what the other corrected measurement should be. For instance, if $\widetilde{F}_{\text{out}}$ is estimated from the model and the other corrected measurements,

$$\widetilde{F}_{\text{out}} = \widetilde{F}_{\text{in}} - A\frac{d\widetilde{h}}{dt} = (F_{\text{in}} + b_{\text{in}}) - A\frac{d(h + b_h)}{dt} = (F_{\text{in}} + b_{\text{in}}) - A\frac{dh}{dt} \tag{45.4}$$

or if $\widetilde{F}_{\text{out}}$ is calculated from the measurement and the bias,

$$\widetilde{F}_{\text{out}} = (F_{\text{out}} + b_{\text{out}}) \tag{45.5}$$

The two $\widetilde{F}_{\text{out}}$ values should be identical, and the objective is to determine the bias values to minimize the square of the deviation of the model over the time duration of a data window:

$$\min_{\{b_{\text{in}}, b_{\text{out}}\}} J = \sum_t \left[F_{\text{out},t} + b_{\text{out}} - (F_{\text{in},t} + b_{\text{in}}) + A\frac{dh}{dt}_t \right]^2 \tag{45.6}$$

In general, there are multiple models that are used to adjust each variable, and the data reconciliation method is a bit more complicated. Adjust the b-values to minimize the model-predicted value from its measurement:

$$\min_{\{b\}} J = \sum_{k} \sum_{t} \left[\frac{\widetilde{y}_{k,t} - y'_{k,t}}{\sigma_k} \right]^2 \tag{45.7}$$

The model index is k and this assumes that b_k values are the result of independent perturbations on each sensor. This uses the noise amplitude, σ_k, of the model-predicted online y-value as an equal concern factor that normalizes all OF elements to be dimensionless and places them each as the number of standard deviations. This presumes that the bias values are relatively unchanged over a data window, the time duration of data collection. This presumes that there is adequate redundancy in sensors and model structures to be able to use the data to legitimately adjust the sensor bias values.

There are two desirables in data reconciliation. The expression earlier relates to only one of them: minimizing the residual deviation in the material and energy balance models. The OF represents a soft penalty for a residual in a model. But also one would like the bias corrections to remain reasonably small relative to the noise.

In general, with x representing process inputs and y representing process outcomes, the process models (representing material and energy balances, equilibrium relations, kinetic relations, etc.) are

$$\widetilde{y}_{1,t} = f_1(x_1, x_2, ..., x_i, ..., x_n)_t$$
$$\widetilde{y}_{2,t} = f_2(x_1, x_2, ..., x_i, ..., x_n)_t \tag{45.8}$$

where the subscript, t, represents a particular sampling and i an input, and with j representing an output, $\widetilde{y}_{j,t}$ represents the numerical solution to a modeled output.

With deviations from the measurement indicated as b, the corrected y- and x-values are

$$x'_i = x_i + b_{xi}$$
$$y'_j = y_j + b_{yj} \tag{45.9}$$

Then data reconciliation will seek the b-values to minimize deviations from the measurements, tempered by the expected standard deviation (an equal concern factor):

$$\min_{\{b_x, b_y\}} J = \frac{1}{N_t} \frac{1}{N_k} \sum_t \left[\sum_k \left(\frac{\widetilde{y}_k(x') - y'_k}{\sigma_{yk}} \right)^2 \right] + \frac{1}{N_x} \sum_i \left(\frac{b_{xi}}{\sigma_{xi}} \right)^2 + \frac{1}{N_y} \sum_j \left(\frac{b_{yj}}{\sigma_{yj}} \right)^2 \tag{45.10}$$

The argument of the double sum term represents the "closure" on the N_k number of models and the deviation between data and model. The objective is to reduce this. The residual is scaled by the expected uncertainty in the y-measurement data predicted by the kth model, making it represent the number of standard deviations. It is summed over each of the N_k models predicting a y-variable and summed over N_t time intervals. The reciprocal N scaling normalizes for the number of y-variable models and time samplings. The second and third terms in the RHS of the optimization statement represent the scaled value of the correction, scaled by the expected random uncertainty of the measurement. It is a penalty for large corrections to the measurement. It is summed over all x- and y-values and is independent of time. The reciprocal N scaling accounts for the number of variables. The b_{xi} correction for the x-measurements and b_{yj} correction for the y-measurements are common values over all the time intervals in the data reconciliation period.

In a particular application, the output of one process section may be the input to another. The x–y distinction may not be clear, and it is an inconsequential choice.

Note: In one of the seminal papers on data reconciliation, Mah, Stanley, and Downing do not include the double sum term of the model deviations in the OF. The OF reduction to only consider the b/σ terms can be defended, because when there is only one measurement, or when an average of multiple measurements over a time period is used, the models are constraints that can be used to calculate some of the biases, which are not considered as independent DVs. There are $N_x + N_y$ number of biases and N_k number of models, which are equality constraints. Accordingly there are $N_x + N_y - N_k$ number of DVs in the $\left\{ b_x^-, b_y^- \right\}$ reduced set. In such cases, use

$$\min_{\left\{ b_x^-, b_y^- \right\}} J = \frac{1}{N_x} \sum_i \left(\frac{b_{xi}}{\sigma_{xi}} \right)^2 + \frac{1}{N_y} \sum_j \left(\frac{b_{yj}}{\sigma_{yj}} \right)^2 \tag{45.11}$$

45.2   Exercises

1   Show the optimization statement for data reconciliation for the inventory accumulation process of Figure 45.1.

2   In the in-pipe flow process of Figure 45.2, three inflows all combined to one outflow, with an intermediate F_3 measurement. Assume constant density and derive the two independent material balance models. Show the optimization statement for data reconciliation.

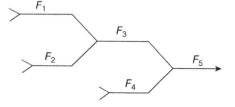

Figure 45.2 Combined flows.

3   The in-pipe mixing of hot and cold water makes an outflow in Figure 45.3. Assume constant density and steady state and derive the material and energy balance equations to model F_3 and T_3. Show the optimization statement for data reconciliation.

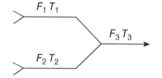

Figure 45.3 Blending flows.

4   Describe the rationale for the sigma and N scaling in the optimization statement of Equation (45.10). Does the scaling meet your equal concern interpretation?

5   If the modeled y-deviation is included in the first term on the RHS, is there a need for the third term on the RHS of the OF?

6   The models could be considered constraints. In this case, the b_{yj} values would not be optimization DVs but would be calculated to make the model residuals be zero. Comment on this version of the optimization statement:

$$\min_{\{b_x\}} J = \frac{1}{N_x} \sum_i \left(\frac{b_{xi}}{\sigma_{xi}}\right)^2 + \frac{1}{N_k} \sum_k \left(\frac{b_{yk}}{\sigma_{yk}}\right)^2$$

$$\text{S.T.:} \quad y_{k,t} + b_{yk} - f_k(x_1, x_2, \dots, x_i, \dots, x_n)_t = 0$$

Consider application simplicity. Consider the impact of the model is an imperfect representation of the process. Suppose measurement error reflected sensor bias, not random sampling noise.

7   Take a simple process that is familiar to you, and show the optimization statement for data reconciliation.

8   Create a simulation for the process in Exercise 1, 2, 3, or 7. Perturb the data with a systematic error, and use the optimization to provide the bias corrections. I think this can be easily done in a spread sheet.

Section 8

Appendices

The six appendices are to provide essential refresher and/or introduction to mathematical, procedural, or programming topics in the book.

Engineering Optimization: Applications, Methods, and Analysis, First Edition. R. Russell Rhinehart.
© 2018 R. Russell Rhinehart. Published 2018 by John Wiley & Sons Ltd.
Companion website: www.wiley.com/go/rhinehart/engineeringoptimization

Appendix A

Mathematical Concepts and Procedures

A.1   Representation of Relations

The representation

$$z = f(x, y) \tag{A1}$$

means that z is a function of x and y. This is a generic statement. The explicit, specific equation might be

$$z = 3x + 5xy^{-1} - 2\sqrt{y} \tag{A2}$$

Here x and y represent the independent variables, which are alternately termed cause, influence, or input variables. And z represents the dependent variable (result, consequence, or output). There are model coefficients (alternately termed parameters) represented by the numerical values of 3, 5, −1, and 2. But these do not need to be integers and could be expressed as the letters a, b, c, and d:

$$z = ax + bxy^c - d\sqrt{y} \tag{A3}$$

In general, this book will use the last letters of the alphabet to refer to decision and response variables in a relation and the first letters to represent model coefficients. However, in regression, the objective is to determine the model coefficient values that best fit the data, in which case the model coefficients are the decision variables.

This example function reveals both linear and nonlinear relations. The response, z, depends on influence x. With other values held constant, the value of z is a linear response to the value of x. However, z also depends on influence y. With other values held constant, the value of z has a nonlinear response to the value of y. Further, when considering finding the best coefficient values, with other variables and coefficients held constant, z is a linear function of a, b, and d. But z has a nonlinear response to the value of c. A relation is classified as linear in variables if $\partial^2 z / \partial x_i \partial x_j = 0$ for any i, j pairing. Similarly, it is termed linear in coefficients if $\partial^2 z / \partial c_i \partial c_j = 0$ for any i, j pairing.

Since the $\sqrt{}$ function restricts y to nonnegative values (choosing to discount imaginary numbers), the function has constraints, which could be expressed as $y \geq 0$. Further, if coefficient c has a negative value, then to prevent a divide by zero, the constraint becomes $y > 0$. However, to prevent a computer overflow error, the constrained reality might be that $> 10^{-100}$.

Engineering Optimization: Applications, Methods, and Analysis, First Edition. R. Russell Rhinehart.
© 2018 R. Russell Rhinehart. Published 2018 by John Wiley & Sons Ltd.
Companion website: www.wiley.com/go/rhinehart/engineeringoptimization

A specific example might represent reaction rate:

$$r = f(k_0, E, R, T, [A], [B]) = k_0 e^{-E/RT} [A][B] \tag{A4}$$

A more generic notation is

$$z = f(c_1, c_2, c_3, c_4, x_1, x_2), \quad x_2 > c_5 \tag{A5}$$

There is no need to use separate symbols for the independent variables. We could represent $\{x, y\}$ as $\{x_1, x_2\}$, but this generic form does not provide the specific rules as to how to calculate the value of z. The procedure, the model, must be specified:

Using vector notation

$$\boldsymbol{c} = \begin{bmatrix} c_1 \\ c_2 \\ c_3 \\ \vdots \end{bmatrix} \quad \boldsymbol{x} = \begin{bmatrix} x_1 \\ x_2 \\ x_3 \\ \vdots \end{bmatrix},$$

then the relation can be stated as

$$z = f(\boldsymbol{c}, \boldsymbol{x}) \tag{A6}$$

Note that the use of vector notation does not imply that the function is linear.

The function of Equation (A3) is explicit, meaning that the value of z is not required to calculate the value of z. The right-hand side can be independently calculated and the value assigned to z. Here is a simple relation that is implicit in z:

$$\ln(z) = \sqrt{z} + a + bx \tag{A7}$$

It could be rearranged to explicitly solve for the value of x given a z-value, but if x is the independent variable, then there is not a way to explicitly solve for z. It might be generically represented as

$$z = f(c_1, c_2, x, z) \tag{A8}$$

We are interested in the function, and there is no need to call it z, because the properties of $f(c_1, c_2, c_3, c_4, x_1, x_2, z)$ or in vector notation $f(\boldsymbol{c}, \boldsymbol{x}, z)$ is our object of study.

Sometimes we search for the best x values (what engine thrust, plane heading, and flap setting are required to arrive on time with minimum fuel consumption?) (what temperature is right to complete a reaction with minimum waste by-product?). In such cases the decision variables (DVs) are elements of the $\boldsymbol{x}$ vector. However, in modeling or system identification, we adjust the coefficient values to get a best fit of model-to-data. Then elements of $\boldsymbol{c}$ are the DVs.

A.2 Taylor Series Expansion (Single Variable)

A Taylor series expands a function to an infinite series of terms referenced to a base value of the argument variable. For a function of a single variable, $f(x)$:

$$f(x) = f(x_0) + \frac{1}{1!} \frac{df}{dx}\bigg|_{x_0} (x - x_0) + \frac{1}{2!} \frac{d^2 f}{dx^2}\bigg|_{x_0} (x - x_0)^2 + \cdots \tag{A9}$$

This can be expressed in alternate ways:

$$f(x) = f(x_0) + \sum_{i=1}^{\infty} \frac{1}{i!} \frac{d^i f}{dx^i}\bigg|_{x_0} (x-x_0)^i \tag{A10}$$

or

$$f(x) = \sum_{i=0}^{\infty} \frac{1}{i!} \frac{d^i f}{dx^i}\bigg|_{x_0} (x-x_0)^i \tag{A11}$$

Also the function deviation can be expressed as its dependence on the x-deviation:

$$\Delta f = f(x) - f(x_0) = \sum_{i=1}^{\infty} \frac{1}{i!} \frac{d^i f}{dx^i}\bigg|_{x_0} (x-x_0)^i = \sum_{i=1}^{\infty} \frac{1}{i!} \frac{d^i f}{dx^i}\bigg|_{x_0} (\Delta x)^i \tag{A12}$$

If $x - x_0$ is less than unity, then $(x-x_0)^2$ is smaller yet, and $(x-x_0)^i$ "vanishes." If the deviation is small enough, then one can reasonably approximate $f(x)$ with a linear model. More generally, if the product of the exponentiated x-deviation with the derivative, scaled by the factorial, is vanishingly small after the first few terms, then $f(x)$ can be reasonably approximated with a linear model:

$$f(x) \cong f(x_0) + \frac{df}{dx}\bigg|_{x_0} (x-x_0) \tag{A13}$$

Note that x_0, $f(x_0)$, and $df/dx|_{x_0}$ are each constants. Although they are dependent on the value of x_0, they are independent of the choice of x. Rearranging to group like terms reveals the conventional linear relation:

$$f(x) \cong \left[f(x_0) - \frac{df}{dx}\bigg|_{x_0} x_0 \right] + \left[\frac{df}{dx}\bigg|_{x_0} \right] x = a + bx \tag{A14}$$

Figure A.1 illustrates a function value w.r.t. the x-value. And the linear relations, which are tangent to function at the several x_0 points, represent the truncated Taylor series relations.

There are many useful rearrangements and interpretations of this locally valid linear model. First, when expressed as a deviation of the function w.r.t. a deviation in x,

$$\Delta f(x) = f(x) - f(x_0) \cong \frac{df}{dx}\bigg|_{x_0} (x-x_0) = \frac{df}{dx}\bigg|_{x_0} (\Delta x) \tag{A15}$$

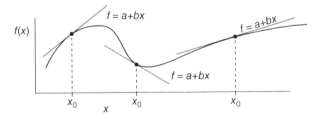

Figure A.1 Illustration of local linear approximating models.

It provides a justification for the numerical approximation to the derivative when x has a small deviation, ε, from x_0:

$$\left.\frac{df}{dx}\right|_{x_0} \cong \left.\frac{\Delta f(x)}{\Delta x}\right|_{x_0} = \frac{f(x)-f(x_0)}{x-x_0} = \frac{f(x_0+\varepsilon)-f(x_0)}{\varepsilon} \tag{A16}$$

Also, one realizes that the analytical derivative is not required in Equation (A13), but the local numerical approximation can be used:

$$f(x) \cong f(x_0) + \left.\frac{\Delta f(x)}{\Delta x}\right|_{x_0}(x-x_0) \tag{A17}$$

Or alternately, any two local $f(x)$ values can be used to estimate the local linear model, the tangent line to the function. Since,

$$f_1 = f(x_1) = a + bx_1 \tag{A18}$$

$$f_2 = f(x_2) = a + bx_2 \tag{A19}$$

When values for x_1 and x_2 are in the close vicinity of x_0, those two equations can be used to solve for the a and b coefficients:

$$b = \frac{f_2 - f_1}{x_2 - x_1} \tag{A20}$$

$$a = f_1 - bx_1 \tag{A21}$$

A.3 Taylor Series Expansion (Multiple Variable)

The analysis for a function of multiple variables is similar. The Taylor series includes the mixed derivatives:

$$f(\boldsymbol{x}) = f(x_1, x_2, x_3 \ldots)$$

$$\cong f(\boldsymbol{x_0}) + \frac{1}{1!}\sum \left.\frac{\partial f}{\partial x_i}\right|_{x_0}(x_i - x_{i_0}) + \frac{1}{2!}\sum\sum \left.\frac{\partial^2 f}{\partial x_i \partial x_j}\right|_{x_0}(x_i - x_{i_0})(x_j - x_{j_0})$$

$$+ \frac{1}{3!}\sum\sum\sum \left.\frac{\partial^3 f}{\partial x_i \partial x_j \partial x_k}\right|_{x_0}(x_i - x_{i_0})(x_j - x_{j_0})(x_k - x_{k_0})$$

$$+ \text{(all higher-order terms)}$$

(A22)

It is usually assumed that $\partial^2 f/\partial x_i \partial x_j = \partial^2 f/\partial x_j \partial x_i$ and $\partial^3 f/\partial x_i \partial x_j \partial x_k = \partial^3 f/\partial x_i \partial x_k \partial x_j$ because the function is assumed continuous (no jumps or discontinuities in value, no jumps in derivative).

Keeping a quadratic model (excluding terms of order 3 and higher—cubic, quartic, etc.), the local approximation to the function is

$$f(\pmb{x}) \cong f(\pmb{x_0}) + \frac{1}{1!}\frac{\partial f}{\partial x_1}\bigg|_{x_0}(x_1 - x_{1_0}) + \frac{1}{2!}\frac{\partial^2 f}{\partial x_1^2}\bigg|_{x_0}(x_1 - x_{1_0})^2 + \cdots$$

$$+ \frac{1}{1!}\frac{\partial f}{\partial x_2}\bigg|_{x_0}(x_2 - x_{2_0}) + \frac{1}{2!}\frac{\partial^2 f}{\partial x_2^2}\bigg|_{x_0}(x_2 - x_{2_0})^2$$

$$+ \vdots \qquad\qquad\qquad\qquad\qquad\qquad\qquad\qquad\qquad (A23)$$

$$+ \frac{\partial^2 f}{\partial x_1 \partial x_2}\bigg|_{x_0}(x_1 - x_{1_0})(x_2 - x_{2_0}) + \frac{\partial^2 f}{\partial x_1 \partial x_3}\bigg|_{x_0}(x_1 - x_{1_0})(x_3 - x_{3_0}) + \cdots$$

Recognizing that the base values for the variables, and derivative evaluations at the base cases are all constants, regrouping terms into like functionalities results in a conventional quadratic local approximation to the function:

$$y \cong a + bx_1 + cx_2 + dx_3 + \cdots + ex_1^2 + fx_2^2 + \cdots + gx_1x_2 + hx_1x_3 + \cdots \qquad (A24)$$

This is often termed a surrogate model. The local quadratic functionality is assumed in many optimization algorithms. This reduced-order representation is just a local convenience; it is not an equivalent alternate for the entire range of the original function.

A.4 Evaluating First Derivatives at x_0

The function may be too difficult to yield analytical derivatives or partial derivatives. So, use finite difference methods to approximate the derivative values. The curved line in Figures A.2a and b illustrate a function response to a single variable. The solid line represents the true slope at x_0, and the dashed lines represent the numerical derivative approximation for a forward difference (Figure A.2a) and central difference (Figure A.2b).

For the forward difference approximation, use the function value at x_0 and at $(x_0 + \Delta x)$:

$$\frac{df}{dx} \approx \frac{\Delta f}{\Delta x} = \frac{f(x_0 + \Delta x) - f(x_0)}{(x_0 + \Delta x) - x_0} = \frac{f(x_0 + \Delta x) - f(x_0)}{\Delta x} \qquad (A25)$$

A backward difference approximation is similar but uses the function value at x_0 and at $(x_0 - \Delta x)$:

$$\frac{df}{dx} \approx \frac{\Delta f}{\Delta x} = \frac{f(x_0) - f(x_0 - \Delta x)}{x_0 - (x_0 - \Delta x)} = \frac{f(x_0) - f(x_0 - \Delta x)}{\Delta x} \qquad (A26)$$

Both forward and backward finite differences use the value of the function at the base value of x_0. Here Δx is presumed to have a positive value. If Δx has a negative value, the backward and forward finite differences are reversed.

By contrast, a central difference approximation uses x-values on either side of the base value:

$$\frac{df}{dx} \approx \frac{f(x_0 + \Delta x) - f(x_0 - \Delta x)}{2\Delta x} \qquad (A27)$$

(a)

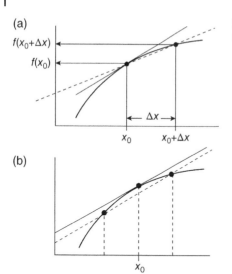

Figure A.2 (a) Illustrating a forward finite difference and (b) a central difference.

(b)

The advantage of the central difference, as Figure A.2 illustrates, is that the numerical estimate of the slope is a better approximation to the true slope. However, the disadvantage is that if the value of the function at $x_0, f(x_0)$, is already known, then the central difference method requires two additional function evaluations while the forward or backward only requires one more.

Each approach is reasonable and provides better values (truer estimates of the slope) as $\Delta x \to 0$. In any case the user must choose Δx. Too large a value for Δx leads to poor approximations. But too small a value for Δx leads to numerical truncation error in the slope estimate. A reasonable Δx value will be small relative to the convergence threshold or precision desired for the x-value, but not smaller than necessary. I recommend that Δx for numerical estimates of derivatives be one or two orders of magnitude smaller than the precision with which the user desires: $\Delta x = 0.1\varepsilon$ or $\Delta x = 0.01\varepsilon$.

A.5 Partial Derivatives: First Order

The partial derivative is represented as $\partial f / \partial x_i$ or more explicitly as $\partial f / \partial x_i|_{x_j}$. The meaning is that the derivative represents the sensitivity of the function to the change in one variable, x_i, when values of all other variables $x_j, j \neq i$ remain constant (unchanged). This notation necessarily indicates that f is a function of more than one variable.

If a function of two variables is

$$f(\mathbf{x}) = a + bx_1 + cx_1x_2 + d\ln(x_2) \tag{A28}$$

an analytical example, in which the partial derivatives are simply the derivative of the function w.r.t. each variable, assuming that the other variable is a constant, would be

$$\frac{\partial f}{\partial x_1} = b + cx_2 \tag{A29}$$

$$\frac{\partial f}{\partial x_2} = cx_1 + \frac{d}{x_2} \tag{A30}$$

Figure A.3 Illustration of the meaning of a partial derivative.

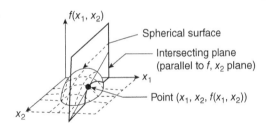

And when evaluated at a base case $\boldsymbol{x}_0 = (x_{10}, x_{20})$,

$$\left.\frac{\partial f}{\partial x_1}\right|_{\boldsymbol{x}_0} = b + cx_{20} \tag{A31}$$

$$\left.\frac{\partial f}{\partial x_2}\right|_{\boldsymbol{x}_0} = cx_{10} + \frac{d}{x_{20}} \tag{A32}$$

What is the meaning of a partial derivative? Figure A.3 illustrates a spherical-like surface poking upward through the x_1, x_2 plane. The vertical plane intersecting the function at the x_{10} value is parallel to the $x_2, f(\boldsymbol{x})$ plane. Every point in the intersecting plane has the $x_1 = x_{10}$ value. The curve in the intersecting x_{10} plane reveals how the function changes with x_2 value, keeping $x_1 = x_{10}$. The point indicated on the curve in the intersecting plane is the function value at the base case (x_{10}, x_{20}), and the tangent to the curve in the intersecting plane is illustrated as the line. The slope of that line in the $x_2, f(\boldsymbol{x})$ plane is the partial derivative of the function w.r.t. x_2 at the (x_{10}, x_{20}) location.

If a function is analytically tractable, then as shown previously the partial derivatives can be analytically derived. However, often the objective function in optimization is not kind to such calculus analysis; and numerical estimation of the partial derivatives is the more convenient, universally applicable, and often only way to obtain a value.

First-order partial derivatives can be estimated by either forward, backward, or central differences:

$$\text{Forward } \frac{\partial f}{\partial x_i} \approx \frac{f\left(x_{1_0}, x_{2_0}, \ldots, x_{i_0} + \Delta x_i, \ldots, x_{N_0}\right) - f\left(\boldsymbol{x}_0\right)}{\Delta x_i} \tag{A33}$$

$$\text{Backward } \frac{\partial f}{\partial x_i} \approx \frac{f\left(\boldsymbol{x}_0\right) - f\left(x_{1_0}, x_{2_0}, \ldots, x_{i_0} - \Delta x_i, \ldots, x_{N_0}\right)}{\Delta x_i} \tag{A34}$$

$$\text{Central } \frac{\partial f}{\partial x_i} \approx \frac{f\left(\ldots x_{i_0} + \Delta x_i \ldots\right) - f\left(\ldots x_{i_0} - \Delta x_i \ldots\right)}{2\Delta x_i} \tag{A35}$$

A.6 Partial Derivatives: Second Order

There are similar representations for second-order partial derivatives, but the central difference approach is usually preferred. First, consider a homogeneous second-order partial derivative, the second derivative w.r.t. one x-variable. Here, the second-order derivative is represented as a sensitivity of

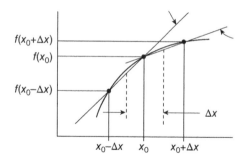

$f(x_0+\Delta x)$

$f(x_0)$

$f(x_0-\Delta x)$

Δx

$x_0-\Delta x \quad x_0 \quad x_0+\Delta x$

Figure A.4 Illustrating second derivative as the difference between sequential derivatives.

the slope w.r.t. the same variable, and the outer (first) derivative is evaluated as a forward difference and the inner with a backward difference. The result is the central difference approximation:

$$\frac{\partial^2 f}{\partial x^2} = \frac{\partial}{\partial x}\frac{\partial f}{\partial x} \cong \frac{\partial f/\partial x|_{x_i+\Delta x_i} - \partial f/\partial x|_{x_0}}{\Delta x} \cong \frac{(f_{x_0+\Delta x}-f_{x_0})/\Delta x - (f_{x_0}-f_{x_0-\Delta x})/\Delta x}{\Delta x} = \frac{f_{x_0+\Delta x}-2f_{x_0}+f_{x_0-\Delta x}}{\Delta x^2}$$

(A36)

The concept is illustrated in Figure A.4.

Although one could use other combinations of finite differences for second-order estimate (such as two central differences or two forward), the result of the forward–backward choice minimizes the number of function evaluations and keeps the function evaluations at localized and balanced locations.

The mixed second-order derivative represents how the slope w.r.t. one variable changes w.r.t another. In Figure A.3 the sensitivity of the function to x_2 is indicated by the tangent line in the intersecting plane. If the question is "How much does that slope change if the intersecting plane moves left or right to a near-by x_1 value?," the answer is the mixed second-order derivative.

A common assumption is that the order of differentiation is not important (that the function and derivatives are continuous in the vicinity). Then expanding each derivative in a central difference form is the most convenient and locally true estimation. The two-stage operation

$$\frac{\partial^2 f}{\partial x_1 \partial x_2} = \frac{\partial^2 f}{\partial x_2 \partial x_1} = \frac{\partial}{\partial x_1}\left(\frac{\partial f}{\partial x_2}\right) \cong \frac{\partial f/\partial x_2|_{x_1+\Delta x_1} - \partial f/\partial x_2|_{x_1-\Delta x_1}}{2\Delta x_1}$$

$$\cong \frac{\left(f\left(x_1^+\ x_2^+\right)-f\left(x_1^+\ x_2^-\right)\right)/2\Delta x_2 - \left(f\left(x_1^-\ x_2^+\right)-f\left(x_1^-\ x_2^-\right)\right)/2\Delta x_2}{2\Delta x_1}$$

(A37)

results in

$$\frac{\partial^2 f}{\partial x_1 \partial x_2} \approx \frac{f\left(x_1^+\ x_2^+\right)-f\left(x_1^+\ x_2^-\right)-f\left(x_1^-\ x_2^+\right)+f\left(x_1^-\ x_2^-\right)}{4\Delta x_1 \Delta x_2}$$

(A38)

where $x_i^+ = x_{i_0} + \Delta x_i$, and $x_i^- = x_{i_0} - \Delta x_i$.

A.7 Linear Algebra Notations

Vector and matrix notation is convenient in representing many of the procedures fundamental to optimization. A vector is simply a listing and ordered array of variables, coefficients, terms, or operators. By contrast, a scalar has one value; it could be considered a one-element vector. Here are vector examples. The $\boldsymbol{x}$ vector would represent variables, and the $\boldsymbol{c}$ vector a collection of coefficient values. The gradient vector, ∇, is an operator vector; it is a set of instructions, but alone its elements have no numerical values. This list of instructions, "take the partial derivatives," is alternately termed the Jacobean or del operator. The partial derivatives must operate on something. If they are operating on the function (a scalar), then the meaning of ∇f is as indicated:

$$
\boldsymbol{x} = \begin{bmatrix} x_1 \\ x_2 \\ x_3 \\ \vdots \end{bmatrix} \quad \boldsymbol{c} = \begin{bmatrix} c_1 \\ c_2 \\ c_3 \\ \vdots \end{bmatrix} \quad \nabla = \begin{bmatrix} \dfrac{\partial}{\partial x_1} \\[2mm] \dfrac{\partial}{\partial x_2} \\[2mm] \dfrac{\partial}{\partial x_3} \\[2mm] \vdots \end{bmatrix} \quad \nabla f = \begin{bmatrix} \dfrac{\partial}{\partial x_1} \\[2mm] \dfrac{\partial}{\partial x_2} \\[2mm] \dfrac{\partial}{\partial x_3} \\[2mm] \vdots \end{bmatrix} f = \begin{bmatrix} \dfrac{\partial f}{\partial x_1} \\[2mm] \dfrac{\partial f}{\partial x_2} \\[2mm] \dfrac{\partial f}{\partial x_3} \\[2mm] \vdots \end{bmatrix} \tag{A39}
$$

Note that scalars can either pre- or post-multiply a vector of scalar elements. However, for an operator vector, such as ∇, the representation $f\nabla$ would mean that elements are $f(\partial/\partial x_i)$, keeping the function that the derivative operates on unspecified. The operator comes prior to its argument, for example, sin (30), not 30sin. Similarly, ∇f, not $f\nabla$.

A vector dot product indicates that corresponding terms in two vectors are multiplied and the products are summed:

$$
\boldsymbol{c} \cdot \boldsymbol{x} = \boldsymbol{c}^T \cdot \boldsymbol{x} = \begin{bmatrix} c_1 & c_2 & c_3 & \cdots \end{bmatrix} \cdot \begin{bmatrix} x_1 \\ x_2 \\ x_3 \\ \vdots \end{bmatrix} = c_1 x_1 + c_2 x_2 + \cdots = \sum c_i x_i \tag{A40}
$$

$$
\boldsymbol{c} \cdot \nabla f = \begin{bmatrix} c_1 & c_2 & c_3 & \cdots \end{bmatrix} \cdot \begin{bmatrix} \dfrac{\partial}{\partial x_1} \\[2mm] \dfrac{\partial}{\partial x_2} \\[2mm] \vdots \end{bmatrix} f(\boldsymbol{x}) = c_1 \frac{\partial f}{\partial x_1} + c_2 \frac{\partial f}{\partial x_2} + \cdots = \sum c_i \frac{\partial f}{\partial x_i} \tag{A41}
$$

Notes: A dot product of two vectors generates a scalar. The number of elements in the two vectors must be the same. Again, operator vectors must have an argument. $f\boldsymbol{c} \cdot \nabla = fc_1(\partial/\partial x_1) + fc_2(\partial/\partial x_2) + \cdots$ makes no sense. The symbol of three vectors in a dot product, $\boldsymbol{c} \cdot \boldsymbol{x} \cdot \boldsymbol{z}$, also does not make sense. The dot product of two vectors is a scalar, and the scalar multiplies each term in a vector, so the notation $\boldsymbol{c} \cdot \boldsymbol{x} \cdot \boldsymbol{z}$ could represent either $\boldsymbol{c}$ times a scalar or $\boldsymbol{z}$ times a scalar. The representation $(\boldsymbol{c} \cdot \boldsymbol{x})\boldsymbol{z}$ is meaningful, representing a scalar pre-multiplying the vector $\boldsymbol{z}$.

A matrix is a two-dimensional array of elements. In optimization it is often the matrix of mixed second derivatives, termed the Hessian matrix:

$$\mathbf{H}f = \begin{bmatrix} \dfrac{\partial^2 f}{\partial x_1{}^2} & \dfrac{\partial^2 f}{\partial x_1 \partial x_2} & \dfrac{\partial^2 f}{\partial x_1 \partial x_3} & \cdots \\[2ex] \dfrac{\partial^2 f}{\partial x_2 \partial x_1} & \dfrac{\partial^2 f}{\partial x_2{}^2} & \dfrac{\partial^2 f}{\partial x_2 \partial x_3} & \cdots \\[2ex] \dfrac{\partial^2 f}{\partial x_3 \partial x_1} & \dfrac{\partial^2 f}{\partial x_3 \partial x_2} & \dfrac{\partial^2 f}{\partial x_3{}^2} & \cdots \\[2ex] \vdots & \vdots & \vdots & \ddots \end{bmatrix} = \left[\dfrac{\partial^2 f}{\partial x_i \partial x_j} \right] \tag{A42}$$

Unfortunately, in optimization, the gradient-squared symbol is often used to represent a matrix of mixed second-order derivatives. But in numerical analysis this is termed the Laplacian and is a scalar:

$$\nabla^2 f = \nabla \cdot \nabla f = \frac{\partial^2 f}{\partial x_1^2} + \frac{\partial^2 f}{\partial x_2^2} + \cdots = \text{Laplacian operator} = L(f) \tag{A43}$$

Accordingly, this book uses "I'll use the **H**, not the ∇^2 notation."

Vectors and matrices can provide short-hand notation for common linear algebra procedures. Defining a matrix **M** as a collection of m_{ij} elements,

$$\mathbf{M} = \begin{bmatrix} m_{11} & m_{12} & m_{13} & \cdots \\ m_{21} & m_{22} & m_{23} & \cdots \\ m_{31} & m_{32} & m_{33} & \cdots \\ \vdots & \vdots & \vdots & \ddots \end{bmatrix} = \left[m_{ij} \right] \tag{A44}$$

The product of a matrix post-multiplied by a vector is another vector:

$$\mathbf{M} \cdot \mathbf{c} = \begin{bmatrix} m_{11} & m_{12} & \cdots \\ m_{21} & m_{22} & \cdots \\ \vdots & \vdots & \ddots \end{bmatrix} \cdot \begin{bmatrix} c_1 \\ c_2 \\ \vdots \end{bmatrix} = \begin{bmatrix} c_1 m_{11} + c_2 m_{12} + c_3 m_{13} + \cdots \\ c_1 m_{21} + c_2 m_{22} + c_3 m_{23} + \cdots \\ c_1 m_{41} + c_2 m_{42} + c_3 m_{43} + \cdots \\ \vdots \end{bmatrix} = \boldsymbol{\nu} \tag{A45}$$

The length of the $\boldsymbol{\nu}$ vector (the number of elements in a column) and the width of the matrix (number of columns in each row) must be identical. The number of elements in the vector $\mathbf{c}$ and the number of elements in each matrix row have a one-to-one correspondence. However, the number of rows in the matrix is independent of the number of its columns.

A.8 Algebra and Assignment Statements

In the computer, $x = x + 1$ is an executable procedure, and each execution of that line increments x by 1. If x starts with a value of 0, the sequence of x-values is $\{0, 1, 2, 3, 4, \ldots\}$.

But, in algebra, $x = x + 1$ looks exactly the same and could be solved as $x = $ infinity: Group like terms, $x - x = 1$, combine coefficients, $x(1 - 1) = 1$, and solve for x, $x = 1/(1 - 1) = 1/0$. Alternately, subtracting x from both sides of $x = x + 1$ results in $0 = 1$, which makes no sense.

The difference is that the subscript is missing in the procedure statement. Algebraically, the equation should be written as

$$x_i = x_{i-1} + 1 \tag{A46}$$

where i is the iteration or sequence counter.

In code it should be

$$x(i) = x(i-1) + 1 \tag{A47}$$

But this requires an array and dedicated storage. If there is no need to remember all the past x-values, and if only the updated x-value is needed, then you can use an assignment statement to update the x-value:

$$x = x + 1 \tag{A48}$$

But this looks like nonsensical algebra, so often I'll indicate assignment statements or computer procedures with the := symbol:

$$x := x + 1 \tag{A49}$$

Appendix B

Root Finding

B.1 Introduction

Root finding is a basic tool for optimization as well as for determining values in mathematical relationships. I'll first develop it as a method to determine dependent variable values in a relation.

Explicit Equation: In an explicit equation, $y(x)$, given a value of x you can arrange the relation so that y is isolated on the LHS. Then given a value for x, you can execute the RHS procedure to explicitly calculate the value for y. For example,

$$y = 17 + 3 \sin(\theta) \tag{B1}$$

Given a value of θ, you can explicitly solve for the value of y. The equation looks like an assignment statement in computer programming. Calculate the RHS value and then assign it to the y storage location.

Implicit Equation: By contrast in an implicit relation, you cannot isolate the dependent variable. For instance, if z is a function of x,

$$z = \ln(z) + 5\sqrt{z} + x \tag{B2}$$

Given an x-value you cannot explicitly isolate z to LHS. The value of z is required to calculate the value of z.

However, either implicit or explicit relations can be rearranged to describe a function that equals zero by subtracting the RHS from both sides of the equation. The RHS becomes zero:

$$f_1(y,\theta) - y - 17 - 3\sin(\theta) = 0 \tag{B3}$$
$$f_2(z,x) = z - \left[\ln(z) + 5\sqrt{z} + x\right] = 0 \tag{B4}$$

Now you have a function of the two variables, and the right values for the variables make the LHS have a value of zero. This is the canonical form of a relation for root finding.

Graph the value of f w.r.t. the dependent value for a range of dependent variable values, such as illustrated in Figure B.1a and b.

Mostly, $f \neq 0$. However, the value of the independent variable that makes $f = 0$ is the right value of the independent variable. Note that $f > 0$ on one side of the root and $f < 0$ on the other side.

Root finding is the procedure of finding the value of dependent variable that makes $f = 0$.

There may be multiple roots. Consider the polynomial

Engineering Optimization: Applications, Methods, and Analysis, First Edition. R. Russell Rhinehart.
© 2018 R. Russell Rhinehart. Published 2018 by John Wiley & Sons Ltd.
Companion website: www.wiley.com/go/rhinehart/engineeringoptimization

(a)

(b)

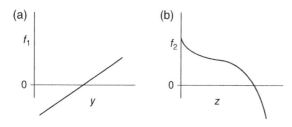

Figure B.1 Illustrations of the function w.r.t. the dependent variable: (a) linear and (b) nonlinear.

$$f(x) = (x-3)(x+1)(x+2) = 0 \tag{B5}$$

In this form it may be obvious that the roots (the values of that make the LHS = 0) are $x = -2, -1, +3$. By multiplying the binomials, the relation can be rearranged to

$$x = \frac{(6-x^3)}{7} \tag{B6}$$

In this form the roots are not obvious, although they are unchanged. With the RHS subtracted from both sides rearranged to the canonical form for root finding,

$$f(x) = x - \frac{(6-x^3)}{7} = 0 \tag{B7}$$

Figure B.2 graphs Equation (B7). Note that the roots are $x = -2, -1, +3$ as expected.
If you cannot use analytical methods to determine the roots, then use numerical methods as follows.

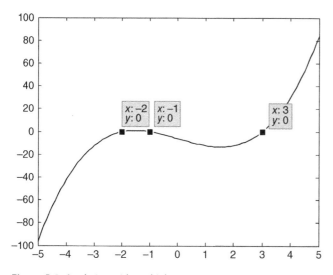

Figure B.2 A relation with multiple roots.

B.2 Interval Halving: Bisection Method

This is a two-part procedure. First, bound the root (determine an interval that contains it) by a marching method. Then once bound use interval halving to find the root. To bound the root, start at x_{low}, and evaluate $f(x)$. Increment x by Δx. Using k as the index for each trial,

$$x_k = x_{k-1} + \Delta x \qquad (B8)$$

Evaluate $f(x_k) = f_k$. If $f_k \cdot f_{k-1} < 0$, then you have bound the root between x_{k-1} and x_k. This is illustrated in Figure B.3.

The root is in either the "+" half or the "−" half of the Δx interval, illustrated as $[x_3, x_4]$ in Figure B.3. Relabel the interval bounds as x_{low}, x_{high}, as illustrated in Figure B.4.

Evaluate the function at the mid-value of x, $x_{mid} = (x_{low} + x_{high})/2$, and determine the value of $f(x_{mid})$. If f_{mid} has the same sign as f_{high}, then the root is in the left half. Reset

$$x_{high} \leftarrow x_m \qquad (B9)$$

$$f_{high} \leftarrow f_m \qquad (B10)$$

Otherwise, the root is in the right half reset x_{low} and f_{low}.

This is an iterative procedure. It follows the same procedure at each iteration or cycle. The relation is termed a recursive relation.

In general, this technique never finds the exact value. However, the number of times the interval is halved relates to the uncertainty about the root value:

$$\text{Uncertainty} = \text{bound} = \text{range} = \Delta x \left(\frac{1}{2}\right)^N \qquad (B11)$$

Figure B.3 Illustration of interval halving stage 1—the marching method.

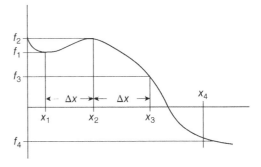

Figure B.4 Illustration of interval halving stage 2—bisection method.

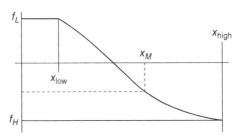

where Δx is the original marching step interval and N is the number of interval halving stages (iterations).

Report the x_{mid} of the Nth iteration.

Choose N for the desired uncertainty on the root value. First, define an acceptable uncertainty range based on the needs of the application. This is the range between x_{high} and x_{low} when the iterative procedure stops because it is close enough to the true value. Epsilon is the half range:

$$\epsilon = \frac{1}{2} \text{range} \qquad (B12)$$

since $x_{mid} \pm \epsilon$ bounds x_{root}.

Choose ϵ and then calculate the required number of stages, N, by desiring $2\epsilon = \Delta x (1/2)^N$, which results in

$$N = \frac{\ln(\Delta x / 2\epsilon)}{\ln(2)} \qquad (B13)$$

B.3   Newton's: Secant Version

For functions that express a well-behaved trend near the root and the initial guess that is near to the root value, the secant method (a variant of Newton's method) is very efficient. Figure B.5 illustrates the concept.

Start with two initial guesses for the root, x_1 and x_2, and evaluate the function at those two points, f_1 and f_2. These are indicated as points A and B on Figure B.5. Point C is the intersection of the f_2 value at the x_1 value. Triangle ABC is a right triangle. Point D is the intersection of the line AB on the $f(x) = 0$ axis and will become the next guess, x_3. Point E is the value of x_2 on the $f(x) = 0$ axis. Triangle BDE is similar to triangle ABC, and using the geometric relations,

$$\frac{ED}{CB} = \frac{BE}{AC} \qquad (B14)$$

Expressed in x and f terms,

$$\frac{x_3 - x_2}{x_2 - x_1} = \frac{f_2 - 0}{f_1 - f_2} \qquad (B15)$$

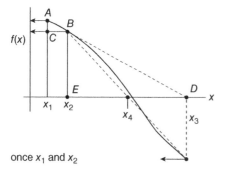

Figure B.5 Illustration of the secant version of Newton's method of root finding.

Figure B.6 Illustrating a misdirection that a Newton type of algorithm can produce.

And rearranged to solve for the next guess, x_3,

$$x_3 = x_2 - \frac{f_2}{(f_2 - f_1)/(x_2 - x_1)} \tag{B16}$$

Then x- and function-values at points 2 and 3 will be used to determine point 4. This is an iterative procedure, outlined as follows:

- Choose x_1 and x_2 (initial human guesses) and evaluate f_1 and f_2.
- Determine the next guess from the recursion formula (exactly the same formula used repeatedly to update or improve the estimate) to continue

$$x_{k+1} = x_k - \frac{f_k}{(f_k - f_{k-1})/(x_k - x_{k-1})} \tag{B17}$$

- Stop when $x_{k+1} - x_k <$ tolerance or $f(x) <$ tolerance.

Note that the denominator in the recursion formula is the finite difference estimate of the derivative of the function w.r.t. x. Using the calculus derivative notation reveals the classical Newton's method for root finding:

$$x_{k+1} = x_k - \frac{f_k}{(df/dx)_k} \tag{B18}$$

If the function has curvature, neither the secant method nor Newton's method jumps to the root. Both are iterative. The advantages of the secant method are that (i) it is not necessary to obtain an analytical derivative and (ii) it saves a function evaluation at each step. Since both are iterative, I prefer the analytically and computationally simpler, more robust secant method.

Near the root, Newton's (or the secant method) converges faster than interval halving. However, far away from the root, or where the function is not rationally directing the search toward the root, a Newton's type of algorithm can send the search in ridiculous directions, as Figure B.6 illustrates.

So, for root finding, start with a marching approach to bound the root. Then use either interval halving or the secant method to converge on the root. Stop the iterations and claim the root has been found when the convergence criterion has been met.

B.4 Which to Choose?

My preference is interval halving over a Newton's method. Interval halving might take a few more computer iterations; but (i) it can guarantee convergence to a tolerance within N iterations, (ii) it is not confounded by shapes that Newton's cannot cope with (poles, discontinuities, inflections), and (iii) it does not require an analytical derivative (extra human and computer work). Optimization

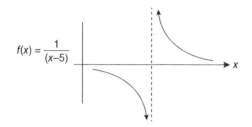

Figure B.7 Illustration of a pole in a function that could misdirect interval halving.

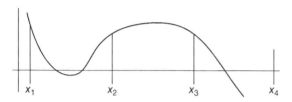

Figure B.8 The marching method may miss a root.

is making the best choices that balance all issues, not just one issue, nor just for a particular class of problems. Analytical sophistication is not an issue that should be considered in an applications context.

However, interval halving may find a "pole" as illustrated in Figure B.7, which is constructed from the simple relation $f(x) = 1/(x-5)$.

So, the declaration of a root should be based on both the function value and the x-range.

Regardless of the root-finding algorithm, if there are, or could be, multiple roots, continue the marching method until you find another function sign change.

However, the marching method may miss roots if Δx is too big. Figure B.8 illustrates a marching method that skips over the first two roots. However, if Δx is too small, it adds computational burden.

User knowledge is required to choose Δx, ϵ, tolerance, and total search range over x values.

Appendix C

Gaussian Elimination

C.1 Linear Equation Sets

Often optimization procedures result in a set of N linear equations with N unknown variable values. Gaussian elimination is one method for determining the variable values.

There are diverse ways to present a system of linear equations, but these are simply notational variations. One is a classic reveal of each equation. Equation set (C1) is an example with three unknowns, x, y, and z. The coefficients a, b, and c, and the right-hand side elements, r, all have known values. Note that the unknowns x, y, and z each appear linearly (to the first power and independent of other unknowns) in the equation set. Remove the third equation and the cz elements, and it is a system of two equations and two unknowns. In general it could be a system of N equations linear in N unknowns. If a variable does not appear in a particular equation, insert it with a coefficient value of zero:

$$a_1x + b_1y + c_1z = r_1$$
$$a_2x + b_2y + c_2z = r_2 \quad\quad\quad (C1)$$
$$a_3x + b_3y + c_3z = r_3$$

Note that the structure of the abovementioned equation set has all of the unknown variables and coefficients on the left-hand side, in the same order, and the constant on the right-hand side. This makes it convenient to represent in matrix–vector notation:

$$\begin{bmatrix} a_1 & b_1 & c_1 \\ a_2 & b_2 & c_2 \\ a_3 & b_3 & c_3 \end{bmatrix} \cdot \begin{bmatrix} x \\ y \\ z \end{bmatrix} = \begin{bmatrix} r_1 \\ r_2 \\ r_3 \end{bmatrix} \quad\quad (C2)$$

Although now there is only one equal sign, for the LHS matrix–vector product (a vector) to equal the RHS vector, each element on the vector must be identical, indicating the same three equations. Note that the matrix is square, the number of rows represents the number of equations, and the number of columns represents the coefficients on each unknown.

A shorthand matrix–vector representation is

$$\mathbf{M} \cdot \mathbf{x} = \mathbf{RHS} \quad\quad\quad (C3)$$

Engineering Optimization: Applications, Methods, and Analysis, First Edition. R. Russell Rhinehart.
© 2018 R. Russell Rhinehart. Published 2018 by John Wiley & Sons Ltd.
Companion website: www.wiley.com/go/rhinehart/engineeringoptimization

There is no need to have separate symbol for each unknown. Instead of x, y, and z, they can be x_1, x_2, and x_3. Similarly the matrix of coefficients could use the symbol $m_{i,j}$ for each element with i representing the row and j representing the column (counting from the upper left element):

$$\begin{bmatrix} m(1,1) & m(1,2) & m(1,3) \\ m(2,1) & m(2,2) & m(2,3) \\ m(3,1) & m(3,2) & m(3,3) \end{bmatrix} \cdot \begin{bmatrix} x(1) \\ x(2) \\ x(3) \end{bmatrix} = \begin{bmatrix} r(1) \\ r(2) \\ r(3) \end{bmatrix} \tag{C4}$$

C.2 Gaussian Elimination

Gaussian elimination is a two-part procedure that solves for the x-values of equation set (C3). The result of the first part of the procedure is to convert the system to one in which the elements of the matrix are unity along the main diagonal and zero in the lower triangle.

$$\begin{bmatrix} 1 & m'(1,2) & m'(1,3) \\ 0 & 1 & m''(2,3) \\ 0 & 0 & 1 \end{bmatrix} \cdot \begin{bmatrix} x(1) \\ x(2) \\ x(3) \end{bmatrix} = \begin{bmatrix} r'(1) \\ r''(2) \\ r'''(3) \end{bmatrix} \tag{C5}$$

With this structure the value of x_3 is easily obtained from the third equation. $x_3 = r'''_3$. With x_3 evaluated, the second to bottom equation permits a direct calculation for x_2. This is the second part of the procedure.

The first part of the procedure is to convert the equation set into the desired form.

C.2.1 Part 1

Stage 1
A) Normalize each of the coefficients in the first column to unity (1.0000...) by dividing all elements in the each equation (coefficients and RHS) by the first coefficient in that equation:

$$\begin{bmatrix} \dfrac{m(1,1)}{m(1,1)} & \dfrac{m(1,2)}{m(1,1)} & \dfrac{m(1,3)}{m(1,1)} \\ \dfrac{m(2,1)}{m(2,1)} & \dfrac{m(2,2)}{m(2,1)} & \cdots \\ \vdots & \vdots & \ddots \end{bmatrix} \cdot \begin{bmatrix} x(1) \\ x(2) \\ x(3) \end{bmatrix} = \begin{bmatrix} \dfrac{r(1)}{m(1,1)} \\ \dfrac{r(2)}{m(2,1)} \\ \vdots \end{bmatrix} \tag{C6}$$

B) Starting with row 2, subtract row 1 from row 2, and then do the same for each subsequent row:

$$\begin{bmatrix} 1 & \dfrac{m(1,2)}{m(1,1)} & \dfrac{m(1,3)}{m(1,1)} \\ 0 & \dfrac{m(2,2)}{m(2,1)} - \dfrac{m(1,2)}{m(1,1)} & \dfrac{m(2,3)}{m(2,1)} - \dfrac{m(1,3)}{m(1,1)} \\ \vdots & \vdots & \ddots \end{bmatrix} \cdot \begin{bmatrix} x(1) \\ x(2) \\ x(3) \end{bmatrix} = \begin{bmatrix} \dfrac{r(1)}{m(1,1)} \\ \dfrac{r(2)}{m(2,1)} - \dfrac{r(1)}{m(1,1)} \\ \vdots \end{bmatrix} \tag{C7}$$

C) For convenience, at the end of this first stage, represent the coefficients and RHS values with a prime:

$$
\begin{bmatrix}
1 & m'(1,2) & m'(1,3) \\
0 & m'(2,2) & m'(2,3) \\
0 & m'(3,2) & m'(3,3)
\end{bmatrix}
\cdot
\begin{bmatrix}
x(1) \\
x(2) \\
x(3)
\end{bmatrix}
=
\begin{bmatrix}
r'(1) \\
r'(2) \\
r'(3)
\end{bmatrix}
\tag{C8}
$$

Note: The upper right element in the matrix has the desired value of unity and all coefficients below it have the desired values of zero.

Stage 2

A) Normalize coefficients in the second column (starting with the second row) to unity by dividing each element in the equation by the second coefficient. Do this for all $N - 1$ rows starting with the second:

$$
\begin{bmatrix}
1 & m'(1,2) & m'(1,3) \\
0 & \dfrac{m'(2,2)}{m'(2,2)} & \dfrac{m'(2,3)}{m'(2,2)} \\
0 & \dfrac{m'(3,2)}{m'(3,2)} & \dfrac{m'(3,3)}{m'(3,2)}
\end{bmatrix}
\cdot
\begin{bmatrix}
x(1) \\
x(2) \\
x(3)
\end{bmatrix}
=
\begin{bmatrix}
r'(1) \\
\dfrac{r'(2)}{m'(2,2)} \\
\dfrac{r'(3)}{m'(3,2)}
\end{bmatrix}
\tag{C9}
$$

B) Subtract the second row from each lower row:

$$
\begin{bmatrix}
1 & m'(1,2) & m'(1,3) \\
0 & 1 & \dfrac{m'(2,3)}{m'(2,2)} \\
0 & 0 & \dfrac{m'(3,3)}{m'(3,2)} - \dfrac{m'(2,3)}{m'(2,2)}
\end{bmatrix}
\cdot
\begin{bmatrix}
x(1) \\
x(2) \\
x(3)
\end{bmatrix}
=
\begin{bmatrix}
r'(1) \\
\dfrac{r'(2)}{m'(2,2)} \\
\dfrac{r'(3)}{m'(3,2)} - \dfrac{r'(2)}{m'(2,2)}
\end{bmatrix}
\tag{C10}
$$

C) And rename the coefficients with a double prime:

$$
\begin{bmatrix}
1 & m'(1,2) & m'(1,3) \\
0 & 1 & m''(2,3) \\
0 & 0 & m''(3,3)
\end{bmatrix}
\cdot
\begin{bmatrix}
x(1) \\
x(2) \\
x(3)
\end{bmatrix}
=
\begin{bmatrix}
r'(1) \\
r''(2) \\
r''(3)
\end{bmatrix}
\tag{C11}
$$

Stage 3 (The Last Stage in an $N = 3$ Equation Set)

A) Normalize remaining coefficient in last row:

$$
\begin{bmatrix}
1 & m'(1,2) & m'(1,3) \\
0 & 1 & m''(2,3) \\
0 & 0 & \dfrac{m''(3,3)}{m''(3,3)}
\end{bmatrix}
\cdot
\begin{bmatrix}
x(1) \\
x(2) \\
x(3)
\end{bmatrix}
=
\begin{bmatrix}
r'(1) \\
r''(2) \\
\dfrac{r''(3)}{m''(3,3)}
\end{bmatrix}
\tag{C12}
$$

B) And relabel the coefficient that has been adjusted a third time:

$$\begin{bmatrix} 1 & m'(1,2) & m'(1,3) \\ 0 & 1 & m''(2,3) \\ 0 & 0 & 1 \end{bmatrix} \cdot \begin{bmatrix} x(1) \\ x(2) \\ x(3) \end{bmatrix} = \begin{bmatrix} r'(1) \\ r''(2) \\ r'''(3) \end{bmatrix} \tag{C13}$$

This is the desired structure that permits a simple step-by-step solution.

Note: If there were N equations, this first part would have N stages.

C.2.2   Part 2

Solve for each variable in reverse order:

$$x(3) = \frac{r'''(3)}{1} \tag{C14}$$

$$x(2) = r''(2) - m''(2,3) \cdot x(3) \tag{C15}$$

$$x(1) = r'(1) - m'(1,2) \cdot x(2) - m'(1,3) \cdot x(3) \tag{C16}$$

C.3   Pivoting

That was easy! However, there could be a problem if a leading coefficient is zero. Then you have a divide by zero in the normalization (A-steps) of each stage. Even if the original coefficient value is not zero, this could happen because it becomes zero after the stages. The solution is to resort the columns to make leading coefficient not zero. Better yet, prior to each A-step, sort the columns so that coefficient with the maximum absolute value is the lead, then this minimizes the effects of truncation error in calculations. This is termed pivoting.

For example, an original equation set might be

$$\begin{aligned} &+ 3x + 5y - 10z = 17 \\ &+ 6x - 20y + 3z = 5 \\ &- 3x + 1y + 2z = 20 \end{aligned} \tag{C17}$$

The A-step would direct that all elements in the first equation are to be divided by the x-coefficient value of "+3." But the z-coefficient value of "−10" has a larger absolute value. Accordingly, rearrange the equation set so that z is the first unknown variable in the list. This does not change the solution to the equations:

$$\begin{aligned} &- 10z + 5y + 3x = 17 \\ &+ 3z - 20y + 6x = 5 \\ &+ 2z + 1y - 3x = 20 \end{aligned} \tag{C18}$$

Note: The elements in the unknown vector have shifted order from (x, y, z) to (z, y, x). In matrix–vector notation,

$$
\begin{bmatrix} +3 & +5 & -10 \\ +6 & -20 & +3 \\ -3 & +1 & +2 \end{bmatrix} \begin{bmatrix} x \\ y \\ z \end{bmatrix} = \begin{bmatrix} 17 \\ 5 \\ 20 \end{bmatrix} \text{ becomes } \begin{bmatrix} -10 & +5 & +3 \\ +3 & -20 & +6 \\ +2 & +1 & -3 \end{bmatrix} \begin{bmatrix} z \\ y \\ x \end{bmatrix} = \begin{bmatrix} 17 \\ 5 \\ 20 \end{bmatrix} \qquad \text{(C19)}
$$

Switching coefficients to make the leading one the maximum requires switching all column elements in the coefficients matrix and row elements in solution column vector.

For simplicity, there is no need to normalize every successive row, just the one you're leaving, and then subtract that row times the leading coefficient:

$$
\begin{bmatrix} 1 & m'(1,2) & m'(1,3) \\ 0 & 1 & m'(2,3) \\ 0 & m'(3,2) & m'(3,3) \end{bmatrix} \begin{bmatrix} x(1) \\ x(2) \\ x(3) \end{bmatrix} = \begin{bmatrix} r'(1) \\ r'(2) \\ r'(3) \end{bmatrix} \qquad \text{(C20)}
$$

$$
\begin{bmatrix} 1 & m'(1,2) & m'(1,3) \\ 0 & 1 & m'(2,3) \\ 0 & m'(3,2)-1\cdot m'(3,2) & m'(3,3)-m'(3,2)\cdot m'(3,2) \end{bmatrix} \begin{bmatrix} x(1) \\ x(2) \\ x(3) \end{bmatrix} = \begin{bmatrix} r'(1) \\ r'(2) \\ r'(3)-r'(2)\cdot m'(3,2) \end{bmatrix} \qquad \text{(C21)}
$$

If a divide by zero occurs, it is because the solution is indeterminate (if two or more equations are linearly dependent). So, if a leading coefficient is zero (or equivalently zero numerically, for instance, its absolute value is less than 0.00000001), then send an "indeterminate" message and stop execution.

Since the rows operated on are at the stage number or higher, this procedure can be expressed in pseudocode.

Stage 1
- Pivot on maximum coefficient in row 1.
- Normalize *row 1* with *m(1, 1)*.
- For each subsequent row (index = 2, 3, 4, ...), subtract *row 1* * *m(index, 2)* from it.

Stage 2
- Pivot on max coefficient value in row 2, but only have to search *m(2, 2), m(2, 3), ...*
- Normalize *row 2* with *m(2, 2)*.
- For each subsequent row (index = 3, 4,...), subtract *row 2* * *m(index, 3)* from it.

In general

Stage *k*
- Pivot on max coefficient value in row *k*, but only need to search *m(k, k), m(k, k + 1), m(k, k + 2), ..., m(k, n)*.
- Normalize *row k* with *m(k, k)*.
- For each subsequent *row i* = *k* + 1 to *n*, subtract *row k* * *m(i, k + 1)* from it.

Since pivoting scrambles the solution position in the solution vector, the answer needs to be unscrambled. In the example, in Equation (C19) the solution $\{x, y, z\}$ became reordered as $\{z, y, x\}$. To keep track of the order of elements in the solution vector, I use a place vector, in which place holds the original element index.

For example, place is initialized as

$$\text{place}(1) = \boxed{1}$$
$$\text{place}(2) = \boxed{2} \qquad \qquad (C22)$$
$$\text{place}(3) = \boxed{3}$$

After the pivot, x and z are switched so the place vector becomes

$$p(1) = \boxed{3}$$
$$p(2) = \boxed{2} \qquad \qquad (C23)$$
$$p(3) = \boxed{1}$$

There is no need to have an M matrix and an M' matrix and an M'' matrix. Since each matrix and RHS vector operations update elements, just update the **m** element.

For example,

$$m(2,3) := m(2,3) - m(1,3)^* m(2,1) \qquad \qquad (C24)$$

The ":=" acknowledges that this is a computer assignment statement, not the algebraic "equals."

This column switching in the coefficient matrix to make the leading coefficient of the top working row be the coefficient with largest magnitude (absolute value) is called partial pivoting. Full pivoting also switches working rows so that the divide by value in the upper left of the working part of the matrix is the largest magnitude of all remaining in the working matrix.

C.4   Code in VBA

```
'  Gauss Elimination with Pivoting
'  R. Russell Rhinehart
'  29 June 2009, 23 Aug 2012
'  Gauss Elimination of a system of N (up to 10) linear equations with pivoting
on lead coefficient

Option Explicit
Dim H(10, 10) As Double       ' Coefficient Matrix - Hessian in Newton-Raphson
Dim V(10) As Double       ' RHS vector - Jacobean or Gradient in Newton-Raphson
Dim X(10) As Double       ' Vector of Unknowns - Change in Decision Variables in
Optimization
Dim XS(10) As Double          ' Scrambled Vector of Unkowns
Dim Place(10) As Integer       ' Position of solution vector
Dim TC As Double        ' Temporary holder for coefficient value for switching
Dim TI As Integer          ' Temporary value of column index when switching or
unscrambling
Dim Nvar As Integer        ' Number of variables, length of solution vector
Dim i As Integer         ' index counter for row
Dim k As Integer         ' index counter for row
```

```
Dim j As Integer           ' index counter for column
Dim MaxCoefficient As Double   ' value for largest coefficient in the row
Dim Index As Integer        ' k-value for largest coefficient
Dim Coefficient As Double    ' value of rearragned equation lead coefficient
Dim sum As Double           ' Sum of values in reconstructing the solution

Sub Gauss_Elimination_w_Pivot()

    Nvar = Cells(17, 13)

    ' Initialize sequence of solution vector elements
    For i = 1 To Nvar
      Place(i) = i
    Next i

    ' Read Data
    For i = 1 To Nvar
      For j = 1 To Nvar
        H(i, j) = Cells(i + 5, j + 1)
      Next j
    Next i
    For i = 1 To Nvar
      V(i) = Cells(i + 5, 13)
    Next i

    ' Perform row operations to get upper-right diagonal matrix
    For k = 1 To Nvar - 1

      ' Search for largest coefficient in row k
      MaxCoefficient = Abs(H(k, k))
      Index = k
      For j = k + 1 To Nvar
        If Abs(H(k, j)) > MaxCoefficient Then
          MaxCoefficient = Abs(H(k, j))
          Index = j
        End If
      Next j

      ' switch Index and k columns
      For i = 1 To Nvar
        TC = H(i, Index)
        H(i, Index) = H(i, k)
        H(i, k) = TC
      Next i
      TI = Place(Index)
      Place(Index) = Place(k)
```

```
      Place(k) = TI

      ' normalize lead coefficient of row k to unity - make H(k,k) = 1
      Coefficient = H(k, k)
      If Abs(Coefficient) < 0.0000000001 Then   'test for linearly
dependent equations.
         For i = 1 To Nvar
            Cells(i + 5, 17) = "Indeterminate"
         Next i
         End
      End If
      For j = 1 To Nvar
         H(k, j) = H(k, j) / Coefficient
      Next j
      V(k) = V(k) / Coefficient

      ' subtract row-k-times-lead-coefficient in each remaining row (i>k)
   - make M(i,k) = 0
      For i = k + 1 To Nvar
         Coefficient = H(i, k)
         For j = 1 To Nvar
            H(i, j) = H(i, j) - H(k, j) * Coefficient
         Next j
         V(i) = V(i) - V(k) * Coefficient
      Next i

   Next k

   ' normalize last row - row i=Nvar
   Coefficient = H(Nvar, Nvar)
   If Abs(Coefficient) < 0.0000000001 Then
      For i = 1 To Nvar
         Cells(i + 5, 17) = "Indeterminate"
      Next i
      End
   End If
   For j = 1 To Nvar
      H(Nvar, j) = H(Nvar, j) / Coefficient
   Next j
   V(Nvar) = V(Nvar) / Coefficient

   ' solve for scrambled solution values
   XS(Nvar) = V(Nvar)
   For j = Nvar - 1 To 1 Step -1
      sum = 0
      For k = j + 1 To Nvar
```

```
      sum = sum + XS(k) * H(j, k)
   Next k
   XS(j) = V(j) - sum
Next j

' Unscramble solution
For i = 1 To Nvar
   TI = Place(i)
   X(TI) = XS(i)
Next i

' Display answer
For i = 1 To Nvar
   Cells(i + 5, 17) = X(i)
Next i

' check if solution vector returns RHS vector
' Re-read original matrix coefficients
For i = 1 To Nvar
   For j = 1 To Nvar
      H(i, j) = Cells(i + 5, j + 1)
   Next j
Next i

' Calculate and display RHS
For i = 1 To Nvar
   sum = 0
   For j = 1 To Nvar
      sum = sum + X(j) * H(i, j)
   Next j
   Cells(i + 5, 14) = sum
Next i

End Sub
Sub GenerateRHS()

   Nvar = Cells(17, 13)

   For i = 1 To Nvar
      For j = 1 To Nvar
         H(i, j) = Cells(i + 5, j + 1)
      Next j
      X(i) = Cells(i + 5, 18)
   Next i

   ' Calculate and display RHS
```

```
For i = 1 To Nvar
   sum = 0
   For j = 1 To Nvar
      sum = sum + X(j) * H(i, j)
   Next j
   Cells(i + 5, 13) = sum
Next i

End Sub
```

Appendix D

Steady-State Identification in Noisy Signals

D.1   Introduction

Identification of both steady state (SS) and transient state (TS) in noisy process signals is important. Steady-state models are widely used in process control, online process analysis, and process optimization; and since manufacturing and chemical processes are inherently nonstationary, selected model parameter values need to be adjusted frequently to keep the models true to the process and functionally useful. But either the use or data-based adjustment of SS models should only be triggered when the process is at SS. Additionally, detection of SS triggers the collection of data for process fault detection, the data reconciliation, the neural network training, the end of an experimental trial (collect data and implement the next set of conditions), etc.

Often, process owners, scientists, and engineers run a sequence of experiments to collect data throughout a range of operating conditions, and process operators sequence the next stage of a trial. Each sampling event is initiated when the operator observes that steady conditions are met. Then the operator implements the new set of operating conditions. Similarly, in batch operations the end of a stage is evidenced by signals reaching their equilibrium or completion values, and when operators observe that the stage is complete, they initiate the next step in the processing sequence. However, this visual method of triggering requires continual human attention, and it is subject to human error in the recognition of steady state. Features that can compromise the visual interpretation include noisy measurements, slow process changes, multiple dynamic trends, scheduled operator duties, upcoming lunch breaks, or change-of-shift timing.

Alternately, the experimental run or batch stage can be scheduled to go to the next set of conditions at preset time intervals. Unfortunately, this method can create inefficiency if the runs are scheduled for an unnecessarily long time, or the data can be worthless if the scheduled time is insufficient for any particular set of conditions to achieve steady state. Since the time to reach steady state varies with operating conditions, it is difficult to accurately predict the necessary hold time.

If SS detection were automated, process sampling or data recording would be initiated, and after, the computer could autonomously implement the next operational stage or set of experimental conditions. But likely on the first sampling after the new signals go to the process, the process will not have responded, and the process output signals will remain at their prior steady state. To prevent this past SS from triggering the next trial, the computer should first seek a TS after implementing new conditions and then seek the resulting SS.

Engineering Optimization: Applications, Methods, and Analysis, First Edition. R. Russell Rinehart.
© 2018 R. Russell Rinehart. Published 2018 by John Wiley & Sons Ltd.
Companion website: www.wiley.com/go/rhinehart/engineeringoptimization

An automated online, real-time SS and TS identification would be useful to trigger the next stage of an experimental plan or process phase.

In the context of optimization, steady-state identification is used several ways:

1) To identify that the optimization in nonlinear regression has converged
2) To identify that the optimization in stochastic applications has converged
3) To identify when an adequate number of realizations have been acquired to provide statistical confidence that the CDF is captured

If a process signal was noiseless, then SS or TS identification would be trivial. At steady state there is no change in data value. Alternately, if there is a change in data value, the process is in a transient state.

However, since process variables are usually noisy, the identification needs to "see" through the noise and would announce probable SS or probable TS situations, as opposed to definitive SS or definitive TS situations. Accordingly, a method also needs to consider more than the most recent pair of samples to confidently make any statement.

Since the noise could be a consequence of autocorrelated trends (of infinite types), varying noise amplitude (including zero), individual spikes, non-Gaussian noise distributions, or spurious events, a useful technique also needs to be robust to such aspects. Fortunately, in applications to determining convergence in optimization, the signal is driven by independent random events (not autocorrelated), and an elementary (easy to understand, and implement, computationally fast) algorithm is applicable.

D.2 Conceptual Model

Begin with this conceptual model of the phenomena: The true process variable (PV) is at a constant value (at SS) and events create "noise," independently distributed fluctuations on the measurement. In optimization, the noise on a measurement (the OF value) may arise from Monte Carlo stochastic functions or from randomized sampling of data. In observing experimental data, such measurement perturbations could be attributed to mechanical vibration, stray electromagnetic interference in signal transmission, electronic noise, flow turbulence, etc. In simulation, the "noise" represents realization-to-realization differences. Alternately, the concept includes a situation in which the true PV is averaging at a constant value, and internal spatial nonuniformity (resulting from nonideal fluid mixing or multiphase mixtures in a boiling situation) or an internal distribution of properties (crystal size, molecular weight) create temporal changes to the local measurement.

If the noise distribution and variance would be unchanging in time, then statistics would call this time series as stationary. However, for a process, the true value, nominal value, or average may be constant in time, but the noise distribution may change. So, SS does not necessarily mean stationary in a statistical sense of the term.

The first null hypothesis of this analysis is the conventional one where the process is at steady state, Ho: SS. For the approach recommended here, the statistic, a ratio of variances, the R-statistic, will be developed. Due to the vagaries of noise, the R-statistic will have a distribution of values at SS. As long as the R-statistic value is within the normal range of a SS process, the null hypothesis cannot be rejected. When the R-statistic has an extreme value, then the null hypothesis can be rejected with a certain level of confidence, and probable TS claimed.

By contrast, there is no single conceptual model of a transient state. A transient condition could be due to a ramp change in the true value, or an oscillation, or a first-order transient to a new value, or a step change, etc. Each is a unique type of transient. Further, each single transient event type has unique characteristics such as ramp rate, cycle amplitude and frequency, and time constant and magnitude of change. Further, a transient could be comprised of any combination or sequence of the not-at-SS events. Since there is no unique model for TS, there can be no null hypothesis or corresponding statistic that can be used to reject the TS hypothesis and claim probable SS. Accordingly, an alternate approach needs to be used to claim probable SS.

The alternate approach used here is to take a transient condition that is barely detectable or decidedly inconsequential (per human judgment) and set the probably SS threshold for the R-statistic as an improbable low value, but not so low as to be improbably encountered when the process is truly at SS.

D.3 Method Equations

This description is extracted from Nonlinear Regression Modeling for Engineering Applications: Modeling, Model Validation, and Enabling Design of Experiments, by R. R. Rhinehart, John Wiley & Sons, Inc., Hoboken, September 2016b.

The method of Cao and Rhinehart (1995) uses a ratio of two variances, as measured on the same set of data by two methods. Figure D.1 illustrates the concept. The value starts at a value of about 20, ramps to a value of about 5 at sample number 50, and then holds steady. The markers about that trend represent the measured data. The true trend is unknowable, only the measurements can be known, and they are infected with noise-like fluctuations, masking the truth.

The method first calculates a filtered value of the process measurements, indicated by the solid line that lags behind the data. Then the variance in the data is measured by two methods. One deviation d_2 is the difference between measurement and the filtered trend. The deviation d_1 is the difference between sequential data measurements.

If the process is at SS, as illustrated in the 80–100 time period, the filtered value, X_f is almost the middle of the data. Then a process variance, σ^2, estimated by d_2 will ideally be equal to σ^2 estimated by d_1. Then the ratio of the variances, $R = \sigma_{d_2}^2/\sigma_{d_1}^2$, will be approximately equal to unity, $R = \sigma_{d_2}^2/\sigma_{d_1}^2 \cong 1$. Alternately, if the process is in a TS, then X_f is not the middle of data, the filtered value lags behind, and the variance as measured by d_2 will be much larger than the variance as estimated by d_1, $\sigma_{d_2}^2 \gg \sigma_{d_1}^2$, and ratio will be much greater than unity, $R = \sigma_{d_2}^2/\sigma_{d_1}^2 \gg 1$.

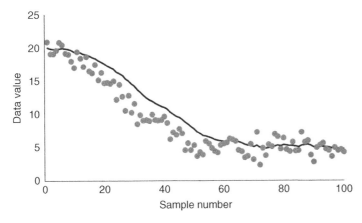

Figure D.1 Illustration of noisy measurements (markers) and filtered data (solid line).

In 2016, I became aware that von Neumann published an analysis of this ratio of variance concept in 1941. But I was not aware of that in the 1980s and seem to have rediscovered the concept and co-analyzed with my PhD candidate Songling Cao. However, contrasting the von Neumann approach, to minimize computational burden, in this method a filtered value (not an average) provides an estimate of the data mean:

$$X_{f,i} = \lambda_1 X_i + (1 - \lambda_1) X_{f,i-1} \tag{D1}$$

where X is the process variable, X_f is the filtered value of X, λ_1 is the filter factor, and i is the time sampling index.

The first method to obtain a measure of the variance uses an exponentially weighted moving "variance" (another first-order filter) based on the difference between the data and the "average":

$$v_{f,i}^2 = \lambda_2 \left(X_i - X_{f,i-1} \right)^2 + (1 - \lambda_2) v_{f,i-1}^2 \tag{D2}$$

where $v_{f,i}^2$ is the filtered value of a measure of variance based on differences between data and filtered values and $v_{f,i-1}^2$ is the previous filtered value.

Equation (D2) is a measure of the variance to be used in the numerator or the ratio statistic. The previous value of the filtered measurement is used instead of the most recently updated value to prevent autocorrelation from biasing the variance estimate, $v_{f,i}^2$, keeping the equation for the ratio simple.

The second method to obtain a measure of variance is an exponentially weighted moving "variance" (another filter) based on sequential data differences:

$$\delta_{f,i}^2 = \lambda_3 (X_i - X_{i-1})^2 + (1 - \lambda_3) \delta_{f,i-1}^2 \tag{D3}$$

where $\delta_{f,i}^2$ is the filtered value of a measure of variance and $\delta_{f,i-1}^2$ is the previous filtered value.

This will be the denominator measure of the variance.

The ratio of variances, the R-statistic, may now be computed by the following simple equation:

$$R = \frac{(2 - \lambda_1) \, v_{f,i}^2}{\delta_{f,i}^2} \tag{D4}$$

The calculated value is to be compared with its critical values to determine SS or TS. Neither Equations (D2) nor (D3) computes the variance. They compute a measure of the variance. Accordingly, the $(2 - \lambda_1)$ coefficient in Equation (D4) is required to scale the ratio and to represent the variance ratio.

The essential assignment statements for Equations (D1)–(D4) are as follows:

```
nu2f := L2 * (measurement - xf ) ^ 2 + cL2 * nu2f
xf := L1 * measurement + cL1 * xf
delta2f := L3 * (measurement - measurement_old) ^ 2 + cL3 * delta2f
measurement_old := measurement
R_Filter := (2 - L1) * nu2f / delta2f
```

The coefficients L1, L2, and L3 represent the lambda values, and the coefficients cL1, cL2, and cL3 represent the complementary values.

The five computational lines of code of this method require direct, no logic, low storage, and low computational operation calculations. In total there are four variables and seven coefficients to be stored, 10 multiplication or divisions, 5 additions, and 2 logical comparisons per observed variable.

Being a ratio of variances, the statistic is scaled by the inherent noise level in the data. It is also independent of the dimensional units chosen for the variable.

If the process is at steady state, then the R-statistic will have a distribution of values near unity. Alternately, if the process is not at steady state, then the filtered value will lag behind the data, making the numerator term larger than the denominator, and the ratio will be larger than unity.

D.4 Coefficient, Threshold, and Sample Frequency Values

For simplicity and for balancing speed of response with surety of decision and robustness to noiseless periods, use filter values of $\lambda_1 = \lambda_2 = \lambda_3 = 0.1$. However, other users have recommended alternate values to optimize speed of response and type-I and type-II errors. The method works well for diverse combinations of filter coefficient values within the range of 0.05–0.2.

If R-calculated > about 2.5, "reject" steady state with fairly high confidence and accept that the process is in a transient state. Alternately, if R-calculated < about 0.85, "accept" that the process is at steady state and reject that it is in a transient state with fairly high confidence. If in-between values for R-calculated, hold the prior SS or TS state, because there is no confidence in changing the declaration.

The filter factors can be related to the number of data (the length of the time window) effectively influencing the average or variance calculation. Ideally, the number of data in a time window for the calculations is effectively $N = 1/\lambda$. However, based on a first-order decay, roughly, the effective number of data in the window of observation that could still have an impact on the filtered value is about $N = 3.5/\lambda$.

Larger λ values mean that fewer data are involved in the analysis, which has a benefit of reducing the time for the identifier to catch up to a process change, reducing the average run length (ARL). But larger λ values have an undesired impact of increasing the variability on the statistic, confounding interpretation. The reverse is true: lower λ values undesirably increase the ARL to detection but increase precision (minimizing statistical errors).

The basis for this method presumes that there is no autocorrelation in the time series process data when at SS. Autocorrelation means that if a measurement is high (or low), the subsequent measurement will be related to it. For example, if a real process event causes a temperature measurement to be a bit high and the event has persistence, then the next measurement will also be influenced by the persisting event and will also be a bit high. Autocorrelation could be related to control action, thermal inertia, noise filters in sensors, etc. Autocorrelation would tend to make all R-statistic distributions shift to the right, requiring a reinterpretation of critical values for each process variable.

For Monte Carlo simulations, the independent realizations in each trial return independent perturbations on the diversity of results (DV*, OF*, 95% limits, etc.). So realization-to-realization or iteration-to-iteration autocorrelation is not an issue. Similarly, when regression data are randomly sampled at each iteration, there is no autocorrelation.

Also, however, for measured process data, autocorrelation is commonly encountered due to persistence of perturbations or signal damping/filtering techniques. Here, it is more convenient to choose a sampling interval that eliminates autocorrelation than to model and compensate for autocorrelation in the test statistic. A plot of the current process measurement versus the previous sampling of the process measurement over a sufficiently long period of time (equaling several time constants) at steady state is required to establish the presence/absence of autocorrelation. To detect autocorrelation, visually choose a segment of data that is at steady state and plot the PV value versus its prior value.

Figure D.2 plots data with a lag of one sample (a measurement vs. the prior measurement) and shows autocorrelation. (Here, "lag" is used in the statistician's sense and means "delay" in the process control sense. For process control, the term lag means a first-order asymptotic response.) Figure D.3 plots the same data but with a lag of five samples and shows zero autocorrelation.

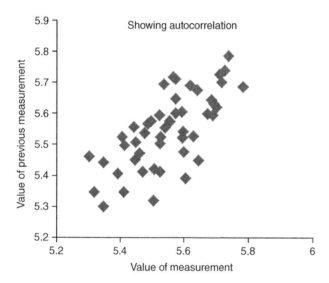

Figure D.2 Data showing autocorrelation.

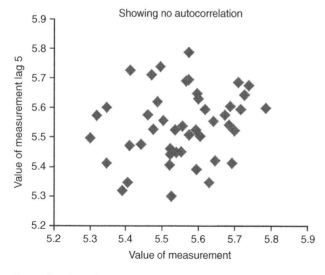

Figure D.3 Data showing no autocorrelation when the interval is five samplings.

However, if the OF of the best player is being observed, it may not change for several iterations until a leap-over finds a better spot. Accordingly, the iteration-to-iteration OF value would have autocorrelation. So, observe the OF of the worst player, not the best.

Generally, hydraulics (flow rates and levels) come to SS faster than gas pressure in large volumes, which is faster than temperature or composition. Further, noise in flow rate and level measurements will have little persistence; however, thermal inertia of temperature sensors may extend autocorrelation. So, a sampling interval for one variable might not be what is needed for another. The user might want to separate the process attributes into a set of hydraulic variables, another set of thermal and composition inventory (T, P, x) variables, and monitor the hydraulic state of the process and the thermal and composition states separately.

Summarizing, use $\lambda_1 = \lambda_2 = \lambda_3 = 0.1$. Use R-critical = 2.5 to reject SS (accept TS) and R-critical = 0.85 to accept SS (reject TS). Choose a sampling interval to eliminate autocorrelation from a visually determined SS period.

Figure D.4 illustrates the method. The process variable, PV, is connected to the left-hand vertical axis (log 10-scale) and is graphed with respect to sample interval. Initially it is at a steady state with a value of about 5. At a sample number 200, the PV begins a first-order rise to a value of about 30. At sample number 700, the PV makes a step rise to a value of about 36. The ratio statistic is attached to the same left-hand axis and shows an initial kick to a high value as variables are initialized and then relaxes to a value that wanders about the unity SS value. When the PV changes at sample 200, the R-statistic value jumps up to values ranging between 4 and 11, which relaxes back to the unity value as the trend hits a steady value at a time of 500. Then when the small PV step occurs at sample 700, the R-value jumps up to about 4 and then decays back to its nominal unity range. The SS value is connected to the right-hand vertical axis and has values of either 0 or 1 that change when the R-value exceeds the two limits of $R_{\beta,\mathrm{TS}}$ and $R_{1-\alpha,\mathrm{SS}}$.

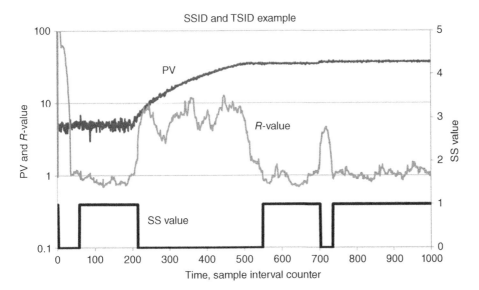

Figure D.4 Example application of SSID and TSID.

D.5 Type-I Error

If the null hypothesis is that a process is at SS, then a type-I error is a claim of not-at-SS when the process is actually at SS. The concept is best understood by considering the distribution of the R-statistic of a SS process. The statistical vagaries in the data create a distribution of the R-statistic values. Figure D.5 represents the statistical distribution of the R-statistic values at SS.

The R-statistic will have some variability because of the random fluctuations in the sequential measured data. If the value of R is less than the upper 95% confidence value, the process may be at steady state, but if it is beyond (larger than) the 95% confidence value, then it is likely that the process is not at steady state. If the process is at steady state, there is a small, $\alpha = 5\%$, chance that $R > R_{0.95}$. The level of significance, α, is the probability of making a type-I error of rejecting the SS hypothesis when it is true.

Five percent is the standard level of significance for economic decisions. However, if $\alpha = 5\%$, and the process is at SS, the identifier will make a false claim 5% of the time, about every 20th sample, which would render it nearly useless for automation.

Figure D.6 includes the distribution of the R-statistic for a process that is not at steady state, one that is in a transient state, with its distribution of R-statistic values greater that unity. For a process that is not at steady state, there is a high chance that $R > R_{0.95}$. As illustrated in Figure D.6, it is about a 70% chance.

So, if $R > R_{0.95}$, the likely explanation is that the process is in a transient state. As illustrated by the shaded areas to the right of the $R_{0.95}$ value, the probability that an excessive R-value could come from the SS or the TS distribution, the odds are about 70–5% or $15:1$ for being in the transient state. Claim TS.

There are several ways to reduce the probability of a T-I error. An obvious one is to choose a smaller value for alpha, and in statistical process control (SPC), the value of $\alpha = 0.27\%$ is normally accepted. But too small a level of significance means that a not-at-SS event might be missed—a T-II error.

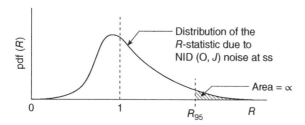

Figure D.5 *R*-Statistic distribution at steady state.

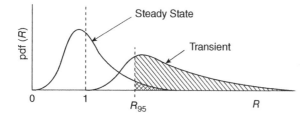

Figure D.6 High chance of not being at steady state.

D.6   Type-II Error

If the null hypothesis is that the process is at SS, then a type-II error is claiming the process is at SS when it is actually in a TS. If $R < R_{0.95}$, this does not indicate the process is at steady state. Figure D.7 provides an illustration using the same two SS and TS distributions. As illustrated, the likelihood of $R < R_{0.95}$ if at steady state is 95%, and if not at steady state it is 30%. Here the odds are about $3:1$ that the steady-state conclusion is true. The $3:1$ odds are not very definitive. So, to be confident that the process is at steady state, consider the left side of the distributions and an alternate T-I error.

D.7   Alternate Type-I Error

For a given transient state, $R_{\beta,\mathrm{TS}}$ is the lower β critical value and for a given steady state, $R_{1-\alpha,\mathrm{SS}}$ is the $1 - \alpha$ upper critical value. If $R > R_{1-\alpha,\mathrm{SS}}$, then it will reject SS (accept TS). And if $R < R_{1-\alpha,\mathrm{SS}}$, it will reject TS (accept SS).

If the process is at a transient state, then β is the probability of $R < R_{\beta,\mathrm{TS}}$. Figure D.7 shows that the TS process has only about a 1% chance that $R < R_{\beta,\mathrm{TS}}$. However, if the process is at steady state, then as illustrated, there is a 40% likelihood that $R < R_{\beta,\mathrm{TS}}$. So, if $R < R_{1-\alpha,\mathrm{SS}}$, the odds are about $40:1$ that process is at steady state. Claim SS.

However, if $R_{\beta,\mathrm{TS}} \leq R \leq R_{1-\alpha,\mathrm{SS}}$, then there is a high likelihood of the process being either at steady or transient state. There is no adequate justification to make either claim. So retain the last claim.

Both type-I and alternate type-I errors are important. A type-I error is the trigger of a "probable not at steady-state" claim when the process is at steady state. An alternate type-I error is the trigger of a "probable at steady-state" claim when the process is in a transient state. In any statistical test, the user needs to choose the level of significance, α, the probability of a T-I error, and power, β, the probability of an alternate T-I error. Once decided, the $R_{1-\alpha,\mathrm{SS}}$ critical value can be obtained from Cao and Rhinehart (1995) and the $R_{\beta,\mathrm{TS}}$ critical value from Shrowti *et al.* (2010).

However, it is more convenient and less dependent on idealizations to visually select data from periods that represent a transient or steady period and to find the R-critical values that make the algorithm agree with the user interpretation. My experience recommends $\lambda_1 = \lambda_2 = \lambda_3 = 0.1$, for simplicity $R_{1-\alpha,\mathrm{SS}} = 3\text{–}4$ and $R_{\beta,\mathrm{TS}} = 0.8\text{–}0.9$, chosen by visual inspection as definitive demarcations for a transient and steady process.

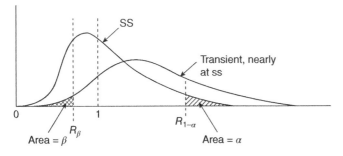

Figure D.7 Critical value for steady-state identification.

D.8 Alternate Array Method

I have been very pleased with the filter method described earlier when applied to detecting SS for convergence in either regression or stochastic functions. The rational for the three first-order filters (for average and variances) is to reduce computational burden.

However, filters have the disadvantage of permitting a past event to linger with an exponentially weighted diminishing factor. This could extend the ARL to detect either SS or TS. Further, the filter structure is not as comfortable for many who are familiar with conventional sums in calculating variances. Further, the window of fixed length N is easier to grasp than the exponential weighted infinite window or the interpretation of the three lambda values. Alternately, in using a window of N data, the event is totally out of the window and out of the analysis after N samples, which provides a clearer user understanding of data window length.

Accordingly, this section describes an alternate calculation method for the R-statistic ratio of variances.

Returning to the base concept with the R-statistic as a ratio of variances, first calculate the average:

$$\bar{X} = \frac{1}{N}\sum_{i=1}^{N} X_i \tag{D5}$$

where i is the sample counter starting at the last sample and is counting back in time N data points in the window. Then the conventional variance can be expanded:

$$\sigma_1^2 = \frac{1}{N-1}\sum_{i=1}^{N}(X_i - \bar{X})^2 = \frac{1}{N-1}\left[\left(\sum_{i=1}^{N} X_i^2\right) - N\bar{X}^2\right] \tag{D6}$$

And substituting for the average,

$$\sigma_1^2 = \frac{1}{N-1}\left[\left(\sum_{i=1}^{N} X_i^2\right) - \frac{1}{N}\left(\sum_{i=1}^{N} X_i\right)^2\right] \tag{D7}$$

Assuming no autocorrelation the variance can also be estimated as the differences between successive data. Since there are $N-1$ data differences in a set of N data, there are $N-1$ terms in the sum of differences in the data window and then $N-2$ in the divisor:

$$\sigma_2^2 = \frac{1}{2(N-2)}\sum_{j=1}^{N-1}(x_j - x_{j-1})^2 \tag{D8}$$

Then the ratio of variances is

$$R = \frac{\sigma_1^2}{\sigma_2^2} = \frac{2(N-2)}{N-1}\left[\frac{\sum_{i=1}^{N} x_i^2 - (1/N)\left(\sum_{i=1}^{N} x_i\right)^2}{\sum_{j=1}^{N-1}(x_j - x_{j-1})^2}\right] = \frac{2(N-2)}{N-1}\left[\frac{S1 - (1/N)(S2)^2}{S3}\right] \tag{D9}$$

Note: Often, when I implement this, I normalize by the number in the sum, not the degrees of freedom (N not $N-1$) and ($N-1$ not $N-2$).

This is essentially the von Neumann et al. (1941) approach. What appears to be a computationally expensive online calculation can be substantially simplified by using an array structure to the window of data and a pointer that indicates the array element to be replaced with the most recent measurement. With this concept, the sums are incrementally updated at each sampling.

Code for the array method with variances incrementally updated, including initialization on first call is revealed at the end of this Appendix. The method is as follows. First increment the counter to the next data storage location. Then decrement the sums with the data point to be removed. Then replace the data array element with the new value. Then increment the sums. Then recalculate the R-statistic. In comparison with the filter method, this requires storage of the N data, the N squared data differences, and the three sums. The pointer updating adds a bit of extra computation. Our investigations indicate that the array approach takes about twice the execution time as the filter approach. Also, since it stores $2N$ data values, it has about that much more RAM requirement. The increased computer burden over the filter method may not be excessive in today's online computing devices.

This approach calculates a true data variance from the average in the data window of N samples and has the additional benefit that the sensitivity of the R-statistic to step changes in a signal is greater with the array method than with the filter method (about twice as sensitive from experience). The array approach can respond to changes of about 1.5σ. Accordingly, this defined window, array approach, is somewhat better in detecting small deviations from SS.

Further, the value of window length, N, explicitly defines the number of data after an event clears to fully pass through the data window and influence on the R-value. By contrast, the filer method exponentially weights the past data, and the persisting influence ranges between about $3/\lambda$ and $4/\lambda$ number of samplings.

Additionally, the array methods seem less conflicted by autocorrelation and discretization flat spots in the time series of data.

However, the array method is computationally more burdensome than the filter method. Nominally, the filter method needs to store and manipulate seven values at each sampling. With $\lambda = 0.1$, this characterizes a window of about 35 samples. For the same characterization, the array method needs to store and manipulate over 180 variables at each sampling.

However, there is only one adjustable variable representing the data window, N. By contrast, the filter approach has three lambda values providing some flexibility in optimizing performance, as done by Bhat and Saraf (2004), but perhaps tripling the user confusion.

Today, my preference is to use the filter method. But if data vagaries (changes in autocorrelation, flat spots, coarse discretization, small shifts relative to noise) make it dysfunctional, then use the array method.

However, for optimization applications, which only need to determine SS (not TS), the filter method is fully adequate, much simpler to implement, and is my preference.

D.9 SSID and TSID VBA Code

D.9.1 Cao–Rhinehart Filter Method

```
Sub SSID_Filter() 'Cao-Rhinehart Filter Method

  If FirstCall THEN
        l1 = 0.1      'filter lambda values
        l2 = 0.1
        l3 = 0.1
        cl1 = 1 - l1
```

```
          cl2 = 1 - l2
          cl3 = 1 - l3
          n2f = 0         'initial values
          d2f = 0
          xf = 0
          xold = 0
          SS = 0.5        'indeterminate
   Else
          x= 'acquire data
          n2f = l2 * (xf - x) ^ 2 + cl2 * n2f
          xf = l1 * x + cl1 * xf
          d2f = l3 * (x - xold) ^ 2 + cl3 * d2f
          xold = x
          If d2f > 0 Then
              r = (2 - l1) * n2f / d2f
          End If
          If (2 - l1) * n2f <= lower * d2f Then SS = 1      'confidently @ SS
          If (2 - l1) * n2f > upper * d2f Then SS = 0      'confidently @TS
    End IF
End Sub
```

D.9.2 Array Method

```
Sub SSID_Array()     'Rhinehart-Gore Array Method
  IF FIrstCall THEN
      Dim y(50)         'up to N=50 values in window
      Dim dy2(50)
      For j = 1 To 50  'initialize array
         y(j) = 0
         dy2(j) = 0
      Next j
      N = 10    'initialize
      sum1 = 0
      sum2 = 0
      sum3 = 0
      yold = 0
      j = 1
      SS = 0.5    'indeterminate
  Else
          j = j + 1     'counter for data array of N elements
          If j > N Then j = 1
          sum1 = sum1 - y(j) ^ 2      'decrement sums
          sum2 = sum2 - y(j)
          y(j) = 'acquire data
          sum1 = sum1 + y(j) ^ 2      'increment sums
          sum2 = sum2 + y(j)
```

```
          jj = jj + 1 'counter for difference array of N-1 elements
          If jj > N - 1 Then jj = 1
          sum3 = sum3 - dy2(jj)          'decrement
          dy2(jj) = (y(j) - yold) ^ 2
          yold = y(j)
          sum3 = sum3 + dy2(jj)          'increment
          If sum3 > 0 Then
            r = 2 * ((N - 1) / N) * (sum1 - (sum2 ^ 2) / N) / sum3
          End If
          If 2 * (sum1 - (sum2 ^ 2) / N) <= 0.9 * sum3 Then SS = 1
          If 2 * (sum1 - (sum2 ^ 2) / N) > 2.5 * sum3 Then SS = 0
   End If
End Sub
```

Appendix E

Optimization Challenge Problems (2-D and Single OF)

E.1 Introduction

There is a need for convenient test functions for creators to test optimizers and for learners to explore issues. Many that have been offered are simple one-line functions. Although these can embody certain aspects or features related to optimization, they misrepresent the complexity of real applications and are often contrived idealizations. This collection of test functions seeks to embody first-principles models of optimization applications to reveal issues, within a step-better reveal of reality, but remaining relatively simple.

This set of objective functions (OF) was created to provide sample challenges for testing optimizers. The examples are all two-dimensional, having two decision variables (DVs) so as to provide visual understanding of the issues that they embody. Most are relatively simple to program and compute rather simply, for user convenience. Most represent physically meaningful situations, for the person who wants to see utility and relevance. Most should be understood by those with a STEM degree. All are presented with minimization as the objective.

In all equations that follow, $x1$ and $x2$ are the DVs, and f_of_x is the OF value. The DVs are programmed for the range [0, 10]. However, not all functions use DV values in that range. So, the DVs are scaled for the appropriate range and labeled $x11$ and $x22$. The OF value f_of_x is similarly scaled for a [0, 10] range. The 0–10 scaling permits a common basis for scaling all functions. It could have been 0–1 or 0–100.

These functions are available in VBA from the Excel file "2-D Optimization Examples" in www.r3eda.com

E.2 Challenges for Optimizers

Classic challenges to optimizers are objective functions that have

1) Non-quadratic behavior
2) Multiple optima
3) Stochastic responses
4) Asymptotic approach to optima at infinity

Engineering Optimization: Applications, Methods, and Analysis, First Edition. R. Russell Rhinehart.
© 2018 R. Russell Rhinehart. Published 2018 by John Wiley & Sons Ltd.
Companion website: www.wiley.com/go/rhinehart/engineeringoptimization

5) Hard inequality constraints, or infeasible regions
6) Slope discontinuities (sharp valleys)
7) A gently sagging channel (effectively slope discontinuities)
8) Level discontinuities (cliffs)
9) Flat spots
10) Nearly flat spots
11) Very thin global optimum in a large surface (an improbable to find, pinhole optima)
12) Discrete, integer, or class DVs mixed with continuous variables
13) Underspecified problems with infinite number of equal solutions
14) Discontinuous response to seemingly continuous DVs because of discretization in a numerical integration
15) Sensitivity of DV or OF solution to givens

Any solution depends on the optimizer algorithm, the coefficients of the algorithm, and the convergence criteria. For instance, a multiplayer optimizer has an increased chance of finding the global optimum. An optimizer based on a quadratic surface assumption (such as successive quadratic or Newton's) will jump to the optimum when near it, but can jump in the wrong direction when not in the proximity. The values for optimizer coefficients (scaling, switching, number of players, number of replicates, initial step size, tempering, or acceleration) can make an optimizer efficient for one application, but with the same values it might be sluggish or divergent in an application with other features. The convergence criteria may be right for one optimizer but stop another for long before arriving at an optimum. When you are exploring optimizers, realize that the results are dependent on your choice of optimizer coefficients and convergence criteria, as well as the optimizer and the features of the test function.

Metrics for evaluating desirability attributes of optimizers were presented in Chapter 33.

What follows is a description of a variety of 2-D optimization applications that I believe are useful test applications to reveal a variety of issues. The number that follows the name is the function number in the 2-D Optimization Examples file.

E.3 Test Functions

E.3.1 Peaks (#30)

This function seems to be a MatLAB creation as a simple representation of several optimization issues. It provides mountains and valleys in the middle of a generally outward sloping surface (Figure E.1). There are two major valleys and, between the three mountains, a small local minima. If the trial solution starts on the north or east side of the mountains, it leads downhill to the north or east boundary. The southern valley has the lowest elevation. The surface is non-quadratic and has multiple optima.

The function is as follows:

```
x11 = 3 * (x1 - 5) / 5    'convert my 0-10 DVs to the -3 to +3 range for the function
x22 = 3 * (x2 - 5) / 5
f_of_x = 3 * ((1 - x11) ^ 2) * Exp(-1 * x11 ^ 2 - (x22 + 1) ^ 2) - _
    10 * (x11 / 5 - x11 ^ 3 - x22 ^ 5) * Exp(-1 * x11 ^ 2 - x22 ^ 2) - _
    (Exp(-1 * (x11 + 1) ^ 2 - x22 ^ 2)) / 3
f_of_x = (f_of_x + 6.75) / 1.5    'to convert to a 0 to 10 f_of_x range
```

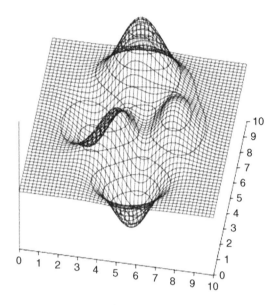

Figure E.1 Peaks surface.

E.3.2 Shortest Time (#47)

This simple appearing function is confounding for successive quadratic or Newton's approaches. It represents a simple situation of a person on sand on one side of a shallow river wanting to minimize the travel time to a point on land on the other side (Figure E.2). It is also kin to light traveling the

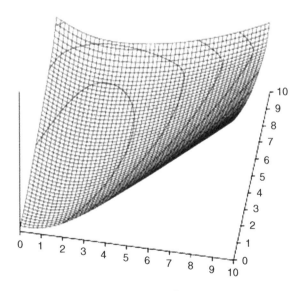

Figure E.2 Shortest-time path surface.

minimum time path through three media. Variables $v1$, $v3$, and $v3$ are the velocities through the near side, water, and far side; and $x1$ and $x2$ represent the E–W distance that the path intersects the near side and far side of the river.

The VBA code is as follows:

```
v1 = 1
v2 = 0.75
v3 = 1.25
xa = 1
ya = 1
xb = 9
yb = 9
a1 = 3
b1 = -0.3
a2 = 5
b2 = 0.4
y1 = a1 + b1 * x1
y2 = a2 + b2 * x2
distance1 = Sqr((x1 - xa) ^ 2 + (y1 - ya) ^ 2)
distance2 = Sqr((x2 - x1) ^ 2 + (y2 - y1) ^ 2)
distance3 = Sqr((xb - x2) ^ 2 + (yb - y2) ^ 2)
timetravel = distance1 / v1 + distance2 / v2 + distance3 / v3
f_of_x = (timetravel - 12) * 10 / (32 - 12)
```

The problem statement from 2012 was: "She was playing in the sandy area across the river from her house when the dinner bell rang. To get home she needed to run across the sand, run through the shallow lazy river, then run across the field to home. She runs faster on land than on sand. Both are faster than running though the knee-deep water. What is the shortest-time path home? On the x-y space of her world, she starts at location (1, 1) and home is at (9, 9). Perhaps the units are deci-kilometers (dKm) (tenths of a kilometer). Her speed through water is 0.75 dKm/min, through sand is 1.0 dKm/min, and on land is 1.25 dKm/min. The boundaries of the river are defined by $y = 3.0 - 0.3x$ and $y = 5.0 + 0.4x$."

E.3.3 Hose Winder (#46)

This function presents discontinuities to the generally well-behaved floor. It represents a storage box that winds up a garden hose on a spool (Figure E.3). When the winding hose gets to the end of the spool, it starts back but at a spool diameter that is larger by two-hose diameters. The diameter jump causes the discontinuities. The objective is to design the storage box size and handle length to minimize the owner's work in winding the hose.

The 2012 problem statement was: "Consider that a hose is 200 feet long, and 1.25 inch in diameter, and is wound on a 6 inch diameter spindle by a gear connection to the handle. The hose goes through a guide, which oscillates side-to-side to make the winding uniform. Each sweep of the guide leads to a new hose layer, making the wind-on diameter 2.5 inches larger.

Figure E.3 Hose winder device.

"Originally the hose is stretched out 200 ft. To reel it in, the human must overcome the drag force of the hose on the ground. Either a smaller spindle diameter or a larger handle radius reduces the handle force required to reel in the hose. As the hose is reeled in, its residual length is less, and the drag force is less. But when the first spindle layer is full and the hose moves to the next layer, the leverage changes, and the windup handle force jumps up.

"After winding 200 ft of hose, the human is exhausted. He had to overcome the internal friction of the device, the hose drag resistance, and move his own body up and down. We wish to design a device that that minimizes the total human work (handle force times distance moving the handle + body work). We also wish to keep the maximum force on the handle less than some excessive value (so that an old man can do the winding). If you increase the handle radius, the force is lessened, but the larger range of motion means more body motion work. There is a constraint on the handle radius—it cannot make the human's knuckles scrape the ground.

"If you make the spindle length longer, so that more hose is wound on each layer, then the hose wind-on diameter does not get as large, and the handle needs less force to counter the drag. But more turns to wind-in the hose means more body motion.

"Further, consider the economics of manufacturing the box. The box volume needs to be large enough to windup the hose on the spindle, and perhaps 20% larger (so that spiders can find space for their webs, it often seems). Use 5 cuft. If the side is square, then defining the length and volume sets the side dimensions. Setting the weight and cost directly proportional to the surface area, the spindle length defines the cost of manufacturing and shipping.

"The objective function is comprised of a penalty for the maximum force, a penalty for the cost, and a penalty for the wind-up work."

The 3-D view indicates a generally smooth approach to a minimum but with local wrinkles on the surface (Figure E.4). The local valleys guide many optimizers to a false minimum.

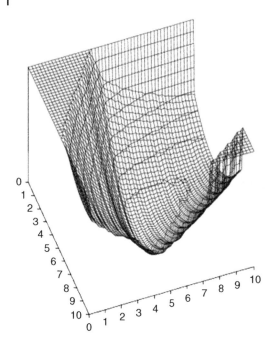

Figure E.4 Hose winder surface.

The VBA code is as follows:

```
If x1 <= 0 Or x2 <= 0 Then
    constraint = "FAIL"
    f_of_x = 15
    Exit Function
End If
'   HandleRadius = x1 / 4        'nominal, scales 0-10 to 0-2.5 ft
'   BoxWidth = x2 / 2            'nominal, scales 0-10 to 0-5 ft
HandleRadius = 1.8 + 0.04 * x1 'to focus on discontinuities'
BoxWidth = 0.6 + 0.02 * x2     'to focus on discontinuities
WindRadius = 0.25              'spindle radius 3 inches as 1/4 ft
BoxVolume = 5                  'cuft
BoxSide = Sqr(BoxVolume / BoxWidth)
If HandleRadius > 0.85 * BoxSide Then
    constraint = "FAIL"
    f_of_x = 15
    Exit Function
End If
Area = 2 * BoxSide ^ 2 + 3 * BoxWidth * BoxSide   'no bottom closure
HoseLength = 200               'ft
```

```
        HoseDiameter = 0.1              '1.25 inches = 0.1 ft
        DragLength = HoseLength
        dcircle = 0.025                 'fraction of circumfrence
        work = 0
        WindLength = 0
        Fmax = 0
        Do Until DragLength < 2         'enough increments to wind up hose, leave
2 feet of hose outside winder
            DragForce = 0.2 * DragLength    '0.2 is coefficient of friction lbf/ft
of hose
                HandleForce = (DragForce * WindRadius + 0.5) / HandleRadius + 1
'hose drag, torque to spin assembly, body motion
                If HandleForce > Fmax Then Fmax = HandleForce
                dwork = HandleForce * HandleRadius * 2 * 3.14159 * dcircle
                work = work + dwork
                dlength = WindRadius * 2 * 3.14159 * dcircle  'incremental length
wound in one circle increment
            WindLength = WindLength + dlength    'total length wound on the layer
            DragLength = DragLength - dlength     'length of hose left unwoound
            If WindLength > (BoxWidth / HoseDiameter) * 2 * 3.14159 * WindRadius
Then   'layer full, move to next
                WindRadius = WindRadius + HoseDiameter    'update winding radius
                WindLength = 0                    'reset wound length on new layer
            End If
        Loop
        ' f_of_x = 10 * ((Fmax / 5) ^ 2 + (Area / 20) ^ 2 + (work / 1000) ^ 2 - 28) /
(70 - 28)  'nominal
        f_of_x = 10 * ((Fmax / 5) ^ 2 + (Area / 20) ^ 2 + (work / 1000) ^ 2 - 28.25) /
(28.45 - 28.25) 'to focus on discontinuities
```

E.3.4 Boot Print in the Snow (#19)

This represents a water reservoir design problem, but with very simple equations. It is a surrogate deterministic model. The objectives of a reservoir are to trap excessive rain water to prevent downstream floods, to release water downstream to compensate for upstream droughts, and to provide water for human recreational and security needs. We also want to minimize the cost of the dam and land. The questions are how tall should the dam be and how full should the reservoir be kept. The taller it is, the more it costs, but the lower will be the probability of flood or drought impact, and the better the recreational and security features. The fuller it is kept, the less it can absorb floods, but the better the drought or recreational performance. On the other hand, if kept nearly empty, it can mitigate any flood, but cannot provide recreation or drought protection. The contour appears as a boot print in the snow (Figure E.5). The contours represent the economic risk (probability of an event times the cost of the event). The minimum is at the toe. The horizontal axis, $x1$, represents the dam height. The vertical axis, $x2$, is the set point portion of full.

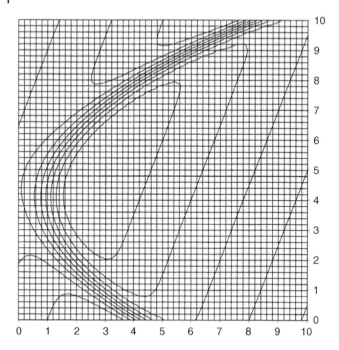

Figure E.5 Boot print contour.

A feature of this application that creates difficulty is that the central portion, in the $(7, 5)$ area of the print, is a plane, and optimization algorithms that use a quadratic model (or second derivatives) cannot cope with the zero values of the second derivative.

The VBA code is as follows:

```
If x1 < 0 Or x2 < 0 Or x1 > 10 Or x2 > 10 Then
      constraint = "FAIL"
      Exit Function
End If
x1line = 1 + 0.2 * (x2 - 4) ^ 2
deviation = (x1line - x1)
penalty = 5 * (1 / (1 + Exp(-3 * deviation)))    'logistic functionality
f_of_x = 0.5 * x1 - 0.2 * x2 + penalty + add_noise
f_of_x = 10 * (f_of_x - 0.3) / 6
```

E.3.5 Boot Print with Pinhole (#21)

This is the same as boot print in the snow, but the global minimum is entered with a small region on the level snow (Figure E.6). Perhaps an acorn fell from a high tree and drilled a hole in the snow. The difficulty is that there is a small probability of starting in the region that would attract the solution to the true global. Nearly everywhere, the trial solution will be attracted to the toe part of the boot print.

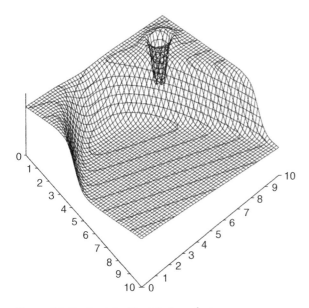

Figure E.6 Boot print with pinhole surface.

The VBA code is as follows:

```
If x1 < 0 Or x2 < 0 Or x1 > 10 Or x2 > 10 Then
      constraint = "FAIL"
      Exit Function
End If
x1line = 1 + 0.2 * (x2 - 4) ^ 2
deviation = (x1line - x1)
penalty = 5 * (1 / (1 + Exp(-3 * deviation)))
'logistic functionality
f_of_x = 0.5 * x1 - 0.2 * x2 + penalty + add_noise
x1mc2 = (x1 - 1.5) ^ 2
x2mc2 = (x2 - 8.5) ^ 2
factor = 1 + (5 * (x1mc2 + x2mc2) - 2) *
Exp(-4 * (x1mc2 + x2mc2))
f_of_x = factor * f_of_x + add_noise
f_of_x = 10 * (f_of_x - 0.3) / 6
```

E.3.6 Stochastic Boot Print (#20)

This represents the same water reservoir design problem as boot print in the snow; however, the surface is stochastic (Figure E.7). This, too, is a surrogate model. The OF value depends on a probability of the flood or drought event. Because of this each realization of the contour will yield a slightly different appearance. One realization of the 3-D view is shown. Note that starting in the middle of the DV space on the planar portion, a downhill optimizer will progressively move toward

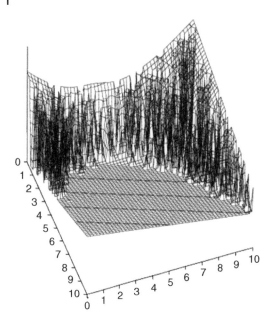

Figure E.7 Stochastic boot print surface.

the far side of the illustration where the spikes are. In that region of a small reservoir kept too full or too empty, there is a probability of encountering a costly flood or drought event that the reservoir cannot mitigate. Moving in the downhill direction, the optimizer may, or may not, encounter a costly event. If it does not, it continues to place trial solutions into a region with a high probability of a disastrous event and continues into the bad region as the vagaries of probability generate fortuitous appearing OF values.

The VBA code is as follows:

```
If x1 < 0 Or x2 < 0 Or x1 > 10 Or x2 > 10 Then
      constraint = "FAIL"
      Exit Function
End If
x1line = 2 + 0.2 * (x2 - 4) ^ 2
penalty = 0
deviation = (x1line - x1)
probability = 1 / (1 + Exp(-1 * deviation)) 'logistic functionality
If probability > Rnd() Then penalty = 5 * probability
f_of_x = 0.5 * x1 - 0.2 * x2 + penalty + add_noise
f_of_x = 10 * (f_of_x + 1.25) / 7.25
```

Two realizations of the contour are shown in Figure E.8, which reveals the stochastic nature of the surface, the non-repeatability of the OF value. Now, in addition to the difficulty of the planar midsection, the optimizer also faces a stochastic surface that could lead to a fortuitous minimum in a high risk section of too small a dam (*x*-axis) or keeping the reservoir too full or empty (*y*-axis).

(a)

(b)

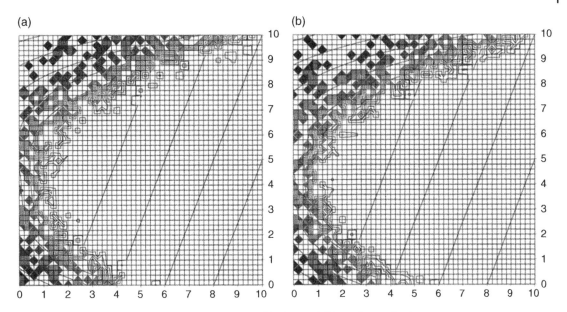

Figure E.8 Stochastic boot print contours: (a) one realization and (b) another realization.

E.3.7 Reservoir (#18)

This is the basis for stochastic boot print, but it is a computationally time-consuming (but not excessively slow) Monte Carlo simulation of a water reservoir. Reservoir capacity and nominal level are the decision variables.

The larger the reservoir, the greater the initial cost. Cost is the objective function. So, superficially, build a small dam and have a small reservoir to reduce cost. However, if the reservoir is too small, and/or it is maintained nearly full, it does not have enough capacity to absorb an upstream flood due to exceptionally heavy rainfall and it will transmit the flood downstream. Downstream flooding incurs a cost of damaged property. But the chance of a flood and the magnitude of the flood depend on the upstream rainfall. So, the simulator models a day-to-day status with a lognormal rainfall distribution for a time period of 20 years. (You can change the simulated time or rainfall distribution.)

Contrasting flooding conditions are drought conditions. If the reservoir is too small, and/or maintained with little reserve of water, an upstream drought will require stopping the water release, which stops downstream river flow. Downstream dwellers, recreationists, or water users will not like this. There is also a cost related to zero downstream flow.

There is a fixed cost of the structure and a probabilistic or stochastic cost associated with extreme flood or drought events.

Since one 20-year simulation will not reveal the confluence of "100-year events," the simulator runs 50 reservoirs for 20 years each—50 realizations of the 20-year period. You can change the number of realizations.

The function is set up to return either the maximum cost for the 50 realizations or the estimated 99% probable upper limit on the cost. You could choose another performance indicator.

An excessively large reservoir kept half full will have ample reserve (to keep water flowing in a drought) and open capacity (to absorb excess rainfall and prevent downstream flooding), but it will cost a lot, both to build and to acquire the land for the lake. A smaller reservoir will have less cost. But too small a reservoir will not prevent problems with drought or flood. So, there is an in-between optimum size.

If the nominal volume is near the full mark, then the reservoir will not be able to absorb floods, but it will have plenty of capacity for a drought. If the nominal level is too low, it will be able to prevent a flood, but not keep water flowing for a drought event. So, there is also an in-between set point capacity that is best. The optimum set point for the level might not be at 50%. It depends on whether the vagaries of rainfall and the associated cost impacts make floods a bigger event than droughts.

For any given sized reservoir, the fuller it is kept, the greater is the fresh water reserve and recreational area. These benefits are added as a negative penalty to the cost.

There is a fixed cost of the structure, a probabilistic or stochastic cost of extreme events, and a negative penalty for reserve and recreation benefits.

Figure E.9 is rotated to provide a good view of the surface. The optimum is in the upper left of the aforementioned figure. The lower axis is $x2$, the water level nominal set point for the reservoir. At zero the reservoir is empty, and at 10 it is completely full. The nearly vertical axis on the right is the reservoir size, where zero is a nonexistent reservoir and 10 is large.

Similar to stochastic boot print, this function presents optimizer difficulties of the planar mid-section and a stochastic surface that could lead to a fortuitous minimum in a high risk section of too small a dam (x-axis) or keeping the reservoir too full or empty (y-axis). It is a more realistic Monte Carlo simulation than stochastic boot print but takes longer to compute and provides the same issues for an optimizer as stochastic boot print.

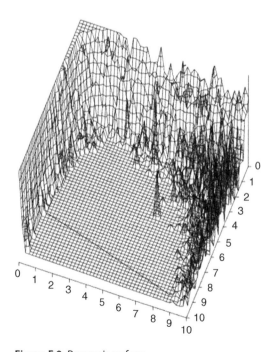

Figure E.9 Reservoir surface.

The VBA code is as follows:

```
twoPi = 2 * 3.1415926535
inchavg = 0.4          'average inches of rainfall/event
prain = 0.25           'daily probability of rain
sy = Log(5 / inchavg) / 1.96   'sigma
for log-normal distribution
 Qinavg = prain * inchavg * 2.54 * 10 ^ 6
   'average water volume
collected/day
 Useravg = 0.4 * Qinavg        'average
daily water demand by users
 Qoutavg = Qinavg - Useravg    'average excess water/day released
downstream
 Vc = (100 + x1 * 100) * 0.25 * inchavg * 2.54 * 10 ^ 6 'reservoir capacity,
from scaled DV x1
 If Vc < 0 Then Vc = 0
 Vset = (0.3 + 0.65 * x2 / 10) * Vc        'reservoir setpoint from scaled DV x2
 Vmin = 0.3 * Vc                'Minimum residual capacity in reservoir, a
constraint
 inchflood = 12          'rain fall amount at one time that causes a flood
downstream
 Qflood = inchflood * 2.54 * 10 ^ 6   'volume of water/day associated with the
flood rain
 v = Vset               'initialize simulation with water at the setpoint
 TotalDays = 1 * 365        'simulation period in days
 NRealizations = 10         'number of realizations simulated
 MaxCost = 0                'initialize total cost for any realization
 For Realization = 1 To NRealizations
   Cost(Realization) = 0.0001 * Vc ^ 0.6      'cost of initial reservoir
   For Daynum = 1 To TotalDays
     If Rnd() > 0.25 Then       'does it rain?
       inches = 0               'if no
     Else
       inches = inchavg * Exp(sy * Sqr(-2 * Log(Rnd())) * Sin(twoPi * Rnd()))
       If inches > 25 Then inches = 25   'log-normal occasionally returns some
excessive numbers. maybe rainfall is not log-normally distributed
     End If
     Qin = inches * 2.54 * 10 ^ 6     'incoming water volume due to rain - level
times area
     If v <= Vmin Then Qout = 0         'control logic for water release
     If Vmin < v And v <= Vset Then Qout = Qoutavg * (v - Vmin) / (Vset - Vmin)
     If Vset < v And v <= Vc Then Qout = Qoutavg + (v - Vset) * (Qflood - Qoutavg) /
(Vc - Vset)
     vnew = v + Qin - Qout - Useravg     'reservoir volume after a day if Qout as
calculated
```

```
          If vnew > Vc Then      'override if reservoir volume would exceed capacity
             Qout = Qin - Useravg - (Vc - v)
             vnew = Vc
          End If
          If vnew < Vmin Then            'override if reservoir volume would fall lower
than Vmin
             Qout = v - Vmin + Qin - Useravg
             If Qout < 0 Then Qout = 0
             vnew = v + Qin - Qout - Useravg
          End If
          v = vnew
          If Qout > Qflood Then               'cost accumulation if a flood event
             discount = (1 + 0.03) ^ Int(Daynum / 365)   'discount factor
           Cost(Realization) = Cost(Realization) + (0.1 * ((Qout - Qflood) / 10 ^
6) ^ 2) / discount
          End If
          If Qout = 0 Then                 'cost accumulation if a drought event
             discount = (1 + 0.03) ^ Int(Daynum / 365)
             Cost(Realization) = Cost(Realization) + 1 / discount
          End If
      Next Daynum
      If Cost(Realization) > MaxCost Then MaxCost = Cost(Realization)
    Next Realization
    costsum = 0          'determine average cost per realizaton
    For Realization = 1 To NRealizations
       costsum = costsum + Cost(Realization)
    Next Realization
    AvgCost = costsum / NRealizations
    cost2sum = 0         'determine variance of realization-to-realization cost
    For Realization = 1 To NRealizations
       cost2sum = cost2sum + (Cost(Realization) - AvgCost) ^ 2
    Next Realization
    SigmaCost = Sqr(cost2sum / (NRealizations - 1))
    f_of_x = AvgCost + 3 * SigmaCost      'Primary OF is 3-sigma, 99.73 probable
upper limit
    f_of_x = f_of_x - 0.2 * x2          'Secondary OF adds a benefit (negative
penalty) for high setpoint level
    f_of_x = 10 * (Sqr(f_of_x) - 2) / (15 - 4) + add_noise       'scale factor for
    display convenience
' f_of_x = MaxCost - 0.2 * x2 - 4      'OF based on max cost for the several
realizations
    If Vset < Vmin Or Vset > 0.95 * Vc Or Vc <= 0 Then 'hard constraint
       constraint = "FAIL"
End If
```

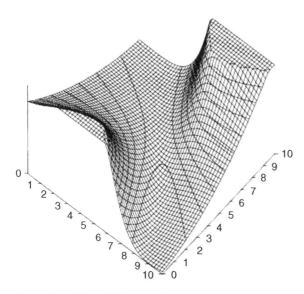

Figure E.10 Chong Vu's normal regression.

E.3.8   Chong Vu's Normal Regression (#11)

Chong Vu was a student exploring various regression objective functions as part of the optimization applications course at OSU and created this test problem for a best linear relation to fit five data points representing contrived noisy data (Figure E.10). The points are (0, 1), (0, 2), (1, 3), (1, 1), and (2, 2). DVs $x1$ and $x2$ are scaled to represent the slope and intercept of the linear model. The OF value is computed as the sum of squared normal distances from the line to the data (as opposed to the traditional vertical deviation least squares that assumes variability in the y-measurement only).

The minimum is at about $x1 = 8$, $x2 = 2$ in the near valley. The function is relatively well behaved, and even though it represents a sum of squared deviations, it is not a quadratic shape. Further, trial solutions in the far portion of the valley send the solution toward infinity. Accordingly, I added a penalty to bring solutions back toward the global best value.

The VBA code is as follows:

```
  m11 = 0.5 * x1 - 3       'coefficients adjusted to fit better on x1, x2
display scale
  b11 = 0.3 * x2 + 0.5
  Sum = 0
  Sum = Sum + (1 - m11 * 0 - b11) ^ 2    'first of 5 pairs of x,y data y=1, x=0
  Sum = Sum + (2 - m11 * 0 - b11) ^ 2
  Sum = Sum + (3 - m11 * 1 - b11) ^ 2
  Sum = Sum + (1 - m11 * 1 - b11) ^ 2
  Sum = Sum + (2 - m11 * 2 - b11) ^ 2
  f_of_x = Sum / (1 + m11 ^ 2) - 2    'sum of vertical distance squared converted
to normal d^2
```

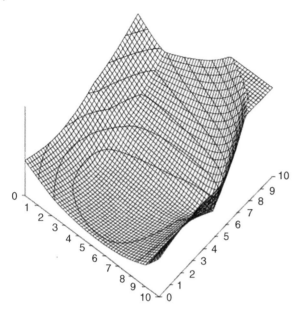

Figure E.11 Windblown surface.

```
' f_of_x = 0.6 * (f_of_x - 1 + 0.02 * (x1 - x2 + 3) ^ 2)  'OF adjusted for display
appearance and to keep
  solution in bounds
    If x1 < 0 Or x2 > 10 Then constraint = "FAIL"           'there is an off-graph
attractor seems to be at -
      infinity, + infinity
```

E.3.9 Windblown (#61)

A person wants to travel from point *A* to point *B* and chooses two linear paths from *A* to point *C* and then *C* to *B* so that the cumulative impact of the wind is minimized. Perhaps he is not wearing his motorcycle helmet, and doesn't want his hair to get messed up. The DVs are the $x1$ and $x2$ coordinates for point *C*. The wind blows in a constant (not stochastic) manner, but the wind velocity and direction change with location. The travel velocity is constant. If point *C* is out of the high windy area but far away, the travel time is high and the low wind experience persists for a long time. If point *C* is on the line between *A* and *B*, representing the shortest distance path and lowest time path, it takes the traveler into the high wind area, and even though the time is minimized, the cumulative wind damage is high. The wind damage impact is due to the square of the difference of the wind and travel velocity. Moving at 25 mph in the same direction as a 25 mph wind is blowing is like being in calm air. But traveling in the opposite direction is like standing in a 50 mph wind. The objective is to minimize the cumulative impact, the integral of the squared velocity difference along the *A–C–B* path (Figure E.11).

The function provides some discontinuities as evidenced by the kinks in the contours. And there are two minima: the global is to the front left of the lower contour, and the secondary is to the back right. Both minima are in relatively flat spots.

Suresh Kumar Jayaraman helped me explore this simulation of a path integral. The VBA code is as follows:

```
xc = x1      'Optimizer chooses point C
yc = x2
xa = 2       'User defines points A and B
ya = 4
xb = 9
yb = 6
N = 200      'Discretization number of intervals
velocity = 1.5       'Velocity of traveller
wind_coefficient = 0.1       'Coefficient for velocity of wind
                     'from A to C
xi = xa      'xi and yi are locations along the path
yi = ya
D1 = Sqr((xa - xc) ^ 2 + (ya - yc) ^ 2)     'total path distance
dD1 = D1 / N                      'path incremental length
time_interval = dD1 / velocity            'travel time along path increment
dxi1 = (xc - xa) / N              'incremental x increments
dyi1 = (yc - ya) / N              'incremental y increments
If xc = xa Then
   vx = 0                         'traveler x velocity
Else
   vx = velocity * (1 + ((yc - ya) / (xc - xa)) ^ 2) ^ -0.5
End If
If yc = ya Then
   vy = 0                         'traveler y velocity
Else
   vy = velocity * (((xc - xa) / (yc - ya)) ^ 2 + 1) ^ -0.5
End If
im1 = 0                          'integral of impact on path 1
For Path_Step = 1 To N
   If D1 = 0 Then Exit For
   xi = xi + dxi1
   yi = yi + dyi1
   wx = (-wind_coefficient * xi ^ 2 * yi) / Sqr((xi + 0.1) ^ 2 + (yi + 0.1) ^ 2)
   wy = (wind_coefficient * xi * yi ^ 2) / Sqr((xi + 0.1) ^ 2 + (yi + 0.1) ^ 2)
   lim = Sqr((vx - wx) ^ 2 + (vy - wy) ^ 2)
   im1 = im1 + lim
Next Path_Step
im1 = im1 * time_interval                'total impact scaled by time
                     'from C to B
xi = xc
yi = yc
D2 = Sqr((xc - xb) ^ 2 + (yc - yb) ^ 2)
dD2 = D2 / N
```

```
time_interval = dD2 / velocity
dxi2 = (xb - xc) / N
dyi2 = (yb - yc) / N
If xc = xb Then
  vx = 0
Else
  vx = velocity * (1 + ((yb - yc) / (xb - xc)) ^ 2) ^ -0.5
End If
If yc = yb Then
  vy = 0
Else
  vy = velocity * (((xb - xc) / (yb - yc)) ^ 2 + 1) ^ -0.5
End If
im2 = 0
For Path_Step = 1 To N
  If D2 = 0 Then Exit For
  xi = xi + dxi2
  yi = yi + dyi2
  wx = (-wind_coefficient * xi ^ 2 * yi) / Sqr((xi + 0.1) ^ 2 + (yi + 0.1) ^ 2)
  wy = (wind_coefficient * xi * yi ^ 2) / Sqr((xi + 0.1) ^ 2 + (yi + 0.1) ^ 2)
  lim = Sqr((vx - wx) ^ 2 + (vy - wy) ^ 2)
  im2 = im2 + lim
Next Path_Step
im2 = im2 * time_interval
f_of_x = im1 + im2
f_of_x = 10 * (f_of_x - 18) / (35 - 18)
f_of_x = f_of_x + add_noise
```

E.3.10 Integer Problem (#33)

This simple example represents a classic manufacturing application (Figure E.12). Minimize a function (perhaps maximize profit) of DVs $x1$ and $x2$ (perhaps the number of items of products A and B to make), subject to constraints (perhaps on capacity) and requiring $x1$ and $x2$ to be integers (you can only sell whole units).

There are many similar examples in textbooks.

The attributes of such applications are the level discontinuities (cliffs) as the integer value changes and the flat spots over the range of DV values that generate the same integer value. The surface is nonanalytic—derivatives are either zero or infinity. This illustration illustrates the constraint regions with a high OF value.

The VBA code is as follows:

```
x11 = Int(x1)
x22 = Int(x2)
f_of_x = 11 - (4 * x11 + 7 * x22) / 10
If 3 * x11 + 4 * x22 > 36 Then constraint = "FAIL"
If x11 + 8 * x22 > 49 Then constraint = "FAIL"
```

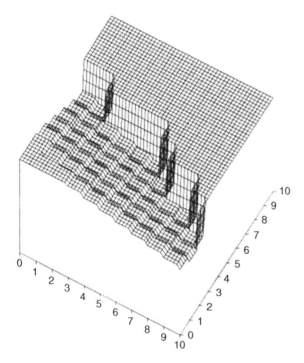

Figure E.12 An integer problem surface.

E.3.11   Reliability (#56)

This application represents a designer's choice of the number and size of parallel items for system success (Figure E.13). Consider a bank of exhaust fans needed to keep building air refreshed. The fans are operating in parallel. If one fails, air quality deteriorates. If there are three operating fans and one spare, when one fails, the spare can be placed online. This increases reliability of the operation but increases the cost of the fan assembly by 4/3. Even so, reliability is not perfect. There is a chance that two fans will fail, or three. Perhaps have three spares. Now, the reliability is high, but the cost is 6/3 of the original fan station. In either case, the cost needs to be balanced by the probability of the fan bank not handling the load. Alternately, it could be balanced by risk (the financial penalty of an event times the probability that an event will happen).

A clever cost reduction option is to use smaller capacity items but more of them. For example, if there are four operating fans, each only has to have 3/4 of the capacity of any one of the original three, each doing 1/3 duty. Using the common 6/10th power law for device cost, the smaller fans each cost $(3/4)^{0.6}$ of the original three, but there are four of them. So, the cost of the zero-spare situation with four smaller fans is higher than the cost of the three larger fans. The ratio is $4 * (3/4)^{0.6}/3 = 1.12$. However, if three spares are adequate, the cost of seven smaller fans is lower than the cost of six larger fans. The ratio is $7 * (3/4)^{0.6}/6 = 0.98$.

The optimization objective is to determine the number of operating units and the number of spare units to minimize cost with a constraint that the system reliability must be greater than 99.99%.

The realizable values of the DVs must be integers. This creates flat surfaces with cliff discontinuities (derivatives are either zero or infinity), and the constraint creates infeasible DV sets.

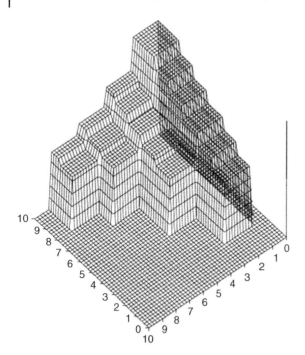

Figure E.13 Reliability surface.

The VBA code is as follows:

```
N = 3 + Int(8 * x1 / 10 + 0.5) 'total number of components
M = Int(5 * x2 / 10 + 0.5) 'number of operating components needed to meet
capacity
p = 0.2          'probability of any one component failing
q = 1 - p        'probability of any one component working
If M > N Then
   f_of_x = 0
   constraint = "FAIL"
   Exit Function
End If
If M > 0 Then
   f_of_x = N * (1 / M) ^ 0.6
Else
   f_of_x = 0
   constraint = "FAIL"
   Exit Function
End If
P_System_Success = 0
For im = 0 To N - M
  P_System_Success = P_System_Success + (factorial(N) / (factorial(im) *
factorial(N - im)))
  * (p ^ im) * (q ^ (N - im))
```

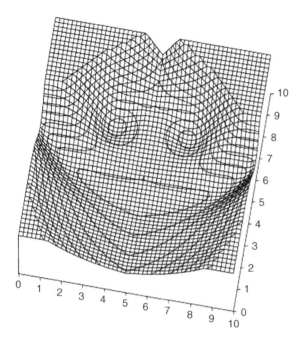

Figure E.14 Frog surface.

```
Next im
If P_System_Success < 0.999 Then
  f_of_x = 0
  constraint = "FAIL"
  Exit Function
End If
```

The factorial function is:

```
factorial = 1
If NUM = 1 Or NUM = 0 Then Exit Function
For iNUM = 2 To NUM
  factorial = factorial * iNUM
Next iNUM
```

E.3.12 Frog (#2)

I generated this function for students to enjoy as a final project for a freshman-level computer programming class at NCSU (Figure E.14). Students would know when they have the right answer!

The eyes represent equal global optima. There are also three minima in the mouth. To add difficulty for an optimizer, there is an oval constraint, a forbidden area, surrounding the eye in the upper left. Downhill-type optimizers get stuck on the constraint northwest of the eye. The face is also relatively flat, tricking some convergence criteria to stop early.

The VBA code is as follows:

```
f_of_x = 2 + 1 * (((x1 - 5) ^ 2) + ((x2 - 5) ^ 2)) * (0.5 - Exp((-((x1 - 3.5) ^ 2)) - _
    ((x2 - 7) ^ 2))) * (0.5 - Exp((-((x1 - 6.5) ^ 2)) - _
    ((x2 - 7) ^ 2))) * (0.5 + (Abs(Sqr(((x1 - 5) ^ 2) / ((x2 - 11) ^ 2)))) - _
    (Exp(-(Sqr(((x1 - 5) ^ 2) + ((x2 - 11) ^ 2)) - 7) ^ 2)))
If (x1 - 3.5) ^ 2 + (x2 - 7) ^ 4 <= 2 Then
    constraint = "FAIL"                              'Hard constraint approach
Else
    constraint = "OK"
End If
```

E.3.13 Hot and Cold Mixing (#36)

This function represents the control action required to meet steady-state mixed temperature and flow rate targets of 70°C and 20 kg/min from the current conditions of 35°C and 8 kg/min (Figure E.15). The DVs are the signals to the hot and cold valves.

Hot and cold fluids are mixed in line, and the objective is to determine the hot and cold flow control valve positions, o_1 and o_2, to produce the desired mixed temperature and total flow rate. The valves have a parabolic inherent characteristic and identical flow rate versus valve position response. The control algorithm is the generic model control (GMC) law with a steady-state model, and the control desire is to target for 20% beyond the biased set point, $K_c = 1.2$. There is uncertainty on the model

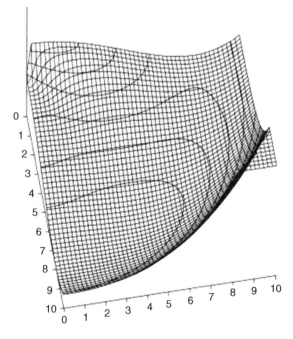

Figure E.15 Hot and cold mixing.

parameters of the supply temperatures, measured flow rate and temperature, and valve C_v. The controller objective is to determine o_1 and o_2 values that minimize the equal-concern-weighted deviations from target at steady state:

$$J = \frac{\left[K_c\left(T_{sp} - T_p\right) + T_p - \left(T_h\, o_1^2 + T_c\, o_2^2\right)/\left(o_1^2 + o_2^2\right)\right]^2}{E_T^2} + \frac{\left[K_c\left(F_{sp} - F_p\right) + F_p - C_v\left(o_1^2 + o_2^2\right)\right]^2}{E_F^2}$$

The first term in the OF relates to the temperature deviation, and the second term the flow rate deviation. The deviations are weighted by the equal-concern factors E_T and E_F. The numerator of each term starts with the calculated target value (beyond the set point) and subtracts from it the modeled value. The OF is the equal-concern-weighted sum of squared deviations.

The VBA code is as follows:

```
If x1 <= 0 Or x1 > 10 Or x2 <= 0 Or x2 > 10 Then
    constraint = "FAIL"
    Exit Function
  End If
  o1 = 10 * x1       'hot valve position, %
  o2 = 10 * x2       'cold valve position, %
  SetpointT = 70              'Celsius
  add_noise = Worksheets("Main").Cells(8, 12) * Sqr(-2 * Log(Rnd())) * Sin(2
* 3.14159 * Rnd())
  FromT = 35 * (1 + add_noise)    'Celsius
  SetpointF = 20              'm^3/min
  add_noise = Worksheets("Main").Cells(8, 12) * Sqr(-2 * Log(Rnd())) * Sin(2 *
3.14159 * Rnd())
  FromF = 8 * (1 + add_noise)     'm^3/min
  add_noise = Worksheets("Main").Cells(8, 12) * Sqr(-2 * Log(Rnd())) * Sin(2 *
3.14159 * Rnd())
  HotTin = 80 * (1 + add_noise)    'Celsius
  add_noise = Worksheets("Main").Cells(8, 12) * Sqr(-2 * Log(Rnd())) * Sin(2 *
3.14159 * Rnd())
  ColdTin = 20 * (1 + add_noise)   'Celsius
  add_noise = Worksheets("Main").Cells(8, 12) * Sqr(-2 * Log(Rnd())) * Sin(2 *
3.14159 * Rnd())
  ValveCv = 0.0036 * (1 + add_noise) 'm^3/min/%^2
  EC4T = 0.15                 'Celsius^(-2)
  EC4F = 1                    '(m^3/min)^(-2)
  f_of_x = EC4T * (1.2 * (SetpointT - FromT) + FromT - (HotTin * o1 ^ 2 + ColdTin
* o2 ^ 2) / (o1 ^ 2 + o2 ^ 2)) ^ 2 + EC4F * (1.2 * (SetpointF - FromF) + FromF -
ValveCv * (o1 ^ 2 + o2 ^ 2)) ^ 2
  f_of_x = f_of_x / 150
```

This is a simple function but provides substantial misdirection to a steepest descent optimizer that starts in the far side. It has steep walls, but a low slope at the proximity of the minimum. Some optimizers starting in the proximity of the optimum do not make large enough DV changes, and convergence criteria can stop them where they start.

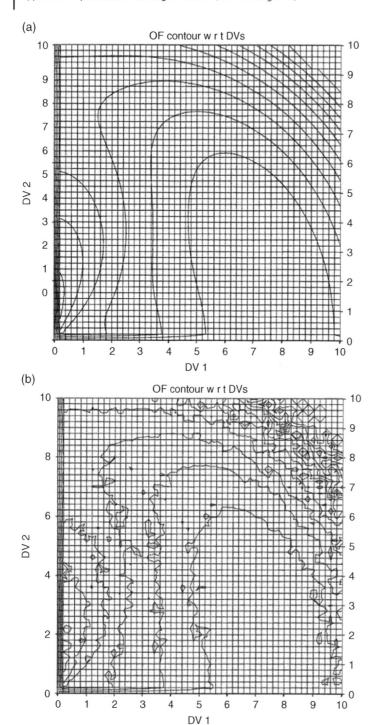

Figure E.16 Hot and cold mixing: (a) deterministic contour, (b) one contour realization with uncertainty, and (c) one surface realization with uncertainty.

(c)

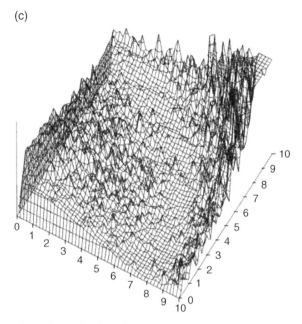

Figure E.16 (Continued)

With no uncertainty on model values, the contour of the 2-D search or o_1 and o_2 appears as Figure E.16a, which is interesting enough as a test case for nonlinear optimization. However, with a 5% nominal uncertainty (Figure E.16b), the contours are obviously irregular, and the 3-D plot of OF versus DVs (Figure E.16c) reveals the irregular surface.

E.3.14 Curved Sharp Valleys (#37)

This is a contrivance to provide a simple function with a slope discontinuity at the global minimum (in the valley near the lower right of Figure E.17, but in the interior, at about the point $x1 = 8$, $x2 = 3$). At the minimum the slope of the valley floor is low compared with the side walls. This means that from any point in the bottom of the valley, there is only a small directional angle to move to a lower spot. Nearly all directions point uphill. Also the valley is curved, so that once the right direction is found, it is not the right direction for the next move. Most optimizers will "think" they have converged when they are in the steep valley, and multiple runs will lead to multiple ending points that trace the valley bottom. The surface has another minimum in a valley in the upper right and a well-behaved local minimum up on the hill in the far right. Both the parameter correlation and the ARMA regression function have a similar feature. This exaggerates it and has a very simple formulation.

The VBA code is as follows:

```
f_of_x = 0.015 * (((x1 - 8) ^ 2 + (x2 - 6) ^ 2) + _
    15 * Abs((x1 - 2 - 0.001 * x2 ^ 3) * (x2 - 4 + 0.001 * x1 ^ 3)) - _
    500 * Exp(-((x1 - 9) ^ 2 + (x2 - 9) ^ 2)))
```

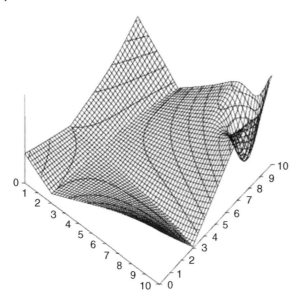

Figure E.17 Curved sharp valleys surface.

E.3.15 Parallel Pumps (#41)

This represents a redundancy design application (Figure E.18). A company has three identical centrifugal pumps in parallel in a single process stream. The pumps run at a constant impeller speed. They wish to have a method that chooses how many pumps should be operating and what flow rate should

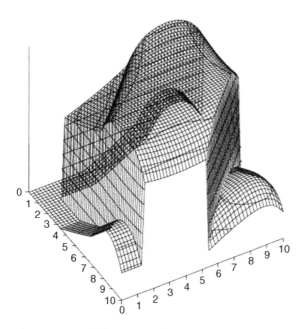

Figure E.18 Parallel pumps surface.

go through each operating pump to minimize energy consumption for a given total flow rate. The inlet and outlet pressures on the overall system of pumps remain a constant, but not on each individual pump. The individual flow rates out of each pump are controlled by a flow control valve.

The pump characteristic curves (differential pressure vs. flow rate) are described as

$$\Delta P_i = 1.22 \Delta P_{\text{nominal}} \left(1 - 0.017 e^{2.37 F_i / F_{\text{nominal}}} \right)$$

That means that at high flow rates, the outlet pressure will be low. The choice of the pump flow rate must create a pressure that is greater than the downstream line pressure.

The efficiency of each pump is roughly a quadratic relation to flow rate:

$$\epsilon_i = 0.7 r_i (2 - r_i), \quad r_i = \frac{F_i}{F_{\text{nominal}}}$$

The nominal flow rate rating is at the peak efficiency, but the pumps can operate at a flow rate that is 50% above nominal, $F_{\text{maximum}} = 1.5 F_{\text{nominal}}$. The energy consumption for any particular pump is the flow rate times the pressure drop divided by efficiency. And the total power is the sum for all operating pumps:

$$\dot{E} = \sum \dot{E}_i = \sum \frac{\Delta P_i F_i}{\epsilon} = \frac{1.22 \Delta P_{\text{nominal}} F_n{}^2}{0.7} \sum \frac{\left(1 - 0.017 e^{2.37 F_i / F_{\text{nominal}}} \right)}{(2 F_n - F_i)}$$

We prefer to not run a pump if its flow rate is less than 20% of nominal, because of the very low energy efficiency. We also wish to not use all three pumps unless absolutely needed—we have a fairly strong desire to keep a spare pump off-line for maintenance so that we always have one in peak condition. Although it might be best for energy efficiency when the target total flow rate is $3 * F_{\text{nominal}}$ to run all three pumps at F_{nominal}, the desire to keep a spare would prefer that we operate two pumps at $1.5 * F_{\text{nominal}}$.

This OF reveals surface discontinuities, cliffs, due to the switching of pumps and multiple optima.

The VBA code provides an option for either hard or soft constraints on the issue of running pumps when they would be less than 20% of nominal. The DVs, $x1$ and $x2$, represent the fraction of maximum of any two pumps. The third pump makes up the balance.

The code is as follows:

```
Qnominal = 6
Qmaximum = 1.5 * Qnominal
Qminimum = 0.2 * Qnominal
If x1 < 0 Or x2 < 0 Or x1 > 10 Or x2 > 10 Then constraint = "FAIL"
Qtotal = 15
o1 = x1 * Qmaximum / 10    'Transfer to permit an override if Q<min
o2 = x2 * Qmaximum / 10
penalty = 0
ConstraintType = "Hard"    ' "Soft"  ' Preference on running pumps less than
Qminimum
If ConstraintType = "Soft" And o1 < 0.1 * Qminimum Then o1 = 0 'Absolutely off
if a trivial flow rate
If ConstraintType = "Hard" And o1 < Qminimum Then o1 = 0
If ConstraintType = "Soft" And o1 < Qminimum Then penalty = penalty + 2 * (o1 -
0) ^ 2    'What to use for equal concern factors?
```

```
If ConstraintType = "Soft" And o2 < 0.1 * Qminimum Then o2 = 0
If ConstraintType = "Hard" And o2 < Qminimum Then o2 = 0
If Constraint Type = "Soft" And o2 < Qminimum Then penalty = penalty + 2 * (o2 -
0) ^ 2
o3 = Qtotal - o1 - o2
If o3 < 0 Or o3 > Qmaximum Then constraint = "FAIL"
If o3 > Qmaximum Then o3 = Qmaximum
If ConstraintType = "Soft" And o3 < 0.1 * Qminimum Then o3 = 0
If ConstraintType = "Hard" And o3 < Qminimum Then o3 = 0
If ConstraintType = "Soft" And o3 < Qminimum Then penalty = penalty + 2 * (o3 -
0) ^ 2
If Abs(o1 + o2 + o3 - Qtotal) > 0.0001 * Qtotal Then constraint = "FAIL" 'hard
constraint on production
If ConstraintType = "Hard" And 1.22 * (1 - 0.017 * Exp(2.37 * o1 / Qnominal)) <
0.7 Then constraint = "FAIL"     'Constraint on output pressure
If ConstraintType = "Hard" And 1.22 * (1 - 0.017 * Exp(2.37 * o2 / Qnominal)) <
0.7 Then constraint = "FAIL"
If ConstraintType = "Hard" And 1.22 * (1 - 0.017 * Exp(2.37 * o3 / Qnominal)) <
0.7 Then constraint = "FAIL"
If ConstraintType = "Soft" And 1.22 * (1 - 0.017 * Exp(2.37 * o1 / Qnominal)) <
0.7 Then penalty = penalty + 10 * (1.22 * (1 - 0.017 * Exp(2.37 * o1 / Qnominal))
- 0.7) ^ 2
If ConstraintType = "Soft" And 1.22 * (1 - 0.017 * Exp(2.37 * o2 / Qnominal)) <
0.7 Then penalty = penalty + 10 * (1.22 * (1 - 0.017 * Exp(2.37 * o2 / Qnominal))
- 0.7) ^ 2
If ConstraintType = "Soft" And 1.22 * (1 - 0.017 * Exp(2.37 * o3 / Qnominal)) <
0.7 Then penalty = penalty + 10 * (1.22 * (1 - 0.017 * Exp(2.37 * o3 / Qnominal))
- 0.7) ^ 2
f1 = 0
f2 = 0
f3 = 0
If o1 > 0 Then f1 = (1 - 0.017 * Exp(2.37 * o1 / Qnominal)) / (2 * Qnominal - o1)
'Calculate scaled energy rate
If o2 > 0 Then f2 = (1 - 0.017 * Exp(2.37 * o2 / Qnominal)) / (2 * Qnominal - o2)
If o3 > 0 Then f3 = (1 - 0.017 * Exp(2.37 * o3 / Qnominal)) / (2 * Qnominal - o3)
If o1 * o2 * o3 > 0 Then penalty = penalty + 1     'soft penalty if all three pumps
are running
```

E.3.16 Underspecified (#43)

There are an infinite number of simple functions to create that are underspecified that permit infinite solutions with the same OF value (Figure E.19). These arise in optimization applications that are missing one (or more) of the impacts of a DV, when the OF is incomplete. In this one contrivance, the code is as follows:

```
f_of_x = 2 * (x1 * x2 / 20 - Exp(-x1 * x2 / 20)) ^ 2
```

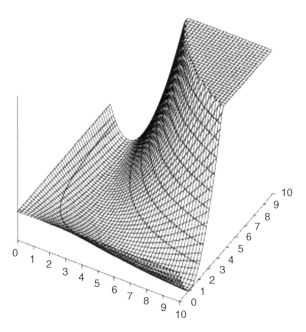

Figure E.19 Underspecified surface.

Since $x1$ and $x2$ always appear as a product, the function does not have independent influence by the DVs. Regardless of the value of $x1$, as long as the value of $x2$ makes the product equal 11.343, the function is at its optimum.

E.3.17 Parameter Correlation (#54)

This is similar to the underspecified case but brought about by an extreme condition that makes what appears to be independent functionalities reduce to a joint influence. I encountered this in modeling viscoelastic material using a hyper-elastic model for a nonlinear spring. The model computes stress, σ, given strain, ϵ:

$$\sigma = A\left(e^{B\epsilon} - 1\right)$$

The DVs, A and B, appear independent. But if ϵ is low, the product $B\epsilon$ is small and, in the limit, the exponential becomes $e^{B\epsilon} \cong 1 + B\epsilon$, resulting in $\sigma \cong AB\epsilon$ in which the value of coefficients A and B are dependent. Run many trials and plot the values of A w.r.t. B, and there will be a strong correlation. The 3-D plot happens to be very similar to that of the underspecified example, but there is a true minimum in the valley (Figure E.20).

The VBA code is as follows:

```
epsilon1 = 0.01        'no problem if epsilons = 0.1 and 0.2.  Then it doesn't
degenerate to a product
epsilon2 = 0.02
' the data are generated by three true values of x1=5 and x2=5.
'minimize the model deviations from data.
```

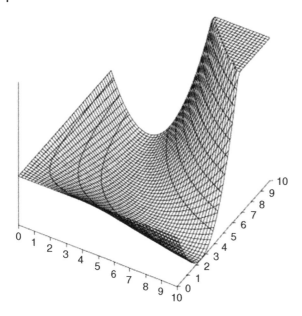

Figure E.20 Parameter correlation surface.

```
f_of_x = (x1 * (Exp(x2 * epsilon1) - 1) - 5 * (Exp(5 * epsilon1) - 1)) ^ 2 + (x1 *
(Exp(x2 * epsilon2) - 1) - 5 * (Exp(5 * epsilon2) - 1)) ^ 2
f_of_x = 10 * f_of_x   'minimize normal equations deviation from zero
```

E.3.18   Robot Soccer (#35)

The Automation Society at OSU (the Student Section of the International Society of Automation at OSU) creates an automation contest each year. In 2011 it was to create the rules for a simulated soccer defender to intercept the offensive opponent dribbling the ball toward the goal. The players start in randomized locations on their side of the field, and the offensive player has randomized evasive rules. Both players are subject to limits on speed and momentum changes. The ball-carrying opponent moves slightly slower than the defensive player. Figure E.21 shows one realization of the path of the two players and the final interception of the offensive player by the defender. The contest objective was to create motion rules (vertical and horizontal speed targets) for the defender to intercept the opponent. The OF was to maximize the fraction of successful stops in 100 games. If a perfect score, then the objective was to minimize the average playing time to intercept the opponent.

   Much thanks to graduate student Solomon Gebreyohannes for creating the code and simulator.

   Simple motion rules for the defender might be as follows: Always run as fast as possible. And run directly toward the offensive player. In that case, with too much defender momentum, an evasive move by the offensive player may permit the ball-carrier to get to the goal when momentum carries the defender past.

   The defender motion rules in the succeeding code are as follows: Extrapolate the opponent speed and direction to anticipate where it will be in "slead" seconds. Run toward that spot at a speed that gets you there in "stemper" reciprocal seconds. The scaled variables for "slead" and "stemper" are $x1$ and $x2$. The optimizer objective is to determine $x1$ and $x2$ to maximize fraction of successful stops and, if 100%, then to minimize playtime. This is a stochastic problem, and Figure E.22 shows the OF surface.

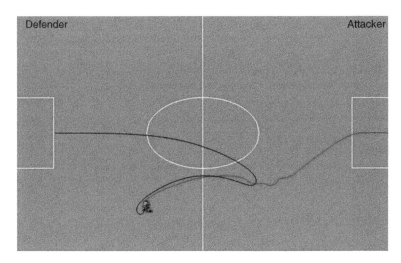

Figure E.21 Robot soccer field.

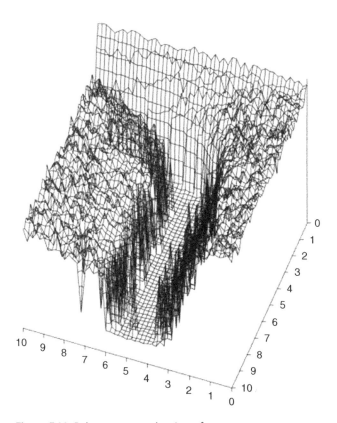

Figure E.22 Robot soccer stochastic surface.

$x1$, the lead time, is on the right-hand axis, and $x2$, the speed temper factor, is on the near axis. The rotation is chosen to provide a best view of the surface. The successful region is seen as a curving and relatively flat valley within steep walls.

The high surface is stochastic, and the value of any point changes with each sampling; and optimizers that start there will wander around in confusion seeking a local fortuitous best.

However, the floor of the valley is also stochastic. It represents the game time for speed rules that are 100% successful, and replicate DV values do not produce identical OF values. Accordingly, optimizers based on surface models will be confounded.

Further, the motion rules limit the defender speed, so that regardless of combinations of desired motion rules, the player cannot move any faster, and the floor of the valley is truly flat. It would be underspecified, except for the stochastic OF fluctuations.

The VBA code is as follows:

```
Dim GameTimeCounter As Integer
If x1 < 0 Or x1 > 10 Or x2 < 0 Or x2 > 10 Then
    constraint = "FAIL"
    Exit Function
End If
slead = 3 * (x1 - 3) 'DV 1
stemper = (x2 - 0) / 10 'DV 2
NumGames = 50   '500
GameTimeCounterMax = 10000
Distance = 0
SuccessCount = 0
undercount = 0
sdt = 0.03
Lamdad = 0.2          'momentum constraints on speed rate of change
CLamdad = 1 - Lamdad
Lamdaa = 0.2
CLamdaa = 1 - Lamdaa
AAngle = 3.14159
For GameNumber = 1 To NumGames
  Randomize
  'Player defender initialize
  xd = 1
  yd = 5
  vxd = 0
  vyd = 0
  'Player attacker initialize
  xa = 8 + 2 * Rnd()   'Int(Rnd() * 3)
  ya = 10 * Rnd()      'Int(Rnd() * 11)
  vxa = -2 * Rnd()
  vya = 2 * (Rnd() - 0.5)
  For GameTimeCounter = 1 To GameTimeCounterMax
    'Changing the attacker vx and vy speed
    If xa > 3 Then
      separation = Sqr((xa - xd) ^ 2 + (ya - yd) ^ 2)
```

```
      If separation < 3 And xa > xd Then
         LoAngle = 2 * 3.14159 * 45 / 360
         HiAngle = 2 * 3.14159 * 315 / 360
         If GameTimeCounter = 3 * Int(GameTimeCounter / 3) Then AAngle =
         LoAngle + Rnd() * (HiAngle - LoAngle)
      Else
         LoAngle = 2 * 3.14159 * 120 / 360
         HiAngle = 2 * 3.14159 * 250 / 360
         If GameTimeCounter = 10 * Int(GameTimeCounter / 10) Then AAngle =
         LoAngle + Rnd() * (HiAngle - LoAngle)
      End If
   Else
      LoAngle = 3.14159 + Atn((6.5 - ya) / (-xa))
      HiAngle = 3.14159 + Atn((3.5 - ya) / (-xa))
      If GameTimeCounter = 3 * Int(GameTimeCounter / 3) Then AAngle =
   LoAngle + Rnd() * (HiAngle - LoAngle)
   End If
   ASpeed = 4 + Rnd() * (5 - 4)
   vxa = CLamdaa * vxa + Lamdaa * ASpeed * Cos(AAngle)
   vya = CLamdaa * vya + Lamdaa * ASpeed * Sin(AAngle)
 xa = xa + vxa * sdt      'Calculate current position of player 2 (Attacker)
   ya = ya + vya * sdt
   'Field constraints
   If ya > 10 Then ya = 10
   If ya < 0 Then ya = 0
   If xa > 10 Then xa = 10
   If xa < 0 Then xa = 0
  ' Defender rules - Anticipate where target will be at next interval and
seek to get there
      xdtarget = xa + slead * sdt * vxa         'anticipated attacker location
      leadnumber of dts intot he future
      ydtarget = ya + slead * sdt * vya
      Ux = stemper * (xdtarget - xd) / sdt   'speed to get defender there -
tempered faster or slower
      Uy = stemper * (ydtarget - yd) / sdt
      Utotal = Sqr(Ux ^ 2 + Uy ^ 2)
      If Utotal > 5 Then
         Ux = Ux * 5 / Utotal               'constrained defender target speed, 5 is
the max
         Uy = Uy * 5 / Utotal
      End If
      'Calculate new speed and position of the defender using calculated Ux
and Uy
      vxd = CLamdad * vxd + Lamdad * Ux
      vyd = CLamdad * vyd + Lamdad * Uy
      xd = xd + vxd * sdt
```

```
      yd = yd + vyd * sdt
      'Field constraints
      If yd > 10 Then yd = 10
      If yd < 0 Then yd = 0
      If xd > 10 Then xd = 10
      If xd < 0 Then xd = 0
      If xa <= 0 And ya >= 3.5 And ya <= 6.5 Then Exit For   'Goal scored
      'Check if both players are in 0.2 radius region
      If Sqr((xd - xa) ^ 2 + (yd - ya) ^ 2) < 0.2 Then
        SuccessCount = SuccessCount + 1
        PlayTime = PlayTime + sdt * GameTimeCounter            'play time to
intercept
        Distance = Distance + Sqr((xa - 0) ^ 2 + (ya - 5) ^ 2)    'stopped this
distance from goal
        Exit For
      End If
   Next GameTimeCounter
 Next GameNumber
 AvgDistance = Distance / NumGames       'larger is better
 AvgPlayTime = PlayTime / NumGames       'smaller is better
 SuccessRatio = SuccessCount / NumGames
 If SuccessRatio = 1 Then                'Only important if zero goals scored
   f_of_x = AvgPlayTime
 Else
   f_of_x = 10 - 5 * SuccessRatio   'if a goal is scored, then playtime and
distance are unimportant
 End If
```

E.3.19   ARMA(2, 1) Regression (#62)

This optimization seeks the coefficients in an equation to model a second-order process. The true process is $y_i = ay_{i-1} + by_{i-2} + u_{i-1}$. Here y is the process response; u is the process independent variable, influence; and i is the time counter. The term u, the forcing function, is expressed as the variable labeled "push" in the VBA assignment statements. The model has the same functional form as the process. The optimizer objective is to determine model coefficient values that minimize the sum of squared differences between data and model. A good optimizer will find the right values of $a = 0.3$ and $b = 0.5$.

The difficulty of this problem is that the true solution is in the middle of a long valley of steep walls and very gentle longitudinal slope. Like the steep valley and parameter correlation functions, when the trial solution is in the bottom of the valley, optimizers have difficulty in moving it along the bottom to the optimum (see Figure E.23).

The VBA code is as follows:

```
x11 = 0.3 + (x1 - 5) / 100
x22 = 0.5 + (x2 - 5) / 100
y_trueOld = 0
```

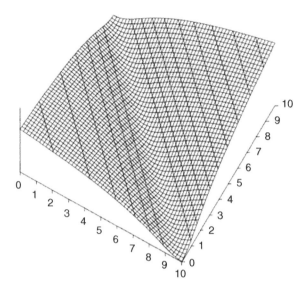

Figure E.23 ARMA (2, 1) regression surface.

```
y_trueOldOld = 0
y_modelOld = 0
y_modelOldOld = 0
Sum = 0
For i_stage = 1 To 100
   If i_stage = 1 Then push = 5
   If i_stage = 20 Then push = 0
   If i_stage = 40 Then push = -2
   If i_stage = 60 Then push = 7
   If i_stage = 80 Then push = 3
   y_true = 0.3 * y_trueOld + 0.5 * y_trueOldOld + push
   y_model = x11 * y_modelOld + x22 * y_modelOldOld + push
   y_trueOldOld = y_trueOld
   y_trueOld = y_true
   y_modelOldOld = y_modelOld
   y_modelOld = y_model
   Sum = Sum + (y_true - y_model) ^ 2
Next i_stage
rms = Sum / (i_stage - 1)
f_of_x = 2 * Log(rms + 1)
```

E.3.20   Algae Pond (#63)

This seeks to determine an economic optimization of an algae farm. Algae are grown in a pond, in a batch process, and in wastewater from a fertilizer plant that contains needed nutrients. The algae biomass is initially seeded in the pond, grows naturally in the sunlight, and eventually is harvested. The

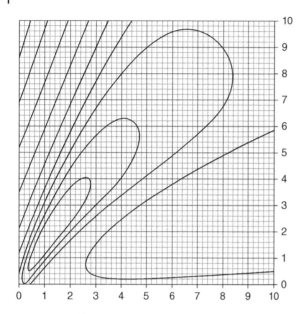

Figure E.24 Algae pond contour.

questions are "How deep should the pond be?" and "How long a time between harvestings?" Considering depth, a deeper pond has greater volume and can grow more algae. But sunlight attenuates with pond depth; so, a too deep pond does not add growing volume. Instead, a too deep pond creates more water that must be processed in harvesting, which increases costs.

Considering growth time, initially, there are no algae in the water. The initial seeded batch grows in time, but the mass asymptotically approaches a steady-state concentration when growth rate matches death rate. Growth is limited by CO_2 infusion and sunlight, and high concentration of biomass shadows the lower levels. If you harvest on a short time schedule, you have a low algae concentration and produce low mass on a batch but still have to process all the water. Alternately, if you harvest on a long time schedule, the excessive weeks don't contribute biomass and you lose batches of production per year.

The simulation uses a simple model of the algae growth, and I appreciate Suresh Kumar Jayaraman's role in devising the simulator. The objective function is to maximize annual profit, which is based on value of the mass harvested per year less harvest costs and fixed costs.

The DVs are scaled to represent 2–25 batch growth days (horizontal axis) and 0.1–2 m of pond depth (vertical axis). The minimum is in the lower left of the contour, a steep valley. The feature in the lower right of Figure E.24 is a broad maximum that sends downhill searches either upward to the valley or downward toward the axis representing a very shallow pond.

There are three difficulties with this problem. First, the steep valley causes difficulty for some gradient-based optimizers. Second, the nearly flat plane in the lower right can lead to premature convergence. Third, the surface has slope discontinuities due to the time discretization of the simulation. These can be made more visible with a larger simulation time increment and appear as irregularities on the contour lines on the left. Ideally, the function is continuous in time, and the contours are smooth. Larger simulation discretization time intervals permit the simulation to run faster, which is desirable; but this has the undesired impact of creating discontinuities. Even with the small discretization used to generate the contours on the right, where there are no visible discontinuities, the surface undulations are present and will confound gradient- and Hessian-based searches.

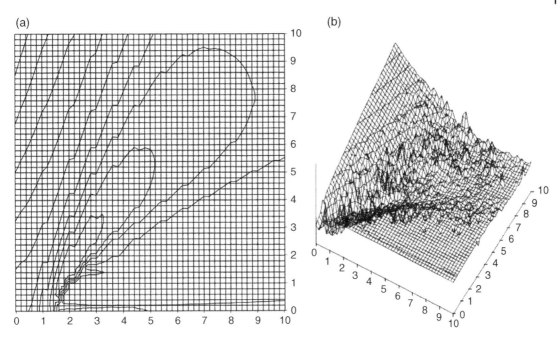

(a) (b)

Figure E.25 Algae pond: (a) stochastic contour and (b) stochastic surface.

One can add a fourth difficulty, uncertainty on parameter and coefficient values in the model. Which creates a stochastic "live" fluctuations to the surface, as illustrated in Figure E.25.
The VBA code is as follows:

```
If x2 < 0 Or x1 < 0 Or x1 > 10 Or x2 > 10 Then
    constraint = "FAIL"
    Exit Function
End If
Dim C68(5000) As Single
Dim p68(5000) As Single
Dim n68(5000) As Single
Dim profit68 As Single
add_noise = Worksheets("Main").Cells(8, 12) * Sqr(-2 * Log(Rnd())) * Sin
(2 * 3.14159 * Rnd())
    Izero68 = 6 * (1 + add_noise)
    add_noise = Worksheets("Main").Cells(8, 12) * Sqr(-2 * Log(Rnd())) * Sin
(2 * 3.14159 * Rnd())
    K168 = 0.01 * (1 + add_noise) 'growth rate constant
    add_noise = Worksheets("Main").Cells(8, 12) * Sqr(-2 * Log(Rnd())) * Sin
(2 * 3.14159 * Rnd())
    K268 = 0.03 * (1 + add_noise) 'death rate constant
    add_noise = Worksheets("Main").Cells(8, 12) * Sqr(-2 * Log(Rnd())) * Sin
(2 * 3.14159 * Rnd())
```

```
    alpha68 = 5 * (1 + add_noise)
    add_noise = Worksheets("Main").Cells(8, 12) * Sqr(-2 * Log(Rnd())) * Sin
(2 * 3.14159 * Rnd())
    tr68 = 288 * (1 + add_noise)  'pond temperature
    add_noise = Worksheets("Main").Cells(8, 12) * Sqr(-2 * Log(Rnd())) * Sin
(2 * 3.14159 * Rnd())
    topt68 = 293 * (1 + add_noise)   'optimum temperature
    add_noise = Worksheets("Main").Cells(8, 12) * Sqr(-2 * Log(Rnd())) * Sin
(2 * 3.14159 * Rnd())
    kt68 = 0.001 * (1 + add_noise)  'temperature rate const
    add_noise = Worksheets("Main").Cells(8, 12) * Sqr(-2 * Log(Rnd())) * Sin
(2 * 3.14159 * Rnd())
    kp68 = 0.02 * (1 + add_noise)  'P growth rate const
    add_noise = Worksheets("Main").Cells(8, 12) * Sqr(-2 * Log(Rnd())) * Sin
(2 * 3.14159 * Rnd())
    kn68 = 0.02 * (1 + add_noise)  'N growth rate const
    add_noise = Worksheets("Main").Cells(8, 12) * Sqr(-2 * Log(Rnd())) * Sin
(2 * 3.14159 * Rnd())
    dt68 = 0.01 * (1 + add_noise)  'time increment (hr) 0.1 reveals
discretization discontinuities
    simtime68 = 2 + x1 * (25 - 2) / 10     'time: min is 2 days and max is 25 days
    Depth68 = 0.1 + x2 * (2 - 0.1) / 10     'depth: min is 0.1 m and max is 2 m
    M68 = simtime68 / dt68
    rp68 = 0.025  'gP/gAlgae
    np68 = 0.01  'gN/gAlgae
    pc68 = 0.0001  'critical P conc
    nc68 = 0.0001  'critical N conc
    hn68 = 0.0145  'half saturation N conc
    hp68 = 0.0104  'half saturation P conc
    C68(1) = 0.1  'scaled value - starts at 1% of maximum possible
    p68(1) = 0.1  'initial P conc
    n68(1) = 0.1  'initial N conc
    For i68 = 1 To M68
     'dependence of the biomass growth on temperature f(T), intensity f(I),
     phosphorus f(P), Nitrogen F(N)
      If (p68(i68) - pc68) > 0 Then
        fp68 = ((p68(i68) - pc68)) / (hp68 + (p68(i68) - pc68))
      Else
        fp68 = 0
      End If
      If (n68(i68) - nc68) > 0 Then
        fn68 = ((n68(i68) - nc68)) / (hn68 + (n68(i68) - nc68))
      Else
        fn68 = 0
      End If
      ft68 = Exp(-kt68 * (tr68 - topt68) ^ 2)
      fi68 = (Izero68 / (alpha68 * Depth68)) * (1 - Exp(-alpha68 * Depth68))
```

```
      p68(i68 + 1) = p68(i68) + dt68 * (kp68 * (-fi68 * ft68 * fp68 * fn68 * rp68)
* C68(i68))
       If p68(i68 + 1) < 0 Then p68(i68 + 1) = 0
       n68(i68 + 1) = n68(i68) + dt68 * (kn68 * (-fi68 * ft68 * fp68 * fn68 * np68)
* C68(i68))
       If n68(i68 + 1) < 0 Then n68(i68 + 1) = 0
       C68(i68 + 1) = C68(i68) + dt68 * (((fi68 * ft68 * fp68 * fn68) - K268 *
C68(i68)) * C68(i68))
       If C68(i68 + 1) < 0 Then C68(i68 + 1) = 0
    Next i68
    avg68 = C68(i68 - 1)
    sales68 = 1 * avg68 * 25 * Depth68 * 365 / (simtime68 + 5)   '$/year  5 days to
process pond
    expenses68 = 10 * Depth68 * 25 * 365 / (simtime68 + 5)       '$/year  cost to
process and stock
    profit68 = sales68 - expenses68
    f_of_x = -profit68
    f_of_x = 10 * (f_of_x + 16700) / 43000   'for 25 days and 2 meters
```

E.3.21   Classroom Lecture Pattern (#73)

Lectures are effective in the beginning when students are paying attention. But as time drags on, student attention wanders, and fewer and fewer students are sufficiently intently focused to make the continued lecture be of any use. A "logistic" model that represents the average attention is

$$f = \frac{1}{1 + e^{(t-15)}}$$

where f is the fraction of class attention and t is the lecture time in minutes. Graph this, and you'll see that after 15 min $f = 0.5$, half the class is not paying attention. After about 20 min, effectively no one is attentive to the lecturer. The average f over a 55-min period is about 0.26. Students only "get" 26% of the material. That is not very efficient. A teacher wants to maximize learning, as measured by maximizing the average f over a lecture. He/she plans on doing this by taking 3-min breaks in which the students stand up, stretch, and slide over to the adjacent seat and then resuming the lecture with fully renewed student attention ($t = 0$ again). If a teacher lectured 2 min and took a 3 min break, on a 5-min cycle, then the value of f would be nearly 1.00 for the 2 min, but there would only be 2 * (55/5) = 22 min of lecture per session. But 22 min at an f of 0.9999 over the 55 min is an average f of 0.40. This is an improvement in student focus on the lecture material. Is there a better lecture-stretch cycle period?

   It can be argued that the lecture segments should have an equal duration. If there is an optimal duration, then exceeding it in one segment loses more than what is not lost in the other. There would be n equal duration lecture segments and $(n-1)$ 3-min breaks within in a class period of p total minutes:

$$nx + (n-1)3 = p$$

How long should a class period be to maximize learning? Say, a school wants to have seven lecture periods in a day so that students have a one-period lunch break, when they are enrolled in six classes each semester. And the schedule needs to permit 10 min for between class transitions. Longer classes would permit more learning. The work-day length would be

$$l = 7p + (7-1)^*10$$

But an excessive day would receive objection from the faculty member's home. So, let's maximize learning by choosing p and n, but penalize an excessive day.

The code is as follows:

```
n_segments = Int(1 * x1)    '# segments between 3-min breaks, integer value
Period = 10 * x2       'lecture duration, continuous, minutes
If n_segments > Period / 3 + 1 Or n_segments < 1 Then
        constraint = "FAIL"
        Exit Function
End If
Sum = 0    'initialize integral of student attention
For x11 = 0 To ((Period + 3) / n_segments - 3) Step 0.001
        Sum = Sum + 0.001 / (1 + Exp(x11 - 15))
Next x11
Duration = 7 * Period + 6 * 10    'work-day length planning 6 class periods
per day
penalty = 0
If Duration > 10 * 60 Then penalty = 0.05 * (Duration - 10 * 60) ^ 2
f_of_x = -7 * n_segments * Sum + penalty
f_of_x = 10 * (f_of_x + 458) / (1617)
```

This is a mixed integer continuous DV problem. The front axis in Figure E.26 represents the number of lecture segments within a class period, and the RHS axis represents classroom period in 10-min intervals.

The horizontal direction in each segment is flat and becomes discontinuous when the n integer value changes.

Not visible in this view is the additional problem of the time discretization in using the rectangle rule of integration to solve for the average f-value. With a time step of 0.001 min, it is not visible. But with a step size of 1 min, the discontinuity along the "continuous" "p" axis becomes visible.

E.3.22 Mountain Path (#29)

This uses the peaks function to represent the land contour of a mountain and valley region. City A is at location $(1, 1)$ and city B at location $(9, 9)$ in Figure E.27 (see Figure E.1 for a 3-D view). The objective is to find a path from A to B that minimizes distance and avoids excessive steepness along the path. The path is described as a cubic relation of y (the vertical axis in the peaks contour) and x (the horizontal axis in the peaks contour):

$$y = a + bx + cx^2 + dx^3$$

If the path follows a straight line on the x–y projection surface, from $(1, 1)$, the path goes over the mountain at $(4, 4)$, down the valley at $(5.5, 5.5)$, over the side of the mountain at $(7, 7)$, and down to $(9, 9)$. Alternately, if the path goes from $(1, 1)$ to $(5, 3)$, it stays on relatively level ground. Then to $(7, 6.5)$ it does not go into the middle valley and only rises to the lowest point (the pass) between the two mountains in the upper right. While the projection of the curved path might be longer, it does not rise or fall and is actually a shorter path.

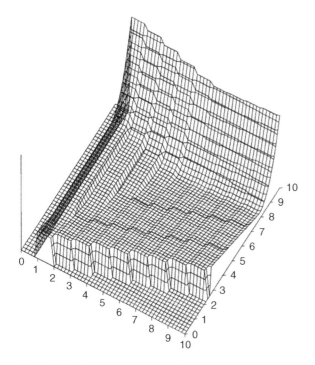

Figure E.26 Classroom lecture pattern.

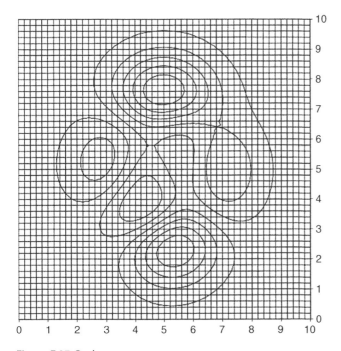

Figure E.27 Peaks contour.

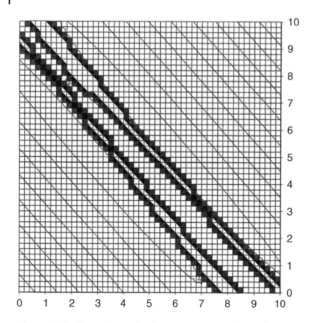

Figure E.28 Mountain path of contour w.r.t. path coefficients c (horizontal) and d (vertical).

A section of an alternate path might go from (6, 6) to (8, 7), taking the shortest distance across the mountain pass, but this might have a steep climb, about (6.5, 6.5). An alternate path through the pass, from (6, 4) to (8, 8), sideways up the mountain would not have the steep climb.

In the peaks function the DVs are the x- and y-direction, and the objective is to find the min (or max) point. By contrast, here, the DVs are coefficients c and d in the $y(x)$ relation. Once the optimizer chooses trial values for c and d, a and b are determined by the equality constraints of $y(\text{at } xA) = yA$ and $y(\text{at } xB) = yB$. The path length includes the x, y, and z (elevation) distances, and the function calculates path length by discretizing the path into 100 Δx increments and summing the 100 $\Delta s = \sqrt{\Delta x^2 + \Delta y^2 + \Delta z^2}$ elements. The steepness of the path is evaluated at each of the 100 increments, and if $|dz/ds|$ exceeds a threshold, the constraint is violated.

The objective statement is

$$\min_{\{c,d\}} J = S = \sum \Delta s$$

$$\text{S.T.} \quad \left| \frac{dz}{ds} \right| \le \text{slope}$$

The function $S(c, d)$ results in multiple local optima comprised of steep, narrow valleys with low sloping floor. The dark bands in Figure E.28 indicate steep rises to a maximum value at constrained conditions. The minimum is in the (DV1 = 7, DV2 = 2) area. However, if trial solutions start outside of that area, constraints would prevent the trial solution from migrating to the optimum. Many optimizers migrate to local optima at (4, 4) and (5, 6).

The VBA code is as follows:

```
CoeffC = 0.2 * x1 - 1.8 'scaled for best reveal of the function detail
CoeffD = 0.012 * x2 + 0
xa = 1          'start location - E-W direction
ya = 1          'start location - N-S direction
xb = 9          'end location
yb = 9
CoeffB = ((yb - ya) - CoeffC * (xb ^ 2 - xa ^ 2) - CoeffD * (xb ^ 3 - xa ^ 3)) / (xb
- xa)
CoeffA = ya - CoeffB * xa - CoeffC * xa ^ 2 - CoeffD * xa ^ 3
N = 100
deltax = (xb - xa) / N
PathS = 0
MaxSteepness = 0
For xstep = 1 To N
    x = xa + xstep * deltax
    y = CoeffA + CoeffB * x + CoeffC * x ^ 2 + CoeffD * x ^ 3
    dydx = CoeffB + 2 * CoeffC * x + 3 * CoeffD * x ^ 2 'derivative of y w.r.t. x
    x11 = 3 * (x - 5) / 5    'convert my 0-10 DVs to the -3 to +3 range for the
function
    x22 = 3 * (y - 5) / 5
    zbase = 3 * ((1 - x11) ^ 2) * Exp(-1 * x11 ^ 2 - (x22 + 1) ^ 2) - _
        10 * (x11 / 5 - x11 ^ 3 - x22 ^ 5) * Exp(-1 * x11 ^ 2 - x22 ^ 2) - _
        (Exp(-1 * (x11 + 1) ^ 2 - x22 ^ 2)) / 3
    zbase = (zbase + 6.75) / 1.5
    x11 = 3 * (x + 0.0001 - 5) / 5 'convert my 0-10 DVs to the -3 to +3 range for the
function
    Z = 3 * ((1 - x11) ^ 2) * Exp(-1 * x11 ^ 2 - (x22 + 1) ^ 2) - _
        10 * (x11 / 5 - x11 ^ 3 - x22 ^ 5) * Exp(-1 * x11 ^ 2 - x22 ^ 2) - _
        (Exp(-1 * (x11 + 1) ^ 2 - x22 ^ 2)) / 3
    Z = (Z + 6.75) / 1.5
    pzpx = (Z - zbase) / 0.0001    'partial derivative of elevation w.r.t. x
    x11 = 3 * (x - 5) / 5    'convert my 0-10 DVs to the -3 to +3 range for the
function
    x22 = 3 * (y + 0.0001 - 5) / 5
    Z = 3 * ((1 - x11) ^ 2) * Exp(-1 * x11 ^ 2 - (x22 + 1) ^ 2) - _
        10 * (x11 / 5 - x11 ^ 3 - x22 ^ 5) * Exp(-1 * x11 ^ 2 - x22 ^ 2) - _
        (Exp(-1 * (x11 + 1) ^ 2 - x22 ^ 2)) / 3
    Z = (Z + 6.75) / 1.5
    pzpy = (Z - zbase) / 0.0001    'partial derivative of elevation w.r.t. y
    dzdx = pzpx + pzpy * dydx    'total derivative of elevation w.r.t. x
    dsdx = Sqr(1 + dydx ^ 2 + dzdx ^ 2) 'total derivative of path length w.r.t. x
    dPathS = deltax * dsdx    'Path segment length
    PathS = PathS + dPathS    'Cumulative path length
```

```
       dzds = dzdx / dsdx        'Path slope
       Steepness = Abs(dzds)
       If Steepness > MaxSteepness Then MaxSteepness = Steepness
       badness = 0
       '  Soft Constraint
'        If Steepness > 0.8 Then    '  .85=tan(40), .60=tan(30), .35=tan(20)
'          badness = (Steepness - 0.8)
'          PathS = PathS + 2 * badness ^ 2
'        End If
       '  Hard Constraint
       If Steepness > 0.95 Then    '  .85=tan(40), .60=tan(30), .35=tan(20)
         constraint = "FAIL"
         f_of_x = 10          ' visually acknowledge violation on contour
         Exit Function
       End If
     Next xstep
   f_of_x = PathS
   f_of_x = 10 * (PathS - 13.28) / 52.36
```

E.3.23   Chemical Reaction with Phase Equilibrium (#75)

A reversible homogeneous reaction of $A + B$ goes to C seems simple; but there are two immiscible phases, oil and water, and the A, B, and C species are dispersed in both. The A and B species are preferentially soluble in the water, and C is preferentially soluble in the oil. The reaction is slow relative to mass transfer and diffusion, so the A, B, and C species are considered in thermodynamic equilibrium between the oil and water phases:

1) The reaction in the oil layer of volume v is $a + b \leftrightarrow c$, $r = -k1ab + k2c$. Lower case letters represent oil.
2) In the water layer of volume, V is $A + B \leftrightarrow C$, $R = -k3AB + k4C$. Uppercase letters represent water.
3) Each component is in concentration equilibrium between the two phases $KA = a/A$, $KB = b/B$, $KC = c/C$, because the reactions are relatively slow compared with the rate of mass transfer between phases.
4) The total number of moles of each species are NA ($=va + VA$), NB, and NC.
5) Given an initial number of moles, $NA0$, and phase equilibria, the number of moles in each phase are determined by $a0 = NA0/(v + V/KA)$ and $A0 = NA0/(vKA + V)$. bo, co, Bo, and Co are similarly calculated.
6) Combined mass balances on species in the individual oil and water phases, eliminate the unknowable terms for the interphase rates, and result in

$$\frac{dNA}{dt} = vr + VR = -vk1ab + vk2c - Vk3AB + Vk4C$$

7) Equilibrium relates the a and A concentrations to NA as in line 5. Stoichiometry and equilibrium relate the b, c, B, and C concentrations to NA. Concentrations a, b, c, A, B, and C can be

reformulated in terms of NA. Stoichiometry gives $NB = NB0 - (NA - NA0)$. Equilibrium gives $b = NB/(v + V/KB)$. Combined, $(NB0 - (NA - NA0))/(v + V/KB)$. Etc.
8) There are four rate coefficients ($k1, k2, k3, k4$); however, only two are independent. The reaction is the same whether in the oil phase or the water phase. But the reaction is not really dependent on concentrations. It is dependent on activities of the species in the oil and water medium. Further, the equilibrium of a to A (b to B and c to C) are defined as equal activities. This means that the activity coefficients are related to the partition coefficient. $KA = \gamma a/\gamma A$. Etc. Then $k3 = k1 * KA * KB$, and $k4 = k2 * KC$. So,

$$\frac{dNA}{dt} = vr + VR = -vk1ab + vk2c - V(k1KAKB)AB + V(k2KC)C$$

9) This differential equation represents a phase-equilibrium constrained model.
10) Only the one nonlinear ODE needs to be solved, because all species concentrations are related to the single NA and the initial conditions on $NA0$, $NB0$, and $NC0$.
11) This can be expressed in terms of reaction extent:

$$\xi = NAo - NA$$

$$\frac{d\xi}{dt} = vk1ab - vk2c + V(k1KAKB)AB - V(k2KC)C$$

12) which is solved with a simple Euler's method.

The optimization seeks to find $k1$ and $k2$ values that make the dynamic model best match the experimental data in a least squares sense. The function in Figure E.29 has the "steep valley" property that makes many optimizers converge along the bottom, with parameter correlation.

The VBA code is as follows:

```
k1075 = 200 + 200 * (x1 - 5) / 5     'oil forward reaction rate coefficient
k2075 = 0.015 + 0.015 * (x2 - 5) / 5  'oil backward reaction rate coefficient
If k1075 < 0 Or k2075 < 0 Then
   constraint = "FAIL"
   Exit Function
End If
sum75 = 0    'sum of penalties for deviations from measured values
dt75 = 0.5   'time increment for Euler's method, min
vo75 = 100   'volume of oil, liters, human supposition
VW75 = 200   'volume of Water, liters, human supposition
NA75 = 5     'initial moles of A, combined, in oil and water, human
supposition
NB75 = 3     'initial moles of B, human supposition
NC75 = 1     'initial moles of C, human supposition
KA75 = 0.1   'partition coefficient for A, human supposition
KB75 = 0.05   'for B, human supposition
KC75 = 5.5   'for C, human supposition
k1W75 = k1075 * KA75 * KB75   'water forward reaction rate coefficient
k2W75 = k2075 * KC75          'water backward reaction rate coefficient
e75 = 0                      'reaction extent sum of oil and water
```

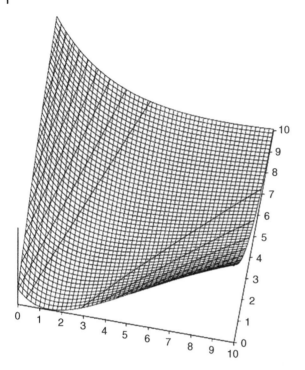

Figure E.29 Chemical reaction with phase-equilibrium surface.

```
For i75 = 1 To 80          '80 time steps for dynamic model
   a075 = (NA75 - e75) / (v075 + VW75 / KA75)  'concentration of A in oil
   b075 = (NB75 - e75) / (v075 + VW75 / KB75)  'B in oil
   c075 = (NC75 + e75) / (v075 + VW75 / KC75)  'C in oil
   AW75 = (NA75 - e75) / (v075 * KA75 + VW75)  'A in water
   BW75 = (NB75 - e75) / (v075 * KB75 + VW75)  'B in water
   CW75 = (NC75 + e75) / (v075 * KC75 + VW75)  'C in water
   r075 = k1075 * a075 * b075 - k2075 * c075   'reaction rate in oil
   RW75 = k1W75 * AW75 * BW75 - k2W75 * CW75   'reaction rate in water
   e75 = e75 + dt75 * (v075 * r075 + VW75 * RW75)          'Euler's method
   t75 = i75 * dt75                        'simulation time
        'data generated by Excel simulator - "Kinetic Model with Phase
        Equilibrium Sheet 1"
   If i75 = 8 Then sum75 = sum75 + (e75 - 0.235176) ^ 2      'penalty for
deviation from data
      If i75 = 16 Then sum75 = sum75 + (e75 - 0.361498) ^ 2
      If i75 = 24 Then sum75 = sum75 + (e75 - 0.453315) ^ 2
      If i75 = 32 Then sum75 = sum75 + (e75 - 0.590891) ^ 2
      If i75 = 40 Then sum75 = sum75 + (e75 - 0.658838) ^ 2
```

```
    If i75 = 48 Then sum75 = sum75 + (e75 - 0.637532) ^ 2
    If i75 = 56 Then sum75 = sum75 + (e75 - 0.717511) ^ 2
    If i75 = 64 Then sum75 = sum75 + (e75 - 0.701781) ^ 2
    If i75 = 72 Then sum75 = sum75 + (e75 - 0.70742) ^ 2
    If i75 = 80 Then sum75 = sum75 + (e75 - 0.763436) ^ 2
  Next i75
  f_of_x = 10 * (sum75 - 0.0059) / 18
```

E.3.24 Cork Board (#1)

This seeks to determine the number of rows and columns for a bulletin board made of wine corks. The length of a cork is between about 1.7 and 1.8 inches, which is nearly double the width. So a pair of adjacent corks makes a square, and squares tile a surface. I used 1.73 inches for the length in this design application. Artistically, the aspect ratio of a picture (the cork region) should be equal to the golden ratio, $\gamma = \left(-1 + \sqrt{5}\right)/2 \cong 0.61803$.

Functionally, it needs a minimum of about 150 corks to provide adequate posting area, but more than that number takes assembly time and uses up an endangered commodity. (Real corks are being replaced with plastic stoppers or screw tops.) There are many ways to tile the corks—rectangular, chevron, etc.—and align with the frame or on the diagonal. Figure E.30 shows a rectangular arrangement along the diagonal. In this pattern, choose an integer number of rows and columns, so the unused portion of corks on one side fill the complementary space on the other side, conserving material. The DVs are the number of rows and columns, and the OF is a combination of the deviations from the ideal γ and 150 corks. In the combined OF, I choose EC factors of 0.05 deviation from the golden ratio as creating concern equivalent to 50 corks deviation from the target.

This has integer DVs, which create flat surfaces with discontinuities between (see Figure E.31). Although there is a general trend to the minimum, there are local traps (minima surrounded by worse spots). Far from the minimum, the large OF values make the entire minimum region seem featureless. So to better visualize the region of interest, I log transformed the OF.

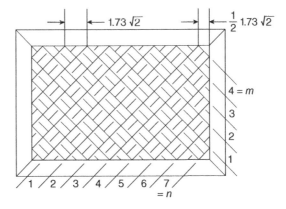

Figure E.30 Cork board sketch.

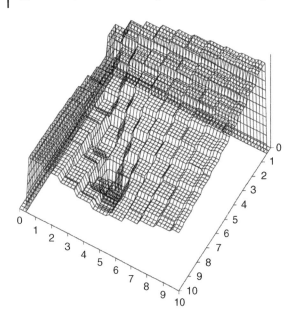

Figure E.31 Cork board surface.

The VBA code is as follows:

```
The VBA Code is:
  If Fchoice = 1 Then  'Cork Board dimension
    n = Int(x1 + 0.5)  'number of full rows on the bias for length count
    m = Int(x2 + 0.5)  'number of full rows on the bias for height count
    If n < 1 Or m < 1 Then
       constraint = "FAIL"
       Exit Function
    End If
    corks = 4 * (n + 1) * (m + 1)   'number of corks needed
    L1 = (n + 1) * Sqr(2) * 1.73 '+ 2 * 2.5   'frame inside length add 2 times
    thickness for exterior length
    L2 = (m + 1) * Sqr(2) * 1.73 '+ 2 * 2.5   'frame inside height
    ratio = L2 / L1
    golden = (-1 + Sqr(5)) / 2    'ideal aspect ratio
    badness1 = (ratio - golden) ^ 2    'deviation from ideal
    badness2 = (150 - corks) ^ 2      'deviation from ideal
    ec1 = 0.05                'Equal Concern factor
    ec2 = 50                  'Equal Concern factor
    f_of_x = badness1 / ec1 ^ 2 + badness2 / ec2 ^ 2  'the OF
    f_of_x = Log(f_of_x)        'transform to make the features in low values
    more visible
    f_of_x = 10 * (f_of_x + 2.83) / 12    'scaling on a 0 to 10 basis
    Exit Function
  End If
```

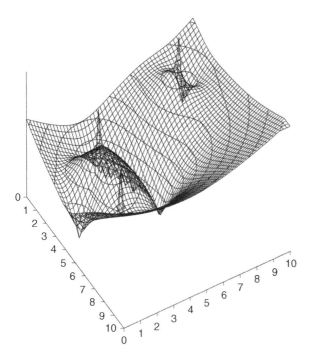

Figure E.32 Three springs surface.

E.3.25 Three Springs (#12)

In this application three springs are joined at a common but floating node 4. They are fixed at the other ends, nodes 1, 2, and 3. Initially all springs are at their resting length. There is no tension or compression in any spring. Then fixed node 3 is moved, which stresses the springs. When they move to equilibrium, node 4 will be relocated and springs will retain some residual stress. This application seeks to determine the location of the common floating node 4 that balances the *x*- and *y*-force on node 4. Springs have a linear stress–strain response, and compression and tension have the same *k*-values. Springs are not bendable. Interestingly, there are several unstable equilibria locations. The code has an option to minimize energy in the springs. Shown in Figure E.32 is the force balance version.

There are spikes at nodes 1, 2, and 3. If node 4 is at any of those points, then the compression of a spring is infinity. Thanks to MAE MS Candidate Romit Maulik for the idea, summer 2015, who was actually seeking to minimize the collective energy in all springs. A reader might want to convert this function to use energy as the OF or to make tension and compression have alternate *k*-values.

The VBA code is as follows:

```
If Fchoice = 12 Then      'Three Springs, unbendable, linear
'   Thanks to Romit Maulik for the basic idea - June 2015
'   All three springs are joined at the common, and free to move (y4, x4) node
'   Originally one end of Spring 1 is fixed at position (y1,x1), Spring 2 at
(y2,x2), and 3 at (y3o, x3o)
'   Originally all three springs are at rest
'   Node 3 moves to (y3, x3). Where does Node 4 move to?
```

```
'  OF from Case 1: At rest the y and x forces on Node 4 must be balanced. Then
'    minimize the squared sum of component forces.  This works, but permits
'   unstable equilibrium points of high tension when the forces are balanced.
'  OF from Case 2: A more rational surface happens when finding the Node 4
position
'      that minimizes the energy in all springs.
  F12y1 = 8   'you can initialize springs in any locations
  F12x1 = 2
  F12y2 = 2
  F12x2 = 4
  F12y3o = 7
  F12x3o = 7
  F12y4o = 5
  F12x4o = 5
  F12y3 = 1   'choose a variety of interesting move to points
  F12x3 = 8
  F12y4 = x2
  F12x4 = x1
  F12k1 = 1   'choose a variety of values for the spring constants
  F12k2 = 2
  F12k3 = 0.05
  F12L1o = Sqr((F12y1 - F12y4o) ^ 2 + (F12x1 - F12x4o) ^ 2)  'spring original at
rest lengths
  F12L2o = Sqr((F12y2 - F12y4o) ^ 2 + (F12x2 - F12x4o) ^ 2)
  F12L3o = Sqr((F12y3o - F12y4o) ^ 2 + (F12x3o - F12x4o) ^ 2)
  F12L1 = Sqr((F12y1 - F12y4) ^ 2 + (F12x1 - F12x4) ^ 2)    'spring lengths after
Node 3 moves
  F12L2 = Sqr((F12y2 - F12y4) ^ 2 + (F12x2 - F12x4) ^ 2)
  F12L3 = Sqr((F12y3 - F12y4) ^ 2 + (F12x3 - F12x4) ^ 2)
'  Force Analysis (more interesting surface, permits unstable equilibrium)
  If F12L1 = 0 Or F12L2 = 0 Or F12L3 = 0 Then  'to prevent a divide by zero
    constraint = "FAIL"
    f_of_x = 10
    Exit Function
  End If
  F12F1 = F12k1 * (F12L1o - F12L1)    'negative if tension, positive if
compression
  F12F2 = F12k2 * (F12L2o - F12L2)
  F12F3 = F12k3 * (F12L3o - F12L3)
  F12F1y = F12F1 * (F12y4 - F12y1) / F12L1
  F12F1x = F12F1 * (F12x4 - F12x1) / F12L1
  F12F2y = F12F2 * (F12y4 - F12y2) / F12L2
  F12F2x = F12F2 * (F12x4 - F12x2) / F12L2
  F12F3y = F12F3 * (F12y4 - F12y3) / F12L3
  F12F3x = F12F3 * (F12x4 - F12x3) / F12L3
  F12Fy = F12F1y + F12F2y + F12F3y
```

```
  F12Fx = F12F1x + F12F2x + F12F3x
  f_of_x = F12Fy ^ 2 + F12Fx ^ 2 + add_noise
  f_of_x = 2.5 * f_of_x ^ 0.25    'this power scaling emphasizes visibility of
  features in the low OF region
''  Energy Analysis (more rational surface, and simpler to compute)
'   F12E1 = F12k1 * (F12L1o - F12L1) ^ 2
'   F12E2 = F12k2 * (F12L2o - F12L2) ^ 2
'   F12E3 = F12k3 * (F12L3o - F12L3) ^ 2
'   f_of_x = F12E1 + F12E2 + F12E3
'   f_of_x = Sqr(f_of_x)
  Exit Function
End If
```

E.3.26   Retirement (#15)

This investigates two decisions: when to retire and what portion of your salary you should set aside each year while working to finance retirement.

The objective is to maximize cumulative joy in life. Joy is not income. Joy is based on income but is as much determined by the discretionary time that permits one to pursue personal happiness and by the affirmation pleasure of working.

The model has one start working at age 25 and die at 85. While working, a portion of the salary is invested in savings for retirement. In retirement, withdrawing from the savings provides income. But since the person might live to 90, the retirement income is allocated to last until age 90. One DV is the age to retire: retiring at an early age would leave a long retirement period to pursue individual choices, but there would not be much $ saved to finance the pursuit. Retiring at a late age would mean that a lot of $ had accumulated to finance retirement activities, but then there is little time left to enjoy it. The other DV is the portion of salary to invest: Don't invest much, and there is more $ to enjoy life while working but little in retirement. Invest a high portion of work salary, and there will be more $ to enjoy retirement, but it means a low-joy life while working. If retirement income is too high, then it is excessive to support joy and means that the tax rate is very high.

The equations are described in detail in Case Study 3, Chapter 38. The code accounts for the time value of $ and the compounding of investment $. The code also accounts for tax rate and has joy as a diminishing returns consequence of salary. Further the code accounts for an age factor that diminishes the ability to enjoy with increasing age. As complicated as that seems, it is still a simple representation, not accounting for partner activity, pensions, inheritances, postretirement employment, etc. I modeled this as how I perceive life with 50% of my time at work, leading to personal joys (a professor's view), which also has a salary-capped academic position, a 50 percentile death age of 85, and an age of half function at 80. But you will make alternate career choices and have differing life expectancy. You can adapt the code to better suit your choices.

The 3-D response is shown in Figure E.33. DV1 is scaled age, from 40 to 90, and is on the lower-left axis. DV2 is scaled portion of salary, from 0 to 100%, on the right-hand axis. The discontinuity of the surface at DV1 = 8 represents the impact of attempting to work after the age factor has diminished and led to forced retirement.

The surface seems continuous (smooth), but a detailed look in Figure E.34, over the 67–72-year period, reveals the impact of age discretization.

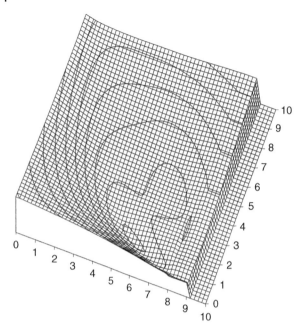

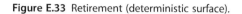

Figure E.33 Retirement (deterministic surface).

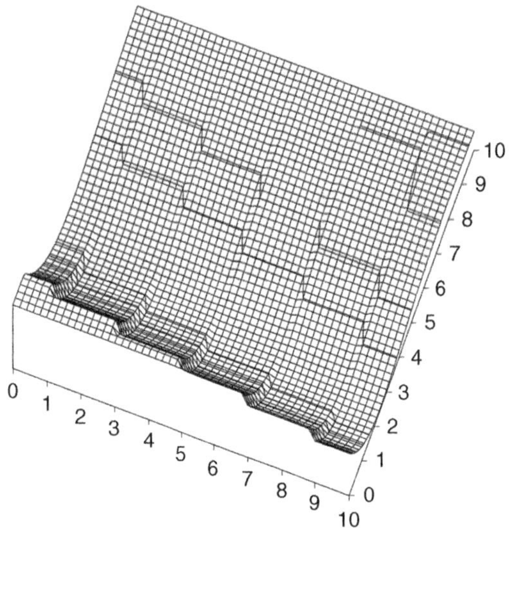

Figure E.34 Detail of retirement surface.

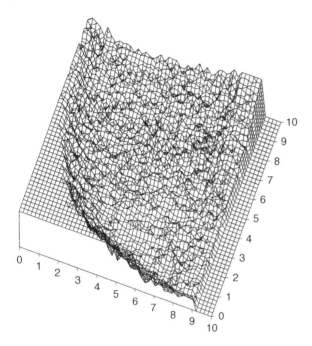

Figure E.35 Retirement stochastic surface.

In addition, the nominal values of interest rate, inflation rate, half age, and death age are uncertain. If selected these can be a stochastic factor. One realization of the stochastic surface is revealed in Figure E.35, and the reality of uncertainty obscures the apparently rational surface of Figure E.33.

The VBA code is as follows:

```
If Fchoice = 15 Then
'  Seeks optimum retirement age and portion of annual salary to
'     set aside for retirement.
'  The discretization associated with year-to-year age of retirement makes
'     discontinuities that sibstantially confound gradient based techniques.
   salary15begin = 12
   salary15end = 150
   basesalary15 = salary15begin
   savings15 = 0
   joy15 = 0
   discount15 = 0.0375
   interest15 = 0.07
   halfage15 = 80
   deathage15 = 90
```

```
  plandeathage15 = 85
  retireage15 = Int (40 + x1 * 5 + 0.5)
'  retireage15 = Int (66 + 0.5 * x1 + 0.5)   'to see discretization detail near
optimum
  portion15 = x2 / 10
'  portion15 = 0.06 + x2 / 40      'to see discretization detail near optimum

'  Stochastic? If Yes, uncomment these lines
'  discount15 = discount15 + 0.002 * Sqr (-2 * Log (Rnd())) * Sin (2 * 3.14159 *
Rnd()) '0.02 * (Rnd() - 0.5)
'  interest15 = interest15 + 0.002 * Sqr (-2 * Log (Rnd())) * Sin (2 * 3.14159 *
Rnd()) '0.02 * (Rnd() - 0.5)
'  halfage15 = halfage15 + Int (2 * Sqr (-2 * Log (Rnd())) * Sin (2 * 3.14159 * Rnd
())) 'Int (20 * (Rnd() - 0.5))
'  deathage15 = deathage15 + Int (2 * Sqr (-2 * Log (Rnd())) * Sin (2 * 3.14159 *
Rnd()))  ' Int (30 * (Rnd() - 0.5))

  If retireage15 > plandeathage15 Then
    constraint = "FAIL"
'    f_of_x = 10
    Exit Function
  End If

  If retireage15 < plandeathage15 - 5 Then
    withdrawportion15 = 1 / (plandeathage15 + 5 - retireage15)
  Else
    withdrawportion15 = 0.2
  End If

  For age15 = 25 To deathage15 Step 1    'calculate cumulative joy over the
lifetime
    agefactor15 = 0.1 + 0.9 / (1 + Exp (0.2 * (age15 - halfage15)))
'functionality to work or enjoy
    If agefactor15 < 0.5 And retireage15 > age15 Then retireage15 = age15
'forced retirement
    If age15 = retireage15 Then capitalwithdraw15 = savings15 *
withdrawportion15
    If age15 <= retireage15 Then      'working
    savings15 = savings15 * (1 + interest15) + portion15 * salary15 'compound
retirement savings15
      basesalary15 = basesalary15 * (1 + 0.15 * (salary15end - basesalary15) /
(salary15end - salary15begin)) 'my model for how my salary increased over my
career
        salary15 = agefactor15 * basesalary15    'The base salary is for a 100%
```

functioning person. The actual income could be reduced by days on the job, etc.

```
        discountedsalary15 = salary15 / (1 + discount15) ^ (age15 - 25)
'normalized to the start, at age 25, by the inflation rate
        tax15 = 0.85 / (1 + Exp(0.1 * (35 - discountedsalary15 * (1 -
portion15)))) 'an approximate model for the graduated income tax, based on
income after investing in tax-deferred savings15
        joyfactor15 = 1 - 2.5 * Exp(-discountedsalary15 * (1 - portion15) * (1 -
tax15) / 10) 'my approximation for how joy scales with discretionary $
        joy15 = joy15 + joyfactor15 * agefactor15 * (1 * 52 + 3 * 5) / 364 'joy of
personal time per year - based on a 52 week year of 364 days
        joyfactor15 = 1 - 2.5 * Exp(-discountedsalary15 / 10)      'now based on
salary as a measure of value to humanity
        joy15 = joy15 + joyfactor15 * agefactor15 * 0.5 * (5 * 45) / 364  'joy of
the job per year - only 20% of the time generates joy on average. My 50% of job
is joy might not match most jobs.
    Else      'retired
        salary15 = capitalwithdraw15 + savings15 * interest15  'take all
interest plus capital draw from savings15
        savings15 = (savings15 - salary15) * (1 + interest15)  'residual $ in
savings15 compounds
        discountedsalary15 = salary15 / (1 + discount15) ^ (age15 - 25)
        tax15 = 0.85 / (1 + Exp(0.1 * (35 - discountedsalary15)))
        joyfactor15 = 1 - 2.5 * Exp(-discountedsalary15 * (1 - tax15) / 10)
        joy15 = joy15 + joyfactor15 * agefactor15 * (6 * 52) / 364 'joy of
personal time less one day per week for bills, taxes etc.
    End If
  Next age15
  age15 = age15 - 1
  If savings15 > 0 Then
    discountedsavings15 = savings15 / (1 + discount15) ^ (age15 - 25)
    joyfactor15 = 1 - 2.5 * Exp(-discountedsavings15 * (1 - tax15) / 10)
    joy15 = joy15 + joyfactor15
  Else
    joy15 = joy15 - 3
  End If
'   f_of_x = 10 * (-joy15 + 20) / 70 'normal scaling - misses floor detail
'   f_of_x = 10 * (-joy15 + 20) / 20 'normal scaling - sees floor detail but
misses high detail
  f_of_x = 10 * Sqr(Sqr((-joy15 + 20) / 20)) - 3.5 'normal, transformed to
reveal
'   f_of_x = 10 * (-joy15 + 20) / 20            'for stochastic function
'   f_of_x = 10 * (-joy15 + 19.4) / 4        'to see discretization detail
  Exit Function
End If
```

E.3.27 Solving an ODE (#23)

Some differential equations can be solved exactly analytically. But some are too complicated. This investigates the use of a polynomial relation to approximate the solution. (This approach is similar to that of Case Study 7, Chapter 42, but the model types are distinctly different.) Consider the ODE $(dy/dx) = f(x,y), y(x_0) = y_0$. If there is a solution, $y(x)$, then it can be expanded in a Taylor series and terms rearranged to make it a power series, $y(x) = a + bx + cx^2 + dx^3 + \cdots$. If the truncated power series (e.g., a cubic) is a reasonable approximation, then its derivative, $dy/dx = b + 2cx + 3dx^2$, is a reasonable approximation to $dy/dx = f(x,y) \cong f(x, a + bx + cx^2 + dx^3) = b + 2cx + 3dx^2$ for all x-values. The objective is to find the b- and c-values that make $b + 2cx + 3dx^2$ best match $f(x, a + bx + cx^2 + dx^3)$ for many x-values.

With a quadratic approximation (appropriate for a 2-D exercise), and b and c chosen by the optimizer, the value of coefficient "a" is determined from the initial condition, $y_0 = a + bx_0 + cx_0^2$. The generic optimization statement is

$$\min_{\{b,c\}} J = \sum_{i=1}^{N} \left[f\left(x, a + bx + cx^2\right) - \left(b + 2cx\right) \right]^2$$

$$\text{S.T.:} \quad \Delta x = \frac{(x_{end} - x_0)}{N}$$

$$x_i = x_0 + i\Delta x$$

$$a = y_0 - bx_0 - cx_0{}^2$$

My 2-D example uses the quadratic polynomial to approximate the solution to a first-order linear ODE $\tau dy/dx + y = k$, $y(x_0) = y_0$ where $f(x,y) = (k-y)/\tau$. The specific optimization statement is

$$\min_{\{b,c\}} J = \sum_{i=1}^{N} \left[\frac{k - (a + bx + cx^2)}{\tau} - (b + 2cx) \right]^2$$

$$\text{S.T.:} \quad \Delta x = \frac{(x_{end} - x_0)}{N}$$

$$x_i = x_0 + i\Delta x$$

$$a = y_0 - bx_0 - cx_0^2$$

The function is a steep valley type, as seen in Figure E.36.

Feel free to change the ODE coefficients and initial conditions and its functionality. How would you test the optimal solution to determine if the approximation is adequate?

The VBA code is as follows:

```
If Fchoice = 23 Then            'RRR power series model solution to an ODE
'  Best fit of a polynomial y=a+bx+cx^2 as a solution to the
'  linear first order ODE  tau* (dy/dx) +y=k
'  By minimizing the difference between the polynomial derivative =b+2cx and
'  The ODE derivative = (k-y)/tau
'   If x1 < 0 Or x2 < 0 Or x1 > 10 Or x2 > 10 Then
'      constraint = "Fail"
'      Exit Function
```

```
'  End If
   x023 = 1
   xend23 = 10
   y023 = 1
   k23 = 10
   tau23 = 4
   b23 = 2 * (x1 - 5)
   c23 = 0.2 * (x2 - 5)
   a23 = y023 - b23 * x023 - c23 * (x023) ^ 2
   sum23 = 0
   deltax23 = (xend23 - x023) / 100
   For i23 = 1 To 100
      x23 = x023 + i23 * deltax23
      y23 = a23 + b23 * x23 + c23 * (x23) ^ 2
      sum23 = sum23 + (((k23 - y23) - tau23 * (b23 + 2 * c23 * x23)) * deltax23) ^ 2
   Next i23
   f_of_x = 10 * (Sqr(sum23) - 2) / 150
   Exit Function
End If
```

E.3.28 Space Slingshot (#39)

Determine (i) the drawback on the slingshot and (ii) the angle to aim it, and send the projectile along its free-fall trajectory in 2-D space to accumulate points, as illustrated in Figure E.37.

In this example there are three planets. The home planet is at the origin and the main target planet is centered at 4-up and 7-over. The third, smaller planet is at 9-up, 2-over. All three planets have gravitational pull on the projectile, but they remain stationary. The two larger planets have an atmosphere

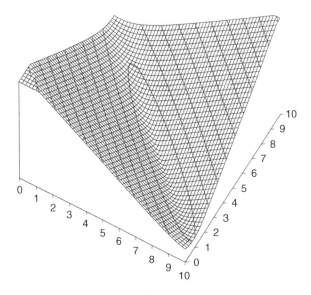

Figure E.36 Solving an ODE.

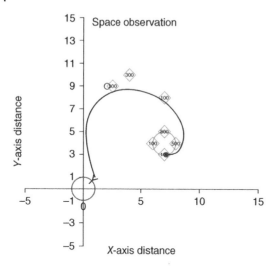

Figure E.37 Space slingshot illustration.

that adds drag to the projectile when in the proximity. Otherwise translational *x*- and *y*-motion is governed by the gravitational force balance in this 2-D space.

The diamond markers indicate modest prize locations in space and larger prize locations on the planets. Get near or crash into them to gather target points. The OF is the target points accumulated and the DVs are pullback and angle. Figure E.38 shows a 3-D plot of OF w.r.t. DVs. There are many local optima and very narrow pinholes. There are flat spots. It was a surprisingly difficult surface for optimization.

Figure E.38 Space slingshot surface.

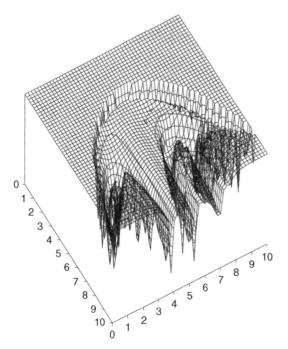

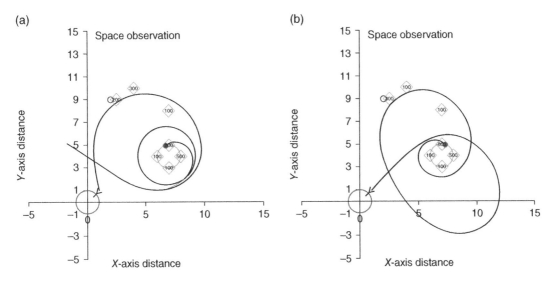

Figure E.39 (a) A best path and (b) a nearly as good path.

One of the best paths (Figure E.39a) sends the projectile upward, gravity pulls it to the right passing near the in-space prize points, and then overshoots the target planet and moves to the upper left; but the projectile is prevented from going into deep space by the combined gravitational pull (at about 5 on the vertical axis). It then falls back toward the target planet, encircles it, and eventually crashes into the North Pole target post.

There are many nearly as good (point-accumulating) patterns. Figure E.39b shows a very distinctly different, but nearly as good, path.

The VBA code is as follows:

```
If Fchoice = 26 Then       'Projectiles in Space
  If x1 < 0 Or x1 > 10 Or x2 < 0 Or x2 > 10 Then
     constraint = "Fail"
     Exit Function
  End If
  abx1 = 0
  aby1 = 0
  abr1 = 1
  abGm1 = 1.25
  abx2 = 7
  aby2 = 4
  abr2 = 1
  abGm2 = 1
  abx3 = 2
  aby3 = 9
  abr3 = 0.5
  abGm1 = 0.25
```

```
abcdr2m = 1
abalpha2 = -3
abalpha1 = -2
abx = 1
aby = 1
abd = 0.06 + x1 * (0.2 - 0.06) / 10 '0.06 + (x1 - 3) / 50 'slingshot draw
abangle = 36 * (-2 + x2 * 6 / 10) '(x2 - 4) * 36
abv = 40 * abd ^ 2
abvx = abv * Cos(2 * 3.14159 * abangle / 360)
abvy = abv * Sin(2 * 3.14159 * abangle / 360)
abn = 50 '5000
abdt = 1
f_of_x = 0
Value1 = 300
Value2 = 100
Value3 = 300
Value4 = 800
Value5 = 100
Value6 = 100
value7 = 500
For abtcount = 1 To 500
    For abncount = 1 To abn
        abx = abx + abvx * abdt / abn
        aby = aby + abvy * abdt / abn
        abrb1 = Sqr((abx - abx1) ^ 2 + (aby - aby1) ^ 2)
        abrb2 = Sqr((abx - abx2) ^ 2 + (aby - aby2) ^ 2)
        abrb3 = Sqr((abx - abx3) ^ 2 + (aby - aby3) ^ 2)
        If abrb1 <= abr1 Then Exit For 'crashed
        If abrb2 <= abr2 Then Exit For 'crashed
        If abrb3 <= abr3 Then Exit For 'crashed
        abax = abGm1 * (abx1 - abx) / abrb1 ^ 3 + abGm2 * (abx2 - abx) / abrb2 ^ 3 +
abGm3 * (abx3 - abx) / abrb3 ^ 3 - abcdr2m * abv * abvx * Exp(abalpha1 * (abrb1 -
abr1)) - abcdr2m * abv * abvx * Exp(abalpha2 * (abrb2 - abr2))
        abay = abGm1 * (aby1 - aby) / abrb1 ^ 3 + abGm2 * (aby2 - aby) / abrb2 ^ 3 +
abGm3 * (aby3 - aby) / abrb3 ^ 3 - abcdr2m * abv * abvy * Exp(abalpha1 * (abrb1 -
abr1)) - abcdr2m * abv * abvy * Exp(abalpha2 * (abrb2 - abr2))
        abvx = abvx + abdt * abax / abn
        abvy = abvy + abdt * abay / abn
        abv = Sqr(abvx ^ 2 + abvy ^ 2)
        abdragx = abcdr2m * abv * abvx * Exp(abalpha * (abrb2 - abr2)) / abax

        distance = Sqr((abx - 4) ^ 2 + (aby - 10) ^ 2)
        If Value1 > 0 Then
            increment = 3 * Exp(-2 * distance)
            f_of_x = f_of_x + increment
            Value1 = Value1 - increment
        End If
```

```
        distance = Sqr((abx - 7) ^ 2 + (aby - 8) ^ 2)
        If Value2 > 0 Then
           increment = 1 * Exp(-2 * distance)
           f_of_x = f_of_x + increment
           Value2 = Value2 - increment
        End If
        distance = Sqr((abx - 2.5) ^ 2 + (aby - 9) ^ 2)
        If Value3 > 0 Then
           increment = 3 * Exp(-2 * distance)
           f_of_x = f_of_x + increment
           Value3 = Value3 - increment
        End If
        distance = Sqr((abx - 7) ^ 2 + (aby - 5) ^ 2)
        If Value4 > 0 Then
           increment = 8 * Exp(-3 * distance)
           f_of_x = f_of_x + increment
           Value4 = Value4 - increment
        End If
        distance = Sqr((abx - 7) ^ 2 + (aby - 3) ^ 2)
        If Value5 > 0 Then
           increment = 1 * Exp(-2 * distance)
           f_of_x = f_of_x + increment
           Value5 = Value5 - increment
        End If
        distance = Sqr((abx - 6) ^ 2 + (aby - 4) ^ 2)
        If Value6 > 0 Then
           increment = 1 * Exp(-2 * distance)
           f_of_x = f_of_x + increment
           Value6 = Value6 - increment
        End If
        distance = Sqr((abx - 8) ^ 2 + (aby - 4) ^ 2)
        If value7 > 0 Then
           increment = 5 * Exp(-2 * distance)
           f_of_x = f_of_x + increment
           value7 = value7 - increment
        End If

     Next abncount
     If abrb1 <= abr1 Then Exit For  'crashed
     If abrb2 <= abr2 Then Exit For  'crashed
     If abrb3 <= abr3 Then Exit For  'crashed
     If abx < -10 Or abx > 30 Or aby < -10 Or aby > 30 Then Exit For
   Next abtcount
   f_of_x = -Sqr(f_of_x)
   f_of_x = 10 * (f_of_x + 30) / 30
   Exit Function
End If
```

E.3.29   Goddard Problem (#79)

There are many variations on this application named to honor Robert H. Goddard, rocket scientist. The objective is to schedule the thrust to minimize fuel consumption to achieve a desired rocket height. If the initial thrust is not high enough, then the rocket never lifts off the launch pad and burns all the fuel. The best initial thrust is 100% to get the rocket moving. But, once moving upward, the faster it goes, the greater is the air drag, and continuing full thrust leads to excessive velocity and excessive drag, which wastes fuel. In this simple version of the exercise, the initial thrust is 100%, and the objective is to find the elevation to cut the thrust to 20% and then the next elevation to cut it to 0%. Even at 0% thrust, the upward velocity of the rocket continues until gravitational pull and air drag counter the momentum. The two elevations are the DVs. Although the DVs are continuum variables, the thrust discretization makes the OF response have discontinuities, as revealed in Figure E.40. The OF has flat spots, cliff discontinuities, and hard constraints. There is small sensitivity to the second stage of thrust, and numerical time discretization of the dynamic model has a visible impact on the OF surface.

There are many variations on the OF and scheduling basis for the thrust. See Case Study 4, Chapter 39, for an alternate. Functions #77 through #80 in the 2-D Optimization Examples file explore alternate 2-DV options.

Thanks to Thomas Hays for pointing me to this in the fall of 2013. The model and coefficient values are from http://bocop.saclay.inria.fr/?page_id=465, and the model equations are explained in detail in Case Study 4, Chapter 39.

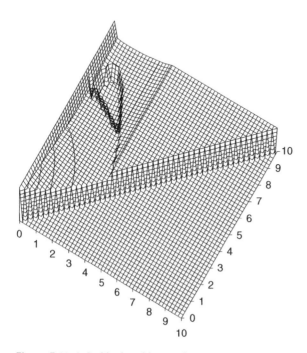

Figure E.40 A Goddard problem surface.

The VBA code is as follows:

```
If Fchoice = 79 Then    'Goddard Problem 3.  Start with u1 thrust = 1, then
    'find elevation to switch to u2=0.2 and then elevation to switch fuel off
    'to maximize remaining fuel content to reach desired height
    'time is i79*dtprime*7915, sec
  'elevation is (rprime-1)*20170, thousands of feet (rocket can get to 243k
feet, 46 miles)

    If x1 > 10 Or x2 > 10 Or x1 < 0 Or x2 < 0 Then
      constraint = "FAIL"
      Exit Function
    End If
    rprimeend = 1.005     'about 100k feet
    rprime1 = 1 + (rprimeend - 1) * x1 / 10
    rprime2 = 1 + (rprimeend - 1) * x2 / 10
    If rprime1 > rprime2 Then
      constraint = "FAIL"
      Exit Function
    End If

    Stage79 = 1
    mprime0 = 1         'initial scaled mass of rocket with fuel
    rprime0 = 1         'initially on the Earth's surface, scaled radius
    vprime0 = 0         'initial scaled velocity
   dtprime = 0.0001     'small dt is required to make numerical discretization
inconsequential, should use 0.00001
    For i79 = 1 To 100000     'takes about a quarter this many iterations for a
simulation
    If Stage79 = 1 Then influence = 1
    If Stage79 = 2 Then influence = 0.2
    If Stage79 = 3 Then influence = 0
      mprime = mprime0 - dtprime * 24.5 * influence
     vprime = vprime0 + dtprime * (3.5 * influence / mprime0 - 1 / rprime0 ^ 2 -
310 * vprime0 ^ 2 * Exp(-500 * (rprime0 - 1)) / mprime0)
      rprime = rprime0 + dtprime * vprime0
      If vprime < 0 Then Exit For    'falling
      If mprime < 0.2 Then Exit For  'fuel out not a good trial
      If rprime > rprime1 And Stage79 = 1 Then  'entered stage 2
        dtincrement = dtprime * (rprime - rprime1) / (rprime - rprime0)
'recalculate last dt
        mprime = mprime0 - dtincrement * 24.5 * influence
      vprime = vprime0 + dtincrement * (3.5 * influence / mprime0 - 1 / rprime0 ^
2 - 310 * vprime0 ^ 2 * Exp(-500 * (rprime0 - 1)) / mprime0)
        rprime = rprime0 + dtincrement * vprime0
        Stage79 = 2
```

```
            End If
            If rprime > rprime2 And Stage79 = 2 Then  'entered stage 3
               dtincrement = dtprime * (rprime - rprime2) / (rprime - rprime0)
    'recalculate last dt
               mprime = mprime0 - dtincrement * 24.5 * influence
           vprime = vprime0 + dtincrement * (3.5 * influence / mprime0 - 1 / rprime0 ^
    2 - 310 * vprime0 ^ 2 * Exp(-500 * (rprime0 - 1))) / mprime0)
               rprime = rprime0 + dtincrement * vprime0
               Stage79 = 3
            End If
            mprime0 = mprime
            vprime0 = vprime
            rprime0 = rprime
            If rprime > rprimeend Then Exit For  'made it 1.005 is about 100k ft.
         Next i79
         f_of_x = -mprime '+ 10 * vprime ^ 2 'plus penalty for excessive velocity
    (unnecessary)
    '    If rprime < rprimeend Then f_of_x = f_of_x + (1000 * (rprimeend - rprime))
    ^ 2      'soft penalty for not making height
    '      f_of_x = 10 * (f_of_x + 0.353) / 24
         If rprime < rprimeend Then f_of_x = (100 * (rprimeend - rprime)) ^ 2 'change
    OF penalty for not meeting height
         f_of_x = 10 * (f_of_x + 0.25826) / 0.50824
         Exit Function
      End If
```

E.3.30 Rank Model (#84)

The objective is to create a formula that can determine competitiveness of players from their attributes. Perhaps you need a method to invite people to be junior members of your school team and want to know who has the best promise to develop into a winner at the varsity level. Perhaps you want to know whether your 6-year-old child has talent and whether you should invest time and money in developing them for a possible place on the Olympics team. Perhaps you want to know if you should join the tennis ladder at work.

The hope is that there are fundamental physical attributes of the individuals that can indicate their competitive fitness. For instance, reflex rate could be easily measured in a laboratory test of individual's delay in hand response to visual signals, and such reflex rate seems to be a relevant metric to imply fast responses on the tennis court, which seems important to winning. Smaller would be better. Another measureable attribute might be how far a person can reach when standing. Larger would be better. There are several such attributes, and the hope is that a model of overall fitness can be developed from the following power-law function: $F = f_1^a f_2^b f_3^c \ldots$, where F is the overall player fitness, f_i are the lab-measured attributes, and coefficients a, b, c, etc., are adjustable model parameters (these are the DVs in the optimization).

Contrasting additive fitness functions, typical of the "cost functions" models, this composite fitness model indicates that if one factor is very low (perhaps $f_2 = 0.02$), even if two are very high (perhaps

$f_1 = 0.93$ and $f_3 = 0.89$), the player is not as fit for the game as one with mediocre values for all attributes (perhaps $f_1 = f_2 = f_3 = 0.7$).

The model will be developed by taking N number of high-level players, testing them for their attribute values, and have them play in tournaments under various conditions (indoor, outdoor, clay court, morning, evening, etc.). The model coefficients will be adjusted so that the rank predicted by the overall fitness model best matches the average rank found from the tournaments.

Typical of rank models, there are cliff discontinuities and flat spots. If overall fitness values for players A, B, and C are 10, 7, and 2, respectively, then their ranks are first, second, and third. If the fitness values were 8, 7.1, and 2, the ranking does not change. If 7.3, 7.29, and 2, it is still not changed. However, with a tiny change to that last set of values, to 7.3, 7.301, and 2, the ranking abruptly changes to second, first, and third (see Figure E.41).

In this exercise, I choose to use three metrics, a geometric mean for the overall fitness, and to set the power of the first factor to unity. As long as $a > 0$, the $F = f_1^a f_2^b f_3^c$ functionality can be raised to the $1/a$ power without changing ranking. Or equivalently, set $a = 1$. I choose to use a geometric mean of the fitness function to keep ranges on F and f similar. As long as the value of $(1 + a + b) > 0$, this is a strictly positive transformation that does not change the DV values:

$$F = \sqrt[1+b+c]{f_1 f_2^b f_3^c}$$

This is a 2-DV application. The example uses three contestants, with three attributes each, and seeks to best match the ranking from three contests between the three players.

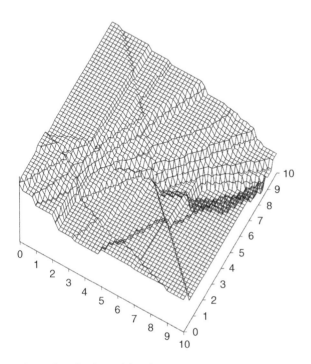

Figure E.41 Rank model surface.

The VBA code is as follows:

```
If Fchoice = 84 Then   'rank exploration - 6 sets of 3 contestants, 3
attributes each - what is model that best matches ranks
    x11 = -2 + x1 / 2   'model parameter 1
    x22 = -2 + x2 / 2   'model parameter 2
    If x1 < 0 Or x1 > 10 Or x2 < 0 Or x2 > 10 Then   'prevent extreme values
        constraint = "FAIL"
        Exit Function
    End If
                    '  Set 1
    FA184 = 0.4      'Player attributes - Player A, Attribute 1
    FA284 = 0.3
    FA384 = 0.9
    FB184 = 0.4
    FB284 = 0.6
    FB384 = 0.5
    FC184 = 0.7
    FC284 = 0.7
    FC384 = 0.8
    fA84 = (FA184 * FA284 ^ x11 * FA384 ^ x22) '^ (1 / (x11 + x22 + 1))  'composite
fitness
    fB84 = (FB184 * FB284 ^ x11 * FB384 ^ x22) '^ (1 / (x11 + x22 + 1))
    fC84 = (FC184 * FC284 ^ x11 * FC384 ^ x22) '^ (1 / (x11 + x22 + 1))
    minf84 = fA84             ' find worst
    If fB84 < minf84 Then minf84 = fB84
    If fC84 < minf84 Then minf84 = fC84
    maxf84 = fA84             ' find best
    If fB84 > maxf84 Then maxf84 = fB84
    If fC84 > maxf84 Then maxf84 = fC84
    RankA = 2                 ' Assign rank - very primitive approach
    RankB = 2
    RankC = 2
    If fA84 = maxf84 Then RankA = 1
    If fB84 = maxf84 Then RankB = 1
    If fC84 = maxf84 Then RankC = 1
    If fA84 = minf84 Then RankA = 3
    If fB84 = minf84 Then RankB = 3
    If fC84 = minf84 Then RankC = 3
    f_of_x = ((RankC - 1) ^ 2) / 1 + ((RankB - 2) ^ 2) / 2 + ((RankA - 3) ^ 2) / 3
'Assess - compare model to actual
                '  Set 2
    FA184 = 0.3      'Player attributes
    FA284 = 0.5
    FA384 = 0.5
    FB184 = 0.7
    FB284 = 0.8
```

```
    FB384 = 0.2
    FC184 = 0.5
    FC284 = 0.5
    FC384 = 0.6
  fA84 = (FA184 * FA284 ^ x11 * FA384 ^ x22) '^ (1 / (x11 + x22 + 1))  'composite
   fB84 = (FB184 * FB284 ^ x11 * FB384 ^ x22) '^ (1 / (x11 + x22 + 1))
   fC84 = (FC184 * FC284 ^ x11 * FC384 ^ x22) '^ (1 / (x11 + x22 + 1))
   minf84 = fA84                  ' find worst
   If fB84 < minf84 Then minf84 = fB84
   If fC84 < minf84 Then minf84 = fC84
   maxf84 = fA84                   ' find best
   If fB84 > maxf84 Then maxf84 = fB84
   If fC84 > maxf84 Then maxf84 = fC84
   RankA = 2                  ' Assign rank
   RankB = 2
   RankC = 2
   If fA84 = maxf84 Then RankA = 1
   If fB84 = maxf84 Then RankB = 1
   If fC84 = maxf84 Then RankC = 1
   If fA84 = minf84 Then RankA = 3
   If fB84 = minf84 Then RankB = 3
   If fC84 = minf84 Then RankC = 3
  f_of_x = f_of_x + ((RankC - 1) ^ 2) / 1 + ((RankB - 2) ^ 2) / 2 + ((RankA - 3) ^ 2)
/ 3 'Assess
                   '  Set 3
    FA184 = 0.6       'Player attributes
    FA284 = 0.5
    FA384 = 0.6
    FB184 = 0.7
    FB284 = 0.6
    FB384 = 0.5
    FC184 = 0.8
    FC284 = 0.7
    FC384 = 0.5
  fA84 = (FA184 * FA284 ^ x11 * FA384 ^ x22) '^ (1 / (x11 + x22 + 1))  'composite
   fB84 = (FB184 * FB284 ^ x11 * FB384 ^ x22) '^ (1 / (x11 + x22 + 1))
   fC84 = (FC184 * FC284 ^ x11 * FC384 ^ x22) '^ (1 / (x11 + x22 + 1))
   minf84 = fA84                  ' find worst
   If fB84 < minf84 Then minf84 = fB84
   If fC84 < minf84 Then minf84 = fC84
   maxf84 = fA84                   ' find best
   If fB84 > maxf84 Then maxf84 = fB84
   If fC84 > maxf84 Then maxf84 = fC84
   RankA = 2                  ' Assign rank
   RankB = 2
   RankC = 2
   If fA84 = maxf84 Then RankA = 1
```

```
    If fB84 = maxf84 Then RankB = 1
    If fC84 = maxf84 Then RankC = 1
    If fA84 = minf84 Then RankA = 3
    If fB84 = minf84 Then RankB = 3
    If fC84 = minf84 Then RankC = 3
  f_of_x = f_of_x + ((RankC - 1) ^ 2) / 1 + ((RankB - 2) ^ 2) / 2 + ((RankA - 3) ^ 2)
/ 3 'Assess
                              '  Set 4
    FA184 = 0.6        'Player attributes
    FA284 = 0.5
    FA384 = 0.5
    FB184 = 0.7
    FB284 = 0.6
    FB384 = 0.5
    FC184 = 0.8
    FC284 = 0.8
    FC384 = 0.4
  fA84 = (FA184 * FA284 ^ x11 * FA384 ^ x22) '^ (1 / (x11 + x22 + 1)) 'composite
  fB84 = (FB184 * FB284 ^ x11 * FB384 ^ x22) '^ (1 / (x11 + x22 + 1))
  fC84 = (FC184 * FC284 ^ x11 * FC384 ^ x22) '^ (1 / (x11 + x22 + 1))
    minf84 = fA84              ' find worst
    If fB84 < minf84 Then minf84 = fB84
    If fC84 < minf84 Then minf84 = fC84
    maxf84 = fA84              ' find best
    If fB84 > maxf84 Then maxf84 = fB84
    If fC84 > maxf84 Then maxf84 = fC84
    RankA = 2                  ' Assign rank
    RankB = 2
    RankC = 2
    If fA84 = maxf84 Then RankA = 1
    If fB84 = maxf84 Then RankB = 1
    If fC84 = maxf84 Then RankC = 1
    If fA84 = minf84 Then RankA = 3
    If fB84 = minf84 Then RankB = 3
    If fC84 = minf84 Then RankC = 3
  f_of_x = f_of_x + ((RankC - 1) ^ 2) / 1 + ((RankB - 2) ^ 2) / 2 + ((RankA - 3) ^ 2)
/ 3 'Assess
                              '  Set 5
    FA184 = 0.1        'Player attributes
    FA284 = 0.6
    FA384 = 0.2
    FB184 = 0.5
    FB284 = 0.2
    FB384 = 0.2
    FC184 = 0.3
    FC284 = 0.3
    FC384 = 0.3
```

```
    fA84 = (FA184 * FA284 ^ x11 * FA384 ^ x22) '^ (1 / (x11 + x22 + 1)) 'composite
    fB84 = (FB184 * FB284 ^ x11 * FB384 ^ x22) '^ (1 / (x11 + x22 + 1))
    fC84 = (FC184 * FC284 ^ x11 * FC384 ^ x22) '^ (1 / (x11 + x22 + 1))
    minf84 = fA84                ' find worst
    If fB84 < minf84 Then minf84 = fB84
    If fC84 < minf84 Then minf84 = fC84
    maxf84 = fA84                ' find best
    If fB84 > maxf84 Then maxf84 = fB84
    If fC84 > maxf84 Then maxf84 = fC84
    RankA = 2                    ' Assign rank
    RankB = 2
    RankC = 2
    If fA84 = maxf84 Then RankA = 1
    If fB84 = maxf84 Then RankB = 1
    If fC84 = maxf84 Then RankC = 1
    If fA84 = minf84 Then RankA = 3
    If fB84 = minf84 Then RankB = 3
    If fC84 = minf84 Then RankC = 3
  f_of_x = f_of_x + ((RankC - 1) ^ 2) / 1 + ((RankA - 2) ^ 2) / 2 + ((RankB - 3) ^ 2)
/ 3 'Assess
                          '  Set 6
    FA184 = 0.7       'Player attributes
    FA284 = 0.4
    FA384 = 0.8
    FB184 = 0.8
    FB284 = 0.3
    FB384 = 0.9
    FC184 = 0.8
    FC284 = 0.5
    FC384 = 0.6
    fA84 = (FA184 * FA284 ^ x11 * FA384 ^ x22) '^ (1 / (x11 + x22 + 1)) 'composite
    fB84 = (FB184 * FB284 ^ x11 * FB384 ^ x22) '^ (1 / (x11 + x22 + 1))
    fC84 = (FC184 * FC284 ^ x11 * FC384 ^ x22) '^ (1 / (x11 + x22 + 1))
    minf84 = fA84                ' find worst
    If fB84 < minf84 Then minf84 = fB84
    If fC84 < minf84 Then minf84 = fC84
    maxf84 = fA84                ' find best
    If fB84 > maxf84 Then maxf84 = fB84
    If fC84 > maxf84 Then maxf84 = fC84
    RankA = 2                    ' Assign rank
    RankB = 2
    RankC = 2
    If fA84 = maxf84 Then RankA = 1
    If fB84 = maxf84 Then RankB = 1
    If fC84 = maxf84 Then RankC = 1
    If fA84 = minf84 Then RankA = 3
    If fB84 = minf84 Then RankB = 3
```

```
      If fC84 = minf84 Then RankC = 3
      f_of_x = f_of_x + ((RankC - 1)^2) / 1 + ((RankA - 2)^2) / 2 + ((RankB - 3)^2)
/ 3 'Assess
      f_of_x = Sqr(f_of_x)          ' Transformed to visualize contours
      f_of_x = 10 * (f_of_x - 0) / 5.066
      Exit Function
   End If
```

E.3.31   Liquid–Vapor Separator (#27)

The objective is to design a horizontal cylindrical tank to disengage liquid and vapor. Two-phase flow enters the top of the tank on one side, and the liquid drops out to the lower portion of the tank and the vapor to the upper portion. They exit at the other end of the tank, liquid at the bottom and vapor at the top. As the vapor flows from entrance to exit, mist droplets gravity-fall from the vapor to the liquid. The higher the vertical dimension of the head space (the vapor space), or the higher the vapor velocity, the longer the tank must be for droplets to have time to fall out of the vapor. This places a constraint relating tank length to diameter. There are two additional constraints: The vapor flow rate must be small enough so that it does not make waves on the liquid surface and re-entrain droplets. And the liquid holdup must provide a reservoir to provide continuity of downstream flow rate in spite of inflow pulses. The process tank designer can choose tank aspect ratio (L/D) and liquid level in the tank, and then tank diameter will be a consequence of compliance to the limiting constraint. The objective is to minimize the capital cost of the tank.

I modified this problem from one given to me by Josh Ramsey, ChE design instructor, who modified it from an example that Jan Wagner had used in the class, who created it based on the publication by W. Y. Svrcek and W. D. Monnery, "Design Two-Phase Separators within the Right Limits," Chemical Engineering Progress, October 1993, pp. 53–60. There are actually three stages of separation in a tank, the initial impingement plate to deflect incoming liquid, a settling zone to let large droplets gravity-fall from vapor to liquid, and then demisters to coalesce small droplets. For simplicity, this exercise only considers the second-stage mechanisms of droplet settling and idealizes some aspects to keep the equations easily recognized and understood by the general reader.

What are the undesirables that need to be balanced? If liquid level is high in the tank, then the area for vapor flow is low, vapor velocity is high, and mist is re-entrained. To prevent this, tank D must be large, which is costly. By contrast, if the level is low, then liquid residence time is too low to either degas or provide surge capacity. To prevent this, either tank D or L/D must be large. There is an intermediate level that minimizes cost. Similarly, if L/D is low, then D needs to be high to provide low gas velocity and high enough droplet settling and degas times and liquid holdup. This increases tank cost. Alternately, if L/D is large, the diameter is limited to ensure gas velocity constraints, and the long tank is expensive.

Tanks come in standard sizes; for this exercise L/D ratios will be restricted to half-integer values (1, 1.5, 2, 2.5, 3, 3.5, etc.). This means that one of the DVs, the L/D ratio, is discretized, which creates ridges in the OF response. Further, the tank diameter is limited to standard sizes here based on 3-inch (quarter-ft) increments. This means that although the DV representing the choice for liquid level in the tank is continuum valued, the OF will have flat spots. Further, the three conditions on tank diameter (ensure settling time, ensure liquid holdup time, and ensure that vapor velocity is low enough to not re-entrain droplets) create a discontinuity in the surface, intersecting sharp valleys, even if the DV and OF were continuum valued.

For simplicity, I consider that the cost of the vessel is directly related to the mass of the metal needed to make it (which will be the surface area times thickness times density), and I modeled the vessel as a right circular cylinder with flat ends. Since density, thickness, and cost/mass are constants, the OF can be stated as $J = (0.5 + L/D)D^2$. Additionally to reveal detail of the lower values of the OF surface, I have the function as the square root of J (a strictly monotonic function transformation does not change the DV^* values).

The detailed derivation of the equations is revealed in Case Study 8, Chapter 43.

The liquid level can be stated as a fraction of the tank diameter, and the geometry/trigonometry of a segment of a circle is used to calculate the cross-sectional area faction for liquid and vapor flow. Again, for simplicity, this model assumes plug flow of both liquid and vapor, immediately developed at the entrance. Then the liquid residence time is $t = A_l L / \dot{Q}_l$, which must be greater than the holdup time desired. This can be rearranged to calculate a required tank D. Further, the vapor flow rate Reynolds number must be low enough to prevent re-entrainment $\mathrm{Re} = D_{ch} u \rho / \mu \leq \mathrm{Re}_{threshold}$. This can be rearranged to determine the required tank D. Finally, the vapor residence time must be longer than the time for droplets to fall the largest vapor distance, $A_v L / \dot{Q}_v \leq D(1-f)/v_{terminal}$, which also can be rearranged to determine the required tank D. The required tank D is the maximum of the D determined by each of the three constraints and then rounded to the next higher 3-inch increment.

Without the increments on either L/D or D, the surface has discontinuities due to the choice of diameter-controlling mechanism. This can be seen as discontinuities in either the net lines or contour lines at the bottom of the three valleys in Figure E.42a. With discretized L/D and D values, the surface has additional ridges and many flat spots as Figure E.42b reveals.

Even if there is no discretization (Figure E.43a), the discontinuity in the valleys is a problem for most optimizers. Gradient-based optimizers stop anywhere along the discontinuity, as shown in

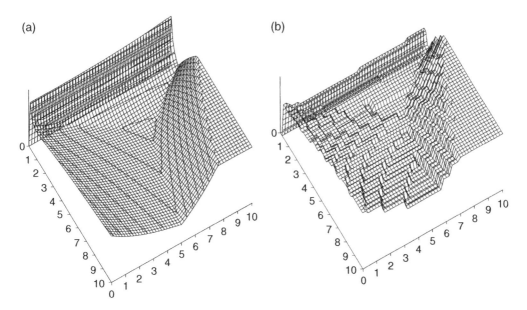

Figure E.42 Liquid–vapor separator surfaces: (a) continuum size variables and (b) discretized size.

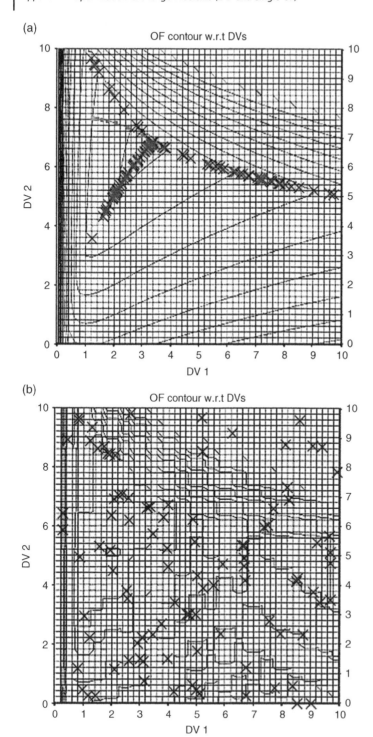

Figure E.43 Liquid–vapor separator issues for optimizers: (a) effect of ridges and (b) effect of flat spots.

Figure E.43a. You can easily comment out the two lines of VBA code that discretize *L/D* or *D* values and generate the left-hand graph and apply the optimizers to see the issues in their solutions.

With the flat spots from discretization (Figure E.43b), most single TS optimizers stop where they are initialized (on a flat spot, no direction is downhill). The optimum from Figure E.43a (continuum version of the application) can be rounded to the nearest *L/D* and tank *D* values; but this is not the optimum when the application considers the discretization. This is a great application example!

The VBA code is as follows:

```
If Fchoice = 27 Then      'vapor/liquid disengagement tank (British Units)
  'based on design example created by Jan Wagner and Josh Ramsey for process
Design course
  'RRR added the gas Re constraint, and rounded the "givens"
  ld27 = x1             'L/D ratio for the horizontal cylinder, 0 to 10
  ld27 = Round(2 * ld27) / 2   'L/D ratio discretized to halves, comment this
line to make the DV continuous
  f27 = 0.2 + 0.4 * x2 / 10     'height of liquid as a fraction of cylinder
diameter, 20% to 80% nominal range
  If ld27 <= 0 Or f27 <= 0 Or f27 >= 1 Then
    constraint = "FAIL"
    Exit Function
  End If
  '  Here are the "givens", the process basis
  QV27 = 2            'vapor volumetric flow rate, cuft/s
  QL27 = 0.05         'liquid volumetric flow rate, cuft/s
  vt27 = 0.1          'mist droplet in vapor settling velocity, ft/s
  tL27 = 500          'liquid degas miminum time, sec
  densV27 = 0.6       'gas density, lbm/cuft
  viscV27 = 0.0001    'gas viscosity, lbm/ft-s
  If f27 <= 0.5 Then
    y27 = 1 - 2 * f27     'vertical scaled distance from surface up to center
    beta27 = Atn(y27 / Sqr(1 - y27 ^ 2))  'angle in radians, of surface-edge to
center
    alpha27 = 3.1415926536 - 2 * beta27   'angle in radians, of the segment end-
to-end
    ALf27 = (alpha27 - Sin(alpha27)) / (2 * 3.1415926536)  'fraction of cross
section area that is liquid
    AVf27 = 1 - ALf27               'fraction of cross section area that is vapor
  Else
    y27 = 2 * f27 - 1     'vertical scaled distance from surface down to center
    beta27 = Atn(y27 / Sqr(1 - y27 ^ 2))   'angle in radians
    alpha27 = 3.1415926536 - 2 * beta27    'angle in radians
    AVf27 = (alpha27 - Sin(alpha27)) / (2 * 3.1415926536) 'fraction of cross
section area that is vapor
    ALf27 = 1 - AVf27               'fraction of cross section area that is liquid
  End If
  '  effective d of the vapor segment is taken as the height of the head space
```

```
   dVRe27 = 4 * QV27 * densV27 / (3.1415926536 * 20000 * AVf27 * viscV27 * (1 -
f27))  'diameter based on moderately turbulent, ft
   dV27 = Sqr((1 - f27) * QV27 * 4 / (vt27 * 3.1415926536 * AVf27 * ld27))
'diameter based on droplet settling time, ft
   dL27 = (tL27 * QL27 * 4 / (3.1415926536 * ALf27 * ld27)) ^ (1 / 3)   'diameter
based on liquid degassing time, ft
   d27 = dL27
   If dV27 > d27 Then d27 = dV27   'maximum of the two diameters - this creates a
sharp valley discontinuity
   If dVRe27 > d27 Then d27 = dVRe27  'maximum of the three diameters
   d27 = Int(4 * (d27 + 0.49999)) / 4   'discretizing to next larger in 3-inch
increments, comment this line to make the OF continuous
   f_of_x = (0.5 + ld27) * d27 ^ 2   'OF is porportional to capital based on mass
of tank
   f_of_x = Sqr(f_of_x)        'transformation to see detail in low OF values
   f_of_x = 10 * (f_of_x - 5) / 5     'scaled for display on 0-10 range
   Exit Function
End If
```

E.4   Other Test Function Sources

There are numerous compilations of 2-D sample problems for optimization. Here is a list of sites that were reported on ResearchGate during the summer of 2015:

http://www.mat.univie.ac.at/~neum/glopt/test.html
http://www.gamsworld.org/performance/selconglobal/selconglobállib.htm
http://www.geatbx.com/docu/fcnindex-01.html
http://coco.gforge.inria.fr/doku.php?id=bbob-2013-downloads
https://en.wikipedia.org/wiki/Test_functions_for_optimization
http://www.mat.univie.ac.at/~vpk/math/funcs.html
http://www.cs.bham.ac.uk/~xin/papers/published_tec_jul99.pdf
http://www.ntu.edu.sg/home/EPNSugan/index_files/CEC2013/CEC2013.htm
https://en.wikipedia.org/wiki/Test_functions_for_optimization

Appendix F

Brief on VBA Programming

Excel in Office 2013

F.1 Introduction

This is intended as a basic how-to program in Excel VBA. It presumes that you can program, in some computer language, and just need to know how to do it in VBA. I'm currently running Office 2013, but the operations are very similar in other versions.

F.2 To Start

- Open an Excel worksheet (and save it as macro-enabled, as .xlsm).
- Set the Macro Security Setting to Medium. Click on the Office Button in the upper left of the menu bar, then the Excel Options button in bottom of window, then Trust Center in the left hand menu, then Trust Center Settings in middle right of window, and then choose your security preference ("disable with notification", probably). Finally click OK.
- Press ALT-F11 to open the VBA editor. Alternately, use the "Launch VBA" icon on the Developer Toolbar. [To install the Developer Toolbar, click on the "Office Button" (upper left), click on "Excel Options" (lower border of new window), and in the "Popular" set, check the "Show Developer Tab in the Ribbon" box. Close the "Office Button" window, and you should see the "Developer" tab on the upper ribbon.]
- In the new open window "VBA Editor", you should see three windows (Project, Properties, and Immediate) and a grayed space for another window. If the "Project" window does not appear, use the "View" menu item to add "Project Explorer".
- Right-click in the Project window, and then click on "Insert", then "Module". This will open a window in the grayed space for writing VBA code. Alternately, you can double click on the spread sheet name in the Project window to open a code window. Either way allows you to write VBA code. However, code in a code window attached to a spread sheet (for instance, "Sheet1") can only interface with that sheet.
- Add the Edit tool to the Menu: Right-click in the Menu area, check the Edit box, and Enter. Drag the Edit tool to the menu ribbon.
- There are two categories of code—subroutine and function—with attributes similar to that of any language. To create a subroutine, type "SUB", spacebar, subname, and "()". For example, "Sub

Engineering Optimization: Applications, Methods, and Analysis, First Edition. R. Russell Rhinehart.
© 2018 R. Russell Rhinehart. Published 2018 by John Wiley & Sons Ltd.
Companion website: www.wiley.com/go/rhinehart/engineeringoptimization

Practice()". Then hit enter, and "End Sub" will appear. Your code goes in between these two lines. To create a function, type "FUNCTION", spacebar, functionname, "(", argument list, and ")". For example, "Function F_Multiply(x,y)". Then hit Enter, and "End Function" will appear. Your code goes in between these two lines. Subs do not need an argument list, but can take one. There can be no spaces in the name of either the subroutine or the function, but you can use the underscore. Such as "Sub Help_Me()".

F.3 General

- There are no rules about the use of columns (unlike Fortran columns 1–6, 7, and 8–72).
- To continue a long statement on the next line, break it with a blank_underscore at the end of the line. " _" acts like a hyphen in English.
- Names for variables, functions, and subroutines cannot include spaces or characters. They must start with a letter and can contain numbers. However, you can use the underscore character to join two_words into a single name. This is an exception to the no symbols rule.
- Some combinations of letters are reserved for functions or values. These include log, sin, sqr, rand, name, TRUE, single, INT, Time, rem, etc. You cannot use these for variable names. If you inadvertently choose a reserved name, VBA will capitalize it for you and change the color, which will be your signal to change the name.
- The assignment symbol is simply "=", not ":=".
- There is no end-of-line symbol, like ";".
- Unless the user declares "Option Explicit", variables types do not need to be explicitly declared.
- Comments are indicated by the single quote mark, " ' ", and whether starting a new line or following code on the same line, anything on the line after the " ' " mark is ignored. The term "Rem", short for "Remark", does the same.
- Run a subroutine by (i) placing the cursor within the range of the subroutine and (ii) pressing the F5 key, or by clicking on the "Run arrow", which looks like a sideways triangle on the VBA toolbar.
- If an Excel cell is in the edit mode, you cannot run a VBA program. Press the "Esc" key or click on another cell to close the cell that was in the edit mode.
- If you get the message "Procedure Too Long" when you attempt to run a VBA program, it means that your code has too many lines. If you keep adding test functions or optimizer subroutines, this will happen. On some versions the system only allocates so much space for the code, and this will happen. Just delete unnecessary code from the active file. Save it first, under a unique name! Preserve the currently "unnecessary" code. You may want to revisit it later.

F.4 I/O to Excel Cells

- Cells are addressed by the row and column number, as Cells(4,2) representing the B4 cell. The operator name is "Cells", not "Cell". For your convenience, switch the Excel display from the default "A1" notation (column-as-a-letter row-as-a-number) to "R1C1" notation (row-as-a-number, column-as-a-number). To do this, open an Excel workbook; click on the "Office Button", then "Excel Options" at the bottom of the window, then "Formulas" in the left hand menu; and then check the R1C1 notation box in the "Working with Formulas" group of items. Click OK out of the sequence.

- Like an array, "Cells" is the name (not "Cell") and the row and column numbers are the indices. The VBA assignment statement "A = Cells(5,6)" reads the value in cell "F5" (R5C6) and assigns it to the variable "A." "Cells(i,j) = q" assigns the value of "q" for display in the cell in the ith row and jth column.

F.5 Variable Types and Declarations

- Variables that store text are called "string" variables. Those storing integer values are called "integers". Double-precision integers are called "long." Variables storing single precision real values are called "single". And double-precision reals are called "double". They do not have to be explicitly declared, but it is good practice. They are declared using a "dimension" statement, "Dim" for short, which sets up storage space as follows:
  - Dim Name As String 'sets a location called "Name" for text
  - Dim I As Integer
  - Dim J As Long
  - Dim a As Single
  - Dim b As Double
- Dimensions of vectors and arrays have to be declared, but the variable type does not. Examples include:
  - Dim List_One(25) as String 'declares both number and type
  - Dim Table(10, 10) 'only declares the array size
  - Dim Intensity(20, 20, 20)
  - Dim Population(100, 3) As Double
- Assignment statements for String Variables require double quotes. The others are conventional. Examples include:
  - List_One(k) = "Paul Taylor"
  - Table(I , J) = I * J
- It is good practice to organize all declarations in the header of the program, above all subroutine and function declarations, but VBA does not require this structure. If you declare variables within each subprogram, they are locally used, but variable values are forgotten when exiting the subprogram. Declaration in the header space above all subprograms makes the variable commonly available, with its value retained when control is switched from one subprogram to another.
- The statement "Option Explicit" requires every variable to be declared with a "Dim" statement. This is good for subsequent users, if you also include the definition and units as a comment line. It is also good to be sure that you did not inadvertently type JELLO for JELL0, or line for 1ine, or dimond2 for diamond2, or H0G for HOG.

F.6 Operations

- Precedence of mathematical operators is conventional. Parenthesis first, in to out and then left to right. The parenthesis symbols are "(" and ")" whether nested or not. Exponentiation is second in precedence, in to out and then left to right. The VBA symbol for exponentiation is "^", in Fortran it is "**". Next are by multiplication and division, left to right. Then addition and subtraction, left to right. These symbols are "*", "/", "+", and "−".

- Functions have parenthesis and return the value that results from operating on their argument value. For example,
  - A = fun_fun(b)
- VBA can concatenate. The operator symbol is " & " (space-&-space). For example, the statements:

```
Name2 = "Rhinehart"
Name1 = "Russ"
Cells( 5, 9 ) = Name2 & ", " & Name1
N = 3
Cells( 4, N ) = "R^" & N
```

would write "Rhinehart, Russ" in cell I5, and "R^3" in cell C4.

F.7 Loops

- The loop-type closest in function to the Fortran Do loop is the FOR–NEXT loop. The loop is started with a FOR statement and an initialization of the loop index, an extreme limit, and an optional step increment. The loop ends with a NEXT statement, at which place the loop index is incremented and tested whether it is beyond the extreme to either exit the loop or reenter. Here is an example of a loop that counts by 2s:

```
For k = 0 to 20 Step 2
        Cells(k, 1) = k
Next k
```

- If the "Step" is not defined, the default is "Step 1". "Step -5" would decrement.
- The "Next k" operation first adds the step increment to the loop index then tests if the value exceeds the limit. If not, execution returns to the loop range. If it does, execution proceeds to the next stage. In this case, the first value of k is 0. Then it will have the values 2, 4, 6, etc. Upon exiting the loop, the value of k will be 22 not the loop limit of 20.

F.8 Conditionals

- The conditional statement closest in function to the Fortran IF is the IF–THEN–ELSE conditional. It starts with "IF" followed by the condition and the word "THEN". If the condition is true, code following the THEN is executed; otherwise not. The <, =, and > keyboard symbols are used for the comparison. Here are two examples, one with and one without an "ELSE:"

```
IF name < 'C' THEN Cells (17, I) = name
IF Abs(x) > = Abs(y) THEN
        Cells(k, 1) = 'Great! x is at least as large as y''
ELSE
        Cells(k, 1) = 'Sorry, x had a smaller magnitude than y.''
END IF
```

- Compound antecedent tests require each comparison to be explicitly stated and joined with either the "AND" or "OR" or other conjunction operator. For example, to test if the value of x is between "a" and "b," as "Is $a < x < b$?" you would write:

$$IF \ a < x \ AND \ x < b \ THEN \ ...$$

- DO WHILE or DO UNTIL. These are preferred over the GOTO statement. The end of the WHILE or UNTIL range is the statement LOOP. The form is as follows:

```
Constraint = 'INDETERMINED'
DO UNTIL Constraint = 'PASS'
 x = Rnd()             'Assigns a random value to variable x
 Constraint = Constraint_Test(x) 'Calls function
LOOP
```

F.9   Debugging

- If VBA detects an error during the compile stage (actually an interpreter but it does look at the entire code for syntax errors such as FOR-without-NEXT or a GOTO-Label-not-defined) (whether pre-execution or during your edit stage), or if it encounters an impossible operation during the execution stage, it will make a "BONK" sound (so that all of your friends can hear) and open an error message window. If you click "debug" it will highlight the problem line in yellow. This is termed being in "Break Mode". If the break mode occurs during execution, you can place the cursor on any variable in the VBA editor window, and it will display its value. This is a great convenience for debugging.
- Click on the square "Reset" button on the toolbar to exit "Break Mode". You must do this to be able to run after fixing the code.
- If you want the execution to stop at a particular line so that you can mouse-over variables to see their values, then click in the slightly gray column just to the left of the programming text area. It will create a red dot. The program will stop, in break mode, when it gets to that line. Reclick on the red dot to eliminate it. After encountering a red dot, to take the computer out of break mode, click on the square "Reset" button on the toolbar.
- Alternately, you can step through the program during execution, line by line, by pressing the F8 key. VBA will highlight each line that is about to be executed. In "Break mode" you can observe variable values from formerly executed lines by a mouse-over (placing the cursor over) the variable symbol. Use the Escape key or the reset button to exit this "Step Into" mode.
- Instead of printing intermediate values to the Excel cells for display, you can add a watch window to the VBA editor. Use the "Debug" drop down menu, click on "Add Watch", and follow the directions.

F.10   Run Buttons (Commands)

- You can place a button (called a command) on the Excel spreadsheet to run a subroutine. In Excel, open the "Developer" toolbar, and click on the "Insert" icon in the "Controls" category. Then click on the "Button" icon in the upper left. Move the cursor to the location on the spreadsheet where you want the button (the cursor will have a + shape) and left-click. It will open a window that provides a list of subroutines and create a button in the grayed-boundary edit mode. Choose the sub that you want the button to start, and edit the button name. If the button is not in the active edit mode, right-click the button to change the text or to reassign the macro.

F.11   Objects and Properties

- The Excel workbook is an object. Objects contain objects, and objects have properties (attributes). For example, consider "The red car has a flat tire." Tire is an object of the car object. The car object has a property (color) and a property value (red). The tire object has a property (air pressure) and a value (0 psig). The workbook object contains worksheet objects, which contain cell objects, which might contain a number. The number object has a numerical value, but it also has other properties such as font type, font size, and font color.
- You can use VBA commands to do any formatting or data processing operation that you can do using the Excel toolbar items. Formatting includes change font, create cell borders, set display characteristics, set font and background colors, set cell dimensions, etc. Data processing includes recalculate, sort, clear contents, etc. Use the VBA help menu for details.

F.12   Keystroke Macros

- Often it is easier to perform an operation such as sort, plot, clear contents, and change font and color with keystroke/mouse operations in an Excel worksheet than to figure out how to program it in VBA. You can record the keystroke/mouse sequence as a macro (as a subroutine) in VBA and then call the subroutine from VBA programs.
- Click on the "Developer" tool tab on the Excel ribbon, and then click on "Record Macro". This opens a window. Fill in the name you want to assign to the keystroke sequence and location to write it. I usually accept the default name and location. Perform your keystroke/mouse operations, and then click on "Stop Recording" where you found the "Record Macro". The VBA code that the keystroke/mouse sequence represents will be in a new module in the VBA project window.
- Your VBA code can call that Macro subroutine.
- You can read the Macro code to see what the VBA instructions are.
- You can copy/paste that code, or appropriate sequences from it, into your VBA program.

F.13   External File I/O

- To open an external file for input or output, state "Open" the file path including name, the purpose, then the number you choose for the file. If the path is not explicitly defined, the default is the directory of the open Excel workbook. The numbers are your choice. For example:

```
OPEN "C:/My Documents/VBA Programs/filename.txt" FOR OUTPUT AS #4
OPEN "source data.txt" FOR INPUT AS #1
```

- You should close a file, when execution is finished, with the CLOSE #n statement, for example, CLOSE #7.
- INPUT and PRINT (read and write) from and to text files. Specify the file number followed by a comma and then the variable list. When the Input or Print list is complete, the next read or write "call" starts at the beginning of the next line in the file. Examples are as follows:

```
INPUT #8, a, b, c(2)
PRINT #2, Name2 & ", " & Name1, age
```

F.14   Solver Add-In

Solver is an optimization add-in to Excel. To install Solver, open Excel, and click on file and then options. Click on Add-Ins in the left column and the view changes. In the lower section, there is a "Manage" window, choose the "Excel Add-Ins", and click on the "Go" button. This opens a selection list. Click on the Solver box to check it, and then click "OK". Now Solver should appear on the far right of the Excel Data menu.

To use Solver, click on the Menu icon. Then select the cell to be optimized (choose Objective), choose min or max or value of, and choose the decision variable cells with the values that are adjusted to optimize the objective.

Solver has several optimization algorithms. In the 2013 version, I prefer GRG Nonlinear. If you click on Options, you can change the convergence criterion and choice of forward or central difference derivatives. GRG stands for "generalized reduced gradient." I think it is a Cauchy sequential line search approach that converts inequality constraints to equality with slack variables, locally linearizes the equality constraints, and solves for some of the DVs from the slack variable values.

F.15   Calling Solver from VBA

To Call Solver in VBA:
1) Make sure the Solver add-in is installed to VBA (one step more than just having it available in Excel). In the VBA editor, go to the Tools Menu and then References, and add Solver as a reference to the project. See Solver help in VBA editor for more details.
2) Record a Solver Macro from a keyboard implementation of Solver.
3) Modify the arguments for Solver, if needed for other functions or DVs.

```
SolverOk SetCell:="$J$" & nI & "", MaxMinVal:=3, ValueOf:=0, _
 ByChange:="$D$" & nI & "", Engine _
 :=1, EngineDesc:="GRG Nonlinear"            'a typical call
```

The following two lines are needed to close Solver and to save results. For some reason the mouse clicks on the "accept Solver solution" are not recorded in the macro. You will have to add these lines to the macro.

```
    SolverSolve UserFinish:=True
    SolverFinish KeepFinal:=1
```

Section 9

References and Index

The references are provided for those who are interested in more information about the topics or seeing the origins of some landmark publications.

References and Additional Resources

Books on Optimization

Beveridge, G. S. G. and R. S. Schechter, *Optimization: Theory and Practice*, McGraw-Hill, New York, 1970.

Chong, E. P. K. and S. H. Zak, *An Introduction to Optimization*, 2nd Edition, John Wiley & Sons, Inc., New York, 2001.

Edgar, T. F., D. M. Himmelblau, and L. S. Lasdon, *Optimization of Chemical Processes*, McGraw-Hill, New York, 2001.

Hillier, F. S. and G. J. Lieberman, *Introduction to Operations Research*, McGraw-Hill, New York, 2001.

Nocedal, J. and S. J. Wright, *Numerical Optimization*, Springer-Verlag, New York, 1999.

Rao, S. S., *Engineering Optimization: Theory and Practice*, 4th Edition, John Wiley & Sons, Inc., Hoboken, 2009.

Ravindran, A., K. M. Ragsdell, and G. V. Reklaitis, *Engineering Optimization—Methods and Applications*, John Wiley & Sons, Inc., Hoboken, 2006.

Snyman, J. A., *Practical Mathematical Optimization*, Springer, New York, 2005.

Books on Probability and Statistics

Bethea, R. M. and R. R. Rhinehart, *Applied Engineering Statistics*, Taylor & Francis, Boca Raton, FL, 1991. ISBN 0-8247-8503-7.

Rhinehart, R. R., *Instrument and Automation Engineers' Handbook, Vol I, Process Measurement and Analysis*, 5th Edition, B. Liptak and K. Venczel, Editors, Section 1.10, "Uncertainty—Estimation, Propagation, & Reporting," Taylor & Francis, CRC Press, Boca Raton, FL, 2016a.

Rhinehart, R. R., *Nonlinear Regression Modeling for Engineering Applications: Modeling, Model Validation, and Enabling Design of Experiments*, John Wiley & Sons, Inc., Hoboken, 2016b. ISBN 9781118597965.

Engineering Optimization: Applications, Methods, and Analysis, First Edition. R. Russell Rhinehart.
© 2018 R. Russell Rhinehart. Published 2018 by John Wiley & Sons Ltd.
Companion website: www.wiley.com/go/rhinehart/engineeringoptimization

Books on Simulation

Law, A. M. and W. D. Kelton, *Simulation Modeling and Analysis*, 2nd Edition, McGraw Hill, New York, 1991.

Specific Techniques

Akaho's Approximation for Normal Least Squares—S. Akaho, "Curve Fitting That Minimizes the Mean Square of Perpendicular Distances from Sample Points," Proc. SPIE 2060, Vision Geometry II, 237, December 1, 1993. doi:10.1117/12.164998; 10.1117/12.164998.

Best-of-N Method—Iyer, M. S. and R. R. Rhinehart, "A Method to Determine the Required Number of Neural Network Training Repetitions," *IEEE Transactions on Neural Networks*, Vol. **10**, No. 2, 1999, pp. 427–432.

Dynamic Programming—Rhinehart, R. R. and J. D. Beasley, "Dynamic Programming for Chemical Engineering Applications," *Chemical Engineering*, Vol. **94**, No. 18, 1987, pp. 113–119. That article was the basis for the chapter, Rhinehart, R. R. and J. D. Beasley, *Encyclopedia of Chemical Processing and Design*, Vol. **44**, J.J. McKetta, Editor, "Dynamic Programming," 1993, pp. 411–424.

Generating Gaussian Noise—Box, G. E. P. and M. E. Muller, "A Note on the Generation of Random Normal Deviates," *The Annals of Mathematical Statistics*, Vol. **29**, No. 2, 1958, pp. 610–611.

Initialization of Players—Manimegalai-Sridhar, U., A. Govindarajan, and R. R. Rhinehart, "Improved Initialization of Players in Leapfrogging Optimization," *Computers & Chemical Engineering*, Vol. **60**, 2014, pp. 426–429.

Leapfrogging—Rhinehart, R. R., M. Su, and U. Manimegalai-Sridhar, "Leapfrogging and Synoptic Leapfrogging: A New Optimization Approach," *Computers & Chemical Engineering*, Vol. **40**, 2012, pp. 67–81.

Snyman, J. A. and L. P. Fatti, "A Multi-Start Global Minimization Algorithm with Dynamic Search Trajectories," *Journal of Optimization Theory and Applications*, Vol. **54**, 1987, pp. 121–141.

Steady State Identification—Bhat, S. A. and D. N. Saraf, "Steady-State Identification, Gross Error Detection, and Data Reconciliation for Industrial Process Units," *Industrial and Engineering Chemistry Research*, Vol. **43**, No. 15, 2004, pp. 4323–4336.

Steady State Identification—Cao, S. and R. R. Rhinehart, "An Efficient Method for On-Line Identification of Steady-State," *Journal of Process Control*, Vol. **5**, No. 6, 1995, pp. 363–374.

Steady State Identification—Shrowti, N., K. Vilankar, and R. R. Rhinehart, "Type-II Critical Values for a Steady-State Identifier," *Journal of Process Control*, Vol. **20**, No. 7, pp. 885–890, 2010.

Steady State Identification—von Neumann, J., R. Kent, H. Bellison, and B. Hart, "The Mean Square Successive Difference," *The Annals of Mathematical Statistics*, Vol. **12**, 1941, pp. 153–162.

Stochastic Convergence—Rhinehart, R. R., "Convergence Criterion in Optimization of Stochastic Processes," *Computers & Chemical Engineering*, Vol. **68**, 2014, pp. 1–6.

Selected Landmark Papers

Data Reconciliation—Mah, R. S., G. M. Stanley, and D. M. Downing, "Reconciliation and Rectification of Process Flow and Inventory Data," *Industrial & Engineering Chemistry, Process Design and Development*, Vol. **15**, No. 1, 1976, pp. 175–183.

Differential Evolution—Storn, R. and K. Price, "Differential Evolution—A Simple and Efficient Heuristic for Global Optimization over Continuous Spaces," *Journal of Global Optimization*, Vol. **11**, 1997, pp. 341–359. doi:10.1023/A:1008202821328.

GRG—Lasdon, L. S., R. L. Fox, and M. W. Ratner, "Nonlinear Optimization Using the Generalized Reduced Gradient Method," AD-774 723, Prepared for the Office of Naval Research, National Technical information Service, US Department of Commerce, Springfield, VA, 1973, Technical Memorandum No. 325.

Hooke–Jeeves—Hooke, R. and T. A. Jeeves, "'Direct Search' Solution of Numerical and Statistical Problems," *Journal of the Association for Computing Machinery (ACM)*, Vol. **8**, No. 2, 1961, pp. 212–229. doi:10.1145/321062.321069.

Levenberg–Marquardt—Levenberg, K., "A Method for the Solution of Certain Non-Linear Problems in Least Squares," *Quarterly of Applied Mathematics*, Vol. **2**, 1944, pp. 164–168. Marquardt, D., "An Algorithm for Least-Squares Estimation of Nonlinear Parameters," *SIAM Journal on Applied Mathematics*, Vol. **11**, No. 2, 1963, 431–441. doi:10.1137/0111030.

Particle Swarm—Kennedy, J. and R. Eberhart, "Particle Swarm Optimization," *Proceedings of IEEE International Conference on Neural Networks*, Vol. **4**, 1995, pp. 1942–1948. doi:10.1109/ICNN.1995.488968.

Simplex—Nelder–Mead—Nelder, J. A. and R. Mead, "A Simplex Method for Function Minimization," *Computer Journal*, Vol. **7**, 1965, pp. 308–313. doi:10.1093/comjnl/7.4.308.

Simplex—Spendley–Hext–Himsworth—Spendley, W., G. R. Hext, and F. R. Himsworth, "Sequential Application of Simplex Designs in Optimization and Evolutionary Operation," *Technometrics*, Vol. **4**, 1962, pp. 441–461.

Selected Websites Resources

https://www.wikipedia.org/
www.r3eda.com
https://sourceforge.net/projects/leapfrog-optimizer/ (accessed November 13, 2017).
http://www.mat.univie.ac.at/~neum/glopt/test.html (accessed November 13, 2017).
http://www.gamsworld.org/performance/selconglobal/selcongloballib.htm (accessed November 13, 2017).
http://www.geatbx.com/docu/fcnindex-01.html (accessed November 13, 2017).
http://coco.gforge.inria.fr/doku.php?id=bbob-2013-downloads (accessed November 13, 2017).
https://en.wikipedia.org/wiki/Test_functions_for_optimization (accessed November 13, 2017).
http://www.mat.univie.ac.at/~vpk/math/funcs.html (accessed November 13, 2017).
http://www.cs.bham.ac.uk/~xin/papers/published_tec_jul99.pdf (accessed November 13, 2017).
http://www.ntu.edu.sg/home/EPNSugan/index_files/CEC2013/CEC2013.htm (accessed November 13, 2017).

Index

Engineering Optimization: Applications, Methods, and Analysis, First Edition. R. Russell Rhinehart.
© 2018 R. Russell Rhinehart. Published 2018 by John Wiley & Sons Ltd.
Companion website: www.wiley.com/go/rhinehart/engineeringoptimization

Printed and bound by CPI Group (UK) Ltd, Croydon, CR0 4YY

16/04/2025

14658395-0001